Limit Distributions for Sums of Independent Random Vectors

WILEY SERIES IN PROBABILITY AND STATISTICS
PROBABILITY AND STATISTICS SECTION

A complete list of the titles in this series appears at the end of this volume.

Limit Distributions for Sums of Independent Random Vectors

Heavy Tails in Theory and Practice

Mark M. Meerschaert
Hans-Peter Scheffler

A Wiley-Interscience Publication
JOHN WILEY & SONS, INC.
New York / Chichester / Weinheim / Brisbane / Singapore / Toronto

This text is printed on acid-free paper. ♾

Published simultaneously in Canada.

For ordering and customer service, call 1-800-CALL-WILEY.

Library of Congress Cataloging-in-Publication Data:

Meerschaert, Mark M., 1955–
Limit distributions for sums of independent random vectors : heavy tails in theory and practice / Mark M. Meerschaert, Hans-Peter Scheffler.
p. cm. — (Wiley series in probability and statistics)
Includes bibliographical references and index.
ISBN 0-471-35629-8 (cloth: acid-free paper)
1. Central limit theorem. 2. Random variables. I. Scheffler, Hans-Peter. II. Title. III. Series

QA273.67 .M45 2001
519.2—dc21 2001017844

Printed in the United States of America

10 9 8 7 6 5 4 3 2 1

Contents

Preface

Limit Distributions for Sums of Independent Random Variables was published (in Russian) by B. V. Gendenko and A. N. Kolmogorov in 1949 and translated into English by K. L. Chung [44] in 1968. This book provided an accessible but serious introduction to the central limit theory of random variables, which lies at the heart of probability and statistics. It required only a knowledge of analysis, yet it took the reader to the frontiers of current research. Fifty years later, there is no better reference for much of this material. Just as important, the exposition has provided a framework for research based on these fundamental principles. We set out to write another book in the same spirit, except that now the basic theory can be presented in a multivariable setting. There is an old adage in mathematics: The right proof is the proof that generalizes. History has validated the work of Gnedenko and Kolmogorov because their approach to the one variable problem extends to the multivariable case.

The central limit theory in probability and statistics treats a fundamental scientific question. If repeated experiments under controlled conditions produce different answers, how can we draw useful conclusions from our data? The answer lies in the fact that there is a regular, nonrandom pattern to the variations in repeated experiments. The most familiar example is the bell-shaped curve. The central limit theorem presented in most basic probability and statistics courses justifies use of the bell-shaped curve. It states that sums of large numbers of independent random variables must approximately fit this distribution under a broad range of realistic conditions, an approximation that becomes more precise as the quantity of data increases. This limit theorem applies as long as the tails of the distribution are not too heavy, so that the variance is finite. This means, roughly speaking, that the probability of experimental outcomes very far from average is very small. Other central limit theorems may apply when the distributional tails are so heavy that the variance is infinite. The resulting limit distributions are called stable laws. Real world problems often involve several related measurements, so that the results of repeated experiments may usefully be represented as random vectors. Then the requisite central limit theory needs to be multidimensional. In the multivariable central limit theorem the corresponding analogue of the bell-shaped curve is called a multivariate normal or Gaussian distribution. There is also a richer limit theory that pertains when the data has heavy tails. Here the limit distributions include multivariable stable and operator stable laws, along with the semistable and operator semistable laws.

Stable laws were once considered just a mathematical curiosity. Around 1960 researchers began to discover evidence of heavy tail fluctuations in financial data [39, 99]. This line of research led to the discovery of fractals [100]. By now, stable models are firmly established in the area of finance. There is extensive empirical evidence of heavy tail price fluctuations in stock markets, futures markets, and currency exchange rates [38, 72, 90, 107]. Heavy tail probability distributions are also used in electrical engineering [140] and hydrology [4, 12, 63]. Several additional applications to economics and computer science appear in [2]. Applications in other areas of science are emerging rapidly, and the subject continues to gain momentum. Multivariable heavy tail models are used in finance for portfolio analysis involving several different stock issues or mutual funds [138, 142], and in hydrology to describe certain multivariable diffusion models [130].

This book is intended as an accessible reference that treats the general central limit theory in detail. In our own collaborations with graduate students and faculty in economics and hydrogeology, as well as researchers in probability and statistics, it has become clear that such a book could be useful. Following the example of Gnedenko and Kolmogorov [44], this book starts at the beginning and carefully develops the central limit theory in detail. Starting with the basic constructions of modern probability theory, we develop the fundamental tools of infinitely divisible distributions and regular variation. Then, we lay out the general central limit theory for independent random vectors. Finally, we provide a number of extensions and applications to probability and statistics. Our regular variation approach mirrors that of Feller [40], which has become a standard reference for theoreticians as well as practicing engineers and scientists. We hope that our text will also serve as a handy reference for multivariable regular variation theory.

Probability and Statistics is a diverse field of study. The central limit theory, aside from being fundamental, provides an accessible starting point. Because of the fundamental nature of the subject, we can take the motivated reader all the way from the basic foundations of probability theory to cutting-edge research, all in one volume. Anyone with a working knowledge of analysis and linear algebra should find the book accessible. Although we do not intend our book as a textbook, it could easily form the basis for a Ph.D. level course or seminar in probability theory. For researchers, we hope that this book will provide an efficient and logical path through a large collection of results with many possible applications to real world phenomena. Heavy tail models are rapidly gaining acceptance and importance in many fields of science and engineering, yet some of the results that are used are not easily accessible in their original form as published research articles. This situation has led to numerous mistakes and misunderstandings. We hope that our exposition will help to clarify and unify this important and fundamental body of research.

Acknowledgments

The material in this book is taken from hundreds of research papers by a large number of talented authors, who in turn relied on the work of their predecessors. The credit for these results is shared by such a large number of mathematicians that it seems impossible to acknowledge them all. Although we did attempt to point out connections with the existing literature, it seems inevitable that we should leave out some important contributors, for which we can only beg their forgiveness.

The book in its present form owes a great deal to those generous colleagues who agreed to carefully review portions of the book for accuracy and content. Thanks especially to Peter Becker-Kern, University of Dortmund, who read and corrected the most technical portions of the manuscript. Thanks also go out to our colleagues Inmaculada Aban, Boris Baeumer, and Rina Schumer at the University of Nevada; Paul Anderson at Albion College; Piotr Kokoszka at Utah State University; Makoto Maejima at Keio University; V. Mandrekar at Michigan State University; and Jerry Alan Veeh at Auburn University. Any errors that remain are, of course, solely the responsibility of the authors.

M.M.M. and H.P.S.

Part I

INTRODUCTION

In this part of the book, we lay out the mathematical foundations for the study of limit theorems for sums of independent random vectors. The three main ingredients are probability theory for random vectors, linear operator theory, and the theory of infinitely divisible laws and convergence of triangular arrays.

1 Random Vectors

In this chapter, we collect some well-known results on random vectors in $\mathbb{R}^d$, their probability distributions, and their characteristic functions. This summary is to provide a convenient reference, and to establish notation. The reader who is familiar with measure theoretic probability can proceed directly to Chapter 2.

1.1 PROBABILITY DISTRIBUTIONS

In this section, we recall some basic ideas from probability theory. We state our results on the space $\mathbb{R}^d$ of d-dimensional column vectors, but the same is true for any finite dimensional real vector space. Modern probability theory is based on measure theory, and we assume the reader is familiar with basic measure theory as laid out in Billingsley [18] or Royden [155].

Definition 1.1.1. Given a set Ω we say that the collection $\mathcal{F}$ of subsets $B \subseteq \Omega$ is a *σ-algebra* if

(a) $\Omega \in \mathcal{F}$

(b) if $B \in \mathcal{F}$, then $B^c \in \mathcal{F}$

(c) If $B_i \in \mathcal{F}$, then $\bigcup_{i=1}^{\infty} B_i \in \mathcal{F}$

We call $(\Omega, \mathcal{F})$ a *measure space.* If $(\Omega', \mathcal{F}')$ is another measure space, then $f : \Omega \to \Omega'$ is *measurable* if the *preimage* $f^{-1}(B) = \{\omega \in \Omega : f(\omega) \in B\}$ belongs to $\mathcal{F}$ for every $B \in \mathcal{F}'$. A *measure* $P : \mathcal{F} \to [0, \infty)$ satisfies $P(\emptyset) = 0$

and $P(\bigcup_{i=1}^{\infty} B_i) = \sum_{i=1}^{\infty} P(B_i)$ for any disjoint sequence $(B_i) \subset \mathcal{F}$; meaning that $B_i \cap B_j = \emptyset$ for all $i \neq j$. If $P(\Omega) < \infty$, we say that P is a *finite measure.* If $P(A_i) < \infty$ and $\Omega = \bigcup_{i=1}^{\infty} A_i$, then P is a *σ-finite measure.* Given a class of subsets $\mathcal{A}$ we define $\sigma(\mathcal{A})$ to be the smallest σ-algebra that contains $\mathcal{A}$. If P, P' are σ-finite measures on the measure spaces $(\Omega, \mathcal{F})$ and $(\Omega', \mathcal{F}')$, then the *product measure* $P \times P'$ is the unique measure on the space $(\Omega \times \Omega', \sigma(\mathcal{F} \times \mathcal{F}'))$ such that $P \times P'(A \times B) = P(A)P'(B)$ for all $A \in \mathcal{F}, B \in \mathcal{F}'$. If $P(\Omega) = 1$, we call P a *probability measure* and we call $(\Omega, \mathcal{F}, P)$ an *abstract probability space.*

Definition 1.1.2. A *random vector* X is a measurable mapping from an abstract probability space $(\Omega, \mathcal{F}, P)$ to $(\mathbb{R}^d, \mathcal{B})$, where $\mathcal{B} = \sigma(\mathcal{O})$ is the σ-algebra of *Borel sets,* i.e., the smallest σ-algebra of subsets of $\mathbb{R}^d$ that contains the class $\mathcal{O}$ of open subsets of $\mathbb{R}^d$. Then X is a random element of $\mathbb{R}^d$ such that $P(X \in B)$ makes sense for Borel sets $B \subseteq \mathbb{R}^d$.

Theorem 1.1.3. *Let $\mathcal{A}$ be a subclass of the class $\mathcal{B}$ of all Borel subsets of $\mathbb{R}^d$ with $\mathcal{B} = \sigma(\mathcal{A})$ such that if $A, B \in \mathcal{A}$, then $A \cap B \in \mathcal{A}$. If μ, ν are σ-finite Borel measures on $\mathbb{R}^d$ and $\mu(A) = \nu(A)$ for all $A \in \mathcal{A}$, then $\mu = \nu$.*

Proof. This is a special case of Theorem 10.3 in Billingsley [18]. □

Definition 1.1.4. Let $\mathcal{M}^1$ denote the set of Borel probability measures on $\mathbb{R}^d$, i.e., the set of positive measures μ on $(\mathbb{R}^d, \mathcal{B})$ for which $\mu(\mathbb{R}^d) = 1$. If X is a random vector on $\mathbb{R}^d$ and $\mu(B) = P(X \in B)$ for all Borel sets $B \subseteq \mathbb{R}^d$, we call μ the *probability distribution* of X. In the special case where $X = a$ with probability 1, we will also write $\mu = \varepsilon_a$, the *point mass* at $a \in \mathbb{R}^d$. If Y is another random vector on $\mathbb{R}^d$ with probability distribution ν, we say that X, Y (or μ, ν) are *independent* if $P(X \in A, Y \in B) = P(X \in A)P(Y \in B)$ for all Borel sets $A, B \subseteq \mathbb{R}^d$. In other words, the probability distribution of the random vector (X, Y) on $\mathbb{R}^d \times \mathbb{R}^d$ is the product measure $\mu \times \nu$. The *convolution* of $\mu * \nu$ is defined by

$$\mu * \nu(B) = \int \mu(B - y)\nu\{dy\}$$

for all Borel sets $B \subseteq \mathbb{R}^d$, so that $\mu * \nu$ is the probability distribution of $X + Y$ when X, Y are independent.

Definition 1.1.5. Given a Borel measure μ on $\mathbb{R}^d$ and $T : \mathbb{R}^d \to \mathbb{R}^m$ Borel measurable, we define the Borel measure $T\mu$ on $\mathbb{R}^m$ by letting $T\mu(B) = \mu(T^{-1}B)$ for all Borel subsets $B \subseteq \mathbb{R}^m$, where the $T^{-1}B$ is the preimage. If X is a random vector on $\mathbb{R}^d$ with probability distribution μ, then the random vector TX on $\mathbb{R}^m$ has probability distribution $T\mu$.

Definition 1.1.6. We say that $A : \mathbb{R}^d \to \mathbb{R}^d$ is a *linear operator* on $\mathbb{R}^d$ if

(a) $Ax + Ay = A(x + y)$ for all $x, y \in \mathbb{R}^d$

(b) $A(\lambda x) = \lambda A x$ for all $x \in \mathbb{R}^d$ and $\lambda \in \mathbb{R}$

Let $\mathrm{L}(\mathbb{R}^d)$ denote the set of linear operators $A : \mathbb{R}^d \to \mathbb{R}^d$.

Proposition 1.1.7. *Suppose that $\mu, \nu \in \mathcal{M}^1$, $A \in \mathrm{L}(\mathbb{R}^d)$, $a \in \mathbb{R}^d$, and n is a positive integer. Then*

(a) $\mu * \nu = \nu * \mu$

(b) $A(\mu * \nu) = (A\mu) * (A\nu)$

(c) $(\mu * \nu)^n = (\mu^n) * (\nu^n)$

(d) $(A\mu)^n = A(\mu^n)$

(e) $A\varepsilon_b = \varepsilon_{Ab}$

(f) $(\varepsilon_b)^n = \varepsilon_{nb}$

Proof. If X, Y are random vectors with distribution μ, ν respectively, then $X + Y = Y + X$ and $A(X + Y) = AX + AY$ which proves (a) and (b). The rest of the proof is similar, and is left to the reader. □

Definition 1.1.8. A closed set $A \subseteq \mathbb{R}^d$ *supports* a Borel measure μ on $\mathbb{R}^d$ if $\mu(A^c) = 0$. The *support* $\mathrm{supp}(\mu)$ is the intersection of all closed sets that support μ.

Lemma 1.1.9. *Suppose that ϕ is a Borel measure on $\mathbb{R}^d$ and $T \in \mathrm{L}(\mathbb{R}^d)$. Then* $\mathrm{supp}(T\phi) = T\,\mathrm{supp}(\phi)$.

Proof. If $y \notin \mathrm{supp}(T\phi)$, then there exists a $A \subset \mathbb{R}^d$ closed with $y \notin A$ and $T\phi(A^c) = 0$. Let $B = T^{-1}A$. Since T is continuous B is closed, and $y \notin TB = TT^{-1}A = A$. Then $\phi(B^c) = \phi((T^{-1}A)^c) = \phi(T^{-1}(A^c)) = T\phi(A^c) = 0$, so $y \notin T\,\mathrm{supp}(\phi)$. Conversely, if $y \notin T\,\mathrm{supp}(\phi)$, then for some $B \subset \mathbb{R}^d$ closed, $y \notin TB$ and $\phi(B^c) = 0$. Let $A = TB$. Since T is an open mapping, A is closed and $y \notin A$. Then $B \subseteq T^{-1}TB = T^{-1}A$ so $T^{-1}(A^c) = (T^{-1}A)^c \subseteq B^c$. Then $T\phi(A^c) = \phi(T^{-1}(A^c)) \leq \phi(B^c) = 0$ so $y \notin \mathrm{supp}(T\phi)$. □

Definition 1.1.10. If μ is a Borel measure on $\mathbb{R}^d$ and $f : \mathbb{R}^d \to \mathbb{R}^n$ is Borel measurable, then we denote by

$$\int f(x)\mu\{dx\} = \Big(\int f_1(x)\mu\{dx\}, \ldots, \int f_n(x)\mu\{dx\}\Big)$$

the *Lebesgue integral* of f with respect to μ. We say that f is *integrable* with repect to μ if this integral exists. Note that f is integrable with respect to μ if and only if $\int |f_i(x)|\mu\{dx\} < \infty$ for $i = 1, \ldots, d$.

The following standard results can be found, e.g., in [18].

Lemma 1.1.11. *(Fatou) If μ is a Borel measure on $\mathbb{R}^d$ and $f_n : \mathbb{R}^d \to [0, \infty)$ are Borel measurable, then*

$$\int \liminf_{n\to\infty} f_n(x)\mu\{dx\} \leq \liminf_{n\to\infty} \int f_n(x)\mu\{dx\}.$$

Theorem 1.1.12. *(Monotone Convergence) If μ is a Borel measure on $\mathbb{R}^d$, $f_n : \mathbb{R}^d \to [0, \infty)$ are Borel measurable, $f_n(x)$ is monotone nondecreasing in n and $f_n(x) \to f(x)$ off a set of μ-measure zero, then*

$$\int f_n(x)\mu\{dx\} \to \int f(x)\ \mu\{dx\} \quad \text{as } n \to \infty.$$

Theorem 1.1.13. *(Lebesgue Dominated Convergence) If μ is a Borel measure on $\mathbb{R}^d$, $f_n, g : \mathbb{R}^d \to \mathbb{R}$ are Borel measurable, $\int g(x)\mu\{dx\}$ exists, and $|f_n(x)| \leq g(x)$ and $f_n(x) \to f(x)$ off a set of μ-measure zero, then*

$$\int f_n(x)\mu\{dx\} \to \int f(x)\ \mu\{dx\} \quad \text{as } n \to \infty.$$

Definition 1.1.14. Given a finite Borel measure μ in $\mathbb{R}$ we define the *distribution function* $F : \mathbb{R} \to [0, \infty)$ of μ by letting $F(x) = \mu(-\infty, x]$. Then $F(x)$ is bounded, nondecreasing, and continuous from the right. Conversely, any such function defines a finite Borel measure on the real line. In this case, we will also write

$$\int g(x)\mu\{dx\} = \int g(x)dF(x)$$

the *Lebesgue-Stieltjes* integral.

Theorem 1.1.15. *(Integration by parts) Given $F, G : [a, b] \to \mathbb{R}$ nondecreasing and continuous from the right, if F, G have no common points of discontinuity, then*

$$\int_{(a,b]} G(x)dF(x) = F(b)G(b) - F(a)G(a) - \int_{(a,b]} F(x)dG(x).$$

Theorem 1.1.16. *(Change of Variables) If μ is a Borel measure on $\mathbb{R}^d$ and if $T : \mathbb{R}^d \to \mathbb{R}^m$ and $f : \mathbb{R}^m \to \mathbb{R}^n$ are Borel measurable, then*

$$\int f(Tx)\mu\{dx\} = \int f(y)\ T\mu\{dy\}.$$

Theorem 1.1.17. *(Fubini) If μ, ν are σ-finite Borel measures on $\mathbb{R}^d, \mathbb{R}^m$ then*

$$\begin{aligned}\int\int f(x,y)\mu\{dx\}\nu\{dy\} &= \int f(x,y)\mu \times \nu\{dx, dy\} \\ &= \int\int f(x,y)\nu\{dy\}\mu\{dx\}\end{aligned}$$

for any Borel measurable $f \geq 0$, or for any $f : \mathbb{R}^d \times \mathbb{R}^m \to \mathbb{R}^n$ integrable with respect to $\mu \times \nu$.

Definition 1.1.18. We say that $<\cdot, \cdot> : \mathbb{R}^d \times \mathbb{R}^d \to [0, \infty)$ is an *inner product* on $\mathbb{R}^d$ if

(a) $<x + y, z> = <x, z> + <y, z>$ for all $x, y, z \in \mathbb{R}^d$

(b) $<x, y> = <y, x>$ for all $x, y \in \mathbb{R}^d$

(c) $<\lambda x, y> = \lambda <x, y>$ for all $x, y \in \mathbb{R}^d$ and $\lambda \in \mathbb{R}$

(d) $<x, x> = 0$ implies $x = 0$

Let $e_1, \ldots, e_d$ denote the standard basis for $\mathbb{R}^d$. Given $x, y \in \mathbb{R}^d$ we can write $x = \sum_i x_i e_i$ and $y = \sum_i y_i e_i$ in this basis, where $x_i, y_i \in \mathbb{R}$ are unique. The *standard inner product* is defined by $<x, y> = \sum_i x_i y_i$. Let x' denote the *transpose* of the column vector $x \in \mathbb{R}^d$. Then we can also write the standard inner product in the form $<x, y> = x'y$. Given $x, y \in \mathbb{R}^d$ we also define the *outer product* xy', a $d \times d$ matrix with ij entry $x_i y_j$. We say that a basis $b_1, \ldots, b_d$ for $\mathbb{R}^d$ is *orthonormal* with respect to a given inner product if $<b_i, b_j> = I(i = j)$ for all i, j.

Definition 1.1.19. Given an inner product $<\cdot, \cdot>$ on $\mathbb{R}^d$ and a linear operator $A \in \mathrm{L}(\mathbb{R}^d)$ we define the *transpose* $A^* \in \mathrm{L}(\mathbb{R}^d)$ by requiring that $<Ax, y> = <x, A^*y>$ for all $x, y \in \mathbb{R}^d$, so that the ji entry of the matrix of A^* in any orthonormal basis equals the ij entry of A. We say that A is *orthogonal linear operator* if $A^{-1} = A^*$. We say that A is *symmetric* if $A^* = A$. If A is symmetric then $Q(x) = <x, Ax>$ is called a *quadratic form.* If $Q(x) \geq 0$ for all $x \in \mathbb{R}^d$ we say that Q (or A) is *nonnegative definite.* If $Q(x) > 0$ for all nonzero $x \in \mathbb{R}^d$, we say that Q (or A) is *positive definite.*

Proposition 1.1.20. *Let $e_1, \ldots, e_d$ denote the standard basis for $\mathbb{R}^d$ and $<\cdot, \cdot>$ the standard inner product on $\mathbb{R}^d$. Then*

(a) Any inner product $(\cdot, \cdot)$ on $\mathbb{R}^d$ is of the form $(x, y) = <x, Ay>$, where A is a symmetric, positive definite linear operator on $\mathbb{R}^d$. The matrix of A in the standard basis has ij entry $a_{ij} = (e_i, e_j)$.

(b) Given any basis $b_1, \ldots, b_d$ for $\mathbb{R}^d$ there exists a unique inner product $(\cdot, \cdot)$ on $\mathbb{R}^d$ with respect to which this basis is orthonormal.

(c) If $Q(x) = <x, Ax>$ is a quadratic form then $a_{ij} = B(e_i, e_j)$, where $B(x, y) = \frac{1}{2}[Q(x + y) - Q(x) - Q(y)]$. If Q is positive definite then B is an inner product.

Proof. Given $(\cdot, \cdot)$ define A to be the matrix with ij entry $a_{ij} = (e_i, e_j)$. Then $(x, y) = \sum_i \sum_j x_i y_j (e_i, e_j) = \sum_i \sum_j x_i a_{ij} y_j = <x, Ay>$. Then A is symmetric since $a_{ij} = a_{ji}$ and A is positive definite since $(\cdot, \cdot)$ is an inner product. Given a basis $b_1, \ldots, b_d$ define $(b_i, b_j) = I(i = j)$ and extend by linearity to obtain an inner product $(\cdot, \cdot)$. Given a quadratic form $Q(x) = <x, Ax>$ compute that $a_{ij} = <e_i, Ae_j> = B(e_i, e_j)$ and that B is symmetric and linear in each entry. □

Definition 1.1.21. We say that $|| \cdot ||$ is a *norm* on $\mathbb{R}^d$ if

(a) $||x + y|| \leq ||x|| + ||y||$ for all $x, y \in \mathbb{R}^d$ (the *triangle inequality*)

(b) $||\lambda x|| = |\lambda| \cdot ||x||$ for all $x \in \mathbb{R}^d$ and $\lambda \in \mathbb{R}$

(c) $||x|| = 0$ implies $x = 0$

For any inner product $<\cdot, \cdot>$ on $\mathbb{R}^d$, we define the associated *Euclidean norm* $||x|| = \sqrt{<x, x>}$. For any basis $b_1, \ldots, b_d$ the Euclidean norm in the inner product of Proposition 1.1.20 (b) that makes this basis orthonormal will be called the *norm associated with this basis.*

Proposition 1.1.22. *If $<\cdot, \cdot>$ is any inner product on $\mathbb{R}^d$, and if $||x|| = \sqrt{<x, x>}$ is the associated Euclidean norm, then*

$$<x, y> \leq ||x|| \cdot ||y|| \tag{1.1}$$

for all $x, y \in \mathbb{R}^d$.

Proof. If $y = 0$ the result is obvious. Otherwise $<y, y> \neq 0$ and for any $\lambda \in \mathbb{R}$ we have $<x + \lambda y, x + \lambda y> \geq 0$ and hence $<x, x> + 2\lambda <x, y> + \lambda^2 <y, y> \geq 0$. Substitute $\lambda = -<x, y>/<y, y>$. □

We call (1.1) the *Cauchy-Schwarz Inequality* on $\mathbb{R}^d$. Given a Borel measure μ on $\mathbb{R}^d$, let $L^2_\mu(\mathbb{R}^n)$ denote the collection of all Borel measurable functions $f : \mathbb{R}^d \to \mathbb{R}^n$ such that $\int f(x)' f(x) \mu\{dx\} < \infty$. Then $L^2_\mu(\mathbb{R}^n)$ is a real vector space with the inner product $<f, g> = \int f(x)' g(x) \mu\{dx\}$ and the associated norm $||f|| = \sqrt{<f, f>}$. The following result generalizes (1.1) to the vector space $L^2_\mu(\mathbb{R}^n)$. The proof is the same as before.

Theorem 1.1.23. *(Cauchy-Schwarz Inequality) If μ is a Borel measure on $\mathbb{R}^d$ and $f, g : \mathbb{R}^d \to \mathbb{R}^n$ are Borel measurable then*

$$\left(\int f(x)' g(x) \mu\{dx\} \right)^2 \leq \left(\int ||f(x)||^2 \mu\{dx\} \right) \cdot \left(\int ||g(x)||^2 \mu\{dx\} \right) \tag{1.2}$$

provided that the integrals on the right exist.

If $f, g : \mathbb{R}^d \to \mathbb{R}$ then (1.2) reduces to

$$\left(\int f(x) g(x) \mu\{dx\} \right)^2 \leq \left(\int f(x)^2 \mu\{dx\} \right) \cdot \left(\int g(x)^2 \mu\{dx\} \right).$$

If we take $g(x) = 1$ and $\mu \in \mathcal{M}^1$, we obtain the useful inequality

$$\left(\int f(x)\mu\{dx\} \right)^2 \leq \int f(x)^2 \mu\{dx\}.$$

Definition 1.1.24. Given a random vector X on $\mathbb{R}^d$ with probability distribution μ we define the *mean* or *expectation*

$$EX = \int x \ \mu\{dx\}$$

whenever the integral exists. If X, Y are random vectors on $\mathbb{R}^d$ and $m_1 = EX$, $m_2 = EY$ exist, we define the *covariance* $\text{Cov}(X, Y) = E[(X - m_1)(Y - m_2)']$ if this expectation exists. We will also call $\text{Cov}(X, X)$ the *covariance matrix of* X. If $d = 1$, we will also define the *variance* of X as $\text{Var}(X) = \text{Cov}(X, X)$.

Proposition 1.1.25. *Suppose that X is a random vector on $\mathbb{R}^d$ with probability distribution μ and that $g : \mathbb{R}^d \to \mathbb{R}^n$ is Borel measurable. Then:*

(a) $E[g(X)] = \int g(x) \ \mu\{dx\}$

(b) If $d = 1$ and $m = EX$ exists then $\text{Var}(X) = \int (x - m)^2 \mu\{dx\}$

If X, Y are random vectors on $\mathbb{R}^d$ such that $m_X = EX$ and $m_Y = EY$ exist, and if ν is the probability distribution of the random vector (X, Y) on $\mathbb{R}^d \times \mathbb{R}^d$ then the covariance matrix

$$C = \text{Cov}(X, Y) = \int (x - m_X)(y - m_Y)' \nu\{dx, dy\}.$$

If we write $X = (X_1, \ldots, X_d)$ and $Y = (Y_1, \ldots, Y_d)$, then the ij entry of the matrix C is $\text{Cov}(X_i, Y_j)$.

Proof. Apply Theorem 1.1.16 to Definition 1.1.24. □

Remark 1.1.26. Using Theorem 1.1.23 and Proposition 1.1.25, it is easy to show that if $E\|X\|^2 < \infty$ and $E\|Y\|^2 < \infty$ then $\text{Cov}(X, Y)$ exists.

Remark 1.1.27. If the probability distribution μ of X has distribution function $F(x)$, then we can also write Proposition 1.1.25 in terms of the Lebesgue-Stieltjes integral, e.g., $E[g(X)] = \int g(x) \ dF(x)$. If the random vector X has a density $f(x)$, then $\int g(x) \ dF(x) = \int g(x) f(x) dx$. If X is discrete with $P(X = x_i) = p_i$, then $\int g(x) \ dF(x) = \sum_i g(x_i) p_i$.

Proposition 1.1.28. *(Markov Inequality) If X is a random vector on $\mathbb{R}^d$ then*

$$P(\|X\| \geq a) \leq \frac{1}{a^k} E[\|X\|^k]$$

for any $a > 0$ and any positive integer k.

Proof. Since $||x||^k/a^k \geq 1$ when $||x|| \geq a$ we have

$$P(||X|| \geq a) = \int_{||X|| \geq a} \mu\{dx\} \leq \frac{1}{a^k} \int_{||X|| \geq a} ||x||^k \mu\{dx\} \leq \frac{1}{a^k} E[||X||^k]$$

where μ is the distribution of X. □

Corollary 1.1.29. *If X, Y are random variables on $\mathbb{R}$, then*

$$E[|XY|]^2 \leq E[X^2] \cdot E[Y^2]$$

Proof. The proof follows immediately from Theorem 1.1.23. □

Corollary 1.1.30. *(Chebyshev Inequality) If X is a random variable on $\mathbb{R}$ such that* Var(X) *exists, then $m = EX$ exists and*

$$P(|X - m| \geq a) \leq \frac{1}{a^2} \operatorname{Var}(X)$$

for any $a > 0$.

Proof. Combine Proposition 1.1.28 with Corollary 1.1.29, where $Y \equiv 1$. □

1.2 CONVERGENCE IN DISTRIBUTION

In this section, we collect some basic properties of weak convergence for sequences of random vectors. The following are standard results, see, e.g., Billingsley [17].

Definition 1.2.1. If X_n, X are random vectors on $\mathbb{R}^d$ with probability distribution μ_n, μ we say that $X_n \Rightarrow X$ (or $\mu_n \Rightarrow \mu$) if $\mu_n(A) \to \mu(A)$ for Borel sets $A \subset \mathbb{R}^d$ such that $\mu(\partial A) = 0$. In other words, $P(X_n \in A) \to P(X \in A)$ for all such Borel sets. This is called *weak convergence* on the class $\mathcal{M}^1$ of Borel probability measures on $\mathbb{R}^d$. Here ∂A denotes the topological boundary of A, i.e. $\partial A = \bar{A} \setminus A^o$ where $\bar{A}$ is the closure of A, defined as the intersection of all closed sets containing A, and A^o is the interior of A, defined as the union of all open sets contained in A. A Borel set A such that $\mu(\partial A) = 0$ will be called a *μ-continuity set.* In the special case where $\mu = \varepsilon_a$, the point mass at $a \in \mathbb{R}^d$, we will also say that *$X_n \to a$ in probability.*

Theorem 1.2.2. *(Portmanteau Theorem for probability measures) Suppose that $\mu_n, \mu \in \mathcal{M}^1$. The following are equivalent:*

(a) $\mu_n \Rightarrow \mu$

(b) $\int f(x)\mu_n\{dx\} \to \int f(x)\mu\{dx\}$ *for bounded, continuous $f : \mathbb{R}^d \to \mathbb{R}$*

(c) $\limsup \mu_n(A) \leq \mu(A)$ *for every closed set $A \subseteq \mathbb{R}^d$*

(d) $\liminf \mu_n(A) \geq \mu(A)$ *for every open set* $A \subseteq \mathbb{R}^d$

Proof. This is Theorem 2.1 in [17]. □

Definition 1.2.3. Given $\varepsilon > 0$ we call $B_\varepsilon(x) = \{y \in \mathbb{R}^d : \|x - y\| < \varepsilon\}$ an *ε-neighborhood* of $x \in \mathbb{R}^d$. We call $B_\varepsilon(A) = \cup_{x \in A} B_\varepsilon(x)$ an *ε-neighborhood* of $A \subseteq \mathbb{R}^d$.

Proposition 1.2.4. *If* $\mu_n \Rightarrow \mu$ *and* $A \subseteq \mathbb{R}^d$ *is a* μ*-continuity set, then*

$$\int_A f(x)\mu_n\{dx\} \to \int_A f(x)\mu\{dx\}$$

for any bounded continuous function $f : B_\varepsilon(A) \to \mathbb{R}$.

Proof. Let $\|x - \bar{A}\| = \min\{\|x - y\| : y \in \bar{A}\}$ be the distance between $x \in \mathbb{R}^d$ and the closed set $\bar{A} \subseteq \mathbb{R}^d$. Let $\varphi(t) = \max\{0, 1 - t\}$ and $g_k(x) = f(x)\varphi(k\|x - \bar{A}\|)$ so that $g_k(x)$ is continuous, $f(x) = g_k(x)$ for $x \in A$ and $h_k(x) = g_k(x) - f(x)I(x \in A) \to 0$ as $k \to \infty$. For any k and any $\varepsilon > 0$, Proposition 1.2.2 (b) implies that

$$\left| \int g_k(x)\mu_n\{dx\} - \int g_k(x)\mu\{dx\} \right| < \varepsilon$$

for all n large. Let $B_k = \{x \in \mathbb{R}^d : \|x - \bar{A}\| \leq 1/k\}$. If $|f(x)| \leq M$ for all $x \in B_k$ then, since $h_k(x) = 0$ off the closed set $B_k \setminus A^o$, we have

$$\left| \int g_k(x)\mu\{dx\} - \int_A f(x)\mu\{dx\} \right| \leq M\mu(B_k \setminus A^o) < \varepsilon$$

for all k large, since $B_k \setminus A^o$ decreases to $\bar{A} \setminus A^o = \partial A$ and $\mu(\partial A) = 0$. For any such k, Proposition 1.2.2 (c) implies that

$$\left| \int g_k(x)\mu_n\{dx\} - \int_A f(x)\mu_n\{dx\} \right| \leq M\mu_n(B_k \setminus A^o) < 2\varepsilon$$

for all n large. Then

$$\left| \int_A f(x)\mu_n\{dx\} - \int_A f(x)\mu\{dx\} \right| < 4\varepsilon$$

where $\varepsilon > 0$ is arbitrary, and the result follows. □

Theorem 1.2.5. *Let* $\mathcal{A}$ *be a subclass of the class* $\mathcal{B}$ *of all Borel subsets of* $\mathbb{R}^d$ *such that:*

(a) If $A, B \in \mathcal{A}$, *then* $A \cap B \in \mathcal{A}$

(b) If $S \subseteq \mathbb{R}^d$ *is open, then* $S = \bigcup_{i=1}^\infty A_i$ *for some* $A_i \in \mathcal{A}$

If $\mu_n, \mu \in \mathcal{M}^1$ and $\mu_n(A) \to \mu(A)$ for all $A \in \mathcal{A}$, then $\mu_n \Rightarrow \mu$

Proof. If $A_1, \ldots, A_n$ belong to $\mathcal{A}$, then so do their intersections, hence the inclusion-exclusion formula of set theory yields

$$\begin{aligned}\mu_n\left(\bigcup_{i=1}^{n} A_i\right) &= \sum_{i=1}^{n} \mu_n(A_i) - \sum_{i=1}^{n}\sum_{j=1}^{n} \mu_n(A_i \cap A_j) + \cdots \\ &\to \sum_{i=1}^{n} \mu(A_i) - \sum_{i=1}^{n}\sum_{j=1}^{n} \mu(A_i \cap A_j) + \cdots \\ &= \mu\left(\bigcup_{i=1}^{n} A_i\right)\end{aligned}$$

If $S \subseteq \mathbb{R}^d$ is open, then $S = \bigcup_{i=1}^{\infty} A_i$ for some $A_i \in \mathcal{A}$ so for all $\varepsilon > 0$ we have $\mu(\bigcup_{i=1}^{n} A_i) > \mu(S) - \varepsilon$ for all n large. Then $\mu(S) - \varepsilon < \mu(\bigcup_{i=1}^{n} A_i) = \lim_m \mu_m(\bigcup_{i=1}^{n} A_i) \leq \liminf_m \mu_m(S)$ and it follows that condition (d) of the Portmanteau Theorem 1.2.2 holds, so that $\mu_n \Rightarrow \mu$. □

Example 1.2.6. We now give some useful examples of the class of sets $\mathcal{A}$ in Theorem 1.2.5.

(a) Write $x = (x_1, \ldots, x_d) \in \mathbb{R}^d$. Given $a \in \mathbb{R}$ and $i = 1, \ldots, d$ define the $d-1$ dimensional hyperplane $H_i(a) = \{x \in \mathbb{R}^d : x_i = a_i\}$. Let $\mathcal{A}$ denote the set of rectangles

$$R = \{x \in \mathbb{R}^d : a_i < x_i \leq b_i \ \forall\, i = 1, \ldots, d\} \tag{1.3}$$

where a_i, b_i are any real numbers such that the hyperplanes $H_i(a_i)$ and $H_i(b_i)$ have μ measure zero. Since μ is a finite measure, the set of $a \in \mathbb{R}$ such that $\mu(H_i(a)) > 0$ is countable. Then it is easy to see that Theorem 1.2.5 holds for this class of sets.

(b) We can also take $a_i \leq x_i \leq b_i$ in (1.3), or $a_i < x_i < b_i$ or $a_i \leq x_i < b_i$.

(c) Given $x \in \mathbb{R}^d$ and $\delta > 0$ define

$$B_\delta(x) = \{y \in \mathbb{R}^d : \|x - y\| < \delta\} \tag{1.4}$$

and let $\mathcal{A}$ denote the collection of all finite intersections of μ-continuity sets of the form (1.4). Since μ is a finite measure, the set of $\delta > 0$ such that $\mu(\partial B_\delta(x)) > 0$ is countable. Then it is easy to see that Theorem 1.2.5 also holds for this class of sets.

(d) We can also take $\|x - y\| \leq \delta$ in (1.4).

Given $\mu \in \mathcal{M}^1$ and Borel measurable function $h : \mathbb{R}^d \to \mathbb{R}^n$, recall from Definition 1.1.5 that we can define another probability measure $h\mu$ on $\mathbb{R}^n$ by setting $h\mu(A) = \mu(h^{-1}(A))$ for all Borel sets $A \subseteq \mathbb{R}^d$. If X is a random vector on $\mathbb{R}^d$ with probability distribution μ then $h(X)$ is a random vector on $\mathbb{R}^n$ with probability distribution $h\mu$.

Theorem 1.2.7. *(Continuous Mapping Theorem) If $\mu_n \Rightarrow \mu$ in $\mathcal{M}^1$ and $T : \mathbb{R}^d \to \mathbb{R}^n$ is Borel measurable and continuous off a set of μ-measure zero, then $T\mu_n \Rightarrow T\mu$.*

Theorem 1.2.8. *If $\mu_n \Rightarrow \mu$ in $\mathcal{M}^1$, $T_n, T : \mathbb{R}^d \to \mathbb{R}^n$ are Borel measurable and if the set of $x \in \mathbb{R}^d$ such that $T_n(x_n) \not\to T(x)$ for some $x_n \to x$ has μ-measure zero, then $T_n\mu_n \Rightarrow T\mu$.*

Definition 1.2.9. We say that a sequence of probability measures (μ_n) in $\mathcal{M}^1$ is *weakly relatively compact* if every subsequence contains a further subsequence along which $\mu_n \Rightarrow \mu$ for some $\mu \in \mathcal{M}^1$. We say that (μ_n) is *tight* if for every $\varepsilon > 0$ there exists a compact set $K \subset \mathbb{R}^d$ such that $\mu_n(K) \geq 1 - \varepsilon$ for all n.

Theorem 1.2.10. *(Prohorov Theorem) A sequence of probability measures $\mu_n \in \mathcal{M}^1$ is tight if and only if it is weakly relatively compact.*

Remark 1.2.11. We can also write the above results in terms of random vectors. If X_n, X are random vectors on $\mathbb{R}^d$ then $X_n \Rightarrow X$ is equivalent to the condition that $E[f(X_n)] \to E[f(X)]$ for every function $f : \mathbb{R}^d \to \mathbb{R}$ that is bounded and continuous. If $h_n, h : \mathbb{R}^d \to \mathbb{R}^d$ are Borel measurable, then $h_n(X_n) \Rightarrow h(X)$ provided that $P(X \in E_h) = 0$, where E_h is the set of all $x \in \mathbb{R}^n$ such that $h_n(x_n) \not\to h(x)$ for some $x_n \to x$. A sequence of random vectors (X_n) on $\mathbb{R}^d$ is weakly relatively compact if every subsequence contains a further subsequence along which $X_n \Rightarrow X$ for some random vector X on $\mathbb{R}^d$, (X_n) is tight if for every $\varepsilon > 0$ there exists a compact set $K \subset \mathbb{R}^d$ such that $P(X_n \in K) \geq 1 - \varepsilon$ for all n, and (X_n) is weakly relatively compact if and only if it is tight.

Convergence results for various other classes of measures can be easily obtained by reduction to the case of probability measures. We present some basic results here that will be useful later in this book.

Definition 1.2.12. Denote by $\mathcal{M}_0$ the collection of finite Borel measures on $\mathbb{R}^d$. For $\mu_n, \mu \in \mathcal{M}_0$ we will write $\mu_n \to \mu$ if $\mu_n(A) \to \mu(A)$ for Borel sets $A \subseteq \mathbb{R}^d$ such that $\mu(\partial A) = 0$. This is called *vague convergence* on $\mathcal{M}_0$. A Borel set A such that $\mu(\partial A) = 0$ will be called a *μ-continuity set.*

Proposition 1.2.13. *(Portmanteau Theorem for finite measures) Suppose that $\mu_n, \mu \in \mathcal{M}_0$. The following are equivalent:*

(a) $\mu_n \to \mu$

(b) $\int f(x)\mu_n\{dx\} \to \int f(x)\mu\{dx\}$ *for bounded, continuous* $f : \mathbb{R}^d \to \mathbb{R}$

(c) $\limsup \mu_n(A) \leq \mu(A)$ *for every closed set* $A \subseteq \mathbb{R}^d$

(d) $\liminf \mu_n(A) \geq \mu(A)$ *for every open set* $A \subseteq \mathbb{R}^d$

Proof. Let $c_n = \mu_n(D)$ and $c = \mu(D)$. If $c > 0$, then (a) holds if and only if $c_n \to c$ and $\nu_n \Rightarrow \nu$, where $\nu_n = c_n^{-1}\mu_n$ for n large and $\nu = c^{-1}\mu$. Condition (b) holds if and only if $c_n \to c$ and the integrals converge with μ_n, μ replaced by ν_n, ν. Taking complements shows that (c) and (d) are equivalent. Recall that $\mathbb{R}^d$ is both open and closed. Then condition (c) holds if and only if $c_n \to c$ and the lim sup condition holds with μ_n, μ replaced by ν_n, ν and likewise for (d). Then the theorem follows immediately from Theorem 1.2.2 in this case. Finally, if $\mu(\mathbb{R}^d) = 0$, then (a) implies $\mu_n(\mathbb{R}^d) \to 0$ so that $|\int f(x)\mu_n\{dx\}| \leq M\mu_n(\mathbb{R}^d) \to 0$ for $|f| \leq M$ and (b) holds. Then we can take $f \equiv 1$ to see that $\mu_n(A) \leq \int f(x)\mu_n\{dx\} \to 0$ for any Borel set A so that (c) holds. If (c) holds, then $\mu_n(A) \leq \mu_n(\mathbb{R}^d) \to 0$ for all Borel sets A so that (a) holds. Since (c) and (d) are equivalent, the proof is complete. □

Proposition 1.2.14. *If* $\mu_n \to \mu$ *in* $\mathcal{M}_0$ *and* $A \subseteq \mathbb{R}^d$ *is a* μ*-continuity set, then*

$$\int_A f(x)\mu_n\{dx\} \to \int_A f(x)\mu\{dx\}$$

for any bounded continuous function $f : B_\varepsilon(A) \to \mathbb{R}$.

Proof. The proof is identical to Proposition 1.2.4, using Proposition 1.2.13 in place of Proposition 1.2.2. □

Theorem 1.2.15. *Let* $\mathcal{A}$ *be a subclass of the class* $\mathcal{B}$ *of all Borel subsets of* $\mathbb{R}^d$ *such that conditions (a) and (b) of Theorem 1.2.5 hold. If* $\mu_n, \mu \in \mathcal{M}_0$ *and* $\mu_n(A) \to \mu(A)$ *for all* $A \in \mathcal{A}$, *then* $\mu_n \to \mu$ *in* $\mathcal{M}_0$.

Proof. The proof is identical to Theorem 1.2.5, using Theorem 1.2.13 instead of Theorem 1.2.2. □

Proposition 1.2.16. *If* $\mu_n \to \mu$ *in* $\mathcal{M}_0$, $T_n, T : \mathbb{R}^d \to \mathbb{R}^n$ *are Borel measurable and* $T_n(x_n) \to T(x)$ *whenever* $x_n \to x$ *in* $\mathbb{R}^d$, *then* $T_n\mu_n \Rightarrow T\mu$.

Proof. Let $c_n = \mu_n(D)$ and $c = \mu(D)$. If $c > 0$, then $\mu_n \to \mu$ if and only if $c_n \to c$ and $\nu_n \Rightarrow \nu$, where $\nu_n = c_n^{-1}\mu_n$ for n large and $\nu = c^{-1}\mu$. Proposition 1.2.8 shows that $T_n\nu_n \Rightarrow T\nu$ so that $T_n\mu_n \Rightarrow T\mu$ in this case. If $c = 0$, then $T\mu$ is the zero measure and $T_n\mu_n(A) \leq T_n\mu_n(\mathbb{R}^n) \leq \mu_n(\mathbb{R}^d) \to 0$ for all Borel sets $A \subseteq \mathbb{R}^n$. □

Definition 1.2.17. A set $A \subset \mathbb{R}^d$ is *bounded away from zero* if for some $\varepsilon > 0$ we have $\|x\| \geq \varepsilon$ for all $x \in A$. Denote by $\mathcal{M}$ the collection of Borel measures on $\Gamma = \mathbb{R}^d \setminus \{0\}$ that assign finite measure to sets bounded away from zero.[1]

[1]For convenience we will identify a measure $\mu \in \mathcal{M}$ with a σ-finite Borel measure on $\mathbb{R}^d$ by setting $\mu\{0\} = 0$

A Borel set $A \subseteq \Gamma$ such that $\mu(\partial A) = 0$ will be called a *μ-continuity set.* For $\mu_n, \mu \in \mathcal{M}$ we will write $\mu_n \to \mu$ if $\mu_n(A) \to \mu(A)$ for all μ-continuity sets A that are bounded away from zero.

Definition 1.2.18. A closed set $A \subseteq \mathbb{R}^d$ *supports* a Borel measurable function $f : \mathbb{R}^d \to \mathbb{R}$ if $f(x) \neq 0$ implies $x \in A$. The *support* $\mathrm{supp}(f)$ is the intersection of all closed sets that support f.

Proposition 1.2.19. *(Portmanteau Theorem on $\mathcal{M}$) Suppose $\mu_n, \mu \in \mathcal{M}$. The following are equivalent:*

(a) $\mu_n \to \mu$

(b) $\int f(x)\mu_n\{dx\} \to \int f(x)\mu\{dx\}$ *for every bounded, continuous function* $f : \mathbb{R}^d \to \mathbb{R}$ *for which supp(f) is bounded away from zero*

(c) $\limsup \mu_n(A) \leq \mu(A)$ *for every closed set* $A \subseteq \Gamma$ *that is bounded away from zero*

(d) $\liminf \mu_n(A) \geq \mu(A)$ *for every open set* $A \subseteq \Gamma$ *that is bounded away from zero*

Proof. Suppose (a) holds and $f(x) = 0$ on D^c, where $D = \{x : \|x\| \geq \varepsilon\}$ and $\mu(\partial D) = 0$. Let ν_n, ν denote the restriction of μ_n, μ to D. Then $\nu_n \to \nu$ in $\mathcal{M}_0$ and (b) follows immediately from Proposition 1.2.13. If (b) holds and $A \subset D$ is closed, then $\limsup \nu_n(A) \leq \nu(A)$ by Proposition 1.2.13 and (c) follows. If $A \subset D$ is open, then $D \setminus A$ is closed and since $\mu_n(D) \to \mu(D)$ it follows that (c) and (d) are equivalent. If (c) and (d) hold, then (a) follows as in the proof of Proposition 1.2.13. □

Proposition 1.2.20. *If $\mu_n \to \mu$ in $\mathcal{M}$ and $A \subseteq \mathbb{R}^d$ is a μ-continuity set that is bounded away from zero, then*

$$\int_A f(x)\mu_n\{dx\} \to \int_A f(x)\mu\{dx\}$$

for any bounded continuous function $f : B_\varepsilon(A) \to \mathbb{R}$.

Proof. The proof is identical to Proposition 1.2.4, using Proposition 1.2.19 in place of Proposition 1.2.2. □

Theorem 1.2.21. *Let $\mathcal{A}$ be a subclass of the class $\mathcal{B}$ of all Borel subsets of $\mathbb{R}^d$ such that conditions (a) and (b) of Theorem 1.2.5 hold. If $\mu_n, \mu \in \mathcal{M}$ and $\mu_n(A) \to \mu(A)$ for all $A \in \mathcal{A}$, then $\mu_n \to \mu$ in $\mathcal{M}$.*

Proof. This proof is identical to Theorem 1.2.5, using Theorem 1.2.19 instead of Theorem 1.2.2. □

Example 1.2.22. The classes of sets in Example 1.2.6 also determine convergence in $\mathcal{M}_0$. They also determine convergence in $\mathcal{M}$, even if we omit those sets that contain the origin. Let $S_r = \{x \in \mathbb{R}^d : \|x\| = r\}$ denote the sphere of radius $r > 0$ in some norm. Then Theorem 1.2.21 also holds for the class of sets

$$\{r\theta : a < r \leq b, \theta \in B\} \tag{1.5}$$

where $\mu(S_a) = \mu(S_b) = 0$ and B is a Borel subset of the unit sphere S_1 in this norm. We can also take $a \leq r \leq b$ in (1.5), or $a < r < b$ or $a \leq r < b$. We can also restrict B to a subclass of Borel sets that satisfies the conditions (a) and (b) of Theorem 1.2.5.

Proposition 1.2.23. *If $\mu_n \to \mu$ in $\mathcal{M}$, $T_n, T : \mathbb{R}^d \to \mathbb{R}^n$ are Borel measurable and $T_n(x_n) \to T(x)$ whenever $x_n \to x$ in $\mathbb{R}^d$, then $T_n\mu_n \Rightarrow T\mu$.*

Proof. It suffices to show that condition (b) of Proposition 1.2.19 holds. Choose $D = \{x : \|x\| \geq \varepsilon\}$ such that $\mu(\partial D) = 0$ and $f(x) = 0$ on D^c. Let ν_n, ν denote the restriction of μ_n, μ to D. Then the desired result follows immediately from Proposition 1.2.16. □

1.3 CHARACTERISTIC FUNCTIONS

Characteristic functions are a kind of Fourier transform. As in other branches of applied mathematics, Fourier transforms are useful in probability because they transform convolutions into products. Since the distribution of the sum of independent random variables is a convolution, we will often use these transforms.

Definition 1.3.1. For any $\mu \in \mathcal{M}^1$, we define the *characteristic function*

$$\hat{\mu}(t) = \int e^{i<t,x>} \mu\{dx\} \tag{1.6}$$

where $t \in \mathbb{R}^d$. Note that $\hat{\mu}(t)$ always exists, and that $|\hat{\mu}(t)| \leq 1 = \hat{\mu}(0)$ since μ is a probability measure.

If μ is a probability distribution on $\mathbb{R}^1$ with density $f(x)$, then $\hat{\mu}(-t) = \int e^{-itx} f(x)dx$ is the best known Fourier transform. Formula (1.6) is preferred in probability theory because it simplifies the formula for the moments of μ in terms of the derivatives of $\hat{\mu}$. The following properties of the characteristic function are well known, see, e.g., Feller [40].

Definition 1.3.2. If X is a random vector on $\mathbb{R}^d$ with distribution μ and $\nu\{dx\} = \mu\{-dx\}$ is the distribution of $-X$, then we say that X (or μ) is *symmetric* if $\mu = \nu$. If X_1, X_2 are i.i.d. with X, then we call the distribution $\omega = \mu * \nu$ of $X_1 - X_2$ the *symmetrization* of μ.

Lemma 1.3.3. *Suppose that $\mu \in \mathcal{M}^1$ and that $\hat{\mu}$ is the corresponding characteristic function. Then μ is symmetric if and only if $\hat{\mu} = \operatorname{Re} \hat{\mu}$.*

Proposition 1.3.4. *Suppose that $\mu \in \mathcal{M}^1$ and that $\hat{\mu}$ is the corresponding characteristic function. Then*

$$1 - \operatorname{Re}\hat{\mu}(t_1 + t_2) \le 2\big(1 - \operatorname{Re}\hat{\mu}(t_1) + 1 - \operatorname{Re}\hat{\mu}(t_2)\big) \tag{1.7}$$

for all $t_1, t_2 \in \mathbb{R}^d$.

Proposition 1.3.5. *Suppose that $\mu, \nu \in \mathcal{M}^1$ and that $\hat{\mu}, \hat{\nu}$ are the corresponding characteristic functions. Then*

(a) *$\hat{\mu}(t) = \hat{\nu}(t)$ for all $t \in \mathbb{R}^d$ if and only if $\mu = \nu$*

(b) *$\hat{\mu}(t)$ is a uniformly continuous function of $t \in \mathbb{R}^d$*

(c) *If $\omega = \mu * \nu$, then $\hat{\omega}(t) = \hat{\mu}(t)\hat{\nu}(t)$*

Theorem 1.3.6. *(Lévy Continuity Theorem) Suppose that $\mu_n, \mu \in \mathcal{M}^1$ and that $\hat{\mu}_n, \hat{\mu}$ are the corresponding characteristic functions. Then*

(a) *If $\mu_n \Rightarrow \mu$, then $\hat{\mu}_n \to \hat{\mu}$ uniformly on compact subsets of $\mathbb{R}^d$*

(b) *If $\hat{\mu}_n(t) \to \phi(t)$ for all $t \in \mathbb{R}^d$, where $\phi(t)$ is continuous at $t = 0$, then $\phi = \hat{\mu}$ for some $\mu \in \mathcal{M}^1$ and $\mu_n \Rightarrow \mu$*

Theorem 1.3.7. *(Fourier Inversion Theorem) Suppose $\mu \in \mathcal{M}^1$ has characteristic function $\hat{\mu}$. Then μ has a density*

$$f(x) = (2\pi)^{-d} \int e^{-i<t,x>} \hat{\mu}(t) dt$$

with respect to Lebesgue measure on $\mathbb{R}^d$ provided that $\hat{\mu}$ is integrable with respect to Lebesgue measure.

Proposition 1.3.8. *Suppose that $\mu \in \mathcal{M}^1$, $A \in \mathrm{L}(\mathbb{R}^d)$ and $a \in \mathbb{R}^d$. If $\omega = A\mu * \varepsilon_a$, then $\hat{\omega}(t) = \hat{\mu}(A^*t)e^{i<a,t>}$.*

Proof. This proof is an easy calculation using Definition 1.3.1. □

Definition 1.3.9. We say that a probability measure on $\mathbb{R}^1$ is *nondegenerate* if it is not a point mass. Suppose that X is a random vector on $\mathbb{R}^d$ with probability distribution μ. We say that X (or μ) is *full* if $<t, X>$ is nondegenerate for every $t \neq 0$ in $\mathbb{R}^d$.

Lemma 1.3.10. *A probability measure μ on $\mathbb{R}^1$ is degenerate if and only if $|\hat{\mu}(t)| = 1$ for all real t.*

Proof. If μ is degenerate, then $\mu = \varepsilon_a$ for some $a \in \mathbb{R}^1$ so that $|\hat{\mu}(t)| = |e^{ita}| = 1$. Conversely, if $|\hat{\mu}(t)| = 1$ for some $t \neq 0$ then $\hat{\mu}(t) = e^{ita}$ for some $a \in \mathbb{R}^1$ and then $1 = e^{-ita}\hat{\mu}(t) = \int \cos(tx)(\mu * \varepsilon_{-a})\{dx\}$ so that $\mu * \varepsilon_{-a}$ is supported on $\{x \in \mathbb{R}^1 : tx = k\pi \ \exists\ k \in \mathbb{Z}\}$. If this is true for all real t, then $\mu = \varepsilon_a$ so that μ is degenerate. □

Lemma 1.3.11. *A probability measure $\mu \in \mathcal{M}^1$ is full if and only if there is no $x \neq 0$ in $\mathbb{R}^d$ such that $|\hat{\mu}(tx)| = 1$ for all real t.*

Proof. Note that $\hat{\mu}(tx)$ is the characteristic function of the probability distribution $\pi_x\mu$ on $\mathbb{R}^1$, where $\pi_x y = <x, y>$. If μ is not full, then $\pi_x\mu$ is degenerate for some nonzero $x \in \mathbb{R}^d$, and then Lemma 1.3.10 shows that $|\hat{\mu}(tx)| = 1$ for all real t. Conversely, if there is some nonzero $x \in \mathbb{R}^d$ such that $|\hat{\mu}(tx)| = 1$ for all real t, then $\pi_x\mu$ is degenerate, so that μ is not full. □

Lemma 1.3.12. *If $\mu, \nu \in \mathcal{M}^1$ are full, then $\mu * \nu$ is full.*

Proof. Let $\omega = \mu * \nu$ so that $\hat{\omega} = \hat{\mu} \cdot \hat{\nu}$. If ω is not full, then Lemma 1.3.11 implies that for some $x \neq 0$ in $\mathbb{R}^d$ we have $|\hat{\omega}(tx)| = 1$ for all real t. Then $|\hat{\mu}(tx)| \cdot |\hat{\nu}(tx)| = 1$ for all real t and since $|\hat{\mu}| \leq 1, |\hat{\nu}| \leq 1$ this implies that $|\hat{\mu}(tx)| = 1$ and $|\hat{\nu}(tx)| = 1$ for all real t so that neither μ nor ν are full. This is a contradiction, hence $\mu * \nu$ is full. □

Lemma 1.3.13. *If $\mu \in \mathcal{M}^1$, then μ^n is full for each integer $n \geq 1$ if and only if μ is full.*

Proof. Lemma 1.3.11 shows that μ is not full if and only if there exists some $x \neq 0$ in $\mathbb{R}^d$ such that $|\hat{\mu}(tx)| = 1$ for all real t. This is true if and only if $|\hat{\mu}(tx)|^n = |\hat{\mu}^n(tx)| = 1$ for all real t, and this equivalent to μ^n not full. □

Lemma 1.3.14. *If $\mu \in \mathcal{M}^1$, then μ is full if and only if the symmetrization of μ is full.*

Proof. Let $\omega = \mu * \nu$ denote the symmetrization of μ as in Definition 1.3.2, so that $\hat{\nu}(x) = \hat{\mu}(-x)$ and $\hat{\omega} = \hat{\mu} \cdot \hat{\nu}$. Lemma 1.3.11 shows that μ is not full if and only if there exists some $x \neq 0$ in $\mathbb{R}^d$ such that $|\hat{\mu}(tx)| = 1$ for all real t. This is true if and only if $|\hat{\nu}(tx)| = |\hat{\mu}(-tx)| = 1$ for all real t, so that μ is full if and only if ν is full. Hence if μ is full, then ν is full and Lemma 1.3.12 implies that $\omega = \mu * \nu$ is full. If μ is not full, then $|\hat{\omega}(tx)| = |\hat{\mu}(tx)| \cdot |\hat{\mu}(-tx)| = 1$ for all real t so that ω is not full. □

Lemma 1.3.15. *If $\mu \in \mathcal{M}^1$ is full, then for some $\varepsilon > 0$ we have $|\hat{\mu}(t)| < 1$ for all $0 < \|t\| < \varepsilon$.*

Proof. Suppose that for some $t_n \neq 0$ we have $t_n \to 0$ and $|\hat{\mu}(t_n)| = 1$ for all n. Let μ_0 be the symmetrization of μ, so that $\hat{\mu}_0(t_n) = 1$ for all n. Then

$$0 = \int (1 - \cos<x, t_n>)\mu_0\{dx\}$$

so that $\text{supp}(\mu_0) \subseteq \{x : <x, t_n> = 2\pi k \ \exists\ k \in \mathbb{Z}\}$. Write $t_n = r_n\theta_n$ where $r_n > 0$ and $\|\theta_n\| = 1$. Since the unit sphere in $\mathbb{R}^d$ is compact there exist $\|\theta\| = 1$ and $n' \to \infty$ such that $\theta_{n'} \to \theta$. Let $\omega_{n'} = \theta_{n'} - \theta$ so that $\|\omega_{n'}\| \to 0$. Given $\delta > 0$ and $R < \infty$ choose n_0 such that $\|\omega_n\| \leq \delta/R$ and $2\pi r_n^{-1} \geq R$ for all $n = n' \geq n_0$. Then for all such n we have that $\|x\| \leq R$ and $<x, \theta_n> = 0$

implies $|\langle x, \theta\rangle| = |\langle x, \omega_n\rangle| \le R\|\omega_n\| \le \delta$, and also that $\langle x, \theta_n\rangle = 2\pi k r_n^{-1}$ for some integer $k \ne 0$ implies $\|x\| \ge R$. Hence for all $n = n' \ge n_0$ we have

$$\{x : |\langle x, \theta_n\rangle| = 2\pi k r_n^{-1} \ \exists\ k \in \mathbb{Z}\} \subseteq \{x : |\langle x, \theta\rangle| \le \delta\} \cup \{x : \|x\| \ge R\}$$

so that $\{x : \delta < |\langle x, \theta\rangle|, \|x\| < R\}$ is contained in a set of μ_0-measure zero. Since δ, R are arbitrary this implies that $\mu_0\{x : |\langle x, \theta\rangle| > 0\} = 0$ so μ_0 is not full, and then Lemma 1.3.14 implies that μ is not full. □

Example 1.3.16. The *standard normal distribution* μ_1 on $\mathbb{R}^1$ has density $f(x) = (2\pi)^{-1/2}\exp(-x^2/2)$ and characteristic function $\hat{\mu}_1(t) = \exp(-t^2/2)$. If $X = (X_1, \ldots, X_d)$ has i.i.d. standard normal components, then the distribution μ of X has the characteristic function

$$\hat{\mu}(t) = \prod_{i=1}^{n} \hat{\mu}_1(t_i) = \exp\left(-\tfrac{1}{2}\sum_{i=1}^{n} t_i^2\right) = \exp\left(-\tfrac{1}{2}\|t\|^2\right)$$

in the standard Euclidean norm, and $Y = AX + a$ has distribution $\nu = A\mu * \varepsilon_a$, where

$$\hat{\nu}(t) = \mu(A^*t)e^{i\langle a,t\rangle} = \exp\left(i\langle a, t\rangle - \tfrac{1}{2}\|A^*t\|^2\right) = \exp\left(i\langle a, t\rangle - \tfrac{1}{2}Q(t)\right)$$

where $Q(t) = \langle A^*t, A^*t\rangle = \langle t, Ct\rangle$ and $C = AA^*$ is symmetric so that Q is a quadratic form on $\mathbb{R}^d$. Lemma 1.3.11 implies that Y is full if and only if Q is positive definite. Taking first and second derivatives shows that $a = EY$ and C is the covariance matrix of Y.

Definition 1.3.17. If $\mu \in \mathcal{M}_0$, we call (1.6) the *Fourier transform* of the finite Borel measure μ. Note that $\hat{\mu}(t)$ always exists, and that $|\hat{\mu}(t)| \le \mu(\mathbb{R}^d) = \hat{\mu}(0)$ since μ is a finite measure.

Proposition 1.3.18. *Suppose that $\mu, \nu \in \mathcal{M}_0$ and that $\hat{\mu}, \hat{\nu}$ are the corresponding Fourier transforms. Then*

(a) *Inequality* (1.7) *holds for all $t_1, t_2 \in \mathbb{R}^d$*

(b) *$\hat{\mu}(t) = \hat{\nu}(t)$ for all $t \in \mathbb{R}^d$ if and only if $\mu = \nu$*

(c) *$\hat{\mu}(t)$ is a uniformly continuous function of $t \in \mathbb{R}^d$*

Proof. Let $c = \mu(D)$, and note that μ is the zero measure if and only if $\hat{\mu}(t) = 0$ for all $t \in \mathbb{R}^d$. If $c > 0$, then $\nu = c^{-1}\mu$ is a probability measure, $c\hat{\nu} = \hat{\mu}$, and the result follows from Propositions 1.3.4 and 1.3.5. □

Theorem 1.3.19. *(Lévy Continuity Theorem on $\mathcal{M}_0$) Suppose $\mu_n, \mu \in \mathcal{M}_0$ and that $\hat{\mu}_n, \hat{\mu}$ are the corresponding Fourier transforms. Then*

(a) *If $\mu_n \to \mu$, then $\hat{\mu}_n \to \hat{\mu}$ uniformly on compact subsets of $\mathbb{R}^d$*

(b) If $\hat{\mu}_n(t) \to \phi(t)$ *for all* $t \in \mathbb{R}^d$ *where* $\phi(t)$ *is continuous at* $t = 0$*, then* $\phi = \hat{\mu}$ *for some* $\mu \in \mathcal{M}_0$ *and* $\mu_n \to \mu$

Proof. Let $c_n = \mu_n(D)$ and $c = \mu(D)$. If $c > 0$, then $\mu_n \to \mu$ if and only if $c_n \to c$ and $\nu_n \Rightarrow \nu$, where $\nu_n = c_n^{-1}\mu_n$ for n large and $\nu = c^{-1}\mu$. In this case, the result follows from Theorem 1.3.19. If $\mu_n \to \mu$ and $c = 0$, then $|\hat{\mu}_n(t)| \le \mu_n(\mathbb{R}^d) \to 0 = \hat{\mu}(t)$ uniformly on $t \in \mathbb{R}^d$. If $|\hat{\mu}_n(t)| \to 0$ for all $t \in \mathbb{R}^d$, then $\mu_n(\mathbb{R}^d) = \hat{\mu}_n(0) \to 0$. □

Remark 1.3.20. In Chapter 3, we will use Fourier transform methods to prove convergence of certain σ-finite measures in $\mathcal{M}$. Because the Fourier transform of a σ-finite measure may not exist, the methods used there are significantly more difficult. The main idea will be to modify the integrand in (1.6) so that the integral converges for certain measures in $\mathcal{M}$ that satisfy a growth condition, see Theorem 3.1.11.

2

Linear Operators

Suppose that S_n, Y are random vectors on $\mathbb{R}^d$ and that $A_n S_n \Rightarrow Y$, where A_n are linear operators on $\mathbb{R}^d$. If A_n is invertible, then for large n the distribution of S_n is close to that of $A_n^{-1}Y$. If S_n is the sum of n independent random vectors on $\mathbb{R}^d$, then we have the basic setup for this book. In order to make use of this convergence, we will use some well-known properties of linear operators on $\mathbb{R}^d$. We collect those results in this chapter for the convenience of the reader, and to establish the notation to be used throughout the rest of the book. We also establish a convergence of types theorem that characterizes the class of possible norming operators A_n and limits Y.

2.1 OPERATOR NORMS

In this section, we lay out the basic definitions and properties of linear operators and operator norms. For simplicity, we restrict our attention to the space $\mathbb{R}^d$, but any finite dimensional real vector space has the same properties.

Definition 2.1.1. Let $\mathrm{GL}(\mathbb{R}^d) \subset \mathrm{L}(\mathbb{R}^d)$ denote the group of invertible linear operators, i.e., the set of all linear operators $A : \mathbb{R}^d \to \mathbb{R}^d$ for which A^{-1} exists, along with the group operations $(A, B) \mapsto AB$ and $A \mapsto A^{-1}$.

Since $A \in \mathrm{L}(\mathbb{R}^d)$ is invertible if and only if the determinant $\det(A) \neq 0$, $\mathrm{GL}(\mathbb{R}^d)$ is the preimage of the open set $\mathbb{R} \backslash \{0\}$ under this continuous mapping, and hence $\mathrm{GL}(\mathbb{R}^d)$ is an open subset of $\mathrm{L}(\mathbb{R}^d)$.

Definition 2.1.2. Fix any norm $||\cdot||$ for $\mathbb{R}^d$ and define the associated *operator norm* on $\mathrm{L}(\mathbb{R}^d)$ by letting

$$||A|| = \sup\{||Ax|| : ||x|| = 1\}. \tag{2.1}$$

Note that $||A|| = \sup\{||Ax||/||x|| : x \neq 0\}$. We say that $A_n \to A$ in $\mathrm{L}(\mathbb{R}^d)$ if $||A_n - A|| \to 0$. If $||\cdot||$ is the standard norm on $\mathbb{R}^d$ (i.e., the Euclidean norm associated with the standard basis for $\mathbb{R}^d$) we will call (2.1) the *standard norm* on $\mathrm{L}(\mathbb{R}^d)$. If $||\cdot||$ is the Euclidean norm associated with a given inner product on $\mathbb{R}^d$, we will call (2.1) the *Euclidean operator norm* associated with that inner product.

Proposition 2.1.3. *For any norm on $\mathbb{R}^d$ the associated operator norm is a norm on the real vector space $\mathrm{L}(\mathbb{R}^d)$ that satisfies*

(a) $||Ax|| \le ||A|| \cdot ||x||$ *for all* $A \in \mathrm{L}(\mathbb{R}^d)$ *and all* $x \in \mathbb{R}^d$

(b) *If* $A \in \mathrm{GL}(\mathbb{R}^d)$ *then* $||Ax|| \ge ||x||/||A^{-1}||$ *for all* $x \in \mathbb{R}^d$

(c) $||AB|| \le ||A|| \cdot ||B||$ *for all* $A, B \in \mathrm{L}(\mathbb{R}^d)$

(d) *If* $A \in \mathrm{GL}(\mathbb{R}^d)$ *then* $1/||A^{-1}|| = \inf\{||Ax|| : ||x|| = 1\}$

Proof. It is easy to check that $||A||$ is a norm on $\mathrm{L}(\mathbb{R}^d)$ such that (a) holds. If A^{-1} exists then $||x|| = ||A^{-1}Ax|| \le ||A^{-1}|| \cdot ||Ax||$ by (a), and then (b) follows. If $||x|| = 1$ then $||ABx|| \le ||A|| \cdot ||Bx||$ by (a), and then (c) follows from the definition. If $A \in \mathrm{GL}(\mathbb{R}^d)$ then $||A^{-1}|| = \sup\{||A^{-1}y||/||y|| : y \neq 0\} = \sup\{||A^{-1}Ax||/||Ax|| : x \neq 0\} = \sup\{||x||/||Ax|| : x \neq 0\}$ and (d) follows. □

Proposition 2.1.4. *Every norm $||\cdot||_0$ on $\mathbb{R}^d$ is continuous, and equivalent to the standard norm in the sense that for some $0 < C_1 < C_2 < \infty$ we have*

$$C_1||x||_0 \le ||x|| \le C_2||x||_0$$

for all $x \in \mathbb{R}^d$. Hence every operator norm gives $\mathrm{L}(\mathbb{R}^d)$ the same topology.

Proof. Given $x = \sum_j x_j e_j \in \mathbb{R}^d$ let $||x||_1 = \max_j |x_j|$ and note that $||x||_1 \le ||x|| \le \sqrt{d}||x||_1$. Then

$$||x||_0 \le \sum_{j=1}^{d} |x_j|\ ||e_j||_0 \le M \sum_{j=1}^{d} |x_j| \le Md||x||_1 \le Md||x||$$

where $M = \max_j ||e_j||_0$. If $x_n \to x$ in $\mathbb{R}^d$, then $||x_n - x||_0 \le Md||x_n - x|| \to 0$ so that $f(x) = ||x||_0$ is continuous. Then the unit sphere $S_0 = \{x : f(x) = 1\}$ in this norm is compact, hence $g(x) = ||x||$ must attain its maximum $C_2 < \infty$ and minimum $C_1 > 0$ on S_0. Given $x \in \mathbb{R}^d$ write $x = r\theta$, where $r = ||x||_0$ and $||\theta||_0 = 1$. Then $C_1||x||_0 \le r||\theta|| = ||x|| \le C_2||x||_0$. If we let $||A||, ||A||_0$ denote the associated operator norms, then $||A||_0 = \sup\{||Ax||_0 : ||x||_0 \le 1\} \le$

$\sup\{C_1^{-1}||Ax|| : C_2^{-1}||x|| \le 1\} = (C_2/C_1)||A||$. Similarly, $||A||_0 \ge (C_1/C_2)||A||$ so that $||A_n - A||_0 \to 0$ if and only if $||A_n - A|| \to 0$. □

Remark 2.1.5. If A^* is the transpose of A with respect to some inner product $<\cdot,\cdot>$ on $\mathbb{R}^d$ and $||\cdot|| = \sqrt{<\cdot,\cdot>}$ is the Euclidean norm associated with this inner product, then $||A||^2 = \max\{<Ax, Ax> : ||x|| = 1\} = \max\{<A^*x, A^*x> : ||x|| = 1\} = ||A^*||^2$ so that $||A|| = ||A^*||$ in the Euclidean operator norm associated with this inner product. Then for any other norm $||\cdot||_0$ we have that $||A^*||_0 \le (C_2/C_1)||A^*|| = (C_2/C_1)||A|| \le (C_2/C_1)^2||A||_0$ and likewise $||A^*||_0 \ge (C_1/C_2)^2||A||_0$. Hence $||A_n^*||_0 \to 0$ if and only if $||A_n||_0 \to 0$, and $||A_n^*||_0 \to \infty$ if and only if $||A_n||_0 \to \infty$.

Corollary 2.1.6. *If $<\cdot,\cdot>$ is any inner product on $\mathbb{R}^d$, then $<x_n, y_n> \to <x, y>$ whenever $x_n \to x$ and $y_n \to y$ in $\mathbb{R}^d$.*

Proof. Let $||x|| = \sqrt{<x,x>}$ be the norm associated with this inner product. Then $|<x_n, y_n> - <x, y>| \le |<x_n - x, y_n>| + |<x, y_n - y>| \le ||x_n - x|| \cdot ||y_n|| + ||x|| \cdot ||y_n - y|| \to 0$ by Proposition 1.1.22. □

Now we want to show that convergence in $\mathrm{L}(\mathbb{R}^d)$ is equivalent to uniform convergence $A_n x \to Ax$ on compact subsets of $\mathbb{R}^d$. We begin with a more general result, which will also be useful later. The argument in the proof of (b) will be used repeatedly throughout this book, where typically K is the unit sphere in some norm.

Proposition 2.1.7. *Suppose that $K \subset \mathbb{R}^d$ is compact. Then*

(a) If $f : K \to \mathbb{R}$ is continuous, then f is uniformly continuous

(b) If $f_n, f : K \to \mathbb{R}$ and f is continuous, then $f_n \to f$ uniformly if and only if $f_n(x_n) \to f(x)$ whenever $x_n \to x$ in $\mathbb{R}^d$

Proof. Given $\varepsilon > 0$ for each $x \in K$ there exists some $\delta > 0$ such that $||f(y) - f(x)|| < \varepsilon$ for all $y \in B_\delta(x) = \{y \in K : ||x - y|| < \delta\}$. Then the collection $(B_{\delta/2}(x) : x \in K)$ covers K compact. Choose a finite subcover $B_{\delta_1/2}(x_1), \ldots, B_{\delta_n/2}(x_n)$ and let $\delta = \min\{\delta_1, \ldots, \delta_n\}$. For any $z \in K$, we have $z \in B_{\delta_i/2}(x_i)$ for some i, and hence for any $y \in K$ such that $||y - z|| < \delta/2$ we have $y \in B_\delta(x_i)$ so that $||f(z) - f(y)|| \le ||f(z) - f(x_i)|| + ||f(x_i) - f(y)|| < 2\varepsilon$, which proves (a).

If $f_n \to f$ uniformly and $x_n \to x$ then $||f_n(x_n) - f(x)|| \le ||f_n(x_n) - f(x_n)|| + ||f(x_n) - f(x)|| \to 0$ since $||f_n(x_n) - f(x_n)|| \to 0$ by uniform convergence and $||f(x_n) - f(x)|| \to 0$ by continuity. Conversely, suppose by way of contradiction that $f_n(x_n) \to f(x)$ whenever $x_n \to x$ in K but that $f_n \to f$ is not uniform. Then there exists a sequence (x_n) in K for which $||f_n(x_n) - f(x_n)||$ is bounded away from zero. Since K is compact the sequence (x_n) contains a convergent subsequence $x_{n'} \to x \in K$. But then $||f_{n'}(x_{n'}) - f(x_{n'})|| \le ||f_{n'}(x_{n'}) - f(x)|| + ||f(x) - f(x_{n'})|| \to 0$, giving a contradiction. This proves (b). □

Proposition 2.1.8. *Suppose $A_n, A \in \mathrm{L}(\mathbb{R}^d)$. The following are equivalent:*

(a) $A_n \to A$ *in* $\mathrm{L}(\mathbb{R}^d)$

(b) $A_n x \to Ax$ *for every* $x \in \mathbb{R}^d$

(c) $A_n x \to Ax$ *uniformly on compact subsets of* $x \in \mathbb{R}^d$

(d) $A_n x_n \to Ax$ *whenever* $x_n \to x$ *in* $\mathbb{R}^d$

(e) The matrix representations for A_n with respect to some fixed basis for $\mathbb{R}^d$ converge elementwise to the matrix representation for A with respect to that same basis

Proof. If (a) holds, then $\|A_n - A\| \to 0$. Then $\|A_n x - Ax\| \leq \|A_n - A\| \cdot \|x\| \to 0$ by Proposition 2.1.3 (a) so that (b) holds. Proposition 2.1.7 (b) shows that (c) and (d) are equivalent. If $x_n \to x$ in $\mathbb{R}^d$, then $\|x_n\| \leq M < \infty$ for all n so $\|A_n x_n - Ax\| \leq \|A_n x_n - Ax_n\| + \|Ax_n - Ax\| \leq M\|A_n - A\| + \|A\| \cdot \|x_n - x\| \to 0$ so that (c) and (d) hold. If $b_1, \ldots, b_d$ is some basis for $\mathbb{R}^d$ then the ij entry of the matrix A in that basis is $a_{ij} = <e_i, Ae_j>$. Since $A_n e_j \to Ae_j$ by (b) we have $a_{ij}(n) = <e_i, A_n e_j> \to <e_i, Ae_j> = a_{ij}$ by Corollary 2.1.6. Given $\|x\| = 1$ in the norm associated with this basis, write $x = \sum_j x_j b_j$, where $|x_j| \leq 1$. Then $\|A_n x - Ax\|^2 \leq \sum_i (\sum_j (a_{ij}^n - a_{ij}) x_j)^2$ so that $\|A_n - A\|^2 \leq \sum_i (\sum_j (a_{ij}^n - a_{ij}))^2 \to 0$ as $n \to \infty$, and then (a) holds. □

Corollary 2.1.9. *If $Q_n(t) = <t, A_n t>$ are quadratic forms on $\mathbb{R}^d$ and if $Q_n(t) \to Q(t)$ for all $t \in \mathbb{R}^d$ then $A_n \to A$ in $\mathrm{L}(\mathbb{R}^d)$ and $Q(t) = <t, At>$ is a quadratic form on $\mathbb{R}^d$. Furthermore, $Q_n(t_n) \to Q(t)$ whenever $t_n \to t$ in $\mathbb{R}^d$.*

Proof. Proposition 1.1.20 (c) implies that the matrix A_n has ij entry $a_{ij}(n) = <e_i, A_n e_j> = B_n(e_i, e_j)$, where $B_n(x, y) = \frac{1}{2}[Q_n(x + y) - Q_n(x) - Q_n(y)]$. If we define $B(x, y) = \frac{1}{2}[Q(x + y) - Q(x) - Q(y)]$, then it follows that $a_{ij}(n) \to a_{ij} = B(e_i, e_j)$ for all i, j. Since B_n is linear in each argument and since $B_n(x, y) \to B(x, y)$ for all $x, y \in \mathbb{R}^d$ it follows that B is also linear in each argument. Then $B(x, y) = <x, Ay>$ for all $x, y \in \mathbb{R}^d$, where A has ij entry a_{ij}, so $Q(x) = B(x, x)$ is a quadratic form on $\mathbb{R}^d$. Then Proposition 2.1.8 yields $A_n \to A$ in $\mathrm{L}(\mathbb{R}^d)$ and $A_n t_n \to At$ whenever $t_n \to t$ in $\mathbb{R}^d$, and Corollary 2.1.6 implies that $Q_n(t_n) \to Q(t)$. □

Example 2.1.10. Proposition 2.1.8 implies that if $\|A_n x\| \to 0$ for all $x \in \mathbb{R}^d$, then this convergence is automatically uniform on compact subsets. However, if $\|A_n x\| \to \infty$ for all nonzero $x \in \mathbb{R}^d$, then this convergence is *not* automatically uniform on compact subsets of $\Gamma = \mathbb{R}^d \setminus \{0\}$. Consider the case $d = 2$ and let

$$A_n = \begin{pmatrix} 1 + n^2 & n \\ n & 1 \end{pmatrix} \quad \text{and} \quad x_n = \begin{pmatrix} -1/n \\ 1 \end{pmatrix} \quad \text{so that} \quad A_n x_n = \begin{pmatrix} -1/n \\ 0 \end{pmatrix}.$$

Then $\|A_n x\| \to \infty$ for all $x \in \Gamma$ but $\|A_n x_n\| \to 0$ so that the convergence $\|A_n x\| \to \infty$ is not uniform on compact subsets of Γ.

Remark 2.1.11. If $A_n \to A$ in $\mathrm{L}(\mathbb{R}^d)$ and $a_n \to a$ in $\mathbb{R}^d$, then it follows from Proposition 2.1.8 that the mappings $T_n x = A_n x + a_n$ converge to $Tx = Ax + a$ uniformly on compact subsets of $x \in \mathbb{R}^d$. If X_n, X are random vectors on $\mathbb{R}^d$ and $X_n \Rightarrow X$, then the Theorem 1.2.8 implies that $T_n X_n \Rightarrow TX$. If $\mu_n \to \mu$ in $\mathcal{M}$, then Theorem 1.2.23 implies that $T_n \mu_n \to T\mu$.

Definition 2.1.12. Given $A_n, A \in \mathrm{GL}(\mathbb{R}^d)$ we will write $A_n \to A$ provided that $\|A_n - A\| \to 0$ in some operator norm. Since $\mathrm{GL}(\mathbb{R}^d)$ is an open subset of $\mathrm{L}(\mathbb{R}^d)$, we also have $A_n \to A$ in $\mathrm{L}(\mathbb{R}^d)$. Since the group operations $A \mapsto A^{-1}$ and $(A, B) \mapsto AB$ in $\mathrm{GL}(\mathbb{R}^d)$ are continuous, $\mathrm{GL}(\mathbb{R}^d)$ is a *topological group* with this topology.

The following decomposition theorem forms the basis for many important representation theorems, both in the theory of regular variation and in our study of limit theorems for sums of independent random vectors.

Definition 2.1.13. If $A \in \mathrm{L}(\mathbb{R}^d)$, then we can also consider A as a linear operator on the complex vector space $\mathbb{C}^d = \{x + iy : x, y \in \mathbb{R}^d\}$ by setting $A(x + iy) = Ax + iAy$. If $Ax = \lambda x$ for some $x \in \mathbb{R}^d$ and some $\lambda \in \mathbb{C}$, then λ is an *eigenvalue* of A and x is the corresponding *eigenvector* of A. The *spectrum* of A is the set of eigenvalues, and the *real spectrum* of A consists of the real parts of the eigenvalues of A. The *spectral radius* $r(A) = \max\{|\lambda|\}$, where $\lambda = a + ib$ ranges over the eigenvalues of A and $|\lambda|^2 = a^2 + b^2$ is the complex absolute value. The *minimal polynomial* $f(x)$ of a linear operator $A \in \mathrm{L}(\mathbb{R}^d)$ is a polynomial of lowest degree such that the linear operator $f(A) = 0$. The eigenvalues of A are the roots of f. We say that $\mathbb{R}^d$ is the *direct sum* of subspaces $V_1, \ldots, V_p$ and we write $\mathbb{R}^d = V_1 \oplus \cdots \oplus V_p$ if for every $x \in \mathbb{R}^d$ there exist unique $x_i \in V_i$ such that $x = x_1 + \cdots + x_p$. Then for any linear operators A_i on V_i we define the *direct sum* $A = A_1 \oplus \cdots \oplus A_p$ to be the linear operator on $\mathbb{R}^d$ defined by $Ax = A_1 x_1 + \cdots + A_p x_p$.

Theorem 2.1.14. *(Spectral Decomposition) Given $A \in \mathrm{L}(\mathbb{R}^d)$, factor the minimal polynomial of A into $f_1(x) \cdots f_p(x)$ where all roots of f_i have real part a_i and $a_i < a_j$ for $i < j$. Define $V_i = \mathrm{Ker}(f_i(A))$. Then $V_1 \oplus \cdots \oplus V_p$ is a direct sum decomposition of $\mathbb{R}^d$ into A-invariant subspaces, and we may write $A = A_1 \oplus \cdots \oplus A_p$, where $A_i : V_i \to V_i$ and every eigenvalue of A_i has real part equal to a_i. The matrix for A in an appropriate basis is then block-diagonal with p blocks, the ith block corresponding to the matrix for A_i.*

Proof. This result is a special case of the primary decomposition theorem of linear algebra (see, e.g., [31]). □

Definition 2.1.15. We will call the subspaces $V_1, \ldots, V_p$ in Theorem 2.1.14 the *spectral decomposition* of $\mathbb{R}^d$ with respect to A. We will also call the operators $A_1, \ldots, A_p$ the *spectral decomposition* of A. Note that $a_1 < \cdots < a_p$

is the real spectrum of A. We say that A_i is *spectrally simple*, since every eigenvalue of A_i has the same real part. In other words, the real spectrum of A_i consists of a single value $a_i \in \mathbb{R}$.

Theorem 2.1.16. *(Jordan decomposition) For any $A \in \mathrm{L}(\mathbb{R}^d)$, there exists a basis for $\mathbb{R}^d$ with respect to which the matrix of A is block diagonal, where every block is of the form*

$$\begin{pmatrix} a & 0 & 0 & \cdots & 0 \\ 1 & a & 0 & \cdots & 0 \\ 0 & 1 & a & & \vdots \\ \vdots & & \ddots & \ddots & \\ 0 & \cdots & & 1 & a \end{pmatrix} \quad \text{or} \quad \begin{pmatrix} B & 0 & 0 & \cdots & 0 \\ I & B & 0 & \cdots & 0 \\ 0 & I & B & & \vdots \\ \vdots & & \ddots & \ddots & \\ 0 & \cdots & & I & B \end{pmatrix} \tag{2.2}$$

where a is a real eigenvalue of A in the first case, and in the second case

$$B = \begin{pmatrix} a & -b \\ b & a \end{pmatrix} \quad \text{and} \quad I = \begin{pmatrix} 1 & 0 \\ 0 & 1 \end{pmatrix} \tag{2.3}$$

where $a \pm ib$ is a complex conjugate pair of eigenvalues of A.

Proof. This is a standard result (see, e.g., [60] p. 129). □

Definition 2.1.17. A linear operator $S \in \mathrm{L}(\mathbb{R}^d)$ is said to be *semisimple* if its minimal polynomial is the product of distinct prime factors. This is equivalent to the statement that the linear operator is diagonalizable over the complex numbers. In this case, we will say that the eigenvalues of S are *simple*. A linear operator $N \in \mathrm{L}(\mathbb{R}^d)$ is *nilpotent* if $N^k = 0$ for some k.

Theorem 2.1.18. *($S + N$ decomposition) Any linear operator $A \in \mathrm{L}(\mathbb{R}^d)$ can be written uniquely in the form $A = S + N$ where S is semisimple, N is nilpotent, and $NS = SN$. Furthermore, both S and N are polynomials in A.*

Proof. See, e.g., [61]. □

Theorem 2.1.19. *If A is orthogonal, then every eigenvalue $\lambda = a + ib$ of A is simple with $a^2 + b^2 = 1$.*

Proof. See, e.g., [31] p. 286. □

Proposition 2.1.20. *For A symmetric there exists an orthonormal basis of eigenvectors of A, and the corresponding eigenvalues are all real. A is nonnegative definite if and only if every eigenvalue is nonnegative, and in this case A has a unique nonnegative definite square root.*

Proof. This result follows immediately from [31] p. 286. □

Remark 2.1.21. In Theorem 2.1.14, the component A_i in the spectral decomposition is the direct sum of the Jordan blocks with $a = a_i$. For a real eigenvalue a, the Jordan block can be written in the form $S + N$ where $S = aI$.

For complex eigenvalues, we can take $S = \mathrm{diag}(B, \ldots, B)$. Theorem 2.1.19 shows that each block in the Jordan decomposition of an orthogonal linear operator is a 1×1 block $a = \pm 1$ or a 2×2 block B with $a^2 + b^2 = 1$.

2.2 EXPONENTIAL OPERATORS AND POWERS

In this section, we summarize the relevant facts about exponential operators. We also discuss power operators, which are related by a logarithmic transformation, just as in the scalar case. At the end of this section, we discuss exponential operators as mappings from the tangent space.

Definition 2.2.1. Given a linear operator $A \in \mathrm{L}(\mathbb{R}^d)$ we define the *exponential operator*

$$\exp(A) = \sum_{k=0}^{\infty} \frac{A^k}{k!} \tag{2.4}$$

where $A^0 = I$ the *identity operator* on $\mathbb{R}^d$. We also define the family of *power operators*

$$t^A = \exp(A \log t) \tag{2.5}$$

for any $A \in \mathrm{L}(\mathbb{R}^d)$ and $t > 0$.

Proposition 2.1.3 implies that

$$\| \exp(A) \| \leq \sum_{k=1}^{\infty} \frac{\|A\|^k}{k!} = \exp(\|A\|)$$

so that the infinite series in (2.4) always converges. Hence t^A is well defined for any $t > 0$. In applications to limit theorems in probability, we will be primarily concerned with t^A rather than the exponential operator. Exponential operators are important in the study of differential equations because $x(s) = x(0)\exp(As)$ solves the initial value problem $dx/ds = Ax$. Most of the results in this section can be obtained directly from that literature by a simple reparameterization $s = \log t$, (see, e.g., [60]).

Proposition 2.2.2. *If $A, B \in \mathrm{L}(\mathbb{R}^d)$ and $s, t > 0$, then*

(a) $t^A = I$ *when* $t = 1$

(b) If $AB = BA$, *then* $t^A t^B = t^{A+B}$

(c) $s^A t^A = (st)^A$

(d) $t^{-A} = (1/t)^A = (t^A)^{-1}$

(e) *If $C = ABA^{-1}$, then $t^C = At^BA^{-1}$*

(f) *$Ax = \lambda x$ for some $x \in \mathbb{R}^d$ and some $\lambda \in \mathbb{C}$ if and only if $t^A x = t^\lambda x$ for all $t > 0$*

Proof. If $t = 1$, then $t^A = \exp(A \log 1) = \exp(0) = I$ from (2.4). If $XY = YX$, then $\exp(X + Y) = \exp(X)\exp(Y)$ follows directly from (2.4). Substitute $X = A \log t$, $Y = B \log t$ to get (b), and $X = A \log s$, $Y = A \log t$ to get (c). If $Y = -X$, then $\exp(X + Y) = \exp(0) = I$. Substitute $X = A \log t$ and $Y = -A \log t = A \log(1/t)$ to get (d). Compute $\exp(AXA^{-1}) = A\exp(X)A^{-1}$ directly from (2.4), and substitute $X = B \log t$ to get (e). If $Ax = \lambda x$, then $\exp(As)x = e^{\lambda s}x$ for all real s by (2.4). Conversely, if $\exp(As)x = e^{\lambda s}x$ for all real s, then $(I + As + o(s))x = (1 + \lambda s + o(s))x$ as $s \to 0$ implies $Ax = \lambda x$. Substitute $s = \log t$ to get (f). □

Lemma 2.2.3. *For any $A \in \mathrm{L}(\mathbb{R}^d)$, there exists a basis for $\mathbb{R}^d$ in which every coordinate of $y = t^A x$ is a linear combination of terms of the form*

$$t^a (\log t)^j \sin(b \log t) \quad \text{and} \quad t^a (\log t)^j \cos(b \log t) \tag{2.6}$$

where $a + ib$ is an eigenvalue of A, $0 \le j < d$, and the coefficients in this linear combination are independent of both $x \in \mathbb{R}^d$ and $t > 0$.

Proof. Theorem 2.1.16 yields a basis in which the matrix of A is block diagonal, where each block A_i is of the form (2.2). Write $A_i = S + N$ where S is semisimple and N is nilpotent, and $SN = NS$. In the first case, $t^S = t^a I$ and

$$t^N = \begin{pmatrix} 1 & 0 & 0 & \cdots & 0 \\ \log t & 1 & 0 & \cdots & 0 \\ (\log t)^2/2! & \log t & 1 & & \vdots \\ \vdots & & \ddots & \ddots & \\ (\log t)^k/k! & \cdots & & \log t & 1 \end{pmatrix}$$

so that

$$t^{A_i} = t^S t^N = \begin{pmatrix} t^a & 0 & 0 & \cdots & 0 \\ t^a \log t & t^a & 0 & \cdots & 0 \\ t^a (\log t)^2/2! & t^a \log t & t^a & & \vdots \\ \vdots & & \ddots & \ddots & \\ t^a (\log t)^k/k! & \cdots & & t^a \log t & t^a \end{pmatrix}$$

where $k+1$ is the rank of this block. If $x_i = (x_1, \ldots, x_{k+1})$ are the coordinates for this block and $z(t) = t^{A_i}x = (z_1(t), \ldots, z_{k+1}(t))$, then

$$z_j(t) = \sum_{n=0}^{j-1} \frac{t^a (\log t)^n}{n!} x_{j-n} \tag{2.7}$$

for all $j = 1, \ldots, k+1$. In the second case, $B = aI + bR$ where I is the 2×2 identity matrix and R represents the counterclockwise rotation through an angle $\pi/2$, with $R^2 = -I$, $R^3 = -R$ and $R^4 = I$. Then $t^{aI} = t^a I$ and

$$t^{bR} = \begin{pmatrix} \cos(b \log t) & -\sin(b \log t) \\ \sin(b \log t) & \cos(b \log t) \end{pmatrix}$$

by a Taylor series expansion. Then $t^A = t^S t^N$, where t^S is block diagonal with 2×2 blocks of this form, and

$$t^N = \begin{pmatrix} I & 0 & 0 & \cdots & 0 \\ N_1(t) & I & 0 & \cdots & 0 \\ N_2(t) & N_1(t) & I & & \vdots \\ \vdots & & \ddots & \ddots & \\ N_k(t) & \cdots & & N_1(t) & I \end{pmatrix}$$

where

$$N_j(t) = \begin{pmatrix} (\log t)^j/j! & 0 \\ 0 & (\log t)^j/j! \end{pmatrix}$$

and $2k+2$ is the rank of this block. If $u = (x_1, y_1, \ldots, x_{k+1}, y_{k+1})$ are the coordinates for this block and $t^{A_i}u = (z_1(t), w_1(t), \ldots, z_{k+1}(t), w_{k+1}(t))$, then

$$\begin{aligned} z_j(t) &= \sum_{n=0}^{j-1} \frac{t^a (\log t)^n}{n!} (\cos(b \log t) x_{j-n} - \sin(b \log t) y_{n-j}) \\ w_j(t) &= \sum_{n=0}^{j-1} \frac{t^a (\log t)^n}{n!} (\sin(b \log t) x_{j-n} + \cos(b \log t) y_{n-j}) \end{aligned} \tag{2.8}$$

for all $j = 1, \ldots, k+1$. □

Theorem 2.2.4. *Suppose that $A \in \mathrm{L}(\mathbb{R}^d)$, and that $\|\cdot\|$ is any norm on $\mathbb{R}^d$. If all eigenvalues of A have real part greater than α, then*

$$t^{-\alpha} \|t^A x\| \to \infty \tag{2.9}$$

as $t \to \infty$ uniformly on compact subsets of $x \neq 0$ in $\mathbb{R}^d$. If all eigenvalues of A have real part less than β, then

$$t^{-\beta} \|t^A x\| \to 0 \tag{2.10}$$

as $t \to \infty$ uniformly on compact subsets of $x \in \mathbb{R}^d$.

Proof. Since the proof of both parts is similar we will only prove (a). Let $(\cdot, \cdot)$ denote the inner product from Proposition 1.1.20 (b) that makes the basis from Theorem 2.1.16 orthonormal. Let $\|u\|$ denote the norm associated with this inner product, so that $\|u\|^2 = \sum_i u_i^2$ in these coordinates. Given $x(t) \to x \neq 0$ in $\mathbb{R}^d$ let $y(t) = t^{-\alpha} t^A x(t)$ and use Lemma 2.2.3 to write $y_j(t) = \sum_i c_{ij}(t) x_i(t)$ in these coordinates, where each $c_{ij}(t)$ is a linear combination of terms of the form (2.6) and $a > \alpha$ ranges over the real spectrum of A. If (2.7) holds, then

$$|z_j(t)|^2 = \sum_{n=0}^{j-1} \frac{t^{2(a-\alpha)} (\log t)^{2n}}{(n!)^2} |x_{j-n}(t)|^2 \tag{2.11}$$

and if $x_{j-n}(t) \to x_{j-n} \neq 0$, then $|z_j(t)|^2 \to \infty$. If (2.8) holds, then

$$|z_j(t)|^2 + |w_j(t)|^2 = \sum_{n=0}^{j-1} \frac{t^{2a} (\log t)^{2n}}{(n!)^2} \left(|x_{j-n}(t)|^2 + |y_{n-j}(t)|^2\right) \tag{2.12}$$

and if $(x_{j-n}(t), y_{j-n}(t)) \to (x_{j-n}, y_{j-n}) \neq (0,0)$, then $|z_j(t)|^2 + |w_j(t)|^2 \to \infty$. Since $\|y(t)\|^2$ is the sum of terms of these two forms, with each term nonnegative, and since $x \neq 0$, it follows that (2.9) holds for this norm. Then in view of Proposition 2.1.4 it also holds for any norm. □

Now we apply the spectral decomposition of Theorem 2.1.14 to sharpen the results of Theorem 2.2.4.

Corollary 2.2.5. *Suppose that $A \in \mathrm{L}(\mathbb{R}^d)$, and that $\|\cdot\|$ is any norm on $\mathbb{R}^d$. Let $V_1, \ldots, V_p$ be the spectral decomposition of $\mathbb{R}^d$ with respect to A, and $A = A_1 \oplus \cdots \oplus A_p$ the spectral decomposition of A.*

(a) If $\alpha < a_i$, then

$$t^{-\alpha} \|t^A x\| \to \infty \tag{2.13}$$

as $t \to \infty$ uniformly on compact subsets of $x \neq 0$ in $V_i \oplus \cdots \oplus V_p$

(b) If $\beta > a_i$, then

$$t^{-\beta} \|t^A x\| \to 0 \tag{2.14}$$

as $t \to \infty$ uniformly on compact subsets of $x \in V_1 \oplus \cdots \oplus V_i$

Proof. The linear operator $B = A_i \oplus \cdots \oplus A_p$ on the vector space $V = V_i \oplus \cdots \oplus V_p$ has eigenvalues with real parts $a_i < \cdots < a_p$ and if $x \in V$ then $Ax = Bx$ so $t^A x = t^B x$. Then part (a) follows immediately from Theorem 2.2.4 applied to $t^B x$. The proof of part (b) is similar. □

Definition 2.2.6. Given $x \in \mathbb{R}^d$ and a closed set $C \subset \mathbb{R}^d$ we will write $\|x - C\| = \min\{\|x - y\| : y \in C\}$. For a sequence (x_n) in $\mathbb{R}^d$, we will write $x_n \rightharpoonup C$ if $\|x_n - C\| \to 0$. This is equivalent to saying that (x_n) is limit point compact and every limit point lies in C.

Corollary 2.2.7. *Suppose that $A \in \mathrm{L}(\mathbb{R}^d)$, and that $\|\cdot\|$ is any norm on $\mathbb{R}^d$. Let $L_0 = \{0\}$ and $L_i = V_1 \oplus \cdots \oplus V_i$, where $\mathbb{R}^d = V_1 \oplus \cdots \oplus V_p$ is the spectral decomposition of $\mathbb{R}^d$ relative to A. Then*

(a) if $x \in L_i$, then $t^{-\rho}\|t^A x\| \to 0$ for all $\rho > a_i$

(b) if $x \notin L_i$, then $t^{-\rho}\|t^A x\| \to \infty$ for all $\rho < a_{i+1}$

(c) if $x \in L_i \setminus L_{i-1}$, then $t^A x/\|t^A x\| \rightharpoonup V_i$

Furthermore, each convergence is uniform on compact subsets.

Proof. Part (a) follows immediately from Corollary 2.2.5. Let $b_1, \ldots, b_d$ denote the basis from Theorem 2.1.16 and let $(\cdot,\cdot)$ denote the inner product from Proposition 1.1.20 (b) that makes this an orthonormal basis. Then $V_1, \ldots, V_p$ are mutually orthogonal in this inner product. Let $\|x\|$ denote the norm associated with this inner product. Write $A = A_1 \oplus \cdots \oplus A_p$ and $x = x_1 + \cdots + x_p$ with respect to this direct sum decomposition. If $x(t) \to x \in \mathbb{R}^d \setminus L_i$, then $x(t) \in \mathbb{R}^d \setminus L_i$ for all large t since this is an open set. Define $j = j(x) = \max\{i : x_i \neq 0\}$ and choose $a_{j-1} < \alpha < a_j$. Corollary 2.2.5 (a) implies that $t^{-\alpha}\|t^{A_j}x_j(t)\| \to \infty$, and Corollary 2.2.5 (b) implies that $t^{-\alpha}\|t^{A_i}x_i(t)\| \to 0$ for all $i < j$. Then $\|t^A x(t)\|^2 = \sum_i \|t^{A_i}x_i(t)\|^2 \sim \|t^{A_j}x_j(t)\|^2$ as $t \to \infty$. If $\rho < a_{i+1} \leq a_j$ we have $t^{-\rho}\|t^A x(t)\| \sim t^{-\rho}\|t^{A_j}x_j(t)\| \to \infty$ and this along with Proposition 2.1.4 proves (b). If $x(t) \to x \in L_j \setminus L_{j-1}$, then $x(t) \in L_j \setminus L_{j-1} = \{x \in \mathbb{R}^d : j(x) = j\}$ for all t large, so

$$\frac{t^A x(t)}{\|t^A x(t)\|} \sim \frac{\sum_{i=1}^{j} t^{A_i}x_i(t)}{\|t^{A_j}x_j(t)\|} \rightharpoonup V_j$$

since $t^{A_i}x_i(t)/\|t^{A_j}x_j(t)\| \to 0$ for $i < j$, and (c) follows. □

Remark 2.2.8. Corollary 2.2.7 shows that $\|t^A x\|$ grows like the power function t^{a_i} and $t^A x/\|t^A x\| \rightharpoonup V_i$ when $j(x) = i$. In Section 4.3, we will generalize these results to *regularly varying* sequences of linear operators, whose growth behavior is asymptotically equivalent to the power operators t^A.

Now we consider exponential and power mappings from a different point of view. The following material will be important in Chapters 5 and 6.

Definition 2.2.9. Use the matrix representations of linear operators to identify $\mathrm{L}(\mathbb{R}^d)$ with $\mathbb{R}^{d\times d}$. A *smooth curve* $x(t)$ on $\mathcal{H} \subseteq \mathrm{GL}(\mathbb{R}^d)$ is a differentiable function $x : \mathbb{R} \to \mathcal{H}$. If $\mathcal{H}$ is a closed subgroup of $\mathrm{GL}(\mathbb{R}^d)$ the *tangent space* $T\mathcal{H}$ at the identity consists of all derivatives $x'(0)$ of smooth curves $x(t)$ on $\mathcal{H}$ for which $x(0) = I$.

Proposition 2.2.10. *Suppose that $\mathcal{H}$ is a closed subgroup of $\mathrm{GL}(\mathbb{R}^d)$ and $T\mathcal{H}$ is the tangent space.*

(a) $T\mathcal{H}$ is a linear subspace of $\mathrm{L}(\mathbb{R}^d)$

(b) $\mathrm{L}(\mathbb{R}^d)$ *is the tangent space of* $\mathrm{GL}(\mathbb{R}^d)$

(c) $t^A \in \mathcal{H}$ *for all* $A \in T\mathcal{H}$ *and all* $t > 0$

(d) there exists an open set $U \subset \mathcal{H}$ *such that* $I \in U$ *and every* $B \in U$ *satisfies* $B = t^A$ *for some* $A \in T\mathcal{H}$ *and some* $t > 0$

Proof. If $A \in T\mathcal{H}$, then $A = x'(0)$ for some smooth curve on $\mathcal{H}$. Given $\lambda \in \mathbb{R}$ let $y(t) = x(\lambda t)$. Then y is a smooth curve on $\mathcal{H}$ with $y(0) = I$ and $y'(0) = \lambda A$. If $A, B \in T\mathcal{H}$, then $A = x'(0)$ and $B = y'(0)$, so if $z = x \cdot y$ then $z'(t) = x'(t)y(t) + x(t)y'(t)$, and hence $A + B = z'(0)$ so that (a) holds. If $A \in \mathrm{L}(\mathbb{R}^d)$, then it follows from Proposition 2.2.2 that $x(t) = \exp(tA)$ is a smooth curve on $\mathrm{GL}(\mathbb{R}^d)$ with $x(0) = I$ and a computation shows that $x'(0) = A$, so that (b) holds. If $A \in T\mathcal{H}$, then $A = x'(0)$ for some smooth curve on $\mathcal{H}$. Let $X_n = x(1/n)$ so that $n(X_n - I) \to A$ as $n \to \infty$. Then $G_n = n(X_n - I) - A \to 0$ and $X_n = I + (A + G_n)/n$. As $n \to \infty$ we have $X_n^n = (I + (A + G_n)/n)^n \to \exp(A)$ since $\mathcal{H}$ is a closed group, which proves (c). Define

$$\log(I - A) = -\sum_{k=1}^{\infty} \frac{A^k}{k}$$

and note that if $\|A\| < 1$, then

$$\|\log(I - A)\| \leq \sum_{k=1}^{\infty} \frac{\|A\|^k}{k}$$

converges, so that $\log A$ is well defined whenever $\|I - A\| < 1$. A computation shows that if $X = \log A$, then $\exp(X) = A$. Hence if $A \in \mathcal{H}$ then (c) implies that $X \in T\mathcal{H}$, which proves (d). □

Proposition 2.2.11. *The mapping* $\exp : \mathrm{L}(\mathbb{R}^d) \to \mathrm{GL}(\mathbb{R}^d)$ *is continuous. Furthermore*

(a) $(t, A) \mapsto t^A$ *is jointly continuous*

(b) $(t, A, x) \mapsto t^A x$ *is jointly continuous*

(c) If every eigenvalue of A *has real part greater than* α*, then for any* $t_0 > 0$ *there exists* $C > 0$ *such that* $\|t^A x\| \geq C t^{\alpha} \|x\|$ *for all* $t \geq t_0$ *and all* $x \in \mathbb{R}^d$

(d) If every eigenvalue of A *has real part less than* β*, then for any* $t_0 > 0$ *there exists* $C > 0$ *such that* $\|t^A x\| \leq C t^{\beta} \|x\|$ *for all* $t \geq t_0$ *and all* $x \in \mathbb{R}^d$

Proof. If $X_n \to X$ in $\mathrm{L}(\mathbb{R}^d)$, then $\|X_n\| \to \|X\|$, and hence for some $M > 0$ we have $\|X\| \le M$ and $\|X_n\| \le M$ for all n. Since

$$\begin{aligned}\|X_n^k - X^k\| &= \|X_n^k - X_n^{k-1}X + X_n^{k-1}X - X_n^{k-2}X^2 + \cdots - X^k\| \\ &= \|X_n^{k-1}(X_n - X) + X_n^{k-2}(X_n - X)X + \cdots + (X_n - X)X^{k-1}\| \\ &\le \|X_n\|^{k-1}\|X_n - X\| + \cdots + \|X_n - X\| \cdot \|X\|^{k-1} \\ &\le kM^{k-1}\|X_n - X\|\end{aligned}$$

we have

$$\|\exp(X_n) - \exp(X)\| \le \sum_{k=1}^{\infty} \frac{\|X_n^k - X^k\|}{k!} \le e^M \|X_n - X\| \to 0$$

so that exp is continuous. Hence if $t_n \to t > 0$ and $A_n \to A$ in $\mathrm{L}(\mathbb{R}^d)$, then $t_n^{A_n} = \exp(A_n \log t_n) \to \exp(A \log t) = t^A$ so that (a) holds. Then Proposition 2.1.8 shows that (b) holds. If every eigenvalue of A has real part greater than α, then Theorem 2.2.4 implies that $t^{-\alpha}\|t^A x\| \to \infty$ uniformly on $\|x\| = 1$. Choose t_1 such that $\|t^A x\| \ge t^\alpha$ for all $t \ge t_1$ and all $\|x\| = 1$. Part (b) along with Definition 1.1.21 and Proposition 2.1.4 shows that $(t, x) \to \|t^A x\|$ is positive and continuous, and hence $\|t^A x\| \ge m > 0$ for $t_0 \le t \le t_1$ and $\|x\| = 1$. Then (c) holds with $C = \min\{1, mt_0^{-\alpha}, mt_1^{-\alpha}\}$. The proof of (d) is similar. □

Remark 2.2.12. If we use the matrix representations of linear operators to identify $\mathrm{L}(\mathbb{R}^d)$ with $\mathbb{R}^{d\times d}$, then the group operations on $\mathrm{GL}(\mathbb{R}^d)$ are infinitely differentiable so that $\mathrm{GL}(\mathbb{R}^d)$ is a *Lie group*. The tangent space $\mathrm{L}(\mathbb{R}^d)$ together with the *Lie bracket* $[A, B] = AB - BA$ is a *Lie algebra*. The exponential maps the tangent space into the Lie group. See [28, 68, 149, 186] for more details and properties.

2.3 CONVERGENCE OF TYPES

Convergence of types is a simple but powerful limit theorem that is useful in many branches of probability theory. In this section, we present the multivariable version, which is less well known than convergence of types for probability distributions on the real line.

Definition 2.3.1. If X, Y are random variables on $\mathbb{R}$ with probability distributions μ, ν respectively, we say that X, Y (or μ, ν) are the same *type* if $\nu = a\mu * \varepsilon_b$ for some $a > 0$ and some $b \in \mathbb{R}$. In other words, Y and $aX + b$ have the same distribution. If X, Y are random vectors on $\mathbb{R}^d$ with probability distributions μ, ν respectively, we say that X, Y (or μ, ν) are the same *type* if $\nu = A\mu * \varepsilon_b$ for some $A \in \mathrm{GL}(\mathbb{R}^d)$ and some $b \in \mathbb{R}^d$. In other words, Y and $AX + b$ have the same distribution.

Theorem 2.3.2. *(Convergence of Types in $\mathbb{R}$) Suppose μ_n, μ, ν are probability measures on $\mathbb{R}$ with μ, ν nondegenerate. If $a_n\mu_n * \varepsilon_{b_n} \Rightarrow \mu$ for some $a_n > 0$ and $b_n \in \mathbb{R}$, then $a'_n\mu_n * \varepsilon_{b'_n} \Rightarrow \nu$ for some $a'_n > 0$ and $b'_n \in \mathbb{R}$ if and only if $\nu = a\mu * \varepsilon_b$ and $a'_n/a_n \to a$ for some $a > 0$ and $b \in \mathbb{R}^d$.*

Proof. See, e.g., Billingsley [18] Theorem 14.2. □

Suppose X_n, X, Y are random variables on $\mathbb{R}$ and that X, Y are nondegenerate. If $a_nX_n + b_n \Rightarrow X$ and $a'_nX_n + b'_n \Rightarrow Y$, then Theorem 2.3.2 shows that X, Y are of the same type. Now we derive a similar result for random vectors. Convergence of types in $\mathbb{R}^d$ is more complicated because of possible symmetries of the limit.

Definition 2.3.3. We say that $A \in \mathrm{L}(\mathbb{R}^d)$ is a *symmetry* for $\mu \in \mathcal{M}^1$ if $\mu = A\mu * \varepsilon_a$ for some $a \in \mathbb{R}^d$. We denote the collection of all symmetries of μ by $\mathcal{S}^1(\mu)$.

Proposition 2.3.4. *For any $\mu \in \mathcal{M}^1$ and any $T \in \mathrm{GL}(\mathbb{R}^d)$, $b \in \mathbb{R}^d$ we have $\mathcal{S}^1(T\mu * \varepsilon_b) = T\mathcal{S}^1(\mu)T^{-1}$.*

Proof. Let $\mu' = T\mu * \varepsilon_b$. If $\mu = A\mu * \varepsilon_a$ and $s = Ta + b - TAT^{-1}b$, then $TAT^{-1}\mu' * \varepsilon_s = T(A\mu * \varepsilon_a) * \varepsilon_b = \mu'$ so $T\mathcal{S}^1(\mu)T^{-1} \subseteq \mathcal{S}^1(\mu')$. Since $\mu = T^{-1}\mu' * \varepsilon_{-T^{-1}b}$ it follows from this inclusion that $T^{-1}\mathcal{S}^1(\mu')T \subseteq \mathcal{S}^1(\mu)$ as well, and the result follows easily. □

Lemma 2.3.5. *Suppose that $\mu \in \mathcal{M}^1$, $A \in \mathrm{L}(\mathbb{R}^d)$ and $a \in \mathbb{R}^d$. Then the following are equivalent:*

(a) $A\mu$ is full

(b) $A\mu * \varepsilon_a$ is full

(c) A is invertible and μ is full

Proof. It follows immediately from Definition 1.3.9 that $A\mu * \varepsilon_a$ is full if and only if $A\mu$ is full. Proposition 1.3.8 shows that $\hat{\mu}(A^*t)$ is the characteristic function of $A\mu$, and Lemma 1.3.11 shows that μ is full if and only if there is no $x \neq 0$ such that $|\hat{\mu}(tx)| = 1$ for all real t. If A is invertible, then A^* is also invertible, hence for any $y \neq 0$ we have $y = A^*x$ for some $x \neq 0$. If $A\mu$ is not full, then for some $x \neq 0$ we have $|\hat{\mu}(A^*tx)| = |\hat{\mu}(ty)| = 1$ for all real t, so μ is not full. Conversely, if $A\mu$ is full and A is singular then $A^*x = 0$ for some $x \neq 0$, and then $\hat{\mu}(A^*tx) = 1$ for all real t, which is a contradiction, hence A is invertible. If $y \neq 0$, then we cannot have $|\hat{\mu}(ty)| = |\hat{\mu}(A^*tx)| = 1$ for all real t, so that μ is full. □

Lemma 2.3.6. *If $\mu_n \Rightarrow \mu$ in $\mathcal{M}^1$ and μ is full, then μ_n is full for all n large.*

Proof. If each $\mu_{n'}$ is not full, where $n' \to \infty$, then Lemma 1.3.11 shows that there exist unit vectors $\theta_{n'}$ such that $|\hat{\mu}_{n'}(t\theta_{n'})| = 1$ for all t real. Choose a further subsequence (n'') along which $\theta_{n''} \to \theta$. Then Theorem 1.3.6 yields $|\hat{\mu}_{n''}(t\theta_{n''})| \to |\hat{\mu}(t\theta)| = 1$ for all real t, and then Lemma 1.3.11 implies that μ is not full. This is a contradiction, hence μ_n is full for all n large. $\square$

Lemma 2.3.7. *Suppose $\mu_n, \mu \in \mathcal{M}^1$, μ is full, $A_n \in \mathrm{L}(\mathbb{R}^d)$ and $a_n \in \mathbb{R}^d$. If $A_n\mu_n * \varepsilon_{a_n} \Rightarrow \mu$, then for all n large A_n is invertible and μ_n is full.*

Proof. If $A_n\mu_n * \varepsilon_{a_n} \Rightarrow \mu$, then Lemma 2.3.6 shows that $A_n\mu_n * \varepsilon_{a_n}$ is full for n large, and then Lemma 2.3.5 shows that $A_n \in \mathrm{GL}(\mathbb{R}^d)$ and μ_n is full. $\square$

Lemma 2.3.8. *Suppose $\mu_n, \mu, \nu \in \mathcal{M}^1$, where μ, ν are full and $A_n \in \mathrm{L}(\mathbb{R}^d)$. If $\mu_n \Rightarrow \mu$ and $A_n\mu_n \Rightarrow \nu$, then $\sup \|A_n\| < \infty$, and hence (A_n) is relatively compact.*

Proof. Suppose by way of contradiction that there exists a subsequence along which $\|A_n\| \to \infty$. Then there exist unit vectors $x_n \in \mathbb{R}^d$ such that $\|A_n x_n\| \to \infty$ along this subsequence. Write $A_n^* x_n = r_n \theta_n$, where $r_n \to \infty$ and $\|\theta_n\| = 1$. Choose a further subsequence along which $\theta_n \to \theta$, a unit vector in $\mathbb{R}^d$. Then for all real t we have $\hat{\mu}(t\theta_n) \to \hat{\mu}(t\theta)$ along this subsequence. But $\theta_n = A_n^*(x_n/r_n)$, where $(x_n/r_n) \to 0$, and then it follows from Theorem 1.3.6 that $\hat{\mu}(t\theta_n) = \hat{\mu}_n(A_n^*(t x_n/r_n)) \to \hat{\nu}(0) = 1$ so that $\hat{\mu}(t\theta) = 1$ for all real t. In view of Lemma 1.3.11, this contradicts the assumption that μ is full, hence for some $M > 0$ we have $\|A_n\| \le M$ for all n. Then (A_n) is contained in a compact subset of $\mathrm{L}(\mathbb{R}^d)$, so it is relatively compact. $\square$

Lemma 2.3.9. *Suppose $\mu_n, \mu, \nu \in \mathcal{M}^1$, where μ, ν are full and $a_n \in \mathbb{R}^d$. If $\mu_n \Rightarrow \mu$ and $\mu_n * \varepsilon_{a_n} \Rightarrow \nu$, then $\sup \|a_n\| < \infty$, and hence (a_n) is relatively compact.*

Proof. Choose $R > 0$ such that $D = \{x : \|x\| \le R\}$ is both a μ-continuity set and a ν-continuity set, and such that $\mu(D) > 2/3$ and $\nu(D) > 2/3$. Then $\mu_n(D) > 1/2$ and $\mu_n * \varepsilon_{a_n}(D) = \mu_n(D - a_n) > 1/2$ for all large n. If $\|a_n\| \to \infty$ along a subsequence then $D, D - a_n$ are disjoint for some large n, which gives a contradiction. $\square$

Theorem 2.3.10. *If $\mu \in \mathcal{M}^1$ is full, then $\mathcal{S}^1(\mu)$ is a compact subgroup of $\mathrm{GL}(\mathbb{R}^d)$, and a compact subset of $\mathrm{L}(\mathbb{R}^d)$.*

Proof. If $A \in \mathcal{S}^1(\mu)$, then $\mu = A\mu * \varepsilon_a$ is full for some $a \in \mathbb{R}^d$, so $A\mu$ is full, and then Lemma 2.3.5 shows that A is invertible, hence $\mathcal{S}^1(\mu)$ is a subset of $\mathrm{GL}(\mathbb{R}^d)$. Proposition 1.1.7 implies that $\mu = A^{-1}\mu * \varepsilon_{-A^{-1}a}$, and if $\mu = B\mu * \varepsilon_b$ as well, then $\mu = AB\mu * \varepsilon_{Ab+a}$, so that $\mathcal{S}^1(\mu)$ is a subgroup of $\mathrm{GL}(\mathbb{R}^d)$. Now suppose that $\mu = A_n\mu * \varepsilon_{a_n}$, where $A_n \to A$ in $\mathrm{L}(\mathbb{R}^d)$ and $a_n \to a$ in $\mathbb{R}^d$. Then Theorem 1.2.8 implies that $\mu = \alpha_n\mu \Rightarrow \alpha\mu$, where $\alpha_n(x) = A_n x + a_n$ and $\alpha(x) = Ax + a$. Hence $A \in \mathcal{S}^1(\mu)$, so that $\mathcal{S}^1(\mu)$ is a closed subset of $\mathrm{L}(\mathbb{R}^d)$. Lemma 2.3.8 implies that $\|A_n\|$ remains bounded, hence $\mathcal{S}^1(\mu)$ is a compact subset of $\mathrm{L}(\mathbb{R}^d)$ and a compact subgroup of $\mathrm{GL}(\mathbb{R}^d)$. $\square$

Here we pause for a moment to indicate the structure of compact subgroups of $\mathrm{GL}(\mathbb{R}^d)$, and to explore the possible symmetry groups.

Theorem 2.3.11. *Any compact subgroup of* $\mathrm{GL}(\mathbb{R}^d)$ *is of the form* $L\mathcal{K}L^{-1}$, *where* $\mathcal{K}$ *is a compact subgroup of the group* $\mathcal{O}$ *of orthogonal linear operators on* $\mathbb{R}^d$ *and* $L \in \mathrm{GL}(\mathbb{R}^d)$ *is symmetric and positive definite.*

Proof. This is a classical result (see, e.g., Corollary 2.4.2 in [75]). □

Remark 2.3.12. The group of rotations on $\mathbb{R}^2$ cannot be the symmetry group of any probability measure, because if every rotation is a symmetry, then so is every reflection. Define an equivalence relation $\sim$ on the class of all subgroups of $\mathrm{GL}(\mathbb{R}^d)$ by saying that $\mathcal{K}_1 \sim \mathcal{K}_2$ if the orbits $\mathcal{K}_1 x = \mathcal{K}_2 x$ for all $x \in \mathbb{R}^d$. Let $[\mathcal{K}]$ denote the equivalence class of $\mathcal{K}$, and partial order the subsets of $\mathrm{GL}(\mathbb{R}^d)$ by set inclusion. Then a compact subgroup $\mathcal{K}$ of $\mathrm{GL}(\mathbb{R}^d)$ satisfies $\mathcal{K} = \mathcal{S}^1(\mu)$ for some $\mu \in \mathcal{M}^1$ if and only if $\mathcal{K}$ is a maximal element of $[\mathcal{K}]$ (see [119]).

Remark 2.3.13. If $\mathcal{S}^1(\mu) = L\mathcal{K}L^{-1}$, where $\mathcal{K}$ is a compact subgroup of $\mathcal{O}$, then Proposition 2.3.4 shows that $\mathcal{S}^1(L^{-1}\mu) = \mathcal{K}$. Then every symmetry of $L^{-1}\mu$ is orthogonal, so that $<Ax, Ay> = <x, y>$ for all $x, y \in \mathbb{R}^d$ and $A \in \mathcal{S}^1(L^{-1}\mu)$ in the standard inner product. By using the inner product $(x, y) = <L^{-1}x, L^{-1}y>$, it is easy to check that every $A \in \mathcal{S}^1(\mu)$ satisfies $(Ax, Ay) = (x, y)$ for all $x, y \in \mathbb{R}^d$, so that every symmetry of μ is orthogonal with respect to this inner product. Then $\|Ax\| = \|x\|$ for all $A \in \mathcal{S}^1(\mu)$ and $x \in \mathbb{R}^d$ in the norm associated with this inner product.

Definition 2.3.14. For $A \in \mathrm{L}(\mathbb{R}^d)$ and $\mathcal{K} \subset \mathrm{L}(\mathbb{R}^d)$ closed we define

$$\|A - \mathcal{K}\| = \inf\{\|A - B\| : B \in \mathcal{K}\}$$

the distance between A and $\mathcal{K}$. For a sequence (A_n) in $\mathrm{L}(\mathbb{R}^d)$ we will write $A_n \rightharpoonup \mathcal{K}$ to mean that $\|A_n - \mathcal{K}\| \to 0$ as $n \to \infty$. This is equivalent to saying that (A_n) is limit point compact and every limit point lies in $\mathcal{K}$.

Lemma 2.3.15. *Suppose* $\mu_n, \mu \in \mathcal{M}^1$, μ *is full and* $A_n \in \mathrm{L}(\mathbb{R}^d)$. *If* $\mu_n \Rightarrow \mu$, *then* $A_n\mu_n * \varepsilon_{a_n} \Rightarrow \mu$ *for some* $a_n \in \mathbb{R}^d$ *if and only if* $A_n \rightharpoonup \mathcal{S}^1(\mu)$.

Proof. If $\mu_n \Rightarrow \mu$ full and $A_n\mu_n * \varepsilon_{a_n} \Rightarrow \mu$, then Lemmas 2.3.8 and 2.3.9 show that (A_n, a_n) is relatively compact, so any subsequence contains a further subsequence along which $(A_n, a_n) \to (A, a)$ in $\mathrm{L}(\mathbb{R}^d) \times \mathbb{R}^d$. Then Theorem 1.2.8 implies that $A_n\mu_n * \varepsilon_{a_n} \Rightarrow A\mu * \varepsilon_a$ along this subsequence, hence $\mu = A\mu * \varepsilon_a$ and $A \in \mathcal{S}^1(\mu)$. Then every limit point of the sequence (A_n) lies in $\mathcal{S}^1(\mu)$, i.e., $A_n \rightharpoonup \mathcal{S}^1(\mu)$. Conversely, if $A_n \rightharpoonup \mathcal{S}^1(\mu)$, then (A_n) is relatively compact, and any subsequence has a further subsequence along which $A_n \to A \in \mathcal{S}^1(\mu)$. Then $A\mu * \varepsilon_a = \mu$ for some $a \in \mathbb{R}^d$, and Theorem 1.2.8 implies that $A_n\mu_n * \varepsilon_a \Rightarrow A\mu * \varepsilon_a = \mu$ along this subsequence. Since any subsequence has a further subsequence with this property, we have $A_n\mu_n * \varepsilon_a \Rightarrow \mu$. □

Lemma 2.3.16. *Given $\mu_n, \mu, \nu \in \mathcal{M}^1$ with μ, ν full and $A_n \in \mathrm{L}(\mathbb{R}^d)$, if $\mu_n \Rightarrow \mu$ and $A_n\mu_n * \varepsilon_{a_n} \Rightarrow \nu$ for some $a_n \in \mathbb{R}^d$, then $\nu = A\mu * \varepsilon_a$ for some $A \in \mathrm{GL}(\mathbb{R}^d)$ and some $a \in \mathbb{R}^d$.*

Proof. If $\mu_n \Rightarrow \mu$ and $A_n\mu_n * \varepsilon_{a_n} \Rightarrow \nu$, then Lemmas 2.3.8 and 2.3.9 show that (A_n, a_n) is relatively compact, so any subsequence contains a further subsequence along which $(A_n, a_n) \to (A, a)$. Theorem 1.2.8 implies that $A_n\mu_n * \varepsilon_{a_n} \Rightarrow A\mu * \varepsilon_a$ along this subsequence, so that $A\mu * \varepsilon_a = \nu$. Since ν is full, Lemma 2.3.5 implies that A is invertible. □

Now we come to the main result of this section. The following theorem is originally due to Fisz [42] and Billingsley [16]. See also Weissman [197] and Sharpe [181].

Theorem 2.3.17. *(Convergence of Types in $\mathbb{R}^d$) Given $\mu_n, \mu, \nu \in \mathcal{M}^1$ where μ, ν are full and $A_n, B_n \in \mathrm{L}(\mathbb{R}^d)$, suppose that $B_n\mu_n * \varepsilon_{b_n} \Rightarrow \mu$ for some $b_n \in \mathbb{R}^d$. Then $A_n\mu_n * \varepsilon_{a_n} \Rightarrow \nu$ for some $a_n \in \mathbb{R}^d$ if and only if $\nu = A\mu * \varepsilon_a$ and $A_nB_n^{-1} \rightharpoonup A\mathcal{S}^1(\mu)$ for some $A \in \mathrm{GL}(\mathbb{R}^d)$ and $a \in \mathbb{R}^d$.*

Proof. Lemma 2.3.5 implies that B_n^{-1} exists for all large n. Write $\omega_n = B_n\mu_n * \varepsilon_{b_n}$ and use Proposition 1.1.7 to compute that $\mu_n = B_n^{-1}\omega_n * \varepsilon_{s_n}$ where $s_n = -B_n^{-1}b_n$. If $A_n\mu_n * \varepsilon_{a_n} \Rightarrow \nu$, then $\omega_n \Rightarrow \mu$ and $(A_nB_n^{-1})\omega_n * \varepsilon_{s_n'} \Rightarrow \nu$, where $s_n' = A_ns_n + a_n$. Then Lemma 2.3.16 implies that $\nu = A\mu * \varepsilon_a$ for some $A \in \mathrm{GL}(\mathbb{R}^d)$ and some $a \in \mathbb{R}^d$. Then $\mu = A^{-1}\nu * \varepsilon_{a'}$ where $a' = -A^{-1}a$, and Theorem 1.2.7 implies that $A^{-1}(A_nB_n^{-1}\omega_n * \varepsilon_{s_n'}) * \varepsilon_{a'} \Rightarrow A^{-1}\nu * \varepsilon_{a'} = \mu$. Then Lemma 2.3.15 implies that $A^{-1}A_nB_n^{-1} \rightharpoonup \mathcal{S}^1(\mu)$. Multiply both sides by A on the left to obtain $A_nB_n^{-1} \rightharpoonup A\mathcal{S}^1(\mu)$.

Conversely, if $\nu = A\mu * \varepsilon_a$ and $A_nB_n^{-1} \rightharpoonup A\mathcal{S}^1(\mu)$, then since $\mathcal{S}^1(\mu)$ is compact by Theorem 2.3.10, any subsequence contains a further subsequence along which $A_nB_n^{-1} \to AB$ for some $B \in \mathrm{GL}(\mathbb{R}^d)$ such that $B\mu * \varepsilon_b = \mu$ for some $b \in \mathbb{R}^d$. Define $b' = Ab + a$ and $a_n = A_nB_n^{-1}b_n + b'$, and let $\omega_n = B_n\mu_n * \varepsilon_{b_n}$. Then $A_n\mu_n * \varepsilon_{a_n} = A_nB_n^{-1}\omega_n * \varepsilon_{b'} \Rightarrow AB\mu * \varepsilon_{b'} = A(B\mu * \varepsilon_b) * \varepsilon_a = A\mu * \varepsilon_a = \nu$ along this subsequence, and since any subsequence contains a further subsequence with this property we have $A_n\mu_n * \varepsilon_{a_n} \Rightarrow \nu$. □

In the remainder of this book, we will often consider the case where μ_n is the distribution of a sum $S_n = X_1 + \cdots + X_n$ of independent random variables. In this case, we seek norming operators A_n and centering constants a_n such that $A_nS_n + a_n \Rightarrow Y$ full. Theorem 2.3.17 characterizes the possible norming operators and limits. In particular, every such limit is of the same type.

3

Infinitely Divisible Distributions and Triangular Arrays

The main limit theorems for sums of independent random variables are proved using characteristic functions, which were introduced in Section 1.3. In this chapter, we use characteristic function to establish the Lévy representation for infinitely divisible laws, and convergence criteria for triangular arrays of random vectors. Our first applications involve some general limit theorems for sums of i.i.d. random vectors.

3.1 INFINITELY DIVISIBLE DISTRIBUTIONS

Infinitely divisible distributions are studied almost exclusively in terms of their characteristic functions. Recall from Section 1.3 that the characteristic function of a probability distribution $\mu \in \mathcal{M}^1$ is defined by

$$\omega(t) = \int e^{i<t,x>} \mu\{dx\} \tag{3.1}$$

for all $t \in \mathbb{R}^d$. If X_1, X_2 are independent random vectors with probability distribution μ_1, μ_2 and characteristic function ω_1, ω_2, then the distribution of $X_1 + X_2$ is the convolution $\mu_1 * \mu_2$, which has characteristic function of $\omega_1 \cdot \omega_2$.

Definition 3.1.1. Suppose that Y is a random vector on $\mathbb{R}^d$ with probability distribution μ and characteristic function ω. Let $\mu^n = \mu * \cdots * \mu$ denote the *n-fold convolution* of μ with itself. We say that Y (or μ or ω) is *infinitely divisible* if for each $n = 1, 2, 3, \ldots$ there exist $Y_{n1}, \ldots, Y_{nn}$ i.i.d. such that $Y_{n1} + \cdots + Y_{nn}$ is identically distributed with Y. Hence if Y_{ni} has distribution

μ_n and characteristic function ω_n, then $\mu_n^n = \mu$ and $\omega = \omega_n^n$. We call μ_n the *nth root* of μ. If for some continuous function $\psi(t)$ with $\psi(0) = 0$ we have $\omega(t) = e^{\psi(t)}$ for all $t \in \mathbb{R}^d$, we call ψ the *log-characteristic function* of μ.

Our first result shows that every infinitely divisible law has a unique log-characteristic function.

Theorem 3.1.2. *Let $\mu \in \mathcal{M}^1(\mathbb{R}^d)$ be infinitely divisible with characteristic function ω as in (3.1). Then*

(a) $\omega(t) \neq 0$ for every $t \in \mathbb{R}^d$

(b) There exist one and only one continuous function $\varphi : \mathbb{R}^d \to \mathbb{R}$ with $\varphi(0) = 0$ and

$$\omega(t) = |\omega(t)|e^{i\varphi(t)} \tag{3.2}$$

hence there exists a unique continuous function $\psi(t)$ with $\psi(0) = 0$ such that $\omega(t) = e^{\psi(t)}$ for all $t \in \mathbb{R}^d$, so that $\psi(t)$ is the log-characteristic function of μ[1]

Proof. By Definition 3.1.1, for each $n = 1, 2, \ldots$ we have $|\omega_n(t)| = |\omega(t)|^{1/n}$, and hence

$$\lim_{n\to\infty} |\omega_n(t)| = \begin{cases} 1 & \text{if } \omega(t) \neq 0 \\ 0 & \text{if } \omega(t) = 0 \end{cases}$$

Proposition 1.3.5 (b) shows that $\omega(t)$ is continuous, and since $\omega(0) = 1$ we know that $\omega(t) \neq 0$ in some neighborhood of the origin, so $\lim |\omega_n(t)| = 1$ in that neighborhood. Then the function $\lim |\omega_n|^2 = \lim |\omega_n|$ is continuous at the origin. But $|\omega_n|^2$ is the characteristic function of the probability measure $\mu_n * \tilde{\mu_n}$, where μ_n has characteristic function ω_n and $\tilde{\mu}_n(B) = \mu_n(-B)$. Then Levy's Continuity Theorem 1.3.6 implies that $\lim |\omega_n|^2$ is the characteristic function of a probability measure, and hence continuous. But $\lim |\omega_n|^2 = \lim |\omega_n|$ has only values zero and one with one for $t = 0$, and hence $\lim |\omega_n(t)| = 1$ for all t implying $\omega(t) \neq 0$ for all $t \in \mathbb{R}^d$. Part (b) follows from a general result for continuous functions from $\mathbb{R}^d \to \mathbb{C}$ derived in Theorem 3.4.1. □

Corollary 3.1.3. *The nth roots of any infinitely divisible $\mu \in \mathcal{M}^1(\mathbb{R}^d)$ are uniquely determined. Using the notation of (3.2), we get for the characteristic function ω_n of μ_n*

$$\omega_n = |\omega|^{1/n} e^{i\varphi/n} \tag{3.3}$$

[1] The log-characteristic function of μ does not necessarily correspond to any analytic determination of the complex logarithm function.

or equivalently $\omega_n = e^{\psi/n}$, *where* ψ *is the log-characteristic function of* μ.

Proof. Let ω_n be any characteristic function with $(\omega_n)^n = \omega$. Since by Theorem 3.1.2 $\omega_n(t) \neq 0$ for all $t \in \mathbb{R}^d$, by Theorem 3.4.1 there exists a uniquely determined continuous real valued function φ_n such that $\varphi_n(0) = 0$ and $\omega_n = |\omega_n|e^{i\varphi_n}$. Then (3.2) implies $|\omega_n|^n = |\omega|$ and since $\omega(t) \neq 0$ for all $t \in \mathbb{R}^d$, $e^{in\varphi_n} = e^{i\varphi}$. Now each side of this last equation satisfies the hypothesis of Theorem 3.4.1 and thus by the uniqueness statement in that theorem it follows that $n\varphi_n = \varphi$. Thus, for any characteristic function ω_n for which $(\omega_n)^n = \omega$, $|\omega_n| = |\omega|^{1/n}$ and $\varphi_n = \varphi/n$, so ω_n is unique. □

Corollary 3.1.4. *For every infinitely divisible law* $\mu \in \mathcal{M}^1(\mathbb{R}^d)$, *the sequence* (μ_n) *of roots satisfies* $\mu_n \Rightarrow \varepsilon_0$.

Proof. Let ω_n denote the characteristic function of μ_n. Then part (a) of Theorem 3.1.2 along with Corollary 3.1.3 imply that $\omega_n(t) \to 1$ as $n \to \infty$ for all $t \in \mathbb{R}^d$, and hence $\mu_n \Rightarrow \varepsilon_0$ by Levy's Continuity Theorem 1.3.6. □

We need two more properties of the class of infinitely divisible distributions. Recall from Definition 1.1.5 that if X is a random vector on $\mathbb{R}^d$ with distribution μ and A is a linear operator on $\mathbb{R}^d$, then AX has distribution $A\mu$.

Lemma 3.1.5. *For any infinitely divisible measures* $\mu, \nu \in \mathcal{M}^1(\mathbb{R}^d)$ *and any linear operator* $A \in \mathrm{L}(\mathbb{R}^d)$, *the measures* $\mu * \nu$ *and* $A\mu$ *are infinitely divisible.*

Proof. By Corollary 3.1.3, for any natural n there exist unique μ_n, ν_n such that $\mu = \mu_n^n$ and $\nu = \nu_n^n$. Proposition 1.1.7 (a) yields $\mu * \nu = (\mu_n * \nu_n)^n$ so that $\mu_n * \nu_n$ is the nth root of $\mu * \nu$. If $\hat{\mu}_n, \hat{\mu}$ are the characteristic functions of μ_n, μ, then $\hat{\mu}_n^n = \hat{\mu}$ and Proposition 1.3.8 implies that $\omega_n(t) = \hat{\mu}_n(A^*t)$ and $\omega(t) = \hat{\mu}(A^*t)$ are the characteristic functions of $A\mu_n, A\mu$ respectively. Then $(\omega_n(t))^n = (\hat{\mu}_n(A^*t))^n = \hat{\mu}(A^*t) = \omega(t)$, so $A\mu_n$ is the nth root of $A\mu$. □

Lemma 3.1.6. *Suppose* $\mu_n, \mu \in \mathcal{M}^1$ *and* $\mu_n \Rightarrow \mu$. *If each* μ_n *is infinitely divisible then* μ *is infinitely divisible.*

Proof. Let ω_m denote the characteristic function of μ_m. Then $\omega_m = (\omega_{m,n})^n$ for every natural m and n, where $\omega_{m,n}$ is the characteristic function of the nth root of μ_m. By Theorem 1.3.6 (a), $\omega_{m,n} = (\omega_m)^{1/n} \to (\omega)^{1/n}$ as $m \to \infty$, where ω is the characteristic function of μ. The limit is continuous at the origin and hence by Theorem 1.3.6 (b) the characteristic function of some probability measure ρ_n. Hence $\omega = (\hat{\rho}_n)^n$ so μ is infinitely divisible. □

Definition 3.1.7. The *compound Poisson distribution*

$$\Pi(c, \mu) = e^{-c} \sum_{n=0}^{\infty} \frac{c^n}{n!} \mu^n \tag{3.4}$$

where $\mu \in \mathcal{M}^1$ and $c > 0$. This probability distribution evidently has characteristic function $e^{c(\omega-1)}$, where ω is the characteristic function of μ and hence is infinitely divisible with nth root $\Pi(c/n, \mu)$.

Corollary 3.1.8. *Every infinitely divisible distribution μ on $\mathbb{R}^d$ is the weak limit of the sequence of compound Poisson distributions $\Pi(n, \mu_n)$, where μ_n is the nth root of μ.*

Proof. Let μ be infinitely divisible with characteristic function e^{ψ}. Then by Corollary 3.1.3 the roots μ_n of μ have characteristic function $e^{\psi/n}$. We claim that $\Pi(n, \mu_n) \Rightarrow \mu$. In fact, the characteristic function of $\Pi(n, \mu_n)$ is $\exp(n(e^{\psi/n} - 1))$. Since $\psi/n \to 0$ it follows from Taylor's expansion that $e^{\psi/n} - 1 = \psi/n + O((\psi/n)^2)$, and hence $\exp(n(e^{\psi/n} - 1)) \to e^{\psi}$ proving the claim. □

Remark 3.1.9. Since the compound Poisson distribution has a physical meaning that is easy to interpret, this gives a concrete way in which to understand the more ethereal class of infinitely divisible distributions. In particular, suppose that Y_n are i.i.d. random vectors with common distribution μ, and let N denote a Poisson random variable with mean $c > 0$ independent of (Y_n) . Then the random sum

$$S_N = \sum_{n=1}^{N} Y_n$$

has the compound Poisson distribution (3.4). One typical application of a compound Poisson distribution is the following. Let N denote the number of vehicles that arrive at a popular tourist spot on a given day, and let Y_n denote the number of passengers in the nth vehicle. Then S_N is the total number of visitors to the tourist spot on a given day.

Lemma 3.1.10. *Let μ_n, μ be infinitely divisible with log-characteristic function ψ_n, ψ respectively. Then $\mu_n \Rightarrow \mu$ if and only if $\psi_n(t) \to \psi(t)$ for all $t \in \mathbb{R}^d$, and in this case the log-characteristic functions also converge uniformly on compact subsets of $\mathbb{R}^d$.*

Proof. Let ω_n, ω denote the characteristic functions of μ_n, μ respectively. If $\psi_n(t) \to \psi(t)$ for all $t \in \mathbb{R}^d$, then $e^{\psi_n(t)} = \omega_n(t) \to \omega(t) = e^{\psi(t)}$ for all $t \in \mathbb{R}^d$, and it follows from Levy's Continuity Theorem 1.3.6 that $\mu_n \Rightarrow \mu$. Conversely, if $\mu_n \Rightarrow \mu$ then it follows from Lévy's Continuity Theorem that $\omega_n(t_n) \to \omega(t)$ whenever $t_n \to t$ in $\mathbb{R}^d$, and Theorem 3.1.2 (a) shows that $\omega(t) \neq 0$. Use Theorem 3.1.2 to write $\psi_n(t) = \log|\omega_n(t)| + i\varphi_n(t)$ and $\psi(t) = \log|\omega(t)| + i\varphi(t)$. Since $\omega_n(t_n) \to \omega(t)$ we also have $\log|\omega_n(t_n)| \to \log|\omega(t)|$ whenever $t_n \to t$ in $\mathbb{R}^d$, and then $e^{i\varphi_n(t_n)} = \omega_n(t_n)/|\omega_n(t_n)| \to \omega(t)/|\omega(t)| = e^{i\varphi(t)}$. Hence $e^{i(\varphi_n(t_n)-\varphi(t))} \to 1$. Since $\omega(t)$ is continuous we also have

$$e^{i(\varphi(t)-\varphi(t_n))} = \frac{\omega(t)}{|\omega(t)|} \cdot \frac{|\omega(t_n)|}{\omega(t_n)} \to 1.$$

It follows that $e^{if(t)} \to 1$ uniformly on compact subsets of $t \in \mathbb{R}^d$, where $f(t) = \varphi_n(t) - \varphi(t)$. Then Theorem 3.4.6 implies that $f(t) \to 0$ uniformly on compact subsets of $\mathbb{R}^d$, and this completes the proof. □

One of the basic tools we will use to study limit theorems for sums of independent random vectors is the Lévy representation for infinitely divisible laws, due to P. Lévy [88].

Theorem 3.1.11. *(Lévy representation) A probability measure μ on $\mathbb{R}^d$ is infinitely divisible if and only if we can write the characteristic function of μ in the form* $\exp(\psi(t))$, *where*

$$\psi(t) = i\langle a,t\rangle - \tfrac{1}{2}Q(t) + \int_{x\neq 0}\Big(e^{i\langle t,x\rangle} - 1 - \frac{i\langle t,x\rangle}{1+\|x\|^2}\Big)\phi\{dx\}, \tag{3.5}$$

where $a \in \mathbb{R}^d$, $Q(t)$ is a nonnegative definite quadratic form on $\mathbb{R}^d$, and ϕ is a σ-finite Borel measure on $\mathbb{R}^d \setminus \{0\}$ such that

$$\int_{x\neq 0} \min\{1, \|x\|^2\}\phi\{dx\} < \infty. \tag{3.6}$$

The triple $[a, Q, \phi]$ is unique.

Definition 3.1.12. The triple $[a, Q, \phi]$ in Theorem 3.1.11 is called the *Lévy representation* of μ. A σ-finite Borel measure on $\mathbb{R}^d \setminus \{0\}$ such that (3.6) holds is called a *Lévy measure.* The first term in (3.5) is just a shift. The second term is called the *normal component* since it is the log-characteristic function of a centered normal law, see Remark 1.3.16. The remaining term is called the *Poissonian* component, because when ϕ is a finite measure, it is the log-characteristic function of a shifted compound Poisson distribution. A normal law has no Poissonian component. An infinitely divisible law with no normal component will be called *Poissonian.* Any infinitely divisible law is the sum of two independent components, one normal and one Poissonian, and these components are unique up to a shift.

Proof of Theorem 3.1.11. We first show that a probability distribution with log-characteristic function (3.5) is infinitely divisible. It suffices to show that the Poissonian component

$$f_0(t) = \int_{x\neq 0}\Big(e^{i\langle t,x\rangle} - 1 - \frac{i\langle t,x\rangle}{1+\langle x,x\rangle}\Big)\phi\{dx\}$$

defines the log-characteristic function of some infinitely divisible probability distribution. For natural n let ϕ_n be the restriction of ϕ to $\{x : \|x\| > 1/n\}$, let $c_n = \phi\{x : \|x\| > 1/n\}$, and let $\nu_n = c_n^{-1} \cdot \phi_n$. Then ν_n is a probability measure and we can write

$$\begin{aligned} f_n(t) &= \int_{x\neq 0}\Big(e^{i\langle t,x\rangle} - 1 - \frac{i\langle t,x\rangle}{1+\langle x,x\rangle}\Big)\phi_n\{dx\} \\ &= I_n^{(1)} + I_n^{(2)} \end{aligned}$$

where

$$I_n^{(1)} = c_n \int_{x \neq 0} \left(e^{i<t,x>} - 1\right) \nu_n\{dx\}$$

is the log-characteristic function of the compound Poisson probability distribution $\Pi(c_n, \nu_n)$ and

$$I_n^{(2)} = -i<t, \int_{x \neq 0} \frac{x}{1+<x,x>} \phi_n\{dx\}> = i<t, b_n>$$

for some $b_n \in \mathbb{R}^d$ is the log-characteristic function of ε_{b_n}. Hence f_n is the log-characteristic function of the infinitely divisible distribution $\Pi(c_n, \nu_n) * \varepsilon_{b_n}$. Rewrite

$$f_n(t) = \int_{x \neq 0} \left(e^{i<t,x>} - 1 - \frac{i<t,x>}{1+<x,x>}\right) I(\|x\| > 1/n) \phi\{dx\}$$

and note that the integrand is bounded by

$$C_1 \|x\|^2 I(\|x\| \leq 1) + C_2 I(\|x\| > 1)$$

for some positive constants C_1 and C_2 depending on t. By (3.6), this function is integrable with respect to ϕ and then dominated convergence implies that $f_n(t) \to f_0(t)$ for all t where f_0 is a continuous function. Hence $\exp(f_n(t)) \to \exp(f_0(t))$, so $\Pi(c_n, \nu_n) * \varepsilon_{b_n} \Rightarrow \mu_0$ with characteristic function $\exp(f_0)$. By Lemma 3.1.6, μ_0 is infinitely divisible and hence f_0 is a log-characteristic function, proving the sufficiency.

For the proof of the necessity, suppose that μ is infinitely divisible with log-characteristic function ψ and let μ_n denote the nth root of μ. Then $\psi_n = n(\hat{\mu}_n - 1)$ is the log-characteristic function of the compound Poisson distribution $\Pi(n, \mu_n)$ and it follows from Corollary 3.1.8 and Lemma 3.1.10 together with Lévy Continuity Theorem 1.3.6 that $\psi_n \to \psi$ uniformly on compact sets. We associate to ψ a function

$$\chi(t) = \psi(t) - 2^{-d} \int_B \psi(t+h)dh \tag{3.7}$$

where $B = \{(x_1, \ldots, x_d) \in \mathbb{R}^d : -1 \leq x_i \leq 1 \text{ for all } i = 1, \ldots, d\}$ and likewise χ_n to ψ_n. Now Fubini's Theorem 1.1.17 yields

$$\begin{aligned}
\chi_n(t) &= \int (e^{i<t,x>} - 1) n\mu_n\{dx\} - 2^{-d} \int_B \int (e^{i<t+h,x>} - 1) n\mu_n\{dx\}dh \\
&= \int \left((e^{i<t,x>} - 1) - 2^{-d} \int_B (e^{i<t+h,x>} - 1)dh\right) n\mu_n\{dx\} \\
&= \int \left(e^{i<t,x>} - 2^{-d} \int_B e^{i<t+h,x>} dh\right) n\mu_n\{dx\} \\
&= \int e^{i<t,x>} K(x) n\mu_n\{dx\}
\end{aligned}$$

where

$$K(x) = 1 - \prod_{j=1}^{d} \frac{1}{2} \int_{-1}^{1} e^{ih_j x_j} dh_j = 1 - \prod_{j=1}^{d} \frac{\sin(x_j)}{x_j} \tag{3.8}$$

with the convention $u^{-1} \sin u = 1$ when $u = 0$. Note that K is a bounded continuous function on $\mathbb{R}^d$ that is positive for $x \neq 0$ and that $K(x) = (1/6)||x||^2 + o(||x||^2)$ as $x \to 0$. Since $\psi_n \to \psi$ uniformly on compact sets, $\chi_n \to \chi$ uniformly on compact sets. Observe that χ is continuous at the origin and that χ_n is the Fourier transform of the finite Borel measure $H_n\{dx\} = K(x) n\mu_n\{dx\}$. By Theorem 1.3.19, there exists a finite Borel measure H on $\mathbb{R}^d$ such that $H_n \to H$ in $\mathcal{M}_0$, in the sense of Definition 1.2.12.

Define $q(x) = ||x||^2/((1 + ||x||^2)K(x))$ for $x \neq 0$ and $q(0) = 6$. Then q is a bounded, continuous, positive function on $\mathbb{R}^d$. Define $G_n\{dx\} = q(x)H_n\{dx\}$ and $G\{dx\} = q(x)H\{dx\}$ so that G_n and G are finite Borel measures on $\mathbb{R}^d$ with $G_n \to G$ in $\mathcal{M}_0$, and also note that

$$G_n\{dx\} = \frac{||x||^2}{1 + ||x||^2} n\mu_n\{dx\}.$$

We need some more notation. Let

$$g(t, x) = \left(e^{i<t,x>} - 1 - \frac{i<t, x>}{1 + ||x||^2} \right) \frac{1 + ||x||^2}{||x||^2}.$$

Then, for every fixed $t \in \mathbb{R}^d$, $g(t, x)$ is a bounded continuous function on $x \neq 0$ and Taylor's expansion yields

$$g(t, x) = -\frac{1}{2} \frac{<t, x>^2}{||x||^2} + O(||x||) \quad \text{as } ||x|| \to 0 \tag{3.9}$$

so that we cannot extend $g(t, x)$ to a continuous function on $x \in \mathbb{R}^d$. Furthermore, let

$$\gamma_n(t) = \int \frac{<t, x>}{1 + ||x||^2} n\mu_n\{dx\}$$

and

$$\begin{aligned} \psi^{(1)}(t) &= \lim_{\eta \to 0} \int_{||x|| \geq \eta} g(t, x) G\{dx\} \\ &= \int_{x \neq 0} \left(e^{i<t,x>} - 1 - \frac{i<t, x>}{1 + ||x||^2} \right) \phi\{dx\} \end{aligned}$$

where

$$\phi\{dx\} = \frac{1 + ||x||^2}{||x||^2} G\{dx\} \quad \text{for } x \neq 0.$$

Note that since G is a bounded measure ϕ satisfies (3.6). Compute that

$$\begin{aligned} -i\gamma_n(t) + \psi_n(t) &= \int_{x\neq 0} g(t,x) G_n\{dx\} \\ &= \int_{\|x\|\geq\eta} g(t,x) G_n\{dx\} + \int_{0<\|x\|<\eta} g(t,x) G_n\{dx\} \end{aligned}$$

where $\eta > 0$ is chosen such that $G\{x : \|x\| = \eta\} = 0$.

Since $G_n \to G$ and $g(t,x)$ is bounded, Proposition 1.2.14 implies

$$-i\gamma_n(t) + \psi_n(t) \to \psi^{(1)}(t) \quad \text{as } n \to \infty.$$

Now

$$\begin{aligned} \psi^{(2)}(t) &= \psi(t) - \psi^{(1)}(t) \\ &= \lim_{\eta\to 0} \lim_{n\to\infty} \left(\psi_n(t) - \int_{\|x\|\geq\eta} g(t,x) G_n\{dx\} \right) \\ &= \lim_{\eta\to 0} \lim_{n\to\infty} \left(\int_{0<\|x\|<\eta} g(t,x) G_n\{dx\} + i\gamma_n(t) \right). \end{aligned}$$

In view of (3.9) along with the fact that $G_n \to G$, it follows that

$$\psi^{(2)}(t) = \lim_{\eta\to 0} \lim_{n\to\infty} \left(-\frac{1}{2} \int_{0<\|x\|<\eta} \frac{<t,x>^2}{\|x\|^2} G_n\{dx\} + i\gamma_n(t) \right). \tag{3.10}$$

By definition γ_n is linear, and then $G_n \to G$ along with the fact that the integrand in (3.10) is bounded implies that

$$\lim_{\lambda\to 0} \frac{\psi^{(2)}(\lambda t)}{\lambda} = \lim_{n\to\infty} i\gamma_n(t) = i\gamma(t)$$

exists and is linear. Hence, by (3.10)

$$Q(t) = \lim_{\eta\to 0} \limsup_{n\to\infty} \int_{\|x\|<\eta} \frac{<t,x>^2}{\|x\|^2} G_n\{dx\}$$

exists, and defines a nonnegative definite quadratic form on $\mathbb{R}^d$. Putting things together we have shown that

$$\begin{aligned} \psi(t) &= \psi^{(2)}(t) + \psi^{(1)}(t) \\ &= i\gamma(t) - \frac{1}{2} Q(t) + \int_{x\neq 0} \left(e^{i<t,x>} - 1 - \frac{i<t,x>}{1+\|x\|^2} \right) \phi\{dx\} \end{aligned}$$

showing the existence of the representation (3.5).

For the proof of the uniqueness, assume that we are given two representations ψ and ψ' as in (3.5) of the log characteristic function of μ with triples

$[a, Q, \phi]$ and $[a', Q', \phi']$ respectively. As in (3.7) associate to ψ the function χ and note that direct integration using Fubini's theorem yields

$$\chi(t) = \frac{1}{6}\text{trace}(Q) + \int_{x \neq 0} e^{i<t,x>} K(x)\phi\{dx\} \tag{3.11}$$

so χ is the Fourier transform of the finite Borel measure $M\{dx\} = K(x)\phi\{dx\}$ extended to all $\mathbb{R}^d$ by setting $M\{0\} = (1/6)\,\text{trace}(Q)$. Since $\chi = \chi'$ [χ' defined by (3.7) with ψ'] the Proposition 1.3.18 (b) implies $M = M'$ and since $K(x) > 0$ for $x \neq 0$, we obtain $\phi = \phi'$. Hence

$$i{<}a, t{>} - \frac{1}{2}Q(t) = i{<}a', t{>} - \frac{1}{2}Q'(t)$$

for all $t \in \mathbb{R}^d$. But both sides of the equation above represent the log characteristic function of a normal law with mean a, covariance function Q and mean a' and covariance function Q'. Then the uniqueness of the mean and covariance matrix for normal laws implies $a = a'$ and $Q = Q'$. □

Remark 3.1.13. In one variable, the Lévy representation is written

$$f(t) = iat - \frac{1}{2}\sigma^2 t^2 + \int_{x \neq 0} \left(e^{itx} - 1 - \frac{itx}{1 + x^2}\right)\phi\{dx\} \tag{3.12}$$

where σ is the standard deviation of the normal component. In this case, one can also employ the *Khintchine representation*

$$f(t) = iat + \int \left(e^{itx} - 1 - \frac{itx}{1 + x^2}\right)\frac{1 + x^2}{x^2} M\{dx\} \tag{3.13}$$

where $M\{dx\} = x^2\phi\{dx\}/(1 + x^2)$ is a finite Borel measure on $\mathbb{R}^1 \setminus \{0\}$, which we extend to a finite Borel measure on all of $\mathbb{R}^1$ by setting $M\{0\} = \sigma^2$. In (3.13), we also extend the integrand continuously through the origin by assigning the value $-t^2/2$ at $x = 0$. The Khintchine formula is useful in $\mathbb{R}^1$ because it treats the normal and Poissonian components together, and because it only involves finite measures. Unfortunately, the Khintchine formula has no useful vector analogue. We can certainly write the integral part of (3.5) in the form

$$\int_{x \neq 0} \left(e^{i<t,x>} - 1 - \frac{i<t, x>}{1 + <x, x>}\right)\frac{1 + \|x\|^2}{\|x\|^2} M\{dx\} \tag{3.14}$$

where $M\{dx\} = \|x\|^2\phi\{dx\}/(1 + \|x\|^2)$ is a finite Borel measure on $\mathbb{R}^d \setminus \{0\}$. However, the integrand in (3.14) cannot be extended continuously through the origin. Also, if we extend M to all of $\mathbb{R}^d$ by assigning an atom at the origin, this real number $M\{0\}$ cannot possibly represent the general nonnegative definite quadratic form on $\mathbb{R}^d$ for $d > 1$.

The integrand in (3.5) is somewhat arbitrary. In particular, the role of the last term is to make the integrand $O(\|x\|^2)$ so that the integral exists near the origin. In some applications, it is useful to employ alternate forms.

Theorem 3.1.14. *If μ is an infinitely divisible probability measure on $\mathbb{R}^d$ with Lévy representation $[a, Q, \phi]$, then we can also write the characteristic function of μ in the form $\exp(\psi_R(t))$, where*

$$\psi_R(t) = i\langle a_R, t\rangle - \frac{1}{2}Q(t) + \int_{x\neq 0}\left(e^{i\langle t,x\rangle} - 1 - i\langle t, x\rangle I(0 < \|x\| < R)\right)\phi\{dx\} \quad (3.15)$$

for any $R > 0$. The triple (a_R, Q, ϕ) is unique. Furthermore

(a) If we have

$$\int_{0<\|x\|<R} \|x\|\phi\{dx\} < \infty \quad (3.16)$$

then we can also write the characteristic function of μ in the form $\exp(\psi_1(t))$, where

$$\psi_1(t) = i\langle a_1, t\rangle - \frac{1}{2}Q(t) + \int_{x\neq 0}\left(e^{i\langle t,x\rangle} - 1\right)\phi\{dx\} \quad (3.17)$$

for some unique triple (a_1, Q, ϕ)

(b) If we have

$$\int_{\|x\|\geq R} \|x\|\phi\{dx\} < \infty \quad (3.18)$$

then we can also write the characteristic function of μ in the form $\exp(\psi_2(t))$, where

$$\psi_2(t) = i\langle a_2, t\rangle - \frac{1}{2}Q(t) + \int_{x\neq 0}\left(e^{i\langle t,x\rangle} - 1 - i\langle t, x\rangle\right)\phi\{dx\} \quad (3.19)$$

for some unique triple (a_2, Q, ϕ)

Proof. If μ is infinitely divisible with Lévy representation $[a, Q, \phi]$, then μ has characteristic function $\exp(\psi(t))$, where $\psi(t)$ is given by (3.5). Define

$$\tilde{a} = \int_{x\neq 0}\left(\frac{x}{1 + \|x\|^2} - xI(0 < \|x\| < R)\right)\phi\{dx\} \quad (3.20)$$

and note that $\tilde{a}$ exists since the integrand is bounded and $O(\|x\|^3)$ as $x \to 0$. It follows that $\psi(t) = \psi_R(t)$, where $a_R = a - \tilde{a}$, and the uniqueness of the triple (a_R, Q, ϕ) follows from Theorem 3.1.11. If (3.16) holds, then $\psi_R(t) = \psi_1(t)$, where

$$a_1 = a_R - \int_{0<\|x\|<R} x\phi\{dx\}$$

and if (3.18) holds, then $\psi_R(t) = \psi_2(t)$, where

$$a_2 = a_R + \int_{\|x\|\geq R} x\phi\{dx\}$$

so that in either case, the representation is unique. □

Remark 3.1.15. An infinitely divisible law with log-characteristic function (3.19) has mean a_2. Suppose the distribution ν of Y is infinitely divisible with Lévy representation $[a, Q, \phi]$ and (3.18) holds. Then Theorem 1 in [183] implies that $E(\|Y\|) < \infty$, and hence $E(Y)$ exists. Let $Y = (Y_1, \ldots, Y_d)$ and $a_2 = (a_{12}, \ldots, a_{d2})$. Then it follows from the definition of the characteristic function of Y that for $j = 1, \ldots, d$

$$E(Y_j) = \frac{1}{i}\frac{\partial}{\partial t_j}\hat{\nu}(t)|_{t=0}$$

In view of (3.19), we have

$$\frac{\partial}{\partial t_j}\hat{\nu}(t) = \hat{\nu}(t)\Big(ia_{j2} - \frac{1}{2}\frac{\partial}{\partial t_j}Q(t) + \int_{x\neq 0} ix_j\big(e^{i<t,x>} - 1\big)\phi\{dx\}\Big)$$

and since $\hat{\nu}(0) = 1$ and Q is a quadratic form, we get $E(Y_j) = a_{j2}$ for $j = 1, \ldots, d$.

Suppose that μ_n is an infinitely divisible probability distribution on $\mathbb{R}^d$ with Lévy representation $[a_n, Q_n, \phi_n]$ for each n. If $\mu_n \Rightarrow \mu$, then Corollary 3.1.6 shows that μ is infinitely divisible, so it has a unique Lévy representation $[a, Q, \phi]$. The following result, due to Rvačeva [158], describes the convergence of infinitely divisible distributions in terms of the Lévy representation. Recall from Definition 1.2.17 that $\mathcal{M}$ is the collection of Borel measures on $\Gamma = \mathbb{R}^d \setminus \{0\}$ that assign finite measure to sets bounded away from zero, and that $\mu_n \to \mu$ in $\mathcal{M}$ if $\mu_n(A) \to \mu(A)$ for all μ-continuity sets A that are bounded away from zero.

Theorem 3.1.16. *A sequence of infinitely divisible laws (μ_n) on $\mathbb{R}^d$ with Lévy representations $[a_n, Q_n, \phi_n]$ converges weakly to a probability measure μ on $\mathbb{R}^d$ if and only if:*

(a) For some $\phi \in \mathcal{M}$ we have

$$\phi_n \to \phi \tag{3.21}$$

(b) $a_n \to a$ for some $a \in \mathbb{R}^d$; and

(c) For some nonnegative definite quadratic form $Q(t)$ on $\mathbb{R}^d$ we have

$$\lim_{\varepsilon \to 0} \lim_{n \to \infty} \left[\int_{0<\|x\|<\varepsilon} <t, x>^2 \phi_n\{dx\} + Q_n(t) \right] = Q(t) \tag{3.22}$$

for all $t \in \mathbb{R}^d$

In this case, μ is infinitely divisible with Lévy representation $[a, Q, \phi]$.

Proof. The original proof in [158] contains an error that we have been unable to repair.[2] The following original proof is modeled after the corresponding one variable result in [40]. As in the proof of Theorem 3.1.11 we associate to any infinitely divisible distribution on $\mathbb{R}^d$ with log characteristic function ψ of the form (3.5) the function χ as in (3.7) so that (3.11) holds, where K is defined by (3.8). Then χ is the Fourier transform of the finite Borel measure $M\{dx\} = K(x)\phi\{dx\}$ extended to all $\mathbb{R}^d$ by setting $M\{0\} = (1/6)\,\mathrm{trace}(Q)$.

Now we prove necessity. Suppose that $\mu_n \Rightarrow \mu$ and obtain χ_n, χ and M_n, M by the above construction, where ψ_n, ψ are the log characteristic functions of μ_n, μ respectively. It follows from Lemma 3.1.10 that $\psi_n \to \psi$ uniformly on compact sets, and hence $\chi_n(t) \to \chi(t)$ for all $t \in \mathbb{R}^d$. Thus $M_n \to M$ in $\mathcal{M}$, and then part (a) follows since $K(x) > 0$ for $x \neq 0$.

Let $f(t,x)$ denote the integrand in the Lévy representation (3.5). For any $\varepsilon > 0$ such that $\phi\{x : \|x\| = \varepsilon\} = 0$, Proposition 1.2.20 implies

$$\int_{\|x\|>\varepsilon} f(t,x)\phi_n\{dx\} \to \int_{\|x\|>\varepsilon} f(t,x)\phi\{dx\} \tag{3.23}$$

since $f(t,x)$ is a bounded, continuous function of x. Since $f(t,x)$ is integrable with respect to $\phi\{dx\}$ for all t, we have

$$\begin{aligned}\psi(t) = \lim_{\varepsilon \to 0} \lim_{n \to \infty} &\left[i<t, a_n> - \frac{1}{2} Q_n(t) + \int_{0<\|x\|<\varepsilon} f(t,x)\phi_n\{dx\} \right] \\ &+ \int_{x \neq 0} f(t,x)\phi\{dx\}.\end{aligned}$$

By a Taylor expansion,

$$f(t,x) = -\frac{1}{2}<t, x>^2 + o(<t, x>^2) \text{ as } x \to 0. \tag{3.24}$$

[2] The original proof in [158] uses a multivariable Khintchine formula, the impossibility of which was discussed in Remark 3.1.13.

It follows easily that

$$\psi(t) = \lim_{n\to\infty} i\langle t, a_n\rangle - \frac{1}{2}\lim_{\varepsilon\to 0}\lim_{n\to\infty}\left[Q_n(t) + \int_{0<\|x\|<\varepsilon} \langle t, x\rangle^2 \phi_n\{dx\}\right] + \int_{x\neq 0} f(t,x)\phi\{dx\}$$

and now conditions (b) and (c) follow by virtue of the uniqueness of the Lévy representation.

Conversely, suppose that conditions (a–c) hold. By part (a), we again obtain (3.23) whenever $\phi\{\|x\| = \varepsilon\} = 0$. By part (b) and the integrability of f with respect to ϕ, we have

$$\lim_{n\to\infty} \psi_n(t) = i\langle a, t\rangle + \lim_{\varepsilon\to 0}\lim_{n\to\infty}\left[-\frac{1}{2}Q_n(t) + \int_{0<\|x\|<\varepsilon} f(t,x)\phi_n\{dx\}\right] + \int_{x\neq 0} f(t,x)\phi\{dx\}. \tag{3.25}$$

It follows from part (c) and (3.24) that the limit in (3.25) is $\psi(t)$. Then $e^{\psi_n} \to e^{\psi}$ and so $\mu_n \Rightarrow \mu$, which completes the proof. □

Corollary 3.1.17. *Theorem 3.1.16 remains true if (3.22) is replaced by*

$$\begin{aligned} Q(t) &= \lim_{\varepsilon\to 0}\liminf_{n\to\infty}\left[\int_{0<\|x\|<\varepsilon} \langle t, x\rangle^2\phi_n\{dx\} + Q_n(t)\right] \\ &= \lim_{\varepsilon\to 0}\limsup_{n\to\infty}\left[\int_{0<\|x\|<\varepsilon} \langle t, x\rangle^2\phi_n\{dx\} + Q_n(t)\right] \end{aligned} \tag{3.26}$$

Proof. Since (3.22) implies (3.26) we only need to prove sufficiency. Assume that (a) and (b) of Theorem 3.1.16 hold, and that (c) holds with (3.26) in place of (3.22). By part (a), we again obtain (3.23) whenever $\phi\{\|x\| = \varepsilon\} = 0$. By part (b) and the integrability of f with respect to ϕ, we have

$$\begin{aligned} \limsup_{n\to\infty} \psi_n(t) &= i\langle a, t\rangle + \int_{x\neq 0} f(t,x)\phi\{dx\} \\ &\quad + \lim_{\varepsilon\to 0}\limsup_{n\to\infty}\left[-\frac{1}{2}Q_n(t) + \int_{0<\|x\|<\varepsilon} f(t,x)\phi_n\{dx\}\right]. \end{aligned} \tag{3.27}$$

It follows from part (c) and (3.24) that $\limsup_n \psi_n(t) = \psi(t)$. A similar argument shows that $\liminf_n \psi_n(t) = \psi(t)$, so $\psi_n(t) \to \psi(t)$. Then $e^{\psi_n} \to e^{\psi}$ and so $\mu_n \Rightarrow \mu$, which completes the proof. □

Remark 3.1.18. Theorem 3.1.16 can also be used to give a concrete interpretation to the Lévy measure for a Poissonian infinitely divisible law. Suppose that ν is infinitely divisible with Lévy representation $[0, 0, \phi]$. For every positive integer n define $D_n = \{x \in \mathbb{R}^d : \|x\| > n^{-1}\}$, $c_n = \phi(D_n)$ and $\mu_n(A) = c_n^{-1}\phi(A \cap D_n)$ a probability measure on $\mathbb{R}^d$. The compound Poisson distribution $\Pi(c_n, \mu_n)$ has Lévy representation $[a_n, 0, \phi_n]$ where $\phi_n = c_n\mu_n$ is the restriction of ϕ to D_n and $a_n = \int x/(1+\|x\|^2)\phi_n\{dx\}$. If (3.16) holds, then $a_n \to 0$ and $\Pi(c_n, \mu_n) \Rightarrow \nu$. Hence the infinitely divisible law ν is the weak limit of compound Poisson random vectors of the form

$$Y_n = \sum_{i=1}^{N_n} X_{in}$$

where N_n is Poisson with mean c_n and X_{in} are i.i.d. with distribution μ_n. In the general case where (3.16) fails, we have $\Pi(c_n, \mu_n) * \varepsilon_{-a_n} \Rightarrow \nu$ so that ν is the weak limit of shifted Poisson random vectors of the same description. In this sense, any Poissonian infinitely divisible law is the sum of a random number of jumps, and the Lévy measure determines both the number and the magnitude of the jumps.

Definition 3.1.19. We say that $\phi \in \mathcal{M}$ is $\mathcal{M}$-full if ϕ is not supported on any $d-1$ dimensional subspace of $\mathbb{R}^d$.

Proposition 3.1.20. *Let ν be infinitely divisible with Lévy triple $[a, 0, \phi]$. Then ν is full if and only if ϕ is $\mathcal{M}$-full.*

Proof. If ν is not full, then Proposition 1.3.11 implies that for some $x_0 \in \mathbb{R}^d$ we have $|\hat{\nu}(tx_0)| = 1$ for all $t \in \mathbb{R}$. In view of (3.5), we have for all real t

$$1 = \left(\hat{\nu}(tx_0)\overline{\hat{\nu}(tx_0)}\right)^{1/2} = \exp\left\{\int_{x\neq 0}[\cos<x, tx_0> - 1]\phi\{dx\}\right\}$$

and hence

$$\int_{x\neq 0}[\cos<x, tx_0> - 1]\phi\{dx\} = 0 \quad \text{for all } t \in \mathbb{R}.$$

This implies that for all $t \in \mathbb{R}$ we have $\cos<x, tx_0> = 1$ for ϕ-almost all $x \in \mathbb{R}^d$. Therefore ϕ is supported on $\{x \in \mathbb{R}^d : <x, tx_0> = 2\pi k \;\exists\; k \in \mathbb{Z}\}$. Since this is true for all real t it follows that ϕ is supported on the $d-1$ dimensional subspace $\{x \in \mathbb{R}^d : <x, x_0> = 0\}$, and hence ϕ is not $\mathcal{M}$-full.

If ϕ is not $\mathcal{M}$-full, then ϕ is supported on some lower dimensional subspace W of $\mathbb{R}^d$, and hence there exists a nonzero $x_0 \in \mathbb{R}^d$ such that $<x, x_0> = 0$ for all $x \in W$. Then, using (3.5) again, we get for $t \in \mathbb{R}$

$$\begin{aligned}
|\hat{\nu}(tx_0)| &= \exp\left\{\int_{x\neq 0}\big[\cos<x, tx_0> - 1\big]d\phi(x)\right\} \\
&= \exp\left\{\int_{W\setminus\{0\}}\big[\cos<x, tx_0> - 1\big]d\phi(x)\right\} \\
&= \exp(0) = 1
\end{aligned}$$

and hence Proposition 1.3.11 implies that ν is not full. □

Proposition 3.1.21. *Suppose that ν_1, ν_2 are infinitely divisible with Lévy representation $[a_i, Q_i, \phi_i]$ for $i = 1, 2$, $A \in \mathrm{L}(\mathbb{R}^d)$ and $b \in \mathbb{R}^d$. Then*

(a) *The convolution $\nu_1 * \nu_2$ is infinitely divisible with Lévy representation $[a_1 + a_2, Q_1 + Q_2, \phi_1 + \phi_2]$*

(b) *$A\nu_1 * \varepsilon_b$ is infinitely divisible with Lévy representation $[Aa_1 + b', Q', A\phi]$ where $Q'(t) = Q_1(A^*t)$ for all $t \in \mathbb{R}^d$, $A\phi$ is as in Definition 1.1.5, and*

$$b' = b + \int_{x \neq 0} \Big(\frac{x}{1 + \|x\|^2}\Big) A\phi\{dx\} - \int_{x \neq 0} \Big(\frac{Ax}{1 + \|x\|^2}\Big) \phi\{dx\} \tag{3.28}$$

Proof. Part (a) follows from Theorem 3.1.11 along with the fact that the log-characteristic function of $\nu_1 * \nu_2$ is the sum of the individual log-characteristic functions. Part (b) follows from Theorem 3.1.11 along with Proposition 1.3.8 by a simple substitution in the integral portion of (3.5). □

The following result relates the symmetry group $\mathcal{S}^1(\nu)$ of a Poissonian infinitely divisible law ν with the *symmetry group*

$$\mathcal{S}(\phi) = \{A \in \mathrm{L}(\mathbb{R}^d) : A\phi = \phi\}$$

of its Lévy measure ϕ.

Proposition 3.1.22. *Let ν be infinitely divisible with Lévy triple $[a, 0, \phi]$. Then $\mathcal{S}^1(\nu) = \mathcal{S}(\phi)$.*

Proof. By Definition 2.3.3, if $A \in \mathcal{S}^1(\nu)$, then $\nu = A\nu * \varepsilon_a$ for some $a \in \mathbb{R}^d$. Proposition 3.1.21 shows that $A\nu * \varepsilon_a$ has Lévy measure $A\phi$ and in view of the uniqueness of the Lévy triple it follows that $\phi = A\phi$, and hence $A \in \mathcal{S}(\phi)$. On the other hand, if $A \in \mathcal{S}(\phi)$ another application of Proposition 3.1.21 yields $\nu = A\nu * \varepsilon_a$ for a suitable $a \in \mathbb{R}^d$, so $A \in \mathcal{S}^1(\nu)$ and the proof is complete. □

Definition 3.1.23. Suppose that ν is infinitely divisible with Lévy representation $[a, Q, \phi]$ and $c > 0$. Then we define ν^c to be the unique infinitely divisible law with Lévy representation $[c \cdot a, c \cdot Q, c \cdot \phi]$. We will call ν^c the cth *convolution power* of ν. Note that when c is a positive integer this agrees with Definition 1.1.4.

Proposition 3.1.24. *Suppose that μ, ν are infinitely divisible, $A \in \mathrm{L}(\mathbb{R}^d)$, $a \in \mathbb{R}^d$ and $c > 0$. Then*

(a) $(\mu * \nu)^c = (\mu^c) * (\nu^c)$

(b) $A(\mu^c) = (A\mu)^c$

(c) $(\varepsilon_b)^c = \varepsilon_{cb}$

Proof. This is an easy calculation, using Definition 3.1.23 along with Proposition 3.1.21. □

3.2 CONVERGENCE OF TRIANGULAR ARRAYS

Suppose that (X_n) are independent random vectors on $\mathbb{R}^d$ and $S_n = \sum_{j=1}^{n} X_j$. We are interested in the situation where $A_n S_n - a_n \Rightarrow Y$ for some norming operators $A_n \in \mathrm{L}(\mathbb{R}^d)$ and some shifts $a_n \in \mathbb{R}^d$. The collection of random vectors $Z_{nj} = A_n X_j$ is an example of a triangular array. In this section, we establish necessary and sufficient conditions for the row sums of a triangular array to converge in distribution after an appropriate centering. The proof relies heavily on the Lévy representation developed in Section 3.1. Most of the important limit theorems in this book are based on these convergence criteria for triangular arrays.

Definition 3.2.1. A *triangular array* is a collection of random vectors on $\mathbb{R}^d$ $\{Z_{nj} : 1 \leq j \leq k_n,\ n = 1, 2, 3, \ldots\}$ such that $Z_{n1}, \ldots, Z_{nk_n}$ are independent for each n, and $k_n \to \infty$. A triangular array is *uniformly asymptotically constant* if there exist constants $b_{nj} \in \mathbb{R}^d$ such that

$$\lim_{n\to\infty} \sup_{1\leq j\leq k_n} P(\|Z_{nj} - b_{nj}\| > \varepsilon) = 0 \tag{3.29}$$

for all $\varepsilon > 0$. A triangular array is *infinitesimal* (or uniformly asymptotically negligible) if (3.29) holds for all $\varepsilon > 0$ with all $b_{nj} = 0$.

Necessary and sufficient conditions for the convergence in distribution of an infinitesimal triangular array of random vectors are given in the following important theorem, due to Rvačeva [158].

Theorem 3.2.2. *Given an infinitesimal triangular array of random vectors* $\{Z_{nj} : 1 \leq j \leq k_n,\ n = 1, 2, 3, \ldots\}$ *on* $\mathbb{R}^d$, *where* μ_{nj} *is the distribution of* Z_{nj}, *there exists a random vector* Y *on* $\mathbb{R}^d$ *and* $a_n \in \mathbb{R}^d$ *such that*

$$(Z_{n1} + \cdots + Z_{nk_n}) - a_n \Rightarrow Y \tag{3.30}$$

if and only if:

(a) For some σ*-finite Borel measure* ϕ *on* $\mathbb{R}^k \setminus \{0\}$ *we have*

$$\sum_{j=1}^{k_n} \mu_{nj}(S) \to \phi(S) \text{ as } n \to \infty \tag{3.31}$$

for all Borel subsets S *of* $\mathbb{R}^k \setminus \{0\}$ *for which* $\phi(\partial S) = 0$*; and*

(b) For some nonnegative definite quadratic form Q on $\mathbb{R}^d$ we have

$$
\begin{aligned}
Q(t) &= \lim_{\varepsilon\to 0}\limsup_{n\to\infty}\sum_{j=1}^{k_n}\left[\int_{\|x\|<\varepsilon} <t,x>^2\mu_{nj}\{dx\} - \left(\int_{\|x\|<\varepsilon} <t,x>\mu_{nj}\{dx\}\right)^2\right] \\
&= \lim_{\varepsilon\to 0}\liminf_{n\to\infty}\sum_{j=1}^{k_n}\left[\int_{\|x\|<\varepsilon} <t,x>^2\mu_{nj}\{dx\} - \left(\int_{\|x\|<\varepsilon} <t,x>\mu_{nj}\{dx\}\right)^2\right] \qquad (3.32)
\end{aligned}
$$

for all $t \in \mathbb{R}^d$

In this case, the distribution of the limit Y is infinitely divisible with Lévy representation $[a, Q, \phi]$, where $a \in \mathbb{R}^d$ depends on the choice of the centering constants a_n. The centering constants a_n may be chosen according to the formula

$$
a_n = \sum_{j=1}^{k_n}\int_{\|x\|<R} x\ \mu_{nj}\{dx\} \qquad (3.33)
$$

where $R > 0$ is chosen so that $\phi\{x : \|x\| = R\} = 0$. In that case, Y has characteristic function $\exp(\psi_R(t))$ where $\psi_R(t)$ is given by (3.15) with $a_R = 0$.

We follow the proof of Rvačeva [158].

Definition 3.2.3. The *median* of a random vector $X = (X_1, \ldots, X_d)$ on $\mathbb{R}^d$ is a nonrandom vector $m = (m_1, \ldots, m_d)$ such that m_i is a median of X_i for each $i = 1, \ldots, d$, i.e., $P(X_i \le m_i) \ge 1/2$ and $P(X_i \ge m_i) \ge 1/2$.

Lemma 3.2.4. *For a uniformly asymptotically constant triangular array, we may always take b_{nj} in (3.29) to be the median of the array element Z_{nj}.*

Proof. Given $\varepsilon > 0$ choose n_0 such that for all $n \ge n_0$ and all $1 \le j \le k_n$ we have $P(\|Z_{nj} - b_{nj}\| > \varepsilon) < 1/2$. Then for all such n, j the median m_{nj} of Z_{nj} lies between $b_{nj} - \varepsilon$ and $b_{nj} + \varepsilon$, hence $\|m_{nj} - b_{nj}\| < \varepsilon$ for all such n, j. Then $\sup_j \|m_{nj} - b_{nj}\| \to 0$ and the lemma follows easily. □

Lemma 3.2.5. *A triangular array is infinitesimal if and only if*

$$
\lim_{n\to\infty}\sup_{1\le j\le k_n}\int\left(\frac{\|x\|^2}{1+\|x\|^2}\right)\mu_{nj}\{dx\} = 0 \qquad (3.34)
$$

where μ_{nj} is the probability distribution of the array element Z_{nj}.

Proof. If $\{Z_{nj}\}$ is infinitesimal, then for all $0 < \varepsilon < 1/2$, for some n_0, for all $n \geq n_0$ we have

$$\begin{aligned}
&\sup_{1\leq j\leq k_n} \int \left(\frac{||x||^2}{1+||x||^2}\right) \mu_{nj}\{dx\} \\
\leq\ & \sup_{1\leq j\leq k_n} \left\{ \int_{||x||<\varepsilon} ||x||^2 \mu_{nj}\{dx\} + \int_{||x||\geq\varepsilon} \mu_{nj}\{dx\} \right\} \\
\leq\ & \varepsilon^2 + \sup_{1\leq j\leq k_n} P(||Z_{nj}|| > \varepsilon) \\
\leq\ & \frac{\varepsilon}{2} + \frac{\varepsilon}{2} = \varepsilon.
\end{aligned}$$

Conversely, if (3.34) holds, then since

$$\begin{aligned}
\int \left(\frac{||x||^2}{1+||x||^2}\right) \mu_{nj}\{dx\} &\geq \int_{||x||\geq\varepsilon} \left(\frac{||x||^2}{1+||x||^2}\right) \mu_{nj}\{dx\} \\
&\geq \frac{\varepsilon^2}{1+\varepsilon^2} P(||Z_{nj}|| \geq \varepsilon)
\end{aligned}$$

we get

$$\frac{\varepsilon^2}{1+\varepsilon^2} \sup_{1\leq j\leq k_n} P(||Z_{nj}|| \geq \varepsilon) \to 0$$

as $n \to \infty$, and the lemma follows easily. □

Lemma 3.2.4 shows that $\{Z_{nj}\}$ is uniformly asymptotically constant if and only if $\{Z_{nj} - m_{nj}\}$ is infinitesimal, where m_{nj} is a median of Z_{nj}. Then Lemma 3.2.5 shows that $\{Z_{nj}\}$ is uniformly asymptotically constant if and only if

$$\lim_{n\to\infty} \sup_{1\leq j\leq k_n} \int \Big(\frac{||x||^2}{1+||x||^2}\Big) \mu_{nj}\{dx + m_{nj}\} = 0. \tag{3.35}$$

Lemma 3.2.6. *If a triangular array is infinitesimal, then*

$$\sup_{1\leq j\leq k_n} |\varphi_{nj}(t) - 1| \to 0 \tag{3.36}$$

uniformly on compact subsets of $t \in \mathbb{R}^d$, where φ_{nj} is the characteristic function of the array element Z_{nj}.

Proof. For any $\varepsilon > 0$,

$$\begin{aligned}
|\varphi_{nj}(t) - 1| &= \left| \int (e^{i<t,x>} - 1) \mu_{nj}\{dx\} \right| \\
&\leq \int_{||x||<\varepsilon} |e^{i<t,x>} - 1| \ \mu_{nj}\{dx\} + 2\int_{||x||\geq\varepsilon} \mu_{nj}\{dx\} \\
&\leq \varepsilon||t|| + 2P(||Z_{nj}|| \geq \varepsilon)
\end{aligned}$$

using the inequality $|e^{iz} - 1| \leq |z|$ for z real. The result follows easily using the definition of infinitesimal. □

The next step in the proof of Theorem 3.2.2 is to establish necessary and sufficient conditions for (3.30) in the special case where the limit $Y = 0$, i.e., where the left-hand side of (3.30) converges in probability to zero. Since convergence in probability of a sequence of random vectors on $\mathbb{R}^d$ is equivalent to convergence in probability of each component, most of the work will be to establish results in the scalar case $d = 1$.

Lemma 3.2.7. *If $X, \bar{X}$ are i.i.d. random variables with median m, then*

$$E[s(X - \bar{X})] \geq \frac{1}{2} E[s(X - m)] \tag{3.37}$$

where $s(x) = \min\{1, |x|\}$.

Proof. If $m \neq 0$, we can replace $X, \bar{X}$ by $X - m, \bar{X} - m$ respectively, so that $m = 0$ without loss of generality. Let μ denote the probability distribution of $X, \bar{X}$. Note that $s(x) = s(|x|)$ is a monotone nondecreasing function of $|x|$, so that if x, y are of opposite sign, then $s(x - y) \geq s(x)$. Now

$$\begin{aligned}
& E[s(X - \bar{X})] \\
= & \int \int s(x - y) \mu\{dx\} \mu\{dy\} \\
\geq & \int_{\{x \geq 0, y < 0\}} s(x - y) \mu\{dx\} \mu\{dy\} + \int_{\{x < 0, y \geq 0\}} s(x - y) \mu\{dx\} \mu\{dy\} \\
\geq & \int_{y<0} \mu\{dy\} \int_{x \geq 0} s(x) \mu\{dx\} + \int_{y \geq 0} \mu\{dy\} \int_{x<0} s(x) \mu\{dx\} \\
\geq & \frac{1}{2} \int s(x) \mu\{dx\} = \frac{1}{2} E[s(X)]
\end{aligned}$$

which finishes the proof. □

Theorem 3.2.8. *Given a triangular array of random variables $\{Z_{nj}\}$ on $\mathbb{R}^1$ with probability distributions μ_{nj} and medians m_{nj}, if there exist $a_n \in \mathbb{R}^1$ such that*

$$(Z_{n1} + \cdots + Z_{nk_n}) - a_n \xrightarrow{P} 0 \tag{3.38}$$

then

(a) $\displaystyle\lim_{n \to \infty} \sum_{j=1}^{k_n} \int_{|x|>1} \mu_{nj}\{x + m_{nj}\} = 0$; *and*

(b) $\displaystyle\lim_{n \to \infty} \sum_{j=1}^{k_n} \int_{|x| \leq 1} x^2 \mu_{nj}\{x + m_{nj}\} = 0$

Proof. Let $S_n = Z_{n1} + \cdots + Z_{nk_n}$. If $S_n - a_n \to 0$ in probability, then $S_n - a_n \Rightarrow \varepsilon_0$ and Levy's Continuity Theorem 1.3.6 implies that

$$e^{-ita_n} \prod_{j=1}^{k_n} \varphi_{jn}(t) \to 1$$

uniformly on compact subsets of $t \in \mathbb{R}^1$, where φ_{jn} is the characteristic function of Z_{jn}. Taking absolute values on both sides we get

$$\prod_{j=1}^{k_n} |\varphi_{jn}(t)| \to 1 \tag{3.39}$$

since $|e^{-ita_n}| = 1$. Since $|\varphi_{jn}(t)| \leq 1$ this implies that

$$\sup_{1 \leq j \leq k_n} \{1 - |\varphi_{jn}(t)|\} \to 0, \tag{3.40}$$

in other words, $|\varphi_{jn}(t)| \to 1$ uniformly in $1 \leq j \leq k_n$ and uniformly on compact subsets. Suppose that all Z_{nj} are symmetric, so that $\varphi_{jn}(t)$ is always a real number. Then for some n_0, t_0 we have

$$1 - |\varphi_{jn}(t)| < 1/2$$

for all $n \geq n_0$, all $1 \leq j \leq k_n$, and all $|t| \leq t_0$. Since $\varphi_{jn}(0) = 1$, and since φ_{jn} is continuous, we cannot have $\varphi_{jn}(t) \leq 0$ for such j, n, t. Then (3.39) implies $\prod_j \varphi_{jn}(t) \to 1$, and so $\sum_j \log \varphi_{jn}(t) \to 0$. Using the fact that $\log(1 - x) \leq x$ for $0 < x \leq 1$, we obtain $\sum_j \log \varphi_{jn}(t) \leq \sum_j [1 - \varphi_{jn}(t)] \leq 0$, and so

$$\sum_{j=1}^{k_n} [1 - \varphi_{jn}(t)] \to 0 \tag{3.41}$$

uniformly on compact subsets. Since φ_{nj} is real we have

$$\begin{aligned}
\int_{-1}^{1} [1 - \varphi_{jn}(t)]dt &= \int_{-1}^{1} \int [1 - \cos(tx)]\mu_{jn}\{dx\}dt \\
&= \int \int_{-1}^{1} [1 - \cos(tx)]dt \ \mu_{jn}\{dx\} \\
&= 2 \int \left(1 - \frac{\sin x}{x}\right) \mu_{jn}\{dx\}.
\end{aligned}$$

This integrand is positive, and a Taylor expansion shows that we can choose a, b positive so that the $1 - (\sin x)/x \geq a$ for $|x| > 1$ and $1 - (\sin x)/x \geq bx^2$

for $|x| \leq 1$ (e.g., we can let $a = 1/10$ and $b = 1/8$). Then we have

$$\begin{aligned}
&\int_{-1}^{1} [1 - \varphi_{jn}(t)]dt \\
&= 2\int_{|x|\leq 1} \left(1 - \frac{\sin x}{x}\right)\mu_{jn}\{dx\} + 2\int_{|x|>1} \left(1 - \frac{\sin x}{x}\right)\mu_{jn}\{dx\} \\
&\geq 2b\int_{|x|\leq 1} x^2 \mu_{jn}\{dx\} + 2a\ \mu_{jn}\{|x| > 1\} \geq 0.
\end{aligned}$$

Thus

$$\begin{aligned}
0 &\leq 2b\sum_{j=1}^{k_n} \int_{|x|\leq 1} x^2 \mu_{jn}\{dx\} + 2a\sum_{j=1}^{k_n} \int_{|x|>1} \mu_{jn}\{dx\} \\
&\leq \sum_{j=1}^{k_n} \int_{-1}^{1} [1 - \varphi_{jn}(t)]dt \\
&= \int_{-1}^{1} \sum_{j=1}^{k_n} [1 - \varphi_{jn}(t)]\ dt
\end{aligned}$$

which tends to zero as $n \to \infty$ because (3.41) holds uniformly on $-1 \leq t \leq 1$. This implies (a) and (b) in the case where Z_{nj} are symmetric, in which case $m_{nj} = 0$.

If Z_{nj} are not necessarily symmetric, introduce $\bar{Z}_{nj}$ i.i.d. with Z_{nj}, and let $Z^*_{nj} = Z_{nj} - \bar{Z}_{nj}$, $\sigma_n = \sum_j Z^*_{nj}$. The characteristic function of Z^*_{nj} is $\varphi_{jn}(t)\bar{\varphi}_{jn}(t) = |\varphi_{jn}(t)|^2$ and from (3.39) we get $|\varphi_{jn}(t)|^2 \to 1$, so that $\sigma_n \to 0$ in probability. Letting μ^*_{nj} be the distribution of X_{nj}, from the symmetric case we have that (a) and (b) hold with μ_{nj} replaced by μ^*_{nj}. Now apply Lemma 3.2.7 to see that

$$\begin{aligned}
&\sum_{j=1}^{k_n} \left(\int_{|x|>1} \mu_{nj}\{dx + m_{nj}\} + \int_{|x|\leq 1} x^2 \mu_{nj}\{dx + m_{nj}\} \right) \\
&= \sum_{j=1}^{k_n} E[s(Z_{jn} - m_{jn})] \\
&\leq 2\sum_{j=1}^{k_n} E[s(Z_{jn} - \bar{Z}_{jn})] \\
&= 2\sum_{j=1}^{k_n} \left(\int_{|x|>1} \mu^*_{nj}\{dx\} + \int_{|x|\leq 1} x^2 \mu^*_{nj}\{dx\} \right)
\end{aligned}$$

which tends to zero as $n \to \infty$ by the symmetric case, and it follows immediately that (a) and (b) hold. □

Remark 3.2.9. The conditions (a) and (b) in Theorem 3.2.8 are actually necessary and sufficient (see [44] p. 106), but we only need the result in one direction.

Lemma 3.2.10. *If (3.30) holds for some uniformly asymptotically constant triangular array of random vectors on $\mathbb{R}^d$, then for some constant $C > 0$, for all n*

$$\sum_{j=1}^{k_n} \int \Big(\frac{\|x\|^2}{1+\|x\|^2}\Big)\mu_{nj}\{dx + m_{nj}\} < C \tag{3.42}$$

Proof. First, we prove the corresponding result for triangular arrays of random variables. By Theorem 3.2.8 we have $\sum_j E[s(Z_{nj} - m_{nj})] \to 0$ where $s(x) = \min\{1, |x|\}$ as before. Since

$$\frac{x^2}{1+x^2} \le x^2 \le 2\frac{x^2}{1+x^2}$$

for $|x| \le 1$ and

$$\frac{x^2}{1+x^2} \le 1 \le 2\frac{x^2}{1+x^2}$$

for $|x| > 1$, we have

$$\frac{x^2}{1+x^2} \le s(x) \le 2\frac{x^2}{1+x^2}$$

for all x, and so Theorem 3.2.8 also implies that

$$\sum_{j=1}^{k_n} E\Big[\frac{(Z_{nj} - m_{nj})^2}{1 + (Z_{nj} - m_{nj})^2}\Big] \to 0.$$

Now if $\alpha_n \to 0$ is any sequence of real numbers, then $\alpha_n(S_n - a_n) \to 0$ in probability, so we also have

$$\sum_{j=1}^{k_n} E\Big[\frac{\alpha_n^2(Z_{nj} - m_{nj})^2}{1 + \alpha_n^2(Z_{nj} - m_{nj})^2}\Big] \to 0.$$

For n sufficiently large, we have $\alpha_n^2 < 1$ and so

$$\alpha_n^2 \sum_{j=1}^{k_n} E\Big[\frac{(Z_{nj} - m_{nj})^2}{1 + (Z_{nj} - m_{nj})^2}\Big] \le \alpha_n^2 \sum_{j=1}^{k_n} E\Big[\frac{(Z_{nj} - m_{nj})^2}{1 + \alpha_n^2(Z_{nj} - m_{nj})^2}\Big] \to 0.$$

Let s_n denote the sum on the left-hand side above, so that $\alpha_n^2 s_n \to 0$ for any sequence $\alpha_n \to 0$. Suppose by way of contradiction that $s_{n_i} \to \infty$ along some subsequence $n_i \to \infty$. Choose $\alpha_n^2 = 1/s_n$ for $n = n_i$ and $\alpha_n^2 = 1/n$ otherwise. Then $\alpha_n \to 0$ but $\limsup \alpha_n^2 s_n = 1$, which is a contradiction. Thus for some real number $C > 0$ we have $s_n < C$ for all n, and this is the same thing as (3.42) in the case $d = 1$.

Since (3.42) holds in the one variable case, and recalling the definition of the median of a random vector, we have

$$\sum_{j=1}^{k_n} \int \Big(\frac{x_i^2}{1+x_i^2}\Big) \mu_{nj}\{dx + m_{nj}\} < C$$

for each $i = 1, \ldots, d$. Now (3.42) follows from

$$\sum_{i=1}^{d} \Big(\frac{x_i^2}{1+x_i^2}\Big) \geq \sum_{i=1}^{d} \left(\frac{x_i^2}{1+\sum_{r=1}^{d} x_r^2}\right) = \frac{\|x\|^2}{1+\|x\|^2}$$

which finishes the proof. □

Lemma 3.2.11. *If (3.30) holds for some infinitesimal triangular array, then there exists a constant $C > 0$ such that*

$$\sum_{j=1}^{k_n} \int \left(\frac{\|x\|^2}{1+\|x\|^2}\right) \mu_{nj}\{dx + b_{nj}\} < C \tag{3.43}$$

for all n, where

$$b_{nj} = \int_{\|x\|<R} x \; \mu_{nj}\{dx\} \tag{3.44}$$

and $R > 0$ is an arbitrary number.

Proof. Since Z_{nj} is infinitesimal, for any $\varepsilon > 0$ there exists an n_0 such that $P(\|Z_{nj}\| < \varepsilon) \geq 1/2$ for all $n \geq n_0$ and all $1 \leq j \leq k_n$. Then the medians satisfy $\|m_{nj}\| \leq \varepsilon$ for all such j, n, and so

$$\lim_{n\to\infty} \sup_{1\leq j\leq k_n} \|m_{nj}\| = 0.$$

We also have

$$\begin{aligned} \|b_{nj}\| &= \left\| \int_{\|x\|<R} x \; \mu_{nj}\{dx\} \right\| \\ &\leq \varepsilon \; \mu_{nj}\{\|x\| \leq \varepsilon\} + R \; \mu_{nj}\{\|x\| > \varepsilon\} \\ &< \varepsilon \cdot 1 + R \cdot \varepsilon \end{aligned}$$

for all j and all n sufficiently large, using again the fact that Z_{nj} is infinitesimal. Then we also have

$$\lim_{n\to\infty} \sup_{1\leq j\leq k_n} \|b_{nj}\| = 0. \tag{3.45}$$

Using the inequality $||a+b||^2 \leq 2||a||^2 + 2||b||^2$, we have

$$\begin{aligned}
&\int \Big(\frac{||x||^2}{1+||x||^2}\Big)\mu_{nj}\{dx+b_{nj}\} \\
= \;& \int \Big(\frac{||x+m_{nj}-b_{nj}||^2}{1+||x+m_{nj}-b_{nj}||^2}\Big)\mu_{nj}\{dx+m_{nj}\} \\
\leq \;& 2\int \Big(\frac{||x||^2}{1+||x+m_{nj}-b_{nj}||^2}\Big)\mu_{nj}\{dx+m_{nj}\} \\
&+2\int \Big(\frac{||m_{nj}-b_{nj}||^2}{1+||x+m_{nj}-b_{nj}||^2}\Big)\mu_{nj}\{dx+m_{nj}\}.
\end{aligned}$$

For n sufficiently large we have $||m_{nj} - b_{nj}|| < 1/2$, which ensures that the integrand in the first term is at most $2||x||^2/(1+||x||^2)$, so the first term is bounded above by

$$4\int \left(\frac{||x||^2}{1+||x||^2}\right) \mu_{nj}\{dx+m_{nj}\}.$$

The second term is

$$2||m_{nj}-b_{nj}||^2 \int \left(\frac{1}{1+||x+m_{nj}-b_{nj}||^2}\right) \mu_{nj}\{dx+m_{nj}\} \leq 2||m_{nj}-b_{nj}||^2$$

where

$$\begin{aligned}
||m_{nj}-b_{nj}||^2 &= \left\| \int_{||x||<R} x\ \mu_{nj}\{dx\} - m_{nj} \right\|^2 \\
&= \left\| \int_{||x||<R} (x-m_{nj})\ \mu_{nj}\{dx\} - m_{nj}\int_{||x||\geq R} \mu_{nj}\{dx\} \right\|^2 \\
&\leq 2\left\| \int_{||x||<R} (x-m_{nj})\ \mu_{nj}\{dx\} \right\|^2 \\
&\quad +2\left\| m_{nj}\int_{||x||\geq R} \mu_{nj}\{dx\} \right\|^2 \\
&\leq 2\Big(\int_{||x+m_{nj}||<R} ||x||\ \mu_{nj}\{dx+m_{nj}\}\Big)^2 \\
&\quad +2\ ||m_{nj}||^2 \Big(\int_{||x+m_{nj}||\geq R} \mu_{nj}\{dx+m_{nj}\}\Big)^2 \\
&\leq 2\int_{||x+m_{nj}||<R} ||x||^2 \mu_{nj}\{dx+m_{nj}\} \\
&\quad +2\ ||m_{nj}||^2 \int_{||x+m_{nj}||\geq R} \mu_{nj}\{dx+m_{nj}\}
\end{aligned}$$

by the Cauchy-Schwarz inequality. Thus if n is large enough to ensure that $||m_{nj}|| \leq R/2$, then

$$\begin{aligned} ||m_{nj} - b_{nj}||^2 &\leq 2\int_{||x||<2R} ||x||^2 \mu_{nj}\{dx + m_{nj}\} \\ &\quad +2R^2 \int_{||x||\geq R/2} \mu_{nj}\{dx + m_{nj}\} \\ &\leq 2(1+4R^2)\int_{||x||<2R} \left(\frac{||x||^2}{1+||x||^2}\right) \mu_{nj}\{dx + m_{nj}\} \\ &\quad +2R^2\left(\frac{4+R^2}{R^2}\right)\int_{||x||\geq R/2} \left(\frac{||x||^2}{1+||x||^2}\right) \mu_{nj}\{dx + m_{nj}\}. \end{aligned}$$

Now sum over j and apply Lemma 3.2.10 to complete the proof. □

Theorem 3.2.12. *Given an infinitesimal triangular array whose array elements Z_{nj} have probability distributions μ_{nj}, the convergence (3.30) holds if and only if the sequence of certain infinitely divisible random vectors Y_n converges to the same limit, where Y_n has characteristic function $e^{\psi_n(t)}$ with*

$$\psi_n(t) = -i\langle t, a_n\rangle + \sum_{j=1}^{k_n}\left(i\langle t, b_{nj}\rangle + \int \left[e^{i\langle t,x\rangle} - 1\right]\mu_{nj}\{dx + b_{nj}\}\right) \tag{3.46}$$

where

$$b_{nj} = \int_{||x||<R} x \ \mu_{nj}\{dx\} \tag{3.47}$$

and $R > 0$ is an arbitrary number.

Proof. Let ω_n denote the characteristic function of the left-hand side of (3.30). Levy's Continuity Theorem 1.3.6 implies that (3.30) holds if and only if

$$\omega_n(t) = e^{-i\langle t,a_n\rangle} \prod_{j=1}^{k_n} \varphi_{nj}(t) \to \omega(t) \tag{3.48}$$

where φ_{nj} is the characteristic function of Z_{nj} and ω is the characteristic function of the limit Y. Let $\mu'_{nj}\{dx\} = \mu_{nj}\{dx + b_{nj}\}$ denote the distribution of $Z_{nj} - b_{nj}$, and let $\varphi'_{nj}(t) = e^{-i\langle t,b_{nj}\rangle}\varphi_{nj}(t)$ denote the characteristic function of $Z_{nj} - b_{nj}$. Then (3.48) can be rewritten in the form

$$e^{-i\langle t,a_n\rangle} e^{i\sum_{j=1}^{k_n}\langle t,b_{nj}\rangle} \prod_{j=1}^{k_n} \varphi'_{nj}(t) \to \omega(t).$$

Fix t and set

$$\beta_{nj} = \varphi'_{nj}(t) - 1 = \int \left[e^{i<t,x>} - 1\right]\mu_{nj}\{dx + b_{nj}\}.$$

In the proof of Lemma 3.2.11, we showed that $\sup_j \|b_{nj}\| \to 0$ and thus

$$\lim_{n\to\infty} \sup_{1\le j\le k_n} |1 - e^{-i<t,b_{nj}>}| = 0.$$

Since Z_{nj} are infinitesimal, Lemma 3.2.6 yields

$$\lim_{n\to\infty} \sup_{1\le j\le k_n} |1 - \varphi_{nj}(t)| = 0.$$

Combining these, we find that $\sup_j |\beta_{nj}| \to 0$. Thus for all large n we have for all $1 \le j \le k_n$ that

$$\log \varphi'_{nj} = \log(1 + \beta_{nj}) = \beta_{nj} - \frac{1}{2}\beta_{nj}^2 + \frac{1}{3}\beta_{nj}^3 + \cdots$$

and so by the formula for a geometric series

$$\begin{aligned}
\left|\sum_{j=1}^{k_n}[\log \varphi'_{nj}(t) - \beta_{nj}]\right| &\le \sum_{j=1}^{k_n}\sum_{i=2}^{\infty}\frac{1}{i}|\beta_{nj}|^i \\
&\le \sum_{j=1}^{k_n}\frac{1}{2}\left(\frac{|\beta_{nj}|^2}{1 - |\beta_{nj}|}\right) \\
&\le \sup_{1\le j\le k_n} |\beta_{nj}| \cdot \sum_{j=1}^{k_n} |\beta_{nj}|. \qquad (3.49)
\end{aligned}$$

Now, using the fact that $|e^{iu} - 1 - iu| \le u^2$ for all $u \in \mathbb{R}^1$, we have

$$\begin{aligned}
|\beta_{nj}| &= \left|\int (e^{i<t,x>} - 1)\mu'_{nj}\{dx\}\right| \\
&= \left|\int_{\|x\|<R} (e^{i<t,x>} - 1 - i<t,x>)\mu'_{nj}\{dx\}\right. \\
&\quad + \int_{\|x\|\ge R} (e^{i<t,x>} - 1)\mu'_{nj}\{dx\} \\
&\quad \left. + i<t, \int_{\|x\|<R} x\ \mu'_{nj}\{dx\}>\right| \\
&\le \|t\|^2 \int_{\|x\|<R} \|x\|^2\mu'_{nj}\{dx\} + 2\int_{\|x\|\ge R} \mu'_{nj}\{dx\} \\
&\quad + \|t\| \cdot \left\|\int_{\|x\|<R} x\ \mu'_{nj}\{dx\}\right\|.
\end{aligned}$$

If n is large enough so that $\|b_{nj}\| \leq R/2$, then

$$
\begin{aligned}
&\left\| \int_{\|x\|<R} x\ \mu'_{nj}\{dx\} - \int_{\|x-b_{nj}\|<R} x\ \mu'_{nj}\{dx\} \right\| \\
&\leq \int_{\{R/2<\|x\|<2R\}} \|x\| \mu'_{nj}\{dx\} \\
&\leq 2R \int_{\|x\|>R/2} \mu'_{nj}\{dx\}
\end{aligned}
$$

and

$$
\begin{aligned}
&\left\| \int_{\|x-b_{nj}\|<R} x\ \mu'_{nj}\{dx\} \right\| \\
=& \left\| \int_{\|x-b_{nj}\|<R} x\ \mu_{nj}\{dx + b_{nj}\} \right\| \\
=& \left\| \int_{\|x\|<R} (x - b_{nj}) \mu_{nj}\{dx\} \right\| \\
=& \left\| \int_{\|x\|<R} x\ \mu_{nj}\{dx\} - b_{nj} \int_{\|x\|<R} \mu_{nj}\{dx\} \right\| \\
=& \left\| b_{nj} \left(1 - \int_{\|x\|<R} \mu_{nj}\{dx\} \right) \right\| = \left\| b_{nj} \int_{\|x\|\geq R} \mu_{nj}\{dx\} \right\| \\
\leq& \frac{R}{2} \int_{\|x\|\geq R} \mu_{nj}\{dx\} = \frac{R}{2} \int_{\|x-b_{nj}\|\geq R} \mu_{nj}\{dx + b_{nj}\} \\
\leq& \frac{R}{2} \int_{\|x\|\geq R/2} \mu'_{nj}\{dx\}
\end{aligned}
$$

and thus

$$
\left\| \int_{\|x\|<R} x\ \mu'_{nj}\{dx\} \right\| \leq \frac{5}{2} R \int_{\|x\|\geq R/2} \mu'_{nj}\{dx\}.
$$

Now suppose that (3.30) holds. Then it follows from Lemma 3.2.11 that, for any $\tau > 0$,

$$
\begin{aligned}
\sum_{j=1}^{k_n} \int_{\|x\|\leq\tau} \|x\|^2 \mu'_{nj}\{dx\} &< C \\
\sum_{j=1}^{k_n} \int_{\|x\|>\tau} \mu'_{nj}\{dx\} &< C
\end{aligned}
$$

which shows that $\sum_j |\beta_{nj}|$ is bounded. Then by (3.49) we have

$$
\lim_{n\to\infty} \sum_{j=1}^{k_n} [\log \varphi'_{nj}(t) - \beta_{nj}] = 0. \tag{3.50}
$$

Since

$$\psi_n(t) = -i\langle t, a_n\rangle + \sum_{j=1}^{k_n} i\langle t, b_{nj}\rangle + \sum_{j=1}^{k_n} \beta_{nj}$$

and

$$\omega_n(t) = \exp\left(-i\langle t, a_n\rangle + \sum_{j=1}^{k_n} i\langle t, b_{nj}\rangle + \sum_{j=1}^{k_n} \log \varphi'_{nj}(t)\right)$$

we have

$$e^{\psi_n(t)} = \omega_n(t) \cdot \exp\left(-\sum_{j=1}^{k_n} [\log \varphi'_{nj}(t) - \beta_{nj}]\right) \tag{3.51}$$

and by (3.50) the second term tends to $e^0 = 1$ as $n \to \infty$. Thus $\omega_n(t) \to \omega(t)$ implies that $e^{\psi_n(t)} \to \omega(t)$, completing the proof of necessity.

Conversely, if the infinitely divisible random vectors $Y_n \Rightarrow Y$, then Theorem 3.1.16 applies. Note that Y_n has Lévy measure $\sum_j \mu'_{nj}$, and apply (3.22) with $t = e_1, \ldots, e_d$ the standard basis vectors for $\mathbb{R}^d$. Summing over $i = 1, \ldots, d$ yields that

$$\lim_{n\to\infty} \sum_{j=1}^{k_n} \int_{\|x\|\le\delta} \sum_{i=1}^{d} x_i^2 \mu'_{nj}\{dx\} = \lim_{n\to\infty} \sum_{j=1}^{k_n} \int_{\|x\|\le\delta} \|x\|^2 \mu'_{nj}\{dx\}$$

exists and is finite. Also (3.21) implies that

$$\lim_{n\to\infty} \mu'_{nj}\{x : \|x\| \ge \delta\}$$

exists and is finite. Together these two facts imply that

$$\sum_{j=1}^{k_n} \int \left(\frac{\|x\|^2}{1+\|x\|^2}\right) \mu'_{nj}\{dx\} < C$$

for all n. As in the proof of necessity, this implies that (3.50) holds, and then the equivalence (3.51) implies that (3.30) must hold, with the same limit Y. □

Remark 3.2.13. Theorem 3.2.12 allows us to reduce the question of the convergence of an infinitesimal triangular array (3.30) to the simpler case of the convergence of infinitely divisible laws. It allows us to replace the array elements Z_{nj} with random vectors Z^*_{nj} which are shifted compound Poisson, without affecting the limiting distribution. To be specific, the random variable $Z^*_{nj} = b_{nj} + S_N$ where $S_N = X_1 + \cdots + X_N$ with N Poisson with mean one, and X_i i.i.d. with the same distribution as $Z_{nj} - b_{nj}$.

Theorem 3.2.14. *The class of infinitely divisible distributions equals the class of possible limiting distributions in (3.30) for infinitesimal triangular arrays.*

Proof. It follows from Theorem 3.2.12 that the probability distribution of the limit Y in (3.30) is the weak limit of a sequence of infinitely divisible laws, and then Theorem 3.1.16 implies that this limit is infinitely divisible. Conversely, if Y is infinitely divisible then, by definition, for each $n = 1, 2, 3, \ldots$ there exist $Y_{n1}, \ldots, Y_{nn}$ i.i.d. such that $Y_{n1} + \cdots + Y_{nn}$ is identically distributed with Y. Then $Y_{n1} + \cdots + Y_{nn} \Rightarrow Y$ and $\{Y_{nj}\}$ is an infinitesimal triangular array. □

Proof of Theorem 3.2.2. Apply Theorem 3.2.12 to see that (3.30) holds if and only if $\psi_n(t) \to \psi(t)$ for all $t \in \mathbb{R}^d$, where $\psi_n(t)$ is defined by (3.46), and e^{ψ} is the characteristic function of the limit Y in (3.30). Comparing (3.46) with (3.5) shows that the infinitely divisible law with characteristic function e^{ψ_n} has Lévy representation $[c_n, 0, \phi_n]$ where $\phi_n = \sum_j \mu'_{nj}$ and

$$c_n = a_n - \sum_{j=1}^{k_n} \left\{ b_{nj} - \int \left(\frac{x}{1 + <x, x>} \right) \mu'_{nj}\{dx\} \right\}.$$

Now Theorem 3.1.16 together with Corollary 3.1.17 shows that (3.30) holds if and only if:

(i) For some σ-finite Borel measure ϕ on $\mathbb{R}^k \setminus \{0\}$ we have

$$\sum_{j=1}^{k_n} \mu'_{nj}(S) \to \phi(S) \tag{3.52}$$

for all Borel subsets S of $\mathbb{R}^k \setminus \{0\}$ for which $\phi(\partial S) = 0$;

(ii) As $n \to \infty$ we have

$$a_n - \sum_{j=1}^{k_n} \left\{ b_{nj} - \int \left(\frac{x}{1 + <x, x>} \right) \mu'_{nj}\{dx\} \right\} \to a \tag{3.53}$$

for some $a \in \mathbb{R}^d$; and

(iii) For some nonnegative definite quadratic form $Q(t)$ on $\mathbb{R}^d$ we have

$$\begin{aligned} Q(t) &= \lim_{\varepsilon \to 0} \limsup_{n \to \infty} \sum_{j=1}^{k_n} \left[\int_{\|x\| < \varepsilon} <t, x>^2 \mu_{nj}\{dx\} \right] \\ &= \lim_{\varepsilon \to 0} \liminf_{n \to \infty} \sum_{j=1}^{k_n} \left[\int_{\|x\| < \varepsilon} <t, x>^2 \mu_{nj}\{dx\} \right] \end{aligned} \tag{3.54}$$

for all $t \in \mathbb{R}^d$

Suppose that these three conditions hold, for some $a_n \in \mathbb{R}^d$. To show that (3.30) still holds if we substitute $\tilde{a}_n = \sum_j b_{nj}$ for a_n, we need to show that for some $\tilde{a} \in \mathbb{R}^d$ we have

$$\lim_{n\to\infty} \sum_{j=1}^{k_n} \int \left(\frac{x}{1 + <x, x>} \right) \mu'_{nj}\{dx\} = \tilde{a}. \tag{3.55}$$

Let $g(x) = x/(1 + <x, x>)$ and $h(x) = xI\{\|x\| < R\}$, for some $R > 0$ such that $\phi\{x : \|x\| = R\} = 0$. By condition (i), we have

$$\sum_{j=1}^{k_n} \int_{\|x\|\geq\delta} (g(x) - h(x))\mu'_{nj}\{dx\} \to \int_{\|x\|\geq\delta} (g(x) - h(x))\phi\{dx\} \tag{3.56}$$

as $n \to \infty$ for any $\delta > 0$ such that $\phi\{x : \|x\| = \delta\} = 0$. Since $g(x) - h(x) = O(\|x\|^3)$ as $x \to 0$, and since by (iii) we have

$$\lim_{n\to\infty} \sum_{j=1}^{k_n} \int_{\|x\|<\delta} \|x\|^2 \mu'_{nj}\{dx\} < C$$

for all small $\delta > 0$, it follows that for some $K > 0$

$$\begin{aligned} &\left\| \sum_{j=1}^{k_n} \int_{\|x\|<\delta} (g(x) - h(x))\mu'_{nj}\{dx\} \right\| \\ &\leq K \sum_{j=1}^{k_n} \int_{\|x\|<\delta} \|x\|^3 \mu'_{nj}\{dx\} \\ &\leq K\delta \int_{\|x\|<\delta} \|x\|^2 \mu'_{nj}\{dx\} \leq K\delta(C + 1) \end{aligned} \tag{3.57}$$

for all large n. Thus

$$\lim_{n\to\infty} \sum_{j=1}^{k_n} \int (g(x) - h(x))\mu'_{nj}\{dx\}$$

exists and is finite. Now in order to complete the proof of (3.55) we will show that

$$\lim_{n\to\infty} \sum_{j=1}^{k_n} \int h(x)\mu'_{nj}\{dx\} = 0. \tag{3.58}$$

Write the expression under the limit in the form

$$\begin{aligned}
&\sum_{j=1}^{k_n} \int_{\|x\|<R} x\ \mu'_{nj}\{dx\} \\
&= \sum_{j=1}^{k_n} \int_{\|x-b_{nj}\|<R} (x - b_{nj})\mu_{nj}\{dx\} \\
&= \sum_{j=1}^{k_n} \int_{\|x-b_{nj}\|<R} x\ \mu_{nj}\{dx\} - \sum_{j=1}^{k_n} b_{nj}(1 - \mu'_{nj}\{x : \|x\| \geq R\}) \\
&= \sum_{j=1}^{k_n} \left(\int_{\|x-b_{nj}\|<R} x\ \mu_{nj}\{dx\} - \int_{\|x\|<R} x\ \mu_{nj}\{dx\} \right) \\
&\quad + \sum_{j=1}^{k_n} b_{nj}\mu'_{nj}\{x : \|x\| \geq R\}.
\end{aligned}$$

In the proof of Lemma 3.2.11, we showed that $\sup_j \|b_{nj}\| \to 0$ for any infinitesimal array. This, together with the fact that $\sum_j \mu'_{nj}\{\|x\| \geq R\} \to \phi\{\|x\| \geq R\}$, shows that the second term tends to zero. For $\varepsilon > 0$ such that $\sup_j \|b_{nj}\| < \varepsilon$, the first term is bounded above by

$$\begin{aligned}
&(R+\varepsilon) \sum_{j=1}^{k_n} \mu'_{nj}\{R - \varepsilon \leq \|x\| \leq R + \varepsilon\} \\
\to\ &(R+\varepsilon)\ \phi\{R - \varepsilon \leq \|x\| \leq R + \varepsilon\}
\end{aligned}$$

as $n \to \infty$. Since $\phi\{\|x\| = R\} = 0$ this limit can be made arbitrarily small by choice of $\varepsilon > 0$, which shows that (3.58) holds, and (3.55) follows immediately. This shows that we may always choose the centering constants a_n in (3.30) according to the formula (3.55). Furthermore, if we choose the centering constants in this manner, then it follows from (3.53) that $a = \tilde{a}$ in the Lévy representation of the limit Y. In view of (3.56)–(3.58), we have that

$$\tilde{a} = \int_{x \neq 0} (g(x) - h(x))\phi\{dx\}$$

which is the same constant as (3.20), so that Y also has characteristic function $\exp(\psi_R(t))$, where $\psi_R(t)$ is given by (3.15) and $a_R = a - \tilde{a} = 0$.

We can complete the proof of Theorem 3.2.2 by showing that conditions (i) and (iii) are equivalent to conditions (a) and (b) of the theorem. Since $\mu'_{nj}\{dx\} = \mu\{dx + b_{nj}\}$ where $\sup_j \|b_{nj}\| \to 0$ it is immediate that (i) is the same as (a). To show that (iii) is equivalent to (b), first note that since

$\sup_j \|b_{nj}\| \to 0$ we have

$$\begin{aligned}
&\lim_{\varepsilon\to 0} \limsup_{n\to\infty} \sum_{j=1}^{k_n} \int_{\|x\|<\varepsilon} <t,x>^2 \mu'_{nj}\{dx\} \\
= \quad &\lim_{\varepsilon\to 0} \limsup_{n\to\infty} \sum_{j=1}^{k_n} \int_{\|x+b_{nj}\|<\varepsilon} <t,x>^2 \mu'_{nj}\{dx\} \\
= \quad &\lim_{\varepsilon\to 0} \limsup_{n\to\infty} \sum_{j=1}^{k_n} \int_{\|x\|<\varepsilon} <t,(x-b_{nj})>^2 \mu_{nj}\{dx\}.
\end{aligned}$$

Since

$$<t, b_{nj}> = \int_{\|x\|<R} <t,x> \mu_{nj}\{dx\}$$

we have

$$\int_{\|x\|<\varepsilon} <t,(x-b_{nj})>^2 \mu_{nj}\{dx\} = I_1 + I_2 + I_3$$

where

$$\begin{aligned}
I_1 &= \int_{\|x\|<\varepsilon} <t,x>^2 \mu_{nj}\{dx\} \\
I_2 &= -2\Big(\int_{\|x\|<R} <t,x> \mu_{nj}\{dx\}\Big) \cdot \Big(\int_{\|x\|<\varepsilon} <t,x> \mu_{nj}\{dx\}\Big) \\
&= -2\Big(\int_{\|x\|<\varepsilon} <t,x> \mu_{nj}\{dx\}\Big)^2 \\
&\quad -2\Big(\int_{\varepsilon\le\|x\|<R} <t,x> \mu_{nj}\{dx\}\Big) \cdot \Big(\int_{\|x\|<\varepsilon} <t,x> \mu_{nj}\{dx\}\Big) \\
I_3 &= \Big(\int_{\|x\|<R} <t,x> \mu_{nj}\{dx\}\Big)^2 \Big(1 - \int_{\|x\|\ge\varepsilon} \mu_{nj}\{dx\}\Big) \\
&= \Big(\int_{\|x\|<\varepsilon} <t,x> \mu_{nj}\{dx\}\Big)^2 \\
&\quad +2\Big(\int_{\varepsilon\le\|x\|<R} <t,x> \mu_{nj}\{dx\}\Big) \cdot \Big(\int_{\|x\|<\varepsilon} <t,x> \mu_{nj}\{dx\}\Big) \\
&\quad +\Big(\int_{\varepsilon\le\|x\|<R} <t,x> \mu_{nj}\{dx\}\Big)^2 \\
&\quad -<t, b_{nj}>^2 \int_{\|x\|\ge\varepsilon} \mu_{nj}\{dx\}
\end{aligned}$$

so that

$$I_1 + I_2 + I_3 = \int_{\|x\|<\varepsilon} <t,x>^2 \mu_{nj}\{dx\} - \left(\int_{\|x\|<\varepsilon} <t,x> \mu_{nj}\{dx\}\right)^2 + I_{nj}$$

where

$$I_{nj} = \left(\int_{\varepsilon\le\|x\|<R} <t,x> \mu_{nj}\{dx\}\right)^2 - <t,b_{nj}>^2 \int_{\|x\|\ge\varepsilon} \mu_{nj}\{dx\}.$$

Then

$$\begin{aligned}\sum_{j=1}^{k_n} I_{nj} &\le \sum_{j=1}^{k_n} \left| \left(\int_{\varepsilon\le\|x\|<R} <t,x> \mu_{nj}\{dx\}\right)^2 - <t,b_{nj}>^2 \int_{\|x\|\ge\varepsilon} \mu_{nj}\{dx\} \right| \\ &\le R^2 \|t\|^2 \sup_{1\le j\le k_n} \mu_{nj}\{\|x\|\ge\varepsilon\} \sum_{j=1}^{k_n} \mu_{nj}\{\|x\|\ge\varepsilon\} \\ &\quad + \|t\|^2 \big(\sup_{1\le j\le k_n} \|b_{nj}\|\big)^2 \sum_{j=1}^{k_n} \mu_{nj}\{\|x\|\ge\varepsilon\}\end{aligned}$$

where $\sup_j \|b_{nj}\| \to 0$ for infinitesimal arrays, and $\sup_j \mu_{nj}\{\|x\| \ge \varepsilon\} \to 0$ by the definition of infinitesimal. If condition (a) of the theorem holds, then $\sum_j \mu_{nj}\{\|x\| \ge \varepsilon\} \to \phi\{\|x\| \ge \varepsilon\}$ and so $\sum_j I_{nj} \to 0$. Since (i) and (a) have already been shown to be equivalent, condition (i) also implies that $\sum_j I_{nj} \to 0$. In either case, it follows that

$$\begin{aligned}&\lim_{\varepsilon\to 0} \limsup_{n\to\infty} \sum_{j=1}^{k_n} \int_{\|x\|<\varepsilon} <t,x>^2 \mu'_{nj}\{dx\} \\ =\ &\lim_{\varepsilon\to 0} \limsup_{n\to\infty} \sum_{j=1}^{k_n} \left[\int_{\|x\|<\varepsilon} <t,x>^2 \mu_{nj}\{dx\} - \left(\int_{\|x\|<\varepsilon} <t,x> \mu_{nj}\{dx\}\right)^2 \right]\end{aligned}$$

and a similar argument shows that the same equivalence holds for liminf. Then (iii) holds if and only (b) holds, which completes the proof. □

Corollary 3.2.15. *In the preceeding Theorem 3.2.2, we have convergence to a normal limit Y if and only if:*

(a) We have

$$\sum_{j=1}^{k_n} \mu_{nj}(S) \to 0 \text{ as } n \to \infty \tag{3.59}$$

for all Borel subsets S of $\mathbb{R}^k \setminus \{0\}$; and

(b) For some nonnegative definite quadratic form Q on $\mathbb{R}^d$ we have

$$Q(t) = \lim_{n\to\infty} \sum_{j=1}^{k_n} \left[\int_{\|x\|<\varepsilon} <t,x>^2 \mu_{nj}\{dx\} - \left(\int_{\|x\|<\varepsilon} <t,x> \mu_{nj}\{dx\} \right)^2 \right] \tag{3.60}$$

for all $\varepsilon > 0$ and all $t \in \mathbb{R}^d$

In this case, the distribution of the limit Y has Lévy representation $[a, Q, 0]$, where $a \in \mathbb{R}^d$ depends on the choice of the centering constants a_n as in Theorem 3.2.2. Furthermore, if (a) holds then it is sufficient that (b) holds for one $\varepsilon = \varepsilon(t) > 0$ for each $t \in \mathbb{R}^d$.

Proof. If we have convergence to a normal limit, then the conditions in Theorem 3.2.2 hold where ϕ is the zero measure, so (3.31) implies (3.59). Now

$$\begin{aligned}
&\sum_{j=1}^{k_n} \left[\int_{\|x\|<\varepsilon} <t,x>^2 \mu_{nj}\{dx\} - \left(\int_{\|x\|<\varepsilon} <t,x> \mu_{nj}\{dx\} \right)^2 \right] \\
&= \sum_{j=1}^{k_n} \int_{\delta \le \|x\|<\varepsilon} <t,x>^2 \mu_{nj}\{dx\} + \sum_{j=1}^{k_n} \int_{\|x\|<\delta} <t,x>^2 \mu_{nj}\{dx\} \\
&\quad - \sum_{j=1}^{k_n} \left(\int_{\|x\|<\delta} <t,x> \mu_{nj}\{dx\} \right)^2 - \sum_{j=1}^{k_n} \left(\int_{\delta \le \|x\|<\varepsilon} <t,x> \mu_{nj}\{dx\} \right)^2 \\
&\quad -2 \sum_{j=1}^{k_n} \left[\left(\int_{\|x\|<\delta} <t,x> \mu_{nj}\{dx\} \right) \cdot \left(\int_{\delta \le \|x\|<\varepsilon} <t,x> \mu_{nj}\{dx\} \right) \right].
\end{aligned}$$

Proposition 1.2.20 implies that

$$\sum_{j=1}^{k_n} \int_{\delta \le \|x\|<\varepsilon} <t,x>^p \mu_{nj}\{dx\} \to \int_{\delta \le \|x\|<\varepsilon} <t,x>^p \phi\{dx\} = 0$$

for $p = 1, 2$. Then the Cauchy-Schwarz inequality implies

$$\sum_{j=1}^{k_n} \left(\int_{\delta \le \|x\|<\varepsilon} <t,x> \mu_{nj}\{dx\} \right)^2 \le \sum_{j=1}^{k_n} \left(\int_{\delta \le \|x\|<\varepsilon} <t,x>^2 \mu_{nj}\{dx\} \right) \to 0.$$

Also, (3.59) implies

$$\begin{aligned}
&\sum_{j=1}^{k_n} \left| \int_{\|x\|<\delta} <t,x>\mu_{nj}\{dx\} \right| \cdot \left| \int_{\delta\le\|x\|<\varepsilon} <t,x>\mu_{nj}\{dx\} \right| \\
\le\ & \delta\|t\| \sum_{j=1}^{k_n} \left| \int_{\delta\le\|x\|<\varepsilon} <t,x>\mu_{nj}\{dx\} \right| \\
\le\ & \delta\|t\|^2\varepsilon \sum_{j=1}^{k_n} \mu_{nj}\{\|x\|\ge\delta\} \to 0.
\end{aligned}$$

Then the limsup equals the liminf in (3.32) and neither depends on $\varepsilon > 0$, so we can dispense with the outer limit, leaving (3.60). Conversely, if (a) and (b) of the corollary hold then the array convergence follows immediately from Theorem 3.2.2. □

Corollary 3.2.16. *Theorem 3.2.2 remains true if the domain of integration* $\{\|x\| < \varepsilon\}$ *in (3.32) is replaced by the set* $\{|<t,x>| < \varepsilon\|t\|\}$.

Proof. It suffices to show that the two versions of (3.32) are equivalent in the case where (3.31) holds. Fix $t \neq 0$ and let $\theta = t/\|t\|$. Let $S_1 = \{\|x\| < \varepsilon\}$ and $S_2 = \{|<x,\theta>| < \varepsilon\}$ denote the two domains of integration, and note that $S_1 \subset S_2$. For all $\varepsilon > 0$ such that S_1, S_2 are ϕ-continuity sets, we have

$$\sum_{j=1}^{k_n} \int_{S_2\setminus S_1} <t,x>^2\mu_{nj}\{dx\} \to \int_{S_2\setminus S_1} <t,x>^2\phi\{dx\} \tag{3.61}$$

and the limit in (3.61) tends to zero as $\varepsilon \to 0$, since $<t,x>^2$ is integrable with respect to $\phi\{dx\}$. By the Cauchy-Schwarz inequality,

$$\begin{aligned}
\lim_{\varepsilon\to0} \limsup_{n\to\infty} \sum_{j=1}^{k_n} \Bigg[& \int_{S_2\setminus S_1} <t,x>^2\mu_{nj}\{dx\} \\
& - \Big(\int_{S_2\setminus S_1} <x,t>\mu_{nj}\{dx\} \Big)^2 \Bigg] = 0.
\end{aligned}$$

Also, in view of (3.29) with $b_{nj} = 0$ we have that

$$\begin{aligned}
&\left| \sum_{j=1}^{k_n} \int_{S_2\setminus S_1} <t,x>\mu_{nj}\{dx\} \int_{S_1} <t,x>\mu_{nj}\{dx\} \right| \\
\le\ & \max_{1\le j\le k_n} \left| \int_{S_1} <t,x>\mu_{nj}\{dx\} \right| \cdot \|t\|\varepsilon \sum_{j=1}^{k_n} \mu_{nj}(S_2\setminus S_1) \qquad (3.62)
\end{aligned}$$

tends to zero as $n \to \infty$, and the desired result follows. □

Corollary 3.2.17. *Corollary 3.2.15 remain true if the domain of integration* $\{\|x\| < \varepsilon\}$ *in (3.60) is replaced by the set* $\{|<t,x>| < \varepsilon\|t\|\}$.

Proof. The proof is similar to Corollary 3.2.16. Again (3.61) holds whenever S_1, S_2 are ϕ-continuity sets, and now the limit in (3.61) is equal to zero, since ϕ is the zero measure. By the Cauchy-Schwarz inequality,

$$\sum_{j=1}^{k_n}\left[\int_{S_2\setminus S_1} <t,x>^2 \mu_{nj}\{dx\} - \left(\int_{S_2\setminus S_1} <x,t>\mu_{nj}\{dx\}\right)^2\right] \to 0.$$

Also (3.62) again holds, and so we can substitute S_2 for S_1 in (3.60) given that (3.59) holds. □

3.3 DOMAINS OF ATTRACTION

In Section 3.2, we derived necessary and sufficient conditions for the row sums of an infinitesimal triangular array to converge weakly to some distribution. As shown in Theorem 3.2.14, the limit is necessarily infinitely divisible. In this section, we apply these convergence criteria to the special case of centered and normalized sums of independent and identically distributed random vectors on $\mathbb{R}^d$.

Let $X_1, X_2, \ldots$ be independent and identically distributed random vectors in $\mathbb{R}^d$ with distribution μ. For $k_n \geq 1$ and $1 \leq j \leq k_n$, let $Z_{nj} = A_n X_j$, where $A_n \in \mathrm{L}(\mathbb{R}^d)$ for $n \geq 1$. Then the collection

$$\{Z_{nj} : 1 \leq j \leq k_n,\ n = 1, 2, 3, \ldots\} \tag{3.63}$$

forms a triangular array with independent and identically distributed rows $Z_{n1}, \ldots, Z_{nk_n}$. For $n \geq 1$, the partial sum $S_n = X_1 + \cdots + X_{k_n}$ has distribution μ^{k_n}, and the sum $A_n S_{k_n}$ of the nth row of the triangular array (3.63) has distribution $A_n \mu^{k_n}$.

Definition 3.3.1. Suppose that $X, X_1, X_2, \ldots$ are independent and identically distributed random vectors on $\mathbb{R}^d$ with common distribution μ and that Y is a random vector on $\mathbb{R}^d$ with distribution ν. We say that X (or equivalently μ) belongs to the *generalized domain of partial attraction* of Y (or ν), if there exist linear operators $A_n \in \mathrm{L}(\mathbb{R}^d)$, nonrandom vectors $a_n \in \mathbb{R}^d$ and an increasing sequence (k_n) of natural numbers tending to infinity such that

$$A_n(X_1 + \cdots + X_{k_n}) + a_n \Rightarrow Y. \tag{3.64}$$

In terms of measures, we can rewrite (3.64) as

$$A_n \mu^{k_n} * \varepsilon_{a_n} \Rightarrow \nu \quad \text{as } n \to \infty. \tag{3.65}$$

We write $\mu \in \mathrm{GDOPA}(\nu)$ in this case. We call (A_n) the sequence of *norming operators*, (a_n) the sequence of *shifts* or *centering constants* and (k_n) the *sampling sequence.*

The following examples illustrate that, unless we assume that the distribution of Y is full, the mere fact that X belongs to the generalized domain of partial attraction of Y says very little about either the norming operators A_n or the distribution of X.

Example 3.3.2. Let $d = 2$ and let N be a real valued random variable with a standard normal distribution and let

$$Y = \begin{pmatrix} N \\ 0 \end{pmatrix}.$$

Note that the distribution of Y is not full. Let $U_1, U_2, \ldots$ be independent and identically distributed random variables with uniform distribution on $[-\sqrt{3}, \sqrt{3}]$ so that $E(U_i^2) = 1$ and define

$$X_i = \begin{pmatrix} U_i \\ 1 \end{pmatrix}.$$

Let $S_n = X_1 + \cdots + X_{k_n}$. Then for the same sampling sequence $k_n = n$ we can choose different sequences of norming operators (A_n) to get the same limiting distribution.

(a) Let

$$A_n = \begin{pmatrix} n^{-1/2} & 0 \\ 0 & 0 \end{pmatrix}.$$

Then $\|A_n\| \to 0$ and the classical central limit theorem implies

$$A_n S_n = \begin{pmatrix} n^{-1/2} \sum_{i=1}^{n} U_i \\ 0 \end{pmatrix} \Rightarrow \begin{pmatrix} N \\ 0 \end{pmatrix} = Y.$$

Furthermore, $A_n X_i \to 0$ in probability as $n \to \infty$, so the triangular array (3.63) is infinitesimal.

(b) Let

$$A_n = \begin{pmatrix} n^{-1/2} & 0 \\ 0 & 1 \end{pmatrix}.$$

Then $\|A_n\| \to 1$ and if we choose shifts

$$a_n = \begin{pmatrix} 0 \\ -n \end{pmatrix}$$

then

$$A_n S_n + a_n = \begin{pmatrix} n^{-1/2} \sum_{i=1}^n U_i \\ 0 \end{pmatrix} \Rightarrow \begin{pmatrix} N \\ 0 \end{pmatrix} = Y.$$

But here

$$A_n X_i = \begin{pmatrix} n^{-1/2} U_i \\ 1 \end{pmatrix} \to \begin{pmatrix} 0 \\ 1 \end{pmatrix}$$

in probability, so the triangular array (3.63) is not infinitesimal.

If we assume that the distribution of Y is full, then the fact that X belongs to the domain of partial attraction of Y implies that the triangular array (3.63) is infinitesimal, and that the distribution of Y is infinitely divisible. Then we can employ the theory of infinitely divisible distributions and convergence of triangular arrays developed in Sections 3.1 and 3.2.

Lemma 3.3.3. *Suppose that $\mu \in \mathrm{GDOPA}(\nu)$ and (3.65) holds. If ν is full then μ is full, A_n is invertible for all large n, $\|A_n\| \to 0$ as $n \to \infty$, and the triangular array (3.63) is infinitesimal.*

Proof. Let μ_1, ν_1 denote the symmetrization of μ, ν as in Definition 1.3.2. It follows easily from (3.65) that $A_n \mu_1^{k_n} \Rightarrow \nu_1$. Lemma 1.3.3 shows that $\hat{\nu}_1$ is real, and Proposition 1.3.5 (b) implies that $\hat{\nu}_1$ is continuous. Since $\hat{\nu}_1(0) = 1$, there exists some $a > 0$ such that $\hat{\nu}_1(x) > 0$ for $\|x\| \leq a$. Hence for $\|x\| \leq a$ we have $\hat{\mu}_1(A_n^* x) \to 1$. Since $\hat{\mu}_1$ is real, it follows from Proposition 1.3.4 that $\hat{\mu}_1(A_n^* x) \to 1$ for all $x \in \mathbb{R}^d$. Then Lévy's Continuity Theorem 1.3.6 implies that

$$A_n \mu_1 \Rightarrow \varepsilon_0 \quad \text{as } n \to \infty. \tag{3.66}$$

Now we want to show that $\|A_n\| \to 0$ in the Euclidean operator norm on $\mathrm{L}(\mathbb{R}^d)$. We argue by contradiction. Assume (without loss of generality, by passing to a subsequence) that $\|A_n\| \to \delta > 0$ as $n \to \infty$. Recall from Remark 2.1.5 that $\|A_n^*\| = \|A_n\|$, and select vectors $\|z_n\| = 1$ with $\|A_n^* z_n\| = \|A_n\|$ for all $n \geq 1$. Set $u_n = \|A_n^*\|^{-1} A_n^* z_n$. Then $\|u_n\| = 1$, so without loss of generality we assume that $u_n \to u$ for some $\|u\| = 1$. Since the sequence $(\|A_n^*\|^{-1} z_n)_n$ is bounded (3.66) along with Lévy's Continuity Theorem implies that for all $t \in \mathbb{R}$

$$\lim_{n \to \infty} (A_n \mu_1)\hat{}\ (t \|A_n^*\|^{-1} z_n) = 1.$$

On the other hand,

$$(A_n \mu_1)\hat{}\ (t \|A_n^*\|^{-1} z_n) = \hat{\mu}_1(t u_n) \to \hat{\mu}_1(t u).$$

Thus $\hat{\mu}_1(tu) = 1$ for all $t \in \mathbb{R}$ and some $\|u\| = 1$. Consequently, $(\hat{\mu}_1(tu))^{k_n} = 1$ for all $t \in \mathbb{R}$ and all n.

Since ν is full, Lemma 1.3.14 implies that ν_1 is full. Then it follows from Lemma 2.3.7 that A_n is invertible and $\mu_1^{k_n}$ is full for all large n. Then Lemma 1.3.13 yields that μ_1 is full, and Lemma 1.3.14 shows that μ is full. Let $x_n = (A_n^*)^{-1}u$ and note that $x_n \neq 0$ since $u \neq 0$. Let $\bar{x}$ be an arbitrary limit point of the sequence $(x_n/\|x_n\|)$. Then, since $A_n\mu_1^{k_n} \Rightarrow \nu_1$ we obtain, by passing to a subsequence if necessary, that

$$(A_n\mu_1^{k_n})^\wedge\left(t\frac{x_n}{\|x_n\|}\right) \to \hat{\nu}_1(t\bar{x}).$$

But

$$(A_n\mu_1^{k_n})^\wedge\left(t\frac{x_n}{\|x_n\|}\right) = \left[\hat{\mu}_1\left(\frac{t}{\|x_n\|}A_n^*(A_n^*)^{-1}u\right)\right]^{k_n} = \hat{\mu}_1\left(\frac{t}{\|x_n\|}u\right)^{k_n} = 1$$

for all n. Thus we obtain $\hat{\nu}_1(t\bar{x}) = 1$ for all $t \in \mathbb{R}$ and some $\|\bar{x}\| = 1$, and then Proposition 1.3.11 implies that ν_1 is not full, which is a contradiction. Hence $\|A_n\| \to 0$.

Since $\|A_n\| = \|A_n^*\|$ we also have $A_n^*x \to 0$ for all $x \in \mathbb{R}^d$, and since $\hat{\mu}$ is continuous this implies that $\hat{\mu}(A_n^*x) \to \hat{\mu}(0) = 1$ for all $x \in \mathbb{R}^d$. Then Lévy's Continuity Theorem yields $A_n\mu \to \varepsilon_0$ so that $A_nX_i \to 0$ in probability for $1 \leq i \leq k_n$, hence the triangular array (3.63) is infinitesimal. □

Theorem 3.3.4. *If ν is full, then* GDOPA(ν) *is not empty if and only if ν is infinitely divisible.*

Proof. Suppose that ν is full, $\mu \in$ GDOPA(ν) and (3.65) holds. Then Lemma 3.3.3 implies that the triangular array (3.63) is infinitesimal, and it follows from Theorem 3.2.14 that ν is infinitely divisible. For the proof of the converse implication, we need the following technical result, which we also need for the construction of the so-called universal laws below. □

Lemma 3.3.5. *Let (ν_n) be a sequence of infinitely divisible laws on $\mathbb{R}^d$ with log-characteristic functions ψ_n such that $\|\psi_n\|_\infty \leq 2n$. Then there exist an increasing sequence (n_k) of positive integers tending to infinity and constants $a_k > 0$ such that the series*

$$\lambda(t) = \sum_{k=1}^{\infty} \psi_k(a_k t)/n_k \tag{3.67}$$

converges for all $t \in \mathbb{R}^d$ and is the log-characteristic function of some infinitely divisible law. Furthermore, we have

$$n_r\lambda(t/a_r) - \psi_r(t) \to 0 \quad \text{as } r \to \infty \tag{3.68}$$

for all $t \in \mathbb{R}^d$.

Proof. Let $n_1 = 1$ and define n_k inductively by $n_k = 2^{k+2} \cdot k \cdot n_{k-1}$. Since $\|\psi_k\|_\infty \le 2k$ we then get that $n_k/n_{k-1} > 2^k\|\psi_k\|_\infty$. Furthermore, since the series $\sum_{k=1}^\infty (2k)/n_k < \infty$, it follows that the series in (3.67) is absolutely converging for every $t \in \mathbb{R}^d$ and any choice of the constants a_k. Note that by Lemma 3.1.5 $\sum_{k=1}^N \psi_k(a_k t)/n_k$ is the log-characteristic function of some infinitely divisible law. Then an application of Lemma 3.1.6 together with Lévy's Continuity Theorem 1.3.6 shows that $\lambda(t)$ is the log-characteristic function of some infinitely divisible law for any choice of $a_k > 0$.

Now we construct $a_k > 0$ such that (3.68) holds. Let $a_1 = 1$ and assume that for $r \ge 2$ $a_1, \ldots, a_{r-1}$ are given. Consider the expression

$$n_r \sum_{k=1}^{r-1} |\psi_k(ta_k/a_r)| + \sum_{k=r+1}^{\infty} 2^{-k}. \tag{3.69}$$

Since every ψ_k is continuous with $\psi_k(0) = 0$ we can choose $a_r > 0$ large enough so that the expression (3.69) is smaller than $1/r$ for all $\|t\| < r$.

By our choice of the sequence (n_k), it follows that for $k \ge r+1$ we have $n_r/n_k \le n_{k-1}/n_k < 2^{-k}\|\psi_k\|_\infty^{-1}$. Hence, using the monotonicity of n_k and (3.69) we obtain

$$\begin{aligned} |n_r\lambda(t/a_r) - \psi_r(t)| &\le \sum_{k=1}^{r-1} |\psi_k(ta_k/a_r)| \frac{n_r}{n_k} + \sum_{k=r+1}^{\infty} \|\psi_k\|_\infty \frac{n_r}{n_k} \\ &< n_r \sum_{k=1}^{r-1} |\psi_k(ta_k/a_r)| + \sum_{k=r+1}^{\infty} 2^{-k} < \frac{1}{r} \end{aligned}$$

for all $\|t\| < r$ and (3.68) follows using Proposition 1.3.4. □

Proof of Theorem 3.3.4(Continued). If ν is a full infinitely divisible law, then Corollary 3.1.8 shows that ν is the weak limit of the compound Poisson distributions $\Pi(n, \mu_n)$, where μ_n is the nth root of ν. Note that $\Pi(n, \mu_n)$ has the log-characteristic function $\psi_n = n(\hat{\mu}_n - 1)$ and that by Lemma 3.1.10 we have $\psi_n(t) \to \psi(t)$ as $n \to \infty$ for all $t \in \mathbb{R}^d$, where ψ is the log-characteristic function of ν. Since $|\hat{\mu}_n| \le 1$ it follows that $\|\psi_n\|_\infty \le 2n$. Apply Lemma 3.3.5 to obtain a log-characteristic function λ and define μ by $\hat{\mu} = e^\lambda$. Then it follows from (3.68) that

$$\begin{aligned} \lim_{r\to\infty} \left(\frac{1}{a_r}\mu^{n_r}\right)^{\wedge}(t) &= \lim_{r\to\infty} \exp\big(n_r\lambda(t/a_r)\big) \\ &= \lim_{r\to\infty} \exp(\psi_r(t)) \\ &= \exp(\psi(t)) = \hat{\nu}(t) \end{aligned}$$

for all $t \in \mathbb{R}^d$. By Lévy's Continuity Theorem 1.3.6, this is equivalent to $(1/a_r)\mu^{n_r} \Rightarrow \nu$ so $\mu \in \mathrm{GDOPA}(\nu)$, which concludes the proof. □

Recall from Definition 2.3.1 that two probability distributions $\nu, \nu' \in \mathcal{M}^1$ are the same type if $\nu' = A\nu * \varepsilon_a$ for some $A \in \mathrm{GL}(\mathbb{R}^d)$ and some $a \in \mathbb{R}^d$.

Lemma 3.3.6. *If two full infinitely divisible laws ν, ν' are of the same type, then* $\mathrm{GDOPA}(\nu) = \mathrm{GDOPA}(\nu')$.

Proof. If ν, ν' are the same type, then $\nu' = \alpha\nu$, where $\alpha(x) = Ax + a$ for some $A \in \mathrm{GL}(\mathbb{R}^d)$ and some $a \in \mathbb{R}^d$. Suppose that $\mu \in \mathrm{GDOPA}(\nu)$ so that (3.65) and (3.64) hold. Write $\beta_n(x) = A_n(x) + a_n$ and $\mu_n = \mu^{k_n}$, so that $\beta_n\mu_n \Rightarrow \nu$. Then the Convergence of Types Theorem 2.3.17 implies that $\alpha\beta_n\mu_n \to \alpha\nu = \nu'$, which shows that $\mathrm{GDOPA}(\nu) \subset \mathrm{GDOPA}(\nu')$. Interchange the roles of ν, ν' to obtain the reverse inclusion. Then $\mathrm{GDOPA}(\nu) = \mathrm{GDOPA}(\nu')$. □

The following technical lemma will be useful in Chapter 7 to interpolate between integer convolution powers.

Proposition 3.3.7. *If $\mu \in \mathrm{GDOPA}(\nu)$ and (3.65) holds for some full ν, then ν is infinitely divisible and*

$$A_n\mu^{[k_n t_n]} * \varepsilon_{a_n(t_n)} \Rightarrow \nu^t \tag{3.70}$$

whenever $t_n \to t > 0$, where

$$a_n(t) = \frac{[k_n t]}{k_n} a_n. \tag{3.71}$$

Proof. Theorem 3.3.4 implies that ν is infinitely divisible, and hence ν^t is well defined for every $t > 0$. Lévy's Continuity Theorem 1.3.6 implies that (3.65) is equivalent to

$$\hat{\mu}(A_n^*\xi)^{k_n} e^{i<\xi, a_n>} \to \hat{\nu}(\xi) \quad \text{as } n \to \infty \tag{3.72}$$

for all $\xi \in \mathbb{R}^d$. Using (3.71) the characteristic function of the left-hand side of (3.70) can be written as

$$\hat{\mu}(A_n^*\xi)^{[k_n t_n]} e^{i<\xi, a_n(t_n)>} = \left(\hat{\mu}(A_n\xi)^{k_n} e^{i<\xi, a_n>}\right)^{[k_n t_n]/k_n} \to \hat{\nu}(\xi)^t$$

whenever $t_n \to t > 0$ by (3.72). Applying Theorem 1.3.6 again yields the assertion. □

A characterization of generalized domains of partial attraction for full measures is given by the following special case of the general convergence criteria for infinitesimal triangular arrays, Theorem 3.2.2. Recall from Definition 1.1.5 that for any Borel measure μ on $\mathbb{R}^d$ and any $T : \mathbb{R}^d \to \mathbb{R}^m$ Borel measurable we define the Borel measure $T\mu$ on $\mathbb{R}^m$ by letting $T\mu(B) = \mu(T^{-1}B)$ for all Borel subsets $B \subseteq \mathbb{R}^d$, where the $T^{-1}B$ is the preimage. If the random vector X has probability distribution μ then and $A \in \mathrm{L}(\mathbb{R}^d)$ then $A\mu$ is the probability distribution of the random vector AX. Recall from Definition 1.2.17 that

$\mathcal{M}$ is the class of σ-finite Borel measures on $\Gamma = \mathbb{R}^d \setminus \{0\}$ that assign finite measure to sets bounded away from zero, and that $\mu_n \to \mu$ if $\mu_n(B) \to \mu(B)$ for all μ-continuity sets B that are bounded away from zero.

Theorem 3.3.8. *If ν is full and $\mu \in$ GDOPA(ν) then ν is infinitely divisible. Furthermore, if $[a, Q, \phi]$ denotes the Lévy triple of ν, then (3.65) holds if and only if*

$$k_n \cdot A_n\mu \to \phi \quad in\ \mathcal{M} \tag{3.73}$$

and

$$\begin{aligned} Q(t) &= \lim_{\varepsilon\to 0} \limsup_{n\to\infty} k_n \Big[\int_{\|x\|<\varepsilon} <t,x>^2 A_n\mu\{dx\} - \Big(\int_{\|x\|<\varepsilon} <t,x> A_n\mu\{dx\}\Big)^2\Big] \\ &= \lim_{\varepsilon\to 0} \liminf_{n\to\infty} k_n \Big[\int_{\|x\|<\varepsilon} <t,x>^2 A_n\mu\{dx\} - \Big(\int_{\|x\|<\varepsilon} <t,x> A_n\mu\{dx\}\Big)^2\Big] \end{aligned} \tag{3.74}$$

for all $t \in \mathbb{R}^d$.

The shifts a_n may be chosen according to the formula

$$a_n = -k_n \int_{\|x\|<R} x A_n\mu\{dx\} \tag{3.75}$$

where $R > 0$ is chosen so that $\phi\{x : \|x\| = R\} = 0$. In that case, ν has characteristic function $\exp(\psi_R(t))$ where $\psi_R(t)$ is given by (3.15) with $a_R = 0$.

Proof. Note that by Lemma 3.3.3 the triangular array (3.63) is infinitesimal and the distribution of Z_{nj} is $\mu_{nj} = A_n\mu$. Then the result is just a reformulation of Theorem 3.2.2. □

In case of attraction to a normal law, we get the following special version of Corollary 3.2.15

Corollary 3.3.9. *Let ν be a full normal law with Lévy triple $[a, Q, 0]$. Then $\mu \in$ GDOPA(ν) and (3.65) holds if and only if*

$$k_n \cdot (A_n\mu) \to 0 \quad in\ \mathcal{M} \tag{3.76}$$

and

$$Q(t) = \lim_{n\to\infty} k_n \left[\int_{\|x\|<\varepsilon} <t,x>^2 A_n\mu\{dx\} - \left(\int_{\|x\|<\varepsilon} <t,x> A_n\mu\{dx\}\right)^2\right] \tag{3.77}$$

for all $\varepsilon > 0$ and all $t \in \mathbb{R}^d$. In this case, $a \in \mathbb{R}^d$ depends on the choice of the centering constants a_n as in Theorem 3.3.8. Furthermore, if (3.76) holds, then it is sufficient that (3.77) holds for one $\varepsilon = \varepsilon(t) > 0$ for each $t \in \mathbb{R}^d$.

Proof. This Corollary is just a reformulation of Corollary 3.2.15. □

We also have the following alternative characterizations of GDOPA that we will use in most of our applications.

Corollary 3.3.10. *Theorem 3.3.8 and Corollary 3.3.9 remain true if the we replace the domain of integration* $\{x : \|x\| < \varepsilon\}$ *in* (3.74) *and* (3.77) *by* $\{x : |<t, x>| < \varepsilon\|t\|\}$.

Proof. This Corollary is just a special case of Corollaries 3.2.16 and 3.2.17. □

In Chapter 8, we will consider the two separate cases of a normal and a nonnormal limit. In the case of a nonnormal limit, we will apply the following sufficient condition.

Proposition 3.3.11. *Suppose that for a probability measure* μ *there exist linear operators* $A_n \in \mathrm{L}(\mathbb{R}^d)$ *and an increasing sequence* (k_n) *of natural numbers tending to infinity such that*

$$k_n \cdot A_n\mu \to \phi \tag{3.78}$$

for some full $\phi \in \mathcal{M}$, *and*

$$\lim_{\varepsilon\to 0} \limsup_{n\to\infty} k_n \int_{|<x,t>|<\varepsilon\|t\|} <x, t>^2 A_n\mu\{dx\} = 0 \tag{3.79}$$

for all $t \in \mathbb{R}^d$. *Then there exists a full infinitely divisible probability measure* ν *with Lévy representation* $[a, 0, \phi]$ *for some* $a \in \mathbb{R}^d$ *such that* $\mu \in \mathrm{GDOPA}(\nu)$, *e.g.* (3.65) *holds. The shifts can be chosen according to the formula* (3.75), *and then* $a \in \mathbb{R}^d$ *depends on the choice of the centering constants* a_n *as in Theorem 3.3.8.*

Proof. In view of Theorem 3.3.8 and Corollary 3.3.10, we have to show that (3.79) implies (3.74) with domain of integration equal to $\{x : |<x, t>| < \varepsilon\|t\|\}$. Note that since $Q = 0$ it is enough to consider the limsup. Since $A_n\mu$ is a probability measure, $A_n\mu\{x : |<x, t>| < \varepsilon\|t\|\} \leq 1$, and hence by the Cauchy-Schwarz Inequality we have

$$\left(\int_{|<x,t>|<\varepsilon\|t\|} <x, t> A_n\mu\{dx\}\right)^2 \leq \int_{|<x,t>|<\varepsilon\|t\|} <x, t>^2 A_n\mu\{dx\}$$

and the equivalent version of (3.74) follows from (3.79). It follows from Proposition 3.1.20 that since ϕ is $\mathcal{M}$-full ν is full. □

In the case of a normal limit, we will show in Chapter 8 that we can always reduce to the case where $EX = 0$. In this case, we can apply the following results. First we show that, for the mean zero case, the squared term in Corollary 3.3.9 can be deleted.

Corollary 3.3.12. *Let ν be a full normal law with Lévy triple $[a, Q, 0]$, and suppose that $EX = 0$, where X has distribution μ. Then $\mu \in \mathrm{GDOPA}(\nu)$ and (3.65) holds if and only if*

$$k_n \cdot (A_n \mu) \to 0 \quad \text{as } n \to \infty \tag{3.80}$$

and

$$Q(t) = \lim_{n\to\infty} k_n \int_{\|x\|<\varepsilon} <t,x>^2 A_n\mu\{dx\} \tag{3.81}$$

for all $\varepsilon > 0$ and all $t \in \mathbb{R}^d$. The shifts can be chosen according to the formula (3.75), and then $a \in \mathbb{R}^d$ depends on the choice of the centering constants a_n as in Theorem 3.3.8. Furthermore, if (3.80) holds, then it is sufficient that (3.81) holds for one $\varepsilon = \varepsilon(t) > 0$ for each $t \in \mathbb{R}^d$.

Proof. Suppose that $EX = 0$ and that for all $\varepsilon > 0$ we have

$$k_n \left[\int_{\|x\|<\varepsilon} <t,x>^2 A_n\mu\{dx\} - \left(\int_{\|x\|<\varepsilon} <t,x> A_n\mu\{dx\} \right)^2 \right] \to Q(t) \tag{3.82}$$

for all $t \in \mathbb{R}^d$. Fix $t \in \mathbb{R}^d$ and write $A_n^* t = r_n \theta_n$ where $\|\theta_n\| = 1$ and $r_n > 0$. Lemma 3.3.3 implies that $\|A_n\| = \|A_n^*\| \to 0$, and hence $r_n \to 0$. Then (3.82) is equivalent to

$$k_n r_n^2 \left[I_n - (J_n)^2 \right] \to Q(t) \tag{3.83}$$

as $n \to \infty$, where

$$\begin{aligned} I_n &= \int_{\|A_n x\| \le \varepsilon} <x, \theta_n>^2 \mu\{dx\} \\ J_n &= \int_{\|A_n x\| \le \varepsilon} <x, \theta_n> \mu\{dx\}. \end{aligned}$$

Write $J_n = \int f_n(x)\mu\{dx\}$ where $f_n(x) = <x, \theta_n> I(\|A_n x\| \le \varepsilon)$. Since $\|\theta_n\| = 1$ every subsequence has a further subsequence with $\theta_n \to \theta$. Then $f_n(x) \to f(x) = <x, \theta>$ pointwise along that subsequence, and since $|f_n(x)| \le \|x\|$, which is μ-integrable in view of the fact that EX exists, it follows using dominated convergence that $J_n \to 0$ along this subsequence. Since every subsequence has a further subsequence with this property, it follows that $J_n \to 0$.

Suppose by way of contradiction that $\liminf_{n\to\infty} I_n = 0$. Choose a subsequence along which $I_n \to 0$, and then choose a further subsequence (n') such that $\theta_{n'} \to \theta$. Write $I_{n'} = \int f_{n'}(x)\mu\{dx\}$ where

$$f_{n'}(x) = <x, \theta_{n'}>^2 I(\|A_{n'} x\| \le \varepsilon).$$

Then $f_{n'}(x) \geq 0$ and $f_{n'}(x) \to f(x) = \langle x, \theta \rangle^2$ pointwise, so Fatou's Lemma implies that

$$\int \langle x, \theta \rangle^2 \mu\{dx\} = \int f(x)\mu\{dx\} \leq \liminf_{n' \to \infty} \int f_{n'}(x)\mu\{dx\} = 0$$

which implies that μ is supported on the subspace $\{x \in \mathbb{R}^d : \langle x, \theta \rangle = 0\}$. Then μ is not full. But since ν is full, this contradicts Lemma 3.3.3. Hence there exists some $\delta > 0$ and some $n_0 \geq 1$ such that $I_1^{(n)} \geq \delta$ for all $n \geq n_0$. It follows that

$$\left(J_n\right)^2 = o\left(I_n\right) \quad \text{as } n \to \infty$$

for any $\varepsilon > 0$ and $t \in \mathbb{R}^d$. Then (3.83), and hence (3.82), is equivalent to $k_n r_n^2 I_n \to Q(t)$, which is the same as (3.81). Now the desired result follows from Corollary 3.3.9. □

Next, we show that since the convergence in (3.81) involves quadratic forms, it is automatically uniform convergence on compact subsets.

Lemma 3.3.13. *Suppose ν is a full normal law with Lévy triple $[a, Q, 0]$, and suppose that $EX = 0$, where X has distribution μ. If $\mu \in \mathrm{GDOPA}(\nu)$ and (3.65) holds, then for all $\varepsilon > 0$ we have*

$$\lim_{n \to \infty} k_n \int_{\|x\| \leq \varepsilon} \langle x, t_n \rangle^2 A_n \mu\{dx\} = Q(t) \tag{3.84}$$

whenever $t_n \to t$ in $\mathbb{R}^d$.

Proof. Suppose that ν is a full normal law on $\mathbb{R}^d$ with Lévy triple $[a, Q, 0]$. If $\mu \in \mathrm{GDOPA}(\nu)$ and (3.65) holds with $EX = 0$, then in view of Corollary 3.3.12 we have for any $\varepsilon > 0$ that $Q_n(t) \to Q(t)$ as $n \to \infty$ for all $t \in \mathbb{R}^d$, where $Q_n(t) = \langle t, C_n t \rangle$ and

$$C_n = k_n \int_{\|x\| \leq \varepsilon} x x' A_n \mu\{dx\}$$

where xx' is the outer product, so that Q_n is a quadratic form on $\mathbb{R}^d$. Then Corollary 2.1.9 shows that (3.84) holds. □

Finally we show that, as in Corollary 3.3.10, we can replace the domains of integration in (3.81) and (3.84) by something more convenient for application of the regular variation tools we develop in Part II.

Corollary 3.3.14. *Corollary 3.3.12 remains true if we replace the domain of integration $\{x : \|x\| < \varepsilon\}$ in (3.81) by $\{x : |\langle t, x \rangle| \leq \varepsilon \|t\|\}$.*

Proof. The proof is similar to Corollary 3.2.17. It suffices to show that the two versions of (3.81) are equivalent in the case where (3.80) holds. Fix $t \neq 0$

and let $\theta = t/\|t\|$. Let $S_1 = \{\|x\| < \varepsilon\}$ and $S_2 = \{|<x, \theta>| \leq \varepsilon\}$ denote the two domains of integration, and note that $S_1 \subset S_2$. For all $\varepsilon > 0$ such that S_1, S_2 are ϕ-continuity sets, we have in view of (3.80) that the difference between the two versions of (3.81) is

$$k_n \int_{S_2 \setminus S_1} <t, x>^2 A_n\mu\{dx\} \leq (\varepsilon\|t\|)^2 k_n A_n\mu(S_2 \setminus S_1) \to 0 \tag{3.85}$$

and hence the two versions are equivalent if we assume that (3.80) holds. □

Corollary 3.3.15. *Lemma 3.3.13 remains true if we replace the domain of integration* $\{x : \|x\| < \varepsilon\}$ *in* (3.84) *by* $\{x : |<t_n, x>| \leq 1\}$.

Proof. The proof is similar to Corollary 3.3.14. If $\mu \in \text{GDOPA}(\nu)$, then (3.80) holds. Let $S_1 = \{\|x\| < \varepsilon\}$ and $S_2(n) = \{|<x, t_n>| \leq 1\}$ denote the two domains of integration in the two different versions of (3.84). If $t_n \to t$ in $\mathbb{R}^d$, then $C = \sup\{\|t_n\| : n \geq 1\} < \infty$. Choose $\varepsilon < C^{-1}$. Then $\|x\| < \varepsilon$ implies that $|<x, t_n>| \leq C\|x\| \leq 1$, so that $S_1 \subset S_2(n)$. For all $\varepsilon > 0$ such that S_1 is a ϕ-continuity set, we have in view of (3.80) that the difference between the two versions of (3.84) is

$$k_n \int_{S_2(n) \setminus S_1} <t_n, x>^2 A_n\mu\{dx\} \leq k_n A_n\mu\{x : \|x\| \geq \varepsilon\} \to 0 \tag{3.86}$$

and hence the two versions are equivalent. □

The purpose of a limit theorem for sums of independent and identically distributed random vectors is to estimate the distribution of the sums, and thereby to obtain useful information about the underlying distribution of the summands. Suppose that $\mu \in \text{GDOPA}(\nu)$ according to Definition 3.3.1 and (3.64) holds. Example 3.3.2 illustrates the fact that we must require ν to be full in order to obtain any useful information. If ν is full, then Lemma 3.3.3 implies that the norming operators A_n in (3.64) are invertible for all large n. Then we can estimate the distribution of the sum $(X_1 + \cdots + X_{k_n})$ by that of the centered and normalized limit $A_n^{-1}(Y - a_n)$ for large n, and Theorem 3.3.4 implies that the distribution ν of Y is infinitely divisible. Hence it seems that generalized domains of partial attraction of full infinitely divisible laws are the appropriate limit theorems to study. Unfortunately, this is still far too general to allow us to obtain useful information about the distribution of $\mu \in \text{GDOPA}(\nu)$ for a given full infinitely divisible law ν. In fact, we will show now that there exists a probability measure μ called a *universal law* that belongs to the generalized domain of partial attraction of every infinitely divisible law (including those that are not full).

We first prove a few preliminary results.

Lemma 3.3.16. *Let* (ν_n) *be a sequence of infinitely divisible laws. Then there exists a measure* $\mu \in M^1(\mathbb{R}^d)$ *such that* $\mu \in \text{GDOPA}(\nu_n)$ *for all* $n \geq 1$.

Proof. Let α_n denote the log-characteristic function of ν_n and for $n \geq 1$ and natural $r \geq 1$ let $\psi_r^{(n)} = r(e^{\alpha_n/r} - 1)$. Then $\psi_r^{(n)}$ is the log-characteristic function of the compound Poisson distribution $\Pi(r, \nu_n^{1/r})$, where $\nu_n^{1/r}$ is the r-th root of ν_n, and it follows from Corollary 3.1.8 along with Lemma 3.1.10 that $\psi_r^{(n)} \to \alpha_n$ as $r \to \infty$. Note that $\|\psi_r^{(n)}\|_\infty \leq 2r$.

For $n \geq 1$, let

$$N_n = \{k \geq 1 : k \text{ is divisible by } 2^{n-1} \text{ but not by } 2^n\}$$

so that $\mathbb{N} = \bigcup_{n=1}^\infty N_n$ and the N_n are mutually disjoint. We define a sequence (ψ_r) of log-characteristic functions such that

$$\{\psi_r : r \geq 1\} = \bigcup_{n=1}^{\infty} \{\psi_r^{(n)} : r \in N_n\}$$

such that

$$\psi_r \to \alpha_n \quad \text{as } r \to \infty \text{ and } r \in N_n. \tag{3.87}$$

In fact, we let $\psi_r = \psi_j^{(n)}$ if $r \in N_n$ and

$$j = \min\{i \geq 1 : \psi_i^{(n)} \neq \psi_\ell \text{ for all } 1 \leq \ell \leq r-1\}.$$

Since $\|\psi_r\|_\infty = \|\psi_j^{(n)}\|_\infty \leq 2j \leq 2r$ we can apply Lemma 3.3.5 to obtain a log-characteristic function λ, an increasing sequence (n_r) of natural numbers tending to infinity and constants $a_r > 0$ such that (3.68) holds. Then for any $n \geq 1$ along the subsequence N_n we get from (3.68) and (3.87) that

$$n_r \lambda(t/a_r) \to \alpha_n(t) \quad \text{as } r \to \infty \text{ and } r \in N_n. \tag{3.88}$$

Now define μ by $\hat{\mu} = e^\lambda$ and note that by (3.88)

$$\left(\frac{1}{a_r}\mu\right)^{\wedge}(t) = \exp(n_r\lambda(t/a_r)) \to e^{\alpha_n(t)} = \hat{\nu}_n(t) \quad \text{as } r \to \infty \text{ and } r \in N_n$$

and Lévy's Continuity Theorem implies $\mu \in \mathrm{GDOPA}(\nu_n)$ for all $n \geq 1$. □

Lemma 3.3.17. *Let $\mu \in \mathrm{GDOPA}(\nu_m)$ for all $m \geq 1$ and assume that $\nu_m \Rightarrow \nu$ as $m \to \infty$. Then $\mu \in \mathrm{GDOPA}(\nu)$.*

Proof. Since $\mu \in \mathrm{GDOPA}(\nu_m)$ for all $m \geq 1$ there exist $A_n^{(m)} \in \mathrm{L}(\mathbb{R}^d)$, $b_n^{(m)} \in \mathbb{R}^d$ and sequences $(k_n^{(m)})$ tending to infinity such that for all $m \geq 1$

$$A_n^{(m)}\mu^{k_n^{(m)}} * \varepsilon_{b_n^{(m)}} \Rightarrow \nu_m \quad \text{as } n \to \infty.$$

By Lévy's Continuity Theorem, this implies that for all $m \geq 1$

$$a_n^{(m)}(t) = \left|\hat{\mu}(A_n^{(m)*}t)^{k_n^{(m)}} e^{i<b_n^{(m)},t>} - \hat{\nu}_m(t)\right| \to 0$$

as $n \to \infty$ uniformly on compact subsets of $t \in \mathbb{R}^d$.

Hence, for $m \geq 1$ fixed there exists a $n_m \geq 1$ such that for all $n \geq n_m$ and all $\|t\| \leq m$ we have $a_n^{(m)}(t) \leq 1/m$. We can assume without loss of generality that n_m is increasing to infinity as $m \to \infty$. Then, especially $a_{n_m}^{(m)}(t) \leq 1/m$ for all $\|t\| \leq m$ and $m \geq 1$. Now define $A_m = A_{n_m}^{(m)}$, $k_m = k_{n_m}^{(m)}$ and $b_m = b_{n_m}^{(m)}$. Then, for any fixed $t \in \mathbb{R}^d$ we get

$$\left|\hat{\mu}(A_m^* t)^{k_m} e^{i<b_m,t>} - \hat{\nu}(t)\right| \leq a_{n_m}^{(m)}(t) + |\hat{\nu}_m(t) - \hat{\nu}(t)| \to 0 \quad \text{as } m \to \infty.$$

An application of Lévy's Continuity Theorem yields $\mu \in \text{GDOPA}(\nu)$ and concludes the proof. □

Lemma 3.3.18. *Let $\mathcal{I}(\mathbb{R}^d)$ denote the set of all infinitely divisible laws on $\mathbb{R}^d$. Then there exists a countable subset $\mathcal{D} \subset \mathcal{I}(\mathbb{R}^d)$ that is dense in $\mathcal{I}(\mathbb{R}^d)$ with respect to weak convergence.*

Proof. It follows from Theorem 1.1.6 of [58] that there exists a countable subset $\mathcal{E}$ of probability measures that is dense in $M^1(\mathbb{R}^d)$ with respect to weak convergence. Recall the definition of compound Poisson distributions from (3.4). Then

$$\mathcal{D} = \{\Pi(\lambda, \mu) : \lambda > 0 \text{ rational }, \mu \in \mathcal{E}\}$$

is a countable subset of infinitely divisible laws. Note that if $\lambda_n \to \lambda > 0$ and $\mu_n \Rightarrow \mu$, then by Lévy's Continuity Theorem

$$\Pi(\lambda_n, \mu_n)\hat{}\,(t) = \exp(\lambda_n(\hat{\mu}_n(t) - 1)) \to \exp(\lambda(\hat{\mu}(t) - 1)) = \Pi(\lambda, \mu)\hat{}\,(t)$$

as $n \to \infty$, and hence $\Pi(\lambda_n, \mu_n) \Rightarrow \Pi(\lambda, \mu)$.

Now let $\mu \in \mathcal{I}(\mathbb{R}^d)$ be any infinitely divisible law and let μ_n denote the nth root of μ. Then it follows from the proof of Corollary 3.1.8 and Lévy's continuity theorem that for all $t \in \mathbb{R}^d$

$$\Pi(n, \mu_n)\hat{}\,(t) \to \hat{\mu}(t) \quad \text{as } n \to \infty. \tag{3.89}$$

Since $\mathcal{E}$ is dense in $M^1(\mathbb{R}^d)$, for every $n \geq 1$ there exists a sequence $(\mu_k^{(n)})_k \subset \mathcal{E}$ such that $\mu_k^{(n)} \Rightarrow \mu_n$ as $k \to \infty$, and hence $\Pi(n, \mu_k^{(n)}) \Rightarrow \Pi(n, \mu_n)$ as $k \to \infty$. In view of Lévy's Continuity Theorem, this implies that for any $n \geq 1$ there exists a $k_n \geq 1$ such that

$$e_k^{(n)}(t) = \left|\Pi(n, \mu_k^{(n)})\hat{}\,(t) - \Pi(n, \mu_n)\hat{}\,(t)\right| \leq \frac{1}{n}$$

for all $k \geq k_n$ and all $\|t\| \leq n$. Now let $\bar{\mu}_n = \mu_{k_n}^{(n)}$. Then $\Pi(n, \bar{\mu}_n) \in \mathcal{D}$ and

$$\left|\Pi(n, \bar{\mu}_n)\hat{}\,(t) - \Pi(n, \mu_n)\hat{}\,(t)\right| \leq \frac{1}{n} \tag{3.90}$$

for all $||t|| \leq n$ and $n \geq 1$.

Now, given $\varepsilon > 0$, for a given $t \in \mathbb{R}^d$, chose $n_1 \geq 1$ such that $||t|| \leq n_1$ and $1/n_1 < \varepsilon/2$. Then by (3.89) choose $n_0 \geq n_1$ such that

$$\left|\Pi(n, \mu_n)\hat{}\,(t) - \hat{\mu}(t)\right| < \frac{\varepsilon}{2} \tag{3.91}$$

for all $n \geq n_0$. Then for all $n \geq n_0$ we obtain from (3.90) and (3.91) that

$$\begin{aligned}\left|\Pi(n, \bar{\mu}_n)\hat{}\,(t) - \hat{\mu}(t)\right| \leq \; & \left|\Pi(n, \bar{\mu}_n)\hat{}\,(t) - \Pi(n, \mu_n)\hat{}\,(t)\right| \\ & + \left|\Pi(n, \mu_n)\hat{}\,(t) - \hat{\mu}(t)\right| \\ & < \frac{1}{n} + \frac{\varepsilon}{2} \\ & \leq \frac{\varepsilon}{2} + \frac{\varepsilon}{2} = \varepsilon\end{aligned}$$

so using Lévy's Continuity Theorem again we have shown that $\Pi(n, \bar{\mu}_n) \Rightarrow \mu$, which concludes the proof. □

Theorem 3.3.19. *There exists a probability measure μ belonging to the generalized domain of partial attraction of any infinitely divisible law.*

Proof. Since the set $\mathcal{D}$ obtained in Lemma 3.3.18 is countable, we can write $\mathcal{D} = \{\nu_1, \nu_2, \ldots\}$. By Lemma 3.3.16, there exists a probability measure μ such that $\mu \in \text{GDOPA}(\nu_n)$ for all $n \geq 1$.

Now if $\nu \in \mathcal{I}(\mathbb{R}^d)$ is fixed, since $\mathcal{D}$ is dense in $\mathcal{I}(\mathbb{R}^d)$, there exists a subsequence (n_m) such that $\nu_{n_m} \Rightarrow \nu$ as $m \to \infty$. An application of Lemma 3.3.17 shows that $\mu \in \text{GDOPA}(\nu)$. This concludes the proof since $\nu \in \mathcal{I}(\mathbb{R}^d)$ was arbitrary. □

Suppose that $\mu \in \text{GDOPA}(\nu)$ and (3.65) holds, where ν is full and infinitely divisible. In view of the existence of universal laws, we need some additional restrictions in order to obtain useful limit theorems. Note that for a universal law, the sampling sequence (k_n) in (3.65) always satisfies $k_{n+1}/k_n \to \infty$. It turns out that if we restrict the growth rate of the sequence (k_n), we can obtain useful information about $\mu \in \text{GDOPA}(\nu)$. We first consider the case where the growth rate of the sampling sequence (k_n) is essentially geometrical.

Definition 3.3.20. Suppose $X, X_1, X_2, \ldots$ are independent and identically distributed random vectors on $\mathbb{R}^d$ with common distribution μ and that Y is a random vector whose distribution ν is full.

(a) We say that ν is *operator semistable*, if there exist linear operators $A_n \in \mathrm{L}(\mathbb{R}^d)$, an increasing sequence of positive integers $k_n \to \infty$ with

$$k_{n+1}/k_n \to r \geq 1 \tag{3.92}$$

and nonrandom vectors $b_n \in \mathbb{R}^d$ such that

$$A_n(X_1 + \cdots + X_{k_n}) + b_n \Rightarrow Y. \tag{3.93}$$

In terms of measures, we can rewrite (3.93) as

$$A_n \mu^{k_n} * \varepsilon_{b_n} \Rightarrow \nu. \tag{3.94}$$

(b) We say that X (or equivalently μ) belongs to the *generalized domain of semistable attraction* of Y (or ν), if (3.93) holds for sequences (A_n), (k_n) and (b_n) as in (a). We write $X \in \mathrm{GDOSA}(Y, r)$ (or $\mu \in \mathrm{GDOSA}(\nu, r)$) in this case.

Lemma 3.3.21. *If* $X \in \mathrm{GDOSA}(Y, r)$ *and* (3.93) *holds, then* X *is full and* $A_n \in \mathrm{GL}(\mathbb{R}^d)$ *for all large* n.

Proof. If (3.93) holds then (3.94) holds, and then Lemma 3.3.3 implies that A_n is invertible and μ is full. □

Lemma 3.3.22. *If two full operator semistable laws* ν, ν' *are of the same type, then* $\mathrm{GDOSA}(\nu) = \mathrm{GDOSA}(\nu')$.

Proof. The proof is identical to Lemma 3.3.6, except that now the sampling sequence (k_n) satisfies (3.92). □

Remark 3.3.23. In Chapter 7, we will see that even when the geometrical growth condition on the sequence (k_n) in Definition 3.3.20 is weakened to

$$\limsup_{n \to \infty} \frac{k_{n+1}}{k_n} < \infty$$

we still obtain $\mu \in \mathrm{GDOSA}(\nu, r)$ operator semistable, see Theorem 7.5.6. Hence this is in some sense the most general situation in which a limit theorem says something useful.

In view of the classical central limit theorem, it is natural to also consider the case where $k_n = n$. Then the resulting class of limit distributions is a special case of the operator semistable laws and we define:

Definition 3.3.24. Suppose $X, X_1, X_2, \ldots$ are independent and identically distributed random vectors on $\mathbb{R}^d$ with common distribution μ and that Y is a random vector whose distribution ν is full.

(a) We say that ν is *operator stable*, if there exist linear operators $A_n \in \mathrm{L}(\mathbb{R}^d)$ and nonrandom vectors $b_n \in \mathbb{R}^d$ such that

$$A_n(X_1 + \cdots + X_n) + b_n \Rightarrow Y. \tag{3.95}$$

In terms of measures, we can rewrite (3.95) as

$$A_n \mu^n * \varepsilon_{b_n} \Rightarrow \nu. \tag{3.96}$$

(b) We say that X (or equivalently μ) belongs to the *generalized domain of attraction* of Y (or ν), if (3.95) holds for some sequences (A_n) and (b_n) as in (a). We write $X \in \mathrm{GDOA}(Y)$ (or $\mu \in \mathrm{GDOA}(\nu)$) in this case.

Note that trivially for every full probability measure ν and every $c \geq 1$

$$\begin{aligned} \mathrm{GDOSA}(\nu, c) &\subset \mathrm{GDOPA}(\nu) \quad \text{and} \\ \mathrm{GDOA}(\nu) &\subset \mathrm{GDOPA}(\nu) \end{aligned} \tag{3.97}$$

and hence the characterizations of GDOPA, especially Theorem 3.3.8 and Corollary 3.3.10, apply to GDOSA and GDOA.

Lemma 3.3.25. *If* $X \in \mathrm{GDOA}(Y)$ *and* (3.95) *holds, then* X *is full and* $A_n \in \mathrm{GL}(\mathbb{R}^d)$ *for all large* n.

Proof. If (3.96) holds, then Lemma 3.3.3 implies that A_n is invertible and μ is full. □

Lemma 3.3.26. *Two full operator stable laws* ν, ν' *are of the same type if and only if* $\mathrm{GDOA}(\nu) = \mathrm{GDOA}(\nu')$.

Proof. If ν, ν' are of the same type, the proof that $\mathrm{GDOA}(\nu) = \mathrm{GDOA}(\nu')$ is identical to Lemma 3.3.6, except that now the sampling sequence $k_n = n$. Conversely, if $\mathrm{GDOA}(\nu) = \mathrm{GDOA}(\nu')$, then for any $\mu \in \mathrm{GDOA}(\nu)$ we have $A_n\mu^n * \varepsilon_{a_n} \Rightarrow \nu$ and $B_n\mu^n * \varepsilon_{b_n} \Rightarrow \nu'$ for some $A_n, B_n \in \mathrm{GL}(\mathbb{R}^d)$ and some $a_n, b_n \in \mathbb{R}^d$. Then it follows from Convergence of Types, Theorem 2.3.17, that $\nu' = A\nu * \varepsilon_a$ for some $A \in \mathrm{GL}(\mathbb{R}^d)$ and some $a \in \mathbb{R}^d$. Hence ν, ν' are the same type. □

The following result will be used in the analysis of operator (semi-) stable laws in Section 7.5.

Lemma 3.3.27. *If* ν *is full, then* $\mathrm{GDOA}(\nu) \subset \mathrm{GDOSA}(\nu, c)$ *for any* $c \geq 1$.

Proof. Let $\mu \in \mathrm{GDOA}(\nu)$. Then (3.96) holds. If $c = 1$ there is nothing to show. If $c > 1$, let $k_n = [c^n]$. Then $k_{n+1}/k_n \to c$ and (3.96) implies for $A'_n = A_{k_n}$ and $b_n = a_{k_n}$ that $A'_n\mu^{k_n} * \varepsilon_{b_n} \Rightarrow \nu$, so (3.94) holds, and hence $\mu \in \mathrm{GDOSA}(\nu, c)$. □

Remark 3.3.28. It will be shown in Section 7.5 that for a full operator stable law ν we actually have $\mathrm{GDOA}(\nu) = \mathrm{GDOSA}(\nu, c)$ for any $c \geq 1$ (see Corollary 7.5.11).

Remark 3.3.29. Theorem 3.3.8 shows that

$$k_n \cdot A_n\mu \to \phi \quad \text{in } \mathcal{M} \tag{3.98}$$

is a necessary condition for $\mu \in \mathrm{GDOSA}(\nu, c)$, where k_n satisfies (3.92). A necessary condition for $\mu \in \mathrm{GDOA}(\nu)$ is that (3.98) holds with $k_n = n$. In Part II, we will develop an asymptotic theory of regular variation for probability measures that satisfy (3.98). If $k_n = n$, then (3.98) implies regular variation of μ and if k_n satisfies (3.92), then (3.98) implies R–O variation.

For limit laws ν having no normal component, it turns out that (3.98) is necessary and sufficient for $\mu \in \mathrm{GDOSA}(\nu, c)$ or $\mu \in \mathrm{GDOA}(\nu)$, respectively (see Corollaries 8.2.3 and 8.2.12).

Remark 3.3.30. The convergence (3.98) is also basic to the theory of point processes [139, 153]. Proposition 3.1 of [152] shows that (3.98) with $k_n = n$ (regular variation of μ) is necessary and sufficient for weak convergence of the point processes

$$\xi_n = \sum_{j=1}^{\infty} \varepsilon_{(j/n, A_n X_j)}$$

to a Poisson random measure on $(0, \infty) \times \mathbb{R}^d$ with mean measure $m \times \phi$, where m is Lebesgue measure on the positive real line. This convergence very useful in extreme value theory.

3.4 APPENDIX: CONTINUOUS MAPPINGS INTO THE CIRCLE

In order to define the log-characteristic function of an infinitely divisible law, we use the following general result on continuous functions from $\mathbb{R}^d$ to $\mathbb{C}^* = \mathbb{C} \setminus \{0\}$. We follow the appendix in [10]. The main purpose of this section is to prove:

Theorem 3.4.1. *For every continuous function $f : \mathbb{R}^d \to \mathbb{C}^*$ with $f(0) \in \mathbb{R}_+$, there exists exactly one continuous function $\varphi : \mathbb{R}^d \to \mathbb{R}$ such that $\varphi(0) = 0$ and*

$$f = |f| e^{i\varphi}.$$

Before we prove Theorem 3.4.1, we need to establish some facts about so-called *unessential* mappings. Let $\mathbb{T} = \{z \in \mathbb{C} : |z| = 1\}$ denote the unit circle and let E denote a topological space. A continuous function $f : E \to \mathbb{T}$ is called *unessential* if there is a continuous real function $\varphi : E \to \mathbb{R}$ such that

$$f = e^{i\varphi}. \tag{3.99}$$

Obviously, together with φ, $\varphi + 2\pi k$ is also a function of this kind for every integer k. If E is connected, then any continuous real function on E such that (3.99) holds is of this form. Indeed, for another such function ψ, we have $e^{i(\varphi - \psi)} = 1$ and thus $\phi := 1/(2\pi)(\varphi - \psi)$ is a continuous function of the connected space E into the discrete space $\mathbb{Z}$ of integers. Hence $\phi(E)$ is also connected and thus $\phi(E) = \{k\}$ for a fixed $k \in \mathbb{Z}$, so $\psi = \varphi + 2\pi k$. Therefore the continuous real function φ is uniquely determined by (3.99) and its value at a given point E.

We prepare for the proof of Theorem 3.4.1 by developing some simple properties of unessential functions.

Lemma 3.4.2. *Let E be an arbitrary topological space. Every continuous function $f : E \to \mathbb{T}$ with $f(E) \neq \mathbb{T}$ is unessential.*

Proof. In fact, let $e^{i\alpha} \in \mathbb{T} \setminus f(E)$ for some $\alpha \in \mathbb{R}$. The function $\psi(t) = e^{it}$ is a homeomorphism (i.e., a continuous mapping with a continuous inverse) of $(\alpha, \alpha + 2\pi)$ onto $\mathbb{T} \setminus \{e^{i\alpha}\}$. If $\Psi = \psi^{-1}$, then $z = \psi(\Psi(z)) = e^{i\Psi(z)}$ for all $z \in \mathbb{T} \setminus \{e^{i\alpha}\}$. Thus $\varphi = \Psi \circ f$ is a continuous real function on E with $e^{i\varphi(x)} = e^{i\Psi(f(x))} = f(x)$ for all $x \in E$. □

Lemma 3.4.3. *If f_1 and f_2 are continuous functions of an arbitrary topological space E into $\mathbb{T}$ such that $f_1(x) \neq -f_2(x)$ for all $x \in E$, and if f_1 is unessential, then f_2 is also unessential.*

Proof. Indeed, $f = f_1/f_2$ is a continuous function of E into $\mathbb{T}$ that does not attain the value -1 and thus is unessential by Lemma 3.4.2. Hence there are continuous real functions φ and φ_1 on E such that $f = e^{i\varphi}$ and $f_1 = e^{i\varphi_1}$. Since $f_2 = e^{i(\varphi_1 - \varphi)}$, f_2 is also unessential. □

Lemma 3.4.4. *If E is a compact space and $f : E \times [0,1] \to \mathbb{T}$ is continuous, and if $x \mapsto f(x, 0)$ is unessential, then $x \mapsto f(x, 1)$ is also unessential.*

Proof. Since f is uniformly continuous on the compact space $E \times [0,1]$ there exists a natural n such that $x \in E$, $s, t \in [0,1]$ and $|s - t| \leq 1/n$ imply $|f(x,s) - f(x,t)| \leq 1$. Thus if for every $j = 0, \ldots, n$ we define $f_j(x) = f(x, j/n)$, then $|f_j(x) - f_{j+1}(x)| \leq 1$ for all $x \in E$. If we had $f_j(x) = -f_{j+1}(x)$ for some $x \in E$, this would imply that $2|f_j(x)| \leq 1$ in contrast to $|f_j(x)| = 1$. Thus $f_j(x) \neq -f_{j+1}(x)$ for all $x \in E$ and $j = 0, \ldots, n - 1$. The n-fold application of Lemma 3.4.3 yields the assertion. □

Lemma 3.4.5. *Every continuous mapping $f : \mathbb{R}^d \to \mathbb{T}$ is unessential.*

Proof. We first show that every continuous mapping $f_r : B_r \to \mathbb{T}$ of the compact ball $B_r = \{x \in \mathbb{R}^d : \|x\| \leq r\}$ is unessential. Indeed, $g(x,t) = f_r(tx)$ is a continuous mapping of $B_r \times [0,1] \to \mathbb{T}$ with $g(x,1) = f_r(x)$ and $g(x,0) = f_r(0)$ for all $x \in B_r$. Thus, the fact that f_r is unessential follows from Lemma 3.4.2 and 3.4.4.

For a continuous function $f : \mathbb{R}^d \to \mathbb{T}$, we consider its restriction f_n to B_n $(n \in \mathbb{N})$. Here $f_n(0) = f(0) = e^{i\alpha}$ for a suitable $\alpha \in \mathbb{R}$. Then, by what has just been proved, for every natural n there exists a continuous function $\varphi_n : B_n \to \mathbb{R}$ such that $f_n = e^{i\varphi_n}$. Obviously, $\varphi_n(0) = \alpha + 2\pi k$ for some $k \in \mathbb{Z}$. If we follow the observations following (3.99), we can choose φ_n such that $\varphi_n(0) = \alpha$, and then φ_n is uniquely determined. Hence for every natural n it follows that φ_n is the restriction of φ_{n+1} to B_n. Then if $\varphi : \mathbb{R}^d \to \mathbb{R}$ denotes that function whose restriction to B_n coincides with φ_n, φ is continuous and $f = e^{i\varphi}$; hence f is unessential. □

Proof of Theorem 3.4.1. The function $g = f/|f|$ is unessential by Lemma 3.4.5, and thus there exists a continuous real function ψ on $\mathbb{R}^d$ with $g = e^{i\psi}$.

Since $g(0) = 1$, we have $\psi(0) = 2\pi k$ with suitable $k \in \mathbb{Z}$. Therefore, $\varphi = \psi - 2\pi k$ is a continuous real function on $\mathbb{R}^d$ with $\varphi(0) = 0$ and $f = |f|e^{i\varphi}$. It follows from the introductory discussion that φ is uniquely determined by the given properties. □

Furthermore we show:

Theorem 3.4.6. *Let $(\varphi_n)_n$ be a sequence of continuous real functions on $\mathbb{R}^d$ with $\varphi_n(0) = 0$ for all natural n. Then, if the sequence $(e^{i\varphi_n})_n$ converges on every compact subset of $\mathbb{R}^d$ uniformly to the constant 1, the sequence $(\varphi_n)_n$ converges uniformly to 0 on every compact subset of $\mathbb{R}^d$.*

Proof. It suffices to show the assertion for every compact ball $E = B_R$ with radius $R > 0$ and center 0. We write the sequence being investigated in the form $(2\pi\varphi_n)_n$ without having to change anything in the hypotheses on the function φ_n. For every real number $\gamma \in \mathbb{R}$, let $\{\gamma\} = [\gamma + 1/2]$, an integer satisfying $|\gamma - \{\gamma\}| \leq 1/2$. Then $\{\varphi_n\}$ is a function on E with integer values, that is, $\exp(-2\pi i\{\varphi_n\}) = 1$. Hence $(\exp(-2\pi i(\varphi_n - \{\varphi_n\})))_n$ converges uniformly to 1. Thus for every $\varepsilon > 0$ there exists a natural n_ε with

$$|e^{2\pi i(\varphi_n(x) - \{\varphi_n(x)\})} - 1| < \varepsilon, \quad \text{for all } x \in E \text{ and } n \geq n_\varepsilon.$$

A simple calculation of the square of the above absolute value yields

$$|\sin(\pi(\varphi_n(x) - \{\varphi_n(x)\}))| < \frac{\varepsilon}{2}, \quad \text{for all } x \in E \text{ and } n \geq n_\varepsilon.$$

Now we have $|\varphi_n(x) - \{\varphi_n(x)\}| \leq 1/2$ and thus $|\pi(\varphi_n - \{\varphi_n\})| \leq \pi/2$ for all $n \in \mathbb{N}$. From the well-known inequality $(2/\pi)|\xi| \leq |\sin \xi|$, valid for all $\xi \in \mathbb{R}$ with $|\xi| \leq \pi/2$, it follows that

$$|\varphi_n(x) - \{\varphi_n(x)\}| < \frac{\varepsilon}{4} \quad \text{for all } x \in E \text{ and } n \geq n_\varepsilon. \tag{3.100}$$

As a consequence of the uniform continuity of φ_n on the compact space E for every $n \geq n_\varepsilon$, there is a covering of E by finitely many nonempty open sets $U_1^{(n)}, \ldots, U_{k_n}^{(n)}$ depending in n, such that for arbitrary points x, x' in the same $U_j^{(n)}$ $(j = 1, \ldots, k_n)$,

$$|\varphi_n(x) - \varphi_n(x')| < \frac{\varepsilon}{4}.$$

Hence, from (3.100), we now obviously have

$$|\{\varphi_n(x)\} - \{\varphi_n(x')\}| < \frac{3}{4}\varepsilon. \tag{3.101}$$

If $\varepsilon < 4/3$, then (3.101) yields

$$|\{\varphi_n(x)\} - \{\varphi_n(x')\}| < 1$$

and thus

$$\{\varphi_n(x)\} = \{\varphi_n(x')\}, \tag{3.102}$$

for arbitrary $x, x' \in U_j^{(n)}$ and all $j = 1, \ldots, k_n$. If we decompose $\{1, \ldots, k_n\}$ into two disjoint nonempty sets J and K, then, because E is connected, the set $\bigcup_{j \in J} U_j^{(n)}$ in not disjoint from $\bigcup_{j \in K} U_j^{(n)}$. Therefore, (3.101) implies that

$$\{\varphi_n(x)\} = \text{const.} \quad \text{for all } x \in E.$$

But since $\varphi_n(0) = 0$ by hypotheses and therefore $\{\varphi_n(0)\} = 0$, we have

$$\{\varphi_n(x)\} = 0 \quad \text{for all } x \in E \text{ and } n \geq n_\varepsilon.$$

By (3.100) we finally obtain $|\varphi_n(x)| < \varepsilon/4$ for all $x \in E$ and all $n \geq n_\varepsilon$, provided $\varepsilon < 4/3$. But this is the asserted uniform convergence on E of $(\varphi_n)_n$ to zero. □

3.5 NOTES AND COMMENTS

For the basic properties of infinitely divisible laws in the beginning of Section 3.1, we follow Feller [40]. The remainder of Sections 3.1 and 3.2 follow Rvačeva [158], who extended the arguments in Gnedenko and Kolmogorov [44]. Rvačeva [158] used these results to characterize multivariable stable laws and their domains of attraction (see Chapter 8). The recent book of Sato [162] treats stochastic processes constructed from infinitely divisible laws (see also Chapter 11). The construction of universal laws in Section 3.3 follows in part Feller [40]. Infinitely divisible laws and domains of attraction also generalize to locally compact groups, especially to simply connected nilpotent Lie groups, see Heyer [58] and Hazod and Siebert [55] and the literature cited therein.

Part II

MULTIVARIATE REGULAR VARIATION

Regular variation is an asymptotic property of functions and measures with power law growth. The one variable theory is well-established, and has found numerous applications in both pure and applied mathematics. Feller [40] used regular variation to give an elegant proof of the basic limit theorems for sums of independent random variables. The use of regular variation streamlines earlier proofs of Gnedenko and Kolmogorov [44], and provides a natural setting for the statement of these results. In this part of the book, we develop the corresponding theory of multivariable regular variation. In Part III, we will use multivariable regular variation to simplify both the statement and proof of the most important limit theorems for sums of independent random vectors.

4
Regular Variation for Linear Operators

Regular variation is a flexible description of power law behavior. A Borel measurable function $R : \mathbb{R}^+ \to \mathbb{R}^+$ is said to vary regularly at infinity with index $\rho \in \mathbb{R}$ if

$$\lim_{x\to\infty} \frac{R(\lambda x)}{R(x)} = \lambda^\rho \tag{4.1}$$

for all $\lambda > 0$. A regularly varying function with index zero is called slowly varying. A slowly varying function $L(x)$ fluctuates slower than any power of x. The functions $L(x) = C$ and $L(x) = \log x$ are both slowly varying. If R varies regularly with index ρ, then it is easy to check that $R(x) = x^\rho L(x)$, where L is slowly varying. A comprehensive account of the one variable theory, including a large number of applications, can be found in Bingham, Goldie, and Teugels [19]. Feller [40], Seneta [177] and de Haan [46] are some other useful references for the one variable theory. In this chapter, we develop the corresponding theory of regular variation for linear operators. Regular variation will be used in Part III to characterize the norming operators in some important limit theorems for sums of independent random vectors. It is also the basis for the theory of regularly varying functions and measures that will be discussed in Chapters 5 and 6.

4.1 DEFINITIONS AND BASIC PROPERTIES

Regular variation for linear operators is the most natural extension of the classical one variable theory.

Definition 4.1.1. Suppose $f : \mathbb{R}^+ \to \mathrm{GL}(\mathbb{R}^d)$ is Borel measurable, where $\mathrm{GL}(\mathbb{R}^d)$ is the Lie group of invertible linear operators on $\mathbb{R}^d$. We say that f varies regularly at infinity with index E if

$$\lim_{t\to\infty} f(\lambda t)f(t)^{-1} = \lambda^E \tag{4.2}$$

for all $\lambda > 0$, where λ^E is as in Definition 2.2.1. If $E = 0$ we say that $f(t)$ is slowly varying.

Our first result on regular variation for linear operators is to show that the form of the limit in (4.2) is completely general.

Theorem 4.1.2. *Suppose $f : \mathbb{R}^+ \to \mathrm{GL}(\mathbb{R}^d)$ is Borel measurable and*

$$\lim_{t\to\infty} f(\lambda t)f(t)^{-1} = \psi(\lambda) \in \mathrm{GL}(\mathbb{R}^d) \tag{4.3}$$

for all $\lambda > 0$. Then for some linear operator E we have $\psi(\lambda) = \lambda^E$ for all $\lambda > 0$.

Proof. It follows from (4.3) that ψ is measurable and that $\psi(\lambda\mu) = \psi(\lambda)\psi(\mu)$ for all λ, μ positive. Clearly, $\psi(1)$ is the identity operator, and so ψ is a Borel measurable group homomorphism from the multiplicative group $\mathbb{R}^+$ to $\mathrm{GL}(\mathbb{R}^d)$. Hence $\psi(e^x)$ is a measurable one-parameter subgroup, and it is well known that such a subgroup is automatically continuous (see, e.g., Proposition 9.2 in [8]) and hence takes the form $\exp(xE)$ for some E (see, e.g., [186] p. 10-21). Reparameterize by $x = \log\lambda$. □

Now we provide some examples of regularly varying functions, to develop our intuition about what regular variation means in this setting. We also give some counterexamples, to show where regular variation for linear operators differs from regular variation for functions on $\mathbb{R}^1$.

1. If $R(x)$ varies regularly with index ρ on $\mathbb{R}^+$, then it follows immediately from (4.1) that $1/R(x)$ varies regularly with index $-\rho$. If we take $f(t) = t^E$, then

$$f(\lambda t)f(t)^{-1} = (\lambda t)^E t^{-E} = \lambda^E$$

   so that $f(t)$ varies regularly with index E. In this case, $f^{-1}(t) = t^{-E}$ varies regularly with index $-E$, and $f^*(t) = t^{E^*}$ varies regularly with index E^*.

2. If R_1, R_2 vary regularly on $\mathbb{R}^+$ with index ρ_1, ρ_2 respectively, then it follows immediately from (4.1) that R_1R_2 varies regularly with index $\rho_1 + \rho_2$. If we take $f_i(t) = t^{E_i}$ for $i = 1, 2$ and if $E_1E_2 = E_2E_1$, then f_1f_2 varies regularly with index $E_1 + E_2$ using Proposition 2.2.2 (b).

3. If we take $f(t) = t^E \log(t)$, then

$$f(\lambda t)f(t)^{-1} = \lambda^E \frac{\log(\lambda t)}{\log(t)} \to \lambda^E$$

as $t \to \infty$ so that $f(t)$ varies regularly with index E.

4. Let

$$R(t) = \begin{pmatrix} \cos t & -\sin t \\ \sin t & \cos t \end{pmatrix}$$

denote counterclockwise rotation through an angle t in the plane. The function $R(t)$ is not regularly varying, but $R(\log t)$ is regularly varying, and in fact $R(\log t) = t^Q$, where

$$Q = \begin{pmatrix} 0 & -1 \\ 1 & 0 \end{pmatrix}$$

is skew symmetric.

5. If f varies regularly with index E and $T : \mathbb{R}^d \to \mathbb{R}^d$ is an invertible linear operator, then $g(t) = Tf(t)T^{-1}$ varies regularly with index TET^{-1} since

$$\begin{aligned} g(\lambda t)g(t)^{-1} &= (Tf(\lambda t)T^{-1})(Tf(t)T^{-1})^{-1} \\ &= Tf(\lambda t)T^{-1}Tf(t)^{-1}T^{-1} \\ &= Tf(\lambda t)f(t)^{-1}T^{-1} \to T\lambda^E T^{-1} = \lambda^{TET^{-1}} \end{aligned}$$

by Proposition 2.2.2 (e) for any $\lambda > 0$.

6. If f varies regularly with index E, then $(f^*)^{-1} = (f^{-1})^*$ also varies regularly with index $-E^*$ since

$$[f(\lambda t)^*]^{-1} f(t)^* = ([f(\lambda t)f(t)^{-1}]^*)^{-1} \to ([\lambda^E]^*)^{-1} = \lambda^{-E^*}$$

by continuity of the inverse and transpose operations.

7. Regular variation of f does not imply regular variation of f^{-1} or f^*. The function

$$f(t) = \begin{pmatrix} t \log t & t^2 \\ t & t^2 \log t \end{pmatrix}$$

varies regularly with index

$$E = \begin{pmatrix} 1 & 0 \\ 0 & 2 \end{pmatrix}$$

but f^{-1} is not regularly varying, in fact $f(\lambda t)^{-1} f(t)$ does not converge as $t \to \infty$ for $\lambda \neq 1$. Also, f^* is not regularly varying, since $f^*(\lambda t)(f(t)^*)^{-1}$ does not converge as $t \to \infty$ for $\lambda \neq 1$.

8. For f regularly varying with index E, it is not always possible to write $f(t) = t^E L(t)$ with L slowly varying. For the same regularly varying f as in the previous example, the function

$$L(t) = t^{-E} f(t) = \begin{pmatrix} \log t & t \\ 1/t & \log t \end{pmatrix}$$

is not slowly varying, since $L(\lambda t)L(t)^{-1}$ does not converge for any $\lambda \neq 1$. Similarly, it is not always possible to write $f(t) = L_0(t)t^E$ with L_0 slowly varying. The function

$$L_0(t) = f(t)t^{-E} = \begin{pmatrix} \log t & t \log t \\ 1/t & \log t \end{pmatrix}$$

is not slowly varying, since $L_0(\lambda t)L_0(t)^{-1}$ does not converge for any $\lambda \neq 1$. Since both $f(t)$ and t^{-E} are regularly varying, this example also shows that products of regularly varying linear operators need not vary regularly.

9. The function

$$f(t) = \begin{pmatrix} t \log t & t \\ 0 & t/\log t \end{pmatrix}$$

varies regularly with index

$$E = \begin{pmatrix} 1 & -1 \\ 0 & 1 \end{pmatrix}$$

and f^{-1} also varies regularly, but with index

$$E' = \begin{pmatrix} -1 & 0 \\ 0 & -1 \end{pmatrix} \neq -E.$$

4.2 UNIFORM CONVERGENCE AND PATH BEHAVIOR

Many of the properties of regularly varying functions on $\mathbb{R}^+$ that depend on the commutativity of multiplication do not extend to linear operators. It is no longer true that products or inverses of regularly varying functions are automatically regularly varying, and we can no longer reduce everything to the study of slowly varying functions. In light of these technical difficulties, it is remarkable that some of the deepest results from the one variable theory extend to the operator case. Theorem 4.2.1 is a prime example.

Theorem 4.2.1. *(Uniform Convergence Theorem) If f varies regularly, then the convergence in (4.2) is uniform on compact subsets of $\lambda > 0$.*

Proof. We adapt the proof from Seneta [177], making adjustments for the fact that we cannot reduce to slow variation.[1] It suffices to prove uniform convergence for $\lambda \in I = [a, b]$, where without loss of generality $a < 1/2$ and $b > 2$. Suppose not. Then for some $\varepsilon > 0$ there exists $r_n \to \infty$ and $\lambda_n \in I$ such that

$$\|f(\lambda_n r_n)f(r_n)^{-1} - \lambda_n^E\| \geq \varepsilon \tag{4.4}$$

for all n. Given ε_1 and ε_2 positive define

$$\begin{aligned} U_n &= \{\lambda \in L : \|f(\lambda r_m)f(r_m)^{-1} - \lambda^E\| < \varepsilon_1 \ \forall m \geq n\} \\ V_n &= \{\lambda \in L : \|f(\lambda\lambda_m r_m)f(\lambda_m r_m)^{-1} - \lambda^E\| < \varepsilon_2 \ \forall m \geq n\} \end{aligned} \tag{4.5}$$

where $L = [1/(2b), 2/a]$. As $n \to \infty$ the sets U_n and V_n increase to L in view of the fact that (4.2) holds for each individual $\lambda > 0$. Let $V_n' = \lambda_n V_n$. By construction of L, we must have $[1/2, 2]$ contained in $L \cap \lambda_n L$ for all n, and hence for some N we have $U_N \cap V_N' \neq \emptyset$. If $\lambda \in U_N$ and $\lambda \in V_N'$ then $(\lambda/\lambda_N) \in V_N$, and so

$$\begin{aligned} \|f(\lambda r_N)f(r_N)^{-1} - \lambda^E\| &< \varepsilon_1 \\ \|f(\lambda r_N)f(\lambda_N r_N)^{-1} - (\lambda/\lambda_N)^E\| &< \varepsilon_2. \end{aligned} \tag{4.6}$$

Let $A = f(\lambda r_N)f(r_N)^{-1}$ and $B = f(\lambda r_N)f(\lambda_N r_N)^{-1}$, so that

$$B^{-1}A = f(\lambda_N r_N)f(r_N)^{-1}.$$

By (4.6) and the triangle inequality, we obtain

$$\begin{aligned} \|B^{-1}A - \lambda_N^E\| &= \|B^{-1}A - B^{-1}\lambda^E + B^{-1}\lambda^E - \lambda_N^E\| \\ &\leq \|B^{-1}\| \cdot \|A - \lambda^E\| + \|B^{-1} - (\lambda_N/\lambda)^E\| \cdot \|\lambda^E\|, \end{aligned}$$

and then, by making an appropriate choice of ε_1 and ε_2 in (4.5) and using the continuity of the inversion in $\mathrm{GL}(\mathbb{R}^d)$, we obtain a contradiction to (4.4), and this completes the proof. □

Recall from Definition 2.3.14 that for $A_n \in \mathrm{L}(\mathbb{R}^d)$ and $\mathcal{K}$ a closed subset of $\mathrm{L}(\mathbb{R}^d)$ we write $A_n \rightharpoonup \mathcal{K}$ to mean that (A_n) is limit point compact with every limit point lying in the set $\mathcal{K}$. The following generalization of Theorem 4.2.1 will be used in the study of regularly varying functions and measures.

Lemma 4.2.2. *Suppose $\mathcal{K}$ is a compact subset of* $\mathrm{GL}(\mathbb{R}^d)$ *such that*

$$\mathcal{K} = \lambda^{-E}\mathcal{K}\lambda^E \tag{4.7}$$

[1]For the convenience of the reader, we note here that the proof of Seneta is essentially identical to that of Theorem 5.3.6.

for all $\lambda > 0$, and that

$$f(\lambda t)f(t)^{-1} \rightharpoonup \lambda^E \mathcal{K} \tag{4.8}$$

as $t \to \infty$ for all $\lambda > 0$. Then this convergence is uniform on compact subsets of $\lambda > 0$.

Proof. The proof is similar to Theorem 4.2.1. Suppose by way of contradiction that convergence in (4.8) is not uniform for $\lambda \in I = [a, b]$, where $a < 1/2$ and $b > 2$. Then for some $\varepsilon > 0$ there exists $r_n \to \infty$ and $\lambda_n \in I$ such that

$$\|f(\lambda_n r_n)f(r_n)^{-1} - \lambda_n^E \mathcal{K}\| \geq \varepsilon \tag{4.9}$$

for all n. Given ε_1 and ε_2 positive define

$$\begin{aligned} U_n &= \{\lambda \in L : \|f(\lambda r_m)f(r_m)^{-1} - \lambda^E \mathcal{K}\| < \varepsilon_1 \ \forall m \geq n\} \\ V_n &= \{\lambda \in L : \|f(\lambda \lambda_m r_m)f(\lambda_m r_m)^{-1} - \lambda^E \mathcal{K}\| < \varepsilon_2 \ \forall m \geq n\} \end{aligned} \tag{4.10}$$

where $L = [1/(2b), 2/a]$. Let $V_n' = \lambda_n V_n$, and note that $U_N \cap V_N' \neq \emptyset$ for some N. For $\lambda \in U_N \cap V_N'$, we have

$$\begin{aligned} \|A - \lambda^E G_1\| &< \varepsilon_1 \\ \|B - (\lambda/\lambda_N)^E G_2\| &< \varepsilon_2 \end{aligned} \tag{4.11}$$

for some $G_1, G_2 \in \mathcal{K}$, where $A = f(\lambda r_N)f(r_N)^{-1}$ and $B = f(\lambda r_N)f(\lambda_N r_N)^{-1}$. Define

$$G = \lambda_N^{-E} G_2^{-1} \lambda_N^E G_1$$

and note that $G \in \mathcal{K}$ by virtue of (4.7). By (4.11) and the triangle inequality

$$\begin{aligned} \|B^{-1}A - \lambda_N^E G\| &= \|B^{-1}A - B^{-1}\lambda^E G_1 + B^{-1}\lambda^E G_1 - \lambda_N^E G\| \\ &\leq \|B^{-1}\| \cdot \|A - \lambda^E G_1\| \\ &\quad + \|B^{-1} - G_2^{-1}(\lambda_N/\lambda)^E\| \cdot \|\lambda^E G_1\|, \end{aligned}$$

where both $\|B^{-1}\|$ and $\|\lambda^E G_1\|$ are bounded. By making an appropriate choice of ε_1 and ε_2 in (4.10) we obtain a contradiction to (4.9), and this completes the proof. □

Quite a few important properties of regularly varying functions follow directly from the Uniform Convergence Theorem 4.2.1. Here we collect some of those results. Both the statement and the proof of each of these properties is a direct extension of the analogous one variable result.

Proposition 4.2.3. *If f varies regularly on $\mathrm{GL}(\mathbb{R}^d)$ with index E, then*

(a) $f(t+c)f(t)^{-1} \to I$ for any $c > 0$

(b) *If $R(x)$ varies regularly on $\mathbb{R}^+$ with index $\beta > 0$, then $f(R(x))$ varies regularly on $\mathrm{GL}(\mathbb{R}^d)$ with index βE*

Proof. Given $c > 0$ write $\lambda = \lambda(t) = (t + c)/t$. Since $\lambda \to 1$ as $t \to \infty$ and since $t^E = I$ when $t = 1$, we have $f(t + c)f(t)^{-1} = f(\lambda t)f(t)^{-1} \to I$ by uniform convergence, proving part (a). For part (b), we fix $\lambda > 0$ and define $\delta(x) > 0$ such that $\delta(x)^\beta = R(\lambda x)/(\lambda^\beta R(x))$, so that $\delta(x) \to 1$ by (4.1) as $x \to \infty$. Write

$$f(R(\lambda x))f(R(x))^{-1} = f(R(\lambda x))f(\lambda^\beta R(x))^{-1} \cdot f(\lambda^\beta R(x))f(R(x))^{-1}.$$

Since $R(x) \to \infty$ the second term tends to $(\lambda^\beta)^E$, and the first term tends to the identity by uniform convergence, since $R(\lambda x) = \delta(x)^\beta \lambda^\beta R(x)$. □

The classical uniform convergence theorem for scalar regularly varying functions implies that if R varies regularly with index ρ, then for any $\varepsilon > 0$ we have $x^{\rho-\varepsilon} < R(x) < x^{\rho+\varepsilon}$ for all large x. If L is slowly varying, then $x^{-\varepsilon} < L(x) < x^\varepsilon$ for all large x. An analogous result for linear operators can be expressed as follows. This result also extends Theorem 2.2.4.

Theorem 4.2.4. *Suppose that f varies regularly with index E. If all eigenvalues of E have real part greater than α then*

$$t^{-\alpha}\|f(t)x\| \to \infty \tag{4.12}$$

as $t \to \infty$ uniformly on compact subsets of $x \neq 0$ in $\mathbb{R}^d$. If all eigenvalues of E have real part less than β, then

$$t^{-\beta}\|f(t)x\| \to 0 \tag{4.13}$$

as $t \to \infty$ uniformly on compact subsets of $x \in \mathbb{R}^d$.

Proof. The proof of both assertions is similar, so we only prove the first. Let S denote the unit sphere in $\mathbb{R}^d$ and let $m = \min\{\mathrm{Re}(\lambda)\}$, where λ ranges over the eigenvalues of E. For any $a < m$, Theorem 2.2.4 shows that $\lambda^{-a}\|\lambda^E\theta\| \to \infty$ uniformly on $\theta \in S$ as $\lambda \to \infty$. Select a, b real with $\alpha < b < a < m$. For some $\lambda_0 > 1$, we have $\lambda_0^{-a}\|\lambda_0^E\theta\| > 1$ for all $\theta \in S$ and so $\|\lambda_0^E\theta\| > \lambda_0^a$ for all $\theta \in S$. Now choose $\varepsilon > 0$ such that $\lambda_0^a - \varepsilon \geq \lambda_0^b$, and apply Theorem 4.2.1 to obtain $t_0 > 0$ such that $\|f(\lambda t)f(t)^{-1} - \lambda^E\| < \varepsilon$ whenever $1 \leq \lambda \leq \lambda_0$ and $t \geq t_0$. Now we have $\|f(\lambda_0 t)f(t)^{-1}\theta\| > \lambda_0^b$ for all $\theta \in S$. Write $f(\lambda_0^n t_0)x = r_n\theta_n$ where $r_n > 0$ and $\theta_n \in S$. Then

$$\begin{aligned}\|f(\lambda_0^n t_0)x\| &= \|f(\lambda_0^n t_0)f(\lambda_0^{n-1}t_0)^{-1} \cdot f(\lambda_0^{n-1}t_0)x\| \\ &= r_{n-1}\|f(\lambda_0^n t_0)f(\lambda_0^{n-1}t_0)^{-1}\theta_{n-1}\| \\ &\geq r_{n-1}\lambda_0^b\end{aligned}$$

so that $r_n \geq r_0(\lambda_0^b)^n$ for all n. Lemma 2.2.3 implies that $\|\lambda^E\theta\|$ is bounded away from zero uniformly on $\theta \in S$ and $1 \leq \lambda \leq \lambda_0$. Then another application of Theorem 4.2.1 shows that $\|f(\lambda t)f(t)^{-1}\theta\| \geq A > 0$ for all $\theta \in S$, all $t \geq t_0$, and all $1 \leq \lambda \leq \lambda_0$, where we enlarge t_0 if necessary. For any $t \geq t_0$, there exist unique nonnegative integers $n = n(t)$ and real numbers $\lambda = \lambda(t) \in [1, \lambda_0)$ such that $t = \lambda\lambda_0^n t_0$. Write $f(t)x = r_t\theta_t$ where $r_t > 0$ and $\theta_t \in S$. Then for $t \geq t_0$ we have

$$\begin{aligned} r_t = \|f(\lambda\lambda_0^n t_0)x\| &= \|f(\lambda\lambda_0^n t_0)f(\lambda_0^n t_0)^{-1} \cdot f(\lambda_0^n t_0)x\| \\ &= r_n\|f(\lambda\lambda_0^n t_0)f(\lambda_0^{n-1}t_0)^{-1}\theta_n\| \\ &\geq Ar_n \end{aligned}$$

so that $r_t \geq Ar_0(\lambda_0^b)^n$ for all $t \geq t_0$. Then

$$\begin{aligned} t^{-\alpha}\|f(t)x\| = t^{-\alpha}r_t &\geq (\lambda_0^{n+1}t_0)^{-\alpha}Ar_0(\lambda_0^b)^n \\ &= Ar_0\lambda_0^{-\alpha}(\lambda_0^{b-\alpha})^n \end{aligned}$$

which tends to infinity as $n \to \infty$. This shows that (4.12) holds, and since $r_0 = \|f(t_0)x\|$ is bounded away from zero on compact subsets of $x \neq 0$, this convergence is also uniform on compact subsets. □

Remark 4.2.5. In applications, we often apply the uniform convergence on the unit sphere $\|x\| = 1$. By linearity, this is equivalent to uniform convergence on the set $r \leq \|x\| \leq R$ for any $0 < r < R < \infty$. If $t^{-\beta}\|f(t)x\| \to 0$ for all $x \in \mathbb{R}^d$, then Proposition 2.1.8 implies that $t^{-\beta}f(t)x \to 0$ uniformly on compact subsets of $x \in \mathbb{R}^d$, and $t^{-\beta}f(t) \to 0$ in $\mathrm{L}(\mathbb{R}^d)$. If $t^{-\alpha}\|f(t)x\| \to \infty$ uniformly on compact subsets of $x \neq 0$, then $t^{-\alpha}\|f(t)x_t\| \to \infty$ whenever $x_t \to x \neq 0$ as $t \to \infty$ (see Proposition 2.1.7).

Corollary 4.2.6. *Suppose that f varies regularly with index E. If all eigenvalues of E have real part greater than α, then*

$$t^{\alpha}\|f(t)^{-1}x\| \to 0 \tag{4.14}$$

as $t \to \infty$ uniformly on compact subsets of $x \in \mathbb{R}^d$. If all eigenvalues of E have real part less than β, then

$$t^{\beta}\|f(t)^{-1}x\| \to \infty \tag{4.15}$$

as $t \to \infty$ uniformly on compact subsets of $x \neq 0$ in $\mathbb{R}^d$.

Proof. If $\alpha < m = \min\{\mathrm{Re}(\lambda)\}$, where λ ranges over the eigenvalues of E, it follows from Theorem 4.2.4 that $t^{-\alpha}\min_{\|x\|=1}\|f(t)x\| \to \infty$ as $t \to \infty$. Since by Proposition 2.1.3(d) $\min_{\|x\|=1}\|f(t)x\| = \|f(t)^{-1}\|^{-1}$ it follows that $t^{\alpha}\|f(t)^{-1}\| \to 0$ as $t \to \infty$ and in view of the inequality $\|f(t)^{-1}x\| \leq \|f(t)^{-1}\|\|x\|$ (4.14) follows. If $\beta > M = \max\{\mathrm{Re}(\lambda)\}$, then Theorem 4.2.4 implies $t^{-\beta}\|f(t)\| \to 0$ or equivalently $t^{\beta}\|f(t)\|^{-1} \to \infty$ as $t \to \infty$. Since by Proposition 2.1.3(d) $t^{\beta}\|f(t)^{-1}x\| \geq t^{\beta}\|f(t)\|^{-1}\|x\|$, (4.15) follows. □

Lemma 4.2.7. *Suppose $f : \mathbb{R}^+ \to \mathrm{GL}(\mathbb{R}^d)$ varies regularly with index E.*

(a) *If every eigenvalue of E has real part less than zero, then for any $x_r \to x \neq 0$ in $\mathbb{R}^d$ the function*

$$t(r) = \sup\{t > 0 : \|f(t)rx_r\| \geq 1\} \tag{4.16}$$

is well defined for all r large, and $t = t(r)$ tends to infinity as $r \to \infty$. Furthermore, if we define $y_t = f(t)rx_r$, then $\|y_t\| \to 1$ as $t = t(r) \to \infty$.

(b) *If every eigenvalue of E has real part greater than zero, then for any $x_r \to x \neq 0$ in $\mathbb{R}^d$ the function*

$$t(r) = \sup\{t > 0 : \|f(t)x_r/r\| \leq 1\} \tag{4.17}$$

is well defined for all r large, and $t = t(r)$ tends to infinity as $r \to \infty$. Furthermore, if we define $y_t = f(t)x_r/r$, then $\|y_t\| \to 1$ as $t = t(r) \to \infty$.

Proof. First we will prove part (a). Given $x_r \to x \neq 0$, for some $r_0 > 0$ we have $m \leq \|x_r\|$ for all $r \geq r_0$ by continuity of the norm, where $0 < m < \|x\|$. Then $\|rx_r\| \to \infty$, so for any fixed $t > 0$ we have $\|f(t)rx_r\| \geq c_t\|rx_r\| \geq 1$ for all $r > 0$ large, where $0 < c_t = \min\{\|f(t)\theta\| : \|\theta\| = 1\}$. Hence the set appearing in (4.16) is nonempty for all r large. For $r > 0$ fixed, we also have $\|f(t)rx_r\| \to 0$ as $t \to \infty$ by Theorem 4.2.4 with $\beta = 0$. Thus the set appearing in (4.16) is bounded above for any given r, and so $t(r)$ is well defined for all r large. Suppose by way of contradiction that there exist $r_k \to \infty$ such that $t(r_k) < t_0$ for all k. Then for any $t_1 > t_0$ we have $\|f(t)r_k x_{r_k}\| < 1$ for all k and all $t > t_1$. But this contradicts the fact that $\|f(t)r_k x_{r_k}\| \geq c_t\|r_k x_{r_k}\| \to \infty$ as $k \to \infty$ for any fixed $t > 0$, and so $t(r) \to \infty$ as $r \to \infty$.

We will finish the proof of part (a) by showing that $1 - \varepsilon < \|y_t\| < 1 + \varepsilon$ for all t large, where $\varepsilon > 0$ is arbitrary. From (4.16), it follows immediately that $\|f(t+1)rx_r\| < 1$ when $t = t(r)$, and also that $\|f(t - \delta_t)rx_r\| \geq 1$ for some $0 < \delta_t < 1$. Theorem 4.2.1 shows that $f(t)f(t+1)^{-1} \to I$ as $t \to \infty$, and hence for all t large we have $1 - \varepsilon < \|f(t)f(t+1)^{-1}x\| < 1 + \varepsilon$ for all $\|x\| = 1$. Write $f(t+1)rx_r = \rho_t\theta_t$, where $\rho_t > 0$ and $\|\theta_t\| = 1$, and note that $\rho_t < 1$. Then for $t = t(r)$ large we have $\|y_t\| = \|f(t)f(t+1)^{-1}f(t+1)rx_r\| = \|f(t)f(t+1)^{-1}\rho_t\theta_t\| = \rho_t\|f(t)f(t+1)^{-1}\theta_t\| < 1+\varepsilon$. Write $f(t-\delta_t)rx_r = \sigma_t\mu_t$ where $\sigma_t > 0$ and $\|\mu_t\| = 1$, and note that $\sigma_t \geq 1$. Then for $t = t(r)$ large we have $\|y_t\| = \|f(t)f(t - \delta_t)^{-1}f(t - \delta_t)rx_r\| = \sigma_t\|f(t)f(t - \delta_t)^{-1}\mu_t\| > 1 - \varepsilon$.

The proof of part (b) is similar. Given $x_r \to x \neq 0$, for some $r_0 > 0$ we have $0 < \|x_r\| \leq M$ for all $r \geq r_0$ by continuity of the norm. Then $\|x_r/r\| \to 0$, so for any fixed $t > 0$ we have $\|f(t)x_r/r\| \leq \|f(t)\| \cdot \|x_r/r\| \leq 1$ for all $r > 0$ large, hence the set in (4.17) is nonempty for all r large. For $r \geq r_0$ fixed we also have $\|f(t)x_r/r\| \to \infty$ as $t \to \infty$ by Theorem 4.2.4 with $\alpha = 0$, so that the set in (4.16) is bounded above for any given r, hence $t(r)$ is well defined for all r large. Suppose by way of contradiction that there exist $r_k \to \infty$ such that $t(r_k) < t_0$ for all k. Then for any $t_1 > t_0$ we have $\|f(t)x_{r_k}/r_k\| > 1$ for all k and all $t > t_1$. But this contradicts the fact that $\|f(t)x_{r_k}/r_k\| \le \|f(t)\| \cdot \|x_{r_k}/r_k\| \to 0$ as $k \to \infty$ for any fixed $t > 0$, and so $t(r) \to \infty$ as $r \to \infty$.

From (4.17), it follows immediately that $\|f(t+1)x_r/r\| > 1$ when $t = t(r)$, and also that $\|f(t-\delta_t)x_r/r\| \le 1$ for some $0 < \delta_t < 1$. Write $f(t+1)x_r/r = \rho_t\theta_t$ as before, and note that $\rho_t > 1$. Then for $t = t(r)$ large we have $\|y_t\| = \rho_t\|f(t)f(t+1)^{-1}\theta_t\| > 1 - \varepsilon$ by Theorem 4.2.1. For $f(t-\delta_t)x_r/r = \sigma_t\mu_t$ as before, we have $\sigma_t \le 1$. Then $\|y_t\| = \sigma_t\|f(t)f(t-\delta_t)^{-1}\mu_t\| < 1+\varepsilon$ for $t = t(r)$ large by Theorem 4.2.1, and this completes the proof. □

In some applications it is useful to apply regular variation to sequences of linear operators.

Definition 4.2.8. We say that a sequence of invertible linear operators (A_n) varies regularly with index E if

$$A_{[\lambda n]}A_n^{-1} \to \lambda^E \tag{4.18}$$

as $n \to \infty$ for all $\lambda > 0$. In this case, we will write $(A_n) \in \mathrm{RV}(E)$.

The convergence (4.18) is apparently a weaker condition than the regular variation of $f(t) = A_{[t]}$, but our next result shows these two conditions are actually equivalent.

Theorem 4.2.9. *If the sequence $A_n \in \mathrm{GL}(\mathbb{R}^d)$ varies regularly with index E, then the function $f(t) = A_{[t]}$ varies regularly with the same index E. If f varies regularly with index E, then $A_n = f(n)$ also varies regularly with the same index.*

Proof. Fix $\lambda > 0$. If f varies regularly with index E and $A_n = f(n)$, then

$$A_{[\lambda n]}A_n^{-1} = f([\lambda n])f(n)^{-1} = f([\lambda n])f(\lambda n)^{-1} \cdot f(\lambda n)f(n)^{-1}$$

where the second term tends to λ^E by the regular variation, and the first term tends to I by uniform convergence, since $[\lambda n]/\lambda n \to 1$ as $n \to \infty$. On the other hand, if A_n varies regularly with index E and $f(t) = A_{[t]}$, then

$$f(\lambda t)f(t)^{-1} = A_{[\lambda t]}A_{[t]}^{-1} = A_{[\lambda t]}A_{[\lambda[t]]}^{-1} \cdot A_{[\lambda[t]]}A_{[t]}^{-1}$$

where the second term tends to λ^E by (4.18). To complete the proof, we must show that the first term tends to the identity. For fixed $\lambda > 0$, the difference $[\lambda t] - [\lambda[t]]$ is an integer between zero and $\lambda + 1$, hence we can conclude the proof by repeated application of the following lemma. □

Lemma 4.2.10. *If $A_n \in \mathrm{GL}(\mathbb{R}^d)$ varies regularly, then $A_{n-1}A_n^{-1} \to I$.*

Proof. Define $\psi_n(\lambda) = A_{[\lambda n]}A_n^{-1}$ so that $\psi_{[\lambda n]}(1/\lambda) = A_{[[\lambda n]/\lambda]}A_{[\lambda n]}^{-1}$. By (4.18), we have $\delta_n(\lambda) = \psi_{[\lambda n]}(1/\lambda)\psi_n(\lambda) \to \lambda^{-E}\lambda^E = I$, and in particular

$\delta_n(1/2) \to I$. But when n is odd $\delta_n(1/2) = A_{n-1}A_n^{-1}$ while if n is even $\delta_n(1/2) = I$. □

Corollary 4.2.11. *Suppose that $A_n \in \mathrm{GL}(\mathbb{R}^d)$ for all $n \geq 1$.*

(a) If the sequence A_n varies regularly, then the convergence (4.18) holds uniformly on compact subsets of $\lambda > 0$

(b) If $\mathcal{K}$ is a compact subgroup of $\mathrm{GL}(\mathbb{R}^d)$ such that $\mathcal{K} = \lambda^{-E}\mathcal{K}\lambda^E$ for all $\lambda > 0$ and

$$A_{[\lambda n]}A_n^{-1} \rightharpoonup \lambda^E\mathcal{K} \quad \text{as } n \to \infty \tag{4.19}$$

for all $\lambda > 0$, then this convergence is uniform on compact subsets of $\lambda > 0$

Proof. (a) If (A_n) varies regularly with index E, then Theorem 4.2.9 implies that $f(t) = A_{[t]}$ varies regularly with the same index E. Then Theorem 4.2.1 shows that

$$A_{[\lambda n]}A_n^{-1} = f(\lambda n)f(n)^{-1} \to \lambda^E$$

uniformly on compact subsets of $\lambda > 0$.

(b) Let $f(t) = A_{[t]}$ as before. We show that (4.19) implies (4.8). Write

$$f(\lambda t)f(t)^{-1} = \left(A_{[\lambda t]}A_{[\lambda[t]]}^{-1}\right) \cdot \left(A_{[\lambda[t]]}A_{[t]}^{-1}\right)$$

where the second term tends to $\lambda^E\mathcal{K} = \mathcal{K}\lambda^E$ by (4.19). As in the proof of Theorem 4.2.9, since $\mathcal{K}$ is a group and hence $\mathcal{K}\cdot\mathcal{K} = \mathcal{K}$, it suffices to show that

$$A_{n-1}A_n^{-1} \rightharpoonup \mathcal{K} \quad \text{as } n \to \infty. \tag{4.20}$$

As in the proof of Lemma 4.2.10 let $\psi_n(\lambda) = A_{[\lambda n]}A_n^{-1}$ so that $\psi_{[\lambda n]}(1/\lambda) = A_{[[\lambda n]/\lambda]}A_{[\lambda n]}^{-1}$. Then by (4.19) and the fact that $\lambda^E\mathcal{K} = \mathcal{K}\lambda^E$ we obtain

$$\delta_n(\lambda) = \psi_{[\lambda n]}(1/\lambda)\psi_n(\lambda) \rightharpoonup (\lambda^{-E}\mathcal{K})(\lambda^E\mathcal{K}) = \mathcal{K}(\lambda^{-E}\lambda^E)\mathcal{K} = \mathcal{K}\cdot\mathcal{K} = \mathcal{K}$$

since $\mathcal{K}$ is a group. Then (4.20) follows as in the proof of Lemma 4.2.10, which concludes the proof. □

Finally, the next two results will be the key ingredient in our general theory of regularly varying functions and measures as well as R–O varying measures developed in Chapters 5 and 6.

In the case of regular variation of either functions or measures, we need:

Proposition 4.2.12. *Let $\mathcal{K}$ be a compact subgroup of* $\mathrm{GL}(\mathbb{R}^d)$ *such that* (4.7) *holds for all* $\lambda > 0$. *Assume further that there exists a sequence* $(A_n) \subset \mathrm{GL}(\mathbb{R}^d)$ *such that*

$$A_{[\lambda n]}A_n^{-1} \rightharpoonup \lambda^E \mathcal{K} \quad \text{as } n \to \infty \tag{4.21}$$

for all $\lambda > 0$. *Then there exist* $G_n \in \mathcal{K}$ *such that the sequence defined by* $B_n = G_n A_n$ *is regularly varying with index* E.

Proof. Since $\mathcal{K}$ is a compact subgroup of $\mathrm{GL}(\mathbb{R}^d)$, Remark 2.3.13 shows that there exists an inner product on $\mathbb{R}^d$ that makes every $G \in \mathcal{K}$ orthogonal. We will use the norm associated with this inner product to define the sequence B_n. Note that, in this norm, for $G \in \mathcal{K}$ and $A \in \mathrm{L}(\mathbb{R}^d)$, we have $\|GA\| = \|AG\| = \|A\|$. Let $B_1 = A_1$. For all other n, suppose that $2^k < n \le 2^{k+1}$ for some $k \ge 0$, and let $\lambda = n/2^k$, so that $1 < \lambda \le 2$. Choose $G'_n \in \mathcal{K}$ such that $\|A_n B_{2^k}^{-1} - \lambda^E G'_n\| = \|A_n B_{2^k}^{-1} - \lambda^E \mathcal{K}\|$. Since $\lambda^E \mathcal{K} = \mathcal{K}\lambda^E$, there exists a $G_n \in \mathcal{K}$ such that $\lambda^E G'_n = G_n \lambda^E$. Define $B_n = G_n^{-1} A_n$. Then, by Lemma 4.2.11 (see also Lemma 4.2.2) we have

$$\begin{aligned}
\|B_n B_{2^k}^{-1} - \lambda^E\| &= \|G_n^{-1} A_n B_{2^k}^{-1} - \lambda^E\| \\
&= \|G_n^{-1}(A_n B_{2^k}^{-1} - G_n \lambda^E)\| \\
&= \|A_n B_{2^k}^{-1} - \lambda^E G'_n\| \\
&= \|A_n B_{2^k}^{-1} - \lambda^E \mathcal{K}\| \\
&= \|A_n A_{2^k}^{-1} G_{2^k} - \lambda^E \mathcal{K}\| \\
&= \min\{\|A_n A_{2^k}^{-1} G_{2^k} - \lambda^E G\| : G \in \mathcal{K}\} \\
&= \min\{\|A_n A_{2^k}^{-1} G_{2^k} - \lambda^E G G_{2^k}\| : G \in \mathcal{K}\} \\
&= \min\{\|(A_n A_{2^k}^{-1} - \lambda^E G) G_{2^k}\| : G \in \mathcal{K}\} \\
&= \min\{\|A_n A_{2^k}^{-1} - \lambda^E G\| : G \in \mathcal{K}\} \\
&= \|A_n A_{2^k}^{-1} - \lambda^E \mathcal{K}\| \to 0
\end{aligned}$$

as $n \to \infty$ in view of the uniform convergence. Hence as $n \to \infty$ we have

$$\|B_n B_{2^{k(n)}}^{-1} - \lambda(n)^E\| \to 0 \tag{4.22}$$

where $2^{k(n)} < n \le 2^{k(n)+1}$ and $\lambda(n) = n/2^{k(n)}$, so that $n = \lambda(n)2^{k(n)}$ for some integer $k(n)$ and some $\lambda(n) \in (1, 2]$.

Now we will show that

$$B_{[\lambda 2^k]} B_{2^k}^{-1} \to \lambda^E \tag{4.23}$$

as $k \to \infty$, uniformly on $\lambda \in [1, 2]$. We need to check that for any $\lambda_k \in [1, 2]$ we have

$$\|B_{[\lambda_k 2^k]} B_{2^k}^{-1} - \lambda_k^E\| \to 0.$$

Define $n_k = [\lambda_k 2^k]$, so that $2^k \le n_k \le 2^{k+1}$ and hence $n_k \to \infty$. In view of (4.22), we have

$$\|B_{n_k} B^{-1}_{2^{k(n_k)}} - \lambda(n_k)^E\| \to 0.$$

If $n_k = 2^k$, then $k(n_k) = k - 1$ and $\lambda(n_k) = 2$. Otherwise, $k(n_k) = k$ and $\lambda_k 2^k - 1 < n_k = \lambda(n_k) 2^k = [\lambda_k 2^k] \le \lambda_k 2^k$ which implies that $\lambda_k - 2^{-k} < \lambda(n_k) \le \lambda_k$ so $|\lambda_k - \lambda(n_k)| \to 0$. Given any subsequence $k' \to \infty$, choose a further subsequence along which either $n_k = 2^k$ or $n_k > 2^k$. In the first case, we have $\lambda_k 2^k < 2^k + 1$, so that $\lambda_k < 1 + 2^{-k} \to 0$. Then, since $n_k = [\lambda_k 2^k] = 2^k$ and since $\lambda \mapsto \lambda^E$ is continuous, it follows that

$$\|B_{[\lambda_k 2^k]} B^{-1}_{2^k} - \lambda_k^E\| = \|I - \lambda_k^E\| \to 0.$$

In the second case, using again the fact that $\lambda \mapsto \lambda^E$ is continuous, it follows that

$$\|B_{[\lambda_k 2^k]} B^{-1}_{2^k} - \lambda_k^E\| \le \|B_{n_k} B^{-1}_{2^{k(n_k)}} - \lambda(n_k)^E\| + \|\lambda(n_k)^E - \lambda_k^E\| \to 0.$$

Then (4.23) holds as $k \to \infty$, uniformly on $\lambda \in [1, 2]$.

Now we will show that the sequence $\{B_n\}$ varies regularly with index E. Writing $n = \lambda 2^k$ as before, for all $\alpha \in [1, 2]$ we have

$$B_{[\alpha n]} B_n^{-1} = B_{[\alpha\lambda 2^k]} B^{-1}_{[\lambda 2^k]} = B_{[\alpha\lambda 2^k]} B^{-1}_{2^{k+1}} \cdot B_{2^{k+1}} B^{-1}_{2^k} \cdot B_{2^k} B^{-1}_{[\lambda 2^k]}$$

where the first term tends to $(\alpha\lambda/2)^E$ by (4.23) along with the fact that $1 \le (\alpha\lambda/2) \le 2$, the second term tends to 2^E by (4.23), and the third term tends to λ^{-E} by (4.23) along with the fact $A \mapsto A^{-1}$ is continuous. Then

$$B_{[\alpha n]} B_n^{-1} \to (\alpha\lambda/2)^E \cdot 2^E \cdot \lambda^{-E} = \alpha^E$$

for all $\alpha \in [1, 2]$.

If $\beta > 2$, then $\beta = \alpha 2^p$ for some integer p and some $\alpha \in [1, 2]$, and so

$$\begin{aligned} B_{[\beta n]} B_n^{-1} &= B_{[\alpha 2^p n]} B_n^{-1} \\ &= B_{[\alpha 2^p n]} B^{-1}_{2^p n} \cdot B_{2^p n} B^{-1}_{2^{p-1} n} \cdots B_{2n} B_n^{-1} \\ &\to \alpha^E \cdot 2^E \cdots 2^E \\ &= \beta^E \end{aligned}$$

as $n \to \infty$.

If $\alpha < 1$, let $\lambda = \alpha^{-1}$ and $m = m(n) = [n\alpha]$. We first show that

$$B_n B^{-1}_{[\lambda m(n)]} \to I \quad \text{as } n \to \infty. \tag{4.24}$$

In fact, write $n = \lambda_n 2^{k_n}$ for $2^{k_n} < n \le 2^{k_n + 1}$ and some $1 < \lambda_n \le 2$. Then, if we set $\bar{\lambda}_n = \alpha^{-1}[\lambda_n 2^{k_n} \alpha] 2^{-k_n}$, we easily see that $[\bar{\lambda}_n 2^{k_n}] = [\lambda m(n)]$, and it follows that $\bar{\lambda}_n \leq \lambda_n \leq 2$ as well as $\bar{\lambda}_n > \lambda_n - \alpha^{-1}2^{-k_n}$. Since (λ_n) is relatively compact, every subsequence contains a further subsequence (n') and a $\lambda_0 \in [1,2]$ such that $\lambda_n \to \lambda_0$ along (n'), and hence $\bar{\lambda}_n \to \lambda_0$ along (n') as well. Now we have to consider two cases separately.

If $\bar{\lambda}_n \geq 1$ along a subsequence $(n'') \subset (n')$, then (4.23) implies that

$$B_n B_{[\lambda m(n)]}^{-1} = \left(B_{[\lambda_n 2^{k_n}]} B_{2^{k_n}}^{-1}\right) \cdot \left(B_{2^{k_n}} B_{[\bar{\lambda}_n 2^{k_n}]}^{-1}\right) \to \lambda_0^E \lambda_0^{-E} = I$$

along (n'') using the continuity of the inversion in $\mathrm{GL}(\mathbb{R}^d)$.

On the other hand, if $\bar{\lambda}_n < 1$ along a subsequence $(n'') \subset (n')$, then $\lambda_0 = 1$ and (4.23) implies that

$$\begin{aligned} B_n B_{[\lambda m(n)]}^{-1} &= \left(B_{[\lambda_n 2^{k_n}]} B_{2^{k_n}}^{-1}\right) \cdot \left(B_{2^{k_n}} B_{2^{k_n-1}}^{-1}\right) \cdot \left(B_{2^{k_n-1}} B_{[2\bar{\lambda}_n 2^{k_n-1}]}^{-1}\right) \\ &\to \lambda_0^E 2^E (2\lambda_0)^{-E} = I \end{aligned}$$

along (n'') using the continuity of the inversion in $\mathrm{GL}(\mathbb{R}^d)$ again. Hence (4.24) holds true.

Finally, (4.24) and (4.23) together imply that

$$\begin{aligned} (B_{[\alpha n]} B_n^{-1})^{-1} &= B_n B_{[\lambda m]}^{-1} \cdot B_{[\lambda m]} B_m^{-1} \\ &\to I \cdot \lambda^E \\ &= \alpha^{-E} \end{aligned}$$

as $n \to \infty$. Now use the fact that the inverse operation is continuous to conclude that $B_{[\alpha n]} B_n^{-1} \to \alpha^E$. □

The following results will be essential in studying R–O varying measures as well as generalized domains of semistable attraction.

Theorem 4.2.13. *Let $\mathcal{K} \subset \mathrm{GL}(\mathbb{R}^d)$ be a compact subgroup and let (k_n) be an increasing sequence of integers tending to infinity with $k_{n+1}/k_n \to c > 1$. Assume further that for some $E \in \mathrm{L}(\mathbb{R}^d)$ there exists a sequence (A_n) of invertible linear operators such that*

$$A_{n+1} A_n^{-1} \rightharpoonup c^E \mathcal{K} \tag{4.25}$$

and

$$c^E \mathcal{K} = \mathcal{K} c^E. \tag{4.26}$$

Then there exists a sequence (B_n) varying regularly with index E such that $B_{k_n} = G_n A_n$ for some $G_n \in \mathcal{K}$.

The proof will be preceded by Lemma 4.2.14.

Lemma 4.2.14. *Let $\mathcal{K} \subset \mathrm{GL}(\mathbb{R}^d)$ be a compact subgroup. If for some sequence (A_n) of invertible linear operators and some $c > 1$ (4.25) and (4.26)* *hold, then there exist* $G_n \in \mathcal{K}$ *such that for the sequence defined by* $C_n = G_n A_n$ *we have*

$$C_{n+1} C_n^{-1} \to c^E \quad \textit{as } n \to \infty. \tag{4.27}$$

Proof. As in the proof of Proposition 4.2.12 we can choose an inner product on $\mathbb{R}^d$ that makes every element of $\mathcal{K}$ orthogonal and we denote by $\|A\|$ the corresponding operator norm. Then, if $G \in \mathcal{K}$ and $A \in \mathrm{GL}(\mathbb{R}^d)$ we have $\|GA\| = \|AG\| = \|A\|$.

Let $C_1 = A_1$, where (A_n) are as in (4.25). Assume $C_1, \ldots, C_{n-1}$ have been constructed. Chose $G'_n \in \mathcal{K}$ such that $\|A_n C_{n-1}^{-1} - c^E \mathcal{K}\| = \|A_n C_{n-1}^{-1} - c^E G'_n\|$. By (4.26) there exists a $G_n \in \mathcal{K}$ with $c^E G'_n = G_n c^E$. Define $C_n = G_n^{-1} A_n$. By (4.25) we have

$$\begin{aligned}
\|C_n C_{n-1}^{-1} - c^E\| &= \|G_n^{-1} A_n C_{n-1}^{-1} - c^E\| \\
&= \|G_n^{-1}(A_n C_{n-1}^{-1} - G_n c^E)\| \\
&= \|A_n C_{n-1}^{-1} - c^E G'_n\| \\
&= \|A_n C_{n-1}^{-1} - c^E \mathcal{K}\| \\
&= \|A_n A_{n-1}^{-1} G_{n-1} - c^E \mathcal{K}\| \\
&= \min\{\|A_n A_{n-1}^{-1} G_{n-1} - c^E G\| : G \in \mathcal{K}\} \\
&= \min\{\|A_n A_{n-1}^{-1} G_{n-1} - c^E G G_{n-1}\| : G \in \mathcal{K}\} \\
&= \min\{\|(A_n A_{n-1}^{-1} - c^E G) G_{n-1}\| : G \in \mathcal{K}\} \\
&= \min\{\|A_n A_{n-1}^{-1} - c^E G\| : G \in \mathcal{K}\} \\
&= \|A_n A_{n-1}^{-1} - c^E \mathcal{K}\| \to 0
\end{aligned}$$

as $n \to \infty$, which concludes the proof. □

Proof of Theorem 4.2.13. In view of Lemma 4.2.14, there exist $G_n \in \mathcal{K}$ such that $C_n = G_n A_n$ fulfills (4.27). Now for any $n \geq k_1$ write $n = \lambda(n) k_{p(n)}$, where $k_{p(n)} \leq n < k_{p(n)+1}$, and define

$$B_n = \lambda(n)^E C_{p(n)}. \tag{4.28}$$

Since $\lambda(k_n) = 1$ and $p(k_n) = n$ we have $B_{k_n} = C_n = G_n A_n$. Furthermore, since $k_{n+1}/k_n \to c$ it follows that $(\lambda(n))$ is relatively compact with all limit points in the set $[1, c]$.

It remains to show that (B_n) is regularly varying with index E. We show that

$$B_{[\alpha_n n]} B_n^{-1} \to \alpha^E \quad \text{as } n \to \infty \tag{4.29}$$

whenever $\alpha_n \to \alpha > 0$.

Assume first, that $\alpha \in [1, c]$. Write $[\alpha_n n] = \lambda([\alpha_n n]) k_{p([\alpha_n n])}$. We have to consider several cases separately:

Case I: Assume $p(n) = p([\alpha_n n])$ along a subsequence (n'). Then we have $\lambda([\alpha_n n])/\lambda(n) = [\alpha_n n]/n \to \alpha$ and

$$B_{[\alpha_n n]}B_n^{-1} = \lambda([\alpha_n n])^E C_{p([\alpha_n n])} C_{p(n)}^{-1} \lambda(n)^{-E} = \left(\frac{[\alpha_n n]}{n}\right)^E \to \alpha^E$$

along (n').

Case II: Assume $p([\alpha_n n]) = p(n) + 1$ along a subsequence (n'). Then, along that subsequence $\lambda([\alpha_n n]) = ([\alpha_n n]/n) \cdot (k_{p(n)}/k_{p(n)+1})\lambda(n)$. Since $(\lambda(n))$ is relatively compact every subsequence (n'') of (n') contains a further subsequence (n''') with $\lambda(n) \to \lambda \in [1, c]$ along (n'''). Hence $\lambda([\alpha_n n]) \to \alpha c^{-1}\lambda$ and therefore

$$B_{[\alpha_n n]}B_n^{-1} = \lambda([\alpha_n n])^E C_{p(n)+1} C_{p(n)}^{-1} \lambda(n)^{-E} \to (\alpha c^{-1}\lambda)^E c^E \lambda^{-E} = \alpha^E$$

along (n'''). Since every subsequence (n'') of (n') contains a further subsequence with this property we get (4.29) along the whole subsequence (n').

Case III: Assume $p([\alpha_n n]) = p(n) + 2$ along (n'). This case is similar to case 2.

Case IV: Assume that $p([\alpha_n n]) \geq p(n) + 3$ for infinitely many n, say along (n'). Then for every $\varepsilon > 0$ the relative compactness of the sequence $(\lambda(n))$ yields for all large n'

$$\frac{n}{[\alpha_n n]} = \frac{\lambda(n) k_{p(n)}}{\lambda([\alpha_n n]) k_{p([\alpha_n n])}} \leq (c + \varepsilon) \frac{k_{p(n)}}{k_{p(n)+3}}.$$

But the left-hand side of the inequality above tends to α^{-1} whereas the right-hand side tends to $(c + \varepsilon)c^{-3}$ along (n'). Hence $\alpha \geq (1 + \varepsilon/c)^{-1}c^2$, which contradicts $\alpha \leq c$ if ε is chosen small enough.

Case V: Assume $p([\alpha_n n]) = p(n) - 1$ along (n'). Then we have $\lambda([\alpha_n n]) = ([\alpha_n n]/n)(k_{p(n)}/k_{p(n)-1})\lambda(n)$ and it follows as in case 2 that (4.29) holds along (n')

Case VI: Assume $p([\alpha_n n]) \leq p(n) - 2$ along (n'). Then, using the relative compactness of $(\lambda(n))$, we get for any $\varepsilon > 0$ and all large n' that

$$\frac{[\alpha_n n]}{n} = \frac{\lambda([\alpha_n n]) k_{p([\alpha_n n])}}{\lambda(n) k_{p(n)}} \leq (c + \varepsilon) \frac{k_{p(n)-2}}{k_{p(n)}}.$$

The left-hand side of the inequality above tends to α whereas the right-hand side tends to $c^{-1} + \varepsilon c^{-2}$, which is a contradiction to $\alpha \geq 1$ if ε is small enough.

Therefore every subsequence has a further subsequence (n') such that either case I, II, III or V applies, which shows that (4.29) holds for $\alpha \in [1, c]$.

Now, if $\alpha > c$, write $\alpha = \bar{\alpha}c^j$ for some natural $j \geq 1$ and $\bar{\alpha} \in [1, c)$. Furthermore, write $\alpha_n = \bar{\alpha}_n c^{j_n}$, where eventually $j_n \geq 1$ and $\bar{\alpha}_n \in [1, c)$. Again we have to consider several cases separately:

Case A: If $j_n = j$ along a subsequence (n'), then $\bar{\alpha}_n \to \bar{\alpha}$ along that subsequence. Write

$$B_{[\alpha_n n]} B_n^{-1} = \left(B_{[\bar{\alpha}_n c^j n]} B_{[c^j n]}^{-1}\right)\left(B_{[c^j n]} B_{[c^{j-1} n]}^{-1}\right) \cdots \left(B_{[cn]} B_n^{-1}\right)$$

and note that by the already proved uniform convergence in (4.29) for $\bar{\alpha} \in [1, c]$ we get

$$B_{[\bar{\alpha}_n c^j n]} B_{[c^j n]}^{-1} = B_{\left[\frac{\bar{\alpha}_n c^j n}{[c^j n]}[c^j n]\right]} B_{[c^j n]}^{-1} \to \bar{\alpha}^E$$

and for $k = 1, \ldots, j$

$$B_{[c^k n]} B_{[c^{k-1} n]}^{-1} = B_{\left[\frac{[c^k n]}{[c^{k-1} n]}[c^{k-1} n]\right]} B_{[c^{k-1} n]}^{-1} \to c^E$$

along (n'), and hence (4.29) holds along (n').

Case B: If $j_n = j - 1$ along (n'), then $\alpha_n = \bar{\alpha}_n c^{j-1} \to \bar{\alpha} c^j = \alpha$, and hence $\bar{\alpha}_n \to c\bar{\alpha}$ along (n'). Since $\bar{\alpha}_n < c$ and $\bar{\alpha} \geq 1$, this can only happen if $\bar{\alpha} = 1$, so $\alpha = c^j$. Then

$$B_{[\alpha_n n]} B_n^{-1} = \left(B_{[\bar{\alpha}_n c^{j-1} n]} B_{[c^{j-1} n]}^{-1}\right)\left(B_{[c^{j-1} n]} B_{[c^{j-2} n]}^{-1}\right) \cdots \left(B_{[cn]} B_n^{-1}\right)$$

and as in case A we see that then along (n')

$$B_{[\alpha_n n]} B_n^{-1} \to c^E \cdot (\underbrace{c^E \cdots c^E}_{j-1 \text{ times}}) = (c^j)^E = \alpha^E.$$

Case C: If $j_n = j - k$ along (n') and $k \notin \{0, 1\}$, then $\bar{\alpha}_n \to c^k \bar{\alpha} \in [c^k, c^{k+1})$ along (n'), which contradicts $\bar{\alpha}_n \in [1, c)$ for all n, so this case is void.

These are all possible cases, hence it follows that (4.29) holds for any $\alpha \geq 1$.

Finally, assume $0 < \alpha < 1$ and let $m(n) = [\alpha_n n]$. Then $n/m(n) = n/[\alpha_n n] \to \alpha^{-1} > 1$, and then we get from (4.29) that

$$\left(B_{[\alpha_n n]} B_n^{-1}\right)^{-1} = B_n B_{m(n)}^{-1} = B_{\left[\frac{n}{m(n)} m(n)\right]} B_{m(n)}^{-1} \to (\alpha^{-1})^E = \alpha^{-E}.$$

Taking the inverse on both sides, (4.29) follows in this case too, and the proof is complete. □

We will apply Theorem 4.2.13 in Chapters 6 and 8. There we will also need the following special case.

Corollary 4.2.15. *Let (k_n) be an increasing sequence of integers tending to infinity with $k_{n+1}/k_n \to c > 1$. Assume further that for some $\rho \in \mathbb{R}^1$ there exists a sequence (a_n) of positive real numbers such that*

$$a_{n+1} a_n^{-1} \to c^\rho. \tag{4.30}$$

Then there exists a sequence of positive real numbers (b_n) varying regularly with index ρ such that $b_{k_n} = a_n$.

Proof. Apply Theorem 4.2.13 with $A_n = a_n I$, $E = \rho I$ and $\mathcal{K} = \{I\}$. □

Our next result extends Lemma 4.2.7.

Lemma 4.2.16. *Suppose that $A_n \in \mathrm{GL}(\mathbb{R}^d)$ varies regularly with index E, and that (k_n) is an increasing sequence of integers tending to infinity with $k_{n+1}/k_n \to c \geq 1$.*

(a) If every eigenvalue of E has real part less than zero, then for any $x_r \to x \neq 0$ in $\mathbb{R}^d$ the function

$$n(r) = \sup\{n : \|A_{k_n} r x_r\| \geq 1\} \tag{4.31}$$

is well defined for all r large, and $n = n(r)$ tends to infinity as $r \to \infty$. Furthermore, if we define $y_n = A_{k_n} r x_r$, then (y_n) is limit point compact, and every limit point lies in the compact set $\{y \in \mathbb{R}^d : 1 \leq \|y\| \leq \|c^{-E}\|\}$.

(b) If every eigenvalue of E has real part greater than zero, then for any $x_r \to x \neq 0$ in $\mathbb{R}^d$ the function

$$n(r) = \sup\{n : \|A_{k_n} x_r / r\| \leq 1\} \tag{4.32}$$

is well defined for all r large, and $n = n(r)$ tends to infinity as $r \to \infty$. Furthermore, if we define $y_n = A_{k_n} x_r / r$, then (y_n) is limit point compact, and every limit point lies in the compact set $\{y \in \mathbb{R}^d : 1/\|c^E\| \leq \|y\| \leq 1\}$.

Proof. The proof is similar to that of Lemma 4.2.7. First, we will prove part (a). Given $x_r \to x \neq 0$, for some $r_0 > 0$ we have $m \leq \|x_r\|$ for all $r \geq r_0$ by continuity of the norm, where $0 < m < \|x\|$. Then $\|r x_r\| \to \infty$, so for any fixed $n > 0$ we have $\|A_n r x_r\| \geq c_n \|r x_r\| \geq 1$ for all $r > 0$ large, where $0 < c_n = \min\{\|A_n \theta\| : \|\theta\| = 1\}$. Hence the set appearing in (4.31) is nonempty for all r large. For $r > 0$ fixed, we also have $\|A_n r x_r\| \to 0$ as $n \to \infty$ by Theorem 4.2.4 with $\beta = 0$ along with Theorem 4.2.9. Thus the set appearing in (4.31) is bounded above for any given r, and so $n(r)$ is well defined for all r large. Suppose by way of contradiction that there exist $r_k \to \infty$ such that $n(r_k) < n_0$ for all k. Then $\|A_n r_k x_{r_k}\| < 1$ for all k and all $n > n_0$. But this contradicts the fact that $\|A_n r_k x_{r_k}\| \geq c_n \|r_k x_{r_k}\| \to \infty$ as $k \to \infty$ for any fixed n, and so $n(r) \to \infty$ as $r \to \infty$.

Equation (4.31) implies that $\|A_{k_n} r x_r\| \geq 1$ and $\|A_{k_{n+1}} r x_r\| < 1$ when $n = n(r)$. Rewriting this in terms of y_n yields $\|y_n\| \geq 1$ and $\|A_{k_{n+1}} A_{k_n}^{-1} y_n\| < 1$. Write $A_{k_{n+1}} A_{k_n}^{-1} y_n = A_{k_{n+1}} r x_r = \rho_n \theta_n$ for some $\rho_n > 0$ and $\theta_n \in S = \{\theta \in \mathbb{R}^d : \|\theta\| = 1\}$. Then $\rho_n < 1$. Since $k_{n+1}/k_n \to c$, Corollary 4.2.11 (a) implies that $A_{k_{n+1}} A_{k_n}^{-1} \to c^E$, and since $A \mapsto A^{-1}$ is continuous this implies

$A_{k_n}A_{k_{n+1}}^{-1} \to c^{-E}$. Then $A_{k_n}A_{k_{n+1}}^{-1}\theta \to c^{-E}\theta$ uniformly on $\theta \in S$, so for any $\varepsilon > 0$ we have $\|A_{k_n}A_{k_{n+1}}^{-1}\theta - c^{-E}\theta\| < \varepsilon$ for all n sufficiently large and all $\theta \in S$. It follows that $\|y_n\| = \|A_{k_n}A_{k_{n+1}}^{-1}\rho_n\theta_n\| = \rho_n \cdot \|A_{k_n}A_{k_{n+1}}^{-1}\theta_n\| < 1\cdot(\|c^{-E}\|+\varepsilon)$ for all $n = n(r)$ sufficiently large. Then $1 \le \|y_n\| \le \|c^{-E}\| + \varepsilon$ for all n sufficiently large, and part (a) follows immediately.

The proof of part (b) is similar. Given $x_r \to x \neq 0$, for some $r_0 > 0$ we have $\|x_r\| \le M$ for all $r \ge r_0$ by continuity of the norm. Then $\|x_r/r\| \to 0$, so for any fixed $n > 0$ we have $\|A_n x_r/r\| \le \|A_n\| \cdot \|x_r/r\| \le 1$ for all $r > 0$ large, hence the set in (4.32) is nonempty for all r large. For $r > 0$ fixed we also have $\|A_n x_r/r\| \to \infty$ as $n \to \infty$ by Theorem 4.2.4 with $\alpha = 0$ along with Theorem 4.2.9, so that the set in (4.31) is bounded above for any given r, hence $n(r)$ is well defined for all r large. Suppose by way of contradiction that there exist $r_k \to \infty$ such that $n(r_k) < n_0$ for all k. Then $\|A_n x_{r_k}/r_k\| > 1$ for all k and all $n > n_0$. But this contradicts the fact that $\|A_n x_{r_k}/r_k\| \le \|A_n\| \cdot \|x_{r_k}/r_k\| \to 0$ as $k \to \infty$ for any fixed n, and so $n(r) \to \infty$ as $r \to \infty$.

Equation (4.32) implies that $\|A_{k_n}x_r/r\| \le 1$ and $\|A_{k_{n+1}}x_r/r\| > 1$ when $n = n(r)$. Rewriting this in terms of y_n yields $\|y_n\| \le 1$ and $\|A_{k_{n+1}}A_{k_n}^{-1}y_n\| > 1$. Writing $A_{k_{n+1}}A_{k_n}^{-1}y_n = \rho_n\theta_n$ as before, we have that $\rho_n > 1$. Since $A_{k_n}A_{k_{n+1}}^{-1}\theta \to c^{-E}\theta$ uniformly on $\theta \in S$, for any $\varepsilon > 0$ we have $\|A_{k_n}A_{k_{n+1}}^{-1}\theta\| > \|c^{-E}\theta\| - \varepsilon$ for all n sufficiently large and all $\theta \in S$. Using the general fact that $\|Ax\| \ge \|x\|/\|A^{-1}\|$ it follows that $\|y_n\| = \rho_n \cdot \|A_{k_n}A_{k_{n+1}}^{-1}\theta_n\| > 1\cdot(1/\|c^E\| - \varepsilon)$ for all $n = n(r)$ sufficiently large, and part (b) follows easily. □

4.3 THE SPECTRAL DECOMPOSITION

In this section, we prove a spectral decomposition theorem for regularly varying sequences of linear operators. The spectral decomposition is useful because it reduces to the case where the growth rate is essentially uniform in every radial direction. In Chapters 5 and 6, we will use the spectral decomposition repeatedly to obtain sharp growth estimates for regularly varying functions and measures. It will also be used in Part III, to reduce the study of certain limit theorems for random vectors to the case where the tail behavior is essentially the same in every radial direction.

The growth rate of a regularly varying sequence of linear operators

$$A_{[\lambda n]}A_n^{-1} \to \lambda^E$$

is governed by the exponent E, since $A_{[\lambda n]} \approx \lambda^E A_n$ for n large. If $Ex = \lambda x$ for some nonzero vector $x \in \mathbb{R}^d$ and some complex number $\lambda = a+ib$, then the norm of $\lambda^E x$ grows about as fast as λ^a (see Corollary 2.2.5). The collection of all such a is called the real spectrum of E. In this section, we establish a spectral decomposition that elucidates the connection between the real spectrum of E and the growth rate of the sequence (A_n). This decomposition yields a wealth of useful information about regularly varying functions and measures, through the linear operators which govern their growth. The next result generalizes Corollary 2.2.7. It shows that the growth of a regularly varying sequence with index E is essentially the same as that of the limit function t^E, which is the operator analogue of a power function. Recall from Definition 2.2.6 that for $x_n \in \mathbb{R}^d$ and $\mathcal{S}$ a compact subset of $\mathbb{R}^d$ we write $x_n \rightharpoonup \mathcal{S}$ to mean that (x_n) is limit point compact with every limit point lying in the set $\mathcal{S}$. Let S^{d-1} denote the unit sphere with respect to $\|\cdot\|$ in $\mathbb{R}^d$.

Theorem 4.3.1. *Suppose (B_n) is a regularly varying sequence with index E and let $\mathbb{R}^d = V_1 \oplus \cdots \oplus V_p$ be the spectral decomposition of $\mathbb{R}^d$ relative to E. Then there exists a unique nested sequence of subspaces $L_1 \subset \cdots \subset L_p = \mathbb{R}^d$ such that for each $i = 1, \ldots, p$ we have*

(a) $\dim L_i = \dim(V_1 \oplus \cdots \oplus V_i)$

(b) If $x_n \to x \neq 0$ in L_i, then $B_n x_n / \|B_n x_n\| \rightharpoonup (V_1 \oplus \cdots \oplus V_i) \cap S^{d-1}$

(c) If $x_n \to x$ in $\mathbb{R}^d \setminus L_i$, then $B_n x_n / \|B_n x_n\| \rightharpoonup (V_{i+1} \oplus \cdots \oplus V_p) \cap S^{d-1}$

(d) If $x_n \to x$ in L_i, then $n^{-\rho}\|B_n x_n\| \to 0$ for all $\rho > a_i$

(e) If $x_n \to x$ in $\mathbb{R}^d \setminus L_i$, then $n^{-\rho}\|B_n x_n\| \to \infty$ for all $\rho < a_{i+1}$

Proof. Fix i and let $V' = V_1 \oplus \cdots \oplus V_i$ and $V = V_{i+1} \oplus \cdots \oplus V_p$. We will see the L_i is in some sense the limit of $B_n^{-1}(V')$ as $n \to \infty$. Choose $\rho \in (a_i, a_{i+1})$ and let $C_n = n^{-\rho} B_n$. Then, using the regular variation of (B_n)

$$C_{[\lambda n]} C_n^{-1} \to \lambda^A \quad \text{as } n \to \infty, \tag{4.33}$$

where $A = E - \rho I$. Recall that by Corollary 4.2.11 the convergence in (4.33) is uniform on compact subsets of $\{\lambda > 0\}$.

Now V, V' are A-invariant subspaces and in fact V is the direct sum of the generalized eigenspaces of A whose eigenvalues have positive real part, while V' is the direct sum of the generalized eigenspaces of A whose eigenvalues have negative real part. Corollary 2.2.7 implies that $\|t^A x\| \to \infty$ uniformly on compact subsets of $U = \mathbb{R}^d \setminus V'$ and $\|t^A x\| \to 0$ uniformly on compact subsets of V'.

Then for any $0 < m < 1 < M < \infty$, we have for all $\lambda_0 > 1$ sufficiently large that

$$\begin{aligned} \|\lambda_0^A x\| &\geq M\|x\| \quad \text{for all } x \in V, \text{ and} \\ \|\lambda_0^A x\| &\leq m\|x\| \quad \text{for all } x \in V'. \end{aligned} \tag{4.34}$$

Using the direct sum decomposition, we can write any $y \in \mathbb{R}^d$ in the form $y = r\theta + r'\theta'$, where $r, r' \geq 0$ and $\theta \in V$, $\theta' \in V'$ are unit vectors. Define

$$V_\delta = \{r\theta + r'\theta' \ : \ r > \delta r'\}$$

for any $0 < \delta < 1$. Now choose $\lambda_0 > 1$ so that (4.34) holds for some $0 < m < \delta/2$ and $3 < M < \infty$. Then, by (4.33), given $0 < \varepsilon < \delta/4$, there exists a $n_0 = n_0(\delta)$ such that for all $n \geq n_0$ and all $1 \leq \lambda \leq \lambda_0$,

$$\|C_{[\lambda n]}C_n^{-1} - \lambda^A\| < \varepsilon. \tag{4.35}$$

Now suppose that $y_0 \in V_\delta$ and let $x = C_n^{-1}y_0$ for some $n \geq n_0$. Define $y_j = C_{[\lambda_0^j n]}x$ and write $y_j = r_j\theta_j + r_j'\theta_j'$ using the direct sum decomposition.

Now we outline the remainder of the proof. Let

$$S = \bigcup_{0<\delta<1}\ \bigcup_{n\geq n_0(\delta)} C_n^{-1}(V_\delta)$$

so that if $x \in S$, then $y_0 \in V_\delta$ for $n \geq n_0(\delta)$. First we will prove part (e), using Lemmas 4.3.2 and 4.3.3 to show that $r_j \to \infty$ for $x \in S$, and using Lemma 4.3.4 to interpolate. Next we will prove part (c), using Lemma 4.3.5 to show that r_j dominates r_j' for $x \in S$. Parts (b) and (d) will follow from (c) and (e), using Lemma 4.3.6 to show that $L_i = \mathbb{R}^d \setminus S$ is the linear subspace of $\mathbb{R}^d$ on which $\|C_n x\| \to 0$, and then (a) will follow by construction of the subspace L_i.

Lemma 4.3.2. *Suppose $x = C_n^{-1}y_0$ for some $y_0 \in V_\delta$ and $n \geq n_0(\delta)$, and let $y_j = C_{[\lambda_0^j n]}x$. Then*

$$\begin{aligned} r_{j+1} &\geq Mr_j - \varepsilon(r_j + r_j') \\ r_{j+1}' &\leq mr_j' + \varepsilon(r_j + r_j') \end{aligned}$$

for all $j \geq 0$.

Proof. Observe that $y_{j+1} = C_{[\lambda_0^{j+1}n]}C_{[\lambda_0^j n]}^{-1}y_j$ and write $C_{[\lambda_0^{j+1}n]}C_{[\lambda_0^j n]} = \lambda_0^A + E_j$, where by (4.35) $\|E_j\| < \varepsilon$. Define π, π' to be the natural projections onto V, V' respectively. Then in view of the fact that V and V' are A-invariant we have using (4.34) that

$$\begin{aligned} r_{j+1} &= \|\pi C_{[\lambda_0^{j+1}n]}C_{[\lambda_0^j n]}y_j\| \\ &= \|\pi(\lambda_0^A + E_j)y_j\| \\ &= \|\lambda_0^A \pi y_j + \pi E_j y_j\| \\ &\geq \|\lambda_0^A r_j\theta_j\| - \|\pi E_j y_j\| \\ &\geq Mr_j - \varepsilon(r_j + r_j') \end{aligned}$$

and

$$\begin{aligned} r_{j+1}' &= \|\pi' C_{[\lambda_0^{j+1}n]}C_{[\lambda_0^j n]}^{-1}y_j\| \\ &= \|\pi'(\lambda_0^A + E_j)y_j\| \\ &= \|\lambda_0^A \pi' y_j + \pi' E_j y_j\| \\ &\leq \|\lambda_0^A r_j'\theta_j'\| + \|E_j y_j\| \\ &\leq mr_j' + \varepsilon(r_j + r_j') \end{aligned}$$

for all $j \geq 0$. □

Lemma 4.3.3. *Suppose* $x = C_n^{-1}y_0$ *for some* $y_0 \in V_\delta$ *and* $n \geq n_0(\delta)$, *and let* $y_j = C_{[\lambda_0^j n]}x$. *Then*

$$\begin{aligned} r'_j &\leq r_j/\delta \\ r_{j+1} &\geq \left(M - \frac{2\varepsilon}{\delta}\right) r_j \end{aligned} \tag{4.36}$$

for all $j \geq 0$.

Proof. We will use induction. Obviously, (4.36) is true for $j = 0$ by Lemma 4.3.2 and the definition of V_δ, since $\delta < 1$. Suppose (4.36) is true for $j = 0, \ldots, k-1$. Then we must show it is true for $j = k$. First, compute using Lemma 4.3.2 that

$$\begin{aligned} r'_k &\leq (m+\varepsilon)r'_{k-1} + \varepsilon r_{k-1} \leq (m+\varepsilon)r_{k-1}/\delta + \varepsilon r_{k-1} \\ &= ((m+\varepsilon)/\delta + \varepsilon)r_{k-1} \leq \left(M - \frac{2\varepsilon}{\delta}\right) r_{k-1} \leq r_k \leq r_k/\delta \end{aligned}$$

by induction hypothesis and the choice of m, M, ε, and δ. It follows immediately using Lemma 4.3.2 that

$$r_{k+1} \geq (M-\varepsilon)r_k - \varepsilon r'_k \geq (M-\varepsilon)r_k - \varepsilon r_k/\delta \geq \left(M - \frac{2\varepsilon}{\delta}\right) r_k ,$$

so the claim follows by induction. □

From Lemma 4.3.3, it follows that $r_j \to \infty$, so $\|y_j\| \to \infty$. Since (4.35) holds uniformly in $1 \leq \lambda \leq \lambda_0$ we can interpolate between y_j and y_{j+1} to show that $\|C_m x\| \to \infty$ as $m \to \infty$. In particular, if we write $m = \lambda\lambda_0^j n$ for some $1 \leq \lambda \leq \lambda_0$, $j \geq 0$ and a fixed $n \geq n_0(\delta)$, then

$$C_m x = C_{[\lambda\lambda_0^j n]} C_{[\lambda_0^j n]}^{-1} y_j$$

where

$$C_{[\lambda\lambda_0^j n]} C_{[\lambda_0^j n]}^{-1} = \lambda^A + E_m.$$

Note that by the uniform convergence in (4.33) and by (4.35) $\|E_m\| = \varepsilon_m \to 0$ as well as $\varepsilon_m < \varepsilon$ for all $m \geq n$. Write $C_m x = r_m\theta_m + r'_m\theta'_m$ and $y_j = r_j\theta_j + r'_j\theta'_j$ as before. Then the behavior of $C_m x$ in terms of y_j is given by Lemma 4.3.4.

Lemma 4.3.4. *Suppose* $x = C_n^{-1}y_0$ *for some* $y_0 \in V_\delta$ *and* $n \geq n_0(\delta)$, *and let* $y_j = C_{[\lambda_0^j n]}x$. *Given* $m \geq n$, *let* $j \geq 0$ *be defined by* $m = \lambda\lambda_0^j n$ *for some*

$1 \leq \lambda \leq \lambda_0$. *Then there exist positive real constants B and b independent of m and j such that*

(a) $(b-\varepsilon)\|y_j\| \leq \|C_m x\| \leq (B+\varepsilon)\|y_j\|$

(b) $r_m \geq (b-\varepsilon)r_j - \varepsilon_m r_j'$

(c) $r_m' \leq (B+\varepsilon)r_j' + \varepsilon_m r_j$

Proof. Let $b = \inf\{\|\lambda^A\theta\| : 1 \leq \lambda \leq \lambda_0, \|\theta\| = 1\}$ and let $B = \sup\{\|\lambda^A\theta\| : 1 \leq \lambda \leq \lambda_0, \|\theta\| = 1\}$. Then we have

$$\begin{aligned} \|C_m x\| &= \|C_{[\lambda\lambda_0^j n]} C_{[\lambda_0^j n]}^{-1} y_j\| \\ &= \|(\lambda^A + E_m)y_j\| \\ &\leq \|\lambda^A y_j\| + \|E_m y_j\| \leq (B+\varepsilon)\|y_j\| \end{aligned}$$

and

$$\begin{aligned} \|C_m x\| &= \|\lambda^A y_j + E_m y_j\| \\ &\geq \|\lambda^A y_j\| - \|E_m y_j\| \\ &\geq (b-\varepsilon)\|y_j\|, \end{aligned}$$

which proves (a). Recall the definition of the projections π and π'. Then as in the proof of Lemma 4.3.2 we get

$$\begin{aligned} r_m &= \|\pi(\lambda^A + E_m)y_j\| = \|r_j\lambda^A\theta_j + \pi E_m y_j\| \\ &\geq r_j b - \|\pi E_m y_j\| \geq r_j b - \varepsilon_m(r_j + r_j') \\ &\geq (b-\varepsilon)r_j - \varepsilon_m r_j' \end{aligned}$$

and

$$\begin{aligned} r_m' &= \|\pi'(\lambda^A + E_m)y_j\| = \|r_j'\lambda^A\theta_j' + \pi' E_m y_j\| \\ &\leq r_j' B + \|\pi' E_m y_j\| \leq r_j' B + \varepsilon_m(r_j + r_j') \\ &\leq (B+\varepsilon)r_j' + \varepsilon_m r_j. \end{aligned}$$

This concludes the proof of the lemma. □

Recall that

$$S = \bigcup_{0<\delta<1} \bigcup_{n \geq n_0(\delta)} C_n^{-1}(V_\delta)$$

and $L = L_i = \mathbb{R}^d \setminus S$. Given $x_m \to x$ in S, we have $x \in C_n^{-1}(V_\delta)$ for some $0 < \delta < 1$ and some $n \geq n_0(\delta)$. Then $x_m \in C_n^{-1}(V_\delta)$ for all m large, since $C_n^{-1}(V_\delta)$ is open. Hence $y_0(m) = C_n x_m \in V_\delta$ for all m large. Define $y_j(m), r_j(m), r_j'(m), r_m(m), r_m'(m)$ as before with y_0 replaced by $y_0(m)$. Then

$r_0(m) \to r_0 > 0$. Since the bounds in Lemmas 4.3.3 and 4.3.4 hold for any $x_m \in C_n^{-1}V_\delta$, we have that $r_j(m) \geq (M - 2\varepsilon/\delta)^j r_0(m)$ and $r_m(m) \geq (b - 2\varepsilon/\delta)r_j(m)$ so that $\|C_m x_m\| \to \infty$ as $m \to \infty$, which proves part (e) of Theorem 4.3.1. We will now prove part (c).

Lemma 4.3.5. *Suppose $x \in S$ and let $y_j = C_{[\lambda_0^j n]}x$ for some $n \geq n_0(\delta)$ and $j \geq 0$, where $0 < \delta < 1$ such that $x \in C_n^{-1}(V_\delta)$. Then $r_j/r_j' \to \infty$.*

Proof. Write $C_{[\lambda_0^{j+1}n]}C^{-1}_{[\lambda_0^j n]} = \lambda_0^A + E_j$, where by (4.33) and (4.35) $\|E_j\| = \varepsilon_j \to 0$ as well as $\varepsilon_j < \varepsilon$ for all $j \geq 0$. Then by the proof of Lemma 4.3.2 we have

$$\begin{aligned} r_{j+1} &\geq Mr_j - \varepsilon_j(r_j + r_j') \quad \text{and} \\ r_{j+1}' &\leq mr_j' + \varepsilon_j(r_j + r_j') \quad \text{for all } j \geq 0. \end{aligned}$$

Now we will use induction to show that

$$r_j' \leq (m+\varepsilon)^j \frac{r_0}{\delta} + \sum_{i=0}^{j-1}(m+\varepsilon)^{j-1-i}\varepsilon_i r_i \tag{4.37}$$

for all $j \geq 0$. By Lemma 4.3.3, this is certainly true in the case $j = 0$, where we interpret the sum as being zero. Suppose (4.37) is true for $j = 0, \ldots, k-1$. Then we have

$$\begin{aligned} r_k' &\leq (m+\varepsilon)r_{k-1}' + \varepsilon_{k-1}r_{k-1} \\ &\leq (m+\varepsilon)\left((m+\varepsilon)^{k-1}\frac{r_0}{\delta} + \sum_{i=0}^{k-2}(m+\varepsilon)^{k-2-i}\varepsilon_i r_i\right) + \varepsilon_{k-1}r_{k-1} \\ &= (m+\varepsilon)^k \frac{r_0}{\delta} + \sum_{i=0}^{k-2}(m+\varepsilon)^{k-1-i}\varepsilon_i r_i + \varepsilon_{k-1}r_{k-1} \\ &= (m+\varepsilon)^k \frac{r_0}{\delta} + \sum_{i=0}^{k-1}(m+\varepsilon)^{k-1-i}\varepsilon_i r_i \end{aligned}$$

as required. Lemma 4.3.3 showed that $r_j' \leq r_j/\delta$ and $r_{j+1} \geq (M - 2\varepsilon/\delta)r_j$ for all $j \geq 0$. It follows that $r_{j+1} \geq r_j$ and $r_j \to \infty$. Then

$$\begin{aligned} \frac{r_j'}{r_j} &\leq (m+\varepsilon)^j\delta^{-1}\frac{r_0}{r_j} + \sum_{i=0}^{j-1}(m+\varepsilon)^{j-1-i}\varepsilon_i\left(\frac{r_i}{r_j}\right) \\ &\leq (m+\varepsilon)^j\delta^{-1} + \sum_{i=0}^{j-1}(m+\varepsilon)^{j-i-1}\varepsilon_i \end{aligned}$$

and since the first term above tends to zero as $j \to \infty$, we need only to show that the same is true for the second term. Given $\gamma > 0$ arbitrary small, choose

k such that $\varepsilon_i < \gamma$ for all $i \geq k$. Write the sum in the form $I_1 + I_2$, where I_1 is the sum over $0 \leq i \leq k-1$. Then for any $j \geq k$ we have

$$I_1 = \sum_{i=0}^{k-1} (m+\varepsilon)^{j-1-i} \varepsilon_i \leq \varepsilon \sum_{i=j-k}^{j-1} (m+\varepsilon)^i \leq \sum_{i=j-k}^{\infty} (m+\varepsilon)^i$$

which tends to zero as $j \to \infty$, while

$$\begin{aligned} I_2 &= \sum_{i=k}^{j-1} (m+\varepsilon)^{j-1-i} \varepsilon_i \leq \gamma \sum_{i=k}^{j-1} (m+\varepsilon)^{j-1-i} \\ &= \gamma \sum_{i=0}^{j-1-k} (m+\varepsilon)^i \leq \frac{\gamma}{1-(m+\varepsilon)} \end{aligned}$$

where $\gamma > 0$ is arbitrary small. Thus we have $r'_j/r_j \to 0$, and then $r_j/r'_j \to \infty$ as $j \to \infty$. □

To show that in fact $C_m x/\|C_m x\| \rightharpoonup V$ whenever $x \in S$, as before write $m = \lambda\lambda_0^j n$ for some $1 \leq \lambda \leq \lambda_0$ and $j \geq 0$. Here we have assumed without loss of generality that $m \geq n$ and $x \in C_n^{-1}(V_\delta)$ for some $n \geq n_0(\delta)$ and some $0 < \delta < 1$. Then $y_m = C_{[\lambda\lambda_0^j n]} C^{-1}_{[\lambda_0^j n]} y_j$, where $C_{[\lambda\lambda_0^j n]} C^{-1}_{[\lambda_0^j n]} = \lambda^A + E_m$ with $\|E_m\| = \varepsilon_m \to 0$. Writing $y_m = r_m\theta_m + r'_m\theta'_m$ as before and using $r'_j \leq r_j/\delta$ from Lemma 4.3.3, we get from Lemma 4.3.4 that

$$\begin{aligned} \frac{r'_m}{r_m} &\leq \frac{(B+\varepsilon)r'_j + \varepsilon_m r_j}{(b-\varepsilon)r_j - \varepsilon_m r'_j} \leq \frac{(B+\varepsilon)r'_j + \varepsilon_m r_j}{(b-\varepsilon-\varepsilon_m/\delta)r_j} \\ &= \frac{B+\varepsilon}{b-\varepsilon-\varepsilon_m/\delta} \cdot \frac{r'_j}{r_j} + \frac{\varepsilon_m}{b-\varepsilon-\varepsilon_m/\delta}. \end{aligned} \tag{4.38}$$

Since by Lemma 4.3.5 $r'_j/r_j \to 0$ as $j \to \infty$ and $\varepsilon_m \to 0$ as $m \to \infty$ we get $r'_m/r_m \to 0$. Now suppose that $x_m \to x$ in S. Then as before we have $x_m \in C_n^{-1}V_\delta$ for m large, where $0 < \delta < 1$ and $n \geq n_0(\delta)$. Let $y_0(m) = C_n x_m$ and define $y_j(m), r_j(m), r'_j(m), r_m(m), r'_m(m)$ as before with y_0 replaced by $y_0(m)$. Since the estimates in Lemmas 4.3.5 only depend on the inequalities $r'_j \leq r_j/\delta$ and $r_{j+1} \geq (M - 2\varepsilon/\delta)r_j$, which hold for all $x_m \in C_n^{-1}V_\delta$, it follows that $r_j(m)/r'_j(m) \to \infty$. Since the estimates in Lemmas 4.3.3 and 4.3.4 also hold for all x_m it follows from (4.38) that $r'_m(m)/r_m(m) \to 0$. This proves part (c) of Theorem 4.3.1.

Now we prove part (b) of Theorem 4.3.1. If $x_m \to x$ in L write $C_m x = r_m\theta_m + r'_m\theta'_m$ as before. If $\limsup_{m\to\infty}(r_m/r'_m) > 0$, then for some subsequence $m_n \to \infty$ and some $\delta > 0$ we have $r_{m_n}/r'_{m_n} > \delta$ for all n, so that $C_{m_n}x_{m_n} \in V_\delta$ for all $m_n \geq n_0(\delta)$, and so $x_{m_n} \to x$ in S which is a contradiction. Hence $r_m/r'_m \to 0$, which yields $C_m x_m/\|C_m x_m\| \rightharpoonup V'$ whenever $x_m \to x$ in L, as desired. This proves part (b).

Now we prove part (d) of Theorem 4.3.1. We will actually prove a slightly stronger result, which also shows that $L = L_i$ is a linear subspace of $\mathbb{R}^d$.

Lemma 4.3.6. *$x \in L_i$ if and only if $C_m x \to 0$.*

Proof. Let $x \in \mathbb{R}^d$ and for $j \geq 0$, some $0 < \delta < 1$ and some $n \geq n_0(\delta)$ let $y_j = C_{[\lambda_0^j n]}x$. Then the proof of Lemma 4.3.4(a) shows that $(b-\varepsilon)\|y_j\| \leq \|C_m x\| \leq (B+\varepsilon)\|y_j\|$, where $m = \lambda\lambda_0^j n$ for some $1 \leq \lambda \leq \lambda_0$. Hence it is enough to show that $x \in L_i = L$ if and only if $y_j \to 0$. If $y_j \to 0$, then $x \in L$ because otherwise $x \in L^c = S$, and $\|y_j\| \to \infty$ for all $x \in S$ by Lemma 4.3.3. Conversely, suppose $y_j \not\to 0$. If $r_j/r'_j \to 0$, then $r_j \leq r'_j$ for all j large, and then $r'_{j+1} \leq (m+\varepsilon)r'_j + \varepsilon r_j \leq (m+2\varepsilon)r'_j$ so $r'_j \to 0$, and hence $r_j \to 0$, which contradicts $y_j \not\to 0$. Hence there must exist $j_l \to \infty$ such that $r_{j_l}/r'_{j_l} > \delta_1 > 0$ for all l and some $\delta_1 > 0$, so that $y_{j_l} \in V_{\delta_1}$ for all l. If $m_l = [\lambda_0^{j_l} n] \geq n_0(\delta_1)$ then we have $x = C_{m_l}^{-1} y_{j_l} \in S$, so $x \notin L$. Then $x \in L$ implies $y_j \to 0$. □

Since the restriction of C_n to the subspace L_i is a linear operator, Proposition 2.1.8 together with Lemma 4.3.6 implies that $\|C_n x_n\| \to 0$ whenever $x_n \to x$ in L_i. This proves part (d). Lemma 4.3.6 together with (d) shows that the subspace L_i is unique. Lemma 4.3.6 also implies that the sequence L_i of subsequences is nested, i.e., $L_i \subset L_{i+1}$. In fact, by Lemma 4.3.6, $x \in L_i$ if and only if $C_n x \to 0$ as $n \to \infty$, which is, by the definition of C_n, equivalent to $n^{-\rho} B_n x \to 0$ for any $\rho \in (a_i, a_{i+1})$. Then $n^{-(\rho+\alpha)} B_n x \to 0$ for any $\rho \in (a_i, a_{i+1})$ and any $\alpha > 0$. Choose $\alpha > 0$ such that $\rho' = \rho + \alpha \in (a_{i+1}, a_{i+2})$, so $n^{-\rho'} B_n x \to 0$ which, by Lemma 4.3.6 is equivalent to $x \in L_{i+1}$.

Now we will prove part (a) of Theorem 4.3.1. In the process, we will also show that in some sense $L = \lim C_m^{-1}(V')$ as $m \to \infty$. For each m choose an orthonormal basis $\theta_{m,1}, \ldots, \theta_{m,j}$ for $C_m^{-1}(V')$, where $j = \dim V'$. Choose a subsequence m_k along which $\theta_{m_k,i} \to \theta_i$ for every $i = 1, \ldots, j$. Then $\theta_1, \ldots, \theta_j$ are mutually orthogonal unit vectors. If some $\theta_i \in S$, then $\theta_{m_k,i} \in S$ for all large k, since S is open. (Note that every V_δ is open.) Since $C_m x_m/\|C_m x_m\| \rightharpoonup V$ whenever $x_m \to x$ in S, we have $C_{m_k}\theta_{m_k,i}/\|C_{m_k}\theta_{m_k,i}\| \rightharpoonup V$ as $k \to \infty$, which is a contradiction since $C_{m_k}\theta_{m_k,i} \in V'$ for all k. Thus every $\theta_i \in L$, and so $\dim L \geq \dim V'$. Next note that $L \cap C_n^{-1}(V_\delta) = \emptyset$ for large n, and $V \setminus \{0\} \subset V_\delta$, so $L \cap C_n^{-1}(V) = \{0\}$. Since the sum of the dimensions of these two subspaces is at least d we must have $L \oplus C_n^{-1}(V) = \mathbb{R}^d$ for large n. Then $\dim L = d - \dim C_n^{-1}(V) = d - \dim V = \dim V'$. This concludes the proof of Theorem 4.3.1. □

Remark 4.3.7. We can also state Theorem 4.3.1 in terms of uniform convergence on compact subsets. For example, (c) is equivalent to the assertion that $B_n x/\|B_n x\| \rightharpoonup V_{i+1} \oplus \cdots \oplus V_p$ uniformly on compact subsets of $x \in \mathbb{R}^d \setminus L_i$. In other words, for any compact set $K \subset \mathbb{R}^d \setminus L_i$ and any $\varepsilon > 0$, for some $n_0 > 0$ the distance between $B_n x/\|B_n x\|$ and $V_{i+1} \oplus \cdots \oplus V_p$ is less than ε for all $n \geq n_0$ and all $x \in K$. The simple proof by contradiction is quite similar to Proposition 2.1.7 (b).

Remark 4.3.8. For any $0 < k < d$, the set of all k dimensional linear subspaces $L \subset \mathbb{R}^d$ is called the *Grassman manifold* $\mathcal{G}(k, d)$. If $L_n, L \in \mathcal{G}(k, d)$ and $P_n : \mathbb{R}^d \to L_n$, $P : \mathbb{R}^d \to L$ are orthogonal projections then $L_n \to L$ in $\mathcal{G}(k, d)$ if $P_n \to P$ in $\mathrm{L}(\mathbb{R}^d)$. This is equivalent to $y_n/\|y_n\| \rightharpoonup L$ for all nonzero $y_n \in L_n$. The proof of Theorem 4.3.1 shows that the subspace $L_i = \lim_{n\to\infty} B_n^{-1}(V_1 \oplus \cdots \oplus V_i)$ in this topology.

Our next result is the Spectral Decomposition Theorem for regularly varying sequences of linear operators.

Definition 4.3.9. Given $A_n, B_n \in \mathrm{GL}(\mathbb{R}^d)$ we write $A_n \sim B_n$ if $A_n B_n^{-1} \to I$.

Theorem 4.3.10. *(B_n) varies regularly with index E if and only if $B_n \sim D_n T$ for some $T \in \mathrm{GL}(\mathbb{R}^d)$ and some (D_n) regularly varying with index E such that: each V_i in the spectral decomposition of $\mathbb{R}^d$ with respect to E is D_n-invariant; and $D_n = D_{1n} \oplus \cdots \oplus D_{pn}$ where each $D_{in} : V_i \to V_i$ is regularly varying with index E_i.*

For the proof of Theorem 4.3.10 we need Lemma 4.3.11, which is also of independent interest.

Lemma 4.3.11. *Suppose (B_n) varies regularly with index E and let $\mathbb{R}^d = V_1 \oplus \cdots \oplus V_p$ be the spectral decomposition of $\mathbb{R}^d$ relative to E. Then there exists a nested sequence of linear subspaces $L_1 \subset \cdots \subset L_p = \mathbb{R}^d$ such that for all $x_n \to x$ in $L_i \setminus L_{i-1}$ we have*

$$\frac{B_n x_n}{\|B_n x_n\|} \rightharpoonup V_i. \tag{4.39}$$

Proof. Let L_i be the subspace constructed in Theorem 4.3.1. Then the assertion of the lemma follows directly from Theorem 4.3.1(b) and (c). □

Proof of Theorem 4.3.10. Apply Lemma 4.3.11 and choose $T \in \mathrm{GL}(\mathbb{R}^d)$ so that $T(L_i) = V_1 \oplus \cdots \oplus V_i$ for all $i = 1, \ldots, p$. Let $\pi_i : \mathbb{R}^d \to V_i$ be the projection onto V_i. For $x \in V_i$ define $D_n x = \pi_i B_n T^{-1} x$ and extend by linearity to define $D_n x$ for all x. Then $D_n(V_i) \subset V_i$. for all i. Now we wish to show that D_n is invertible for all large n. Suppose that for some $1 \le i \le p$ and some $\ell_n \to \infty$ and unit vectors $x_n \in V_i$ we have $D_{\ell_n} x_n = 0$ for all n, so that $\pi_i B_{\ell_n} T^{-1} x_n = 0$ for all n. Let $y_n = B_{\ell_n} T^{-1} x_n$ and note that the ith component of y_n with respect to the direct sum decomposition is zero for all n. But by the definition of T we have $T^{-1} x_n \in L_i \setminus L_{i-1}$, so $\pi_i y_n = 0$ for all n contradicts Lemma 4.3.11. Hence there exists a n_0 such that D_n is invertible for all $n \ge n_0$. Redefine $D_n = I$ for $n < n_0$. Then we have $D_n(V_i) = V_i$ and $D_n^{-1}(V_i) = V_i$.

Now since the convergence in (4.39) is uniform on compact subsets of $L_i \setminus L_{i-1}$ (see Remark 4.3.7) it follows easily that

$$\frac{B_nT^{-1}\theta}{||B_nT^{-1}\theta||} - \frac{D_n\theta}{||D_n\theta||} \to 0$$

$$\frac{||B_nT^{-1}\theta||}{||D_n\theta||} \to 1 \tag{4.40}$$

uniformly on compact subsets of V_i for each i. It follows that $B_nT^{-1} \sim D_n$, so D_n also varies regularly with index E. Since each V_i is D_n-invariant, each D_{in} varies regularly with index E_i by projecting (4.18) onto the subspace V_i. Conversely, it follows immediately from the definition (4.18) that if $B_n \sim D_nT$ and D_n varies regularly with index E, then so does B_n. This concludes the proof. □

Corollary 4.3.12. *If (B_n) varies regularly with index E, then $T_0B_n \sim G_n$ for some $T_0 \in \mathrm{GL}(\mathbb{R}^d)$ and some (G_n) regularly varying with index $E_0 = T_0ET_0^{-1}$ such that*

(a) *The subspaces $W_i = T_0(V_i)$ in the spectral decomposition of $\mathbb{R}^d$ with respect to E_0 are mutually orthogonal*

(b) *These subspaces are also G_n invariant*

(c) *If we write $E_0 = E_{10} \oplus \cdots \oplus E_{p0}$ and $G_n = G_{1n} \oplus \cdots \oplus G_{pn}$ with respect to this direct sum decomposition then each $G_{in} : W_i \to W_i$ is regularly varying with index $E_{i0} = T_0E_iT_0^{-1}$*

(d) *Every eigenvalue of E_{i0} has real part equal to a_i*

(e) *Theorem 4.3.1 holds with $L_i = W_1 \oplus \cdots \oplus W_i$ so that $W_1 \cdots W_p$ are unique.*

Proof. Define $W_1 = L_1$ and

$$W_i = \{y \in L_i : <x, y> = 0 \;\forall\; x \in L_{i-1}\}$$

for $1 < i \leq p$, so that W_i is the orthogonal complement of L_{i-1} in L_i. Then $\dim W_i = \dim V_i$ and $L_i = L_{i-1} \oplus W_i = W_1 \oplus \cdots \oplus W_i$, and furthermore the subsets $W_1 \cdots W_p$ are mutually orthogonal. In the proof of Theorem 4.3.10, we may choose any $T \in \mathrm{GL}(\mathbb{R}^d)$ such that $T(L_i) = V_1 \oplus \cdots \oplus V_i$. Now we choose T so that we also have $T(W_i) = V_i$ for all $1 \leq i \leq p$. Let $T_0 = T^{-1}$ so that $T_0(V_i) = W_i$ and $T_0^{-1}(W_i) = V_i$. Since $E(V_i) = V_i$ we also have $E_0(W_i) = T_0ET_0^{-1}(W_i) = T_0E(V_i) = T_0(V_i) = W_i$ so $E_0 = E_{10} \oplus \cdots \oplus E_{p0}$ with respect to the direct sum decomposition $\mathbb{R}^d = W_1 \oplus \cdots \oplus W_p$. Also E_0 has the same real spectrum $a_1 < \cdots < a_p$ as E and furthermore $E_{i0} = T_0E_iT_0^{-1}$. If we define $G_n = T_0D_nT_0^{-1} = T^{-1}D_nT$, where (D_n) is as in Theorem 4.3.10, then

$G_n \sim T_0 B_n$ and G_n varies regularly with index $E_0 = T_0 E T_0^{-1}$. Furthermore, $G_n = G_{1n} \oplus \cdots \oplus G_{pn}$ with respect to the direct sum decomposition $\mathbb{R}^d = W_1 \oplus \cdots \oplus W_p$ where each $G_{in} : W_i \to W_i$ is regularly varying with index E_{i0} spectrally simple with real spectrum a_i. □

Definition 4.3.13. Suppose that (B_n) is a regularly varying sequence with index E and let $\mathbb{R}^d = V_1 \oplus \cdots \oplus V_p$ be the spectral decomposition of $\mathbb{R}^d$ relative to E. If every V_i is a B_n-invariant subspace for each n, then we say that (B_n) is *spectrally compatible with* E.

Proposition 4.3.14. *Suppose that* (B_n) *is a regularly varying sequence with index* E. *Let* $E = E_1 \oplus \cdots \oplus E_p$ *be the spectral decomposition of* E *and* $\mathbb{R}^d = V_1 \oplus \cdots \oplus V_p$ *the spectral decomposition of* $\mathbb{R}^d$ *relative to* E, *so that every eigenvalue of* E_i *has real part equal to* a_i *and* $a_i < a_j$ *for* $i < j$. *If* (B_n) *is spectrally compatible with* E, *then we can write* $B_n = B_{1n} \oplus \cdots \oplus B_{pn}$ *with respect to this direct sum decomposition, and then each* (B_{in}) *varies regularly with index* E_i *on* V_i. *We call the sequences* (B_{in}) *the* spectral decomposition of (B_n) with respect to E. *In this case, the conclusions of Theorem 4.3.1 hold with* $L_i = V_1 \oplus \cdots \oplus V_i$ *for each* $i = 1, \ldots, p$.

Proof. For any $x \in \mathbb{R}^d$, we can write $x = x_1 + \cdots + x_p$ uniquely where $x_i \in V_i$ for each i. Then we can define the unique projections $\pi_i(x) = x_i$ associated with this direct sum decomposition. Since the subspaces V_i are all E-invariant we have $\pi_i E = E_i \pi_i$ in general. Since the subspaces V_i are all B_n-invariant we also have $\pi_i B_n = B_{in} \pi_i$, and since $B_n^{-1} V_i = V_i$ as well we also have $\pi_i B_n^{-1} = B_{in}^{-1} \pi_i$. Now if $B_{[\lambda n]} B_n^{-1} \to \lambda^E$ we can apply π_i to both sides to see that (B_{in}) varies regularly with index E_i on V_i.

Now let $P_i = \pi_1 + \cdots + \pi_i$ denote the natural projection onto $V_1 \oplus \cdots \oplus V_i$. A similar argument shows that $(P_i B_n)$ varies regularly on $V_1 \oplus \cdots \oplus V_i$ with index $E_1 \oplus \cdots \oplus E_i$. Then Theorem 4.3.1 applies to the regularly varying sequence $(P_i B_n)$. If $x \in V_1 \oplus \cdots \oplus V_i$ then $B_n x = P_i B_n x$ and it follows that (b) and (d) of Theorem 4.3.1 hold for all such x. Then it follows from the uniqueness of L_i that $L_i = V_1 \oplus \cdots \oplus V_i$ for each $i = 1, \ldots, p$. □

Remark 4.3.15. The sequence (D_n) in Theorem 4.3.10 and the sequence (G_n) in Corollary 4.3.12 are both spectrally compatible with their index. Hence these results relate any regularly varying sequence with another regularly varying sequence with some nice properties. Throughout the rest of this book, we will apply these results along with Convergence of Types, Theorem 2.3.17, to obtain more convenient regularly varying sequences for our applications.

4.4 NOTES AND COMMENTS

Regular variation theory was introduced by Karamata [77, 78] to simplify the proof of certain Tauberian theorems. Since then the theory has gained importance in probability theory, and other areas of mathematics. A concise history of the one variable theory is included in Seneta [177]. Bingham, Goldie, and Teugels [19] is the most complete and modern reference on the subject, with a large number of diverse applications. Regular variation for linear operators was first treated in Balkema [8], who also proved the Uniform Convergence Theorem 4.2.1. Theorem 4.2.4 appeared in [113], and later Meerschaert [120] showed that regularly varying norming operators can always be chosen for the convergence (3.95) which defines generalized domains of attraction. Hudson, Veeh, and Weiner [67] also proved power law growth bounds for the norming operators in (3.95). In view of convergence of types, Theorem 2.3.17, those results can also be obtained from Theorem 4.2.4. Theorem 4.2.9 on the equivalence between regularly varying functions and regularly varying sequences is adapted from Bojanic and Seneta [21]. The key result in our proof is Lemma 4.2.10, which is based on the argument of Weissman [196]. Theorems 4.2.12 and 4.2.13 extend a result in [120]. Section 4.3 provides a more detailed version of the arguments in [116], see also [124].

5

Regular Variation for Real-Valued Functions

Regular variation is a way to describe power law behavior. When a real-valued function $F(x)$ varies regularly, it means that the tails $F(tx)$ grow like a power function as $t \to \infty$. The rate may vary with the radial direction x, giving a very flexible setup. In this chapter, we lay out the theory of regular variation for real-valued functions. We also introduce the notion of uniform regular variation, which allows the application of some important results from the one variable theory of regular variation to multivariable functions. This is useful when the growth rate of the function F is the same in all directions. When the growth rate varies with direction, we employ in Section 5.3 a generalization of regular variation called R–O variation. Uniform R–O variation will also be used in Chapter 6 to obtain sharp bounds on the growth rate of regularly varying measures.

5.1 REGULARLY VARYING FUNCTIONS

Regular variation for a real-valued function $F(x)$ on $\mathbb{R}^d$ describes power law growth, with a rate that may depend on the radial direction. Regular variation for functions uses the notion of regular variation for linear operators introduced in Chapter 4. Recall that $f : \mathbb{R}^+ \to \mathrm{GL}(\mathbb{R}^d)$ is regularly varying with index $-E$ if $f(\lambda t)f(t)^{-1} \to \lambda^{-E}$ for all $\lambda > 0$. Define $\Gamma = \mathbb{R}^d \setminus \{0\}$. If all eigenvalues of E have positive real part, then it follows from Corollary 4.2.6 (take $\beta = 0$) that $\|f(t)^{-1}x_t\| \to \infty$ as $t \to \infty$ whenever $x_t \to x$ in Γ. If all eigenvalues of E have negative real part, then $\|f(t)^{-1}x_t\| \to 0$ as $t \to \infty$

whenever $x_t \to x$ in Γ. We say that $R : \mathbb{R}^+ \to \mathbb{R}^+$ varies regularly at zero with index $\rho \in \mathbb{R}$ if

$$\lim_{s \to 0} \frac{R(\mu s)}{R(s)} = \mu^\rho \tag{5.1}$$

for all $\mu > 0$. Since (5.1) holds if and only if $R(1/t)$ varies regularly (at infinity) with index $-\rho$, for $R : \mathbb{R}^+ \to \mathbb{R}^+$ it suffices to consider regular variation at infinity. However, for real-valued functions on $\mathbb{R}^d$ it is necessary to treat both cases separately.

Definition 5.1.1. We say that a Borel measurable function $F : \Gamma \to [0, \infty)$ *varies regularly* if there exist $f : \mathbb{R}^+ \to \mathrm{GL}(\mathbb{R}^d)$ regularly varying with index $-E$, and $R : \mathbb{R}^+ \to \mathbb{R}^+$ regularly varying with index $\beta \neq 0$, such that

$$\lim_{t \to \infty} \frac{F(f(t)^{-1} x_t)}{R(t)} = \varphi(x) > 0 \tag{5.2}$$

whenever $x_t \to x$ in $\Gamma = \mathbb{R}^d \setminus \{0\}$. If all eigenvalues of E have positive real part, we will say that F *varies regularly at infinity*, and if all eigenvalues of E have negative real part, we will say that F *varies regularly at zero.*

Regular variation implies that the limit φ in (5.2) is the analogue of a power function for real valued functions on $\mathbb{R}^d$.

Proposition 5.1.2. *If $F : \Gamma \to [0, \infty)$ varies regularly and (5.2) holds for some $f : \mathbb{R}^+ \to \mathrm{GL}(\mathbb{R}^d)$ regularly varying with index $-E$, and $R : \mathbb{R}^+ \to \mathbb{R}^+$ regularly varying with index $\beta \neq 0$, then*

$$\lambda^\beta \cdot \varphi(x) = \varphi(\lambda^E x) \tag{5.3}$$

for all $\lambda > 0$ and all $x \in \Gamma$.

Proof. Fix λ, x and consider the equation

$$\frac{F(f(\lambda r)^{-1} x)}{R(\lambda r)} = \frac{F(f(r)^{-1} x_r)}{R(r)} \cdot \frac{R(r)}{R(\lambda r)} \tag{5.4}$$

where $x_r = f(r) f(\lambda r)^{-1} x$. Since $f(\lambda r) f(r)^{-1} \to \lambda^{-E}$ we can take inverses to obtain $f(r) f(\lambda r)^{-1} \to \lambda^E$ and so $x_r \to \lambda^E x$. Taking the limit of both sides of (5.4) as $r \to \infty$, we obtain $\varphi(x) = \varphi(\lambda^E x) \cdot \lambda^{-\beta}$, which is equivalent to (5.3). □

Remark 5.1.3. If we let $B = (-1/\beta)E$, then it follows from (5.3) that

$$\lambda \cdot \varphi(x) = \varphi(\lambda^{-B} x) \tag{5.5}$$

for all $\lambda > 0$ and all $x \in \Gamma$. We call B an *exponent* of φ. Exponents are not necessarily unique, because of possible symmetries. The structure of exponents and symmetries will be discussed further in Section 5.2.

Definition 5.1.4. Suppose that F varies regularly and β, E are as in Definition 5.1.1. Then $B = (-1/\beta)E$ is an exponent of the limit function φ in (5.2), and for the sake of brevity we will also call B an *exponent* of F. If F varies regularly at zero with exponent B, we will write $F \in \mathrm{RVF}_0(B)$. If F varies regularly at infinity, we will write $F \in \mathrm{RVF}_\infty(B)$.

Remark 5.1.5. If $R : \mathbb{R}^+ \to \mathbb{R}^+$ varies regularly at infinity with index ρ and (4.1) holds, then $F(x) = R(|x|)$ also varies regularly at infinity according to Definition 5.1.1 specialized to $\mathbb{R}^1$. Take $\varphi(x) = |x|^\rho$ and let $f(t) = 1/t$ which varies regularly with index $-1 = -E$ so that $E = 1$. Then uniform convergence implies that

$$\lim_{t\to\infty} \frac{F(f(t)^{-1}x_t)}{R(t)} = \lim_{t\to\infty} \frac{R(t|x_t|)}{R(t)} = |x|^\rho = \varphi(x)$$

whenever $x_t \to x$ in $\Gamma = \mathbb{R}^1 \setminus \{0\}$, so that F varies regularly at infinity. Since $\beta = \rho$ the exponent B of φ is related to the index ρ of regular variation by $B = (-1/\beta)E = (-1/\rho)$. If R varies regularly at zero with index ρ, then $R(1/t)$ varies regularly at infinity with index $-\rho$, and then $F(x) = R(|x|)$ also varies regularly at zero according to Definition 5.1.1 specialized to $\mathbb{R}^1$: Take $\varphi(x) = |x|^\rho$ and let $f(t) = t$, which varies regularly with index $1 = -E$, so that $E = -1$. Then uniform convergence implies that

$$\lim_{t\to\infty} \frac{F(f(t)^{-1}x_t)}{R(1/t)} = \lim_{t\to\infty} \frac{R(|x_t|/t)}{R(1/t)} = |x|^\rho = \varphi(x)$$

whenever $x_t \to x$ in $\Gamma = \mathbb{R}^1 \setminus \{0\}$, so that F varies regularly at zero. Since $\beta = -\rho$ the exponent of φ is again $B = (-1/\beta)E = (-1/\rho)$.

Every regularly varying function $R : \mathbb{R}^+ \to \mathbb{R}^+$ with index $\beta > 0$ has an asymptotic inverse R_1, a regularly varying function with index $1/\beta$ such that $R(R_1(t)) \sim t$ as $t \to \infty$ (see, e.g., Seneta [177] p. 21). Of course a real-valued function on $\mathbb{R}^d$ cannot have an (asymptotic) inverse, but we do have the following result, which serves as a useful analogue.

Proposition 5.1.6. *Suppose that $F : \Gamma \to [0, \infty)$ is regularly varying and (5.2) holds for some $f : \mathbb{R}^+ \to \mathrm{GL}(\mathbb{R}^d)$ regularly varying with index $-E$, and $R : \mathbb{R}^+ \to \mathbb{R}^+$ regularly varying with index $\beta \neq 0$. Then*

(a) If $\beta > 0$, then there exists $g : \mathbb{R}^+ \to \mathrm{GL}(\mathbb{R}^d)$ regularly varying with index $(-1/\beta)E$ such that

$$\lim_{t\to\infty} \frac{F(g(t)^{-1}x_t)}{t} = \varphi(x) > 0 \tag{5.6}$$

whenever $x_t \to x$ in Γ

(b) If $\beta < 0$, then there exists $g : \mathbb{R}^+ \to \mathrm{GL}(\mathbb{R}^d)$ regularly varying with index $(1/\beta)E$ such that

$$\lim_{t\to\infty} tF(g(t)^{-1}x_t) = \varphi(x) > 0 \tag{5.7}$$

whenever $x_t \to x$ *in* Γ

Conversely, if (5.6) *or* (5.7) *hold for some* $F : \Gamma \to [0, \infty)$ *Borel measurable and* $g : \mathbb{R}^+ \to \mathrm{GL}(\mathbb{R}^d)$ *regularly varying, then* F *is regularly varying. Furthermore,*

(c) *If* (5.6) *holds and* g *varies regularly with index* B, *then* F *is regularly varying with exponent* B

(d) *If* (5.7) *holds and* g *varies regularly with index* $-B$, *then* F *is regularly varying with exponent* B

Proof. If $\beta > 0$, then $R(t) \to \infty$ as $t \to \infty$. Let R_1 be an asymptotic inverse of R and define $g(t) = f(R_1(t))$. Then $R_1(t) \to \infty$ as $t \to \infty$, and since R_1 varies regularly with index $1/\beta$ it follows from Theorem 4.2.1 that g varies regularly with index $-\beta^{-1}E$. Then as $t \to \infty$ we have

$$\frac{F(g(t)^{-1}x_t)}{t} \sim \frac{F(f(R_1(t))^{-1}x_t)}{R(R_1(t))} \to \varphi(x)$$

whenever $x_t \to x$ in Γ, which proves part (a).

If $\beta < 0$, then $R(t) \to 0$ as $t \to \infty$ and $1/R$ varies regularly with index $(-\beta)$. Choose an asymptotic inverse R_1 regularly varying with index $(-1/\beta)$ such that $1/R(R_1(t)) \sim t$. Define $g(t) = f(R_1(t))$. Then g is regularly varying with index $(1/\beta)E$, and as $t \to \infty$ we have

$$tF(g(t)^{-1}x_t) \sim \frac{F(f(R_1(t))^{-1}x_t)}{R(R_1(t))} \to \varphi(x)$$

whenever $x_t \to x$ in Γ, which proves part (b).

Conversely, if (5.6) or (5.7) hold for some $F : \Gamma \to [0, \infty)$ Borel measurable and $g : \mathbb{R}^+ \to \mathrm{GL}(\mathbb{R}^d)$ regularly varying, then it follows immediately from Definition 5.1.1 that F is regularly varying.

If (5.6) holds and g varies regularly with index B, then (5.2) holds with $f = g$ regularly varying with index $-E = B$ and $R(t) = t$ regularly varying with index $\beta = 1$. Then $(-1/\beta)E = (-1/1)(-B) = B$ is an exponent of F, so (c) holds.

If (5.7) holds and g varies regularly with index $-B$, then (5.2) holds with $f = g$ regularly varying with index $-E = -B$ and $R(t) = 1/t$ regularly varying with index $\beta = -1$. Then $(-1/\beta)E = (-1/-1)B = B$ is an exponent of F, so (d) holds. □

Corollary 5.1.7. (a) *If* $F \in \mathrm{RVF}_\infty(B)$ *and all eigenvalues of* B *have negative real part, or if* $F \in \mathrm{RVF}_0(B)$ *and all eigenvalues of* B *have positive real part, then there exists* $g : \mathbb{R}^+ \to \mathrm{GL}(\mathbb{R}^d)$ *regularly varying with index* B *such that* (5.6) *holds whenever* $x_t \to x$ *in* Γ

(b) *If* $F \in \mathrm{RVF}_\infty(B)$ *and all eigenvalues of* B *have positive real part, or if* $F \in \mathrm{RVF}_0(B)$ *and all eigenvalues of* B *have negative real part, then there exists* $g : \mathbb{R}^+ \to \mathrm{GL}(\mathbb{R}^d)$ *regularly varying with index* $-B$ *such that* (5.7) *holds whenever* $x_t \to x$ *in* Γ

Proof. If $F \in \mathrm{RVF}_\infty(B)$ and the eigenvalues of $B = (-1/\beta)E$ all have negative real part, then the eigenvalues of E have positive real part, and so $\beta > 0$. If $F \in \mathrm{RVF}_0(B)$ and all eigenvalues of B have positive real part, then the eigenvalues of E have negative real part, and so again $\beta > 0$. In either case, we can apply part (a) of Proposition 5.1.6, and this proves part (a).

If $F \in \mathrm{RVF}_\infty(B)$ and the eigenvalues of $B = (-1/\beta)E$ all have positive real part, then the eigenvalues of E have positive real part, and so $\beta < 0$. If $F \in \mathrm{RVF}_0(B)$ and all eigenvalues of B have negative real part, then the eigenvalues of E have negative real part, and so again $\beta < 0$. In either case, we can apply part (b) of Proposition 5.1.6, and this proves part (b). □

When we use regular variation to investigate limit theorems, in view of (3.98) it is often convenient to consider sequences rather than functions. The following result establishes the equivalence of a sequential definition of regularly varying functions.

Proposition 5.1.8. *Suppose that* $F : \Gamma \to [0, \infty)$ *is regularly varying and (5.2) holds for some* $f : \mathbb{R}^+ \to \mathrm{GL}(\mathbb{R}^d)$ *regularly varying with index* $-E$*, and* $R : \mathbb{R}^+ \to \mathbb{R}^+$ *regularly varying with index* $\beta \neq 0$*. Then*

(a) *If* $\beta > 0$*, then there exists a regularly varying sequence* (A_n) *with index* $(-1/\beta)E$ *such that*

$$\frac{F(A_n^{-1}x_n)}{n} \to \varphi(x) > 0 \tag{5.8}$$

whenever $x_n \to x$ *in* Γ

(b) *If* $\beta < 0$*, then there exists a regularly varying sequence* (A_n) *with index* $(1/\beta)E$ *such that*

$$nF(A_n^{-1}x_n) \to \varphi(x) > 0 \tag{5.9}$$

whenever $x_n \to x$ *in* Γ

Conversely, if (5.8) *or* (5.8) *hold for some* $F : \Gamma \to [0, \infty)$ *Borel measurable and* (A_n) *regularly varying, then* F *is regularly varying. Furthermore,*

(c) *If* (5.8) *holds and* (A_n) *varies regularly with index* B*, then* F *is regularly varying with exponent* B

(d) *If* (5.9) *holds and* (A_n) *varies regularly with index* $-B$*, then* F *is regularly varying with exponent* B

Proof. If F varies regularly according to Definition 5.1.1 with $\beta > 0$, then part (a) of Proposition 5.1.6 shows that (5.6) holds. Then Theorem 4.2.9 shows that $A_n = g(n)$ also varies regularly with the same index, and (5.8) follows immediately. The proof of part (b) is similar, using part (b) of Proposition 5.1.6.

Conversely, if (5.8) holds and (A_n) varies regularly with index B, then Theorem 4.2.9 shows that $g(t) = A_{[t]}$ varies regularly with the same index, and it follows from Theorem 4.2.1 that

$$\frac{F(g(t)^{-1}x_n)}{t} \sim \frac{F(A_n^{-1}x_n)}{n} \to \varphi(x) > 0$$

whenever $x_n \to x$ in Γ. Then F varies regularly with exponent B by part (c) of Proposition 5.1.6. If (5.9) holds and (A_n) varies regularly with index $-B$, then Theorem 4.2.9 shows that $g(t) = A_{[t]}$ varies regularly with the same index, and it follows from Theorem 4.2.1 that

$$tF(g(t)^{-1}x_n) \sim nF(A_n^{-1}x_n) \to \varphi(x) > 0$$

whenever $x_n \to x$ in Γ. Then F varies regularly with exponent B by part (d) of Proposition 5.1.6. □

Corollary 5.1.9. *(a) If $F \in \mathrm{RVF}_\infty(B)$ and all eigenvalues of B have negative real part, or if $F \in \mathrm{RVF}_0(B)$ and all eigenvalues of B have positive real part, then there exists $(A_n) \in \mathrm{RV}(B)$ such that (5.8) holds whenever $x_t \to x$ in Γ*

(b) If $F \in \mathrm{RVF}_\infty(B)$ and all eigenvalues of B have positive real part, or if $F \in \mathrm{RVF}_0(B)$ and all eigenvalues of B have negative real part, then there exists some $(A_n) \in \mathrm{RV}(-B)$ such that (5.9) holds whenever $x_t \to x$ in Γ

Proof. The proof is identical to Corollary 5.1.7, using Proposition 5.1.8 in place of Proposition 5.1.6. □

Remark 5.1.10. If $\varphi : \Gamma \to \mathbb{R}^+$ is continuous and (5.3) holds for all $\lambda > 0$ and all $x \in \Gamma$, then φ varies regularly. From (5.3) and the continuity of φ, it follows that

$$\lim_{t\to\infty} \frac{\varphi(t^E x_t)}{t^\beta} = \varphi(x) \tag{5.10}$$

and

$$\lim_{t\to\infty} \frac{\varphi(t^{-E} x_t)}{t^{-\beta}} = \varphi(x) \tag{5.11}$$

whenever $x_t \to x$ in Γ. If the real parts of the eigenvalues of E are all positive, then (5.10) is of the form (5.2) with $f(t) = t^{-E}$ regularly varying with index $-E$ and $R(t) = t^\beta$ regularly varying with index β, so $\varphi \in \mathrm{RVF}_\infty(B)$, where $B = (-1/\beta)E$. Since (5.11) is of the form (5.2) with $f(t) = t^E$ and $R(t) = t^{-\beta}$, we also have $\varphi \in \mathrm{RVF}_0(B)$. If the real parts of the eigenvalues of E are all negative, then (5.10) shows that $\varphi \in \mathrm{RVF}_0(B)$, and (5.11) shows that $\varphi \in \mathrm{RVF}_\infty(B)$.

Example 5.1.11. Now we provide some examples of regularly varying functions on $\mathbb{R}^d$.

(a) Let $\alpha \neq 0$ and $\varphi(x) = \|x\|^\alpha$. Then φ is continuous, and for $B = (-1/\alpha)I$ we have $t \cdot \varphi(x) = \varphi(t^{-B}x)$ for all $t > 0$ and all $x \in \Gamma$. Then Remark 5.1.10 yields that φ varies regularly at both zero and infinity with exponent B.

(b) Given $0 < \alpha_1 < \cdots < \alpha_d$, define $\varphi(x) = \sum_{i=1}^d |x^{(i)}|^{\alpha_i}$, where $x = (x^{(1)}, \ldots, x^{(d)}) \in \mathbb{R}^d$. Let $B = \mathrm{diag}(-1/\alpha_1, \ldots, -1/\alpha_d)$. Then φ is continuous, and

$$\varphi(t^{-B}x) = \sum_{i=1}^d |t^{1/\alpha_i} x^{(i)}|^{\alpha_i} = t \cdot \varphi(x)$$

for all $t > 0$ and all $x \in \Gamma$. Then Remark 5.1.10 implies that φ varies regularly at both zero and infinity with exponent B.

5.2 EXPONENTS AND SYMMETRIES

If F varies regularly with exponent $B \in \mathrm{GL}(\mathbb{R}^d)$, then B is an exponent of the limit φ in (5.2). Exponents are not necessarily unique, because of symmetry. In this section, we characterize the exponents and symmetries of the limit function φ. We also establish some important technical results for later use.

Definition 5.2.1. A function $\varphi : \Gamma \to (0, \infty)$ is called *admissible* if it is continuous and if either

(i) $\varphi(x_t) \to \infty$ whenever $\|x_t\| \to \infty$ or

(ii) $\varphi(x_t) \to 0$ whenever $\|x_t\| \to \infty$

In order to apply the results of this section to regularly varying functions, we need the following result.

Proposition 5.2.2. *If F varies regularly at zero or infinity and* (5.2) *holds, then the limit φ is admissible.*

Proof. First, we show that φ is continuous. Suppose not. Then for some $\varepsilon > 0$ and some $y_k \to x$ in Γ we have $|\varphi(y_k) - \varphi(x)| > \varepsilon$ for all k. We also have for each fixed k that $F(f(t)^{-1}y_k)/R(t) \to \varphi(y_k)$ as $t \to \infty$, so that for each k there exists $t_k > 0$ such that $|F(f(t)^{-1}y_k)/R(t) - \varphi(y_k)| < \varepsilon/2$ for all $t \geq t_k$.

Choose t_k inductively so that (t_k) is monotone increasing. Now let $x_t = y_k$ for all $t_k < t \leq t_{k+1}$. Then for all $t \geq t_1$ we have

$$\left|\frac{F(f(t)^{-1}x_t)}{R(t)} - \varphi(x)\right| \geq |\varphi(x_t) - \varphi(x)| - \left|\frac{F(f(t)^{-1}x_t)}{R(t)} - \varphi(x_t)\right| \geq \varepsilon/2$$

which is a contradiction. Hence φ is continuous. Furthermore, if F varies regularly at zero or infinity then (5.5) holds and the real parts of the eigenvalues of B are either all negative or all positive. In the first case, if $\|x_t\| \to \infty$ we can write write $x_t = (r_t)^{-B}\theta_t$, where $r_t > 0$ and $\|\theta_t\| = 1$. Then (5.5) yields $\varphi(x_t) = r_t\varphi(\theta_t) \to \infty$ since $r_t \to \infty$ and φ continuous is bounded away from zero on the unit sphere. In the second case, a similar computation shows that $\varphi(x_t) \to 0$. □

Definition 5.2.3. Let $\varphi : \Gamma \to (0, \infty)$. An operator $A \in \mathrm{GL}(\mathbb{R}^d)$ is called a *symmetry* of φ if $\varphi(Ax) = \varphi(x)$ for all $x \in \Gamma$. Let $\mathcal{S}(\varphi)$ denote the set of symmetries of φ.

Theorem 5.2.4. *The symmetries $\mathcal{S}(\varphi)$ of an admissible function φ form a compact subgroup of* $\mathrm{GL}(\mathbb{R}^d)$.

Proof. Obviously, $\mathcal{S}(\varphi)$ is closed under composition and inverses, so $\mathcal{S}(\varphi)$ is a subgroup of $\mathrm{GL}(\mathbb{R}^d)$. If $A_n \in \mathcal{S}(\varphi)$ and $A_n \to A$, then for all $x \in \Gamma$ we have $\varphi(x) = \varphi(A_n x) \to \varphi(Ax)$, which implies that $\varphi(Ax) = \varphi(x)$, and so $\mathcal{S}(\varphi)$ is closed. Finally, if $A_n \in \mathcal{S}(\varphi)$ then $\varphi(x) = \varphi(A_n x)$ implies that $\|A_n x\|$ is bounded [otherwise $\varphi(A_n x)$ tends to either zero or infinity], and so $\mathcal{S}(\varphi)$ is bounded, and hence compact. □

Definition 5.2.5. Let $\mathcal{F}$ denote the set of all Borel measurable functions $F : \Gamma \to \mathbb{R}^+$ that are *monotone on rays*, by which we mean that $t \mapsto F(tx)$ is a monotone function of $t > 0$ for all $x \in \Gamma$. For $F_n \in \mathcal{F}$ and an admissible function $\varphi : \Gamma \to (0, \infty)$ we write $F_n \to \varphi$ if $F_n(x_n) \to \varphi(x)$ whenever $x_n \to x$ in Γ. In this case, we also have $\varphi \in \mathcal{F}$.

Our immediate goal is to establish a *convergence of types* theorem for functions in $\mathcal{F}$. We begin with a few simple lemmas. In what follows, we will use the notation $AF(x) = F(A^{-1}x)$ for $A \in \mathrm{GL}(\mathbb{R}^d)$ and $F : \Gamma \to \mathbb{R}^+$. Recall that $A_n \rightharpoonup \mathcal{S}(\varphi)$ means that the distance between A_n and the compact set $\mathcal{S}(\varphi)$ tends to zero as $n \to \infty$.

Lemma 5.2.6. *Suppose $F_n \in \mathcal{F}$, φ admissible, and $F_n \to \varphi$.*

(i) If φ satisfies (i) of Definition 5.2.1, then for some $n_0 \geq 1$ the function $t \mapsto F_n(tx)$ is monotone increasing for all $x \in \Gamma$ and all $n \geq n_0$

(ii) If φ satisfies (ii) of Definition 5.2.1, then for some $n_0 \geq 1$ the function $t \mapsto F_n(tx)$ is monotone decreasing for all $x \in \Gamma$ and all $n \geq n_0$

Proof. First we prove part (i). Define $m(t) = \inf\{\varphi(x) : \|x\| = t\}$. Since φ is positive and continuous, we have for all $t > 0$ that $m(t) = \varphi(x_t) > 0$ for some $\|x_t\| = t$. If φ satisfies (i) of Definition 5.2.1, then we have that $m(t) = \varphi(x_t) \to \infty$ as $t \to \infty$. Suppose by way of contradiction that for some $x_k \in \Gamma$ and some sequence (n_k) tending to infinity the function $t \mapsto F_{n_k}(tx_k)$ is monotone decreasing for all k. Write $x_k = r_k\theta_k$ where $r_k > 0$ and $\theta_k \in S = \{\theta \in \mathbb{R}^d : \|\theta\| = 1\}$. Then $t \mapsto F_{n_k}(t\theta_k)$ is monotone decreasing for all k, and since S is compact we can choose a subsequence (k') along which $\theta_{k'} \to \theta \in S$. For all $t > t_0$ and all k we have $F_{n_k}(t_0\theta_k) \geq F_{n_k}(t\theta_k)$, so letting $k' \to \infty$ yields $\varphi(t_0\theta) \geq \varphi(t\theta)$. But then $m(t\|\theta\|) \leq \varphi(t\theta) \leq \varphi(t_0\theta) < \infty$ for all $t > t_0$, which is a contradiction.

Next we prove part (ii). Define $M(t) = \sup\{\varphi(x) : \|x\| = t\}$ so that $M(t) = \varphi(x_t) < \infty$ for some $\|x_t\| = t$. If φ satisfies (ii) of Definition 5.2.1, then $M(t) \to 0$ as $t \to \infty$. Suppose by way of contradiction that for some $\theta_k \in S$ and some sequence (n_k) tending to infinity the function $t \mapsto F_{n_k}(t\theta_k)$ is monotone increasing for all k. For all $t > t_0$ and all k, we have $F_{n_k}(t_0\theta_k) \leq F_{n_k}(t\theta_k)$ so $\varphi(t_0\theta) \leq \varphi(t\theta)$, where θ is any limit point of (θ_k). But then $M(t\|\theta\|) \geq \varphi(t\theta) \geq \varphi(t_0\theta) > 0$ for all $t > t_0$ which is a contradiction. □

Lemma 5.2.7. *Suppose $F_n \in \mathcal{F}$ and φ admissible. If $F_n \to \varphi$, then either*

(i) $F_n(x_n) \to \infty$ whenever $\|x_n\| \to \infty$ if φ satisfies (i) of Definition 5.2.1 or

(ii) $F_n(x_n) \to 0$ whenever $\|x_n\| \to \infty$ if φ satisfies (ii) of Definition 5.2.1

Proof. First we prove part (i). Assume by way of contradiction that there exists a sequence $\|x_n\| \to \infty$ and $L > 0$ such that $F_n(x_n) \leq L$ for all n. Write $x_n = r_n\theta_n$, where $r_n > 0$ and $\theta_n \in S = \{\theta \in \mathbb{R}^d : \|\theta\| = 1\}$. We can assume by passing to a subsequence if necessary that r_n is increasing and $\theta_n \to \theta \in S$. As in the proof of Lemma 5.2.6 we define $m(r) = \inf\{\varphi(x) : \|x\| = r\}$ so that $m(t) \to \infty$ as $t \to \infty$. Then there exists $n_0 \geq 1$ such that for all $n \geq n_0$ we have $\varphi(r_n\theta) \geq 2L$. On the other hand, part (i) of Lemma 5.2.6 shows that for all large n the function $t \mapsto F_n(t\theta_n)$ is increasing, so we get for all large $n \geq n_0$ that

$$L \geq F_n(x_n) = F_n(r_n\theta_n) \geq F_n(r_{n_0}\theta_n) \to \varphi(r_{n_0}\theta) \geq 2L$$

which is a contradiction.

Next we prove part (ii). Assume by way of contradiction that there exists a sequence $\|x_n\| \to \infty$ and a $\delta > 0$ such that $F_n(x_n) \geq \delta$ for all n. Writing $x_n = r_n\theta_n$, where $r_n > 0$ and $\|\theta_n\| = 1$, we can assume by passing to a subsequence if necessary that r_n is increasing and $\theta_n \to \theta$. Letting $M(t) = \sup\{\varphi(x) : \|x\| = t\}$ as in the proof of Lemma 5.2.6, we have $M(t) \to 0$ as $t \to \infty$, so there exists $n_0 \geq 1$ such that for all $n \geq n_0$ we have $\varphi(r_n\theta) \leq \delta/2$. On the other hand, since for all large n the function $t \mapsto F_n(t\theta_n)$ is decreasing we get for all large $n \geq n_0$ that

$$\delta \leq F_n(x_n) = F_n(r_n\theta_n) \leq F_n(r_{n_0}\theta_n) \to \varphi(r_{n_0}\theta) \leq \frac{\delta}{2}$$

which is a contradiction. □

Lemma 5.2.8. *Let $F_n \in \mathcal{F}$, φ admissible and $A_n \in \mathrm{GL}(\mathbb{R}^d)$. If $F_n \to \varphi$, then $A_nF_n \to \varphi$ if and only if $A_n \rightharpoonup \mathcal{S}(\varphi)$.*

Proof. If $A_nF_n \to \varphi$, then for all $x \in \Gamma$ we have $F_n(A_n^{-1}x) \to \varphi(x) > 0$. If $||A_n^{-1}x|| \to \infty$, then Lemma 5.2.7 shows that $F_n(A_n^{-1}x)$ tends to either zero or infinity, which is a contradiction. Thus the sequence (A_n^{-1}) is relatively compact. If along a subsequence n' we have $A_{n'}^{-1} \to C$, then it follows that $F_{n'}(A_{n'}^{-1}x) \to \varphi(Cx)$ and so $C \in \mathcal{S}(\varphi)$. Since $\mathcal{S}(\varphi)$ is a compact group, taking inverses yields $A_n \rightharpoonup \mathcal{S}(\varphi)$. Conversely, if $A_n \rightharpoonup \mathcal{S}(\varphi)$, then for any subsequence there is a further subsequence n' such that $A_{n'} \to A$ for some $A \in \mathcal{S}(\varphi)$. Then $A_{n'}F_{n'} \to A\varphi = \varphi$, and it follows that $A_nF_n \to \varphi$. □

Lemma 5.2.9. *Let $F_n \in \mathcal{F}$, φ, φ_0 admissible and $A_n \in \mathrm{GL}(\mathbb{R}^d)$. If $F_n \to \varphi$ and $A_nF_n \to \varphi_0$, then $\varphi_0 = C\varphi$ for some linear operator $C \in \mathrm{GL}(\mathbb{R}^d)$.*

Proof. If $A_nF_n \to \varphi_0$, then for all $x \in \Gamma$ we have $F_n(A_n^{-1}x) \to \varphi_0(x) > 0$. If $||A_n^{-1}x|| \to \infty$ then Lemma 5.2.7 implies that $F_n(A_n^{-1}x)$ tends to either zero or infinity, which is a contradiction. Then the sequence A_n^{-1} is relatively compact. If along a subsequence n' we have $A_{n'}^{-1} \to C_0$, then $F_{n'}(A_{n'}^{-1}x) \to \varphi_0(x) = \varphi(C_0x)$ for all $x \in \Gamma$. Since φ_0 is admissible we cannot have $C_0x = 0$ for any $x \in \Gamma$, so $C = C_0^{-1}$ exists and we may write $\varphi_0 = C\varphi$. □

Theorem 5.2.10. *(Convergence of types) Let $F_n \in \mathcal{F}$, φ, φ_0 admissible and $A_n, B_n \in \mathrm{GL}(\mathbb{R}^d)$. Suppose $B_nF_n \to \varphi$. Then $A_nF_n \to \varphi_0$ if and only if $\varphi_0 = C\varphi$ and $A_nB_n^{-1} \rightharpoonup C\mathcal{S}(\varphi)$ for some $C \in \mathrm{GL}(\mathbb{R}^d)$.*

Proof. If $A_nF_n \to \varphi_0$, then $(A_nB_n^{-1})B_nF_n \to \varphi_0$ and so Lemma 5.2.9 yields $\varphi_0 = C\varphi$. Then $C^{-1}(A_nB_n^{-1})B_nF_n \to \varphi$, and hence by Lemma 5.2.8 we have $C^{-1}(A_nB_n^{-1}) \rightharpoonup \mathcal{S}(\varphi)$. Multiply both sides by C on the left. Conversely, if $A_nB_n^{-1} \rightharpoonup C\mathcal{S}(\varphi)$ then $C^{-1}(A_nB_n^{-1}) \rightharpoonup \mathcal{S}(\varphi)$ and so Lemma 5.2.8 yields $C^{-1}(A_nB_n^{-1})B_nF_n \to \varphi$. Simplify and then multiply both sides by C on the left to obtain $A_nF_n \to \varphi_0$, which concludes the proof. □

Now we establish the basic characterization of regularly varying functions that are monotone on rays. In Part III, we will apply this characterization to obtain necessary and sufficient conditions for the multivariable central limit theorem. We begin with two technical results.

Theorem 5.2.11. *Given $F \in \mathcal{F}$, φ admissible and $A_n \in \mathrm{GL}(\mathbb{R}^d)$, suppose that $a_n > 0$ is a regularly varying sequence with index $\alpha \neq 0$ and $a_n(A_nF) \to \varphi$. Then there exists a linear operator E such that*

$$A_{[\lambda n]}A_n^{-1} \rightharpoonup \lambda^{-E}\mathcal{S}(\varphi) \tag{5.12}$$

for all $\lambda > 0$, and furthermore

$$\lambda \cdot \varphi = \lambda^B \varphi \tag{5.13}$$

for all $\lambda > 0$, with $B = (1/\alpha)E$.

Proof. Let $F_n = a_n \cdot F$ so that $A_n F_n \to \varphi$. Regular variation of the sequence a_n implies that

$$\begin{aligned} A_{[\lambda n]} F_n &= A_{[\lambda n]} \cdot a_n F \\ &= a_n a_{[\lambda n]}^{-1} \cdot A_{[\lambda n]}(a_{[\lambda n]} F) \\ &= (a_{[\lambda n]} a_n^{-1})^{-1} \cdot A_{[\lambda n]} F_{[\lambda n]} \\ &\to \lambda^{-\alpha} \cdot \varphi \end{aligned}$$

as $n \to \infty$.

Since $A_n F_n \to \varphi$ and $A_{[\lambda n]} F_n \to \lambda^{-\alpha} \cdot \varphi$, Theorem 5.2.10 yields a linear operator $C_\lambda \in \mathrm{GL}(\mathbb{R}^d)$ such that $\lambda^{-\alpha} \cdot \varphi = C_\lambda \varphi$ and $A_{[\lambda n]} A_n^{-1} \rightharpoonup C_\lambda \mathcal{S}(\varphi)$ for all $\lambda > 0$. Define $G_\lambda = C_\lambda \mathcal{S}(\varphi)$. If $C \in \mathrm{GL}(\mathbb{R}^d)$ also satisfies $\lambda^{-\alpha} \cdot \varphi = C\varphi$, then $C\varphi = C_\lambda \varphi$ implies $\varphi = C^{-1} C_\lambda \varphi$, and so $C^{-1} C_\lambda \in \mathcal{S}(\varphi)$. Since $\mathcal{S}(\varphi)$ is a group we also have $C_\lambda^{-1} C = (C^{-1} C_\lambda)^{-1} \in \mathcal{S}(\varphi)$, hence $C \in C_\lambda \mathcal{S}(\varphi) = G_\lambda$. Conversely, if $C \in G_\lambda$, then for some $A \in \mathcal{S}(\varphi)$ we have $C\varphi = C_\lambda A \varphi = C_\lambda \varphi = \lambda^{-\alpha} \cdot \varphi$, hence $C\varphi = \lambda^{-\alpha} \cdot \varphi$ if and only if $C \in G_\lambda$.

For any $C \in G_\lambda$, if $D = CA$ for some $A \in \mathcal{S}(\varphi)$ then $D\varphi = CA\varphi = C\varphi = \lambda^{-\alpha} \cdot \varphi$, and so $D \in G_\lambda$. Conversely, if $D \in G_\lambda$, then $D\varphi = C\varphi$ implies $C^{-1} D\varphi = C^{-1} C\varphi = \varphi$, so $C^{-1} D \in \mathcal{S}(\varphi)$ by Lemma 5.2.8, hence $D \in C\mathcal{S}(\varphi)$. Thus $G_\lambda = C\mathcal{S}(\varphi)$ for any $C \in G_\lambda$.

Define G to be the union of all G_λ taken over $\lambda > 0$. If $\lambda^{-\alpha} \cdot \varphi = \mu^{-\alpha} \cdot \varphi$, then obviously $\lambda = \mu$, hence $G_\lambda \cap G_\mu = \emptyset$ for $\lambda \neq \mu$. If $C \in G_\lambda$, then $\varphi = C^{-1} C\varphi = C^{-1}(\lambda^{-\alpha} \cdot \varphi) = \lambda^{-\alpha} \cdot C^{-1}\varphi$, hence $\lambda^{\alpha} \cdot \varphi = C^{-1}\varphi$ and so $C^{-1} \in G_{1/\lambda}$. Then $G_\lambda^{-1} \subset G_{1/\lambda}$. Substitute $t = 1/\lambda$ to obtain $G_{1/t}^{-1} \subset G_t$, and take inverses to obtain $G_{1/t} \subset G_t^{-1}$, which shows that $G_\lambda^{-1} = G_{1/\lambda}$. If $C \in G_\lambda$ and $D \in G_\mu$, then $CD\varphi = \lambda^{-\alpha}\mu^{-\alpha} \cdot \varphi$ implies that $CD \in G_{\mu\lambda}$. Therefore $G_\mu G_\lambda \subset G_{\mu\lambda}$. Substitute $s = 1/\mu$ and $t = \mu\lambda$ to obtain $G_{1/s} G_{st} \subset G_t$, which is equivalent to $G_{st} \subset G_s G_t$ in view of the fact that $G_{1/s} = G_s^{-1}$. Then $G_\mu G_\lambda = G_{\mu\lambda}$, and G is a group.

Define $\zeta(C) = \lambda$ for $C \in G_\lambda$. Since $G_\lambda \cap G_\mu = \emptyset$ for $\lambda \neq \mu$, the mapping $\zeta : G \to \mathbb{R}^+$ is well defined. If $C \in G_\lambda$ and $D \in G_\mu$, then $C^{-1} \in G_{1/\lambda}$ so $\zeta(C^{-1}) = 1/\zeta(C)$ and $CD \in G_{\mu\lambda}$ so $\zeta(CD) = \zeta(C)\zeta(D)$, thus ζ is a group homomorphism.

Suppose $D_n \in G$ and $D_n \to D$ for some $D \in \mathrm{GL}(\mathbb{R}^d)$. Then $D_n \in G_{\lambda_n}$ for some $\lambda_n > 0$. Suppose $\alpha > 0$. If $\lambda_{n'} \to \infty$ along a subsequence, then $D_{n'}\varphi(x) = \lambda_{n'}^{-\alpha} \cdot \varphi(x) \to 0$, and if $\lambda_{n'} \to 0$ along a subsequence, then $D_{n'}\varphi(x) = \lambda_{n'}^{-\alpha} \cdot \varphi(x) \to \infty$. Either one contradicts the fact that $D_n\varphi(x) \to D\varphi(x) > 0$, hence (λ_n) is relatively compact in $\mathbb{R}^+$. The case $\alpha < 0$ is similar. If $\lambda_{n'} \to \lambda$ along a subsequence, then $\lambda > 0$ and $D_n\varphi = \lambda_n^{-\alpha} \cdot \varphi \to \lambda^{-\alpha} \cdot \varphi$, so that $D \in G_\lambda$. Thus G is a closed subgroup of $\mathrm{GL}(\mathbb{R}^d)$ in the relative topology. If $\lambda_{n''} \to \mu \neq \lambda$ along another subsequence, then $D \in G_\lambda \cap G_\mu = \emptyset$, which is a contradiction, hence $\zeta(D_n) = \lambda_n \to \lambda = \zeta(D)$ in this case, so ζ is continuous.

Take $\lambda_n = (n+1)/n$ so that $\lambda_n \to \lambda = 1$ and choose $C_n \in G_{\lambda_n}$. Then $C_n\varphi = \lambda_n^{-\alpha} \cdot \varphi \to \lambda^{-\alpha} \cdot \varphi = \varphi$ and Lemma 5.2.8 implies that $C_n \rightharpoonup \mathcal{S}(\varphi) = G_1$, in other words $\|C_n - \mathcal{S}(\varphi)\| \to 0$. Choose $B_n \in \mathcal{S}(\varphi)$ so that $\|C_n - B_n\| = \|C_n - \mathcal{S}(\varphi)\|$. Then $(C_n - B_n) \to 0$, and $(C_n - B_n)B_n^{-1} \to 0$ in view of the fact that $\mathcal{S}(\varphi)$ is compact, so $C_nB_n^{-1} - I \to 0$ or equivalently $C_nB_n^{-1} \to I$. Letting $K_n = C_nB_n^{-1} \in G_{\lambda_n}$, we have $K_n \to I$ and $\zeta(K_n) = \lambda_n \neq 1$.

Recall from Definition 2.2.9 that the tangent space TG is the set of tangent vectors to G at the identity. Proposition 2.2.10 (d) implies that there exists an open set $U \subset G$ such that $I \in U$ and every $D \in U$ satisfies $D = t^Q$ for some $Q \in TG$ and some $t > 0$. Since $D_n \to I$ we have $D_n \in U$ for all large n. For some such n, let $D = D_n = t^Q$, where $Q \in TG$ and $t > 0$. Then Proposition 2.2.10 (c) implies that $\lambda^Q \in G$ for all $\lambda > 0$. Define $h(\lambda) = \zeta(\lambda^Q)$. Then $h(1) = \zeta(I) = 1$ and $h(\mu\lambda) = \zeta((\mu\lambda)^Q) = \zeta(\mu^Q\lambda^Q) = h(\mu)h(\lambda)$ for all $\lambda > 0$ and $\mu > 0$. Since h is continuous this implies that $h(\lambda) = \lambda^\rho$ for some real ρ, and since $h(t) = \zeta(t^Q) = \zeta(D) \neq 1$ we have $\rho \neq 0$. Define $E = -(1/\rho)Q$. Then $\zeta(\lambda^{-E}) = \zeta((\lambda^{1/\rho})^Q) = h(\lambda^{1/\rho}) = \lambda$ for all $\lambda > 0$, so $\lambda^{-E} \in G_\lambda$ for all $\lambda > 0$. Then $G_\lambda = \lambda^{-E}\mathcal{S}(\varphi)$ for all $\lambda > 0$, and $\lambda^{-E}\varphi = \lambda^{-\alpha} \cdot \varphi$, so that (5.13) holds for all $\lambda > 0$. □

Definition 5.2.12. Let φ be admissible. Then any linear operator B such that (5.13) holds for all $\lambda > 0$ is called an *exponent* of φ. Let $\mathcal{E}(\varphi)$ denote the set of all possible exponents of φ.

Exponents in general are not unique. Theorem 5.2.13 describes the structure of the set $\mathcal{E}(\varphi)$ of exponents of an admissible function φ in terms of the tangent space.

Theorem 5.2.13. *Exponents and symmetries of an admissible function φ are related by*

$$\mathcal{E}(\varphi) = B + T\mathcal{S}(\varphi) \tag{5.14}$$

where $B \in \mathcal{E}(\varphi)$ is arbitrary and $T\mathcal{S}(\varphi)$ denotes the tangent space of the symmetry group $\mathcal{S}(\varphi)$ (see Definition 2.2.9). One can always select an exponent B_0 that commutes with every symmetry $A \in \mathcal{S}(\varphi)$, i.e., $B_0A = AB_0$ for all $A \in \mathcal{S}(\varphi)$.

Proof. Let G and ζ be as in the proof of Theorem 5.2.11. Define a continuous group mapping $L : TG \to \mathbb{R}$ by letting $L(X) = \log \zeta(\exp(X))$. Given $A \in TG$

let $g(t) = L(tA)$. Then

$$\begin{aligned} g(t) + g(s) &= L(tA) + L(sA) \\ &= \log \zeta(\exp(tA)) + \log \zeta(\exp(sA)) \\ &= \log\{\zeta(\exp(tA)) \cdot \zeta(\exp(sA))\} \\ &= \log\{\zeta(\exp(tA) \cdot \exp(sA))\} \\ &= \log\{\zeta(\exp((t+s)A))\} = g(t+s) \end{aligned}$$

and since g is continuous this implies that $g(t) = at$ for some $a \in \mathbb{R}$. Since $g(1) = L(A)$ we have

$$L(tA) = tL(A) \tag{5.15}$$

for all $t \in \mathbb{R}$. Proposition 2.2.10 (d) implies that there exists an open set $U \subset G$ such that $I \in U$ and every $D \in U$ satisfies $D = \exp(X)$ for some $X \in TG$. If $A, B \in TG$ then $\exp(tA)\exp(tB) = I + (A+B)t + o(t) \to I$ as $t \to 0$. Then $\exp(tA)\exp(tB) \in U$ for all t near zero, and hence for some $X(t) \in TG$ we have $\exp(X(t)) = \exp(tA)\exp(tB)$ for all such t. Then

$$\begin{aligned} L(X(t)) &= \log \zeta(\exp(X(t))) \\ &= \log \zeta(\exp(tA) \cdot \exp(tB)) \\ &= \log\{\zeta(\exp(tA) \cdot \zeta(\exp(tB))\} \\ &= \log\{\zeta(\exp(tA)) + \log(\zeta(\exp(tB))\} \\ &= L(tA) + L(tB) \end{aligned}$$

so that by (5.15) we have

$$L(X(t)) = t\{L(A) + L(B)\} \tag{5.16}$$

for all t near zero. Since $h(t) = \exp(X(t)) = I + t(A+B) + o(t)$ it follows that $X'(0) = h'(0) = A + B$. Then

$$\begin{aligned} L(A+B) &= L(X'(0)) \\ &= L\left(\lim_{t\to 0} \frac{X(t)}{t}\right) \\ &= \lim_{t\to 0} L\left(\frac{X(t)}{t}\right) \\ &= \lim_{t\to 0} \frac{1}{t} L(X(t)) \\ &= \lim_{t\to 0} \frac{1}{t}\, t\{L(A) + L(B)\} = L(A) + L(B) \end{aligned}$$

by continuity of L, (5.15) and (5.16), hence L is linear.

Given $X \in TG$, if $L(X) = \log \zeta(e^X) = 0$, then $L(tX) = 0$ for all $t \in \mathbb{R}$ since L is linear. Then $\zeta(e^{tX}) = 1$ so $g(t) = e^{tX} \in G_1 = \mathcal{S}(\varphi)$ for all t, hence $X = g'(0) \in T\mathcal{S}(\varphi)$ by definition of the tangent space. Conversely, if $X \in T\mathcal{S}(\varphi)$, then $e^X \in \mathcal{S}(\varphi) = G_1$ so $\zeta(e^X) = 1$ and $L(X) = 0$. This shows that

$$\mathcal{S}(\varphi) = \{X \in TG : L(X) = 0\}.$$

Given $X \in \mathcal{E}(\varphi)$ we have $\lambda^X\varphi = \lambda\cdot\varphi$, so $\lambda^{-\alpha X}\varphi = \lambda^{-\alpha}\cdot\varphi$, hence $\lambda^{-\alpha X} \in G_\lambda$. Then $\zeta(\lambda^{-\alpha X}) = \lambda$, and it follows that $\zeta(\lambda^X) = \lambda^{-1/\alpha}$, so that $L(X) = -1/\alpha$. Reversing the steps shows that if $L(X) = -1/\alpha$, then $X \in \mathcal{E}(\varphi)$. Hence

$$\mathcal{E}(\varphi) = \{X \in TG : L(X) = -1/\alpha\}.$$

Since L is linear we know that $\{L = -1/\alpha\} = B + \{L = 0\}$ for any B such that $L(B) = -1/\alpha$, so (5.14) holds.

Finally, if $B \in \mathcal{E}(\varphi)$ is arbitrary we may define

$$B_0 = \int_{A\in\mathcal{S}(\varphi)} ABA^{-1}\mathcal{H}(dA) \tag{5.17}$$

where $\mathcal{H}$ is the Haar measure on the compact group $\mathcal{S}(\varphi)$, so that $\mathcal{H}(AdX) = \mathcal{H}(dX)$ for all $A \in \mathcal{S}(\varphi)$ (see, e.g., [56]).

If $A \in \mathcal{S}(\varphi)$ and $B \in \mathcal{E}(\varphi)$, then Proposition 2.2.2 (e) implies $\lambda^{ABA^{-1}}\varphi = A\lambda^B A^{-1}\varphi = A\lambda^B\varphi = A(\lambda\cdot\varphi) = \lambda\cdot(A\varphi) = \lambda\cdot\varphi$ so that $ABA^{-1} \in \mathcal{E}(\varphi)$ as well. Then since (5.14) holds, so that $\mathcal{E}(\varphi)$ is closed and convex, we know that $B_0 \in \mathcal{E}(\varphi)$. Given $A \in \mathcal{S}(\varphi)$ we have

$$\begin{aligned} AB_0A^{-1} &= A\Big(\int_{C\in\mathcal{S}(\varphi)} CBC^{-1}\mathcal{H}(dC)\Big)A^{-1} \\ &= \int_{C\in\mathcal{S}(\varphi)} A(CBC^{-1})A^{-1}\mathcal{H}(dC) \\ &= \int_{C\in\mathcal{S}(\varphi)} (AC)B(AC)^{-1}\mathcal{H}(dC) \\ &= \int_{D\in\mathcal{S}(\varphi)} DBD^{-1}\mathcal{H}(A^{-1}dD) \\ &= \int_{D\in\mathcal{S}(\varphi)} DBD^{-1}\mathcal{H}(dD) = B_0 \end{aligned}$$

and then $AB_0 = B_0A$ for every $A \in \mathcal{S}(\varphi)$. □

Recall from Definition 2.1.17 that a linear operator $A \in \mathrm{L}(\mathbb{R}^d)$ is semisimple if it is diagonalizable over the complex numbers, and nilpotent if $A^k = 0$ for some k. Theorem 2.1.18 shows that any $A \in \mathrm{L}(\mathbb{R}^d)$ can be written uniquely in the form $A = S+N$, where S is semisimple, N is nilpotent, and $SN = NS$. We call S, N the semisimple and nilpotent parts of A. Recall from Definition 2.1.13 that the real spectrum of a linear operator $A \in \mathrm{L}(\mathbb{R}^d)$ consists of the real parts of the eigenvalues of A.

Theorem 5.2.14. *The nilpotent part of every $B \in \mathcal{E}(\varphi)$ is the same, and it commutes with every element of the symmetry group $\mathcal{S}(\varphi)$. Furthermore, every $B \in \mathcal{E}(\varphi)$ has the same real spectrum, and the spectral decomposition of $\mathbb{R}^d$ is the same for every $B \in \mathcal{E}(\varphi)$.*

Proof. Choose a commuting exponent B_0 and use Theorem 2.1.18 to write $B_0 = S_0 + N_0$, where S_0 is semisimple and N_0 is nilpotent. Now both S_0 and N_0 are polynomials in B_0. Hence $AS_0 = S_0A$ and $AN_0 = N_0A$ for every $A \in \mathcal{S}(\varphi)$. Let $B \in \mathcal{E}(\varphi)$ be arbitrary and write $B = S + N$ as before. Since $\mathcal{E}(\varphi) = B_0 + T\mathcal{S}(\varphi)$, we may write $B = B_0 + X$, where $X \in T\mathcal{S}(\varphi)$. Then $B = (S_0 + X) + N_0$. Since S_0 commutes with all $A \in \mathcal{S}(\varphi)$ it also commutes with $X \in T\mathcal{S}(\varphi)$. Proposition 2.2.10 (c) shows that $t^X \in \mathcal{S}(\varphi)$ compact for all $t > 0$, and hence $\|t^X\|$ remains bounded. In the proof of Theorem 2.2.4, we compute $\|t^X y\|$ and it follows that each coefficient in (2.11) and (2.12) must remain bounded. This can only happen if the eigenvalues of X are pure imaginary and X has no nilpotent part, i.e., X is semisimple. If we consider S_0, X as linear operators over the complex vector space $\mathbb{C}^d$, then since S_0 is semisimple there exists a unique direct sum decomposition $\mathbb{C}^d = U_1 \oplus \cdots \oplus U_m$ such that $S_0U_j = U_j$ and $S_0z = \lambda_j z$ for all $z \in U_j$, for each $j = 1, \ldots, m$, where λ_j ranges over the eigenvalues of S_0 and $\lambda_j \neq \lambda_k$ for $j \neq k$. Then $S_0z = \lambda_j z$ implies that $z \in U_j$. If $z \in U_j$, then $S_0Xz = XS_0z = X\lambda_j z = \lambda_j Xz$ so that $Xz \in U_j$ too. Then $XU_j \subseteq U_j$ and since X is invertible this implies that $XU_j = U_j$. Similarly, there exists a unique direct sum decomposition $\mathbb{C}^d = W_1 \oplus \cdots \oplus W_n$ such that $XW_k = W_k$ and $Xz = \omega_k z$ if and only if $z \in W_k$, for each $k = 1, \ldots, n$ where ω_k ranges over the eigenvalues of X and $\omega_j \neq \omega_k$ for $j \neq k$. Then $S_0W_k = W_k$ as well. Let $\pi_k : \mathbb{C}^d \to W_k$ denote the projections associated with this direct sum decomposition. Then $\mathbb{C}^d = \oplus_{j,k}\pi_k U_j$. For any $z \in \pi_k U_j$ we have $Xz = \omega_k z$ since $z \in W_k$, and since $z = \pi_k w$ for some $w \in U_j$ we also have $S_0z = S_0\pi_k w = \pi_k S_0 w = \pi_k \lambda_j w = \lambda_j \pi_k w = \lambda_j z$, hence $(S_0 + X)z = (\lambda_j + \omega_k)z$. If we choose a basis for each nontrivial subspace $\pi_k U_j$, then the union of these bases forms a basis for $\mathbb{C}^d$ in which $S_0, X, S_0 + X$ are all diagonal. This shows that $S_0 + X$ is semisimple. Then the uniqueness of the $S + N$ decomposition, Theorem 2.1.18, shows that $S = S_0 + X$ and $N = N_0$. Furthermore, since every ω_k is purely imaginary, the real spectrum of S_0 and S are the same. The Jordan Decomposition, Theorem 2.1.16, implies that the spectrum of $S + N$ is the spectrum of S. Then B, B_0 have the same real spectrum. Define $R : \mathbb{C}^d \to \mathbb{R}^d$ by $R(x + iy) = x$. The subspace V_i in the spectral decomposition of $\mathbb{R}^d$ with respect to B_0 is RU, where U is the direct sum of the U_j such that $\mathrm{Re}(\lambda_j) = a_i$. The corresponding subspace in the spectral decomposition of $\mathbb{R}^d$ with respect to B is RW, where W is the direct sum of the $\pi_k U_j$ such that $\mathrm{Re}(\lambda_j + \omega_k) = a_i$. Since $\mathrm{Re}(\omega_k) = 0$ both subspaces are the same. □

Now we apply Theorem 5.2.13 to the situation of Theorem 5.2.11, in order to characterize the set of possible exponents E in (5.12).

Corollary 5.2.15. *Given $F \in \mathcal{F}$, φ admissible and $A_n \in \mathrm{GL}(\mathbb{R}^d)$, suppose that $a_n > 0$ is a regularly varying sequence with index $\alpha \neq 0$ and that $a_n(A_nF) \to \varphi$. Then $\mathcal{E}(\varphi)$ is nonempty, and (5.12) holds if and only if $E = \alpha B$ for some $B \in \mathcal{E}(\varphi)$. In particular, we can always choose E so that $EA = AE$ for all $A \in \mathcal{S}(\varphi)$.*

Proof. If $a_n > 0$ is a regularly varying sequence with index $\alpha \neq 0$ and $a_n(A_nF) \to \varphi$ admissible, then Theorem 5.2.11 implies that (5.13) holds for all $\lambda > 0$, with $B = (1/\alpha)E$. Then $B \in \mathcal{E}(\varphi)$, so that $\mathcal{E}(\varphi)$ is nonempty. Apply Theorem 5.2.13 to obtain $B = B_0 + X'$ for some $X' \in T\mathcal{S}(\varphi)$ and some $B_0 \in \mathcal{E}(\varphi)$ such that $B_0A = AB_0$ for all $A \in \mathcal{S}(\varphi)$. Then we also have $B_0X = XB_0$ for all $X \in T\mathcal{S}(\varphi)$, since

$$X = \lim_{n\to\infty} \frac{A_n - I}{n^{-1}}$$

for some $A_n \in \mathcal{S}(\varphi)$. If we let $E_0 = \alpha B_0$ and $X = \alpha X'$, then we have $E = E_0 + X$, and since $E_0X = XE_0$ we have

$$\lambda^{-E}\mathcal{S}(\varphi) = \lambda^{-E_0-X}\mathcal{S}(\varphi) = \lambda^{-E_0}\lambda^{-X}\mathcal{S}(\varphi) = \lambda^{-E_0}\mathcal{S}(\varphi)$$

because $\lambda^X \in \mathcal{S}(\varphi)$ for all $X \in T\mathcal{S}(\varphi)$. Then

$$A_{[\lambda n]}A_n^{-1} \rightharpoonup \lambda^{-E_0}\mathcal{S}(\varphi)$$

where E_0 commutes with every $X \in T\mathcal{S}(\varphi)$. Now take any $B \in \mathcal{E}(\varphi)$ and let $E = \alpha B$. Since $B = B_0 + X'$ for some B_0 commuting and some $X' \in T\mathcal{S}(\varphi)$ we also have $E = E_0 + X$ for some $X \in T\mathcal{S}(\varphi)$ and $E_0X = XE_0$. Then it follows as before that $\lambda^{-E}\mathcal{S}(\varphi) = \lambda^{-E_0}\mathcal{S}(\varphi)$, and so (5.12) also holds for this choice of E. □

Theorem 5.2.16 is our basic characterization of regularly varying functions that are monotone on rays. The one variable version is due to Feller [40].

Theorem 5.2.16. *Let $F \in \mathcal{F}$, φ admissible and $A_n \in \mathrm{GL}(\mathbb{R}^d)$. Suppose $a_n > 0$ is a regularly varying sequence with index $\alpha \neq 0$ and that $a_n(A_nF) \to \varphi$. Then F is regularly varying, and $\mathcal{E}(\varphi)$ is nonempty. Furthermore, for any $B \in \mathcal{E}(\varphi)$, there exist $G_n \in \mathcal{S}(\varphi)$ such that $B_n = G_nA_n$ is regularly varying with index $-\alpha B$ and $a_nB_nF \to \varphi$. Then F varies regularly with exponent B.*

Proof. Corollary 5.2.15 implies that $\mathcal{E}(\varphi)$ is nonempty, and that (5.12) holds for any linear operator of the form $E = \alpha B$ where $B \in \mathcal{E}(\varphi)$. In particular, we can choose $E = E_0$ commuting. Then $\lambda^{-E}\mathcal{S}(\varphi) = \lambda^{-E_0}\mathcal{S}(\varphi) = \mathcal{S}(\varphi)\lambda^{-E_0} = \mathcal{S}(\varphi)\lambda^{-E}$ for all $\lambda > 0$. Theorem 5.2.4 implies that $\mathcal{S}(\varphi)$ is a compact subgroup of $\mathrm{GL}(\mathbb{R}^d)$ and (5.12) holds, and then Proposition 4.2.12 gives the existence of $G_n \in \mathcal{S}(\varphi)$ such that $B_n = G_nA_n$ is regularly varying with index $-E$. Furthermore, if we set $F'_n = a_n(A_nF)$, then $a_n(B_nF) = G_nF'_n$, and hence by Lemma 5.2.8 we have $a_n(B_nF) \to \varphi$ too. We have shown that for any $B \in$

$\mathcal{E}(\varphi)$, there exists a regularly varying sequence (B_n) with index $-E = -\alpha B$ such that $a_n(B_n F) \to \varphi$. The function $f(t) = B_{[t]}$ varies regularly with index $-E$ by Theorem 4.2.9, and also $R_0(t) = a_{[t]}$ varies regularly with index α. Then (5.2) holds with $R(t) = 1/R_0$, where R varies regularly with index $\beta = -\alpha$, so that F is regularly varying with exponent $(-1/(-\alpha))(\alpha B) = B$. $\square$

Remark 5.2.17. In order to apply Theorem 5.2.16, one must verify that the limit function φ is admissible. In our application in Section 8.1 to normal limits for sums of independent random vectors, the limit function is a positive definite quadratic form, so it is automatically admissible. More generally, if $a_n(A_n F) \to \varphi > 0$, then φ must be continuous (see the proof of Proposition 5.2.2). Also Theorem 5.2.11 implies that φ satisfies (5.13), and if the real parts of the eigenvalues of B are either all positive or all negative, then φ is admissible. The proof of these assertions is essentially identical to Proposition 5.2.2.

In Chapter 8, we will also consider the special case of norming by scalars in Definition 5.1.1. Most of the results of this section simplify considerably in this case.

Definition 5.2.18. We say that a Borel measurable function $F : \Gamma \to [0, \infty)$ is *scalar-normed regularly varying* if there exist $a : \mathbb{R}^+ \to \mathbb{R}^+$ regularly varying with index $-\rho$, and $R : \mathbb{R}^+ \to \mathbb{R}^+$ regularly varying with index $\beta \neq 0$, such that

$$\lim_{t\to\infty} \frac{F(a(t)^{-1}x_t)}{R(t)} = \varphi(x) > 0 \tag{5.18}$$

whenever $x_t \to x$ in $\Gamma = \mathbb{R}^d \setminus \{0\}$. If $\rho > 0$ we will say that F is *scalar-normed regularly varying at infinity*, and if $\rho < 0$ we will say that F is *scalar-normed regularly varying at zero.*

For $a > 0$ and $F \in \mathcal{F}$ we define $(aF) = (aI)F$, so that $(aF)(x) = F(a^{-1}x)$. Then we can also write (5.18) in the form $(a(t)F)/R(t) \to \varphi$. Most properties of scalar-normed regularly varying functions follow easily from the results in this section, by specializing to the case where the linear operators are positive scalar multiples of the identity, and using the fact that such linear operators form a closed multiplicative subgroup of $\mathrm{GL}(\mathbb{R}^d)$. We record here the following results, which we will need in Chapter 8.

Proposition 5.2.19. *Suppose that $F \in \mathcal{F}$ and φ is admissible.*

(a) If F is scalar-normed regularly varying and (5.18) *holds, then $\lambda \cdot \varphi = (\lambda^b \varphi)$ for all $\lambda > 0$, where $b = (-\rho/\beta)$. We will call b an exponent of F (or φ).*

(b) If F is scalar-normed regularly varying at zero with exponent $b \neq 0$, then $F \in \mathrm{RVF}_0(bI)$.

(c) *The function $F \in \mathcal{F}$ is scalar-normed regularly varying at zero with exponent $b < 0$ if and only if $na_nF \to \varphi$ for some regularly varying sequence of positive reals (a_n) with index $-b$.*

(d) *If $n(a_nF) \to \varphi$ for some sequence of positive reals (a_n), then (a_n) varies regularly with index $-b$ for some $b \neq 0$. If $b < 0$ then F is scalar-normed regularly varying at zero with exponent b and $\lambda \cdot \varphi = (\lambda^b \varphi)$ for all $\lambda > 0$.*

Remark 5.2.20. It is also useful to consider a notion of regular variation in the special case where the norming operators $f(t)$ are diagonal (with respect to some fixed basis for $\mathbb{R}^d$). See de Haan, et al. [48] and Meerschaert [110, 115] for additional details.

5.3 UNIFORM REGULAR VARIATION

In this section, we introduce a simple extension of the notion of a regularly varying function on $\mathbb{R}^1$ that is useful for functions $F : \mathbb{R}^d \to \mathbb{R}^+$ whose growth rate is essentially the same in every radial direction.

Definition 5.3.1. Suppose that S is a compact topological space, and that $R : \mathbb{R}^+ \times S \to \mathbb{R}^+$ is Borel measurable. We say that $R(r, \theta)$ is *uniformly regularly varying* with index $\rho \in \mathbb{R}$ if for all $\theta \in S$ and all $\lambda > 0$ we have

$$\lim_{r \to \infty} \frac{R(\lambda r, \theta)}{R(r, \theta)} = \lambda^\rho \tag{5.19}$$

and this convergence is uniform with respect to $\theta \in S$.

Remark 5.3.2. Uniform convergence in (5.19) is equivalent to the condition that

$$\lim_{r \to \infty} \frac{R(\lambda r, \theta_r)}{R(r, \theta_r)} = \lambda^\rho$$

whenever $\theta_r \to \theta$ in S. The proof of this equivalence is quite similar to Proposition 2.1.7 (b) and is left to the reader.

Example 5.3.3. Given $A \in \mathrm{L}(\mathbb{R}^d)$ let $b_1, \ldots, b_d$ be the basis obtained from Theorem 2.1.16 in which A has Jordan form, and use the norm associated with this basis, in the sense of Definition 1.1.21. Let $a_1 < \cdots < a_p$ be the real spectrum of A, $V_1 \oplus \cdots \oplus V_p$ the spectral decomposition of $\mathbb{R}^d$ with respect to A and $A = A_1 \oplus \cdots \oplus A_p$ the spectral decomposition of A in the sense of Definition 2.1.15. Using (2.11) and (2.12) from the proof of Theorem 2.2.4, it follows that $R(r, \theta) = \|r^A \theta\|$ for $r > 0$ and $\|\theta\| = 1$ is uniformly regularly varying with index a_i on the unit sphere in V_i. If we write $\theta = \theta_1 + \cdots + \theta_p$ with respect to the spectral decomposition of $\mathbb{R}^d$, then $R(r, \theta)$ varies regularly with index $\alpha(\theta) = \max\{a_i : \theta_i \neq 0\}$ for every $\theta \in S = \{x \in \mathbb{R}^d : \|x\| = 1\}$.

Furthermore, $R(r,\theta)$ is uniformly regularly varying on compact subsets of $S_i = \{\theta \in S : \alpha(\theta) = a_i\}$.

Our first result shows that, for a regularly varying function F whose index is a multiple of the identity, the function $R(r,\theta) = F(r\theta)$ in polar coordinates is uniformly regularly varying.

Theorem 5.3.4. *Suppose that F varies regularly and $E = \alpha I$ in Definition 5.1.1 for some real $\alpha \neq 0$.*

(a) If F varies regularly at infinity then $R(r,x) = F(rx)$ is uniformly regularly varying with index β/α on compact subsets of $x \in \Gamma$.

(b) If F varies regularly at zero then $R(r,x) = F(x/r)$ is uniformly regularly varying with index $-\beta/\alpha$ on compact subsets of $x \in \Gamma$.

Proof. We use the notation of Definition 5.1.1 First we prove part (a). From (5.3) with $E = \alpha I$ it follows that

$$\varphi(\lambda x) = \lambda^{\beta/\alpha} \cdot \varphi(x) \tag{5.20}$$

for all $\lambda > 0$ and all $x \in \Gamma$. We need to show that

$$\frac{R(\lambda r, x_r)}{R(r, x_r)} = \frac{F(\lambda r x_r)}{F(r x_r)} \to \lambda^{\beta/\alpha} \tag{5.21}$$

as $x_r \to x$ in Γ for all $\lambda > 0$. For each $r > 0$ sufficiently large, let $t(r) = \sup\{t : \|f(t) r x_r\| \geq 1\}$, and then define $y_t = f(t) r x_r$, where $t = t(r)$. Part (a) of Lemma 4.2.7 shows that $\|y_t\| \to 1$, so that $\{y_t : t > 0\}$ is limit point compact and $t(r) \to \infty$ as $r \to \infty$. If for some $r_k \to \infty$ we have $y_{t_k} \to y$ in Γ, where $t_k = t(r_k)$, then we have

$$\frac{F(r_k \lambda x_{r_k})}{F(r_k x_{r_k})} = \frac{F(f(t_k)^{-1} \lambda y_{t_k})}{F(f(t_k)^{-1} y_{t_k})} \to \frac{\varphi(\lambda y)}{\varphi(y)} = \lambda^{\beta/\alpha}$$

by (5.20) along with the regular variation of F. Since any sequence has a further subsequence with this property, we see that (5.21) holds.

The argument for part (b) is similar. Given $x_r \to x$ in Γ, we let $t(r) = \sup\{t : \|f(t) x_r / r\| \leq 1\}$ for $r > 0$ sufficiently large and we define $y_t = f(t) x_r / r$ where $t = t(r)$. If for some $r_k \to \infty$ we have $y_{t_k} \to y$ in Γ, then

$$\frac{R(\lambda r_k, x_{r_k})}{R(r_k, x_{r_k})} = \frac{F(x_{r_k}/\lambda r_k)}{F(x_{r_k}/r_k)} = \frac{F(\lambda^{-1} f(t_k)^{-1} y_{t_k})}{F(f(t_k)^{-1} y_{t_k})} \to \frac{\varphi(\lambda^{-1} y)}{\varphi(y)} = \lambda^{-\beta/\alpha}$$

by (5.20) and the regular variation of F. Part (b) of Lemma 4.2.7 implies that any sequence has a further subsequence with this property, and so R is uniformly regularly varying with index $-\beta/\alpha$. $\square$

Remark 5.3.5. If for all $\mu > 0$ we have

$$\lim_{s \to 0} \frac{R(\mu s, \theta)}{R(s, \theta)} = \mu^\rho \tag{5.22}$$

uniformly in $\theta \in S$ compact, we say that R is uniformly regularly varying at zero with index ρ. It is easy to check that (5.22) is equivalent to the uniform regular variation of $R(1/r, \theta)$ (at infinity) with index $-\rho$. Hence we may restate part (b) of Theorem 5.3.4 as follows: If F varies regularly at zero with $E = \alpha I$ for some real $\alpha \neq 0$, then $R(r, x) = F(rx)$ is uniformly regularly varying at zero with index β/α on compact subsets of $x \in \Gamma$.

We will now establish some of the basic properties of uniformly regularly varying functions. If $R(r, \theta)$ is uniformly regularly varying with index ρ, then $R(r, \theta)$ is a regularly varying function of $r > 0$ with index ρ for any fixed $\theta \in S$. Hence all of the properties of regularly varying functions apply to $R(r, \theta)$ for any fixed $\theta \in S$, and many of these properties also hold uniformly in θ. For example, if R_1, R_2 are uniformly regularly varying with index ρ_1, ρ_2 respectively, then it follows immediately from (5.19) that $1/R_1$ is uniformly regularly varying with index $-\rho_1$ and $R_1 R_2$ is uniformly regularly varying with index $\rho_1 + \rho_2$. Also, if we let $L(r, \theta) = r^{-\rho} R(r, \theta)$, then it follows immediately from (5.19) that L is uniformly slowly varying, i.e., $L(\lambda r, \theta)/L(r, \theta) \to 1$ as $r \to \infty$ uniformly in $\theta \in S$ for any $\lambda > 0$. Hence $R(r, \theta) = r^\rho L(r, \theta)$, where L is uniformly slowly varying, and so it is sufficient for most purposes to study uniformly slowly varying functions.

Theorem 5.3.6. *If R is uniformly regularly varying, then the convergence (5.19) holds uniformly with respect to $(\lambda, \theta) \in [a, b] \times S$ for any $0 < a < b < \infty$.*

Proof. We follow the proof of Theorem 1.1 (uniform convergence) in Seneta [177]. It suffices to prove the theorem in the case where R is slowly varying. Then $R(\lambda r, \theta_r)/R(r, \theta_r) \to 1$ whenever $\theta_r \to \theta$ in S. Define $f(r, \theta) = \log R(e^r, \theta)$, and note that for any $\mu \in \mathbb{R}$ we have

$$\begin{aligned} f(r + \mu, \theta_r) - f(r, \theta_r) &= \log R(e^{r+\mu}, \theta_r) - \log R(e^r, \theta_r) \\ &= \log\left(\frac{R(e^\mu e^r, \theta_r)}{R(e^r, \theta_r)} \right) \\ &\to \log 1 = 0 \end{aligned}$$

whenever $\theta_r \to \theta$ in S. Then

$$f(r + \mu, \theta) - f(r, \theta) \to 0 \tag{5.23}$$

as $r \to \infty$ uniformly with respect to $\theta \in S$. Now it suffices to show that (5.23) holds uniformly with respect to $(\mu, \theta) \in I \times S$ for any compact subinterval I of the real line. Suppose $I = [0, 1]$. If the assertion is not true then for some $\varepsilon > 0$ there exists $r_n \to \infty$, $\mu_n \in [0, 1]$, and $\theta_n \in S$ such that

$$|f(r_n + \mu_n, \theta_n) - f(r_n, \theta_n)| \geq \varepsilon \tag{5.24}$$

for all n. Now define

$$\begin{aligned} U_n &= \{\mu \in [0,2] : |f(r_m+\mu,\theta) - f(r_m,\theta)| < \varepsilon/2 \\ &\qquad \text{for all } m \geq n \text{ and all } \theta \in S\} \\ V_n &= \{\lambda \in [0,2] : |f(r_m+\mu_m+\lambda,\theta) - f(r_m+\mu_m,\theta)| < \varepsilon/2 \\ &\qquad \text{for all } m \geq n \text{ and all } \theta \in S\} \end{aligned}$$

and note that both U_n and V_n increase to $[0,2]$ as $n \to \infty$ in view of the fact that (5.23) holds uniformly in θ. Let $m(A)$ denote the Lebesgue measure of a Borel subset $A \subseteq \mathbb{R}^1$. Then for all N sufficiently large we have $m(U_N) > 3/2$ and $m(V_N) > 3/2$. Let $V_N' = V_N + \mu_N$, so that $m(V_N') = m(V_N) > 3/2$. Since $U_N \subseteq [0,2]$ and $V_N' \subseteq [0,3]$ it follows that $U_N \cap V_N' \neq \emptyset$ for all large N. Choose $\mu \in U_N \cap V_N'$ so that $\mu - \mu_N \in V_N$ as well. Then

$$\begin{aligned} |f(r_N+\mu,\theta_N) - f(r_N,\theta_N)| &< \varepsilon/2 \\ |f(r_N+\mu,\theta_N) - f(r_N+\mu_N,\theta_N)| &< \varepsilon/2 \end{aligned}$$

by definition of the sets U_N, V_N and so

$$|f(r_N+\mu_N,\theta_N) - f(r_N,\theta_N)| < \varepsilon$$

by the triangle inequality, which contradicts (5.24). Then (5.23) holds uniformly with respect to $(\mu,\theta) \in [0,1] \times S$.

To finish the proof, it will suffice to show that (5.23) holds uniformly with respect to $(\lambda,\theta) \in [a,b] \times S$, where $a < b$ are arbitrary real numbers. Define $g(r,\theta) = f(a+(b-a)r,\theta)$ and note that

$$g(r+\lambda,\theta) - g(r,\theta) \to 0 \tag{5.25}$$

uniformly in θ as $r \to \infty$ by (5.23), so that (5.25) also holds uniformly with respect to $(\lambda,\theta) \in [0,1] \times S$ by what has already been proven. For any $\mu_r \in [a,b]$ and $\theta_r \in S$, we have that $r/(b-a) \to \infty$ as $r \to \infty$ and that $(\mu_r - a)/(b-a) \in [0,1]$ so

$$g\left(\frac{r}{b-a} + \frac{\mu_r - a}{b-a}, \theta_r\right) - g\left(\frac{r}{b-a}, \theta_r\right) \to 0 \tag{5.26}$$

because (5.25) holds uniformly with respect to $(\lambda,\theta) \in [0,1] \times S$. Furthermore, in view of (5.23)

$$g\left(\frac{r}{b-a}, \theta_r\right) - g\left(\frac{r}{b-a} + \frac{-a}{b-a}, \theta_r\right) = f(a+r,\theta_r) - f(r,\theta_r) \to 0$$

as $r \to \infty$, and hence by (5.26)

$$\begin{aligned} f(r+\mu_r,\theta_r) - f(r,\theta_r) &= g\left(\frac{r}{b-a} + \frac{\mu_r - a}{b-a}, \theta_r\right) - g\left(\frac{r}{b-a}, \theta_r\right) \\ &\quad + g\left(\frac{r}{b-a}, \theta_r\right) - g\left(\frac{r}{b-a} + \frac{-a}{b-a}, \theta_r\right) \\ &\to 0 \end{aligned}$$

as $r \to \infty$. This completes the proof. □

The following results will allow us to apply the theory of uniform regular variation to certain integrals that arise in probability theory. These results extend those of Feller [40] for regularly varying functions of one variable.

Definition 5.3.7. Suppose that S is a compact topological space, and that $R : \mathbb{R}^+ \times S \to \mathbb{R}^+$ is Borel measurable. For $0 \leq A \leq r$, we define

$$R_\beta(r, \theta) = \int_A^r s^\beta R(s, \theta) ds \tag{5.27}$$

and

$$R_\beta^*(r, \theta) = \int_r^\infty s^\beta R(s, \theta) ds \tag{5.28}$$

whenever these integrals exist.

Lemma 5.3.8. *If R is uniformly regularly varying with index ρ, then*

(a) For some $A_0 > 0$, the integral $R_\beta(r, \theta)$ in (5.27) exists for all $r \geq A \geq A_0$ and all $\theta \in S$

(b) For all $r > A_0$ and all $\theta \in S$, the integral $R_\beta^*(r, \theta)$ in (5.28) exists for $\beta < -\rho - 1$ and diverges for $\beta > -\rho - 1$

(c) If $\beta > -\rho - 1$ and if $R_\beta(r, \theta)$ exists for all $r \geq A$ and all $\theta \in S$, then $R_\beta(r, \theta)$ is uniformly regularly varying with index $\beta + \rho + 1 > 0$

(d) If $\beta < -\rho - 1$, then $R_\beta^*(r, \theta)$ is uniformly regularly varying with index $\beta + \rho + 1 < 0$. If $\beta = -\rho - 1$ and if $R_\beta^*(r, \theta)$ exists for all $\theta \in S$ for some $r > 0$, then R_β^* is uniformly slowly varying

Proof. First we prove part (a). Choose $\varepsilon > 0$ and $\lambda_0 > 1$ arbitrary, and use Theorem 5.3.6 to obtain $A_0 > 0$ such that $|R(\lambda r, \theta)/R(r, \theta) - \lambda^\rho| < \varepsilon$ for all $r \geq A_0$, all $\lambda \in [1, \lambda_0]$ and all $\theta \in S$. Then

$$\lambda^\rho - \varepsilon < \frac{R(\lambda r, \theta)}{R(r, \theta)} < \lambda^\rho + \varepsilon \tag{5.29}$$

for all $r \geq A \geq A_0$, all $\lambda \in [1, \lambda_0]$ and all $\theta \in S$. Given $s > A$ write $s = \lambda \lambda_0^n A$ where $n \geq 0$ is an integer and $1 \leq \lambda < \lambda_0$. Writing

$$R(s, \theta) = \frac{R(\lambda \lambda_0^n A, \theta)}{R(\lambda_0^n A, \theta)} \cdot \frac{R(\lambda_0^n A, \theta)}{R(\lambda_0^{n-1} A, \theta)} \cdots \frac{R(\lambda_0 A, \theta)}{R(A, \theta)} \cdot R(A, \theta)$$

it follows from (5.29) that

$$(\lambda_0^\rho - \varepsilon)^{n+1} R(A, \theta) < R(s, \theta) < (\lambda_0^\rho + \varepsilon)^{n+1} R(A, \theta) \tag{5.30}$$

which shows that $R(s,\theta)$ is bounded on $A \le s \le r$ for any $r > A$, for each fixed $\theta \in S$. Then $s^\beta R(s,\theta)$ is bounded and measurable on this interval, which proves part (a). Since $\varepsilon > 0$ is arbitrary, the upper bound in (5.30) also shows that if $\rho < 0$ then $R(r,\theta) \to 0$ as $r \to \infty$ for all $\theta \in S$. The lower bound in (5.30) shows that if $\rho > 0$ then $R(r,\theta) \to \infty$ as $r \to \infty$ for all $\theta \in S$.

Now we follow the proof on p. 280 of Feller [40]. Given $\lambda > 0$ and $\varepsilon > 0$, the definition (5.19) shows that we may choose $r_0 > 0$ such that

$$\left| \frac{R(\lambda r,\theta)}{R(r,\theta)} - \lambda^\rho \right| < \varepsilon$$

for all $\theta \in S$ and all $r \ge r_0$. Then we have

$$(\lambda^\rho - \varepsilon) R(r,\theta) < R(\lambda r,\theta) < (\lambda^\rho + \varepsilon) R(r,\theta) \tag{5.31}$$

for all $\theta \in S$ and all $r \ge r_0$. Suppose that $R^*_\beta(r,\theta)$ exists for some $\theta \in S$ and some $r > 0$. Since

$$R^*_\beta(\lambda r,\theta) = \int_{\lambda r}^\infty s^\beta R(s,\theta)ds = \lambda^{\beta+1} \int_r^\infty s^\beta R(\lambda s,\theta)ds$$

it follows from (5.31) that

$$(\lambda^\rho - \varepsilon)\lambda^{\beta+1} R^*_\beta(r,\theta) < R^*_\beta(\lambda r,\theta) < (\lambda^\rho + \varepsilon)\lambda^{\beta+1} R^*_\beta(r,\theta) \tag{5.32}$$

for all $r \ge r_0$. Since $\varepsilon > 0$ is arbitrary we conclude that

$$\lim_{r\to\infty} \frac{R^*_\beta(\lambda r,\theta)}{R^*_\beta(r,\theta)} = \lambda^{\rho+\beta+1}$$

so that $R^*_\beta(r,\theta)$ varies regularly with index $\rho + \beta + 1$. Since $r \mapsto R^*_\beta(r,\theta)$ is decreasing we cannot have $\rho + \beta + 1 > 0$. Reversing the implication, if $\rho + \beta + 1 > 0$, then $R^*_\beta(r,\theta)$ diverges for all $\theta \in S$.

On the other hand, suppose that $R_\beta(r,\theta)$ exists and that $R_\beta(r,\theta) \to \infty$ as $r \to \infty$ for some $\theta \in S$. Write

$$\begin{aligned} R_\beta(\lambda r,\theta) &= \int_A^{\lambda A} s^\beta R(s,\theta)ds + \int_{\lambda A}^{\lambda r} s^\beta R(s,\theta)ds \\ &= R_\beta(\lambda A,\theta) + \lambda^{\beta+1} \int_A^r s^\beta R(\lambda s,\theta)ds \end{aligned}$$

and apply (5.31) to see that

$$\frac{R_\beta(\lambda A,\theta)}{R_\beta(r,\theta)} + (\lambda^\rho - \varepsilon)\lambda^{\beta+1} < \frac{R_\beta(\lambda r,\theta)}{R_\beta(r,\theta)} < \frac{R_\beta(\lambda A,\theta)}{R_\beta(r,\theta)} + (\lambda^\rho + \varepsilon)\lambda^{\beta+1} \tag{5.33}$$

for all $r \ge r_0$. Since $R_\beta(r,\theta) \to \infty$ and $\varepsilon > 0$ is arbitrary it follows that $R_\beta(r,\theta)$ varies regularly with index $\rho + \beta + 1$, and in view of (5.30) we cannot have $\rho + \beta + 1 < 0$. Reversing the implication, if $\rho + \beta + 1 < 0$ and $R_\beta(r, \theta)$ exists [and part (a) shows that it always exists for $A \geq A_0$], then $R^*_\beta(r, \theta)$ exists for all $\theta \in S$. Thus we have proven part (b).

Next we prove part (c). Suppose that $\beta > -\rho - 1$ so that $\rho + \beta + 1 > 0$. If $R_\beta(r, \theta)$ exists for all $\theta \in S$ and all $r \geq A$, then we know from part (b) that $R_\beta(r, \theta) \to \infty$ as $r \to \infty$ for any $\theta \in S$. Suppose that $\theta_r \to \theta$ in S, so that (5.33) holds with θ_r in place of θ. In order to show that

$$\frac{R_\beta(\lambda r, \theta_r)}{R_\beta(r, \theta_r)} \to \lambda^{\rho+\beta+1}$$

as $r \to \infty$ it suffices to show that $R_\beta(\lambda A, \theta_r)/R_\beta(r, \theta_r) \to 0$. Writing $r = \lambda \lambda_0^n A$ as before and using the fact that $R_\beta(r, \theta)$ is a monotone function of $r > 0$ we have for all $r > \lambda_0 A$ that

$$\begin{aligned}
R_\beta(r, \theta) &\geq \int_{\lambda_0^{n-1}A}^{\lambda_0^n A} s^\beta R(s, \theta)ds \\
&= (\lambda_0^{n-1})^{\beta+1} \int_A^{\lambda_0 A} s^\beta R(\lambda_0^{n-1}s, \theta)ds \\
&> (\lambda_0^{n-1})^{\beta+1} \int_A^{\lambda_0 A} s^\beta (\lambda_0^\rho - \varepsilon)^{n-1} R(s, \theta)ds \\
&= [(\lambda_0^{\beta+1})(\lambda_0^\rho - \varepsilon)]^{n-1} R_\beta(\lambda_0 A, \theta)
\end{aligned}$$

by (5.30) and since $\varepsilon > 0$ is arbitrary this shows that $R_\beta(\lambda A, \theta)/R_\beta(r, \theta) \to 0$ uniformly with respect to $\theta \in S$. Thus, in view of (5.33) $R_\beta(r, \theta)$ is uniformly regularly varying with index $\beta + \rho + 1 > 0$.

Now we prove part (d). If $\rho + \beta + 1 < 0$, then part (b) shows that $R^*_\beta(r, \theta)$ exists for all $r > A_0$ and all $\theta \in S$. If $\theta_r \to \theta$ in S, then (5.32) holds with θ_r in place of θ. Since $\varepsilon > 0$ is arbitrary this implies that $R^*_\beta(r, \theta)$ is uniformly regularly varying. If $\rho + \beta + 1 = 0$, then the same argument applies, assuming that $R^*_\beta(r, \theta)$ exists. Since $R^*_\beta(r, \theta)$ is monotone it suffices that this function exists for all $\theta \in S$ for some $r > 0$. □

Theorem 5.3.9. *(a) If R is uniformly regularly varying with index ρ and $R^*_\beta(r, \theta)$ exists for all $\theta \in S$ for some $r > 0$, then*

$$\frac{r^{\beta+1} R(r, \theta)}{R^*_\beta(r, \theta)} \to \gamma \tag{5.34}$$

as $r \to \infty$ uniformly in $\theta \in S$ where $\gamma = -(\rho + \beta + 1) \geq 0$.

(b) If (5.34) holds uniformly in θ for some $\gamma \geq 0$, then R^*_β is uniformly regularly varying with index $-\gamma$, and if additionally $\gamma > 0$ then R is uniformly regularly varying with index $-\gamma - \beta - 1$.

(c) If R is uniformly regularly varying with index $\rho > -\beta-1$ and if $R_\beta(r,\theta)$ exists for all $\theta \in S$ and all $r \geq A$, then

$$\frac{r^{\beta+1}R(r,\theta)}{R_\beta(r,\theta)} \to \gamma \tag{5.35}$$

as $r \to \infty$ uniformly in $\theta \in S$, where $\gamma = \rho + \beta + 1 > 0$.

(d) If (5.35) holds uniformly in θ for some $\gamma \geq 0$, then R_β is uniformly regularly varying with index γ, and if additionally $\gamma > 0$ then R is uniformly regularly varying with index $\gamma - \beta - 1$.

Proof. First we prove (a). Since

$$\frac{d}{dr}R_\beta^*(r,\theta) = \frac{d}{dr}\int_r^\infty s^\beta R(s,\theta)ds = -r^\beta R(r,\theta)$$

if we denote the left-hand side of (5.34) by

$$\eta(r,\theta) = \frac{r^{\beta+1}R(r,\theta)}{R_\beta^*(r,\theta)}$$

then

$$\begin{aligned}
\log\frac{R_\beta^*(r,\theta)}{R_\beta^*(\lambda r,\theta)} &= \int_r^{\lambda r}\frac{s^\beta R(s,\theta)}{R_\beta^*(s,\theta)}ds \\
&= \int_r^{\lambda r}\eta(s,\theta)\frac{ds}{s} \\
&= \int_1^{\lambda}\eta(rs,\theta)\frac{ds}{s} \qquad (5.36)\\
&= \eta(r,\theta)\int_1^{\lambda}\frac{\eta(rs,\theta)}{\eta(r,\theta)}\frac{ds}{s}
\end{aligned}$$

for any $\lambda > 0$. If R is uniformly regularly varying with index ρ and R_β^* exists then part (b) of Lemma 5.3.8 implies that $-\gamma = \rho + \beta + 1 \leq 0$, and part (d) of that Lemma implies that R_β^* is uniformly regularly varying with index $-\gamma$. Then the left-hand side of (5.36) tends to $\gamma\log\lambda$ uniformly in θ. Since R, R_β^* and the function $R_1(r,\theta) = r^{\beta+1}$ are uniformly regularly varying with index $\rho, \beta+\rho+1, \beta+1$ respectively, the function η is uniformly slowly varying. Then Theorem 5.3.6 implies that $\eta(rs,\theta)/\eta(r,\theta) \to 1$ as $r \to \infty$ uniformly with respect to $(s,\theta) \in [1,\lambda] \times S$, so the integral in the last expression in (5.36) tends to $\log\lambda$ uniformly in θ. Thus $\eta(r,\theta) \to \gamma$ uniformly in θ as $r \to \infty$.

The proof of part (c) is similar. Letting $\eta(r,\theta)$ denote the left-hand side of (5.35), we have as before that

$$\log\frac{R_\beta(\lambda r,\theta)}{R_\beta(r,\theta)} = \eta(r,\theta)\int_1^{\lambda}\frac{\eta(rs,\theta)}{\eta(r,\theta)}\frac{ds}{s} \tag{5.37}$$

for any $\lambda > 0$. If R is uniformly regularly varying with index $\rho > -\beta - 1$ and $R_\beta(r, \theta)$ exists then part (c) of Lemma 5.3.8 implies that R_β is uniformly regularly varying with index $\gamma = \rho + \beta + 1$ and so η is uniformly slowly varying. Then the left-hand side of (5.37) tends to $\gamma \log \lambda$ uniformly in θ, and the integral on the right-hand side tends to $\log \lambda$ uniformly in θ as in the proof of part (a), so that $\eta(r, \theta) \to \gamma$ uniformly in θ as $r \to \infty$.

To prove (b), suppose that (5.34) holds uniformly in θ as $r \to \infty$ for some $\gamma > 0$. Then the left-hand side of (5.34) η is uniformly slowly varying, and it follows from (5.36) that

$$\log \frac{R_\beta^*(r, \theta)}{R_\beta^*(\lambda r, \theta)} \to \gamma \log \lambda$$

uniformly in θ as $r \to \infty$. Then R_β^* is uniformly regularly varying with index $-\gamma$, and it follows from (5.34) that

$$\begin{aligned} \frac{R(\lambda r, \theta)}{R(r, \theta)} &= \frac{1}{\lambda^{\beta+1}} \cdot \frac{(\lambda r)^{\beta+1} R(\lambda r, \theta)}{R_\beta^*(\lambda r, \theta)} \cdot \frac{R_\beta^*(\lambda r, \theta)}{R_\beta^*(r, \theta)} \cdot \frac{R_\beta^*(r, \theta)}{r^{\beta+1} R(r, \theta)} \\ &\to \frac{1}{\lambda^{\beta+1}} \cdot \gamma \cdot \lambda^{-\gamma} \cdot \frac{1}{\gamma} \end{aligned}$$

uniformly in θ as $r \to \infty$, so that R is uniformly regularly varying with index $-\gamma - \beta - 1$. Lastly, if (5.34) holds uniformly in θ as $r \to \infty$ with $\gamma = 0$, apply (5.36) in the form

$$\log \frac{R_\beta^*(r, \theta)}{R_\beta^*(\lambda r, \theta)} = \int_r^{\lambda r} \eta(s, \theta) \frac{ds}{s}$$

and note that, since $\eta(r, \theta) \to 0$ uniformly in θ as $r \to \infty$, the right-hand side tends to zero uniformly in θ as $r \to \infty$, so that R_β^* is uniformly slowly varying.

To prove (d), suppose that (5.35) holds uniformly in θ as $r \to \infty$ for some $\gamma > 0$. Then η, the left-hand side of (5.35), is uniformly slowly varying, and it follows from (5.37) that

$$\log \frac{R_\beta(\lambda r, \theta)}{R_\beta(r, \theta)} \to \gamma \log \lambda$$

uniformly in θ as $r \to \infty$. Then R_β is uniformly regularly varying with index γ, and it follows from (5.35) that

$$\begin{aligned} \frac{R(\lambda r, \theta)}{R(r, \theta)} &= \frac{1}{\lambda^{\beta+1}} \cdot \frac{(\lambda r)^{\beta+1} R(\lambda r, \theta)}{R_\beta(\lambda r, \theta)} \cdot \frac{R_\beta(\lambda r, \theta)}{R_\beta(r, \theta)} \cdot \frac{R_\beta(r, \theta)}{r^{\beta+1} R(r, \theta)} \\ &\to \frac{1}{\lambda^{\beta+1}} \cdot \gamma \cdot \lambda^{\gamma} \cdot \frac{1}{\gamma} \end{aligned}$$

uniformly in θ as $r \to \infty$, so that R is uniformly regularly varying with index $\gamma - \beta - 1$. Finally, if $\eta(r, \theta) \to 0$ uniformly in θ as $r \to \infty$, then (5.37) yields

$$\log \frac{R_\beta(\lambda r, \theta)}{R_\beta(r, \theta)} = \int_r^{\lambda r} \eta(s, \theta) \frac{ds}{s} \to 0$$

uniformly in θ as $r \to \infty$, so that R_β is uniformly slowly varying. □

Definition 5.3.10. Suppose that for every $\theta \in S$ the function $r \mapsto R(r, \theta)$ is monotone decreasing, continuous from the right, and not constant. Define

$$U_\zeta(r, \theta) = \int_0^r s^\zeta R(-ds, \theta) \tag{5.38}$$

and

$$V_\eta(r, \theta) = \int_r^\infty s^\eta R(-ds, \theta) \tag{5.39}$$

whenever these integrals exist.

Theorem 5.3.11. *Suppose that $\zeta > \eta$ and that the integrals $U_\zeta(r, \theta), V_\eta(r, \theta)$ exist for all $r > 0$ and all $\theta \in S$. Then*

(a) If U_ζ is uniformly regularly varying with index ρ, then $\rho = \zeta - \alpha$ for some $\alpha \in [\eta, \zeta]$ and

$$\frac{r^{\zeta - \eta} V_\eta(r, \theta)}{U_\zeta(r, \theta)} \to \frac{\zeta - \alpha}{\alpha - \eta} \tag{5.40}$$

uniformly in θ as $r \to \infty$, where we interpret the limit in (5.40) as infinity when $\alpha = \eta$

(b) If V_η is uniformly regularly varying with index $\rho > \eta - \zeta$, then $\rho = \eta - \alpha$ for some $\alpha \in [\eta, \zeta)$, and (5.40) holds uniformly in θ

(c) If (5.40) holds uniformly in θ for some $\alpha \in (\eta, \zeta]$, then U_ζ is uniformly regularly varying with index $\zeta - \alpha$. If (5.40) holds uniformly in θ for some $\alpha \in [\eta, \zeta)$, then V_η is uniformly regularly varying with index $\eta - \alpha$

Proof. First we prove part (a). Integrate by parts to obtain

$$\begin{aligned} V_\eta(r, \theta) - V_\eta(t, \theta) &= \int_r^t s^\eta R(-ds, \theta) \\ &= \int_r^t s^{\eta - \zeta} U_\zeta(ds, \theta) \\ &= t^{\eta - \zeta} U_\zeta(t, \theta) - r^{\eta - \zeta} U_\zeta(r, \theta) \\ &\quad - (\eta - \zeta) \int_r^t s^{\eta - \zeta - 1} U_\zeta(s, \theta) ds \end{aligned}$$

and then rearrange to get

$$\begin{aligned} V_\eta(r,\theta) + r^{\eta-\zeta}U_\zeta(r,\theta) &= V_\eta(t,\theta) + t^{\eta-\zeta}U_\zeta(t,\theta) \\ &\quad +(\zeta-\eta)\int_r^t s^{\eta-\zeta-1}U_\zeta(s,\theta)ds. \end{aligned}$$

Since each term is nonnegative, the integral is bounded above by the left-hand side, and so the integral term converges to a finite nonnegative limit as $t \to \infty$. Since $V_\eta(t,\theta) \to 0$ as $t \to \infty$ it follows that $t^{\eta-\zeta}U_\zeta(t,\theta) \to L \geq 0$ as $t \to \infty$. If $L > 0$, then for some $t_0 > 0$ we have $t^{\eta-\zeta}U_\zeta(t,\theta) > L/2$ for all $t > t_0$, so that

$$\int_r^\infty s^{\eta-\zeta-1}U_\zeta(s,\theta)ds \geq \int_{t_0}^\infty \frac{L}{2}\frac{ds}{s} = \infty$$

which is a contradiction. Hence $L = 0$ and we have shown that

$$V_\eta(r,\theta) = -r^{\eta-\zeta}U_\zeta(r,\theta) + (\zeta-\eta)\int_r^\infty s^{\eta-\zeta-1}U_\zeta(s,\theta)ds. \tag{5.41}$$

Multiply by $r^{\zeta-\eta}/U_\zeta(r,\theta)$ to obtain

$$\frac{r^{\zeta-\eta}V_\eta(r,\theta)}{U_\zeta(r,\theta)} = -1 + \frac{(\zeta-\eta)}{r^{\eta-\zeta}U_\zeta(r,\theta)}\int_r^\infty s^{\eta-\zeta-1}U_\zeta(s,\theta)ds. \tag{5.42}$$

If U_ζ is uniformly regularly varying with index ρ then, since the integral in (5.42) converges, we can apply part (a) of Theorem 5.3.9 with $\beta = \eta - \zeta - 1$ to see that

$$\frac{r^{\eta-\zeta}U_\zeta(r,\theta)}{\int_r^\infty s^{\eta-\zeta-1}U_\zeta(s,\theta)ds} \to \gamma \geq 0 \tag{5.43}$$

uniformly in θ as $r \to \infty$, where $\gamma = -(\rho + \beta + 1) = \zeta - \eta - \rho$. Since U_ζ is monotone nondecreasing we must have $\rho \geq 0$, so we can write $\rho = \zeta - \alpha$ for some $\alpha \leq \zeta$. Then we also have $\gamma = \alpha - \eta \geq 0$, so $\eta \leq \alpha \leq \zeta$. Now it follows from (5.42) and (5.43) that

$$\frac{r^{\zeta-\eta}V_\eta(r,\theta)}{U_\zeta(r,\theta)} \to -1 + \frac{\zeta-\eta}{\alpha-\eta} = \frac{\zeta-\alpha}{\alpha-\eta}$$

uniformly in θ as $r \to \infty$, which finishes the proof of part (a). If $\gamma = \alpha - \eta = 0$, then we get uniform convergence to infinity in (5.40), since by (5.42) and (5.43) the reciprocal converges uniformly to zero.

Now we prove part (b). Integrate by parts to obtain

$$\begin{aligned} U_\zeta(r,\theta) - U_\zeta(t,\theta) &= \int_t^r s^\zeta R(-ds,\theta) \\ &= -\int_t^r s^{\zeta-\eta} V_\eta(ds,\theta) \\ &= t^{\zeta-\eta}V_\eta(t,\theta) - r^{\zeta-\eta}V_\eta(r,\theta) \\ &\quad +(\zeta-\eta)\int_t^r s^{\zeta-\eta-1}V_\eta(s,\theta)ds \end{aligned}$$

and then rearrange to get

$$\begin{aligned} U_\zeta(r,\theta) + r^{\zeta-\eta}V_\eta(r,\theta) &= U_\zeta(t,\theta) + t^{\zeta-\eta}V_\eta(t,\theta) \\ &\quad +(\zeta-\eta)\int_t^r s^{\zeta-\eta-1}V_\eta(s,\theta)ds. \end{aligned}$$

Since each term is nonnegative, the integral is bounded above by the left-hand side, and so the integral term converges to a finite nonnegative limit as $t \to 0$. Since $U_\zeta(t,\theta) \to 0$ as $t \to 0$ it follows that $t^{\zeta-\eta}V_\eta(t,\theta) \to L \geq 0$ as $t \to 0$, and since the integral converges for $t = 0$ we must have $L = 0$. Taking $t = 0$, we have

$$U_\zeta(r,\theta) = -r^{\zeta-\eta}V_\eta(r,\theta) + (\zeta-\eta)\int_0^r s^{\zeta-\eta-1}V_\eta(s,\theta)ds.$$

Divide by $r^{\zeta-\eta}V_\eta(r,\theta)$ to obtain

$$\frac{U_\zeta(r,\theta)}{r^{\zeta-\eta}V_\eta(r,\theta)} = -1 + \frac{(\zeta-\eta)}{r^{\zeta-\eta}V_\eta(r,\theta)}\int_0^r s^{\zeta-\eta-1}V_\eta(s,\theta)ds. \tag{5.44}$$

Let $\beta = \zeta - \eta - 1$. If V_η is uniformly regularly varying with index $\rho > -\beta - 1 = \eta - \zeta$, then part (c) of Theorem 5.3.9 implies that

$$\frac{U_\zeta(r,\theta)}{r^{\zeta-\eta}V_\eta(r,\theta)} \to -1 + \frac{\zeta-\eta}{\gamma} \tag{5.45}$$

uniformly in θ as $r \to \infty$, where $\gamma = \rho + \beta + 1 = \rho + \zeta - \eta$. Since V_η is monotone nonincreasing we must have $\rho \leq 0$, and we can write $\rho = \eta - \alpha$ for some $\alpha \geq \eta$. Then $\gamma = \zeta - \alpha > 0$ and the limit in (5.45) reduces to $(\alpha - \eta)/(\zeta - \alpha)$ for some $\eta \leq \alpha < \zeta$, which finishes the proof of part (b).

Now we prove part (c). Suppose that (5.40) holds uniformly in θ as $r \to \infty$ for some $\alpha \in (\eta, \zeta]$. Rewrite (5.42) in the form

$$\frac{\int_r^\infty s^{\eta-\zeta-1}U_\zeta(s,\theta)ds}{r^{\eta-\zeta}U_\zeta(r,\theta)} = \frac{1}{\zeta-\eta}\left(1 + \frac{r^{\zeta-\eta}V_\eta(r,\theta)}{U_\zeta(r,\theta)}\right) \tag{5.46}$$

and apply (5.40) to see that

$$\frac{\int_r^\infty s^{\eta-\zeta-1}U_\zeta(s,\theta)ds}{r^{\eta-\zeta}U_\zeta(r,\theta)} \to \frac{1}{\zeta-\eta}\left(1+\frac{\zeta-\alpha}{\alpha-\eta}\right) = \frac{1}{\alpha-\eta}$$

uniformly in θ as $r \to \infty$. Then part (b) of Theorem 5.3.9 with $\gamma = \alpha - \eta > 0$ and $\beta = \eta - \zeta - 1$ implies that U_ζ is uniformly regularly varying with index $-\gamma - \beta - 1 = \zeta - \alpha \geq 0$. If (5.40) holds uniformly in θ as $r \to \infty$ for some $\alpha \in [\eta, \zeta)$, we rewrite (5.44) in the form

$$\frac{\int_0^r s^{\zeta-\eta-1}V_\eta(s,\theta)ds}{r^{\zeta-\eta}V_\eta(r,\theta)} = \frac{1}{\zeta-\eta}\left(1+\frac{U_\zeta(r,\theta)}{r^{\zeta-\eta}V_\eta(r,\theta)}\right) \tag{5.47}$$

and apply (5.40) to see that

$$\frac{\int_0^r s^{\zeta-\eta-1}V_\eta(s,\theta)ds}{r^{\zeta-\eta}V_\eta(r,\theta)} \to \frac{1}{\zeta-\eta}\left(1+\frac{\alpha-\eta}{\zeta-\alpha}\right) = \frac{1}{\zeta-\alpha}$$

uniformly in θ as $r \to \infty$. Then part (d) of Theorem 5.3.9 with $\gamma = \zeta - \alpha$ and $\beta = \zeta - \eta - 1$ implies that V_η is uniformly regularly varying with index $\gamma - \beta - 1 = \eta - \alpha \leq 0$, and this finishes the proof. □

If $F : \Gamma \to \mathbb{R}^+$ varies regularly at infinity with exponent $B = aI$ for some $a \neq 0$, then by Proposition 5.1.6 and Theorem 5.3.4 $R(r,x) = F(rx)$ varies regularly with index $-1/a$. This implies that for any $\varepsilon > 0$, for all $r_0 > 0$ sufficiently large we have

$$r^{-1/a-\varepsilon} < F(rx) < r^{-1/a+\varepsilon}$$

for all $r \geq r_0$. In other words, $F(rx)$ grows about as fast as $r^{-1/a}$. In the general case, where B is not necessary a multiple of the identity we can obtain bounds on the growth rate of a regularly varying function by using a generalization of regular variation called R–O variation.

Definition 5.3.12. Suppose that S is a compact topological space, and that $R : \mathbb{R}^+ \times S \to \mathbb{R}^+$ is Borel measurable. We say that $R(r,\theta)$ is uniformly R–O varying if there exist $r_0 > 0$, $\lambda_0 > 1$, $0 < m < 1$ and $M > 1$ such that

$$m \leq \frac{R(\lambda r,\theta)}{R(r,\theta)} \leq M \tag{5.48}$$

for all $r \geq r_0$, $1 \leq \lambda \leq \lambda_0$ and $\theta \in S$.

If R is uniformly regularly varying with index ρ, then $R(\lambda r,\theta) \sim \lambda^\rho R(r,\theta)$ as $r \to \infty$, so $R(r,\theta)$ grows like r^ρ as $r \to \infty$. If R is uniformly R–O varying we have instead that $R(\lambda r,\theta) = O(R(r,\theta))$ as $r \to \infty$. Our next result shows that an R–O varying function changes slower than some power law.

Lemma 5.3.13. *If $R : \mathbb{R}^+ \times S \to \mathbb{R}^+$ is uniformly R-O varying and r_0, λ_0, m, M are as in Definition 5.3.12, then*

$$m\lambda^h \leq \frac{R(\lambda r, \theta)}{R(r, \theta)} \leq M\lambda^H \tag{5.49}$$

for all $\lambda \geq 1$, $r \geq r_0$ and $\theta \in S$, where $h = \log m/\log \lambda_0$ and $H = \log M/\log \lambda_0$.

Proof. For $\lambda > 1$ fixed, choose a nonnegative integer p such that $\lambda_0^p \leq \lambda < \lambda_0^{p+1}$. Then for all $r \geq r_0$ and all $\theta \in S$ we get from (5.48)

$$m \leq \frac{R(\lambda_0^k r, \theta)}{R(\lambda_0^{k-1} r, \theta)} \leq M \quad \text{for } k = 1, \dots, p$$

and

$$m \leq \frac{R(\lambda r, \theta)}{R(\lambda_0^p r, \theta)} \leq M.$$

Hence, upon a multiplication of these $p + 1$ inequalities we get

$$m^{p+1} \leq \frac{R(\lambda r, \theta)}{R(r, \theta)} \leq M^{p+1}$$

for all $r \geq r_0$ and all $\theta \in S$. But by definition of p we know $p \leq \log \lambda/\log \lambda_0$ and since $m < 1$ we have

$$\begin{aligned} m^p \geq m^{\log \lambda/\log \lambda_0} &= e^{(\log \lambda/\log \lambda_0)\log m} \\ &= e^{(\log m/\log \lambda_0)\log \lambda} = \lambda^{\log m/\log \lambda_0} = \lambda^h. \end{aligned}$$

Similarly, since $M > 1$ we have $M^p \leq M^{\log \lambda/\log \lambda_0} = \lambda^H$, and then (5.49) follows. □

Theorem 5.3.14 is a sweeping generalization of Theorem 5.3.4. It provides limits on the growth rate of a regularly varying function $F : \Gamma \to \mathbb{R}^+$.

Theorem 5.3.14. *Suppose that F varies regularly at zero or infinity and (5.3) holds for all $\lambda > 0$ and all $x \in \Gamma = \mathbb{R}^d \setminus \{0\}$. Define*

$$\begin{aligned} q &= \min\{0, \beta/\operatorname{Re}(\alpha_1), \dots, \beta/\operatorname{Re}(\alpha_j)\} \\ Q &= \max\{0, \beta/\operatorname{Re}(\alpha_1), \dots, \beta/\operatorname{Re}(\alpha_j)\} \end{aligned} \tag{5.50}$$

where $\alpha_1 \cdots \alpha_j$ are the eigenvalues of E.

(a) *If F varies regularly at infinity, then $R(r, x) = F(rx)$ is uniformly R–O varying on compact subsets of $x \in \Gamma$. In particular, for any $S \subset \Gamma$*

compact, and any $\delta > 0$ sufficiently small there exist $r_0 > 0$, $0 < m < 1$ and $M > 1$ such that

$$m\lambda^{q-\delta} \leq \frac{R(\lambda r, x)}{R(r, x)} \leq M\lambda^{Q+\delta} \tag{5.51}$$

for all $r \geq r_0$, $\lambda \geq 1$, and $x \in S$.

(b) *If F varies regularly at zero, then $R(r, x) = F(x/r)$ is uniformly R–O varying on compact subsets of $x \in \Gamma$. In particular, for any $S \subset \Gamma$ compact, and any $\delta > 0$ sufficiently small there exist $r_0 > 0$, $0 < m < 1$ and $M > 1$ such that*

$$m\lambda^{-Q-\delta} \leq \frac{R(\lambda r, x)}{R(r, x)} \leq M\lambda^{-q+\delta} \tag{5.52}$$

for all $r \geq r_0$, $\lambda \geq 1$, and $x \in S$.

In order to simplify the proof, we begin with two useful lemmas.

Lemma 5.3.15. *Suppose that $\varphi : \Gamma \to \mathbb{R}^+$ is bounded away from both zero and infinity on compact sets and that*

$$c \cdot \varphi(y) = \varphi(c^A y) \tag{5.53}$$

for some $c > 1$ and all $y \in \Gamma = \mathbb{R}^d \setminus \{0\}$, where every eigenvalue of A has real part greater than or equal to $a_1 > 0$.

(a) *Given $S \subset \Gamma$ compact and $\varepsilon > 0$, for all $\lambda_0 > 1$ sufficiently large we have*

$$\lambda_0^{-\varepsilon} \leq \frac{\varphi(\lambda y)}{\varphi(y)} \leq \lambda_0^{\varepsilon+1/a_1} \tag{5.54}$$

for all $1 \leq \lambda \leq \lambda_0$ and $y \in S$.

(b) *Given $S \subset \Gamma$ compact and $\varepsilon > 0$, for all $0 < \mu_0 < 1$ sufficiently small we have*

$$\mu_0^{\varepsilon+1/a_1} \leq \frac{\varphi(\mu y)}{\varphi(y)} \leq \mu_0^{-\varepsilon} \tag{5.55}$$

for all $\mu_0 \leq \mu \leq 1$ and $y \in S$.

Proof. First we prove part (a). It suffices to prove the lemma in the case where $S = \{x \in \mathbb{R}^d : D_1 \leq \|x\| \leq D_2\}$. For any $x \in \Gamma$, we know that $\|t^A x\|$ tends to infinity as $t \to \infty$ and to zero as $t \to 0$. Proposition 2.1.3 (a) yields $\|c^A x\| \leq \|c^A\| \cdot \|x\|$, and it follows that if $D_2 \geq \|c^A\| D_1$, then we can always write $x = c^{kA} y$ for some integer k and some $y \in S$. In fact, for $x \in \Gamma$ let

$k = k(x) = \sup\{n : \|c^{-nA}x\| \geq D_1\}$ and let $y = y(x) = c^{-k(x)A}x$. Theorem 2.2.4 implies that if $0 < \beta < a_1$ then for any $t_0 > 0$ sufficiently large we have

$$\|t^A y\| > t^\beta > t_0^\beta \tag{5.56}$$

for all $t > t_0$ and all $y \in S$. Let $\lambda_0 = t_0^\beta / D_2$. Enlarge t_0 if necessary to ensure $\lambda_0 > 1$. Given $y \in S$ and $1 \leq \lambda \leq \lambda_0$ choose $k \in \mathbb{Z}$ and $y' \in S$ such that $\lambda y = c^{kA}y'$. Then $\|c^{kA}y'\| = \|\lambda y\| \leq \lambda_0 D_2 = t_0^\beta$, and it follows from (5.56) that $c^k \leq t_0$. For some $C_1 > 0$ and $C_2 < \infty$, we have $C_1 \leq \varphi(y) \leq C_2$ for all $y \in S$. Using (5.53) k times, we see that

$$\frac{\varphi(\lambda y)}{\varphi(y)} = \frac{\varphi(c^{kA}y')}{\varphi(y)} = \frac{c^k \varphi(y')}{\varphi(y)} \leq \frac{t_0 C_2}{C_1} = \frac{C_2}{C_1}(D_2\lambda_0)^{1/\beta} \tag{5.57}$$

whenever $y \in S$ and $1 \leq \lambda \leq \lambda_0$. If we choose β sufficiently close to a_1, then $1/\beta < \varepsilon + 1/a_1$, and so the right-hand side in (5.54) is larger than that in (5.57) for all λ_0 sufficiently large.

Since every eigenvalue of A has positive real part, Theorem 2.2.4 along with Proposition 2.2.2 (d) imply that $t^A y \to 0$ uniformly in $y \in S$ as $t \to 0$. Then we can choose $t_1 > 0$ such that $\|t^A y\| < D_1$ for all $0 < t < t_1$ and all $y \in S$. Given $y \in S$ and $\lambda > 1$ choose $y' \in S$ and $k \in \mathbb{Z}$ such that $c^{kA}y' = \lambda y$. Then $\|c^{kA}y'\| = \|\lambda y\| \geq D_1$, which implies that $c^k \geq t_1$, and so

$$\frac{\varphi(\lambda y)}{\varphi(y)} = \frac{c^k \varphi(y')}{\varphi(y)} \geq \frac{t_1 C_1}{C_2} \tag{5.58}$$

whenever $y \in S$ and $\lambda \geq 1$. Then the right-hand side in (5.54) is smaller than that in (5.58) for all λ_0 sufficiently large.

Next we prove part (b). Theorem 2.2.4 along with Proposition 2.2.2 (d) imply that if $0 < \beta < a_1$ then for any $0 < t_0 < 1$ sufficiently small we have

$$\|t^A y\| < t^\beta < t_0^\beta \tag{5.59}$$

for all $t < t_0$ and all $y \in S$. Let $\mu_0 = t_0^\beta / D_1$. Given $y \in S$ and $\mu_0 \leq \mu \leq 1$ choose $k \in \mathbb{Z}$ and $y' \in S$ such that $\mu y = c^{kA}y'$. Then $\|c^{kA}y'\| = \|\mu y\| \geq \mu_0 D_1 = t_0^\beta$, and then it follows from (5.59) that $c^k \geq t_0$. Then

$$\frac{\varphi(\mu y)}{\varphi(y)} = \frac{\varphi(c^{kA}y')}{\varphi(y)} = \frac{c^k \varphi(y')}{\varphi(y)} \geq \frac{t_0 C_1}{C_2} = \frac{C_1}{C_2}(D_1\mu_0)^{1/\beta} \tag{5.60}$$

whenever $y \in S$ and $\mu_0 \leq \mu \leq 1$. If we choose β sufficiently close to a_1, then $1/\beta < \varepsilon + 1/a_1$, and so the right-hand side in (5.55) is smaller than that in (5.60) for all μ_0 sufficiently small.

Since every eigenvalue of A has positive real part, Theorem 2.2.4 implies that $t^A y \to \infty$ uniformly in $y \in S$ as $t \to \infty$. Then we can choose $t_1 > 0$ such that $\|t^A y\| > D_2$ for all $t > t_1$ and all $y \in S$. Given $y \in S$ and $0 < \mu \leq 1$

choose $y' \in S$ and $k \in \mathbb{Z}$ such that $c^{kA}y' = \mu y$. Then $\|c^{kA}y'\| = \|\mu y\| \le D_2$, which implies that $c^k \le t_1$, and so

$$\frac{\varphi(\mu y)}{\varphi(y)} = \frac{c^k \varphi(y')}{\varphi(y)} \le \frac{t_1 C_2}{C_1} \tag{5.61}$$

whenever $y \in S$ and $\mu \le 1$. Then the right-hand side in (5.55) is larger than that in (5.61) for all μ_0 sufficiently small. □

Lemma 5.3.16. *Suppose that F varies regularly at zero or infinity and (5.3) holds, and define q, Q as in (5.50).*

(a) Given $S \subset \Gamma$ compact and $\varepsilon > 0$, for all $\lambda_0 > 1$ sufficiently large we have

$$\lambda_0^{q-\varepsilon} \le \frac{\varphi(\lambda y)}{\varphi(y)} \le \lambda_0^{Q+\varepsilon} \tag{5.62}$$

for all $1 \le \lambda \le \lambda_0$ and $y \in S$.

(b) Given $S \subset \Gamma$ compact and $\varepsilon > 0$, for all $0 < \mu_0 < 1$ sufficiently small we have

$$\mu_0^{Q+\varepsilon} \le \frac{\varphi(\mu y)}{\varphi(y)} \le \mu_0^{q-\varepsilon} \tag{5.63}$$

for all $\mu_0 \le \mu \le 1$ and $y \in S$.

Proof. First we prove part (a). Note that either $q = 0$ or $Q = 0$. If $q = 0$, then (5.3) implies (5.53), where the eigenvalues of $A = (1/\beta)E$ all have real part greater than or equal to $a_1 = 1/Q$. Then (5.62) follows from Lemma 5.3.15 (a). On the other hand, if $Q = 0$ then $t^{-1}\varphi(y) = \varphi(t^A y)$, where $A = (-1/\beta)E$ has eigenvalues whose real parts are all greater than or equal to $a_1 = -1/q$, and then (5.62) follows from applying Lemma 5.3.15 (a) to the function $1/\varphi$. The proof of part (b) is similar. □

Proof of Theorem 5.3.14. First we will prove part (a). Given $\varepsilon > 0$, since F varies regularly at infinity we can apply Lemma 5.3.16 (a) to see that for all λ_0 sufficiently large, (5.62) holds for all $x \in S = \{x \in \mathbb{R}^d : \|x\| = 1\}$ and all $1 \le \lambda \le \lambda_0$. Given $x_r \to x$ in Γ, apply Lemma 4.2.7 (a) to obtain $t = t(r) \to \infty$ as $r \to \infty$ such that $y_t = f(t)(rx_r)$ satisfies $\|y_t\| \to 1$ as $r \to \infty$. For any sequence $r_n \to \infty$ there exists a further subsequence $r_{n'} \to \infty$ along which $y_t \to y \in S$. If $1 \le \lambda \le \lambda_0$, then

$$\frac{R(\lambda r, x_r)}{R(r, x_r)} = \frac{F(f(t)^{-1}\lambda y_t)}{F(f(t)^{-1} y_t)} \to \frac{\varphi(\lambda y)}{\varphi(y)} \tag{5.64}$$

along this subsequence. Since $y \in S$ we know that the right-hand side in (5.64) satisfies (5.62) for any limit point $y \in S$. Since this is true along any subsequence it follows that

$$m = \frac{1}{2}\lambda_0^{q-\varepsilon} \le \frac{R(\lambda r, x)}{R(r, x)} \le 2\lambda_0^{Q+\varepsilon} = M$$

for all $r \ge r_0$, $1 \le \lambda \le \lambda_0$ and $x \in S$. Now apply Lemma 5.3.13 to see that (5.49) holds with

$$\begin{aligned} h &= \frac{\log m}{\log \lambda_0} = q - \varepsilon - \frac{\log 2}{\log \lambda_0} \\ H &= \frac{\log M}{\log \lambda_0} = Q + \varepsilon + \frac{\log 2}{\log \lambda_0} \end{aligned}$$

which implies (5.51) provided that $\varepsilon < \delta$ and λ_0 is chosen sufficiently large.

Now we prove part (b). Given $\varepsilon > 0$, since F varies regularly at zero we can apply Lemma 5.3.16 (b) to see that for all μ_0 sufficiently small, (5.63) holds for all $\mu_0 \le \mu \le 1$ and $x \in S$. Given $x_r \to x$ in Γ apply Lemma 4.2.7 (b) to obtain $t = t(r) \to \infty$ as $r \to \infty$ such that $y_t = f(t)(x_r/r)$ satisfies $\|y_t\| \to 1$ as $r \to \infty$. For any sequence $r_n \to \infty$, there exists a further subsequence $r_{n'} \to \infty$ along which $y_t \to y \in S$. If $1 \le \lambda \le \lambda_0 = \mu_0^{-1}$, then

$$\frac{R(\lambda r, x_r)}{R(r, x_r)} = \frac{F(f(t)^{-1}\lambda^{-1}y_t)}{F(f(t)^{-1}y_t)} \to \frac{\varphi(\lambda^{-1}y)}{\varphi(y)} \tag{5.65}$$

along a subsequence. Since $y \in S$ and $\mu_0 \le \mu = \lambda^{-1} \le 1$ we know that the right-hand side in (5.65) satisfies (5.63) for any limit point $y \in S$. Since this is true along any subsequence it follows that

$$m = \frac{1}{2}\lambda_0^{-Q-\varepsilon} = \frac{1}{2}\mu_0^{Q+\varepsilon} \le \frac{R(\lambda r, x)}{R(r, x)} \le 2\mu_0^{q-\varepsilon} = 2\lambda_0^{-q+\varepsilon} = M$$

for all $r \ge r_0$, $1 \le \lambda \le \lambda_0$ and $x \in S$. Now apply Lemma 5.3.13 to see that (5.49) holds with

$$\begin{aligned} h &= \frac{\log m}{\log \lambda_0} = -Q - \varepsilon - \frac{\log 2}{\log \lambda_0} \\ H &= \frac{\log M}{\log \lambda_0} = -q + \varepsilon + \frac{\log 2}{\log \lambda_0} \end{aligned}$$

which implies (5.52) provided that $\varepsilon < \delta$ and λ_0 is chosen sufficiently large. □

Theorem 5.3.14 always gives an upper bound on the rate at which F changes. Note that $q \le 0 \le Q$ and we always have exactly one of q, Q not equal to zero. Suppose F varies regularly at infinity so that (5.51) holds. If $q = 0$, then $R(\lambda r, x) \le M\lambda^{Q+\delta}R(r, x)$ means that $R(r, x)$ cannot increase as fast as $r^{Q+\delta} \to \infty$. If $Q = 0$, then $R(\lambda r, x) \geq m\lambda^{q-\delta} R(r, x)$ means that $R(r, x)$ cannot decrease as fast as $r^{q-\delta} \to 0$. Now suppose F varies regularly at zero, so that (5.52) holds. If $q = 0$, then $R(\lambda r, x) \geq m\lambda^{-Q-\delta} R(r, x)$ means that $R(r, x)$ cannot decrease as fast as $r^{-Q-\delta} \to 0$. If $Q = 0$, then $R(\lambda r, x) \leq M\lambda^{-q+\delta} R(r, x)$ means that $R(r, x)$ cannot increase as fast as $r^{-q+\delta} \to \infty$. Now we will establish a lower bound on the rate at which F changes. We begin with an extension of Lemma 5.3.13.

Lemma 5.3.17. *Suppose that $R : \mathbb{R}^+ \times S \to \mathbb{R}^+$ is uniformly R–O varying and r_0, λ_0, m, M are as in Definition 5.3.12.*

(a) If there exists $0 < M_1 < 1$ such that

$$\frac{R(\lambda r, \theta)}{R(r, \theta)} \leq M_1 \tag{5.66}$$

for all $r \geq r_0$, $\lambda \geq \lambda_0$ and $\theta \in S$, then

$$\frac{R(\lambda r, \theta)}{R(r, \theta)} \leq \frac{M}{M_1} \lambda^{H_1} \tag{5.67}$$

for all $\lambda \geq 1$, $r \geq r_0$ and $\theta \in S$, where $H_1 = \log M_1 / \log \lambda_0$.

(b) If there exists $m_1 > 1$ such that

$$m_1 \leq \frac{R(\lambda r, \theta)}{R(r, \theta)} \tag{5.68}$$

for all $r \geq r_0$, $\lambda \geq \lambda_0$ and $\theta \in S$, then

$$\frac{m}{m_1} \lambda^{h_1} \leq \frac{R(\lambda r, \theta)}{R(r, \theta)} \tag{5.69}$$

for all $\lambda \geq 1$, $r \geq r_0$ and $\theta \in S$, where $h_1 = \log m_1 / \log \lambda_0$.

Proof. For $\lambda > 1$, let $p = [\log \lambda / \log \lambda_0]$, so that $\lambda = \lambda' \lambda_0^p$ for some $1 \leq \lambda' < \lambda_0$. Then for all $r \geq r_0$ and all $\theta \in S$ we get from (5.66) that

$$\frac{R(\lambda r, \theta)}{R(r, \theta)} = \frac{R(\lambda r, \theta)}{R(\lambda_0^p r, \theta)} \frac{R(\lambda_0^p r, \theta)}{R(\lambda_0^{p-1} r, \theta)} \cdots \frac{R(\lambda_0 r, \theta)}{R(r, \theta)} \leq M \cdot M_1^p = \frac{M}{M_1} M_1^{p+1}$$

and since $M_1 < 1$ we have

$$\begin{aligned} M_1^{p+1} \leq M_1^{\log \lambda / \log \lambda_0} &= e^{(\log \lambda / \log \lambda_0) \log M_1} \\ &= e^{(\log M_1 / \log \lambda_0) \log \lambda} = \lambda^{\log M_1 / \log \lambda_0} = \lambda^{H_1} \end{aligned}$$

which proves part (a).

Similarly, if (5.68) holds for some $m_1 > 1$, then for all $r \ge r_0$ and all $\theta \in S$ we have

$$\frac{R(\lambda r,\theta)}{R(r,\theta)} = \frac{R(\lambda r,\theta)}{R(\lambda_0^p r,\theta)}\frac{R(\lambda_0^p r,\theta)}{R(\lambda_0^{p-1} r,\theta)}\cdots\frac{R(\lambda_0 r,\theta)}{R(r,\theta)} \ge m\cdot m_1^p = \frac{m}{m_1}m_1^{p+1}$$

and since $m_1 > 1$ we have $m_1^{p+1} \ge m_1^{\log\lambda/\log\lambda_0} = \lambda^{h_1}$, which proves (b). □

Our next result complements Theorem 5.3.14.

Theorem 5.3.18. *Suppose that F varies regularly at zero or infinity and (5.3) holds for all $\lambda > 0$ and all $x \in \Gamma$. Define*

$$\begin{aligned} q_1 &= \min\{\beta/\operatorname{Re}(\alpha_1),\ldots,\beta/\operatorname{Re}(\alpha_j)\} \\ Q_1 &= \max\{\beta/\operatorname{Re}(\alpha_1),\ldots,\beta/\operatorname{Re}(\alpha_j)\} \end{aligned} \tag{5.70}$$

where $\alpha_1\cdots\alpha_j$ are the eigenvalues of E.

(a) *If F varies regularly at infinity, $R(r,x) = F(rx)$ and $q_1 > 0$, then for any $S \subset \Gamma$ compact, and any $\delta > 0$ sufficiently small there exist $r_0 > 0$ and $0 < C < 1$ such that*

$$C\lambda^{q_1-\delta} \le \frac{R(\lambda r,x)}{R(r,x)} \tag{5.71}$$

for all $r \ge r_0$, $\lambda \ge 1$, and $x \in S$.

(b) *If F varies regularly at infinity, $R(r,x) = F(rx)$ and $Q_1 < 0$, then for any $S \subset \Gamma$ compact, and any $\delta > 0$ sufficiently small there exist $r_0 > 0$ and $C > 1$ such that*

$$\frac{R(\lambda r,x)}{R(r,x)} \le C\lambda^{Q_1+\delta} \tag{5.72}$$

for all $r \ge r_0$, $\lambda \ge 1$, and $x \in S$.

(c) *If F varies regularly at zero, $R(r,x) = F(x/r)$ and $q_1 > 0$, then for any $S \subset \Gamma$ compact, and any $\delta > 0$ sufficiently small there exist $r_0 > 0$ and $C > 1$ such that*

$$\frac{R(\lambda r,x)}{R(r,x)} \le C\lambda^{-q_1+\delta} \tag{5.73}$$

for all $r \ge r_0$, $\lambda \ge 1$, and $x \in S$.

(d) *If F varies regularly at zero, $R(r,x) = F(x/r)$ and $Q_1 < 0$, then for any $S \subset \Gamma$ compact, and any $\delta > 0$ sufficiently small there exist $r_0 > 0$ and $0 < C < 1$ such that*

$$C\lambda^{-Q_1-\delta} \le \frac{R(\lambda r,x)}{R(r,x)} \tag{5.74}$$

for all $r \geq r_0$, $\lambda \geq 1$, *and* $x \in S$.

In order to simplify the proof, we begin with two useful results that complement Lemmas 5.3.15 and 5.3.16.

Lemma 5.3.19. *Suppose that* $\varphi : \Gamma \to \mathbb{R}^+$ *is bounded away from both zero and infinity on compact sets and that* (5.53) *holds for some* $c > 1$ *and all* $y \in \Gamma$, *where the real parts of the eigenvalues of* A *are* $0 < a_1 < \cdots < a_p$.

(a) *Given* $S \subset \Gamma$ *compact and* $\varepsilon > 0$, *for all* $\lambda_0 > 1$ *sufficiently large we have*

$$\frac{\varphi(\lambda y)}{\varphi(y)} \geq \lambda_0^{\frac{1}{a_p} - \varepsilon} \tag{5.75}$$

for all $\lambda \geq \lambda_0$ *and* $y \in S$.

(b) *Given* $S \subset \Gamma$ *compact and* $\varepsilon > 0$, *for all* $0 < \mu_0 < 1$ *sufficiently small we have*

$$\frac{\varphi(\mu y)}{\varphi(y)} \leq \mu_0^{\frac{1}{a_p} - \varepsilon} \tag{5.76}$$

for all $0 < \mu \leq \mu_0$ *and* $y \in S$.

Proof. First we prove part (a). As in the proof of Lemma 5.3.15, it suffices to consider $S = \{x \in \mathbb{R}^d : D_1 \leq \|x\| \leq D_2\}$, where $D_2 \geq \|c^A\| D_1$, so that we can always write $x = c^{kA} y$ for some integer k and some $y \in S$. If $\alpha > a_p$ then, since $\|t^A x\| \to 0$ uniformly in $x \in S$ as $t \to 0$ and $t^{-\alpha}\|t^A x\| \to 0$ uniformly in $x \in S$ as $t \to \infty$ in view of Theorem 2.2.4, for any $t_0 > 0$ sufficiently large we have $\|t^A y\| < t_0^\alpha$ for all $0 < t < t_0$ and all $y \in S$. Let $\lambda_0 = t_0^\alpha / D_1 > 1$ for t_0 large. Given $y \in S$ and $\lambda \geq \lambda_0$ choose $k \in \mathbb{Z}$ and $y' \in S$ such that $\lambda y = c^{kA} y'$. Then $\|c^{kA} y'\| = \|\lambda y\| \geq \lambda_0 D_1 = t_0^\alpha$, so $c^k \geq t_0$. For some $C_1 > 0$ and $C_2 < \infty$, we have $C_1 \leq \varphi(y) \leq C_2$ for all $y \in S$. Then

$$\frac{\varphi(\lambda y)}{\varphi(y)} = \frac{c^k \varphi(y')}{\varphi(y)} \geq \frac{t_0 C_1}{C_2} = C \lambda_0^{1/\alpha}$$

whenever $y \in S$ and $\lambda \geq \lambda_0$, where $C = (C_1/C_2) D_1^{1/\alpha}$. If α is sufficiently close to a_p, then $1/\alpha > a_p^{-1} - \varepsilon$, and so (5.75) holds.

Next we prove part (b). If $\alpha > a_p$, then for any $0 < t_0 < 1$ sufficiently small we have $\|t^A y\| > t_0^\alpha$ for all $t > t_0$ and all $y \in S$. Let $\mu_0 = t_0^\alpha / D_2 < 1$ for t_0 sufficiently small. Given $y \in S$ and $0 < \mu \leq \mu_0$ choose $k \in \mathbb{Z}$ and $y' \in S$ such that $\mu y = c^{kA} y'$. Then $\|c^{kA} y'\| = \|\mu y\| \leq \mu_0 D_2 = t_0^\alpha$, so $c^k \leq t_0$. Then

$$\frac{\varphi(\mu y)}{\varphi(y)} = \frac{c^k \varphi(y')}{\varphi(y)} \leq \frac{t_0 C_2}{C_1} = C \mu_0^{1/\alpha}$$

whenever $y \in S$ and $0 < \mu \leq \mu_0$, where $C = (C_2/C_1) D_2^{1/\alpha}$. If α is sufficiently close to a_p, then (5.76) holds. □

Lemma 5.3.20. *Suppose that F varies regularly at zero or infinity and (5.3) holds, and define q_1, Q_1 as in (5.70).*

(a) If $q_1 > 0$, then for any $S \subset \Gamma$ compact and any $\varepsilon > 0$, for all $\lambda_0 > 1$ sufficiently large we have

$$\frac{\varphi(\lambda y)}{\varphi(y)} \geq \lambda_0^{q_1-\varepsilon} \tag{5.77}$$

for all $\lambda \geq \lambda_0$ and $y \in S$, and for all $\mu_0 > 0$ sufficiently small we have

$$\frac{\varphi(\mu y)}{\varphi(y)} \leq \mu_0^{q_1-\varepsilon} \tag{5.78}$$

for all $0 < \mu \leq \mu_0$ and $y \in S$.

(b) If $Q_1 < 0$, then for any $S \subset \Gamma$ compact and any $\varepsilon > 0$, for all $\lambda_0 > 1$ sufficiently large we have

$$\frac{\varphi(\lambda y)}{\varphi(y)} \leq \lambda_0^{Q_1+\varepsilon} \tag{5.79}$$

for all $\lambda \geq \lambda_0$ and $y \in S$, and for all $\mu_0 > 0$ sufficiently small we have

$$\frac{\varphi(\mu y)}{\varphi(y)} \geq \mu_0^{Q_1+\varepsilon} \tag{5.80}$$

for all $0 < \mu \leq \mu_0$ and $y \in S$.

Proof. If $q_1 > 0$, then (5.3) implies (5.53) where the eigenvalues of $A = (1/\beta)E$ all have positive real part less than or equal to $a_p = 1/q_1$. Then (5.77) and (5.78) follow from Lemma 5.3.19. On the other hand, if $Q_1 < 0$ then $t^{-1}\varphi(y) = \varphi(t^A y)$ where $A = (-1/\beta)E$ has eigenvalues whose real parts are all positive and less than or equal to $a_p = -1/Q_1$, and then (5.79) and (5.80) follow from applying Lemma 5.3.19 to the function $1/\varphi$. □

Proof of Theorem 5.3.18. The proof is similar to Theorem 5.3.14. First we will prove part (a). Given $\varepsilon > 0$, since F varies regularly at infinity Lemma 5.3.20 implies that for all λ_0 sufficiently large, (5.77) holds for all $x \in S = \{x \in \mathbb{R}^d : \|x\| = 1\}$ and all $\lambda \geq \lambda_0$. Given $x_r \to x$ in Γ apply Lemma 4.2.7 (a) to obtain $t = t(r) \to \infty$ as $r \to \infty$ such that $y_t = f(t)rx_r$ satisfies $\|y_t\| \to 1$ as $r \to \infty$. For any sequence $r_n \to \infty$, there exists a further subsequence $r_{n'} \to \infty$ along which $y_t \to y \in S$. If $\lambda \geq \lambda_0$, then

$$\frac{R(\lambda r, x_r)}{R(r, x_r)} = \frac{F(f(t)^{-1}\lambda y_t)}{F(f(t)^{-1}y_t)} \to \frac{\varphi(\lambda y)}{\varphi(y)} \geq \lambda_0^{q_1-\varepsilon}$$

along this subsequence. Since this is true along any subsequence it follows that

$$\frac{R(\lambda r, x)}{R(r, x)} \geq \frac{1}{2}\lambda_0^{q_1-\varepsilon} = m_1$$

for all $r \geq r_0$, $\lambda \geq \lambda_0$ and $x \in S$. Now apply part (b) of Lemma 5.3.17 to see that (5.69) holds with $h_1 = \log m_1 / \log \lambda_0 = q_1 - \varepsilon - \log 2/\log \lambda_0$,, which implies (5.71) provided that $\varepsilon < \delta$ and λ_0 is chosen sufficiently large.

The proof of part (b) is quite similar to part (a). Now (5.79) holds and

$$\frac{R(\lambda r, x_r)}{R(r, x_r)} \to \frac{\varphi(\lambda y)}{\varphi(y)} \leq \lambda_0^{Q_1+\varepsilon}$$

along subsequences, which leads to

$$\frac{R(\lambda r, x)}{R(r, x)} \leq 2\lambda_0^{Q_1+\varepsilon} = M_1$$

for all $r \geq r_0$, $\lambda \geq \lambda_0$ and $x \in S$. Then part (a) of Lemma 5.3.17 implies that (5.67) holds with $H_1 = \log M_1 / \log \lambda_0 = Q_1 + \varepsilon + \log 2/\log \lambda_0$, which implies (5.72) provided that $\varepsilon < \delta$ and λ_0 is chosen sufficiently large.

Now we prove part (c). Given $\varepsilon > 0$, since F varies regularly at zero Lemma 5.3.20 implies that for all μ_0 sufficiently small, (5.78) holds for all $x \in S = \{x \in \mathbb{R}^d : \|x\| = 1\}$ and all $0 < \mu \leq \mu_0$. Given $x_r \to x$ in Γ apply Lemma 4.2.7 (b) to obtain $t = t(r) \to \infty$ as $r \to \infty$ such that $y_t = f(t)(x_r/r)$ satisfies $\|y_t\| \to 1$ as $r \to \infty$. For any sequence $r_n \to \infty$, there exists a further subsequence $r_{n'} \to \infty$ along which $y_t \to y \in S$. If $\lambda \geq \lambda_0 = \mu_0^{-1}$, then

$$\frac{R(\lambda r, x_r)}{R(r, x_r)} = \frac{F(f(t)^{-1}\lambda^{-1}y_t)}{F(f(t)^{-1}y_t)} \to \frac{\varphi(\lambda^{-1}y)}{\varphi(y)} \leq \lambda_0^{-q_1+\varepsilon}$$

along a subsequence. Since this is true along any subsequence it follows that

$$\frac{R(\lambda r, x)}{R(r, x)} \leq 2\lambda_0^{-q_1+\varepsilon} = M_1$$

for all $r \geq r_0$, $\lambda \geq \lambda_0$ and $x \in S$. Then part (a) of Lemma 5.3.17 implies that (5.67) holds with $H_1 = -q_1 + \varepsilon + \log 2/\log \lambda_0$, which implies (5.73) provided that $\varepsilon < \delta$ and λ_0 is chosen sufficiently large.

The proof of part (d) is similar to part (c). Now (5.80) holds and

$$\frac{R(\lambda r, x_r)}{R(r, x_r)} \to \frac{\varphi(\lambda y)}{\varphi(y)} \geq \lambda_0^{-Q_1-\varepsilon}$$

along subsequences which leads to

$$\frac{R(\lambda r, x)}{R(r, x)} \geq \frac{1}{2}\lambda_0^{-Q_1-\varepsilon} = m_1$$

for all $r \geq r_0$, $\lambda \geq \lambda_0$ and $x \in S$. Then part (b) of Lemma 5.3.17 implies that (5.69) holds with $h_1 = -Q_1 - \varepsilon - \log 2/\log \lambda_0$, which implies (5.74) provided that $\varepsilon < \delta$ and λ_0 is chosen sufficiently large. □

Remark 5.3.21. Taken together, Theorems 5.3.14 and 5.3.18 give upper and lower bounds on F regularly varying. Suppose F varies regularly at infinity so that (5.51) holds. If $q = 0$, then $Q > q_1 > 0$ and (5.51) together with (5.71) imply that

$$C\lambda^{q_1-\delta} \le \frac{R(\lambda r, x)}{R(r, x)} \le M\lambda^{Q+\delta}$$

for all $\lambda \ge 1$, all $x \in S$ compact, and all r large. This means that as $r \to \infty$, $R(r, x) \to \infty$ slower than $r^{Q+\delta} \to \infty$ but faster than $r^{q_1-\delta} \to \infty$. If $Q = 0$, then $q < Q_1 < 0$ and (5.51) together with (5.72) imply that

$$m\lambda^{q-\delta} \le \frac{R(\lambda r, x)}{R(r, x)} \le C\lambda^{Q_1+\delta}$$

for $\lambda \ge 1$, $x \in S$ and r large. This means that $R(r, x) \to 0$ slower than $r^{q-\delta} \to 0$ but faster than $r^{Q_1+\delta} \to 0$.

Now suppose F varies regularly at zero, so that (5.52) holds. If $q = 0$, then $-Q < -q_1 < 0$ and (5.52) together with (5.73) imply that

$$m\lambda^{-Q-\delta} \le \frac{R(\lambda r, x)}{R(r, x)} \le C\lambda^{-q_1+\delta}$$

for all $\lambda \ge 1$, all $x \in S$ compact, and all r large. This means that as $r \to \infty$, $R(r, x) \to 0$ slower than $r^{-Q-\delta} \to 0$ but faster than $r^{-q_1+\delta} \to 0$. If $Q = 0$, then $-q > -Q_1 > 0$ and (5.52) together with (5.74) imply that

$$C\lambda^{-Q_1-\delta} \le \frac{R(\lambda r, x)}{R(r, x)} \le M\lambda^{-q+\delta}$$

for $\lambda \ge 1$, $x \in S$ and r large. This means that $R(r, x) \to \infty$ slower than $r^{-q+\delta} \to \infty$ but faster than $r^{-Q_1-\delta} \to \infty$.

5.4 NOTES AND COMMENTS

The beginnings of a theory of regular variation in $\mathbb{R}^d$ can be found in Stam [188], Stadtmüller and Trautner [187], Jakimiv [70] and de Haan and Resnick [47]. They assume that the norming operators $f(t)$ in Definition 5.1.1 are scalar multiples of the identity. Later de Haan, Omey, and Resnick [48] considered the case where the norming operators $f(t)$ are all diagonal with respect to the standard basis for $\mathbb{R}^d$. The general case where $f(t) \in \mathrm{GL}(\mathbb{R}^d)$ was first treated in [113], and then generalized to real-valued functions on nilpotent Lie groups in [163]. The Convergence of Types Theorem 5.2.10 for real-valued function on $\mathbb{R}^d$ that are monotone on rays is the analogue of the Convergence of Types Theorem 2.3.17 for probability measures on $\mathbb{R}^d$ due to Billingsley [16], and our proof follows that of Billingsley for the most part.

Theorem 5.2.11 follows Sharpe [181], and Theorem 5.2.13 is modeled after Holmes, Hudson, and Mason [62] and Hudson, Jurek, and Veeh [66]. Theorem 5.2.14 is from [117]. Uniform regular variation is implicit in the work of Jakimiv [70], which includes an application to branching processes. Hahn and Klass [49, 50] characterize the generalized domain of attraction of spherically symmetric operator stable laws in terms of uniform regular variation of a trimmed second moment function. The proof of the uniform convergence theorem, Theorem 5.3.6, is a straightforward extension of the argument in Seneta [177], Theorem 1.1. The proof of Theorems 5.3.9 and 5.3.11 extends the arguments in Section VIII.8 of Feller [40], but we provide more details. The proof of Theorem 5.3.14 follows Appendix A of Seneta [177]. The complementary result of Theorem 5.3.18 seems to be new.

6

Regular Variation for Borel Measures

Regular variation of a measure μ means that the tails fall off like a power function, with possibly different rates in different radial directions. Regular variation will be used to state and prove some important limit theorems for sums of independent random vectors. In particular, we will see that the necessary and sufficient condition for a centered and normalized sum of independent, identically distributed random vectors to converge to a non-Gaussian limit is simply regular variation of the underlying probability measure. Furthermore, the tools of regular variation provide a Calculus of truncated moments and tails, which streamlines the proofs in Part III.

6.1 REGULARLY VARYING MEASURES

Regular variation for linear operators was the subject of Chapter 4. In Chapter 5, we introduced regular variation for real-valued functions on $\mathbb{R}^d$. Now we consider regular variation for σ-finite Borel measures on $\mathbb{R}^d$. Regularly varying measures have power law tails, and the rate at which the tails diminish may depend on radial direction. Recall from Section 1.2 that $\mathcal{M}$ is the set of σ-finite Borel measures on $\Gamma = \mathbb{R}^d \setminus \{0\}$ that are finite outside every neighborhood of the origin, and that $\mu_n \to \mu$ in $\mathcal{M}$ if $\mu_n(U) \to \mu(U)$ for μ-continuity sets $U \subset \Gamma$ that are bounded away from the origin. Recall from Definition 1.1.5 that for $T \in \mathrm{L}(\mathbb{R}^d)$ and $\mu \in \mathcal{M}$ we define $T\mu \in \mathcal{M}$ by $T\mu(U) = \mu(T^{-1}U)$ for Borel subsets $U \subset \Gamma$. Recall from Definition 3.1.19 that $\mu \in \mathcal{M}$ is $\mathcal{M}$-full if it cannot be supported on any $d-1$ dimensional subspace of $\mathbb{R}^d$. The Lévy measures introduced in Section 3.1 belong to $\mathcal{M}$.

Definition 6.1.1. A measure $\mu \in \mathcal{M}$ varies regularly if there exist $f : \mathbb{R}^+ \to \mathrm{GL}(\mathbb{R}^d)$ regularly varying with index $-E$, and $R : \mathbb{R}^+ \to \mathbb{R}^+$ regularly varying with index $\beta \neq 0$, such that

$$\frac{f(t)\mu}{R(t)} \to \phi \quad \text{in } \mathcal{M} \tag{6.1}$$

where ϕ is $\mathcal{M}$*-full.* If all eigenvalues of E have negative real part, we will say that μ *varies regularly at zero*, and if all eigenvalues of E have positive real part, we will say that μ *varies regularly at infinity.*

Proposition 6.1.2. *If $\mu \in \mathcal{M}$ and (6.1) holds for some $f : \mathbb{R}^+ \to \mathrm{GL}(\mathbb{R}^d)$ regularly varying with index $-E$, and $R : \mathbb{R}^+ \to \mathbb{R}^+$ regularly varying with index $\beta \neq 0$, then*

$$\lambda \cdot \phi = \lambda^B \phi \tag{6.2}$$

for all $\lambda > 0$, where $B = (-1/\beta)E$.

Proof. Write

$$\frac{f(\lambda t)\mu}{R(\lambda t)} = f(\lambda t)f(t)^{-1}\frac{f(t)\mu}{R(t)} \cdot \frac{R(t)}{R(\lambda t)} \tag{6.3}$$

where $\lambda > 0$ is fixed. Taking the limit of both sides of (6.3) as $t \to \infty$ we obtain $\phi = \lambda^{-E}\phi \cdot \lambda^{-\beta}$, and then a simple substitution yields (6.2). □

Definition 6.1.3. If $\phi \in \mathcal{M}$ and (6.2) holds for all $\lambda > 0$, we will call B an *exponent* of ϕ. If $\mu \in \mathcal{M}$ varies regularly and (6.1) holds, then (6.2) also holds, and we will also call B an *exponent* of μ. If μ varies regularly at zero with exponent B, we will write $\mu \in \mathrm{RVM}_0(B)$. If μ varies regularly at infinity, we will write $\mu \in \mathrm{RVM}_\infty(B)$.

Since measures $\mu \in \mathcal{M}$ assign finite mass to sets bounded away from the origin, they cannot blow up at infinity. Hence, in contrast to the case of regularly varying functions considered in Chapter 5, only certain exponents B are allowed.

Lemma 6.1.4. *If μ varies regularly with exponent B, then the real part of every eigenvalue of B is greater than or equal to zero. If μ varies regularly at zero or infinity, then the eigenvalues of B all have positive real part.*

Proof. If μ varies regularly with exponent B, then (6.2) holds. Let $a_1 < \cdots < a_p$ denote the real parts of the eigenvalues of B, and recall that B, B^* have the same eigenvalues. Let $B^* = B_1^* \oplus \cdots \oplus B_p^*$ denote the spectral decomposition of B^* and $\mathbb{R}^d = V_1^* \oplus \cdots \oplus V_p^*$ the spectral decomposition of $\mathbb{R}^d$ with respect to B^*, in the sense of Definition 2.1.15. Assume by way of contradiction that $a_1 < 0$. Let $\theta_0 \in V_1^*$ be a unit vector. Then (6.2) implies

$$t \cdot \phi\{x : |<x, \theta_0>| > R\} = \phi\{x : |<x, t^{B^*}\theta_0>| > R\} \tag{6.4}$$

for all $R > 0$ and $t > 0$. Write $t^{B^*}\theta_0 = r_t\theta_t$ for some $\|\theta_t\| = 1$ and $r_t > 0$. Since $a_1 < 0$ and $\theta_0 \in V_1^*$, Corollary 2.2.5 (b) with $\beta = 0$ implies that $r_t \to 0$ as $t \to \infty$. Furthermore, $\phi\{x : |<x, \theta_0>| > R\} \leq \phi\{x : \|x\| > R\} < \infty$. But

$$\phi\{x : |<x, t^{B^*}\theta_0>| > R\} = \phi\{x : |<x, \theta_t>| > R/r_t\} \to 0$$

as $t \to \infty$. Hence the right-hand side of (6.4) tends to zero whereas the left-hand side of (6.4) either tends to infinity as $t \to \infty$, which is a contradiction, or else we have $\phi\{x : |<x, \theta_0>| > R\} = 0$. Since $R > 0$ is arbitrary this implies that $\phi\{x : |<x, \theta_0>| \neq 0\} = 0$, which contradicts the assumption that ϕ is $\mathcal{M}$-full. Finally, if μ varies regularly at zero or infinity then the eigenvalues of E cannot have zero real part, so the eigenvalues of $B = (-1/\beta)E$ all have positive real part. □

Lemma 6.1.5. *Suppose that the real parts of the eigenvalues of B are all positive, and define*

$$\|x\|_0 = \int_0^1 \|t^B x\| \frac{dt}{t} \tag{6.5}$$

where $\|\cdot\|$ is any norm on $\mathbb{R}^d$. Then $\|\cdot\|_0$ is a norm on $\mathbb{R}^d$ such that $t \mapsto \|t^B x\|_0$ is strictly increasing for all nonzero $x \in \mathbb{R}^d$. Furthermore, if we let $S_0 = \{x : \|x\|_0 = 1\}$ denote the unit sphere in $\mathbb{R}^d$ with respect to this norm then the mapping $\Psi : (0, \infty) \times S_0 \to \Gamma$ defined by $\Psi(t, \theta) = t^B\theta$ is a homeomorphism.

Proof. We follow the proof of Jurek and Mason [75] Proposition 3.4.3. Since the real parts of the eigenvalues of B are positive, Proposition 2.2.11 (d) applied to $-B$ implies that there exists a $\delta > 0$ and a $C > 0$ such that $\|t^B x\| \leq Ct^\delta\|x\|$ for all $x \in \mathbb{R}^d$ and all $0 \leq t \leq 1$. Hence $\|\cdot\|_0$ is well defined and it is easy to see that it is a norm on $\mathbb{R}^d$. Compute

$$\|t^B x\|_0 = \int_0^1 \|s^B(t^B x)\| \frac{ds}{s} = \int_0^t \|u^B x\| \frac{du}{u} \tag{6.6}$$

via the substitution $u = st$, showing that $t \mapsto \|t^B x\|_0$ is strictly increasing for all $t > 0$ and $x \in \Gamma$. If $x = t^B\theta = s^B\omega$ for some $s, t > 0$ and some $\omega, \theta \in S$, then $(t/s)^B\theta = \omega$, and since $\|u^B\theta\|_0$ is strictly increasing this implies that $s = t$, and then $\omega = \theta$. Hence Ψ is one-to-one. Theorem 2.2.4 implies that $\|t^B x\|_0 \to 0$ as $t \to 0$, and $\|t^B x\|_0 \to \infty$ as $t \to \infty$. Since $t \mapsto \|t^B x\|_0$ is continuous in view of Propositions 2.2.11 and 2.1.4, it follows that every orbit $t \mapsto t^B x$ must intersect S_0, so for any $x \in \Gamma$ we can write $t^B x = \theta$ for some $\theta \in S_0$. Then $(1/t)^B\theta = x$, showing that Ψ is onto. Obviously, Ψ is continuous.

To finish the proof, we need to show that Ψ^{-1} is continuous. Suppose that $x_n = r_n^B\theta_n \to x = r^B\theta$ so that $\|x_n\|_0 \to \|x\|_0$. Theorem 2.2.4 implies that $\|t^Bx\|_0 \to 0$ as $t \to 0$ and $\|t^Bx\|_0 \to \infty$ as $t \to \infty$ uniformly on compact subsets of $x \in \Gamma$. Then $r_{n'} \to 0$ along a subsequence implies $\|x_{n'}\|_0 = \|r_{n'}^B\theta_{n'}\|_0 \to 0$, which is a contradiction. Similarly, if $r_{n'} \to \infty$ along a subsequence then $\|x_{n'}\|_0 = \|r_{n'}^B\theta_{n'}\|_0 \to \infty$, which is a contradiction. Hence (r_n) is limit point compact, and every limit point is positive. Since (θ_n) is also limit point compact, any subsequence of (r_n, θ_n) contains a further subsequence along which $(r_{n'}, \theta_{n'}) \to (r_0, \theta_0)$. Then $x_{n'} = r_{n'}^B\theta_{n'} \to r_0^B\theta_0 = r^B\theta$, which implies $r = r_0$ and $\theta = \theta_0$. Then $(r_n, \theta_n) \to (r, \theta)$ as required. □

Remark 6.1.6. Suppose that the real parts of the eigenvalues of B are all positive, and apply Theorem 2.1.16 to obtain a basis in which B has Jordan form. Let $\|x\|$ denote the norm associated with this basis, in the sense of Definition 1.1.21. It follows directly from (2.11) and (2.12) that $t \mapsto \|t^Bx\|$ is monotone increasing for each $x \neq 0$, and the remaining conclusions of Lemma 6.1.5 also hold for this norm. In this case, the homeomorphism $R(t, \theta) = t^B\theta$ is a regularly varying function of $t > 0$ for each $\|\theta\| = 1$, see Example 5.3.3. In fact, if $a_1 < \cdots < a_p$ is the real spectrum of B and we write $\theta = \theta_1 + \cdots + \theta_p$ in terms of the spectral decomposition of $\mathbb{R}^d$ with respect to B then $R(t, \theta)$ varies regularly with index $\alpha(\theta) = \max\{a_i : \theta_i \neq 0\}$, uniformly on compact subsets of the level sets of $\alpha(\theta)$ in the unit sphere in this norm. If we use this norm in Lemma 6.1.5, then in view of (6.6) it follows from Lemma 5.3.8 (c) that the homeomorphism $\Psi(t, \theta)$ is also regularly varying with the same index, uniformly on level sets of $\alpha(\theta)$ in S_0.

Theorem 6.1.7. *If $\phi \in \mathcal{M}$ satisfies (6.2) for all $\lambda > 0$, and if every eigenvalue of B has positive real part, then there exists a norm $\|\cdot\|_0$ on $\mathbb{R}^d$ such that for all Borel subsets $A \subset \Gamma$ we have*

$$\phi(A) = \int_{S_0} \int_0^\infty I(t^B\theta \in A) \frac{dt}{t^2} \lambda\{d\theta\} \tag{6.7}$$

where

$$\lambda(D) = \phi\{t^B\theta : \theta \in D; t > 1\} \tag{6.8}$$

is a finite Borel measure on the unit sphere $S_0 = \{x : \|x\|_0 = 1\}$.

Proof. We follow the proof of Jurek and Mason [75], who apply (6.7) to Lévy measures of operator stable laws. Apply Lemma 6.1.5 to obtain a norm $\|\cdot\|_0$ on $\mathbb{R}^d$ such that $t \mapsto \|t^Bx\|_0$ is strictly increasing for all nonzero $x \in \mathbb{R}^d$, and the mapping $\Psi : (0, \infty) \times S_0 \to \Gamma$ defined by $\Psi(t, \theta) = t^B\theta$ is a homeomorphism. For $s > 0$ and a Borel subset $F \subset S_0$, let

$$A(s, F) = \{t^B\theta : t \geq s, \theta \in F\} \tag{6.9}$$

and then define $\lambda(F) = \phi(A(1, F))$, a finite Borel measure on S_0. Then we get from (6.2) and the definition of λ that

$$\begin{aligned}
\int_{S_0}\int_0^\infty I(r^B\theta \in A(s,F))\frac{dr}{r^2}\lambda\{d\theta\} &= \int_F\int_s^\infty \frac{dr}{r^2}\lambda\{d\theta\} \\
&= s^{-1}\lambda(F) \\
&= s^{-1}\phi\{t^B\theta : t \geq 1, \theta \in F\} \\
&= \phi(s^B\{t^B\theta : t \geq 1, \theta \in F\}) \\
&= \phi\{t^B\theta : t \geq s, \theta \in F\} = \phi(A(s,F)).
\end{aligned}$$

Then Theorem 1.1.3 implies that (6.7) holds for all Borel subsets $A \subset \Gamma$. □

Remark 6.1.8. The set S_0 is called a compact cross-section for the continuous one-parameter group $(t^B)_{t>0}$ (see [55]).

By this time, it should be clear that the theory of regularly varying measures parallels that of regularly varying functions in many ways. That being the case, we will not belabor the details of those results whose proofs are essentially the same. Our next two results are the analogues of Propositions 5.1.6 and 5.1.8 and their corollaries. Note that in view on Lemma 6.1.4 the eigenvalues of the exponent of a measure regularly varying at either zero or infinity all have positive real part.

Proposition 6.1.9. *(a) If $\mu \in \mathrm{RVM}_0(B)$ and (6.1) holds, then there exists $g : \mathbb{R}^+ \to \mathrm{GL}(\mathbb{R}^d)$ regularly varying with index B such that*

$$\frac{g(t)\mu}{t} \to \phi \tag{6.10}$$

as $t \to \infty$. Conversely, if (6.10) holds for some $\mu \in \mathcal{M}$, ϕ $\mathcal{M}$-full, and $g : \mathbb{R}^+ \to \mathrm{GL}(\mathbb{R}^d)$ regularly varying with index $B \in \mathrm{GL}(\mathbb{R}^d)$, then μ is regularly varying at zero with exponent B.

(b) If $\mu \in \mathrm{RVM}_\infty(B)$ and (6.1) holds, then there exists $g : \mathbb{R}^+ \to \mathrm{GL}(\mathbb{R}^d)$ regularly varying with index $-B$ such that

$$tg(t)\mu \to \phi \tag{6.11}$$

as $t \to \infty$. Conversely, if (6.11) holds for some $\mu \in \mathcal{M}$, ϕ $\mathcal{M}$-full, and $g : \mathbb{R}^+ \to \mathrm{GL}(\mathbb{R}^d)$ regularly varying with index $-B \in \mathrm{GL}(\mathbb{R}^d)$, then μ is regularly varying at infinity with exponent B.

Proposition 6.1.10. *Suppose that $\mu \in \mathcal{M}$, ϕ is $\mathcal{M}$-full and $B \in \mathrm{GL}(\mathbb{R}^d)$.*

(a) $\mu \in \mathrm{RVM}_0(B)$ if and only if there exists a sequence (A_n) of linear operators varying regularly with index B such that

$$\frac{1}{n} \cdot (A_n\mu) \to \phi \quad \textit{as } n \to \infty. \tag{6.12}$$

(b) $\mu \in \mathrm{RVM}_\infty(B)$ *if and only if there exists a sequence* (A_n) *of linear operators varying regularly with index* $-B$ *such that*

$$n \cdot (A_n \mu) \to \phi \quad \text{as } n \to \infty. \tag{6.13}$$

Probability measures that vary regularly at infinity are very important in the limit theorems to come in Part III. In the remainder of this section, our main goal will be to show that if (6.13) holds for some sequence of linear operators (A_n) and some $\mathcal{M}$-full limit ϕ, then μ is regularly varying. In particular, the assumption that the sequence (A_n) varies regularly is superfluous. We begin by establishing a convergence of types theorem in $\mathcal{M}$, which is analogous to the Convergence of Types Theorem 2.3.17 for probability measures. Theorem 2.3.17 is proved using Fourier transform methods. Since the Fourier transform of a σ-finite measure need not exist, we use a different method of proof for convergence of types in $\mathcal{M}$.

Definition 6.1.11. An operator $A \in \mathrm{L}(\mathbb{R}^d)$ is called a *symmetry* of $\phi \in \mathcal{M}$ if $A\phi = \phi$. We denote by $\mathcal{S}(\phi)$ the set of symmetries of a measure $\phi \in \mathcal{M}$.

Lemma 6.1.12. *Suppose that* $A \in \mathrm{L}(\mathbb{R}^d)$ *and* $\phi \in \mathcal{M}$*. Then* $A\phi$ *is* $\mathcal{M}$*-full if and only if* A *is invertible and* ϕ *is* $\mathcal{M}$*-full.*

Proof. Assume that $A\phi$ is $\mathcal{M}$-full, and suppose by way of contradiction that A is singular. Then A^* is also singular and so there exists some nonzero $x \in \mathbb{R}^d$ such that $A^*x = 0$. Define

$$H_x(r) = \{y \in \mathbb{R}^d : |\langle x, y\rangle| > r\} \tag{6.14}$$

and notice that

$$\begin{aligned} A^{-1}H_x(r) &= \{y : |\langle x, Ay\rangle| > r\} \\ &= \{y : |\langle A^*x, y\rangle| > r\} = H_{A^*x}(r) \end{aligned}$$

is the empty set. Then

$$A\phi\{y : |\langle x, y\rangle| > r\} = A\phi(H_x(r)) = \phi(A^{-1}H_x(r)) = 0$$

for all $r > 0$, which implies that $A\phi\{y : |\langle x, y\rangle| > 0\} = 0$ but this contradicts the assumption that $A\phi$ is $\mathcal{M}$-full. Then A^{-1} exists and $\phi = A^{-1}(A\phi)$ where $A\phi$ is $\mathcal{M}$-full, hence ϕ is also $\mathcal{M}$-full.

Conversely, if A is invertible and ϕ is $\mathcal{M}$-full, then for any nonzero $y \in \mathbb{R}^d$ we have $y = A^*x$ for some nonzero $x \in \mathbb{R}^d$, and so we have $A\phi(H_x(0)) = \phi(H_y(0)) > 0$ so that $A\phi$ is $\mathcal{M}$-full. □

Lemma 6.1.13. *If* $\mu_n \to \phi$ *in* $\mathcal{M}$ *and* ϕ *is* $\mathcal{M}$*-full then* μ_n *is* $\mathcal{M}$*-full for all* n *large.*

Proof. Suppose not. Then there exists a subsequence (n') such that $\mu_{n'}$ is singular for all $n' \to \infty$. Then for some $x_{n'} \in S = \{x \in \mathbb{R}^d : \|x\| = 1\}$ we have $\mu_{n'}(H_{x_{n'}}(\delta)) = 0$ for all n' and all $\delta > 0$, where the sets $H_x(r)$ are defined by (6.14). Given $\varepsilon > 0$ we can choose $R > 0$ such that $\phi\{y : \|y\| > R\} < \varepsilon/2$ and $\phi\{y : \|y\| = R\} = 0$. Then we have $\mu_n\{y : \|y\| > R\} \to \phi\{y : \|y\| > R\}$ and so $\mu_n\{y : \|y\| > R\} < \varepsilon$ for all n large. Since $x_{n'} \in S$ compact, any subsequence of $(x_{n'})$ contains a further subsequence $x_{n''} \to x$ for some $x \in S$. Write $x_{n''} = x + \omega_{n''}$ so that $\omega_{n''} \to 0$. If $|<x_{n''}, y>| \leq \delta/2$ and $\|y\| \leq R$, then

$$\begin{aligned} |<x, y>| &= |<x_{n''}, y> - <\omega_{n''}, y>| \\ &\leq |<x_{n''}, y>| + |<\omega_{n''}, y>| \\ &\leq \delta/2 + \|\omega_{n''}\| \cdot R < \delta \end{aligned}$$

for all large n'', hence for any $\delta > 0$ we have

$$\{y : |<x_{n''}, y>| \leq \delta/2\} \cap \{y : \|y\| \leq R\} \subseteq \{y : |<x, y>| \leq \delta\}$$

for all large n'' or equivalently

$$\{y : |<x, y>| > \delta\} \subseteq H_{x_{n''}}(\delta/2) \cup \{y : \|y\| > R\}$$

where $\delta > 0$ is arbitrary. Then

$$\mu_{n''}\{y : |<x, y>| > \delta\} \leq \mu_{n''}(H_{x_{n''}}(\delta/2)) + \mu_{n''}\{y : \|y\| > R\} < 0 + \varepsilon.$$

For almost all such $\delta > 0$ we have $\phi\{y : |<x, y>| = \delta\} = 0$ so that

$$\mu_{n''}\{y : |<x, y>| > \delta\} \to \phi\{y : |<x, y>| > \delta\}$$

which shows that $\phi\{y : |<x, y>| > \delta\} \leq \varepsilon$ where $\varepsilon > 0$ is arbitrary. Hence $\phi\{y : |<x, y>| > \delta\} = 0$, where $\delta > 0$ can be taken arbitrarily small, so that $\phi\{y : |<x, y>| > 0\} = 0$. But this contradicts the assumption that ϕ is $\mathcal{M}$-full, hence for some $n_0 > 0$ we have μ_n $\mathcal{M}$-full for all $n \geq n_0$. □

Lemma 6.1.14. *Suppose $\mu_n \in \mathcal{M}$, ϕ $\mathcal{M}$-full, and $A_n \in \mathrm{L}(\mathbb{R}^d)$. If $A_n\mu_n \to \phi$ then for all n large A_n is invertible and μ_n is $\mathcal{M}$-full.*

Proof. If $A_n\mu_n \to \phi$, then Lemma 6.1.13 shows that $A_n\mu_n$ is $\mathcal{M}$-full for n large, and then Lemma 6.1.12 shows that $A_n \in \mathrm{GL}(\mathbb{R}^d)$ and μ_n is $\mathcal{M}$-full. □

Lemma 6.1.15. *Let $\mu_n \in \mathcal{M}$, ϕ, ϕ_0 $\mathcal{M}$-full and $A_n \in \mathrm{L}(\mathbb{R}^d)$. If $\mu_n \to \phi$ and $A_n\mu_n \to \phi_0$, then (A_n) is relatively compact.*

Proof. Suppose by way of contradiction that there exists a subsequence (n') along which $\|A_{n'}\| \to \infty$ in the Euclidean operator norm. Then in view of Remark 2.1.5 we have $\|A_{n'}\| = \|A_{n'}^*\| = \|A_{n'}^* x_{n'}\|$ for some vectors $x_{n'} \in S = \{x \in \mathbb{R}^d : \|x\| = 1\}$, so that $\|A_{n'}^* x_{n'}\| \to \infty$. Define

$$B_R = \{y \in \mathbb{R}^d : \|y\| > R\}.$$

For any $\phi \in \mathcal{M}$, we have $\phi(B_R) \to 0$ as $R \to \infty$ and $\phi(\partial B_R) = 0$ for almost all $R > 0$, so that B_R is a ϕ-continuity set. Then for any $\varepsilon > 0$ we can choose $R > 0$ such that $\phi(B_R) < \varepsilon/3$, $\phi_0(B_R) < \varepsilon/3$, and such that B_R is both a ϕ-continuity set and a ϕ_0-continuity set. Then we have $\mu_n(B_R) \to \phi(B_R) < \varepsilon/3$ and so $\mu_n(B_R) < \varepsilon/2$ for all n large. Let $H_x(r)$ be the set defined by (6.14) and note that for any $x \in S$ and any $y \in \mathbb{R}^d$ we have $|<x, y>| \le \|y\|$, so that $H_x(r) \subset B_r$. Then $A_n\mu_{n'}(H_{x_{n'}}(R)) \le A_n\mu_{n'}(B_R) \to \phi_0(B_R) < \varepsilon/3$ and so $A_n\mu_{n'}(H_{x_{n'}}(R)) < \varepsilon/2$ for all n' large.

Write $A^*_{n'}x_{n'} = r_{n'}\theta_{n'}$ where $r_{n'} > 0$ and $\theta_{n'} \in S$. Then we have

$$\begin{aligned} A_{n'}\mu_{n'}(H_{x_{n'}}(R)) &= \mu_{n'}\{y : |<x_{n'}, A_{n'}y>| > R\} \\ &= \mu_{n'}\{y : |<A^*_{n'}x_{n'}, y>| > R\} \\ &= \mu_{n'}\{y : |<r_{n'}\theta_{n'}, y>| > R\} \\ &= \mu_{n'}\{y : |<\theta_{n'}, y>| > r_{n'}^{-1}R\} \end{aligned}$$

where $r_{n'}^{-1} \to 0$ by assumption. Since $\theta_{n'} \in S$ compact, any subsequence of $(\theta_{n'})$ contains a further subsequence $\theta_{n''} \to \theta$ for some $\theta \in S$. Write $\theta_{n''} = \theta + \omega_{n''}$, so that $\omega_{n''} \to 0$. If $|<\theta_{n''}, y>| \le r_{n''}^{-1}R$ and $\|y\| \le R$, then

$$\begin{aligned} |<\theta, y>| &= |<\theta_{n''}, y> - <\omega_{n''}, y>| \\ &\le |<\theta_{n''}, y>| + |<\omega_{n''}, y>| \\ &\le r_{n''}^{-1}R + \|\omega_{n''}\| \cdot R \end{aligned}$$

which tends to zero as $n'' \to \infty$, hence for any $\delta > 0$ we have

$$\{y : |<\theta_{n''}, y>| \le r_{n''}^{-1}R\} \cap \{y : \|y\| \le \mathbb{R}\} \subseteq \{y : |<\theta, y>| \le \delta\}$$

for all large n'' or equivalently

$$\{y : |<\theta, y>| > \delta\} \subseteq \{y : |<\theta_{n''}, y>| > r_{n''}^{-1}R\} \cup B_R$$

where $\delta > 0$ is arbitrary. Hence

$$\begin{aligned} \mu_{n''}\{y : |<\theta, y>| > \delta\} &\le \mu_{n''}\{y : |<\theta_{n''}, y>| > r_{n''}^{-1}R\} + \mu_{n''}(B_R) \\ &< \varepsilon/2 + \varepsilon/2 = \varepsilon \end{aligned}$$

for all large n'', and for almost all such $\delta > 0$ we have $\phi\{y : |<\theta, y>| = \delta\} = 0$ so that

$$\mu_{n''}\{y : |<\theta, y>| > \delta\} \to \phi\{y : |<\theta, y>| > \delta\}$$

which shows that $\phi\{y : |<\theta, y>| > \delta\} \le \varepsilon$ where $\varepsilon > 0$ is arbitrary. Hence $\phi\{y : |<\theta, y>| > \delta\} = 0$, where $\delta > 0$ can be taken arbitrarily small, so that $\phi\{y : |<\theta, y>| > 0\} = 0$. But this contradicts the assumption that ϕ is $\mathcal{M}$-full, hence for some $M > 0$ we have $\|A_n\| < M$ for all n. Then (A_n) is contained in a compact subset of $\mathrm{L}(\mathbb{R}^d)$, so it is relatively compact. □

Theorem 6.1.16. *If ϕ is $\mathcal{M}$-full, then $\mathcal{S}(\phi)$ is a compact subgroup of* $\mathrm{GL}(\mathbb{R}^d)$ *and a compact subset of* $\mathrm{L}(\mathbb{R}^d)$.

Proof. If $A \in \mathcal{S}(\phi)$ and $\phi = A\phi$ is $\mathcal{M}$-full, then Lemma 6.1.12 implies that A^{-1} exists. Then $A^{-1}\phi = A^{-1}(A\phi) = (A^{-1}A)\phi = \phi$ so that $A^{-1} \in \mathcal{S}(\phi)$. If $A, B \in \mathcal{S}(\phi)$, then $(AB)\phi = A(B\phi) = A\phi = \phi$ so $AB \in \mathcal{S}(\phi)$. If $A_n \in \mathcal{S}(\phi)$ and $A_n \to A$ in $\mathrm{L}(\mathbb{R}^d)$, then $A_n\phi \to A\phi$ and $A_n\phi = \phi$ implies that $\phi = A\phi$, so $A \in \mathcal{S}(\phi)$ and A is invertible. Hence $\mathcal{S}(\phi)$ is a subgroup of $\mathrm{GL}(\mathbb{R}^d)$ that is closed in $\mathrm{L}(\mathbb{R}^d)$.

Suppose that for some sequence (A_n) in $\mathcal{S}(\phi)$ we have $\|A_n\| \to \infty$. Since $A_n\phi = \phi$, Lemma 6.1.15 implies that (A_n) is relatively compact. Choose a subsequence (n') along which $A_{n'} \to A$ in $\mathrm{L}(\mathbb{R}^d)$. Then $\|A_{n'}\| \to \|A\| < \infty$ which is a contradiction. Thus for some $M > 0$ we have $\|A_n\| < M$ for all n, so $\mathcal{S}(\phi)$ is closed and bounded, and hence compact. □

Lemma 6.1.17. *Let $\mu_n \in \mathcal{M}$, ϕ $\mathcal{M}$-full and $A_n \in \mathrm{L}(\mathbb{R}^d)$. If $\mu_n \to \phi$, then $A_n\mu_n \to \phi$ if and only if $A_n \rightharpoonup \mathcal{S}(\phi)$ in the sense of Definition 2.3.14.*

Proof. If $\mu_n \to \phi$ and $A_n\mu_n \to \phi$ $\mathcal{M}$-full, then Lemma 6.1.15 shows that (A_n) is relatively compact, so any subsequence contains a further subsequence (n') along which $A_{n'} \to A \in \mathrm{L}(\mathbb{R}^d)$. Then $A_{n'}\mu_{n'} \to A\phi$ and $A_{n'}\mu_{n'} \to \phi$ by assumption, so $A\phi = \phi$, and hence $A \in \mathcal{S}(\phi)$. Since any subsequence has a further subsequence with this property, we have $A_n \rightharpoonup \mathcal{S}(\phi)$. Conversely, if $A_n \rightharpoonup \mathcal{S}(\phi)$, then (A_n) is relatively compact, so any subsequence has a further subsequence (n') along which $A_{n'} \to A \in \mathcal{S}(\phi)$. Then $A_{n'}\mu_{n'} \to A\phi = \phi$, and since any subsequence has a further subsequence with this property, we have $A_n\mu_n \to \phi$. □

Lemma 6.1.18. *Let $\mu_n \in \mathcal{M}$, ϕ, ϕ_0 $\mathcal{M}$-full and $A_n \in \mathrm{L}(\mathbb{R}^d)$. If $\mu_n \to \phi$ and $A_n\mu_n \to \phi_0$, then $\phi_0 = C\phi$ for some $C \in \mathrm{GL}(\mathbb{R}^d)$.*

Proof. If $\mu_n \to \phi$ and $A_n\mu_n \to \phi_0$, then Lemma 6.1.15 shows that (A_n) is relatively compact, so there exists a subsequence (n') along which $A_{n'} \to C$ for some $C \in \mathrm{L}(\mathbb{R}^d)$. If $\mu_n \to \phi$, then $A_{n'}\mu_{n'} \to C\phi$, and $A_{n'}\mu_{n'} \to \phi_0$ by assumption, so that $C\phi = \phi_0$. Since ϕ_0 is $\mathcal{M}$-full, Lemma 6.1.12 implies that C is invertible. □

Theorem 6.1.19. *(Convergence of types) Let $\mu_n \in \mathcal{M}$; ϕ, ϕ_0 $\mathcal{M}$-full and $A_n, B_n \in \mathrm{L}(\mathbb{R}^d)$. Suppose that $B_n\mu_n \to \phi$. Then $A_n\mu_n \to \phi_0$ if and only if $\phi_0 = C\phi$ and $A_nB_n^{-1} \rightharpoonup C\mathcal{S}(\phi)$ for some $C \in \mathrm{GL}(\mathbb{R}^d)$.*

Proof. Since $B_n\mu_n \to \phi$ Lemma 6.1.14 implies that B_n^{-1} exists for all large n. If $A_n\mu_n \to \phi_0$, then $(A_nB_n^{-1})(B_n\mu_n) \to \phi_0$, and Lemma 6.1.18 implies that $\phi_0 = C\phi$ for some $C \in \mathrm{GL}(\mathbb{R}^d)$. Then $C^{-1}(A_nB_n^{-1})(B_n\mu_n) \to \phi$ and Lemma 6.1.17 implies that $C^{-1}(A_nB_n^{-1}) \rightharpoonup \mathcal{S}(\phi)$. Multiply both sides by C on the left. Conversely, if $\phi_0 = C\phi$ and $A_nB_n^{-1} \rightharpoonup C\mathcal{S}(\phi)$, then since $\mathcal{S}(\phi)$ is compact by Theorem 6.1.16, any subsequence contains a further subsequence (n') such that $A_{n'}B_{n'}^{-1} \to CG$ for some $G \in \mathcal{S}(\phi)$. Hence

$$A_{n'}\mu_{n'} = (A_{n'}B_{n'}^{-1})(B_{n'}\mu_{n'}) \to (CG)\phi = C(G\phi) = C\phi = \phi_0,$$

and since this is true for any subsequence we have $A_n\mu_n \to \phi_0$. □

Theorem 6.1.20. *Given $\mu \in \mathcal{M}$, ϕ $\mathcal{M}$-full and $A_n \in \mathrm{L}(\mathbb{R}^d)$, suppose that $a_n > 0$ is regularly varying with index $\alpha \neq 0$ and $a_n(A_n\mu) \to \phi$. Then there exists a linear operator E such that*

$$A_{[\lambda n]}A_n^{-1} \rightharpoonup \lambda^{-E}\mathcal{S}(\phi) \tag{6.15}$$

for all $\lambda > 0$, and furthermore (6.2) holds for all $\lambda > 0$ with $B = (1/\alpha)E$.

Proof. Let $\mu_n = a_n \cdot \mu$ so that $A_n\mu_n \to \phi$ $\mathcal{M}$-full. Then Lemma 6.1.14 implies that $A_n \in \mathrm{GL}(\mathbb{R}^d)$ for all large n. Regular variation of the sequence a_n implies that $A_{[\lambda n]}\mu_n = (a_{[\lambda n]}a_n^{-1})^{-1} \cdot A_{[\lambda n]}\mu_{[\lambda n]} \to \lambda^{-\alpha} \cdot \phi$, and since we also have $A_n\mu_n \to \phi$, Theorem 6.1.19 implies that for some linear operator $C_\lambda \in \mathrm{GL}(\mathbb{R}^d)$ we have $\lambda^{-\alpha} \cdot \phi = C_\lambda\phi$ and $A_{[\lambda n]}A_n^{-1} \rightharpoonup C_\lambda\mathcal{S}(\phi)$ for all $\lambda > 0$. Define $G_\lambda = C_\lambda\mathcal{S}(\phi)$ and let G be the union of all G_λ taken over $\lambda > 0$. The remainder of the proof is the same as Theorem 5.2.11. □

Let $\mathcal{E}(\phi)$ denote the set of all exponents of a measure $\phi \in \mathcal{M}$. The remaining results on regularly varying measures are stated without proof, since the proofs are virtually identical to the corresponding results for regularly varying functions, using the facts about regularly varying measures which we have already established.

Theorem 6.1.21. *The exponents and symmetries of ϕ $\mathcal{M}$-full are related by*

$$\mathcal{E}(\phi) = B + T\mathcal{S}(\phi) \tag{6.16}$$

where $B \in \mathcal{E}(\phi)$ is arbitrary and $T\mathcal{S}(\phi)$ denotes the tangent space. One can always select an exponent B_0 which commutes with every symmetry $A \in \mathcal{S}(\phi)$.

Corollary 6.1.22. *The nilpotent part of every $B \in \mathcal{E}(\phi)$ is the same, and it commutes with every element of the symmetry group $\mathcal{S}(\phi)$. Furthermore, every $B \in \mathcal{E}(\phi)$ has the same real spectrum, and the spectral decomposition of $\mathbb{R}^d$ is the same for every $B \in \mathcal{E}(\phi)$.*

Corollary 6.1.23. *Given $\mu \in \mathcal{M}$, ϕ $\mathcal{M}$-full and $A_n \in \mathrm{L}(\mathbb{R}^d)$, suppose $a_n > 0$ is a regularly varying sequence with index $\alpha \neq 0$ and that $a_n(A_n\mu) \to \phi$. Then $\mathcal{E}(\phi)$ is nonempty, and (6.15) holds if and only if $E = \alpha B$ for some $B \in \mathcal{E}(\phi)$. In particular, we can always choose E so that $EA = AE$ for all $A \in \mathcal{S}(\phi)$.*

Theorem 6.1.24. *Let $\mu \in \mathcal{M}$, ϕ $\mathcal{M}$-full and $A_n \in \mathrm{L}(\mathbb{R}^d)$. Suppose that $a_n > 0$ is regularly varying with index $\alpha \neq 0$ and that $a_n(A_n\mu) \to \phi$. Then μ is regularly varying, and $\mathcal{E}(\phi)$ is nonempty. Furthermore, for any $B \in \mathcal{E}(\phi)$,* *there exist $G_n \in \mathcal{S}(\phi)$ such that $B_n = G_n A_n$ is regularly varying with index $-\alpha B$ and $a_n(B_n\mu) \to \phi$. Then μ varies regularly with exponent B.*

Regular variation at infinity implies that the tails of the measure μ decay like a power function. Regular variation at zero implies that the measure μ blows up at the origin like a power function. In order to establish bounds on the rate at which this happens, we will establish a correspondence between regularly varying functions and measures by setting $F(x) = \mu(H_x)$ where

$$H_x = \{y \in \mathbb{R}^d : |<x, y>| > 1\}. \tag{6.17}$$

Then we will apply the results of Chapter 5.

Lemma 6.1.25. *If $\phi \in \mathcal{M}$ and $\varphi(x) = \phi(H_x)$, then any $x \in \Gamma$ such that $\phi\{y : |<x, y>| = 1\} = 0$ is a point of continuity of φ.*

Proof. Let $\varepsilon > 0$ be arbitrary and $x_n \to x$ in Γ. Since $\phi|_{\{y:\|y\|\geq\delta\}}$ is a finite Borel measure for any $\delta > 0$, there exists $R > 1$ such that for $D = \{y \in \mathbb{R}^d : 0 < \|y\| \leq R\}$ we have $\phi(D^c) < \varepsilon/2$ where $D^c = \Gamma \setminus D$. Let $A \Delta B$ denote the symmetric difference, so that

$$\begin{aligned} |\phi(B) - \phi(A)| &= \big|\big(\phi(B \cap A) + \phi(B \setminus A)\big) - \big(\phi(A \cap B) + \phi(A \setminus B)\big)\big| \\ &= |\phi(B \setminus A) - \phi(A \setminus B)| \\ &\leq \phi(B \setminus A) + \phi(A \setminus B) = \phi(A \Delta B) \end{aligned}$$

since $A \Delta B$ is the disjoint union of $A \setminus B$ and $B \setminus A$. Using $A \subseteq (A \cap D) \cup D^c$ we have

$$\begin{aligned} |\varphi(x_n) - \varphi(x)| &= |\phi(H_{x_n}) - \phi(H_x)| \\ &\leq \phi(H_{x_n} \Delta H_x) \\ &\leq \phi\big((H_{x_n} \Delta H_x) \cap D\big) + \phi(D^c). \end{aligned} \tag{6.18}$$

Write $x_n = x + \delta_n$, and define $\varepsilon_n = R\|\delta_n\|$, so that $|<y, \delta_n>| \leq \varepsilon_n \to 0$ for all $y \in D$. If $y \in (H_{x_n} \setminus H_x) \cap D$, then

$$\begin{aligned} |<y, x>| &= |<y, x_n> - <y, \delta_n>| \\ &\geq |<y, x_n>| - |<y, \delta_n>| \geq 1 - \varepsilon_n \end{aligned}$$

and $|<y, x>| \leq 1$, while $y \in (H_x \setminus H_{x_n}) \cap D$ implies that $|<y, x>| > 1$ and $|<y, x>| \leq |<y, x_n>| + |<y, \delta_n>| \leq 1 + \varepsilon_n$. Then we have

$$(H_{x_n} \Delta H_x) \cap D \subseteq \{y : |<y, x>| \in [1 - \varepsilon_n, 1 + \varepsilon_n]\} \cap D$$

so that as $n \to \infty$ we have

$$\begin{aligned} \phi\big((H_{x_n} \Delta H_x) \cap D\big) &\leq \phi\big(\{y : |<y, x>| \in [1 - \varepsilon_n, 1 + \varepsilon_n]\} \cap D\big) \\ &\to \phi\big(\{y : |<y, x>| = 1\} \cap D\big) = 0. \end{aligned}$$

Then we have $\phi\big((H_{x_n}\triangle H_x)\cap D\big) < \varepsilon/2$ for all n large, and since $\phi(D^c) < \varepsilon/2$ we obtain using (6.18) that $|\varphi(x_n) - \varphi(x)| < \varepsilon$ for all n large, where $\varepsilon > 0$ is arbitrary, and this implies that φ is continuous. □

Lemma 6.1.26. *Suppose $\mu_n \to \phi$ in $\mathcal{M}$, $F_n(x) = \mu_n(H_x)$ and $\varphi(x) = \phi(H_x)$. If $x_n \to x$ in Γ and $\phi\{y : |{<}x, y{>}| = 1\} = 0$, then $F_n(x_n) \to \varphi(x)$.*

Proof. Given $\varepsilon > 0$ arbitrary, choose $D = \{x : 0 < \|x\| < R\}$ so that $\phi(D^c) < \varepsilon/2$. Lemma 6.1.25 implies that φ is continuous at x, so for any $\delta > 0$ sufficiently small we have $\varphi((1+\delta)x) \le \varphi(x) + \varepsilon/2$. Write $x_n = x + \delta_n$, so that $\varepsilon_n = R\|\delta_n\| \to 0$ for all $y \in D$. If $y \in H_{x_n} \cap D$, then

$$\begin{aligned} |{<}y, x{>}| &= |{<}y, x_n{>} - {<}y, \delta_n{>}| \\ &\ge |{<}y, x_n{>}| - |{<}y, \delta_n{>}| > 1 - \varepsilon_n > 1/(1+\delta) \end{aligned}$$

for all large n, hence $H_{x_n} \cap D \subset H_{x(1+\delta)}$. Then $H_{x_n} \subset H_{x(1+\delta)} \cup D^c$, and so $F_n(x_n) = \mu_n(H_{x_n}) \le \mu_n(H_{x(1+\delta)}) + \mu_n(D^c)$. Since H_{tx} is a ϕ-continuity set for almost all $t > 0$, we can choose $\delta > 0$ such that $\phi(\partial H_{x(1+\delta)}) = 0$, and then $\mu_n(H_{x(1+\delta)}) \to \phi(H_{x(1+\delta)})$. Since D^c is closed and bounded away from zero, the Portmanteau Theorem on $\mathcal{M}$ (Proposition 1.2.19) implies that $\limsup_{n\to\infty} \mu_n(D^c) \le \phi(D^c)$, and therefore

$$\begin{aligned} \limsup_{n\to\infty} F_n(x_n) &\le \phi(H_{x(1+\delta)}) + \phi(D^c) \\ &\le \varphi(x(1+\delta)) + \varepsilon/2 \\ &\le \varphi(x) + \varepsilon. \end{aligned}$$

Since $\varepsilon > 0$ is arbitrary we have shown that

$$\limsup_{n\to\infty} F_n(x_n) \le \varphi(x). \tag{6.19}$$

Next, using the fact that φ is continuous at x again, we can choose a $\delta > 0$ such that $\varphi((1-\delta)x) \ge \varphi(x) - \varepsilon/2$. If $y \in H_{(1-\delta)x} \cap D$, then

$$\begin{aligned} |{<}y, x_n{>}| &= |{<}y, x{>} + {<}y, \delta_n{>}| \\ &\ge |{<}y, x{>}| - |{<}y, \delta_n{>}| > 1/(1-\delta) - \varepsilon_n > 1 \end{aligned}$$

for all large n, so that $H_{x_n} \supset H_{(1-\delta)x} \cap D$, hence $F_n(x_n) \ge \mu_n\big(H_{(1-\delta)x} \cap D\big)$. But $H_{(1-\delta)x} \cap D$ is open and therefore the Portmanteau Theorem implies that

$$\begin{aligned} \liminf_{n\to\infty} F_n(x_n) &\ge \phi\big(H_{(1-\delta)x} \cap D\big) \\ &= \phi(H_{(1-\delta)x}) - \phi(H_{(1-\delta)x} \cap D^c) \\ &\ge \phi(H_{(1-\delta)x}) - \phi(D^c) \\ &= \varphi((1-\delta)x) - \varepsilon/2 \\ &\ge \varphi(x) - \varepsilon \end{aligned}$$

so we also have $\liminf_{n\to\infty} F_n(x_n) \geq \varphi(x)$. This together with (6.19) concludes the proof. □

Lemma 6.1.27. *If ϕ is $\mathcal{M}$-full and satisfies (6.2) for all $\lambda > 0$, and if every eigenvalue of B has positive real part, then $\phi\{y : |<x,y>| = 1\} = 0$ for all $x \in \Gamma$, and $\varphi(x) = \phi(H_x)$ is continuous.*

Proof. In view of Lemma 6.1.25 it suffices to show that for all $x \in \Gamma$ we have $\phi\{y : |<x,y>| = 1\} = 0$. Theorem 6.1.7 implies that

$$\phi\{y : |<x,y>| = 1\} = \int_{S_0} \int_0^\infty I(|<x,t^B\theta>| = 1)\frac{dt}{t^2}\lambda\{d\theta\}.$$

The function $h(t) = <x,t^B\theta> - 1$ can be written as a power series in $u = \log t$, which is absolutely convergent for every $t > 0$, so it can be extended to an entire function on the set of complex numbers $u \in \mathbb{C}$. It is well known that an entire function has at most countably many zeroes (see, e.g., Theorem 10.18 in Rudin [157]). Hence $<x,t^B\theta> = 1$ for at most countably many $t > 0$. Analogously, we have $<x,t^B\theta> = -1$ for at most countably many $t > 0$. Then the inner integral on the right-hand side of the equation above is zero for every $\theta \in S_0$, and hence $\phi\{y : |<x,y>| = 1\} = 0$ for every $x \in \Gamma$. □

Lemma 6.1.28. *If ϕ is $\mathcal{M}$-full and $c \cdot \phi = c^B\phi$ for some $c > 1$ and some linear operator B whose eigenvalues all have positive real part, then $\varphi(x) = \phi(H_x) > 0$ for all $x \in \Gamma$.*

Proof. Suppose by way of contradiction that $\varphi(x) = 0$ for some $x \in \Gamma$. Then

$$0 = c^k \cdot \phi\{y : |<y,x>| > 1\} = \phi\{y \in \mathbb{R}^d : |<y, c^{kB^*}x>| > 1\} = 0$$

for any integer k. Write $c^{kB^*}x = r_k\theta_k$ with $r_k > 0$ and θ_k a unit vector. Then $\phi\{y \in \mathbb{R}^d : |<y,\theta_k>| > r_k^{-1}\} = 0$ for all k. Since every eigenvalue of B has positive real part, Theorem 2.2.4 implies that $r_k \to \infty$ as $k \to \infty$. By the compactness of the unit sphere there exists a subsequence (k') with $\theta_k \to \theta$ along (k'). Then we have

$$\{y \in \mathbb{R}^d : |<y,\theta>| > 0\} \subset \bigcup_{k\geq 1} \{y \in \mathbb{R}^d : |<y,\theta_k>| > r_k^{-1}\}. \tag{6.20}$$

In fact, assume that this is not the case. Then for some $y \in \mathbb{R}^d$ we have $|<y,\theta>| > 0$ but $|<y,\theta_k>| \leq r_k^{-1}$ for all $k \geq 1$. Hence

$$0 < |<y,\theta>| = \lim_{k'\to\infty} |<y,\theta_{k'}>| \leq \lim_{k'\to\infty} r_{k'}^{-1} = 0,$$

which is a contradiction. Then by (6.20) we have $\phi\{y \in \mathbb{R}^d : <y,\theta> \neq 0\} = 0$ implying that ϕ is supported on $\theta^\perp$, which contradicts the fullness of ϕ. Hence $\varphi(x) > 0$ for all $x \in \Gamma$. □

Theorem 6.1.29. *Suppose $\mu \in \mathcal{M}$ is regularly varying and thus satisfies (6.1), and H_x is as in (6.17). If μ varies regularly at infinity, then $F(x) = \mu(H_x)$ varies regularly at zero. If μ varies regularly at zero, then $F(x) = \mu(H_x)$ varies regularly at infinity. Furthermore,*

$$\lim_{t\to\infty} \frac{F(f(t)^* x_t)}{R(t)} = \varphi(x) > 0 \tag{6.21}$$

whenever $x_t \to x$ in Γ, where $\varphi(x) = \phi(H_x)$ satisfies

$$\lambda^\beta \cdot \varphi(x) = \varphi(\lambda^{-E^*} x) \tag{6.22}$$

for all $\lambda > 0$ and all $x \in \Gamma$.

Proof. Write (6.1) in the form $\mu_t \to \phi$, where $\mu_t = f(t)\mu/R(t)$ and ϕ is $\mathcal{M}$-full. Since μ varies regularly (6.2) holds, and since μ varies regularly at zero or infinity, Lemma 6.1.4 shows that the eigenvalues of B all have positive real part. Then Lemma 6.1.27 implies that φ is continuous and $\phi(\partial H_x) = 0$ for all $x \in \Gamma$, and Lemma 6.1.28 implies that $\varphi(x) > 0$, so Lemma 6.1.26 implies that $F_t(x_t) \to \varphi(x)$ whenever $x_t \to x$ in Γ, where $F_t = \mu_t(H_x)$. Then

$$\begin{aligned} F_t(x_t) &= \mu\{f(t)^{-1}y : |<x_t, y>| > 1\}/R(t) \\ &= \mu\{y : |<x_t, f(t)y>| > 1\}/R(t) \\ &= \mu\{y : |<f(t)^* x_t, y>| > 1\}/R(t) \\ &= F(f(t)^* x_t)/R(t) \to \varphi(x) \end{aligned}$$

as $t \to \infty$, hence (6.21) holds. Since $(f(t)^*)^{-1}$ varies regularly with index E^*, this shows that F varies regularly. Then Proposition 5.1.2 with $f(t)$ replaced by $(f(t)^*)^{-1}$ shows that (6.22) holds. Since the real parts of the eigenvalues of $E, -E^*$ are opposite, F varies regularly at zero when μ varies regularly at infinity, and vice versa. □

Remark 6.1.30. Theorem 6.1.29 implies that if $\mu \in \mathrm{RVM}_\infty(B)$ then $F \in \mathrm{RVF}_0(-B^*)$, and if $\mu \in \mathrm{RVM}_0(B)$, then $F \in \mathrm{RVF}_\infty(-B^*)$.

Theorem 6.1.31. *Suppose μ varies regularly with exponent $B = aI$ for some $a > 0$.*

(a) If μ varies regularly at infinity then the function

$$R(r, x) = \mu\{y \in \mathbb{R}^d : |<x, y>| > r\} \tag{6.23}$$

is uniformly regularly varying with index $-1/a$ on compact subsets of $x \in \Gamma$.

(b) If μ varies regularly at zero then the function

$$R(r, x) = \mu\{y \in \mathbb{R}^d : |<x, y>| > r^{-1}\} \tag{6.24}$$

is uniformly regularly varying with index $1/a$ *on compact subsets of* $x \in \Gamma$.

Proof. First we prove part (a). If β, E are as in Definition 6.1 then the exponent $B = (-1/\beta)E = aI$ so that $E = -a\beta I$. Apply Theorem 6.1.29 to see that $F(x) = \mu(H_x)$ is regularly varying at zero with E replaced by $-E^* = a\beta I$ in (5.3). Now apply Theorem 5.3.4 to see that $R(r, x) = F(x/r)$ is uniformly regularly varying with index $-\beta/(a\beta) = -1/a$ on compact subsets of $x \in \Gamma$.

The proof of part (b) is similar. Theorem 6.1.29 implies that $F(x) = \mu(H_x)$ is regularly varying at infinity with E replaced by $-E^* = a\beta I$ in (5.3). Then Theorem 5.3.4 implies that $R(r, x) = F(rx)$ is uniformly regularly varying with index $\beta/(a\beta) = 1/a$ on compact subsets of $x \in \Gamma$. □

If μ varies regularly at infinity with exponent $B = aI$, Theorem 6.1.31 shows that the tails of μ diminish at about the same rate in every radial direction. In particular, for any $x \in \Gamma$ and any $\varepsilon > 0$, for some $r_0 > 0$ we have

$$r^{-1/a-\varepsilon} < R(r, x) = \mu\{y \in \mathbb{R}^d : |<x, y>| > r\} < r^{-1/a+\varepsilon}$$

for all $r > r_0$. If μ varies regularly at zero with exponent $B = aI$, Theorem 6.1.31 implies that for any $x \in \Gamma$ and $\varepsilon > 0$, for some $r_0 > 0$ we have

$$r^{1/a-\varepsilon} < R(r, x) = \mu\{y \in \mathbb{R}^d : |<x, y>| > r^{-1}\} < r^{1/a+\varepsilon}$$

for all $r > r_0$. Thus the measure μ blows up at the same rate in every radial direction. When the exponent B is more complicated, the growth rate of a regularly varying measure μ can vary with radial direction. Our next result gives upper and lower bounds on how fast this can occur.

Theorem 6.1.32. *Suppose* μ *varies regularly and* (6.2) *holds for all* $\lambda > 0$, *where the real parts of the eigenvalues of* B *are* $0 < a_1 < \cdots < a_p$.

(a) If μ *varies regularly at infinity, then the function*

$$R(r, x) = \mu\{y \in \mathbb{R}^d : |<x, y>| > r\} \tag{6.25}$$

is uniformly R–O varying on compact subsets of Γ. *Furthermore, for any* $S \subset \Gamma$ *compact, and any* $\delta > 0$ *sufficiently small there exist* $r_0 > 0$, $0 < m < 1$ *and* $1 < M < \infty$ *such that*

$$m\lambda^{-\frac{1}{a_1}-\delta} \leq \frac{R(\lambda r, x)}{R(r, x)} \leq M\lambda^{-\frac{1}{a_p}+\delta} \tag{6.26}$$

for all $r \geq r_0$, $\lambda \geq 1$, *and* $x \in S$.

(b) If μ *varies regularly at zero, then the function*

$$R(r, x) = \mu\{y \in \mathbb{R}^d : |<x, y>| > r^{-1}\} \tag{6.27}$$

is uniformly R–O varying on compact subsets of Γ. *Furthermore, for any* $S \subset \Gamma$ *compact, and any* $\delta > 0$ *sufficiently small there exist* $r_0 > 0$, $0 < m < 1$ *and* $1 < M < \infty$ *such that*

$$m\lambda^{\frac{1}{a_p}-\delta} \le \frac{R(\lambda r, x)}{R(r, x)} \le M\lambda^{\frac{1}{a_1}+\delta} \tag{6.28}$$

for all $r \ge r_0$, $\lambda \ge 1$, *and* $x \in S$.

Proof. First we prove part (a). Apply Theorem 6.1.29 to see that $F(x) = \mu(H_x)$ is regularly varying at zero with E replaced by $-E^*$ in (5.3). Then apply Theorem 5.3.14 to see that $R(r, x) = F(x/r)$ is uniformly R–O varying, and in particular, for any $S \subset \Gamma$ compact, and any $\delta > 0$ sufficiently small there exist $r_0 > 0$, $0 < m < 1$ and $M > 1$ such that

$$m\lambda^{-Q-\delta} \le \frac{R(\lambda r, x)}{R(r, x)} \le M\lambda^{-q+\delta} \tag{6.29}$$

for all $r \ge r_0$, $\lambda \ge 1$, and $x \in S$ where q, Q are given by (5.50) and $\alpha_1 \cdots \alpha_j$ are the eigenvalues of $-E^*$. Since μ varies regularly at infinity, every eigenvalue of E has positive real part and $\beta < 0$. Since the real parts of the eigenvalues of E, E^* are the same, every eigenvalue of $-E^*$ has negative real part, so $q = 0 < Q$. Since $B^* = (1/\beta)(-E^*)$ we have $Q = 1/a_1$. Now apply Theorem 5.3.18 to see that we can replace q by $q_1 = 1/a_p$ and then (6.29) reduces to (6.26).

The proof of part (b) is similar. Theorem 6.1.29 implies that $F(x) = \mu(H_x)$ is regularly varying at infinity with E replaced by $-E^*$ in (5.3). Then Theorem 5.3.14 yields that $R(r, x) = F(rx)$ is uniformly R–O varying, and in particular, for any $S \subset \Gamma$ compact, and any $\delta > 0$ sufficiently small there exist $r_0 > 0$, $0 < m < 1$ and $M > 1$ such that

$$m\lambda^{q-\delta} \le \frac{R(\lambda r, x)}{R(r, x)} \le M\lambda^{Q+\delta} \tag{6.30}$$

for all $r \ge r_0$, $\lambda \ge 1$, and $x \in S$. Since μ varies regularly at zero, every eigenvalue of E has negative real part and $\beta > 0$. Then $q = 0 < Q = 1/a_1$. Use Theorem 5.3.18 to replace q by $q_1 = 1/a_p$. □

Remark 6.1.33. If μ varies regularly at infinity, then Theorem 6.1.32 (a) implies that $R(r, x) \to 0$ slower than $r^{-1/a_1-\delta} \to 0$ but faster than $r^{-1/a_p+\delta} \to 0$ as $r \to \infty$. If μ varies regularly at zero then $R(r, x) \to \infty$ slower than $r^{1/a_1+\delta} \to \infty$ but faster than $r^{1/a_p-\delta} \to \infty$. It is also possible to get sharper bounds. These results will be developed later in Section 6.4 for the more general case of R–O varying measures, since the proofs are essentially the same.

Remark 6.1.34. Since every regularly varying measure gives rise to a corresponding regularly varying function, the reader may suspect that the entire theory of regularly varying measures could be obtained by reference to facts about regularly varying functions. But clearly the transformation $F(x) = \mu(H_x)$ from measures to functions is not one-to-one. Since $H_x = H_{-x}$, if we define $\mu_1\{dx\} = \mu\{-dx\}$ then $\mu_1(H_x) = \mu(H_x)$ so that μ, μ_1 both give rise to the same function F. Therefore the converse to Lemma 6.1.26 is false: $F_n \to \varphi$ does not imply that for the corresponding measures we have $\mu_n \to \phi$. We could also employ the transformation

$$\bar{F}(x) = \phi\{y : <x, y> > 1\} \tag{6.31}$$

without absolute values. This transformation has the drawback that even if μ is a full probability measure, the function $\bar{F}(x)$ may vanish on some rays. For example, if μ is supported on the first quadrant in $\mathbb{R}^2$ then $\bar{F}(x) = 0$ in the entire third quadrant. It is possible to develop a limited theory of regularly varying functions where the limit $\varphi(x) \geq 0$ in (5.2) (see, e.g., [110] Chapter 3). Unfortunately the transformation (6.31) is still not one-to-one. The Cramér-Wold device (see, e.g., [17] pp. 48–49) implies that if ϕ_1, ϕ_2 are finite Borel measures on $\mathbb{R}^d$, then

$$\bar{F}_1(x) = \phi_1\{y : <x, y> > 1\} = \phi_2\{y : <x, y> > 1\} = \bar{F}_2(x)$$

for all $x \in \Gamma$ implies that $\phi_1 = \phi_2$. But this argument fails if the measures involved have a singularity at the origin, as shown in Example 6.1.35.

Example 6.1.35. On $\mathbb{R}^2$ we define for Borel subsets $A \subset \Gamma$

$$\phi(A) = \int_{\|u\|=1} \int_0^\infty I(ru \in A) \frac{dr}{r^2} \lambda\{du\} \tag{6.32}$$

where λ is a finite Borel measure on the unit sphere $S^1 = \{u \in \mathbb{R}^2 : \|u\| = 1\}$. For $t > 0$ and $\|\theta\| = 1$, compute that

$$\begin{aligned} \phi\{y : <y, \theta> > t\} &= \int_{\|u\|=1} \int_0^\infty I(<ru, \theta> > t) \frac{dr}{r^2} \lambda\{du\} \\ &= \int_{\|u\|=1} \int_0^\infty I(<u, \theta> > t/r) \frac{dr}{r^2} \lambda\{du\} \\ &= \int_{<u,\theta>>0} \int_{t/<u,\theta>}^\infty \frac{dr}{r^2} \lambda\{du\} \qquad (6.33) \\ &= \frac{1}{t} G(\theta) \end{aligned}$$

where

$$G(\theta) = \int_{<u,\theta>>0} <u, \theta> \lambda\{du\}. \tag{6.34}$$

Now let λ_1 be the Lebesgue measure on S^1 with $\lambda_1(S^1) = 2\pi$. Since λ_1 is rotation invariant it follows that the corresponding function G_1 defined by

(6.34) is constant, say $G_1(\theta) = D > 0$ for all $\theta \in S^1$ and some constant D. Now let

$$V = \bigcup_{n=0}^{2} \{(\cos t, \sin t) : \frac{2n\pi}{3} \leq t < \frac{(2n+1)\pi}{3}\}$$

and define $\lambda_2(A) = C\lambda_1(A \cap V)$ for some $C > 0$, and let G_2 be the corresponding function defined by (6.34) for λ_2. A simple but tedious calculation shows that $G_2(\theta)$ is constant for all $\theta \in S^1$. Hence an appropriate choice of C yields $G_1(\theta) = G_2(\theta)$ for all $\theta \in S^1$. Now let ϕ_1 and ϕ_2 be defined by (6.32) with λ_1 and λ_2, respectively. Given $x \in \Gamma$, write $x = r\theta$ where $r > 0$ and $\|\theta\| = 1$. In view of (6.33), we have $\bar{F}_1(x) = \phi_1\{y : <x, y> > 1\} = \phi_1\{y : <y, \theta> > r^{-1}\} = r^{-1}G_1(\theta) = r^{-1}G_2(\theta) = \phi_2\{y : <x, y> > 1\} = \bar{F}_2(x)$ for all $x \in \Gamma$, but since $\operatorname{supp} \phi_1 \neq \operatorname{supp} \phi_2$, where supp denotes the support, we have $\phi_1 \neq \phi_2$.

In Chapter 8 we will also consider the special case of norming by scalars in Definition 6.1.1. For $a > 0$ and $\mu \in \mathcal{M}$ we define $a\mu = (aI)\mu$, so that $a\mu\{dx\} = \mu\{a^{-1}dx\}$. To simplify the following, we only consider the case of regular variation at infinity, which is what we need in Chapter 8.

Definition 6.1.36. We will say that μ is *scalar-normed regularly varying* (at infinity) if there exist $a : \mathbb{R}^+ \to \mathbb{R}^+$ regularly varying with index $\rho < 0$, and $R : \mathbb{R}^+ \to \mathbb{R}^+$ regularly varying with index $\beta \neq 0$, such that

$$\frac{a(t)\mu}{R(t)} \to \phi \tag{6.35}$$

as $t \to \infty$ for some ϕ that is $\mathcal{M}$-*full*.

Most of the results of Section 6.1 simplify considerably in this case. It is easy to check, by specializing the arguments in this section, that the following properties hold.

Proposition 6.1.37. *Suppose that $\mu \in \mathcal{M}$ and ϕ is $\mathcal{M}$-full.*

(a) If μ is scalar-normed regularly varying and (6.35) holds, then $\lambda \cdot \phi = \lambda^b \phi$ for all $\lambda > 0$, where $b = (\rho/\beta)$. We will call b an exponent of μ (or ϕ).

(b) If μ is scalar-normed regularly varying with exponent b, then $b > 0$ and $\mu \in \mathrm{RVM}_\infty(bI)$.

(c) The measure μ is scalar-normed regularly varying with exponent b if and only if $n(a_n\mu) \to \phi$ for some regularly varying sequence of positive reals (a_n) with index $-b$.

(d) If $n(a_n\mu) \to \phi$ for some sequence of positive reals (a_n), then (a_n) varies regularly with index $-b$ for some $b > 0$. Hence μ is scalar-normed regularly varying with exponent b and $\lambda \cdot \phi = \lambda^b \phi$ for all $\lambda > 0$.

Remark 6.1.38. Similar results pertain to the special case where the norming operators $f(t)$ are diagonal with respect to some fixed basis for $\mathbb{R}^d$. See Meerschaert [110, 115] for additional details.

Remark 6.1.39. Let $\mathcal{N}$ denote the class of σ-finite Borel measures on $\mathbb{R}^d$ that assign finite measure to bounded sets. For $\mu_n, \mu \in \mathcal{N}$, we will write $\mu_n \to \mu$ if $\mu_n(A) \to \mu(A)$ for all μ-continuity sets A that are bounded. A completely parallel theory of regular variation can be constructed for these measures. For example, Lebesgue measure on $\mathbb{R}^d$ varies regularly in this sense.

6.2 R–O VARYING MEASURES

The theory of regularly varying functions and measures developed above will be used to analyze generalized domains of attraction of operator stable laws. For generalized domains of semistable attraction, the more general theory of R–O variation is needed. Roughly speaking, regular variation describes power law growth, which is a scaling property. If $R(t) = t^\rho$ then $R(\lambda t) = \lambda^\rho R(t)$ for all $\lambda > 0$. Now we weaken this scaling property, so that we only assume $R(ct) = c^\rho R(t)$ for a single $c > 1$. Of course, this implies that $R(c^n t) = c^{n\rho} R(t)$ for all integers n, but the behavior of $R(t)$ for $t \in (1, c]$ is essentially arbitrary. In R–O variation, we generalize this kind of power law growth along a geometric sequence of positive real scaling parameters, in the same way that regular variation generalizes power law growth along the entire set of positive real scales. In this section, we restrict our attention to R–O variation at infinity, which is the relevant case for probability theory.

Definition 6.2.1. Suppose that $\mu \in \mathcal{M}$, ϕ is $\mathcal{M}$-full. We say that μ is *R–O varying* if there exists $f : \mathbb{R}^+ \to \mathrm{GL}(\mathbb{R}^d)$ regularly varying with index $-E$ and $h_t > 0$ increasing with $h_t \to \infty$ and $h_{t+1}/h_t \to c \geq 1$ as $t \to \infty$ such that

$$h_t(f(h_t)\mu) \to \phi \quad \text{as } t \to \infty. \tag{6.36}$$

In this case, we write $\mu \in \mathrm{ROV}(E, c)$. The linear operator E is called an *exponent* of μ and c is called the *scale*. If every eigenvalue of E has positive real part we say that μ is *R–O varying at infinity* and we write $\mu \in \mathrm{ROV}_\infty(E, c)$.

As in the theory of regularly varying functions and measures there is an equivalent definition of R–O variation in terms of sequences.

Lemma 6.2.2. *Suppose $c \geq 1$ and ϕ is $\mathcal{M}$-full. Then $\mu \in \mathrm{ROV}(E, c)$ if and only if there exist a sequence $(B_n) \in \mathrm{RV}(-E)$ and an increasing sequence of natural numbers (k_n) tending to infinity with $k_{n+1}/k_n \to c$ such that*

$$k_n(B_{k_n}\mu) \to \phi \quad \textit{as } n \to \infty. \tag{6.37}$$

Proof. Assume (6.36) holds. Put $B_n = f(n)$ and $k_n = [h_n]$. Then it is easy to see that $k_{n+1}/k_n \to c$ and by Theorem 4.2.9 (B_n) is regularly varying with index $-E$. Furthermore,

$$k_n(B_{k_n}\mu) = \frac{[h_n]}{h_n} \cdot f\left(\frac{[h_n]}{h_n}h_n\right) f(h_n)^{-1} \cdot h_n(f(h_n)\mu) \to \phi$$

by the Uniform Convergence Theorem 4.2.1.

On the other hand, if (6.37) holds, let $f(t) = B_{[t]}$ and $h_t = k_{[t]}$. Then, using Theorem 4.2.9 again, f is varying regularly with index $-E$ and since $[t+1] = [t] + 1$ we have $h_{t+1}/h_t \to c$. Since

$$h_t(f(h_t)\mu) = k_{[t]}(B_{k_{[t]}}\mu) \to \phi$$

as $t \to \infty$, the assertion follows. □

In what follows, when working with sequences, we use (6.37) as an alternative definition of R–O variation for measures $\mu \in \mathcal{M}$.

Roughly speaking, regular variation at infinity for measure means that the tails fall of like a power function, so that any linear rescaling amounts to a power law change in the tails of the measure. R–O varying measures have the same kind of scaling behavior, but only along a geometric subsequence of scales. Then it is to be expected that the class of R–O varying measures is larger, and contains the class of regularly varying measures. The exact relationship between the class RVM_∞ and the class ROV_∞ is given by the following result.

Theorem 6.2.3. *Let $\mu \in \mathcal{M}$.*

(a) If $\mu \in \mathrm{RVM}_\infty(E)$, then $\mu \in \mathrm{ROV}_\infty(E, c)$ for every $c \geq 1$.

(b) If $\mu \in \mathrm{ROV}_\infty(E, 1)$, then $\mu \in \mathrm{RVM}_\infty(E)$.

Proof. Note that by Proposition 6.1.10 $\mu \in \mathrm{RVM}_\infty(E)$ if and only if there exists a sequence (B_n) varying regularly with index $-E$ such that

$$n \cdot (B_n\mu) \to \phi \quad \text{as } n \to \infty. \tag{6.38}$$

(a) Assume $\mu \in \mathrm{RVM}_\infty(E)$ and let $c \geq 1$. If $c = 1$, then trivially $\mu \in \mathrm{ROV}_\infty(E, 1)$ and if $c > 1$ choose (k_n) with $k_{n+1}/k_n \to c$ ($k_n = [c^n]$ will do). Then (6.38) implies (6.37).

(b) If $\mu \in \mathrm{ROV}_\infty(E, 1)$, there exist $k_n \to \infty$ with $k_{n+1}/k_n \to 1$ and (B_n) regularly varying with index $(-E)$ such that (6.37) holds. For any natural number ℓ, define $m = m(\ell) = \sup\{n : k_n \leq \ell\}$. Then $m(\ell) \to \infty$ as $\ell \to \infty$ and $k_m \leq \ell < k_{m+1}$. Now let A be a Borel set bounded away from the origin with $\phi(\partial A) = 0$. Theorem 4.2.1 implies that $B_{(\ell/k_m)k_m} B_{k_m}^{-1} \to I$ and so

$$\ell(B_\ell\mu)(A) \geq k_m(B_\ell B_{k_m}^{-1})(B_{k_m}\mu)(A) \to \phi(A).$$

Similarly,

$$\ell(B_\ell\mu)(A) < k_{m+1}(B_\ell B_{k_{m+1}}^{-1})(B_{k_{m+1}}\mu)(A) \to \phi(A),$$

so (6.38) holds. This concludes the proof. □

Remark 6.2.4. In view of Theorem 6.2.3, we can assume without loss of generality that for $\mu \in \mathrm{ROV}_\infty(E, c)$ we have $c > 1$. In fact, if $\mu \in \mathrm{ROV}_\infty(E, 1)$, then $\mu \in \mathrm{RVM}_\infty(E)$, and hence $\mu \in \mathrm{ROV}_\infty(E, c)$ for any $c > 1$.

We showed in Theorem 6.2.3 that the limiting case $c = 1$ corresponds to regularly varying measures, and hence we restrict our attention to the case $c > 1$. As for regularly varying measures (see Proposition 6.1.2) the limit measure in (6.36) exhibits a certain transformation property.

Lemma 6.2.5. *If $\mu \in \mathrm{ROV}(E, c)$ for some $c > 1$, then*

$$c^E \phi = c \cdot \phi. \tag{6.39}$$

E is also called an exponent of ϕ.

Proof. Write

$$h_{t+1}(f(h_{t+1})\mu) = \frac{h_{t+1}}{h_t} \cdot f(h_{t+1})f(h_t)^{-1} \cdot h_t(f(h_t)\mu) \tag{6.40}$$

where

$$f(h_{t+1})f(h_t)^{-1} = f\left(\frac{h_{t+1}}{h_t} h_t\right) f(h_t)^{-1} \to c^{-E}$$

by Theorem 4.2.1. Then it follows from (6.40) using (6.36) that $\phi = c \cdot (c^{-E}\phi)$, which is equivalent to (6.39). □

Remark 6.2.6. If $\mu \in \mathrm{ROV}(E, c)$ for some $c > 1$ and ϕ is the limit measure in (6.37), then $\phi \in \mathrm{ROV}(E, c)$ too. In fact, we can choose $B_n = n^{-E}$ and $k_n = [c^n]$. Then $(B_n) \in \mathrm{RV}(-E)$ and (6.39) implies that

$$k_n(B_{k_n}\phi) = [c^n]\left(\frac{[c^n]}{c^n}\right)^{-E} (c^{-nE}\phi) = ([c^n]c^{-n}) \cdot \left(\frac{[c^n]}{c^n}\right)^{-E} \phi \to \phi.$$

as $n \to \infty$, so (6.37) holds.

As for regularly varying measures only certain exponents E are possible.

Lemma 6.2.7. *If $\mu \in \mathrm{ROV}(E, c)$ for some $c > 1$, then the real parts of the eigenvalues of E are greater than or equal to zero. If $\mu \in \mathrm{ROV}_\infty(E, c)$ for some $c > 1$ then the real parts of the eigenvalues of E are all greater than zero.*

Proof. Iterating (6.39) yields $c^k \cdot \phi = c^{kE}\phi$ for all integers k. Replacing t by c^k in the proof of Lemma 6.1.4 yields the assertion with a proof almost identical to the proof of that lemma. □

The following result generalizes Theorem 6.1.7 to give a disintegration formula for the limit measure ϕ for R–O varying measures.

Theorem 6.2.8. *Suppose that ϕ is $\mathcal{M}$-full and that (6.39) holds for some $c > 1$ where all real parts of the eigenvalues of E are positive. Then there exists a norm $\|\cdot\|_0$ on $\mathbb{R}^d$ such that for all Borel subsets $B \subset \Gamma$ we have*

$$\phi(B) = \int_{S_0} \int_0^{\infty} I(t^E\theta \in B) H_\theta(-dt)\lambda\{d\theta\} \tag{6.41}$$

where

$$\lambda(D) = \phi\{t^E\theta : \theta \in D; t > 1\} \tag{6.42}$$

is a finite Borel measure on the unit sphere $S_0 = \{x \in \mathbb{R}^d : \|x\|_0 = 1\}$ and $H_\theta(t)$ is a monotone decreasing function of $t > 0$ for all $\theta \in S_0$. Furthermore, we can write $H_\theta(s) = K_\theta(\log s)/s$, where $K_\theta(s)$ is a nonnegative $\log c$*-periodic function such that*

$$K_\theta(t + h)e^{-h} \le K_\theta(t - h)e^h \tag{6.43}$$

for all $t \in \mathbb{R}$, $\theta \in S_0$ and $h > 0$.

Proof. Apply Lemma 6.1.5 to obtain a norm $\|\cdot\|_0$ on $\mathbb{R}^d$ such that $t \mapsto \|t^B x\|_0$ is strictly increasing for all nonzero $x \in \mathbb{R}^d$, and such that the mapping $\Psi : (0, \infty) \times S_0 \to \Gamma$ defined by $\Psi(t, \theta) = t^B\theta$ is a homeomorphism. By iterating (6.39), we get $c^{kE}\phi = c^k \cdot \phi$ for all integers k. Let $\mathcal{B}(S_0)$ denote the class of Borel subsets of S_0. Fix $D \in \mathcal{B}(S_0)$ for a moment and consider the family of sets $B(s) = \{t^E\theta : \theta \in D, t > s\}$ for $s > 0$. Note that $c^{kE}B(s) = B(s \cdot c^k)$ for all integers k and introduce the function $H(s) = \phi(B(s))$. Then H is nonincreasing, nonnegative, and satisfies the equation

$$c^{-1}H(s) = H(s \cdot c) \quad \text{for } s > 0. \tag{6.44}$$

Consider the function $K(u) = e^u H(e^u)$, $u \in \mathbb{R}$. Putting $p = \log c$, by (6.44) we find that $K(u + p) = K(u)$ for any $u \in \mathbb{R}$. Hence we can write $H(s) = K(\log s)/s$, where $K(\cdot)$ is a nonnegative $\log c$-periodic function. Note that since H is nonincreasing, inequality (6.43) holds for K.

Introduce the function

$$N(s, D) = \phi\{t^E\theta : \theta \in D, t > s\} \quad \text{for } s > 0,\ D \in \mathcal{B}(S_0).$$

Note that by (6.42) $\lambda(D) = N(1, D)$. From the above, for any $D \in \mathcal{B}(S_0)$ there exists a nonnegative $\log c$-periodic function $K_D(\cdot)$ such that $N(s, D) = H_D(s) = K_D(\log s)/s$. If $\lambda(D) = 0$, then for $s > 0$ we can chose $n \in \mathbb{Z}$ such that $B(s) \subset B(c^n)$. Since

$$N(s, D) \le \phi(B(c^n)) = (c^{-nE}\phi)(B(1)) = c^{-n}\phi(B(1)) = c^{-n}\lambda(D) = 0$$

it follows that $N(s,\cdot)$ is absolutely continuous with respect to λ for any $s>0$. Hence there exists a function $H_\theta(s)$, $\theta \in S_0$ such that

$$N(s,D) = \int_D H_\theta(s)\lambda\{d\theta\}.$$

For all rational $t > s > 0$, we have

$$0 \le N(s,D) - N(t,D) = \int_D (H_\theta(s) - H_\theta(t))\lambda\{d\theta\}$$

for all $D \in \mathcal{B}(S_0)$. Then it follows that $H_\theta(s) - H_\theta(t) \ge 0$ off a set $U_{s,t}$ with $\lambda(U_{s,t}) = 0$. Taking the countable union of all of these sets, we see that $H_\theta(s) \ge H_\theta(t)$ for all rational $t > s > 0$ and λ-almost all $\theta \in S_0$. For irrational $s > 0$, define $H_\theta(s)$ to be the supremum of $H_\theta(t)$ taken over all rational $t > s$. Then $H_\theta(t)$ is nonnegative and nonincreasing, and well defined off a set of λ-measure zero. In view of monotone convergence, it follows that for irrational $s > 0$ we also have $N(s,D) = \int_D H_\theta(s)\lambda(d\theta)$. Moreover,

$$\begin{aligned}\int_D H_\theta(s\cdot c)\lambda\{d\theta\} &= N(s\cdot c, D) = \phi(B(s\cdot c)) = (c^{-E}\phi)(B(s))\\ &= c^{-1}N(s,D) = \int_D c^{-1}H_\theta(s)\lambda\{d\theta\}\end{aligned}$$

for all $D \in \mathcal{B}(S_0)$. Hence $c^{-1}H_\theta(s) = H_\theta(s\cdot c)$ for λ-almost all $\theta \in S_0$. Arguing as above, $H_\theta(s) = K_\theta(\log s)/s$, where $K_\theta(t)$ is a nonnegative $\log c$-periodic function of t satisfying (6.43) [since $H_\theta(\cdot)$ is nonincreasing]. Using all of the above and putting

$$B = \{t^E\theta : \theta \in D, t \in F\} \quad \text{where } D \in \mathcal{B}(S_0),\ F \in \mathcal{B}(\mathbb{R}_+) \tag{6.45}$$

we obtain, by first considering $F = (a,b]$ for some rational $0 < a < b$ that $\phi(B) = \int_D \int_F H_\theta(-dt)\lambda\{d\theta\}$. This relation is equivalent to

$$\phi(B) = \int_{S_0}\int_0^\infty 1_B(t^E\theta)H_\theta(-dt)\lambda\{d\theta\}$$

for all sets B as in (6.45). Then Theorem 1.1.3 implies that (6.41) holds for all $B \in \mathcal{B}(\mathbb{R}^d \setminus \{0\})$. □

Remark 6.2.9. If the function $K_\theta(s) = K$ for all $s > 0$ and all $\theta \in S_0$, then $H_\theta(-dt) = (dt/t^2)K$, and hence (6.41) is just (6.7) with $\nu = K\cdot\lambda$.

Now suppose that ϕ is $\mathcal{M}$-full and assume that for $\mu \in \mathcal{M}$ there exist an increasing sequence (k_n) of natural numbers tending to infinity with $k_{n+1}/k_n \to c > 1$ and a sequence $(A_n) \subset \mathrm{L}(\mathbb{R}^d)$ such that

$$k_n(A_n\mu) \to \phi \quad \text{as } n \to \infty. \tag{6.46}$$

We say that $\mu \in \Xi(\phi, c)$ if (6.46) holds. Note that by Lemma 6.1.14 for all n large A_n is invertible and μ is $\mathcal{M}$-full. Hence we will assume without loss of generality that $A_n \in \mathrm{GL}(\mathbb{R}^d)$ for all n.

Obviously, if $\mu \in \mathrm{ROV}(E, c)$ and (6.37) holds, then $\mu \in \Xi(\phi, c)$ with $A_n = B_{k_n}$. The following result shows that, surprisingly enough, the reverse implication is also true. It is the analogue of Theorem 6.1.24 for R–O varying measures.

Theorem 6.2.10. *Let ϕ be $\mathcal{M}$-full, $\bar{c} > 1$ and $\mu \in \Xi(\phi, \bar{c})$. Let $c = \bar{c}^2$. Then $\mu \in \mathrm{ROV}(E, c)$ and $c \cdot \phi = c^E \phi$ for some $E \in \mathrm{L}(\mathbb{R}^d)$. Furthermore, for some $(B_n) \in \mathrm{RV}(-E)$ we have*

$$k_n(B_{k_n}\mu) \to \phi \quad \text{as } n \to \infty \tag{6.47}$$

for some increasing sequence (k_n) of natural numbers tending to infinity with $k_{n+1}/k_n \to c$.

The proof of the theorem will be preceded by several lemmas.

Lemma 6.2.11. *Let ϕ be $\mathcal{M}$-full, $\bar{c} > 1$ and $\mu \in \Xi(\phi, \bar{c})$. Then there exists a $B \in \mathrm{GL}(\mathbb{R}^d)$ such that*

$$\bar{c} \cdot \phi = B\phi \tag{6.48}$$

holds.

Proof. Let $\rho_n = k_{n+1}(A_{n+1}\mu)$. Then by (6.46) $\rho_n \to \phi$ as $n \to \infty$. On the other hand, $(A_n A_{n+1}^{-1})\rho_n = k_{n+1}(A_n\mu) = (k_{n+1}/k_n)k_n(A_n\mu) \to \bar{c} \cdot \phi$ using (6.46) again. Since ϕ and $\bar{c} \cdot \phi$ are $\mathcal{M}$-full, Lemma 6.1.18 gives the existence of $B \in \mathrm{GL}(\mathbb{R}^d)$ such that (6.48) holds. □

Iterating (6.48) we obtain

$$\bar{c}^2 \cdot \phi = B^2\phi. \tag{6.49}$$

Lemma 6.2.12. *Let ϕ be $\mathcal{M}$-full and $\bar{c} > 1$. Then $\Xi(\phi, \bar{c}) = \Xi(\phi, \bar{c}^2)$.*

Proof. Assume $\mu \in \Xi(\phi, \bar{c})$ and (6.46) holds. Let $k'_n = k_{2n}$ and $A'_n = A_{2n}$. Then $k'_{n+1}/k'_n \to \bar{c}^2$ and by (6.46) $k'_n(A'_n\mu) \to \phi$, so $\mu \in \Xi(\phi, \bar{c}^2)$. On the other hand, if $\mu \in \Xi(\phi, \bar{c}^2)$ there exist (k_n) with $k_{n+1}/k_n \to \bar{c}^2$ and (A_n) such that (6.46) holds. Put $A'_{2n} = A_n$ and $A'_{2n+1} = B^{-1}A_n$, where B is as in (6.48). Furthermore, let $k'_{2n} = k_n$ and $k'_{2n+1} = \bar{c}k_n$. Then $k'_{n+1}/k'_n \to \bar{c}$ and it follows from (6.48) that

$$k'_n(A'_n\mu) = \begin{cases} k_\ell(A_\ell\mu) & \text{if } n = 2\ell \\ \bar{c}k_\ell(B^{-1}A_\ell\mu) & \text{if } n = 2\ell + 1 \end{cases}$$
$$\to \phi \quad \text{as } n \to \infty,$$

so $\mu \in \Xi(\phi, \bar{c})$. □

Lemma 6.2.13. *Let ϕ be $\mathcal{M}$-full and $\bar{c} > 1$ such that* (6.48) *holds. Then there exists an exponent E such that for $c = \bar{c}^2$ we have*

$$c^E \phi = c \cdot \phi. \tag{6.50}$$

Proof. Let B be as in (6.48). Then there exists a linear operator E_1 such that $B^2 = \exp(E_1)$ (see, e.g., [148] p. 291). Putting $E = E_1/\log c$ we get $B^2 = c^E$, and hence (6.49) is equivalent to (6.50). □

In view of Lemmas 6.2.11–6.2.13, for $\mu \in \Xi(\phi, c)$ we can (and hence will) assume without loss of generality that there exist an increasing sequence of natural numbers (k_n) with $k_{n+1}/k_n \to c$ and $A_n \in \mathrm{GL}(\mathbb{R}^d)$ such that (6.46) holds, and (6.50) also holds for some exponent E. Recall the definition of the symmetry group from Definition 6.1.11 (see also Theorem 6.1.16).

Lemma 6.2.14. *Suppose ϕ is $\mathcal{M}$-full, $c > 1$ and $\mu \in \Xi(\phi, c)$ such that* (6.46) *and* (6.50) *hold. Then*

(a) $c^E \mathcal{S}(\phi) = \mathcal{S}(\phi) c^E$

(b) $A_{n+1} A_n^{-1} \rightharpoonup c^{-E} \mathcal{S}(\phi)$

Proof. (a) Obviously, $\mathcal{S}(\phi) = \mathcal{S}(c \cdot \phi)$. We have to show that $c^E \mathcal{S}(\phi) c^{-E} = \mathcal{S}(\phi)$. Let $D \in c^E \mathcal{S}(\phi) c^{-E}$. Then $D = c^E A c^{-E}$ for some $A \in \mathcal{S}(\phi)$. Using (6.50), we obtain

$$\begin{aligned} D\phi &= (c^E A c^{-E})\phi = (c^E A)(c^{-E}\phi) \\ &= c^{-1} \cdot c^E (A\phi) = c^{-1} \cdot (c^E \phi) = c^{-1} c \cdot \phi = \phi, \end{aligned}$$

so $D \in \mathcal{S}(\phi)$. The proof of the other implication is similar.

(b) Let $\rho_n = k_n(A_n \mu)$. Then, by (6.46) $\rho_n \to \phi$. On the other hand, using (6.46) and (6.50) we obtain

$$(A_{n+1} A_n^{-1})\rho_n = k_n \cdot (A_{n+1}\mu) = \frac{k_n}{k_{n+1}} k_{n+1} \cdot (A_{n+1}\mu) \to c^{-1} \cdot \phi = c^{-E}\phi.$$

Then convergence of types, Theorem 6.1.19 yields the assertion. □

Proof of Theorem 6.2.10. Since ϕ is $\mathcal{M}$-full, by Theorem 6.1.16 we know that $\mathcal{S}(\phi)$ is a compact subgroup of $\mathrm{GL}(\mathbb{R}^d)$. In view of Lemma 6.2.14, we can apply Theorem 4.2.13, and hence we obtain a sequence (B_n) varying regularly with index $-E$ such that $B_{k_n} = G_n A_n$ for some $G_n \in \mathcal{S}(\phi)$. Hence, in view of Lemma 6.1.17 we have

$$k_n(B_{k_n}\mu) = k_n(G_n A_n \mu) = G_n(k_n(A_n \mu)) \to \phi$$

as $n \to \infty$. Then by Lemma 6.2.2 $\mu \in \mathrm{ROV}(E, c)$ as desired. □

Corollary 6.2.15. *If $\mu \in \Xi(\phi, c)$ and (6.46) holds with $c^E\phi = c \cdot \phi$ for some $E \in \mathrm{L}(\mathbb{R}^d)$, then $\mu \in \mathrm{ROV}(E, c)$, and for some $(B_n) \in \mathrm{RV}(-E)$ we have*

$$k_n(B_{k_n}\mu) \to \phi \quad \textit{as } n \to \infty. \tag{6.51}$$

Proof. The proof is identical to Theorem 6.2.10 except that now we are given the exponent E, so that Lemmas 6.2.11–6.2.13 are not needed. This allows us to use the same $c > 1$. □

As for regularly varying measures we also need to consider the case of scalar normed R–O varying measures.

Definition 6.2.16. Suppose that $\mu \in \mathcal{M}$ and $\phi \in \mathcal{M}$ is $\mathcal{M}$-full. We say that μ is *scalar normed R–O varying* if there exist $a : \mathbb{R}^+ \to \mathbb{R}^+$ regularly varying with index $-\rho$ for some $\rho > 0$ and $h_t > 0$ increasing with $h_t \to \infty$ and $h_{t+1}/h_t \to c \geq 1$ as $t \to \infty$ such that

$$h_t \cdot (a(h_t)\mu) \to \phi \quad \text{as } t \to \infty. \tag{6.52}$$

Now it is easy to check, by specializing the arguments in this section, that the following properties hold. Note that in this case we do not need to square the operator B in (6.48) to write $b = c^\rho$ for some $\rho \in \mathbb{R}$. Also note that in this case we can apply Corollary 4.2.15 in place of Theorem 4.2.13.

Proposition 6.2.17. *Suppose that $\mu \in \mathcal{M}$ and $\phi \in \mathcal{M}$ is $\mathcal{M}$-full.*

(a) If μ is scalar normed R–O varying and (6.52) holds where $c > 1$, then $c \cdot \phi = c^\rho\phi$. We will call ρ an exponent of μ (or ϕ).

(b) If μ is scalar normed R–O varying and (6.52) holds, then we also have $\mu \in \mathrm{ROV}_\infty(\rho I, c)$.

(c) The measure μ is scalar normed R–O varying with exponent ρ if and only if there exist a sequence (a_n) varying regularly with index $-\rho$ and an increasing sequence of natural numbers (k_n) tending to infinity with $k_{n+1}/k_n \to c$ such that $k_n(a_{k_n}\mu) \to \phi$ as $n \to \infty$.

(d) If there exist an increasing sequence of natural numbers (k_n) tending to infinity with $k_{n+1}/k_n \to c > 1$ and a sequence (a_n) of positive reals such that $k_n(a_n\mu) \to \phi$, then we have that μ is scalar normed R–O varying, $c \cdot \phi = c^\rho\phi$ for some exponent $\rho > 0$, and $a_n = b_{k_n}$ for some sequence of positive reals (b_n) that varies regularly with index $-\rho$.

6.3 TRUNCATED MOMENTS AND TAIL MOMENTS

In this section, we derive bounds on the truncated moments and tails of an R–O varying measure. We begin by showing that the tails of an R–O varying measure are uniformly R–O varying in the sense of Definition 5.3.12. Then we prove some deep results about uniformly R–O varying functions that are monotone on rays. This allows us to obtain bounds on the truncated moments and tails of an R–O varying measure. We begin by extending Theorem 6.1.32 to R–O varying measures. As usual, we define $H_x = \{y \in \mathbb{R}^d : |<x,y>| > 1\}$ and we let $\varphi(x) = \phi(H_x)$ and $F(x) = \mu(H_x)$. Since F is not regularly varying in this case, we cannot rely on Theorem 5.3.14. Since $\varphi(x)$ is not necessarily continuous, we also need the following result.

Lemma 6.3.1. *Suppose ϕ is $\mathcal{M}$-full and $\varphi(x) = \phi(H_x)$. If* (6.39) *holds for some $c > 1$, where every eigenvalue of E has positive real part, then for any compact $S \subset \Gamma$ there exist constants $m, M > 0$ such that $m \le \varphi(x) \le M$ for all $x \in S$.*

Proof. Since S is compact we have $S \subset B_R = \{y \in \mathbb{R}^d : \|y\| \le R\}$ for some $R > 0$. For any $x \in S$, if $y \in H_x$ then $1 < |<x,y>| \le R\|y\|$, so that $\|y\| > 1/R$, and hence $H_x \subset B^c_{1/R}$. Then for any $x \in S$ we have $\varphi(x) \le \phi(B^c_{1/R}) = M < \infty$ so φ is bounded from above.

Suppose by way of contradiction that for all $m \ge 1$ there exist $x_m \in S$ with $\varphi(x_m) < \frac{1}{m}$. Since S is compact we can choose a subsequence (m_k) so that $x_{m_k} \to x \in S$. Since φ is not necessarily continuous at x we have to argue as follows: Since H_{tx} is a ϕ-continuity set for almost all $t > 0$, Lemma 6.1.25 shows that tx is a continuity point of φ for some $t < 1$. Then $\varphi(tx_{m_k}) \to \varphi(tx)$ as $k \to \infty$ and $\varphi(tx_{m_k}) \le \varphi(x_{m_k}) < 1/m_k \to 0$ as $k \to \infty$ showing that $\varphi(tx) = 0$, which contradicts Lemma 6.1.28. Hence for some $m > 0$ we have $\varphi(x) \ge m$ for all $x \in S$. □

Definition 6.3.2. For $\eta, \zeta \ge 0$, the truncated moments U_ζ and tails V_η of a measure $\mu \in \mathcal{M}$ are defined by

$$\begin{aligned} U_\zeta(r,x) &= \int_{|<x,y>| \le r} |<x,y>|^\zeta \mu\{dy\} \\ V_\eta(r,x) &= \int_{|<x,y>| > r} |<x,y>|^\eta \mu\{dy\} \end{aligned} \tag{6.53}$$

whenever these integrals exist.

Remark 6.3.3. If $\mu \in \mathcal{M}$, then $R(r,x) = \mu\{y \in \mathbb{R}^d : |<x,y>| > r\}$ exists for all $r > 0$ and all $x \in \Gamma = \mathbb{R}^d \setminus \{0\}$. Then (6.53) can be rewritten in the form

$$\begin{aligned} U_\zeta(r,x) &= \int_0^r s^\zeta R(-ds,x), \\ V_\eta(r,x) &= \int_r^\infty s^\eta R(-ds,x). \end{aligned}$$

Hence the results in this section extend the results in Section 5.3 for uniformly regularly varying functions.

Theorem 6.3.4. *Suppose that $\mu \in \mathrm{ROV}(E, c)$ for some $c > 1$ and that the real parts of the eigenvalues of E are $0 < a_1 < \cdots < a_p$. Then*

$$V_0(r, x) = \mu\{y \in \mathbb{R}^d : |\langle x, y\rangle| > r\}$$

is uniformly R–O varying on compact subsets of $x \in \Gamma$. Furthermore, for any compact $S \subset \Gamma$ and $\delta > 0$ there exist $C_1 > 0$, $C_2 < \infty$ and $r_0 > 0$ such that

$$C_1 \lambda^{-\delta - 1/a_1} \leq \frac{V_0(r\lambda, x)}{V_0(r, x)} \leq C_2 \lambda^{\delta - 1/a_p} \tag{6.54}$$

for all $r \geq r_0$, all $\lambda \geq 1$ and all $x \in S$.

Proof. By Lemma 6.2.2 there exist $(B_n) \in \mathrm{RV}(-E)$ and an increasing sequence of positive integers (k_n) with $k_{n+1}/k_n \to c > 1$ such that for $A_n = B_{k_n}$ we have $k_n(A_n \mu) \to \phi$, where ϕ satisfies (6.39). Then Lemma 6.3.1 shows that $\varphi(x) = \phi(H_x)$ is bounded away from zero and infinity on compact subsets of Γ. Since the real parts of the eigenvalues of E, E^* are the same, it follows from Lemma 5.3.15 part (b) that for any $\varepsilon > 0$ sufficiently small and any compact set $S \subset \Gamma$, for all t_0 sufficiently large we have

$$\frac{\varphi(y/t)}{\varphi(y)} \geq t_0^{-1/a_1 - \varepsilon} \tag{6.55}$$

whenever $1 \leq t \leq t_0$ and $y \in S$. Similarly, it follows from part (b) of Lemma 5.3.19 that for any $\varepsilon > 0$ sufficiently small and any compact set $S \subset \Gamma$, for all t_0 sufficiently large we have

$$\frac{\varphi(y/t)}{\varphi(y)} \leq t_0^{-1/a_p + \varepsilon} \tag{6.56}$$

whenever $t \geq t_0$ and $y \in S$. Let $F(x) = \mu(H_x)$. Since $k_n(A_n \mu) \to \phi$ Lemma 6.1.26 shows that

$$k_n F(A_n^* x_n) \to \varphi(x) \tag{6.57}$$

whenever $x_n \to x$ in Γ and H_x is a ϕ-continuity set. Suppose $x_r \to x$ in Γ. Since $(B_n^*)^{-1}$ is regularly varying with index E^*, we can apply part (b) of Lemma 4.2.16 to obtain $n = n(r)$ tending to infinity as $r \to \infty$ such that $y_n = (B_{k_n}^*)^{-1}(x_r/r) = (A_n^*)^{-1}(x_r/r)$ is limit point compact with all limit points in the set $S_1 = \{y \in \mathbb{R}^d : D \leq \|y\| \leq 1\}$ for some $D > 0$. Now let $S_2 = \{y \in \mathbb{R}^d : \frac{D}{2} \leq \|y\| \leq 2\}$ and apply Lemma 6.3.1 to obtain $0 < m_0 < M_0 < \infty$ such that $m_0 \leq \varphi(x) \leq M_0$ for all $x \in S_2$. If $1 \leq s \leq 2$, we have for $y \in S_1$ that $y, sy, y/s$ are all in S_2. Now (6.55) implies that for any λ_0 sufficiently large and any $1 \leq s \leq 2$ we have

$$\frac{\varphi(y/s\lambda)}{\varphi(sy)} = \frac{\varphi(y/s\lambda)}{\varphi(y/s)} \cdot \frac{\varphi(y/s)}{\varphi(sy)} \geq \frac{m_0}{M_0} \lambda_0^{-\varepsilon - 1/a_1} \tag{6.58}$$

whenever $1 \leq \lambda \leq \lambda_0$, $y \in S_1$ and $1 \leq s \leq 2$. Similarly, (6.56) implies that for any λ_0 sufficiently large and any $1 \leq s \leq 2$ we have

$$\frac{\varphi(ys/\lambda)}{\varphi(y/s)} = \frac{\varphi(ys/\lambda)}{\varphi(ys)} \cdot \frac{\varphi(ys)}{\varphi(y/s)} \leq \frac{M_0}{m_0}\lambda_0^{\varepsilon - 1/a_p} \tag{6.59}$$

whenever $\lambda \geq \lambda_0$, $y \in S_1$ and $1 \leq s \leq 2$.

Since (y_n) is limit point compact every subsequence contains a further subsequence along which $y_n \to y \in S_1$. Fix $1 \leq \lambda \leq \lambda_0$ and choose $1 \leq s \leq 2$ such that both $H_{y/s\lambda}$ and H_{ys} are ϕ-continuity sets. Then (6.57) shows that

$$\frac{V_0(\lambda r, x_r)}{V_0(r, x_r)} = \frac{F(x_r/\lambda r)}{F(x_r/r)} = \frac{F(A_n^* y_n/\lambda)}{F(A_n^* y_n)} \geq \frac{F(A_n^* y_n/s\lambda)}{F(A_n^* y_n s)} \to \frac{\varphi(y/s\lambda)}{\varphi(ys)}$$

along such a subsequence. Since every subsequence has a further subsequence with this property, and since $V_0(r, x)$ is monotone in r, it follows from (6.58) by a contradiction argument essentially identical to the proof of Proposition 2.1.7 (b) that for any compact subset S of Γ and any λ_0 sufficiently large, for some $r_0 > 0$ we have

$$m = \frac{m_0}{M_0}\lambda_0^{-2\varepsilon - 1/a_1} \leq \frac{V_0(\lambda r, x)}{V_0(r, x)} \leq 1 = M$$

for all $r \geq r_0$, all $1 \leq \lambda \leq \lambda_0$, and all $x \in S$. Then $V_0(r, x)$ is uniformly R–O varying on compact subsets of $x \in \Gamma$ and Lemma 5.3.13 implies that

$$m\lambda^h \leq \frac{V_0(\lambda r, x)}{V_0(r, x)}$$

for all $\lambda \geq 1$, $r \geq r_0$ and $x \in S$, where

$$h = \frac{\log m}{\log \lambda_0} = -2\varepsilon - 1/a_1 + \frac{\log(m_0/M_0)}{\log \lambda_0} > -1/a_1 - \delta$$

provided that $\varepsilon < \delta/2$ and λ_0 is chosen sufficiently large.

Now suppose again that $y_n \to y \in S_1$. Fix $\lambda \geq \lambda_0$ and choose $1 \leq s \leq 2$ such that both $H_{ys/\lambda}$ and $H_{y/s}$ are ϕ-continuity sets. Then (6.57) shows that

$$\frac{V_0(\lambda r, x_r)}{V_0(r, x_r)} = \frac{F(A_n^* y_n/\lambda)}{F(A_n^* y_n)} \leq \frac{F(A_n^* y_n s/\lambda)}{F(A_n^* y_n/s)} \to \frac{\varphi(ys/\lambda)}{\varphi(y/s)}$$

along such a subsequence. Since every subsequence has a further subsequence with this property, and since $V_0(r, x)$ is monotone in r, it follows from (6.59) that for any compact subset S of Γ and any λ_0 sufficiently large, for some $r_0 > 0$ we have

$$\frac{V_0(\lambda r, x)}{V_0(r, x)} \leq \frac{M_0}{m_0}\lambda_0^{2\varepsilon - 1/a_p} = M_1$$

for all $r \geq r_0$, all $\lambda \geq \lambda_0$, and all $x \in S$. Then part (a) of Lemma 5.3.17 implies that

$$\frac{V_0(\lambda r, x)}{V_0(r, x)} \leq \frac{M}{M_1}\lambda^{H_1}$$

for all $\lambda \geq 1$, $r \geq r_0$ and $x \in S$, where

$$H_1 = \frac{\log M_1}{\log \lambda_0} = 2\varepsilon - 1/a_p + \frac{\log(M_0/m_0)}{\log \lambda_0} < -1/a_p + \delta$$

provided that $\varepsilon < \delta/2$ and λ_0 is chosen sufficiently large. □

Theorem 6.3.4 shows that the tails of an R–O varying measure μ fall off slower than $r^{-1/a_1-\delta}$ and faster than $r^{-1/a_p+\delta}$ for any $\delta > 0$. It also shows that the function

$$R(r, x) = V_0(r, x) = \mu\{y \in \mathbb{R}^d : |<x, y>| > r\}$$

is uniformly R–O varying on compact subsets of $x \in \Gamma$. In order to obtain similar bounds for the truncated moments and tails U_ζ and V_η, we need to develop some additional theory of uniformly R–O varying functions. In Section 5.3, we developed the theory of truncated moments and tails for uniformly regularly varying functions that are monotone on rays. Now we will do the same for uniformly R–O varying functions. Suppose that $R(r, \theta)$ is uniformly R–O varying according to Definition 5.3.12. Then (5.49) holds so that $R(r, \theta)$ grows no faster than r^H and diminishes no faster than r^h (recall that $h < 0 < H$).

Recall from Definition 5.3.7 that for S compact, $R : \mathbb{R}^+ \times S \to \mathbb{R}^+$ Borel measurable and $0 \leq A \leq r$ we define

$$\begin{aligned} R_\beta(r, \theta) &= \int_A^r s^\beta R(s, \theta) ds \\ R_\beta^*(r, \theta) &= \int_r^\infty s^\beta R(s, \theta) ds \end{aligned}$$

whenever these integrals exist. Our first result extends Lemma 5.3.8.

Lemma 6.3.5. *If R is uniformly R–O varying and* (5.49) *holds, then*

(a) For some $A_0 > 0$, the integral $R_\beta(r, \theta)$ exists for all $r \geq A \geq A_0$ and all $\theta \in S$

(b) For all $r > A_0$ and all $\theta \in S$, the integral $R_\beta^*(r, \theta)$ exists for $\beta < -H - 1$ and diverges for $\beta > -h - 1$

(c) If $\beta > -h - 1$ and if $R_\beta(r, \theta)$ exists for all $r \geq A$ and all $\theta \in S$, then $R_\beta(r, \theta)$ is uniformly R–O varying, and for any $\delta > 0$ small there exist $r_0 > 0$ and $M_0 > 0$ such that

$$1 \leq \frac{R_\beta(\lambda r, \theta)}{R_\beta(r, \theta)} \leq M_0 \lambda^{\beta+H+1+\delta} \tag{6.60}$$

for all $r \ge r_0$, $\lambda \ge 1$ *and* $\theta \in S$

(d) *If* $\beta < -H - 1$, *then* $R^*_\beta(r,\theta)$ *is uniformly R–O varying, and for some* $r_0 > 0$ *we have*

$$m\lambda^{\beta+h+1} \le \frac{R^*_\beta(\lambda r,\theta)}{R^*_\beta(r,\theta)} \le 1 \tag{6.61}$$

for all $r \ge r_0$, $\lambda \ge 1$ *and* $\theta \in S$

Proof. First we prove part (a). Apply Lemma 5.3.13 and let $A_0 = r_0$. Then

$$R(r,\theta) = R((r/r_0)r_0,\theta) \le M(r/r_0)^H R(r_0,\theta) \tag{6.62}$$

for all $r \ge r_0$ and all $\theta \in S$. Then $r^\beta R(r,\theta)$ is bounded on $[r_0, r]$, and hence integrable on that interval. This implies that $R_\beta(r,\theta)$ exists for any $r \ge A \ge A_0$.

Next we prove part (b). It follows from (6.62) that

$$r^\beta R(r,\theta) \le M r_0^{-H} R(r_0,\theta)\cdot r^{\beta+H}$$

so that R^*_β exists for $\beta < -H-1$. Similarly

$$r^\beta R(r,\theta) \ge m r_0^{-h} R(r_0,\theta)\cdot r^{\beta+h}$$

so that $R_\beta(r,\theta) \to \infty$ as $r \to \infty$, and hence R^*_β diverges, if $\beta > -h-1$.

Now we prove part (c). Since we can take r_0 as large as we like, we may assume without loss of generality that $r_0 = A$. If $R_\beta(r,\theta)$ exists for all $r \ge A$ and all $\theta \in S$ then

$$\begin{aligned} R_\beta(\lambda r,\theta) &= R_\beta(\lambda r_0,\theta) + \lambda^{\beta+1}\int_{r_0}^{r} s^\beta R(\lambda s,\theta)ds \\ &\le R_\beta(\lambda r_0,\theta) + \lambda^{\beta+1} M\lambda^H \int_{r_0}^{r} s^\beta R(s,\theta)ds \\ &\le R_\beta(\lambda r_0,\theta) + M\lambda^{\beta+1+H} R_\beta(r,\theta) \end{aligned}$$

for all $r \ge r_0$, $\lambda \ge 1$ and $\theta \in S$, and it follows that

$$\frac{R_\beta(\lambda r,\theta)}{R_\beta(r,\theta)} \le \frac{R_\beta(\lambda r_0,\theta)}{R_\beta(r,\theta)} + M\lambda^{\beta+1+H}$$

for all $r \ge r_0$, $\lambda \ge 1$ and $\theta \in S$. Fix $\lambda_0 > 1$ arbitrary. For $r \ge \lambda_0 r_0$, write $r = \lambda\lambda_0^n r_0$ for some $1 \le \lambda < \lambda_0$. Then

$$\begin{aligned} R_\beta(r,\theta) &\ge \int_{\lambda_0^{n-1} r_0}^{\lambda_0^n r_0} s^\beta R(s,\theta)ds \\ &= (\lambda_0^{n-1})^{\beta+1}\int_{r_0}^{\lambda_0 r_0} s^\beta R(\lambda_0^{n-1}s,\theta)ds \\ &\ge (\lambda_0^{n-1})^{\beta+1} m(\lambda_0^{n-1})^h \int_{r_0}^{\lambda_0 r_0} s^\beta R(s,\theta)ds \\ &= m(\lambda_0^{n-1})^{\beta+1+h} R_\beta(\lambda_0 r_0,\theta) \end{aligned}$$

for all $r \geq r_0\lambda_0$ and $\theta \in S$. Then

$$\frac{R_\beta(\lambda r_0, \theta)}{R_\beta(r, \theta)} \leq \frac{1}{m}(\lambda_0^{n-1})^{-(\beta+h+1)} \tag{6.63}$$

for all $r \geq \lambda_0 r_0$, $1 \leq \lambda \leq \lambda_0$, and $\theta \in S$. Given $\varepsilon > 0$ we can choose $\lambda_0 > 1$ large enough so that the right-hand side in (6.63) is bounded above by ε. Then

$$1 \leq \frac{R_\beta(\lambda r, \theta)}{R_\beta(r, \theta)} \leq \varepsilon + M\lambda^{\beta+H+1} \leq M_1\lambda_0^{\beta+H+1}$$

for all $r \geq r_0\lambda_0 = r_1$, $1 \leq \lambda \leq \lambda_0$, and $\theta \in S$. This shows that R_β is uniformly R–O varying. Now apply Lemma 5.3.13 along with monotonicity to see that

$$1 \leq \frac{R_\beta(\lambda r, \theta)}{R_\beta(r, \theta)} \leq M_0\lambda^{H_0}$$

for all $r \geq r_1$, $\lambda \geq 1$, and $\theta \in S$, where

$$H_0 = \beta + H + 1 + \frac{\log M_1}{\log \lambda_0}$$

can be made as close to $\beta + H + 1$ as we like by choosing λ_0 sufficiently large.

Now we prove part (d). If $\beta < -1 - H$, then part (b) shows that $R_\beta^*(r, \theta)$ exists for all $r > A$ and all $\theta \in S$ and

$$\begin{aligned} R_\beta^*(\lambda r, \theta) &= \lambda^{\beta+1} \int_r^\infty s^\beta R(\lambda s, \theta) ds \\ &\geq \lambda^{\beta+1} m\lambda^h \int_r^\infty s^\beta R(s, \theta) ds \\ &= m\lambda^{\beta+h+1} R_\beta^*(r, \theta) \end{aligned}$$

so that

$$m\lambda^{\beta+h+1} \leq \frac{R_\beta^*(\lambda r, \theta)}{R_\beta^*(r, \theta)} \leq 1$$

for all $r \geq r_0$, $\lambda \geq 1$ and $\theta \in S$. □

The following result extends Theorem 5.3.9. A weaker version of the following result also holds without assuming monotonicity, but we require this stronger result.

Theorem 6.3.6. *(a) Suppose that $R(r, \theta)$ is a monotone decreasing function of $r > 0$ for each $\theta \in S$. If R is uniformly R–O varying and (5.49) holds, then for any $\beta > -h - 1$ and any $\delta > 0$ small there exists $r_1 > 0$ such that*

$$m(h + \beta + 1 - \delta) \leq \frac{r^{\beta+1}R(r, \theta)}{R_\beta(r, \theta)} \leq (\beta + 1 + \delta) \tag{6.64}$$

for all $r \geq r_1$ *and* $\theta \in S$.

(b) *Suppose that* $R(r,\theta)$ *is a monotone increasing function of* $r > 0$ *for each* $\theta \in S$, *that* R *is uniformly R–O varying and* (5.49) *holds. Then for any* $\beta < -H - 1$ *we have*

$$\frac{-(\beta + H + 1)}{M} \leq \frac{r^{\beta+1}R(r,\theta)}{R_\beta^*(r,\theta)} \leq -(\beta + 1) \tag{6.65}$$

for all $r \geq r_0$ *and* $\theta \in S$.

(c) *Suppose that* $R(r,\theta)$ *is a monotone decreasing function of* $r > 0$ *for each* $\theta \in S$. *If* $\beta + 1 > 0$ *and if there exist* $m_1 > 0$ *and* $r_1 > 0$ *such that*

$$m_1 \leq \frac{r^{\beta+1}R(r,\theta)}{R_\beta(r,\theta)} \tag{6.66}$$

for all $r \geq r_1$ *and* $\theta \in S$, *then both* R, R_β *are uniformly R–O varying.*

(d) *Suppose that* $R(r,\theta)$ *is a monotone increasing function of* $r > 0$ *for each* $\theta \in S$. *If* $\beta + 1 < 0$ *and* $R_\beta^*(r,\theta)$ *exists and if there exist* $m_1 > 0$ *and* $r_1 > 0$ *such that*

$$m_1 \leq \frac{r^{\beta+1}R(r,\theta)}{R_\beta^*(r,\theta)} \tag{6.67}$$

for all $r \geq r_1$ *and* $\theta \in S$, *then both* R, R_β^* *are uniformly R–O varying.*

Proof. First we prove (a). Lemma 6.3.5 (a) implies that $R(r,\theta)$ exists for all $r \geq A$ and $\theta \in S$. Since $R(r,\theta)$ is a monotone decreasing function of $r > 0$ for all $\theta \in S$, we have

$$\begin{aligned} R_\beta(r,\theta) &= \int_A^r s^\beta R(s,\theta)ds \\ &\geq R(r,\theta)\int_A^r s^\beta ds \\ &= R(r,\theta)\left(\frac{r^{\beta+1} - A^{\beta+1}}{\beta+1}\right) \\ &= r^{\beta+1}R(r,\theta)\left(\frac{1-(A/r)^{\beta+1}}{\beta+1}\right) \end{aligned}$$

so that

$$\frac{r^{\beta+1}R(r,\theta)}{R_\beta(r,\theta)} \leq \frac{\beta+1}{1-(A/r)^{\beta+1}} \tag{6.68}$$

for all $r \geq A$ and all θ. If $\beta > -h - 1$, then $\beta + 1 > -h > 0$ so that for any $\delta > 0$, for some $r_1 > 0$, the upper bound in (6.64) is larger than the upper bound in (6.68) for all $r \geq r_1$.

From (5.49), we see that

$$m(r/s)^h \leq \frac{R(r,\theta)}{R(s,\theta)} \leq M(r/s)^H$$

for all $r \geq s \geq r_0$ and all $\theta \in S$. Then $mR(s,\theta) \leq (s/r)^h R(r,\theta)$ and

$$\begin{aligned} mR_\beta(r,\theta) - mR_\beta(r_0,\theta) &= m\int_{r_0}^r s^\beta R(s,\theta)ds \\ &\leq r^{-h}R(r,\theta)\int_{r_0}^r s^{\beta+h}ds \\ &= r^{-h}R(r,\theta)\left(\frac{r^{\beta+h+1} - r_0^{\beta+h+1}}{\beta+h+1}\right) \\ &= r^{\beta+1}R(r,\theta)\left(\frac{1-(r_0/r)^{\beta+h+1}}{\beta+h+1}\right) \end{aligned}$$

so that

$$\frac{r^{\beta+1}R(r,\theta)}{R_\beta(r,\theta) - R_\beta(r_0,\theta)} \geq \frac{m(\beta+h+1)}{1-(r_0/r)^{\beta+h+1}} \geq m(\beta+h+1) \tag{6.69}$$

for all $r \geq r_0$ and all θ, since $\beta + h + 1 > 0$. It follows from (6.63) that $R_\beta(r_0,\theta)/R_\beta(r,\theta) \to 0$ as $r \to \infty$ uniformly in θ. Then for some $r_1 > 0$ the lower bound in (6.69) is larger than the lower bound in (6.64) for all $r \geq r_1$ and all $\theta \in S$.

The proof of part (b) is similar. Note that by Lemma 6.3.5 (b), $R_\beta^*(r,\theta)$ exists if $\beta < -H - 1$. Since $R(r,\theta)$ is a monotone increasing function of $r > 0$ for all $\theta \in S$, we have

$$\begin{aligned} R_\beta^*(r,\theta) &= \int_r^\infty s^\beta R(s,\theta)ds \\ &\geq R(r,\theta)\int_r^\infty s^\beta ds \\ &= R(r,\theta)\left(\frac{-r^{\beta+1}}{\beta+1}\right) \\ &= r^{\beta+1}R(r,\theta)\left(\frac{-1}{\beta+1}\right) \end{aligned}$$

so that the upper bound in (6.65) holds for all $r \geq A$ and all $\theta \in S$, and note that $\beta < -H - 1$ implies that $\beta + 1 < -H < 0$.

From (5.49), we see that

$$m(s/r)^h \leq \frac{R(s,\theta)}{R(r,\theta)} \leq M(s/r)^H$$

for all $s \geq r \geq r_0$ and all $\theta \in S$. Then $R(s,\theta) \leq M(s/r)^H R(r,\theta)$ and so

$$\begin{aligned}
R_\beta^*(r,\theta) &= \int_r^\infty s^\beta R(s,\theta)ds \\
&\leq Mr^{-H}R(r,\theta)\int_r^\infty s^{\beta+H}ds \\
&= Mr^{-H}R(r,\theta)\left(\frac{-r^{\beta+H+1}}{\beta+H+1}\right) \\
&= Mr^{\beta+1}R(r,\theta)\left(\frac{-1}{\beta+H+1}\right)
\end{aligned}$$

so that the lower bound in (6.65) holds for all $r \geq r_0$ and all θ.

Now we prove part (c). Since the proof of the upper bound in (6.64) depends only on the fact that $R(r,\theta)$ is a monotone decreasing function of $r > 0$ and that $\beta + 1 > 0$ it is still true here. Using (6.66) along with the monotonicity of R_β, we have for some $r_1 > 0$ that

$$\begin{aligned}
1 \leq \frac{R_\beta(\lambda r,\theta)}{R_\beta(r,\theta)} &= \frac{R_\beta(\lambda r,\theta)}{(\lambda r)^{\beta+1}R(\lambda r,\theta)} \cdot \frac{r^{\beta+1}R(r,\theta)}{R_\beta(r,\theta)} \cdot \lambda^{\beta+1} \cdot \frac{R(\lambda r,\theta)}{R(r,\theta)} \\
&\leq \frac{1}{m_1} \cdot (\beta+1+\delta) \cdot \lambda^{\beta+1} \cdot 1
\end{aligned}$$

for all $r \geq r_1$, $\lambda \geq 1$ and $\theta \in S$. Then R_β is uniformly R–O varying. Also for some $r_1 > 0$ we have

$$\begin{aligned}
1 \geq \frac{R(\lambda r,\theta)}{R(r,\theta)} &= \frac{(\lambda r)^{\beta+1}R(\lambda r,\theta)}{R_\beta(\lambda r,\theta)} \cdot \frac{R_\beta(r,\theta)}{r^{\beta+1}R(r,\theta)} \cdot \lambda^{-(\beta+1)} \cdot \frac{R_\beta(\lambda r,\theta)}{R_\beta(r,\theta)} \\
&\geq m_1 \cdot \frac{1}{\beta+1+\delta} \cdot \lambda^{-(\beta+1)} \cdot 1
\end{aligned}$$

for all $r \geq r_1$, $\lambda \geq 1$ and $\theta \in S$, so R is uniformly R–O varying as well.

Now we prove part (d). Since the proof of the upper bound in (6.65) depends only on the fact that $R(r,\theta)$ is a monotone increasing function of $r > 0$ and that $\beta + 1 < 0$ it is still true here. Using (6.67) along with the monotonicity of R_β^*, we have for some $r_1 > 0$ that

$$\begin{aligned}
1 \geq \frac{R_\beta^*(\lambda r,\theta)}{R_\beta^*(r,\theta)} &= \frac{R_\beta^*(\lambda r,\theta)}{(\lambda r)^{\beta+1}R(\lambda r,\theta)} \cdot \frac{r^{\beta+1}R(r,\theta)}{R_\beta^*(r,\theta)} \cdot \lambda^{\beta+1} \cdot \frac{R(\lambda r,\theta)}{R(r,\theta)} \\
&\geq \frac{-1}{\beta+1} \cdot m_1 \cdot \lambda^{\beta+1} \cdot 1
\end{aligned}$$

for all $r \geq r_1$, $\lambda \geq 1$ and $\theta \in S$. Then R_β^* is uniformly R–O varying. Also for some $r_1 > 0$ we have

$$\begin{aligned}
1 \leq \frac{R(\lambda r,\theta)}{R(r,\theta)} &= \frac{(\lambda r)^{\beta+1}R(\lambda r,\theta)}{R_\beta^*(\lambda r,\theta)} \cdot \frac{R_\beta^*(r,\theta)}{r^{\beta+1}R(r,\theta)} \cdot \lambda^{-(\beta+1)} \cdot \frac{R_\beta^*(\lambda r,\theta)}{R_\beta^*(r,\theta)} \\
&\leq -(\beta+1) \cdot \frac{1}{m_1} \cdot \lambda^{-(\beta+1)} \cdot 1
\end{aligned}$$

for all $r \geq r_1$, $\lambda \geq 1$ and $\theta \in S$, so R is uniformly R–O varying as well. □

Corollary 6.3.7. *(a) Suppose that $R(r,\theta)$ is a monotone decreasing function of $r > 0$ for each $\theta \in S$. If $\beta + 1 > 0$, then R is uniformly R–O varying if and only if there exist $m_1 > 0$ and $r_1 > 0$ such that*

$$m_1 \leq \frac{r^{\beta+1} R(r,\theta)}{R_\beta(r,\theta)}$$

for all $r \geq r_1$ and $\theta \in S$.

(b) Suppose that $R(r,\theta)$ is a monotone increasing function of $r > 0$ for each $\theta \in S$. If $\beta+1 < 0$ and $R^*_\beta(r,\theta)$ exists, then R is uniformly R–O varying if and only if there exist $m_1 > 0$ and $r_1 > 0$ such that

$$m_1 \leq \frac{r^{\beta+1} R(r,\theta)}{R^*_\beta(r,\theta)}$$

for all $r \geq r_1$ and $\theta \in S$.

Recall from Definition 5.3.10 that for $r \mapsto R(r,\theta)$ monotone decreasing, continuous from the right, and not constant we define

$$\begin{aligned} U_\zeta(r,\theta) &= \int_0^r s^\zeta R(-ds,\theta) \\ V_\eta(r,\theta) &= \int_r^\infty s^\eta R(-ds,\theta) \end{aligned}$$

whenever these integrals exist. Our next result extends Theorem 5.3.11.

Theorem 6.3.8. *Suppose that $\zeta > \eta$ and that the integrals $U_\zeta(r,\theta), V_\eta(r,\theta)$ exist for all $r > 0$ and all $\theta \in S$. Then*

(a) If U_ζ is uniformly R–O varying and for some $r_0 > 0$, $M > 1$ and $H > 0$ we have

$$\frac{U_\zeta(\lambda r,\theta)}{U_\zeta(r,\theta)} \leq M\lambda^H \tag{6.70}$$

for all $r \geq r_0$, $\lambda \geq 1$ and $\theta \in S$, then for any $\eta < \zeta - H$ there exists $r_1 > 0$ such that

$$\frac{r^{\zeta-\eta} V_\eta(r,\theta)}{U_\zeta(r,\theta)} \leq \frac{(\zeta-\eta)(M-1)+H}{\zeta-\eta-H} \tag{6.71}$$

for all $r \geq r_1$, and $\theta \in S$

(b) If for some $r_0 > 0$ and $M_2 < \infty$ we have

$$\frac{r^{\zeta-\eta} V_\eta(r,\theta)}{U_\zeta(r,\theta)} \leq M_2 \tag{6.72}$$

for all $r \geq r_0$ and all $\theta \in S$, then U_ζ is uniformly R–O varying

(c) *If V_η is uniformly R–O varying and for some $r_0 > 0$, $0 < m < 1$ and $h < 0$ we have*

$$\frac{V_\eta(\lambda r, \theta)}{V_\eta(r, \theta)} \geq m\lambda^h \tag{6.73}$$

for all $r \geq r_0$, $\lambda \geq 1$ and $\theta \in S$, then for any $\zeta > \eta - h$ and any $\delta > 0$ there exists $r_1 > 0$ such that

$$\frac{r^{\zeta-\eta} V_\eta(r, \theta)}{U_\zeta(r, \theta)} \geq \frac{m(\zeta - \eta + h - \delta)}{(\zeta - \eta)(1 - m) + m(\delta - h)} \tag{6.74}$$

for all $r \geq r_1$, and $\theta \in S$

(d) *If for some $r_0 > 0$ and $m_2 > 0$ we have*

$$\frac{r^{\zeta-\eta} V_\eta(r, \theta)}{U_\zeta(r, \theta)} \geq m_2 \tag{6.75}$$

for all $r \geq r_0$ and all $\theta \in S$, then V_η is uniformly R–O varying

Proof. The proof is similar to Theorem 5.3.11 using Theorem 6.3.6 in place of Theorem 5.3.9. First we prove part (a). Lemma 6.3.5 shows that

$$(U_\zeta)^*_{\eta-\zeta-1}(r, \theta) = \int_r^\infty s^{\eta-\zeta-1} U_\zeta(s, \theta) ds$$

exists for all $\theta \in S$ and all $r \geq A_0$ and some $A_0 > 0$. Since U_ζ is uniformly R–O varying and monotone increasing, if $\eta < \zeta - H$ then part (b) of Theorem 6.3.6 with $\beta + 1 = \eta - \zeta$ shows that

$$\frac{r^{\eta-\zeta} U_\zeta(r, \theta)}{\int_r^\infty s^{\eta-\zeta-1} U_\zeta(s, \theta) ds} \geq \frac{-(\eta - \zeta + H)}{M}$$

for all $\theta \in S$ and r sufficiently large. Substitute into (5.42) to see that

$$\frac{r^{\zeta-\eta} V_\eta(r, \theta)}{U_\zeta(r, \theta)} \leq -1 + \frac{M(\zeta - \eta)}{-(\eta - \zeta + H)}$$

for all $\theta \in S$ and r sufficiently large, which is equivalent to (6.71).

To prove part (b), substitute (6.72) into (5.46) to obtain

$$\frac{(U_\zeta)^*_{\eta-\zeta-1}(r, \theta)}{r^{\eta-\zeta} U_\zeta(r, \theta)} = \frac{\int_r^\infty s^{\eta-\zeta-1} U_\zeta(s, \theta) ds}{r^{\eta-\zeta} U_\zeta(r, \theta)} \leq \frac{1 + M_2}{\zeta - \eta}$$

and rearrange to obtain

$$\frac{r^{\eta-\zeta} U_\zeta(r, \theta)}{(U_\zeta)^*_{\eta-\zeta-1}(r, \theta)} \geq \frac{\zeta - \eta}{1 + M_2}.$$

Now part (d) of Theorem 6.3.6 implies that U_ζ is uniformly R–O varying.

Now we prove part (c). Lemma 6.3.5(a) shows that

$$(V_\eta)_{\zeta-\eta-1}(r,\theta) = \int_0^r s^{\zeta-\eta-1} V_\eta(s,\theta)ds$$

exists for all $\theta \in S$ and all $r > 0$. Since V_η is uniformly R–O varying and monotone decreasing, if $\zeta > \eta - h$ then part (a) of Theorem 6.3.6 with $\beta + 1 = \zeta - \eta$ implies that for any $\delta > 0$ sufficiently small we have

$$\frac{r^{\zeta-\eta} V_\eta(r,\theta)}{\int_0^r s^{\zeta-\eta-1} V_\eta(s,\theta)ds} \geq m(h + \zeta - \eta - \delta)$$

for all $\theta \in S$ and r sufficiently large. Substitute into (5.44) to see that

$$\frac{U_\zeta(r,\theta)}{r^{\zeta-\eta} V_\eta(r,\theta)} \leq -1 + \frac{(\zeta - \eta)}{m(h + \zeta - \eta - \delta)}$$

for all $\theta \in S$ and r sufficiently large, which is equivalent to (6.74), and since $\zeta - \eta + h > 0$, $0 < m < 1$ and $h < 0$ this lower bound is strictly positive for $\delta > 0$ sufficiently small.

To prove part (d), substitute (6.75) into (5.47) to obtain

$$\frac{(V_\eta)_{\zeta-\eta-1}(r,\theta)}{r^{\zeta-\eta} V_\eta(r,\theta)} = \frac{\int_0^r s^{\zeta-\eta-1} V_\eta(s,\theta)ds}{r^{\zeta-\eta} V_\eta(r,\theta)} \leq \frac{1}{\zeta - \eta}\left(1 + \frac{1}{m_2}\right)$$

and rearrange to obtain

$$\frac{r^{\zeta-\eta} V_\eta(r,\theta)}{(V_\eta)_{\zeta-\eta-1}(r,\theta)} \geq (\zeta - \eta)\left(\frac{m_2}{1 + m_2}\right).$$

Now part (c) of Theorem 6.3.6 implies that V_η is uniformly R–O varying. □

Corollary 6.3.9. *Suppose that $\zeta > \eta$ and that the integrals $U_\zeta(r,\theta), V_\eta(r,\theta)$ exist for all $r > 0$ and all $\theta \in S$. Then*

(a) U_ζ is uniformly R–O varying if and only if for some $M_2 < \infty$ and $r_0 > 0$ we have

$$\frac{r^{\zeta-\eta} V_\eta(r,\theta)}{U_\zeta(r,\theta)} \leq M_2$$

for all $r \geq r_0$ and all $\theta \in S$

(b) V_η is uniformly R–O varying if and only if for some $m_2 > 0$ and $r_0 > 0$ we have

$$\frac{r^{\zeta-\eta} V_\eta(r,\theta)}{U_\zeta(r,\theta)} \geq m_2$$

for all $r \geq r_0$ *and all* $\theta \in S$

Now we will use uniform R–O variation to characterize the growth rate of the truncated moments U_ζ of an R–O varying measure μ. Instead of the functions $F(x) = \mu(H_x)$ and $\varphi(x) = \phi(H_x)$ we will use the following construction. For $x \in \Gamma$, let

$$K_x = \{y \in \mathbb{R}^d : |\langle x, y\rangle| \leq 1\} \tag{6.76}$$

and define

$$\begin{aligned} G_\zeta(x) &= \int_{K_x} |\langle x, y\rangle|^\zeta \mu\{dy\} \\ \psi_\zeta(x) &= \int_{K_x} |\langle x, y\rangle|^\zeta \phi\{dy\} \end{aligned} \tag{6.77}$$

If U_ζ in (6.3.2) exists, then G_ζ exists and

$$U_\zeta(r, x) = r^\zeta G_\zeta(x/r) \tag{6.78}$$

for all $r > 0$ and $x \in \Gamma$. If (6.39) holds, then using the fact that $A^{-1}K_x = K_{A^*x}$ it follows from (6.77) that

$$c \cdot \psi_\zeta(x) = \psi_\zeta(c^{E^*}x) \tag{6.79}$$

for all $x \in \Gamma$. Before we can show that U_ζ is uniformly R–O varying, we must first establish some properties of G_ζ and ψ_ζ.

Lemma 6.3.10. *Suppose* ϕ *is* $\mathcal{M}$*-full and* (6.39) *holds for some* $c > 1$, *where the real parts of the eigenvalues of* E *are* $0 < a_1 < \cdots < a_p$. *If we define*

$$Q = \{y \in \mathbb{R}^d : a \leq ||y|| \leq b\} \tag{6.80}$$

for some $a > 0$ *and some* $b \geq a||c^E||$ *then*

(a) *Any* $x \in \Gamma$ *can be written in the form* $x = c^{kE}y$ *for some* $y \in Q$ *and some integer* k

(b) *For any integer* k_0 *and any* $\zeta > 1/a_1$, *we have*

$$\sum_{k=k_0}^{\infty} \int_{c^{-kE}(Q)} ||y||^\zeta \phi\{dy\} < \infty \tag{6.81}$$

(c) *For any integer* k_0 *and any* $\eta < 1/a_p$, *we have*

$$\sum_{k=k_0}^{\infty} \int_{c^{kE}(Q)} ||y||^\eta \phi\{dy\} < \infty \tag{6.82}$$

Proof. First we prove part (a). Since the real parts of the eigenvalues of E are all positive, Theorem 2.2.4 implies that $\|c^{-kE}x\| \to 0$ and $\|c^{kE}x\| \to \infty$ as $k \to \infty$. If we define $k(x) = \inf\{k \in \mathbb{Z} : \|c^{kE}x\| \geq a\}$, then for $k' = k(x) - 1$ we have $\|c^{k'E}x\| < a$ and so

$$a \leq \|c^{k(x)E}x\| = \|c^E c^{k'E}x\| < \|c^E\|a \leq b$$

which implies that $y = c^{k(x)E}x \in Q$ and $x = c^{kE}y$ with $k = -k(x)$.

Next we prove part (b). Given any $\zeta > 1/a_1$ we can choose $\delta > 0$ such that $\zeta > 1/(a_1 - \delta)$. Apply Proposition 2.2.11 (d) to $-E$ to see that for any $t_0 > 0$ there exists a positive constant C such that $\|t^E\| \leq Ct^{a_1-\delta}$ for all $0 < t \leq t_0$. If $t_0 \geq c^{-k_0}$, then (6.39) implies

$$\begin{aligned}\sum_{k=k_0}^{\infty} \int_{c^{-kE}(Q)} \|y\|^\zeta \phi\{dy\} &\leq \sum_{k=k_0}^{\infty} \int_{c^{-kE}(Q)} \|c^{-kE}\|^\zeta \|c^{kE}y\|^\zeta \phi\{dy\} \\ &\leq C^\zeta \sum_{k=k_0}^{\infty} c^{-k\zeta(a_1-\delta)} \int_Q \|y\|^\zeta (c^{kE}\phi)\{dy\} \\ &= C^\zeta \sum_{k=k_0}^{\infty} c^{k(1-\zeta(a_1-\delta))} \int_Q \|y\|^\zeta \phi\{dy\}\end{aligned} \tag{6.83}$$

and the series converges since $1 - \zeta(a_1 - \delta) < 0$.

The proof of part (c) is similar. If $\eta < 1/a_p$, choose $\delta > 0$ such that $\eta < 1/(a_p + \delta)$ and apply Proposition 2.2.11 (d) to obtain a positive constant C such that $\|t^E\| \leq Ct^{a_p+\delta}$ for all $t \geq t_0$. If $t_0 \leq c^{k_0}$, then (6.39) implies

$$\begin{aligned}\sum_{k=k_0}^{\infty} \int_{c^{kE}(Q)} \|y\|^\eta \phi\{dy\} &\leq \sum_{k=k_0}^{\infty} \int_{c^{kE}(Q)} \|c^{kE}\|^\eta \|c^{-kE}y\|^\eta \phi\{dy\} \\ &\leq C^\eta \sum_{k=k_0}^{\infty} c^{k\eta(a_p+\delta)} \int_Q \|y\|^\eta (c^{-kE}\phi)\{dy\} \\ &= C^\eta \sum_{k=k_0}^{\infty} c^{k(\eta(a_p+\delta)-1)} \int_Q \|y\|^\eta \phi\{dy\}\end{aligned} \tag{6.84}$$

and the series converges since $\eta(a_p + \delta) < 1$. □

Lemma 6.3.11. *Suppose ϕ is $\mathcal{M}$-full and (6.39) holds for some $c > 1$, where every eigenvalue of E has real part greater than or equal to $a_1 > 0$. Then*

$$\int_{0<\|y\|<R} \|y\|^\zeta \phi\{dy\} < \infty \tag{6.85}$$

for all $R > 0$ and $\zeta > 1/a_1$. Furthermore, if $\zeta > 1/a_1$ then the function $\psi_\zeta(x)$ defined by (6.77) exists for all $x \in \Gamma$.

Proof. Given any $R > 0$, since Theorem 2.2.4 implies that $\|t^E x\| \to \infty$ uniformly in $x \in Q$ as $t \to \infty$, where $Q \subset \Gamma$ is compact , for some k_0 we have $\|c^{kE}x\| > R$ for all $k \geq k_0$ and all $x \in Q$. In view of Lemma 6.3.10 (a), this implies that

$$\{y \in \mathbb{R}^d : 0 < \|y\| < R\} \subset \bigcup_{k=-k_0}^{\infty} c^{-kE}(Q)$$

and then part (b) of Lemma 6.3.10 yields

$$\int_{0<\|y\|<R} \|y\|^\zeta \phi\{dy\} \leq \sum_{k=-k_0}^{\infty} \int_{c^{-kE}(Q)} \|y\|^\zeta \phi\{dy\} < \infty$$

so that (6.85) holds. Now if $x \in \Gamma$ then for any $R > 0$ we have

$$\begin{aligned}
\psi_\zeta(x) = & \int\limits_{|<x,y>|\leq 1} |<x,y>|^\zeta I(0 < \|y\| < R)\phi\{dy\} \\
& + \int\limits_{|<x,y>|\leq 1} |<x,y>|^\zeta I(\|y\| \geq R)\phi\{dy\} \\
& \leq \|x\|^\zeta \int_{0<\|y\|<R} \|y\|^\zeta \phi\{dy\} + \phi\{y \in \mathbb{R}^d : \|y\| \geq R\}
\end{aligned}$$

which is finite in view of (6.85) along with the fact that $\phi \in \mathcal{M}$ assigns finite measure to sets bounded away from the origin. □

The next result is similar to Lemma 6.1.25.

Lemma 6.3.12. *Suppose ϕ is $\mathcal{M}$-full and (6.39) holds for some $c > 1$, where every eigenvalue of E has real part greater than or equal to $a_1 > 0$. Then any $x \in \Gamma$ such that $\phi\{y : |<x,y>| = 1\} = 0$ is a point of continuity of the function $\psi_\zeta(x)$ defined by (6.77) with $\zeta > 1/a_1$.*

Proof. Let $\varepsilon > 0$ be arbitrary. Define $D = \{x : 0 < \|x\| < R\}$ where $R > 0$ is chosen such that $\phi(D^c) < \varepsilon/2$. If $x_n \to x$ in Γ, then it is not hard to check that

$$\begin{aligned}
& |\psi_\zeta(x_n) - \psi_\zeta(x)| \\
& \leq \left| \int_{K_{x_n}\cap D} |<x_n,y>|^\zeta \phi\{dy\} - \int_{K_x \cap D} |<x,y>|^\zeta \phi\{dy\} \right| + \varepsilon \\
& \leq \int_{K_{x_n}\cap D} \left| |<x_n,y>|^\zeta - |<x,y>|^\zeta \right| \phi\{dy\} \\
& \quad + \int_{(K_{x_n} \triangle K_x)\cap D} |<x,y>|^\zeta \phi\{dy\} + \varepsilon.
\end{aligned} \tag{6.86}$$

Write $x_n = x + \delta_n$, and define $\varepsilon_n = R\|\delta_n\|$, so that $|\langle y, \delta_n\rangle| \le \varepsilon_n \to 0$ for all $y \in D$. If $y \in (K_{x_n} \setminus K_x) \cap D$, then

$$1 < |\langle y, x\rangle| \le |\langle y, x_n\rangle| + |\langle y, \delta_n\rangle| \le 1 + \varepsilon_n$$

while $y \in (K_x \setminus K_{x_n}) \cap D$ implies $1 \ge |\langle y, x\rangle| \ge |\langle y, x_n\rangle| - |\langle y, \delta_n\rangle| \ge 1 - \varepsilon_n$. Then

$$\begin{aligned}\int_{(K_{x_n} \triangle K_x)\cap D} |\langle x, y\rangle|^\zeta \phi\{dy\} &\le R^\zeta \|x\|^\zeta \phi\{y : |\langle x, y\rangle| \in [1 - \varepsilon_n, 1 + \varepsilon_n]\} \\ &\to R^\zeta \|x\|^\zeta \phi\{y : |\langle x, y\rangle| = 1\} = 0\end{aligned}$$

as $n \to \infty$. To estimate the first integral on the right-hand side of (6.86) note that $f_n(y) = |\langle x_n, y\rangle|^\zeta - |\langle x, y\rangle|^\zeta \to 0$ as $n \to \infty$ for all $y \in \mathbb{R}^d$. Furthermore, there exists a constant $C > 0$ such that $|f_n(y)| \le C\|y\|^\zeta$, which by Lemma 6.3.11 is integrable with respect to ϕ on the set D. Then Lebesgue's dominated convergence theorem implies

$$\int_{K_{x_n} \cap D} |f_n(y)| \phi\{dy\} \le \int_D |f_n(y)| \phi\{dy\} \to 0$$

as $n \to \infty$. Hence $\lim_{n\to\infty} |\psi_\zeta(x_n) - \psi_\zeta(x)| \le \varepsilon$ arbitrary, so ψ_ζ is continuous at the point x. □

Lemma 6.3.13. *Suppose ϕ is $\mathcal{M}$-full and* (6.39) *holds for some $c > 1$, where every eigenvalue of E has real part greater than or equal to $a_1 > 0$. Then for any $\zeta > 1/a_1$ and any compact set $S \dot{\subset} \Gamma$ there exist constants $M, m > 0$ such that $m \le \psi_\zeta(x) \le M$ for all $x \in S$, where $\psi_\zeta(x)$ is defined by* (6.77).

Proof. The proof is similar to Lemma 6.3.1. There exists a $B \ge 1$ such that $\|x\| \le B$ for all $x \in S$. Then for any $R > 0$ we have

$$\begin{aligned}\psi_\zeta(x) = &\int_{|\langle x,y\rangle| \le 1} |\langle x, y\rangle|^\zeta I(0 < \|y\| < R) \phi\{dy\} \\ &+ \int_{|\langle x,y\rangle| \le 1} |\langle x, y\rangle|^\zeta I(\|y\| \ge R) \phi\{dy\} \\ \le &\, B^\zeta \int_{0<\|y\|<R} \|y\|^\zeta \phi\{dy\} + \phi\{y \in \mathbb{R}^d : \|y\| \ge R\}\end{aligned}$$

and now it follows from Lemma 6.3.11 that $\psi_\zeta(x) \le M$ for all $x \in S$ and some constant $M > 0$.

To prove the lower bound, we first show that $\psi_\zeta(x) \ne 0$ whenever $x \in \Gamma$. Assume there exists $x \in \Gamma$ with $\psi_\zeta(x) = 0$. Writing $c^{-kE^*}x = r_k\theta_k$ for some $r_k > 0$ and $\|\theta_k\| = 1$ it follows from (6.79) that

$$\psi_\zeta(c^{-kE^*}x) = \int_{|\langle y,\theta_k\rangle| \le r_k^{-1}} |\langle y, \theta_k\rangle|^\zeta \phi\{dy\} = 0$$

for all $k \in \mathbb{Z}$. Hence the set $N_k = \{y : 0 < |<y, \theta_k>| \le r_k^{-1}\}$ has ϕ-measure zero for every integer k. Since the real parts of the eigenvalues of E^* are all positive, Theorem 2.2.4 shows that $r_k \to 0$ as $k \to \infty$. Since the unit sphere in $\mathbb{R}^d$ is compact, there exists a subsequence $k' \to \infty$ such that $\theta_{k'} \to \theta$ along (k'). If $<y, \theta> \ne 0$ then $\lim_{k'\to\infty} <y, \theta_{k'}> \ne 0$ as well, and since $r_{k'}^{-1} \to \infty$ we must have $y \in N_{k'}$ for all large k'. Then

$$N = \{y : |<y, \theta>| \ne 0\} \subset \bigcup_{k=1}^{\infty} N_k$$

and hence $\phi(N) = 0$, which contradicts the assumption that ϕ is $\mathcal{M}$-full. Then $\psi_\zeta(x) \ne 0$ whenever $x \in \Gamma$.

Now let S be as above and assume that for all $n \ge 1$ there exists $x_n \in S$ such that $\psi_\zeta(x_n) < 1/n$. Choose a subsequence (n') with $x_{n'} \to x \in S$ and $t > 1$ such that K_{tx} is a ϕ-continuity set. Then Lemma 6.3.12 implies that ψ_ζ is continuous at the point tx, so $\psi_\zeta(tx_{n'}) \to \psi_\zeta(tx)$. But we also have $\psi_\zeta(tx_{n'}) \le t^\zeta \psi_\zeta(x_{n'}) < t^\zeta/n' \to 0$ as $n' \to \infty$, which is a contradiction. Hence, for some $m > 0$ we have $\psi_\zeta(x) > m$ for all $x \in S$. □

Recall from Definition 1.2.12 that $\mathcal{M}_0$ is the collection of finite Borel measures on $\mathbb{R}^d$, and that $\mu_n \to \mu$ in $\mathcal{M}_0$ if $\mu_n(A) \to \mu(A)$ for all μ-continuity sets $A \subseteq \mathbb{R}^d$.

Lemma 6.3.14. *Suppose $\mu \in \mathrm{ROV}(E, c)$ for some $c > 1$ and (6.37) holds, where every eigenvalue of E has real part greater than or equal to $a_1 > 0$, and write $A_n = B_{k_n}$. For $\zeta > 1/a_1$ let $\mu_n\{dx\} = k_n\|x\|^\zeta I(\|x\| < R)(A_n\mu)\{dx\}$ and $\rho\{dx\} = \|x\|^\zeta I(\|x\| < R)\phi\{dx\}$. If U_ζ in (6.3.2) exists and $\phi\{\|x\| = R\} = 0$, then $\mu_n \to \rho$ in $\mathcal{M}_0$.*

Proof. Lemma 6.3.11 implies that ρ is a finite Borel measure on $\mathbb{R}^d$, and if U_ζ exists then $\mu_n \in \mathcal{M}_0$ too. In fact if $\{e_1, \ldots, e_d\}$ denotes the standard basis of $\mathbb{R}^d$ we get for the Euclidean norm $\|\cdot\|_2$ that $\|x\|_2^\zeta \le 2^{\zeta/2} \sum_{i=1}^d |<x, e_j>|^\zeta$. Since all norms on $\mathbb{R}^d$ are equivalent, we are free to use this norm, and hence obtain

$$\begin{aligned}
\mu_n(\mathbb{R}^d) &= k_n \int_{\|x\|<R} \|x\|_2^\zeta (A_n\mu)\{dx\} \\
&\le 2^{\zeta/2} k_n \sum_{j=1}^{d} \int_{|<x,e_j>|\le R} |<x, e_j>|^\zeta (A_n\mu)\{dx\} \\
&= 2^{\zeta/2} k_n \sum_{j=1}^{n} U_\zeta(R, A_n^* e_j) < \infty.
\end{aligned}$$

We first show that

$$\lim_{\varepsilon\to 0} \limsup_{n\to\infty} \mu_n\{\|x\| < \varepsilon\} = 0. \tag{6.87}$$

Apply Lemma 6.3.10 where a, b are chosen so that Q is a ϕ-continuity set. Then, in view of Lemma 6.2.5 $c^{kE}Q$ is a ϕ-continuity set for all k. Let $k(\varepsilon) = \inf\{k : a/\|c^{kE}\| \leq \varepsilon\}$, so that $k(\varepsilon) \to \infty$ as $\varepsilon \to 0$. If $y = c^{-kE}x$ for some $k < k(\varepsilon)$ and $x \in Q$ then $\|c^{-kE}x\| \geq a/\|c^{kE}\| > \varepsilon$ by Proposition 2.1.3 (b), so in view of Lemma 6.3.10 (a) we have

$$\{\|y\| < \varepsilon\} \subset \bigcup_{k=k(\varepsilon)}^{\infty} c^{-kE}(Q).$$

Since $k_n(A_n\mu) \to \phi$ in $\mathcal{M}$ we have $\mu_n\big(c^{-kE}(Q)\big) \to \rho\big(c^{-kE}(Q)\big)$ as $n \to \infty$ for each k in view of Proposition 1.2.20, and hence

$$\begin{aligned}
\limsup_{n\to\infty} \mu_n\{\|x\| < \varepsilon\} &\leq \sum_{k=k(\varepsilon)}^{\infty} \limsup_{n\to\infty} \mu_n\big(c^{-kE}(Q)\big) \\
&= \sum_{k=k(\varepsilon)}^{\infty} \rho\big(c^{-kE}(Q)\big) \\
&= \sum_{k=k(\varepsilon)}^{\infty} \int_{c^{-kE}(Q)} \|x\|^{\zeta} I(\|x\| < R)\phi\{dx\} \\
&\leq \sum_{k=k(\varepsilon)}^{\infty} \int_{c^{-kE}(Q)} \|x\|^{\zeta} \phi\{dx\}.
\end{aligned}$$

Part (b) of Lemma 6.3.10 implies that this series converges, hence it tends to zero as $k(\varepsilon) \to \infty$, which proves (6.87).

Let g be any bounded continuous function on $\mathbb{R}^d$. For $\delta > 0$ arbitrary, choose $\varepsilon > 0$ such that $\limsup_{n\to\infty} \mu_n\{\|x\| < \varepsilon\} < \delta/(2\|g\|_\infty)$ by (6.87), $\rho\{\|x\| = \varepsilon\} = 0$ and $\rho\{\|x\| < \varepsilon\} < \delta/(2\|g\|_\infty)$. Write

$$\begin{aligned}
\left|\int g d\mu_n - \int g d\rho\right| &\leq \|g\|_\infty \mu_n\{\|x\| < \varepsilon\} \\
&\quad + \left|\int_{\|x\|\geq\varepsilon} g d\mu_n - \int_{\|x\|\geq\varepsilon} g d\rho\right| + \|g\|_\infty \rho\{\|x\| < \varepsilon\}.
\end{aligned}$$

Since $k_n(A_n\mu) \to \phi$ in $\mathcal{M}$, Proposition 1.2.20 yields that

$$\int_{\|x\|\geq\varepsilon} g d\mu_n \to \int_{\|x\|\geq\varepsilon} g d\rho$$

and this together with the choice of $R > 0$ and $\varepsilon > 0$ implies

$$\limsup_{n\to\infty} \left|\int g d\mu_n - \int g d\rho\right| < \frac{\delta}{2} + 0 + \frac{\delta}{2}$$

and therefore the assertion. This concludes the proof of Lemma 6.3.14. □

Lemma 6.3.15. *Suppose that $\mu \in \mathrm{ROV}(E,c)$ for some $c>1$ and (6.37) holds, where every eigenvalue of E has real part greater than or equal to $a_1>0$, and write $A_n = B_{k_n}$. Define G_ζ, ψ_ζ as in (6.77) where K_x is given by (6.76). Then for all $\zeta > 1/a_1$ we have $k_n G_\zeta(A_n^* x_n) \to \psi_\zeta(x)$ whenever $x_n \to x \in \Gamma$ and K_x is a ϕ-continuity set.*

Proof. We first show that for $x \in \Gamma$

$$k_n G_\zeta(A_n^* x) \to \psi_\zeta(x) \quad \text{as } n \to \infty \tag{6.88}$$

whenever K_x is a ϕ-continuity set. For $0 \le \rho < 1$, let

$$\begin{aligned} f_n(\rho) &= k_n \int_{\rho<|<x,y>|\le 1} |<x,y>|^\zeta (A_n\mu)\{dy\} \\ f(\rho) &= \int_{\rho<|<x,y>|\le 1} |<x,y>|^\zeta \phi\{dy\}. \end{aligned}$$

Since $k_n(A_n\mu) \to \phi$ in $\mathcal{M}$, Proposition 1.2.20 implies $f_n(\rho) \to f(\rho)$ as $n \to \infty$ for all but countably many $0 \le \rho < 1$. For $0 < \rho \le 1$ define

$$\begin{aligned} h_n(\rho) &= k_n \int_{0<|<x,y>|\le \rho} |<x,y>|^\zeta (A_n\mu)\{dy\} \\ h(\rho) &= \int_{0<|<x,y>|\le \rho} |<x,y>|^\zeta \phi\{dy\}. \end{aligned}$$

Since $h(1) = \psi_\zeta(x) < \infty$ in view of Lemma 6.3.11 it follows that $h(\rho) \to 0$ as $\rho \to 0$. If we write $A_n^* x = r_n\theta_n$ for some $r_n > 0$ and $\|\theta_n\| = 1$, then Theorem 4.2.4 together with Theorem 4.2.9 implies that $r_n \to 0$ as $n \to \infty$.

Write

$$\begin{aligned} h_n(\rho) &= k_n \int_{0<|<x,A_n y>|\le \rho} |<x,A_n y>|^\zeta \mu\{dy\} \\ &= k_n r_n^\zeta \int_{0<|<\theta_n,y>|\le \rho r_n^{-1}} |<\theta_n,y>|^\zeta \mu\{dy\} \\ &= k_n r_n^\zeta U_\zeta(\rho r_n^{-1}, \theta_n) \\ &= \rho^\zeta \frac{U_\zeta(\rho r_n^{-1},\theta_n)}{(\rho r_n^{-1})^\zeta V_0(\rho r_n^{-1},\theta_n)} \cdot \frac{V_0(\rho r_n^{-1},\theta_n)}{V_0(r_n^{-1},\theta_n)} \cdot k_n V_0(r_n^{-1},\theta_n) \end{aligned}$$

where by part (b) of Corollary 6.3.9 together with Theorem 6.3.4 the limsup of the first fraction on the right-hand side of the equation above is bounded by some constant $B_2 > 0$ independent of ρ. Furthermore, Theorem 6.3.4 implies that for $\delta > 0$ there exists a constant $C_1 > 0$ such that

$$\begin{aligned} \limsup_{n\to\infty} \frac{V_0(\rho r_n^{-1},\theta_n)}{V_0(r_n^{-1},\theta_n)} &= \left(\liminf_{n\to\infty} \frac{V_0(\rho^{-1}(\rho r_n^{-1}),\theta_n}{V_0(\rho r_n^{-1},\theta_n)} \right)^{-1} \\ &\le C_1 \rho^{-\delta - 1/a_1} \end{aligned}$$

whenever $0 < \rho \le 1$. But

$$\lim_{n\to\infty} k_n V_0(r_n^{-1},\theta_n) = \lim_{n\to\infty} k_n (A_n\mu)\{y \in \mathbb{R}^d : |<x,y>| > 1\}$$
$$= \phi\{y \in \mathbb{R}^d : |<x,y>| > 1\} < \infty$$

since K_x is a ϕ-continuity set. Hence if we choose $\delta > 0$ such that $\delta_1 = \zeta - \delta - 1/a_1 > 0$ there exists a constant $C > 0$ such that for $0 < \rho < 1$

$$\limsup_{n\to\infty} h_n(\rho) \le C\rho^{\delta_1}.$$

Given $\varepsilon > 0$ choose $0 < \rho < 1$ such that $h(\rho) < \varepsilon/2$, $f_n(\rho) \to f(\rho)$ and $C\rho^{\delta_1} < \varepsilon/2$. Then

$$\begin{aligned}
&\limsup_{n\to\infty} |k_n G_\zeta(A_n^* x) - \psi_\zeta(x)| \\
=&\limsup_{n\to\infty} |f_n(0) - f(0)| \\
\le&\limsup_{n\to\infty} \big(|f_n(0) - f_n(\rho)| + |f_n(\rho) - f(\rho)| + |f(\rho) - f(0)|\big) \\
=&\limsup_{n\to\infty} h_n(\rho) + 0 + h(\rho) \\
<&\frac{\varepsilon}{2} + \frac{\varepsilon}{2} = \varepsilon,
\end{aligned}$$

which proves (6.88).

Finally, assume that $x_n \to x \in \Gamma$ and K_x is a ϕ-continuity set. Since the mapping $t \mapsto t^\zeta \psi_\zeta(x/t)$ is increasing and continuous at $t = 1$, in view of Lemma 6.3.12 given $\varepsilon > 0$ there exists a $0 < \delta < 1$ such that

$$(1+\delta)^\zeta \psi_\zeta(x/(1+\delta)) \le \psi_\zeta(x) + \varepsilon/2$$
$$(1-\delta)^\zeta \psi_\zeta(x/(1-\delta)) \ge \psi_\zeta(x) - \varepsilon/2$$

and $K_{x/(1+\delta)}$ and $K_{x/(1-\delta)}$ are ϕ-continuity sets. Define $D = \{x : \|x\| \le R\}$ where $R > 0$ is chosen so that $\phi(D^c) < (1-\delta)^{-\zeta}\varepsilon/2$ and $\phi\{\|x\| = R\} = 0$. Write $x_n = x + \delta_n$ so that $\varepsilon_n = R\|\delta_n\| \to 0$. If $y \in K_{x_n} \cap D$, then $|<x,y>| \le |<x_n,y>| + |<\delta_n,y>| \le 1 + \varepsilon_n < 1 + \delta$ for all large n, so that $K_{x_n} \cap D \subset K_{x/(1+\delta)}$ for all large n. Then

$$\begin{aligned}
&k_n G_\zeta(A_n^* x_n) \\
=&k_n \int_{K_{x_n}\cap D} |<y,x_n>|^\zeta (A_n\mu)\{dy\} + k_n \int_{K_{x_n}\cap D^c} |<y,x_n>|^\zeta (A_n\mu)\{dy\} \\
=&k_n \int_{K_{x_n}\cap D} g_n(y)(A_n\mu)\{dy\} + k_n \int_{K_{x_n}\cap D} |<x,y>|^\zeta (A_n\mu)\{dy\} \\
&+ k_n \int_{K_{x_n}\cap D^c} |<y,x_n>|^\zeta (A_n\mu)\{dy\}
\end{aligned} \tag{6.89}$$

where $g_n(y) = |\langle x_n, y\rangle|^\zeta - |\langle x, y\rangle|^\zeta \to 0$ as $n \to \infty$ for all $y \in \mathbb{R}^d$. Since $k_n(A_n\mu) \to \phi$ and D^c is a ϕ-continuity set we have

$$k_n \int_{K_{x_n} \cap D^c} |\langle y, x_n\rangle|^\zeta (A_n\mu)\{dy\} \leq k_n(A_n\mu)(D^c) \to \phi(D^c) < (1-\delta)^{-\zeta}\varepsilon/2.$$

Furthermore, by (6.88)

$$\begin{aligned} k_n \int_{K_{x_n} \cap D} |\langle x, y\rangle|^\zeta (A_n\mu)\{dy\} &\leq k_n \int_{K_{x/(1+\delta)}} |\langle x, y\rangle|^\zeta (A_n\mu)\{dy\} \\ &= (1+\delta)^\zeta k_n G_\zeta(A_n^* x/(1+\delta)) \\ &\to (1+\delta)^\zeta \psi_\zeta(x/(1+\delta)) \leq \psi_\zeta(x) + \varepsilon/2. \end{aligned}$$

Since $|g_n(y)| \leq C_n\|y\|^\zeta$, where $C_n = \sup_{\|y\|=1} |g_n(y)| \to 0$ as $n \to \infty$, by Lemma 6.3.14 we have

$$\lim_{n\to\infty} \left| k_n \int_{K_{x_n} \cap D} g_n(y)(A_n\mu)\{dy\} \right| \leq \lim_{n\to\infty} C_n \mu_n(\mathbb{R}^d) = (\lim_{n\to\infty} C_n)\rho(\mathbb{R}^d) = 0.$$

Since $\varepsilon > 0$ is arbitrary, in view of (6.89) we have shown

$$\limsup_{n\to\infty} k_n G_\zeta(A_n^* x_n) \leq \psi_\zeta(x). \tag{6.90}$$

Now we prove the lower bound. Write $x_n = x + \delta_n$ so that $\varepsilon_n = R\|\delta_n\| \to 0$. If $y \in K_{x/(1-\delta)} \cap D$, then $|\langle y, x_n\rangle| \leq |\langle y, x\rangle| + |\langle y, \delta_n\rangle| \leq (1-\delta) + \varepsilon_n < 1$ for all large n. Then $K_{x/(1-\delta)} \cap D \subset K_{x_n}$ for all large n, and so

$$\begin{aligned} &k_n G_\zeta(A_n^* x_n) \\ &\geq k_n \int_{K_{x/(1-\delta)} \cap D} |\langle y, x_n\rangle|^\zeta (A_n\mu)\{dy\} \\ &= k_n \int_{K_{x/(1-\delta)} \cap D} g_n(y)(A_n\mu)\{dy\} + k_n \int_{K_{x/(1-\delta)} \cap D} |\langle x, y\rangle|^\zeta (A_n\mu)\{dy\}. \end{aligned}$$

As before

$$\left| k_n \int_{K_{x/(1-\delta)} \cap D} g_n(y)(A_n\mu)\{dy\} \right| \leq k_n \int_D |g_n(y)|(A_n\mu)\{dy\} \to 0$$

as $n \to \infty$. Furthermore

$$\begin{aligned} &k_n \int_{K_{x/(1-\delta)} \cap D} |\langle x, y\rangle|^\zeta (A_n\mu)\{dy\} \\ &= k_n \int_{K_{x/(1-\delta)}} |\langle x, y\rangle|^\zeta (A_n\mu)\{dy\} - k_n \int_{K_{x/(1-\delta)} \cap D^c} |\langle x, y\rangle|^\zeta (A_n\mu)\{dy\} \\ &\geq k_n(1-\delta)^\zeta G_\zeta(A_n^* x/(1-\delta)) - (1-\delta)^\zeta k_n(A_n\mu)(D^c). \end{aligned}$$

Since $k_n(A_n\mu) \to \phi$ and D^c is a ϕ-continuity set, in view of (6.88) and the choice of $R > 0$ we get

$$\liminf_{n\to\infty} k_n G_\zeta(A_n^* x_n) \geq (1-\delta)^\zeta \psi_\zeta(x/(1-\delta)) - (1-\delta)^\zeta \phi(D^c) \geq \psi_\zeta(x) - \varepsilon$$

and hence

$$\liminf_{n\to\infty} k_n G_\zeta(A_n^* x_n) \geq \psi_\zeta(x).$$

This together with (6.90) concludes the proof. □

Theorem 6.3.16. *Suppose that $\mu \in \mathrm{ROV}(E, c)$ for some $c > 1$ and that the real parts of the eigenvalues of E are $0 < a_1 < \cdots < a_p$. If $\zeta > 1/a_1$ and U_ζ exists, then $U_\zeta(r, x)$ is uniformly R–O varying on compact subsets of $x \in \Gamma$. Furthermore, for any compact $S \subset \Gamma$ and any $\delta > 0$ there exist $C_1 > 0$, $C_2 < \infty$ and $r_0 \geq 1$ such that*

$$C_1 \lambda^{\zeta-\delta-1/a_1} \leq \frac{U_\zeta(\lambda r, x)}{U_\zeta(r, x)} \leq C_2 \lambda^{\zeta+\delta-1/a_p}$$

for all $r \geq r_0$, $\lambda \geq 1$ and $x \in S$.

Proof. The proof is similar to Theorem 6.3.4. By Lemma 6.2.2, there exist $(B_n) \in \mathrm{RV}(-E)$ and an increasing sequence of positive integers (k_n) with $k_{n+1}/k_n \to c > 1$ such that for $A_n = B_{k_n}$ we have $k_n(A_n\mu) \to \phi$ where ϕ is $\mathcal{M}$-full and satisfies (6.39). Then Lemma 6.3.13 shows that the function $\psi_\zeta(x)$ defined by (6.77) is bounded away from zero and infinity on compact subsets of Γ. Since the real parts of the eigenvalues of E, E^* are the same, it follows from Lemma 5.3.15 part (b) along with (6.79) that for any $\varepsilon > 0$ sufficiently small and any compact set $S \subset \Gamma$, for all t_0 sufficiently large we have

$$\frac{\psi_\zeta(y/t)}{\psi_\zeta(y)} \leq t_0^\varepsilon \tag{6.91}$$

whenever $1 \leq t \leq t_0$ and $y \in S$. Lemma 6.3.15 implies that

$$k_n G_\zeta(A_n^* x_n) \to \psi_\zeta(x) \tag{6.92}$$

whenever $x_n \to x$ in Γ and K_x is a ϕ-continuity set, where G_ζ is defined by (6.77) and K_x is defined by (6.76) . Suppose $x_r \to x$ in Γ. Since $(B_n^*)^{-1}$ is regularly varying with index E^*, we can apply part (b) of Lemma 4.2.16 to obtain $n = n(r)$ tending to infinity as $r \to \infty$ such that $y_n = (B_{k_n}^*)^{-1}(x_r/r) = (A_n^*)^{-1}(x_r/r)$ is limit point compact with all limit points in the set $S_1 = \{y \in \mathbb{R}^d : D \leq \|y\| \leq 1\}$ for some $D > 0$. Now let $S_2 = \{y \in \mathbb{R}^d : \frac{D}{2} \leq \|y\| \leq 2\}$ and apply Lemma 6.3.13 to obtain $0 < m_0 < M_0 < \infty$ such that $m_0 \leq \psi_\zeta(y) \leq M_0$

for all $y \in S_2$. If $1/2 \le s \le 1$ we have for $y \in S_1$ that $y, sy, y/s$ are all in S_2. Now apply (6.91) with $S = S_2$ to see for any λ_0 sufficiently large we have

$$\frac{\psi_\zeta(ys/\lambda)}{\psi_\zeta(y/s)} = \frac{\psi_\zeta(ys/\lambda)}{\psi_\zeta(ys)} \cdot \frac{\psi_\zeta(ys)}{\psi_\zeta(y/s)} \le \frac{M_0}{m_0}\lambda_0^\varepsilon \tag{6.93}$$

whenever $1 \le \lambda \le \lambda_0$, $y \in S_1$ and $1/2 \le s \le 1$.

Since (y_n) is limit point compact, every subsequence contains a further subsequence along which $y_n \to y \in S_1$. Fix $1 \le \lambda \le \lambda_0$ and choose $\frac{1}{2} \le s \le 1$ such that both $K_{ys/\lambda}$ and $K_{y/s}$ are ϕ-continuity sets. Using the fact that $U_\zeta(r,x) = r^\zeta G_\zeta(x/r)$ is increasing in r for all $x \in \Gamma$, we get using (6.92) that

$$\begin{aligned}\frac{U_\zeta(\lambda r, x_r)}{U_\zeta(r, x_r)} &= \frac{(\lambda r)^\zeta G_\zeta(x_r/(\lambda r))}{r^\zeta G_\zeta(x_r/r)} \le \frac{(\lambda r/s)^\zeta G_\zeta(x_r/(\lambda r/s))}{(rs)^\zeta G_\zeta(x_r/(rs))} \\ &= \lambda^\zeta s^{-2\zeta}\frac{G_\zeta((x_r/r)\frac{s}{\lambda})}{G_\zeta(x_r/(rs))} = \left(\frac{\lambda}{s^2}\right)^\zeta \frac{k_n G_\zeta(A_n^* y_n \frac{s}{\lambda})}{k_n G_\zeta(A_n^* y_n/s)} \to \left(\frac{\lambda}{s^2}\right)^\zeta \frac{\psi_\zeta(ys/\lambda)}{\psi_\zeta(y/s)}\end{aligned}$$

along this subsequence. Since every subsequence has a further subsequence with this property, and since $U_\zeta(r,x)$ is monotone in r, it follows from (6.93) that for any compact subset S of Γ and any λ_0 sufficiently large, for some $r_0 > 0$ we have

$$m = 1 \le \frac{U_\zeta(\lambda r, x)}{U_\zeta(r, x)} \le \left(\frac{\lambda}{s^2}\right)^\zeta \frac{M_0}{m_0}\lambda_0^{2\varepsilon} \le \frac{M_0}{m_0}4^\zeta \lambda_0^{\zeta+2\varepsilon} = M$$

for all $r \ge r_0$, all $1 \le \lambda \le \lambda_0$, and all $x \in S$. Then $U_\zeta(r,x)$ is uniformly R–O varying on compact subsets of $x \in \Gamma$. Write

$$\frac{U_\zeta(\lambda r, x)}{U_\zeta(r, x)} = \frac{U_\zeta(\lambda r, x)}{(\lambda r)^\zeta V_0(\lambda r, x)} \cdot \frac{r^\zeta V_0(r,x)}{U_\zeta(r,x)} \cdot \lambda^\zeta \cdot \frac{V_0(\lambda r, x)}{V_0(r,x)}. \tag{6.94}$$

Since by Theorem 6.3.4 both V_0 and U_ζ are uniformly R–O varying, Corollary 6.3.9 applies. This together with Theorem 6.3.4 shows that for any compact $S \subset \Gamma$ and $\delta > 0$ there exists $r_0 > 0$ such that

$$\frac{m_2}{M_2} \cdot \lambda^\zeta \cdot C_1 \lambda^{-\delta - 1/a_1} \le \frac{U_\zeta(\lambda r, x)}{U_\zeta(r,x)} \le \frac{M_2}{m_2} \cdot \lambda^\zeta \cdot C_2 \lambda^{\delta - 1/a_p}$$

for all $r \ge r_0$, all $\lambda \ge 1$ and all $x \in S$. The proof is complete. □

Now we will use uniform R–O variation to characterize the growth rate of the tail moments V_η defined by (6.53) of an R–O varying measure μ. For $x \in \Gamma$, we define

$$\begin{aligned} F_\eta(x) &= \int_{H_x} |\langle x, y\rangle|^\eta \mu\{dy\} \\ \varphi_\eta(x) &= \int_{H_x} |\langle x, y\rangle|^\eta \phi\{dy\} \end{aligned} \tag{6.95}$$

where $H_x = \{y \in \mathbb{R}^d : |<x, y>| > 1\}$. If V_η exists, then F_η exists and

$$V_\eta(r, x) = r^\eta F_\eta(x/r) \tag{6.96}$$

for all $r > 0$ and $x \in \Gamma$. If (6.39) holds, then using the fact that $A^{-1}H_x = H_{A^*x}$ it follows from (6.95) that

$$c \cdot \varphi_\eta(x) = \varphi_\eta(c^{E^*} x) \tag{6.97}$$

for all $x \in \Gamma$. Before we can show that V_η is uniformly R–O varying, we must first establish some properties of F_η and φ_η. First we show that φ_η exists for certain values of η.

Lemma 6.3.17. *Suppose ϕ is $\mathcal{M}$-full and* (6.39) *holds for some $c > 1$, where every eigenvalue of E and has positive real part less than or equal to $a_p > 0$. Then*

$$\int_{\|y\|>R} \|y\|^\eta \phi\{dy\} < \infty \tag{6.98}$$

for all $R > 0$ and $\eta < 1/a_p$. Furthermore, if $\eta < 1/a_p$ then the function $\varphi_\eta(x)$ defined by (6.95) *exists for all $x \in \Gamma$.*

Proof. Given any $R > 0$, since Theorem 2.2.4 implies that for any $Q \subset \Gamma$ compact, $\|t^E x\| \to 0$ uniformly in $x \in Q$ as $t \to 0$, for some k_0 we have $\|c^{kE} x\| \le R$ for all $k < k_0$ and all $x \in Q$. In view of Lemma 6.3.10 (a), this implies that

$$\{y \in \mathbb{R}^d : \|y\| > R\} \subset \bigcup_{k=k_0}^{\infty} c^{kE}(Q)$$

and then part (c) of Lemma 6.3.10 yields

$$\int_{\|y\|>R} \|y\|^\eta \phi\{dy\} \le \sum_{k=k_0}^{\infty} \int_{c^{kE}(Q)} \|y\|^\eta \phi\{dy\} < \infty$$

so that (6.98) holds. If $x \in \Gamma$, then

$$\varphi_\eta(x) = \int_{|<x,y>|>1} |<x, y>|^\eta \phi\{dy\} \le \|x\|^\eta \int_{\|y\|>1/\|x\|} \|y\|^\eta \phi\{dy\}$$

which is finite in view of (6.98). □

Lemma 6.3.18. *Suppose ϕ is $\mathcal{M}$-full and* (6.39) *holds for some $c > 1$, where every eigenvalue of E has positive real part less than or equal to $a_p > 0$. Then any $x \in \Gamma$ such that $\phi\{y : |<x, y>| = 1\} = 0$ is a point of continuity of the function $\varphi_\eta(x)$ defined by* (6.95) *with $\eta < 1/a_p$.*

Proof. The proof is quite similar to Lemma 6.3.12. Let $\varepsilon > 0$ be arbitrary and assume that $x_n \to x$ in Γ where $\phi\{y : |<x, y>| = 1\} = 0$. Then $\|x_n\| \le M$ for all $n \ge 1$ and some constant $M > 0$. In view of Lemma 6.3.17, we can choose $R > 0$ such that for the set $D = \{x : 0 < \|x\| < R\}$ we have $\int_{D^c} \|y\|^\eta \phi\{dy\} \le \varepsilon/(2M)$. Then it is not hard to check that

$$\begin{aligned} &|\varphi_\eta(x_n) - \varphi_\eta(x)| \\ &\le \left| \int_{H_{x_n} \cap D} |<x_n, y>|^\eta \phi\{dy\} - \int_{H_x \cap D} |<x, y>|^\eta \phi\{dy\} \right| + \varepsilon \\ &\le \int_{H_{x_n} \cap D} \big| |<x_n, y>|^\eta - |<x, y>|^\varsigma \big| \phi\{dy\} \\ &\quad + \int_{(H_{x_n} \triangle H_x) \cap D} |<x, y>|^\eta \phi\{dy\} + \varepsilon. \end{aligned} \tag{6.99}$$

As in the proof of Lemma 6.1.25 we have for all large n that

$$(H_{x_n} \triangle H_x) \cap D \subseteq \{y : |<y, x>| \in [1 - \varepsilon_n, 1 + \varepsilon_n]\} \cap D$$

where $\varepsilon_n \to 0$. Then

$$\begin{aligned} \int_{(H_{x_n} \triangle H_x) \cap D} |<x, y>|^\eta \phi\{dy\} &\le R^\eta \|x\|^\eta \phi\{y : |<x, y>| \in [1 - \varepsilon_n, 1 + \varepsilon_n]\} \\ &\to R^\eta \|x\|^\eta \phi\{y : |<x, y>| = 1\} = 0 \end{aligned}$$

as $n \to \infty$. To estimate the first integral on the right-hand side of (6.99) note that $f_n(y) = |<x_n, y>|^\eta - |<x, y>|^\eta \to 0$ as $n \to \infty$ for all $y \in \mathbb{R}^d$. Note that $H_{x_n} \cap D \subset \{y : 1/M \le \|y\| \le R\} = S$ and that there exists a constant $C > 0$ such that $|f_n(y)| \le C\|y\|^\eta$, which is integrable with respect to ϕ on the set S. Then Lebesgue's dominated convergence theorem implies

$$\int_{H_{x_n} \cap D} |f_n(y)| \phi\{dy\} \le \int_D |f_n(y)| \phi\{dy\} \to 0$$

as $n \to \infty$. Hence $\lim_{n\to\infty} |\varphi_\eta(x_n) - \varphi_\eta(x)| \le \varepsilon$ arbitrary, so φ_η is continuous at the point x. □

Lemma 6.3.19. *Suppose ϕ is $\mathcal{M}$-full and (6.39) holds for some $c > 1$, where every eigenvalue of E has positive real part less than or equal to $a_p > 0$. Then for any $\eta < 1/a_p$ and any compact set $S \subset \Gamma$ there exist constants $M, m > 0$ such that $m \le \varphi_\eta(x) \le M$ for all $x \in S$, where $\varphi_\eta(x)$ is defined by (6.95).*

Proof. The proof is similar to Lemma 6.3.1. There exists a $B \ge 1$ such that $\|x\| \le B$ for all $x \in S$. Then in view of Lemma 6.3.17, we have

$$\varphi_\eta(x) = \int_{|<x,y>|>1} |<x, y>|^\eta \phi\{dy\} \le B^\eta \int_{\|y\|>1/B} \|y\|^\eta \phi\{dy\} = M < \infty$$

for all $x \in S$.

Assume there exists $x \in \Gamma$ with $\varphi_\eta(x) = 0$. Writing $c^{kE^*}x = r_k\theta_k$ for some $r_k > 0$ and $\|\theta_k\| = 1$, it follows from (6.97) that

$$\varphi_\eta(c^{kE^*}x) = \int_{|<y,\theta_k>|>r_k^{-1}} |<y,\theta_k>|^\eta \phi\{dy\} = 0$$

for all $k \in \mathbb{Z}$. Hence the set $N_k = \{y : |<y,\theta_k>| > r_k^{-1}\}$ has ϕ-measure zero for every integer k. Since the real parts of the eigenvalues of E^* are all positive, Theorem 2.2.4 shows that $r_k \to \infty$ as $k \to \infty$. Since the unit sphere in $\mathbb{R}^d$ is compact, there exists a subsequence $k' \to \infty$ such that $\theta_{k'} \to \theta$ along (k'). If $|<y,\theta>| > 0$, then $\lim_{k'\to\infty} |<y,\theta_{k'}>| > 0$ as well, and since $r_{k'}^{-1} \to 0$ we must have $y \in N_{k'}$ for all large k'. Then

$$N = \{y : |<y,\theta>| \neq 0\} \subset \bigcup_{k=1}^{\infty} N_k$$

and hence $\phi(N) = 0$, which contradicts the assumption that ϕ is $\mathcal{M}$-full. Then $\varphi_\eta(x) \neq 0$ whenever $x \in \Gamma$.

Assume that for all $n \geq 1$ there exists $x_n \in S$ such that $\varphi_\eta(x_n) < 1/n$. Choose a subsequence (n') with $x_{n'} \to x \in S$ and $t < 1$ such that H_{tx} is a ϕ-continuity set. Then Lemma 6.3.18 implies that φ_η is continuous at the point tx, so $\varphi_\eta(tx_{n'}) \to \varphi_\eta(tx)$. But we also have $\varphi_\eta(tx_{n'}) \leq t^\eta\varphi_\eta(x_{n'}) < t^\eta/n' \to 0$ as $n' \to \infty$ which is a contradiction. Hence for some $m > 0$ we have $\varphi_\eta(x) > m$ for all $x \in S$. □

Lemma 6.3.20. *Suppose ϕ is $\mathcal{M}$-full and* (6.39) *holds for some $c > 1$, where every eigenvalue of E has positive real part less than or equal to $a_p > 0$. If $0 \leq \eta < 1/a_p$, then $V_\eta(r,x)$ defined by* (6.53) *exists for all $r > 0$ and all $x \in \Gamma$. Furthermore, the function $F_\eta(x)$ defined by* (6.95) *exist for all $x \in \Gamma$, and*

$$\int_{\|y\|>R} \|y\|^\eta k_n(A_n\mu)\{dy\} < \infty \tag{6.100}$$

for all $n \geq 1$, $R > 0$ and $0 \leq \eta < 1/a_p$.

Proof. Given $\eta < 1/a_p$ choose $\delta > 0$ such that $\eta + \delta - 1/a_p < 0$ and apply Theorem 6.3.4 to obtain $r_0 > 0$ such that $V_0(s,x) \leq C_2(s/r_0)^{\delta-1/a_p}V_0(r_0,x)$ for fixed $x \in \Gamma$ and all $s \geq r_0$. Then for any $r > 0$ we have using integration by parts that

$$\begin{aligned} V_\eta(r,x) &= \int_r^{r_0} s^\eta V_0(-ds,x) + \int_{r_0}^\infty s^\eta V_0(-ds,x) \\ &\le \max\{r^\eta, r_0^\eta\} V_0(\min\{r,r_0\},x) \\ &\quad + \lim_{r_1\to\infty}\left(r_0^\eta V_0(r_0,x) - r_1^\eta V_0(r_1,x) + \int_{r_0}^{r_1} \eta s^{\eta-1} V_0(s,x)ds \right) \\ &\le \max\{r^\eta, r_0^\eta\} V_0(\min\{r,r_0\},x) + r_0^\eta V_0(r_0,x) \\ &\quad + \eta C_2 r_0^{-\delta+1/a_p} V_0(r_0,x) \lim_{r_1\to\infty} \int_{r_0}^{r_1} s^{\eta+\delta-1-1/a_p} ds \end{aligned}$$

and this limit is finite since $\eta + \delta - 1/a_p < 0$. In view of (6.96), the function $F_\eta(x) = V_\eta(1,x)$ also exists for all $x \in \Gamma$.

It suffices to prove (6.100) for $0 < R < 1$ such that $D = \{y : ||y|| > R\}$ is a ϕ-continuity set. Write $y_i = <y, e_i>$ where $e_1, \ldots, e_d$ denotes the standard basis for $\mathbb{R}^d$. Using the elementary inequality $(c_1 + \cdots + c_d)^q \le d^q(c_1^q + \cdots + c_d^q)$ for $c_i > 0$ and $q \ge 0$, we have

$$\begin{aligned} &\int_{||y||>R} ||y||^\eta k_n(A_n\mu)\{dy\} \\ \le\ & \int_{||y||>R} d^{\eta/2}(|y_1|^\eta + \cdots + |y_d|^\eta) k_n(A_n\mu)\{dy\} \\ =\ & d^{\eta/2} \sum_{j=1}^d \left(\int_{D\cap K_{r_j e_j}} |y_j|^\eta k_n(A_n\mu)\{dy\} + \int_{H_{r_j e_j}} |y_j|^\eta k_n(A_n\mu)\{dy\} \right) \end{aligned}$$

where $r_j > 0$ is chosen so that $K_{r_j e_j} = \{y \in \mathbb{R}^d : |y_j| \le r_j^{-1}\}$ is a ϕ-continuity set. Since $k_n(A_n\mu) \to \phi$ in $\mathcal{M}$, Proposition 1.2.20 implies that

$$\int_{D\cap K_{r_j e_j}} |y_j|^\eta k_n(A_n\mu)\{dy\} \to \int_{D\cap K_{r_j e_j}} |y_j|^\eta \phi\{dy\} < \infty$$

and hence

$$\int_{D\cap K_{r_j e_j}} |y_j|^\eta k_n(A_n\mu)\{dy\} \le M < \infty$$

for all $n \ge 1$. The second term is (for $\eta \ne 0$)

$$\int_{|<y,r_j e_j>|>1} |<y,e_j>|^\eta k_n(A_n\mu)\{dy\} = r_j^{-\eta} k_n V_\eta(1, A_n^* e_j r_j)$$

and it follows easily that (6.100) holds. □

Lemma 6.3.21. *Suppose that* $\mu \in \mathrm{ROV}(E,c)$ *for some* $c > 1$ *and* (6.37) *holds, where every eigenvalue of* E *has positive real part less than or equal to*

$a_p > 0$, and write $A_n = B_{k_n}$. Let $\mu_n\{dy\} = k_n\|y\|^\eta I(\|y\| > R)(A_n\mu)\{dy\}$ and $\rho\{dy\} = \|y\|^\eta I(\|y\| > R)\phi\{dy\}$. If $0 < \eta < 1/a_p$ and $\phi\{\|y\| = R\} = 0$ then $\mu_n \to \rho$ in $\mathcal{M}_0$.

Proof. Lemma 6.3.17 implies that ρ is a finite Borel measure on $\mathbb{R}^d$, and Lemma 6.3.20 implies that $\mu_n \in \mathcal{M}_0$ too. We first show that

$$\lim_{t\to\infty} \limsup_{n\to\infty} \mu_n\{\|y\| > t\} = 0. \tag{6.101}$$

Apply Lemma 6.3.10, where a, b are chosen so that Q is a ϕ-continuity set. Let $k(t) = \inf\{k : \|c^{kE}\| \geq t/b\}$, so that $k(t) \to \infty$ as $t \to \infty$. If $y = c^{kE}x$ for some $k < k(t)$ and $x \in Q$, then $\|x\| \leq b$ so $\|c^{kE}x\| \leq \|c^{kE}\|b < t$, so in view of Lemma 6.3.10 (a) we have

$$\{y \in \mathbb{R}^d : \|y\| > t\} \subset \bigcup_{k=k(t)}^{\infty} c^{kE}(Q).$$

Since $k_n(A_n\mu) \to \phi$ in $\mathcal{M}$ we have $\mu_n\big(c^{-kE}(Q)\big) \to \rho\big(c^{-kE}(Q)\big)$ as $n \to \infty$ for each k in view of Proposition 1.2.20, and hence

$$\begin{aligned}
\limsup_{n\to\infty} \mu_n\{\|y\| > t\} &\leq \sum_{k=k(t)}^{\infty} \limsup_{n\to\infty} \mu_n\big(c^{kE}(Q)\big) \\
&= \sum_{k=k(t)}^{\infty} \rho\big(c^{kE}(Q)\big) \\
&= \sum_{k=k(t)}^{\infty} \int_{c^{kE}(Q)} \|y\|^\eta I(\|y\| > R)\phi\{dy\} \\
&\leq \sum_{k=k(t)}^{\infty} \int_{c^{kE}(Q)} \|y\|^\eta \phi\{dy\}
\end{aligned}$$

which tends to zero as $k(t) \to \infty$ in view of Lemma 6.3.10 part (c), and this proves (6.101).

Let g be any bounded continuous function on $\mathbb{R}^d$. For $\delta > 0$ arbitrary, choose $t > 0$ such that $\limsup_{n\to\infty} \mu_n\{\|y\| > t\} < \delta/(2\|g\|_\infty)$ by (6.101), $\rho\{\|y\| = t\} = 0$ and $\rho\{\|y\| > t\} < \delta/(2\|g\|_\infty)$. Write

$$\begin{aligned}
\left|\int g d\mu_n - \int g d\rho\right| &\leq \|g\|_\infty \mu_n\{\|y\| > t\} \\
&\quad + \left|\int_{\|y\|\leq t} g d\mu_n - \int_{\|y\|\leq t} g d\rho\right| + \|g\|_\infty \rho\{\|y\| > t\}.
\end{aligned}$$

Since $k_n(A_n\mu) \to \phi$ in $\mathcal{M}$, Proposition 1.2.20 yields

$$\int_{\|y\|\leq t} g d\mu_n \to \int_{\|y\|\leq t} g d\rho$$

which together with the choice of $R > 0$ and $t > 0$ implies

$$\limsup_{n\to\infty}\left|\int g d\mu_n - \int g d\rho\right| < \frac{\delta}{2} + 0 + \frac{\delta}{2}$$

where $\delta > 0$ is arbitrary, and this completes the proof. □

Lemma 6.3.22. *Suppose that $\mu \in \mathrm{ROV}(E, c)$ for some $c > 1$ and (6.37) holds, where every eigenvalue of E has positive real part less than or equal to $a_p > 0$, and write $A_n = B_{k_n}$. Define F_η, φ_η as in (6.95), where H_x is given by (6.17). Then for all $\eta < 1/a_p$ we have $k_n F_\eta(A_n^* x_n) \to \varphi_\eta(x)$ whenever $x_n \to x \in \Gamma$ and H_x is a ϕ-continuity set.*

Proof. We first show that

$$k_n F_\eta(A_n^* x) \to \varphi_\eta(x) \tag{6.102}$$

whenever H_x is a ϕ-continuity set. Define

$$f_n(t) = -k_n V_0(t, A_n^* x) = -k_n(A_n\mu)\{y : |\langle x, y\rangle| > t\}$$

and $f(t) = -\phi\{y : |\langle x, y\rangle| > t\}$. It follows from Lemma 6.3.17 that

$$\int_1^\infty t^\eta f(dt) = \varphi_\eta(x) < \infty$$

and furthermore, (6.37) implies $f_n \to f$ as $n \to \infty$ almost everywhere on $[1, \infty)$. If we write $A_n^* x = r_n\theta_n$ for some $r_n > 0$, $\|\theta_n\| = 1$ we get for $t \geq 1$

$$|f_n(t)| = k_n V_0(r_n^{-1}t, \theta_n) = \frac{V_0(r_n^{-1}t, \theta_n)}{V_0(r_n^{-1}, \theta_n)} k_n V_0(r_n^{-1}, \theta_n).$$

Since H_x is a ϕ-continuity set, we have in view of (6.37) that $k_n V_0(r_n^{-1}, \theta_n) = k_n(A_n\mu)(H_x) \to \phi(H_x)$, and hence $k_n V_0(r_n^{-1}, \theta_n) \leq M < \infty$ for all $n \geq 1$. Lemma 6.2.7 implies that the real parts of the eigenvalues of E^* are positive, and then Corollary 4.2.6 along with Theorem 4.2.9 implies that $r_n \to 0$. Then Theorem 6.3.4 implies that for any $\delta > 0$ there exists a constant $C > 0$ such that $|f_n(t)| \leq C t^{\delta - 1/a_p}$ for all large n and all $t \geq 1$.

Now integration by parts yields

$$\begin{aligned} k_n F_\eta(A_n^* x) &= k_n \int_{|\langle x,y\rangle|>1} |\langle x, y\rangle|^\eta (A_n\mu)\{dy\} \\ &= \int_1^\infty t^\eta f_n(dt) \\ &= -f_n(1) - \eta \int_1^\infty f_n(t) t^{\eta-1}\, dt. \end{aligned}$$

Since $\eta < 1/a_p$ Lebesgue's dominated convergence theorem implies

$$\int_1^\infty f_n(t)t^{\eta-1}dt \to \int_1^\infty f(t)t^{\eta-1}dt$$

and since H_x is a ϕ-continuity set, $f_n(1) \to f(1)$ too. Integrate by parts again to obtain

$$k_n F_\eta(A_n^* x) \to -f(1) - \eta \int_1^\infty f(t)t^{\eta-1}dt = \int_1^\infty t^\eta f(dt) = \varphi_\eta(x)$$

which proves (6.102).

Suppose that $x_n \to x \in \Gamma$ and H_x is a ϕ-continuity set. Since the mapping $t \mapsto t^\eta \varphi_\eta(x/t)$ is decreasing and by Lemma 6.3.18 is continuous at $t = 1$, given $\varepsilon > 0$ there exists a $\delta > 0$ such that

$$\begin{aligned} (1-\delta)^\eta \varphi_\eta(x/(1-\delta)) &\le \varphi_\eta(x) + \varepsilon/2 \\ (1+\delta)^\eta \varphi_\eta(x/(1+\delta)) &\ge \varphi_\eta(x) - \varepsilon/2 \end{aligned} \tag{6.103}$$

and $H_{x/(1+\delta)}$ and $H_{x/(1-\delta)}$ are ϕ-continuity sets. Choose $B > 0$ such that $\|x_n\| \le B$ for all n. Then $\|x\| \le B$ as well. Next, choose $0 < R_0 < (1-\delta)/B$ such that $\phi\{\|x\| = R_0\} = 0$ and apply Lemma 6.3.21 to see that $\mu_n \to \rho$ in $\mathcal{M}_0$, where $\mu_n\{dy\} = k_n\|y\|^\eta I(\|y\| > R_0)(A_n\mu)\{dy\}$ and $\rho\{dy\} = \|y\|^\eta I(\|y\| > R_0)\phi\{dy\}$. Now define $D = \{x : \|x\| < R\}$, where $R > 0$ is chosen so that $B^\eta \rho(D^c) < \varepsilon/2$ and $\phi\{x : \|x\| = R\} = 0$. Write $x_n = x + \delta_n$ so that $\varepsilon_n = R\|\delta_n\| \to 0$. If $y \in H_{x_n} \cap D$, then $|{<}x, y{>}| \ge |{<}x_n, y{>}| - |{<}\delta_n, y{>}| > 1 - \varepsilon_n > 1 - \delta$ for all large n, hence $H_{x_n} \cap D \subset H_{x/(1-\delta)}$. Then

$$\begin{aligned} &k_n F_\eta(A_n^* x_n) \\ =&k_n \int_{H_{x_n} \cap D} |{<}y, x_n{>}|^\eta (A_n\mu)\{dy\} + k_n \int_{H_{x_n} \cap D^c} |{<}y, x_n{>}|^\eta (A_n\mu)\{dy\} \\ \le&k_n \int_{H_{x/(1-\delta)}} g_n(y)(A_n\mu)\{dy\} + k_n \int_{H_{x/(1-\delta)}} |{<}x, y{>}|^\eta (A_n\mu)\{dy\} \\ &+ B^\eta k_n \int_{D^c} \|y\|^\eta (A_n\mu)\{dy\} \end{aligned} \tag{6.104}$$

where $g_n(y) = |{<}x_n, y{>}|^\eta - |{<}x, y{>}|^\eta \to 0$ as $n \to \infty$ for all $y \in \mathbb{R}^d$. Since $|g_n(y)| \le C_n\|y\|^\eta$ where $C_n = \sup_{\|y\|=1} |g_n(y)| \to 0$ as $n \to \infty$, and since $H_{x/(1-\delta)} \subset \{y \in \mathbb{R}^d : \|y\| > R_0\}$ we have

$$\lim_{n\to\infty} \left| k_n \int_{H_{x/(1-\delta)}} g_n(y)(A_n\mu)\{dy\} \right| \le \lim_{n\to\infty} C_n \mu_n(\mathbb{R}^d) = (\lim_{n\to\infty} C_n)\rho(\mathbb{R}^d) = 0.$$

The second term

$$\begin{aligned} k_n \int_{H_{x/(1-\delta)}} |<x,y>|^\eta (A_n\mu)\{dy\} &= (1-\delta)^\eta k_n F_\eta(A_n^* x/(1-\delta)) \\ &\to (1-\delta)^\eta \varphi_\eta(x/(1-\delta)) \\ &\le \varphi_\eta(x) + \varepsilon/2 \end{aligned}$$

in view of (6.102) and (6.103). The third term

$$B^\eta k_n \int_{D^c} \|y\|^\eta (A_n\mu)\{dy\} = B^\eta \mu_n(D^c) \to B^\eta \rho(D^c) < \varepsilon/2$$

by choice of $R > 0$. Since $\varepsilon > 0$ is arbitrary we have shown that

$$\limsup_{n\to\infty} k_n F_\eta(A_n^* x_n) \le \varphi_\eta(x). \tag{6.105}$$

Now we prove the lower bound. Write $x_n = x + \delta_n$ so that $\varepsilon_n = R\|\delta_n\| \to 0$. If $y \in H_{x/(1+\delta)} \cap D$, then $|<y,x_n>| \ge |<y,x>| - |<y,\delta_n>| \ge (1+\delta) - \varepsilon_n > 1$ for all large n. Then $H_{x/(1+\delta)} \cap D \subset H_{x_n}$ for all large n, and so

$$\begin{aligned} &k_n F_\eta(A_n^* x_n) \ge k_n \int_{H_{x/(1+\delta)}\cap D} |<x_n,y>|^\eta (A_n\mu)\{dy\} \\ =&k_n \int_{H_{x/(1+\delta)}\cap D} g_n(y)(A_n\mu)\{dy\} + k_n \int_{H_{x/(1+\delta)}\cap D} |<x,y>|^\eta (A_n\mu)\{dy\}. \end{aligned}$$

As before

$$\left| k_n \int_{H_{x/(1+\delta)}\cap D} g_n(y)(A_n\mu)\{dy\} \right| \le k_n \int_{H_{x/(1+\delta)}} |g_n(y)|(A_n\mu)\{dy\} \to 0$$

as $n \to \infty$. Furthermore,

$$\begin{aligned} &k_n \int_{H_{x/(1+\delta)}\cap D} |<x,y>|^\eta (A_n\mu)\{dy\} \\ =&k_n \int_{H_{x/(1+\delta)}} |<x,y>|^\eta (A_n\mu)\{dy\} - k_n \int_{H_{x/(1+\delta)}\cap D^c} |<x,y>|^\eta (A_n\mu)\{dy\} \\ \ge&(1+\delta)^\eta k_n F_\eta(A_n^* x/(1+\delta)) - B^\eta \mu_n(D^c). \end{aligned}$$

Using (6.102) and (6.103) together with Lemma 6.3.21, we get

$$\liminf_{n\to\infty} k_n F_\eta(A_n^* x_n) \ge (1+\delta)^\eta \varphi_\eta(x/(1+\delta)) - B^\eta \phi(D^c) \ge \varphi_\eta(x) - \varepsilon$$

for this choice of $R > 0$, and letting $\varepsilon \to 0$ yields

$$\liminf_{n\to\infty} k_n F_\eta(A_n^* x_n) \geq \varphi_\eta(x).$$

Combine with (6.105) to complete the proof. □

Theorem 6.3.23. *Suppose that $\mu \in \mathrm{ROV}(E, c)$ for some $c > 1$ and that the real parts of the eigenvalues of E are $0 < a_1 < \cdots < a_p$. If $\eta < 1/a_p$, then $V_\eta(r, x)$ exists for all $r > 0$ and $x \in \Gamma$, and V_η is uniformly R–O varying on compact subsets of $x \in \Gamma$. Furthermore, for any compact $S \subset \Gamma$ and any $\delta > 0$ there exist $C_1 > 0$, $C_2 < \infty$ and $r_0 \geq 1$ such that*

$$C_1 \lambda^{\eta-\delta-1/a_1} \leq \frac{V_\eta(\lambda r, x)}{V_\eta(r, x)} \leq C_2 \lambda^{\eta+\delta-1/a_p}$$

for all $r \geq r_0$, $\lambda \geq 1$ and $x \in S$.

Proof. The proof is similar to Theorem 6.3.4. By Lemma 6.2.2, there exist $(B_n) \in \mathrm{RV}(-E)$ and an increasing sequence of positive integers (k_n) with $k_{n+1}/k_n \to c > 1$ such that for $A_n = B_{k_n}$ we have $k_n(A_n\mu) \to \phi$ where ϕ satisfies (6.39). Then Lemma 6.3.19 shows that the function $\varphi_\eta(x)$ defined by (6.95) is bounded away from zero and infinity on compact subsets of Γ. Since the real parts of the eigenvalues of E, E^* are the same, it follows from Lemma 5.3.15 part (b) along with (6.97) that for any $\varepsilon > 0$ sufficiently small and any compact set $S \subset \Gamma$, for all t_0 sufficiently large we have

$$\frac{\varphi_\eta(y/t)}{\varphi_\eta(y)} \geq t_0^{-\varepsilon-1/a_1} \tag{6.106}$$

whenever $1 \leq t \leq t_0$ and $y \in S$. Lemma 6.3.22 implies that

$$k_n F_\eta(A_n^* x_n) \to \varphi_\eta(x) \tag{6.107}$$

whenever $x_n \to x$ in Γ and H_x is a ϕ-continuity set, where F_η is defined by (6.95). Suppose $x_r \to x$ in Γ. Since $(B_n^*)^{-1}$ is regularly varying with index E^*, we can apply part (b) of Lemma 4.2.16 to obtain $n = n(r)$ tending to infinity as $r \to \infty$ such that $y_n = (B_{k_n}^*)^{-1}(x_r/r) = (A_n^*)^{-1}(x_r/r)$ is limit point compact with all limit points in the set $S_1 = \{y \in \mathbb{R}^d : D \leq \|y\| \leq 1\}$ for some $D > 0$. Now let $S_2 = \{y \in \mathbb{R}^d : \frac{D}{2} \leq \|y\| \leq 2\}$ and apply Lemma 6.3.19 to obtain $0 < m_0 < M_0 < \infty$ such that $m_0 \leq \varphi_\eta(y) \leq M_0$ for all $y \in S_2$. If $1 \leq s \leq 2$, we have for $y \in S_1$ that $y, sy, y/s$ are all in S_2. Now (6.106) implies that for any λ_0 sufficiently large and any $1 \leq s \leq 2$ we have

$$\frac{\varphi_\eta(y/s\lambda)}{\varphi_\eta(sy)} = \frac{\varphi_\eta(y/s\lambda)}{\varphi_\eta(y/s)} \cdot \frac{\varphi_\eta(y/s)}{\varphi_\eta(sy)} \geq \frac{m_0}{M_0} \lambda_0^{-\varepsilon-1/a_1} \tag{6.108}$$

whenever $1 \leq \lambda \leq \lambda_0$, $y \in S_1$ and $1 \leq s \leq 2$.

Since (y_n) is limit point compact every subsequence contains a further subsequence along which $y_n \to y \in S_1$. Fix $1 \le \lambda \le \lambda_0$ and choose $1 \le s \le 2$ such that both $H_{y/s\lambda}$ and H_{ys} are ϕ-continuity sets. Using the fact that $V_\eta(r,x)$ is decreasing in r for all $x \in \Gamma$, we get using (6.107) together with (6.96) that

$$\begin{aligned}\frac{V_\eta(\lambda r, x_r)}{V_\eta(r, x_r)} &\ge \frac{V_\eta(\lambda rs, x_r)}{V_\eta(r/s, x_r)} = \frac{(\lambda rs)^\eta F_\eta(x_r/(\lambda rs))}{(r/s)^\eta F_\eta(x_r/(r/s))}\\ &= \lambda^\eta s^{2\eta}\frac{F_\eta((x_r/r)/(\lambda s))}{F_\eta((x_r/r)s)} = \lambda^\eta s^{2\eta}\frac{k_n F_\eta(A_n^* y_n/(\lambda s))}{k_n F_\eta(A_n^* y_n s)} \to \lambda^\eta s^{2\eta}\frac{\varphi_\eta(y/(\lambda s))}{\varphi_\eta(sy)}\end{aligned}$$

along such a subsequence. Since every subsequence has a further subsequence with this property, and since $V_\eta(r,x)$ is monotone in r, it follows from (6.108) that for any compact subset S of Γ and any λ_0 sufficiently large, for some $r_0 > 0$ we have

$$m = C\frac{m_0}{M_0}\lambda_0^{-2\varepsilon-1/a_1} \le \frac{V_\eta(\lambda r, x)}{V_\eta(r,x)} \le 1 = M$$

for all $r \ge r_0$, all $1 \le \lambda \le \lambda_0$, and all $x \in S$, where $C = \min\{1, (4\lambda_0)^\eta\}$. Then $V_\eta(r,x)$ is uniformly R–O varying on compact subsets of $x \in \Gamma$. Write

$$\frac{V_\eta(\lambda r, x)}{V_\eta(r,x)} = \frac{(\lambda r)^{\zeta-\eta} V_\eta(\lambda r, x)}{U_\zeta(\lambda r, x)} \cdot \frac{U_\zeta(r,x)}{r^{\zeta-\eta} V_\eta(r,x)} \cdot \lambda^{\eta-\zeta} \cdot \frac{U_\zeta(\lambda r, x)}{U_\zeta(r,x)}. \tag{6.109}$$

Since both V_η and U_ζ are uniformly R–O varying, Corollary 6.3.9 applies. This together with Theorem 6.3.16 shows that for any compact $S \subset \Gamma$ and $\delta > 0$ there exists $r_0 > 0$ such that

$$\frac{m_2}{M_2} \cdot \lambda^{\eta-\zeta} \cdot C_1 \lambda^{\zeta-\delta-1/a_1} \le \frac{V_\eta(\lambda r, x)}{V_\eta(r,x)} \le \frac{M_2}{m_2} \cdot \lambda^{\eta-\zeta} \cdot C_2 \lambda^{\zeta+\delta-1/a_p}$$

for all $r \ge r_0$, all $\lambda \ge 1$ and all $x \in S$. □

We have shown that R–O variation of the functions U_ζ and V_η implies both upper and lower bounds on the growth rate of these functions. Now we will establish upper and lower bounds on the size of these functions.

Lemma 6.3.24. *Suppose that $\mu \in \mathrm{ROV}(E,c)$ for some $c > 1$ and that the real parts of the eigenvalues of E are all positive. Then for any compact $S \subset \Gamma$ and any $r_0 > 0$ there exist $0 < m < M < \infty$ such that $m \le V_0(r_0,x) \le M$ for all $x \in S$.*

Proof. Since S is compact we have $\|x\| \le B$ for all $x \in S$ where $0 < B < \infty$. Then

$$V_0(r_0,x) = \mu\{y \in \mathbb{R}^d : |\langle x,y\rangle| > r_0\} \le M = \mu\{y \in \mathbb{R}^d : \|y\| > r_0/B\}$$

for all $x \in S$. Now suppose by way of contradiction that for each n there exists $x_n \in S$ such that $V_0(r_0, x_n) \le 1/n$. Since S is compact we can choose a subsequence (n') along which $x_{n'} \to x \in S$. Choose $r_1 > r_0$ such that H_{x/r_1} is a μ-continuity set, and apply Lemma 6.1.25 to see that x/r_1 is a continuity point of the function $F(x) = \mu(H_x)$. Then

$$V_0(r_0, x_{n'}) \ge V_0(r_1, x_{n'}) = F(x_{n'}/r_1) \to F(x/r_1)$$

implies that $F(x/r_1) = 0$. Since $F(x/r)$ is monotone this implies $F(x/r) = 0$ for all $r \ge r_1$. Since $k_n(A_n\mu) \to \phi$, Lemma 6.1.26 shows that $k_n F(A_n^* x_n) \to \varphi(x)$ whenever $x_n \to x$ in Γ and H_x is a ϕ-continuity set, where $\varphi(x) = \phi(H_x)$. Apply part (b) of Lemma 4.2.16 to obtain $n = n(r)$ tending to infinity as $r \to \infty$ such that $y_n = (A_n^*)^{-1}(x/r)$ is limit point compact with all limit points in the set $S_1 = \{y \in \mathbb{R}^d : D \le \|y\| \le 1\}$ for some $D > 0$. Since (y_n) is limit point compact every subsequence contains a further subsequence along which $y_n \to y \in S_1$. Choose $s > 1$ such that $H_{y/s}$ is a ϕ-continuity set. Then $k_n F(x/r) \ge k_n F(x/(rs)) = k_n F(A_n^* y_n/s) \to \varphi(y/s)$ and since $F(x/r) = 0$ for all $r \ge r_1$ this implies that $\varphi(y/s) = 0$. But this contradicts Lemma 6.1.28, hence for some $m > 0$ we have that $V_0(r_0, x) \ge m$ for all $x \in S$. □

Theorem 6.3.25. *Suppose that $\mu \in \mathrm{ROV}(E, c)$ for some $c > 1$ and that the real parts of the eigenvalues of E are $0 < a_1 < \cdots < a_p$.*

(a) If $\eta < 1/a_p$, then for any compact $S \subset \Gamma$ and any $\delta > 0$ there exist $C_3 > 0$, $C_4 < \infty$ and $r_0 \ge 1$ such that

$$C_3 r^{\eta - \delta - 1/a_1} \le V_\eta(r, x) \le C_4 r^{\eta + \delta - 1/a_p}$$

for all $r \ge r_0$ and $x \in S$.

(b) If $\zeta > 1/a_1$ and U_ζ exists, then for any compact $S \subset \Gamma$ and any $\delta > 0$ there exist $C_3 > 0$, $C_4 < \infty$ and $r_0 \ge 1$ such that

$$C_3 r^{\zeta - \delta - 1/a_1} \le U_\zeta(r, x) \le C_4 r^{\zeta + \delta - 1/a_p}$$

for all $r \ge r_0$ and $x \in S$.

Proof. First we prove part (a) in the special case where $\eta = 0$. Theorem 6.3.4 implies that for any compact $S \subset \Gamma$ and any $\delta > 0$ there exist $C_1 > 0$, $C_2 < \infty$ and $r_0 \ge 1$ such that

$$C_1 (r/r_0)^{-\delta - 1/a_1} V_0(r_0, x) \le V_0(r, x) \le C_2 (r/r_0)^{\delta - 1/a_p} V_0(r_0, x)$$

for all $r \ge r_0$ and $x \in S$. Lemma 6.3.24 implies that for some $0 < m < M < \infty$ we have $m \le V_0(r_0, x) \le M$ for all $x \in S$, and hence

$$(C_1 r_0^{\delta + 1/a_1} m) r^{-\delta - 1/a_1} \le V_0(r, x) \le (C_2 r_0^{-\delta + 1/a_p} M) r^{\delta - 1/a_p}$$

for all $r \ge r_0$ and $x \in S$.

Next we prove part (b). If $\zeta > 1/a_1$ and U_ζ exists, then Theorems 6.3.4 and 6.3.16 imply that both V_0 and U_ζ are uniformly R–O varying. Then Corollary 6.3.9 implies that

$$\frac{1}{M_2} r^\zeta V_0(r,\theta) \le U_\zeta(r,\theta) \le \frac{1}{m_2} r^\zeta V_0(r,\theta)$$

for all $r \ge r_0$ and $x \in S$. Now it follows from part (a) for $\eta = 0$ that

$$\frac{C_3}{M_2} r^{\zeta-\delta-1/a_1} \le U_\zeta(r,\theta) \le \frac{C_4}{m_2} r^{\zeta+\delta-1/a_p}$$

for all $r \ge r_0$ and $x \in S$.

Finally we prove part (a) in general. If $0 \ne \eta < 1/a_p$, then Theorems 6.3.16 and 6.3.23 imply that both V_η and U_ζ are uniformly R–O varying. Then Corollary 6.3.9 implies that

$$m_2 r^{\eta-\zeta} U_\zeta(r,\theta) \le V_\eta(r,\theta) \le M_2 r^{\eta-\zeta} U_\zeta(r,\theta)$$

for all $r \ge r_0$ and $x \in S$. Now it follows from part (b) that

$$m_2 r^{\eta-\zeta} C_3 r^{\zeta-\delta-1/a_1} \le V_\eta(r,\theta) \le M_2 r^{\eta-\zeta} C_4 r^{\zeta+\delta-1/a_p}$$

for all $r \ge r_0$ and $x \in S$. □

6.4 SHARP SPECTRAL BOUNDS

In Section 6.3, we showed that the truncated moments and tails of an R–O varying measure are uniformly R–O varying. We also showed that the growth rate of these functions is governed by the real parts of the eigenvalues of the exponents of this measure. In this section, we will sharpen these results by using the spectral decomposition developed in Section 4.3. Recall from Theorem 2.1.14 that for any linear operator E on $\mathbb{R}^d$ there exists a direct sum decomposition $\mathbb{R}^d = W_1 \oplus \cdots \oplus W_p$ into E-invariant subspaces such that, if we write $E = E_1 \oplus \cdots \oplus E_p$, where $E_i : W_i \to W_i$, then every eigenvalue of E_i has real part equal to a_i and $a_1 < \cdots < a_p$. This is called the spectral decomposition of $\mathbb{R}^d$ relative to E. Recall from Theorem 6.2.3 that a measure $\mu \in \mathcal{M}$ is R-O varying if and only if there exist a sequence $(B_n) \in \mathrm{RV}(-E)$ and an increasing sequence of natural numbers (k_n) tending to infinity with $k_{n+1}/k_n \to c > 1$ such that

$$k_n(B_{k_n}\mu) \to \phi \tag{6.110}$$

where ϕ is $\mathcal{M}$-full. Our first result applies the spectral decomposition to the limit measure ϕ.

Definition 6.4.1. Suppose that $\phi \in \mathcal{M}$ and $c^E\phi = c \cdot \phi$ for some $c > 1$ and some linear operator $E \in \mathrm{L}(\mathbb{R}^d)$. Let $W_1, \ldots, W_p$ denote the spectral decomposition of $\mathbb{R}^d$ relative to E, and $\pi_i : \mathbb{R}^d \to W_i$ the projections associated with this direct sum decomposition. Then the measures $\phi_i = \pi_i\phi$ for $i = 1, \ldots, p$ are called the *spectral decomposition of ϕ with respect to E*. If every eigenvalue of E has the same real part, then there is only one component in the spectral decomposition. In this case, we say that ϕ is *spectrally simple*.

Lemma 6.4.2. *Suppose that ϕ is $\mathcal{M}$-full and $c^E\phi = c \cdot \phi$ for some $c > 1$ and some linear operator $E \in \mathrm{L}(\mathbb{R}^d)$. Then every ϕ_i in the spectral decomposition of ϕ relative to $E = E_1 \oplus \cdots \oplus E_p$ is $\mathcal{M}$-full on W_i and $c^{E_i}\phi_i = c \cdot \phi_i$. Furthermore, ϕ_i is spectrally simple.*

Proof. If some ϕ_i is not $\mathcal{M}$-full, then $\phi_i = \pi_i\phi$ is supported on a proper subspace $L_i \subset W_i$. Then ϕ is supported on the preimage $\pi_i^{-1}(L_i)$, which is a proper subspace of $\mathbb{R}^d$, and this contradicts the assumption that ϕ is $\mathcal{M}$-full. Hence every ϕ_i is $\mathcal{M}$-full. Since $\pi_i E = E_i \pi_i$ we also have $c \cdot \phi_i = c \cdot \pi_i\phi = \pi_i(c \cdot \phi) = \pi_i c^E\phi = c^{E_i}\pi_i\phi = c^{E_i}\phi_i$. Since every eigenvalue of E_i is equal to a_i, the measure ϕ_i is spectrally simple. □

Definition 6.4.3. Suppose that $\mu \in \mathrm{ROV}(E, c)$ and (6.110) holds. Let $W_1, \ldots, W_p$ denote the spectral decomposition of $\mathbb{R}^d$ relative to E, and let $\pi_i : \mathbb{R}^d \to W_i$ be the projections associated with this direct sum decomposition. If every W_i is a B_n-invariant subspace for each n, then we say that μ is *spectrally compatible with ϕ*, and we call the measures $\mu_i = \pi_i\mu$ the *spectral decomposition of μ with respect to E*.

Lemma 6.4.4. *Suppose that $\mu \in \mathrm{ROV}(E, c)$ and (6.110) holds. If μ is spectrally compatible with ϕ, then the measures $\mu_i = \pi_i\mu \in \mathrm{ROV}(E_i, c)$. Furthermore, we can write $B_n = B_{1n} \oplus \cdots \oplus B_{pn}$ with respect to this direct sum decomposition, and then $k_n(B_{ik_n}\mu_i) \to \phi_i = \pi_i\phi$ for each $i = 1, \ldots, p$.*

Proof. Since each of the subspaces $W_1, \ldots, W_p$ are B_n-invariant we may write $B_n = B_{1n} \oplus \cdots \oplus B_{pn}$ with respect to this direct sum decomposition. Since $\pi_i B_n = B_{in}\pi_i$ we also have $\pi_i(k_n B_{k_n}\mu) \to \pi_i\phi = \phi_i$, where $\pi_i(k_n B_{k_n}\mu) = k_n(\pi_i B_{k_n}\mu) = k_n(B_{ik_n}\pi_i\mu) = k_n(B_{ik_n}\mu_i)$, so that $k_n(B_{ik_n}\mu_i) \to \phi_i$, where (B_{in}) varies regularly with index E_i. Then $\mu_i \in \mathrm{ROV}(E_i, c)$. □

Definition 6.4.5. Suppose that $\mu \in \mathrm{ROV}(E, c)$ and that μ is spectrally compatible with E. Let $W_1, \ldots, W_p$ denote the spectral decomposition of $\mathbb{R}^d$ with respect to E and let $\mu_1, \ldots, \mu_p$ denote the spectral decomposition of μ with respect to E. For each $i = 1, \ldots, p$ and $\eta, \zeta \geq 0$, the truncated moments $U_{i\zeta}$ and tails $V_{i\eta}$ of the measure μ_i are defined by

$$\begin{aligned} U_{i\zeta}(r,\theta) &= \int_{|<x,\theta>| \leq r} |<x,\theta>|^{\zeta} \mu_i\{dx\} \\ V_{i\eta}(r,\theta) &= \int_{|<x,\theta>| > r} |<x,y>|^{\eta} \mu_i\{dx\} \end{aligned} \tag{6.111}$$

for $\theta \in W_i$ whenever these integrals exist.

Since the real parts of the eigenvalues of the exponent E_i are all equal to a_i, the results of Section 6.3 yield sharp bounds on the truncated moments and tails of these measures.

Theorem 6.4.6. *Let $\mu \in \mathrm{ROV}(E, c)$ for some $c > 1$, where the real parts of the eigenvalues of E are $0 < a_1 < \cdots < a_p$. Suppose that μ is spectrally compatible with ϕ and let $\mu_1, \ldots, \mu_p$ denote the spectral decomposition of μ with respect to E, so that $\mu_i \in \mathrm{ROV}(E_i, c)$ for all $i = 1, \ldots, p$. Then*

(a) If $0 \le \eta < 1/a_i$, then $V_{i\eta}(r,\theta)$ exists and is uniformly R–O varying on compact subsets of $\theta \in \Gamma_i = W_i \setminus \{0\}$. Furthermore, for any compact $S \subset \Gamma_i$ and any $\delta > 0$ there exist $C_1 > 0$, $C_2 < \infty$ and $r_0 \ge 1$ such that

$$C_1 \lambda^{\eta-\delta-1/a_i} \le \frac{V_{i\eta}(\lambda r, \theta)}{V_{i\eta}(r,\theta)} \le C_2 \lambda^{\eta+\delta-1/a_i}$$

for all $r \ge r_0$, $\lambda \ge 1$ and $\theta \in S$

(b) For any $0 \le \eta < 1/a_i$, any compact $S \subset \Gamma_i$ and any $\delta > 0$ there exist $C_3 > 0$, $C_4 < \infty$ and $r_0 \ge 1$ such that

$$C_3 r^{\eta-\delta-1/a_i} \le V_{i\eta}(r,\theta) \le C_4 r^{\eta+\delta-1/a_i}$$

for all $r \ge r_0$ and $\theta \in S$

(c) If $\zeta > 1/a_i$ and $U_{i\zeta}$ exists, then $U_{i\zeta}(r,\theta)$ is uniformly R–O varying on compact subsets of $\theta \in \Gamma_i$. Furthermore, for any compact $S \subset \Gamma_i$ and any $\delta > 0$ there exist $C_1 > 0$, $C_2 < \infty$ and $r_0 \ge 1$ such that

$$C_1 \lambda^{\zeta-\delta-1/a_i} \le \frac{U_{i\zeta}(\lambda r, \theta)}{U_{i\zeta}(r,\theta)} \le C_2 \lambda^{\zeta+\delta-1/a_i}$$

for all $r \ge r_0$, $\lambda \ge 1$ and $\theta \in S$

(d) If $\zeta > 1/a_i$ and $U_{i\zeta}$ exists, then for any compact $S \subset \Gamma_i$ and any $\delta > 0$ there exist $C_3 > 0$, $C_4 < \infty$ and $r_0 \ge 1$ such that

$$C_3 r^{\zeta-\delta-1/a_i} \le U_{i\zeta}(r,\theta) \le C_4 r^{\zeta+\delta-1/a_i}$$

for all $r \ge r_0$ and $\theta \in S$

Proof. Lemma 6.4.4 shows that $\mu_i \in \mathrm{ROV}(E_i, c)$ with $c > 1$ where every eigenvalue of E_i has real part equal to a_i. Then the desired results follow immediately from Theorems 6.3.4, 6.3.16, 6.3.23, and 6.3.25. □

Roughly speaking, Theorem 6.4.6 shows that the tails of the component measure μ_i fall off like r^{-1/a_i} as $r \to \infty$ when μ is spectrally compatible with the limit measure ϕ in (6.37). Furthermore, the truncated moments grow like $r^{\zeta-1/a_i}$ and the tail moments fall off like $r^{\eta-1/a_i}$, where of course $\eta < 1/a_i < \zeta$. Our goal in this section is to obtain sharp bounds of this type for any R–O varying measure. The key is to arrange things so that μ is spectrally compatible with ϕ. Our next result describes the class of limit measures ϕ that can occur in (6.110).

Lemma 6.4.7. *Suppose that μ is R–O varying and (6.110) holds for some ϕ $\mathcal{M}$-full. Then for any $T \in \mathrm{GL}(\mathbb{R}^d)$ we have*

$$k'_n(B'_{k_n}\mu) \to \phi' \tag{6.112}$$

with $k'_n = k_n$, $B'_n = TB_n$ regularly varying with index $-TET^{-1}$ and $\phi' = T\phi$ $\mathcal{M}$-full. Conversely, if there exists a regularly varying sequence (B'_n) and an increasing sequence of natural numbers (k'_n) tending to infinity with $k'_{n+1}/k'_n \to c' \geq 1$ such that (6.112) holds for some ϕ' $\mathcal{M}$-full, then $\phi' = T\phi$ for some $T \in \mathrm{GL}(\mathbb{R}^d)$.

Proof. If (6.110) holds and $\phi' = T\phi$ for some $T \in \mathrm{GL}(\mathbb{R}^d)$, then for all $\lambda > 0$ we have

$$B'_{[\lambda n]}(B'_n)^{-1} = TB_{[\lambda n]}B_n^{-1}T^{-1} \to T\lambda^{-E}T^{-1} = \lambda^{-TET^{-1}}$$

so that $B'_n = TB_n$ varies regularly with index $-TET^{-1}$. Then $k_n(B'_{k_n}\mu) = k_n(TB_{k_n}\mu) = T(k_n(B_{k_n}\mu)) \to T\phi = \phi'$ so that (6.112) holds. Conversely, if both (6.110) and (6.112) hold, then it follows from convergence of types, Theorem 6.1.19, that $\phi' = T\phi$ for some $T \in \mathrm{L}(\mathbb{R}^d)$, and since ϕ' is $\mathcal{M}$-full it follows from Lemma 6.1.12 that T is invertible. □

Our next result is the main technical tool for this section. It shows that for any R–O varying measure we can choose a limit measure ϕ in (6.110) so that μ is spectrally compatible with ϕ. This is the key to obtaining sharp bounds on the truncated moments and tails of μ.

Theorem 6.4.8. *For any R–O varying measure μ, there exists an $\mathcal{M}$-full limit measure ϕ, a regularly varying sequence of linear operators (B_n) with index $-E$, and an increasing sequence of natural numbers (k_n) tending to infinity with $k_{n+1}/k_n \to c > 1$ such that*

(a) *$k_n(B_{k_n}\mu) \to \phi$ as $n \to \infty$ so that $\mu \in \mathrm{ROV}(E, c)$*

(b) *The subspaces W_i in the spectral decomposition $\mathbb{R}^d = W_1 \oplus \cdots \oplus W_p$ with respect to E are mutually orthogonal*

(c) *The subspaces W_i are all B_n-invariant for every n, so that μ is spectrally compatible with ϕ*

(d) *If we write $E = E_1 \oplus \cdots \oplus E_p$ and $B_n = B_{1n} \oplus \cdots \oplus B_{pn}$ with respect to this direct sum decomposition, then each $B_{in} : W_i \to W_i$ is regularly*

varying with index $-E_i$ *and every eigenvalue of* E_i *has real part equal to* a_i *where* $0 \le a_1 < \cdots < a_p$

(e) *If we let* $\phi_1, \ldots, \phi_p$ *and* $\mu_1, \ldots, \mu_p$ *denote the spectral decomposition of* ϕ, μ *respectively, then* $k_n(B_{ik_n}\mu_i) \to \phi_i$ *as* $n \to \infty$ *so that* $\mu_i \in \mathrm{ROV}(E_i, c)$ *for each* $i = 1, \ldots, p$

Proof. Since μ is R–O varying there exists an $\mathcal{M}$-full limit measure ϕ', a sequence of linear operators (B'_n) regularly varying with index $-E'$ and an increasing sequence of natural numbers (k_n) tending to infinity with $k_{n+1}/k_n \to c \ge 1$ such that $k_n(B'_{k_n}\mu) \to \phi'$. Apply Corollary 4.3.12 to obtain a linear operator $T \in \mathrm{GL}(\mathbb{R}^d)$ and a regularly varying sequence (B_n) with index $-E = -TE'T^{-1}$ such that $TB'_n \sim B_n$ and parts (b)–(d) of the Theorem are satisfied. Convergence of types, Theorem 6.1.19, implies that $k_n(B_{k_n}\mu) \to \phi = T\phi'$ so that (a) holds. Then (e) follows from Lemma 6.4.4. □

Given an R–O varying measure μ and a nonzero vector $\theta \in \mathbb{R}^d$, we can use the direct sum decomposition of Theorem 6.4.8 to write $\theta = \theta_1 + \cdots + \theta_p$. Theorem 6.4.6 gives sharp bounds on the truncated moments $U_{i\zeta}(r, \theta_i)$ and the tail moments $V_{i\eta}(r, \theta_i)$ of the spectral component measure μ_i whenever $\theta_i \ne 0$. In particular, the tails of μ_i fall off about as fast as r^{-1/a_i}. In order to obtain sharp bounds on the truncated moments U_ζ and tail moments V_η of the parent measure μ, we will compare these functions to the truncated moments $U_{i\zeta}$ and tail moments $V_{i\eta}$ of the component measures μ_i. It is reasonable to believe that the heaviest tail should dominate the asymptotic behavior of these truncated moment and tail functions. If $\theta_i = 0$, then $U_{i\zeta}(r, \theta_i) = V_{i\eta}(r, \theta_i) = 0$ for all $r > 0$, so this component vanishes. Then the largest $i = 1, \ldots, p$ for which $\theta_i \ne 0$ should determine the asymptotics. If this is the case, the tails of the measure μ in the direction θ should fall off like r^{-1/a_i} where i is chosen so that $\theta_i \ne 0$ and $\theta_j = 0$ for all $i < j \le p$.

Definition 6.4.9. Suppose μ is R–O varying at infinity. Given a nonzero vector $\theta \in \mathbb{R}^d$ use the spectral decomposition of Theorem 6.4.8 to write $\theta = \theta_1 + \cdots + \theta_p$. Now define

$$\alpha(\theta) = \min\{1/a_i : \theta_i \ne 0\}. \tag{6.113}$$

We will call this the *spectral index of* θ *with respect to* μ.

If μ is R–O varying at infinity we will show that for any nonzero vector $\theta \in \mathbb{R}^d$ the tails of the measure μ fall off about as fast as $r^{-\alpha(\theta)}$. This requires a few preliminary results.

Lemma 6.4.10. *Suppose that* μ *is R–O varying at infinity and* $\theta \ne 0$. *Write* $\theta = \theta_1 + \cdots + \theta_p$ *with respect to the spectral decomposition* $W_1, \ldots, W_p$ *of Theorem 6.4.8. Then*

$$\begin{aligned} U_{i\zeta}(r, \theta_i) &= U_\zeta(r, \theta) \\ V_{i\eta}(r, \theta_i) &= V_\eta(r, \theta) \end{aligned} \tag{6.114}$$

whenever $\theta \in W_i$. *Furthermore,* $V_\eta(r,\theta)$ *exists whenever* $\eta < 1/a_i$ *and* $\theta \in W_i$.

Proof. Given $x \in \mathbb{R}^d$ write $x = x_1 + \cdots + x_p$ with respect to the spectral decomposition of Theorem 6.4.8. If $\theta \in W_i$, then $\theta_j = 0$ for all $j \neq i$. Furthermore, for any $x \in \mathbb{R}^d$ we have $<x_j, \theta_i> = 0$ for $j \neq i$ since the spaces $W_1, \ldots, W_p$ are mutually orthogonal. Then

$$<x,\theta> = <x,\theta_i> = \sum_{j=1}^{p} <x_j,\theta_i> = <x_i,\theta_i> = <\pi_i x,\theta_i>$$

and so

$$\begin{aligned} U_{i\zeta}(r,\theta_i) &= \int_{|<x,\theta_i>|\le r} |<x,\theta_i>|^\zeta \pi_i\mu\{dx\} \\ &= \int_{|<\pi_i x,\theta_i>|\le r} |<\pi_i x,\theta_i>|^\zeta \mu\{dx\} \\ &= \int_{|<x,\theta>|\le r} |<x,\theta>|^\zeta \mu\{dx\} \\ &= U_\zeta(r,\theta) \end{aligned}$$

and the proof for V_η is similar. Then for $\eta < 1/a_i$ and $\theta \in W_i$ it follows from Theorem 6.4.6 (a) that $V_\eta(r,\theta)$ exists. □

Definition 6.4.11. If $\mu \in \mathcal{M}$ is R–O varying measure and $W_1, \ldots, W_p$ is the spectral decomposition of Theorem 6.4.8 define $L_i = W_1 \oplus \cdots \oplus W_i$ and $P_i : \mathbb{R}^d \to L_i$ the orthogonal projection. Let $\bar{\mu}_i = P_i\mu$. Given $\eta, \zeta \ge 0$ the truncated moments $\bar{U}_{i\zeta}$ and tails $\bar{V}_{i\eta}$ of the measure $\bar{\mu}_i$ are defined by

$$\begin{aligned} \bar{U}_{i\zeta}(r,\theta) &= \int_{|<x,\theta>|\le r} |<x,\theta>|^\zeta \bar{\mu}_i\{dx\} \\ \bar{V}_{i\eta}(r,\theta) &= \int_{|<x,\theta>|> r} |<x,y>|^\eta \bar{\mu}_i\{dx\} \end{aligned} \tag{6.115}$$

for $\theta \in L_i$ whenever these integrals exist.

Theorem 6.4.12. *Let* $\mu \in \mathrm{ROV}(E,c)$ *for some* $c > 1$ *where the real parts of the eigenvalues of* E *are* $0 < a_1 < \cdots < a_p$. *Let* $W_1, \ldots, W_p$ *denote the spectral decomposition of Theorem 6.4.8 and define* $L_i = W_1 \oplus \cdots \oplus W_i$.

(a) *If* $\eta < 1/a_i$, *then* $\bar{V}_{i\eta}(r,\theta)$ *exists and is uniformly R–O varying on compact subsets of* $\theta \in L_i \setminus \{0\}$. *Furthermore, for any compact* $S \subset \bar{\Gamma}_i = L_i \setminus \{0\}$ *and any* $\delta > 0$ *there exist* $C_1 > 0$, $C_2 < \infty$ *and* $r_0 \ge 1$ *such that*

$$C_1 \lambda^{\eta-\delta-1/a_1} \le \frac{\bar{V}_{i\eta}(\lambda r,\theta)}{\bar{V}_{i\eta}(r,\theta)} \le C_2 \lambda^{\eta+\delta-1/a_i}$$

for all $r \ge r_0$, $\lambda \ge 1$ *and* $\theta \in S$.

(b) *For any $\eta < 1/a_i$, any compact $S \subset \bar{\Gamma}_i$ and any $\delta > 0$ there exist $C_3 > 0$, $C_4 < \infty$ and $r_0 \geq 1$ such that*

$$C_3 r^{\eta-\delta-1/a_1} \leq \bar{V}_{i\eta}(r,\theta) \leq C_4 r^{\eta+\delta-1/a_i}$$

for all $r \geq r_0$ and $\theta \in S$.

(c) *If $\zeta > 1/a_1$ and $\bar{U}_{i\zeta}$ exists, then $\bar{U}_{i\zeta}(r,\theta)$ is uniformly R–O varying on compact subsets of $\theta \in \bar{\Gamma}_i$. Furthermore, for any compact $S \subset \bar{\Gamma}_i$ and any $\delta > 0$ there exist $C_1 > 0$, $C_2 < \infty$ and $r_0 \geq 1$ such that*

$$C_1 \lambda^{\zeta-\delta-1/a_1} \leq \frac{\bar{U}_{i\zeta}(\lambda r,\theta)}{\bar{U}_{i\zeta}(r,\theta)} \leq C_2 \lambda^{\zeta+\delta-1/a_i}$$

for all $r \geq r_0$, $\lambda \geq 1$ and $\theta \in S$.

(d) *If $\zeta > 1/a_1$ and $\bar{U}_{i\zeta}$ exists, then for any compact $S \subset \bar{\Gamma}_i$ and any $\delta > 0$ there exist $C_3 > 0$, $C_4 < \infty$ and $r_0 \geq 1$ such that*

$$C_3 r^{\zeta-\delta-1/a_1} \leq \bar{U}_{i\zeta}(r,\theta) \leq C_4 r^{\zeta+\delta-1/a_i}$$

for all $r \geq r_0$ and $\theta \in S$.

Proof. Since each of the subspaces $W_1 \ldots, W_p$ are B_n-invariant, if we define $\bar{B}_n = B_{1n} \oplus \cdots \oplus B_{in}$, then L_i is $\bar{B}_{in}$-invariant for every n, and it follows that $(\bar{B}_{in})$ is regularly varying with index $\bar{E}_i = E_1 \oplus \cdots \oplus E_i$. Since $P_i B_n = \bar{B}_{in} P_i$ we also have as in the proof of Lemma 6.4.4 that $P_i(k_n(B_{k_n}\mu)) = k_n(\bar{B}_{ik_n}\bar{\mu}_i) \to P_i\phi$ $\mathcal{M}$-full on L_i. Then $\bar{\mu}_i \in \mathrm{ROV}(c, \bar{E}_i)$ with $c > 1$, where every eigenvalue of $\bar{E}_i$ has real part greater than or equal to $a_1 > 0$ and less than or equal to a_i. Then the desired results follow immediately from Theorems 6.3.4, 6.3.16, 6.3.23, and 6.3.25. □

Lemma 6.4.13. *Suppose that μ is R–O varying at infinity and $\theta \neq 0$. Write $\theta = \theta_1 + \cdots + \theta_p$ with respect to the spectral decomposition $W_1, \ldots, W_p$ of Theorem 6.4.8 and let $\bar{\theta}_i = \theta_1 + \cdots + \theta_i$. Then*

$$\begin{aligned} \bar{U}_{i\zeta}(r,\bar{\theta}_i) &= U_\zeta(r,\theta) \\ \bar{V}_{i\eta}(r,\bar{\theta}_i) &= V_\eta(r,\theta) \end{aligned} \tag{6.116}$$

whenever $\theta \in L_i = W_1 \oplus \cdots \oplus W_i$. Furthermore, $V_\eta(r,\theta)$ exists whenever $\eta < 1/a_i$ and $\theta \in L_i$.

Proof. If $\theta \in L_i$ then $<P_i x, \bar{\theta}_i> = <x, \bar{\theta}_i> = <x, \theta>$ for all $x \in \mathbb{R}^d$ since $<x_j, \bar{\theta}_i> = 0$ for $j > i$. Then

$$\begin{aligned}\bar{U}_{i\zeta}(r, \bar{\theta}_i) &= \int_{|<P_i x, \bar{\theta}_i>| \le r} |<P_i x, \bar{\theta}_i>|^\zeta P_i \mu\{dx\} \\ &= \int_{|<x, \bar{\theta}_i>| \le r} |<x, \bar{\theta}_i>|^\zeta \mu\{dx\} \\ &= \int_{|<x, \theta>| \le r} |<x, \theta>|^\zeta \mu\{dx\} \\ &= U_\zeta(r, \theta)\end{aligned}$$

and the proof for V_η is similar. If $\eta < 1/a_i$ and $\theta \in L_i$, then it follows from Theorem 6.4.12 (a) that $V_\eta(r, \theta)$ exists. □

If μ is R–O varying at infinity and $W_1, \dots, W_p$ is the spectral decomposition of Theorem 6.4.8, then the spectral index $\alpha(\theta) = 1/a_i$ on the set $L_i \setminus L_{i-1}$, where $L_i = W_1 \oplus \cdots \oplus W_i$. Our next result shows that in this case the truncated moments U_ζ and tail moments V_η of the parent measure μ can be estimated using the truncated moments $U_{i\zeta}$ and tail moments $V_{i\eta}$ of the component measure μ_i.

Proposition 6.4.14. *Let $\mu \in \mathrm{ROV}(E, c)$ for some $c > 1$, where the real parts of the eigenvalues of E are $0 < a_1 < \cdots < a_p$. Write $\theta = \theta_1 + \cdots + \theta_p$ with respect to the spectral decomposition $W_1, \dots, W_p$ of Theorem 6.4.8 and define $L_i = W_1 \oplus \cdots \oplus W_i$.*

(a) *If $0 \le \eta < 1/a_i < \zeta$ and $\bar{U}_{i,\zeta}$ exists, then for any compact $S \subset L_i \setminus L_{i-1}$ there exist constants $C_1 > 0$, $C_2 < \infty$ and $r_0 \ge 1$ such that*

$$C_1 r^{\eta - \zeta} U_{i\zeta}(r, \theta_i) \le V_\eta(r, \theta) \le C_2 V_{i\eta}(r, \theta_i) \tag{6.117}$$

for all $r \ge r_0$ and $\theta \in S$.

(b) *If $0 \le \eta < 1/a_i < \zeta$ and $\bar{U}_{i\zeta}$ exists, then for any compact $S \subset L_i \setminus L_{i-1}$ there exist constants $C_3 > 0$, $C_4 < \infty$ and $r_0 \ge 1$ such that*

$$C_3 U_{i\zeta}(r, \theta_i) \le U_\zeta(r, \theta) \le C_4 r^{\zeta - \eta} V_{i\eta}(r, \theta_i) \tag{6.118}$$

for all $r \ge r_0$ and $\theta \in S$.

Proof. We start with the upper bound in (6.117). Write $\theta = \theta_1 + \cdots + \theta_p$ with respect to the spectral decomposition $W_1, \dots, W_p$ of Theorem 6.4.8 and $\bar{\theta}_i = \theta_1 + \cdots + \theta_i$. If $\theta \in L_i \setminus L_{i-1}$, then $\bar{\theta}_i = \theta = \theta_i + \bar{\theta}_{i-1}$ where $\theta_i \neq 0$, so for any $x \in \mathbb{R}^d$ we have $<x, \theta> = <x, \theta_i> + <x, \bar{\theta}_{i-1}>$. Using the well-known inequality $|x+y|^\eta \le 2^\eta(|x|^\eta + |y|^\eta)$ for $x, y \in \mathbb{R}$, we can bound

$$\begin{aligned} V_\eta(r,\theta) &= \int_{|<x,\theta>|>r} |<x,\theta_i> + <x,\bar{\theta}_{i-1}>|^\eta \mu\{dx\} \\ &\le 2^\eta\Big(\int_{|<x,\theta>|>r} |<x,\theta_i>|^\eta \mu\{dx\} \\ &\quad + \int_{|<x,\theta>|>r} |<x,\bar{\theta}_{i-1}>|^\eta \mu\{dx\}\Big) \\ &= 2^\eta(I_1(r) + I_2(r)). \end{aligned} \tag{6.119}$$

But $\{x : |<x,\theta>| > r\} \subset \{x : |<x,\theta_i>| > r/2\} \cup \{x : |<x,\bar{\theta}_{i-1}>| > r/2\}$, and hence in view of Lemmas 6.4.10 and 6.4.13 we can decompose further to obtain

$$\begin{aligned} I_1(r) &\le \int_{|<x,\theta_i>|>r/2} |<x,\theta_i>|^\eta \mu\{dx\} \\ &\quad + \int_{|<x,\bar{\theta}_{i-1}>|>r/2} |<x,\theta_i>|^\eta \mu\{dx\} \\ &= V_{i\eta}(r/2,\theta_i) + \int\limits_{\substack{|<x,\bar{\theta}_{i-1}>|>r/2 \\ |<x,\theta_i>|\le r/2}} |<x,\theta_i>|^\eta \mu\{dx\} \\ &\qquad + \int\limits_{\substack{|<x,\bar{\theta}_{i-1}>|>r/2 \\ |<x,\theta_i>|>r/2}} |<x,\theta_i>|^\eta \mu\{dx\} \\ &\le 2V_{i\eta}(r/2,\theta_i) + (r/2)^\eta \bar{V}_{i-1,0}(r/2,\bar{\theta}_{i-1}). \end{aligned}$$

On the other hand,

$$\begin{aligned} I_2(r) &\le \int_{|<x,\bar{\theta}_{i-1}>|>r/2} |<x,\bar{\theta}_{i-1}>|^\eta \mu\{dx\} \\ &\quad + \int_{|<x,\theta_i>|>r/2} |<x,\bar{\theta}_{i-1}>|^\eta \mu\{dx\} \\ &= \bar{V}_{i-1,\eta}(r/2,\bar{\theta}_{i-1}) + \int\limits_{\substack{|<x,\theta_i>|>r/2 \\ |<x,\bar{\theta}_{i-1}>|\le r/2}} |<x,\bar{\theta}_{i-1}>|^\eta \mu\{dx\} \\ &\qquad + \int\limits_{\substack{|<x,\theta_i>|>r/2 \\ |<x,\bar{\theta}_{i-1}>|>r/2}} |<x,\bar{\theta}_{i-1}>|^\eta \mu\{dx\} \\ &\le 2\bar{V}_{i-1,\eta}(r/2,\bar{\theta}_{i-1}) + (r/2)^\eta V_{i0}(r/2,\theta_i). \end{aligned}$$

Putting things together, it follows that

$$V_\eta(r,\theta) \le 2^\eta \Big(2V_{i\eta}(r/2,\theta_i) + (r/2)^\eta V_{i0}(r/2,\theta_i) \\ + 2\bar{V}_{i-1,\eta}(r/2,\bar{\theta}_{i-1}) + (r/2)^\eta \bar{V}_{i-1,0}(r/2,\bar{\theta}_{i-1}) \Big) \quad (6.120)$$

for any $\theta \in L_i \setminus L_{i-1}$.

If $\theta \in W_i$, then $\theta = \bar{\theta}_i = \theta_i$, and since $\bar{U}_{i\zeta}$ exists we obtain from Lemmas 6.4.10 and 6.4.13 that $U_{i\zeta}(r,\theta_i) = U_\zeta(r,\theta) = \bar{U}_{i\zeta}(r,\bar{\theta}_i)$, and hence $U_{i\zeta}$ exists. Since $\eta < 1/a_i < \zeta$ we know that $U_{i\zeta}(r,\theta_i)$, $V_{i\eta}(r,\theta_i)$, and $V_{i0}(r,\theta_i)$ are all uniformly R–O varying on compact subsets of $\theta_i \in W_i$. Since S is a compact subset of $L_i \setminus L_{i-1}$, the set $S_i = \pi_i(S)$ is a compact subset of $W_i \setminus \{0\}$. Apply part (a) of Corollary 6.3.9 to obtain $r_0 > 0$ and $M_2 < \infty$ such that

$$\frac{r^\zeta V_{i0}(r,\theta_i)}{U_{i\zeta}(r,\theta_i)} \le M_2$$

for all $r \ge r_0$ and $\theta_i \in S_i$. Then apply part (b) of Corollary 6.3.9 to obtain another $r_0 > 0$ and some $m_2 < \infty$ such that

$$\frac{r^{\zeta-\eta} V_{i\eta}(r,\theta_i)}{U_{i\zeta}(r,\theta_i)} \ge m_2$$

for all $r \ge r_0$ and $\theta_i \in S_i$. Take the larger of these two values of r_0 so that both inequalities hold. Then we have

$$r^\eta V_{i0}(r,\theta_i) = \frac{r^\zeta V_{i0}(r,\theta_i)}{U_{i\zeta}(r,\theta_i)} \cdot \frac{U_{i\zeta}(r,\theta_i)}{r^{\zeta-\eta} V_{i\eta}(r,\theta_i)} \cdot V_{i\eta}(r,\theta_i) \le C V_{i\eta}(r,\theta_i) \quad (6.121)$$

for all $r \ge r_0$ and all $\theta_i \in S_i$, where $C = M_2/m_2 < \infty$. Combining (6.120) with (6.121), we see that

$$V_\eta(r,\theta) \le 2^\eta \Big((2+C) V_{i\eta}(r/2,\theta_i) \\ + 2\bar{V}_{i-1,\eta}(r/2,\bar{\theta}_{i-1}) + (r/2)^\eta \bar{V}_{i-1,0}(r/2,\bar{\theta}_{i-1}) \Big) \quad (6.122)$$

for all $\theta \in S$ and all $r > 0$ sufficiently large.

Since S is a compact subset of $L_i \setminus L_{i-1}$, the set $\bar{S}_{i-1} = P_{i-1}(S)$ is a compact subset of L_{i-1} that may contain the origin. Choose $B > 0$ such that $\|\bar{\theta}_{i-1}\| \le B$ for all $\bar{\theta}_{i-1} \in \bar{S}_{i-1}$. Choose $\delta > 0$ such that $2\delta + 1/a_i < 1/a_{i-1}$. Since $\eta < 1/a_i < 1/a_{i-1}$ we can apply Theorem 6.4.12 (b) to obtain $C_4 < \infty$ and $r_0 \ge 1$ such that

$$\bar{V}_{i-1,\eta}(r,\bar{\theta}_{i-1}) \le C_4 r^{\eta+\delta-1/a_{i-1}} \quad (6.123)$$

for all $r \ge r_0$ and all $\bar{\theta}_{i-1} \in S_0 = \{\theta \in L_{i-1} : 1 \le \|\theta\| \le B\}$. Now for any $\bar{\theta}_{i-1} \in \bar{S}_{i-1}$ with $\|\bar{\theta}_{i-1}\| < 1$ we may write $\bar{\theta}_{i-1} = t\theta'$ for some $0 \le t < 1$ and some $\theta' \in S_0$ and then

$$\begin{aligned}\bar{V}_{i-1,\eta}(r,\bar{\theta}_{i-1}) &= \int_{|<x,t\theta'>|>r} |<x,t\theta'>|^\eta \mu\{dx\} \\ &= t^\eta \int_{|<x,\theta'>|>r/t} |<x,\theta'>|^\eta \mu\{dx\} \\ &\le \int_{|<x,\theta'>|>r} |<x,\theta'>|^\eta \mu\{dx\} = \bar{V}_{i-1,\eta}(r,\theta')\end{aligned}$$

so that (6.123) holds for all $\bar{\theta}_{i-1} \in \bar{S}_{i-1}$ for $r > 0$ sufficiently large. Furthermore, part (b) of Theorem 6.4.6 implies that for some $C_3 > 0$ we have $C_3 r^{\eta-\delta-1/a_i} \le V_{i\eta}(r,\theta_i)$ for all $\theta_i \in S_i$ and all $r > 0$ sufficiently large. Then for any $\varepsilon > 0$ we have using (6.123) that

$$\frac{\bar{V}_{i-1,\eta}(r,\bar{\theta}_{i-1})}{V_{i\eta}(r,\theta_i)} \le \frac{C_4 r^{\eta+\delta-1/a_{i-1}}}{C_3 r^{\eta-\delta-1/a_i}} = \frac{C_4}{C_3} r^{2\delta+1/a_i-1/a_{i-1}} < \varepsilon$$

for all $\theta \in S$ and all $r > 0$ sufficiently large. Since (6.123) also holds for $\eta = 0$ we also have that

$$\frac{r^\eta \bar{V}_{i-1,0}(r,\bar{\theta}_{i-1})}{V_{i\eta}(r,\theta_i)} \le \frac{r^\eta C_4 r^{\delta-1/a_{i-1}}}{C_3 r^{\eta-\delta-1/a_i}} < \varepsilon$$

for all $\theta \in S$ and all $r > 0$ sufficiently large. Then it follows from (6.122) that

$$V_\eta(r,\theta) \le 2^\eta\big((2+C)+3\varepsilon\big)V_{i\eta}(r/2,\theta_i)$$

for all $\theta \in S$ and all $r > 0$ sufficiently large. Since by Theorem 6.4.6(a) we know that $V_{i\eta}(r/2,\theta_i) \le K V_{i\eta}(r,\theta_i)$ for all $r > 0$ sufficiently large and some constant $K > 0$ the upper bound in (6.117) follows.

Now we show the lower bound in (6.118). Assume first that $1/a_i < \zeta < 1/a_{i-1}$. It follows from $|x+y|^\zeta \le 2^\zeta(|x|^\zeta + |y|^\zeta)$ that $|x+y|^\zeta \ge 2^{-\zeta}|x|^\zeta - |y|^\zeta$, and hence

$$\begin{aligned}U_\zeta(r,\theta) &= \int_{|<x,\theta>|\le r} |<x,\theta_i> + <x,\bar{\theta}_{i-1}>|^\zeta \mu\{dx\} \\ &\ge 2^{-\zeta} \int_{|<x,\theta>|\le r} |<x,\theta_i>|^\zeta \mu\{dx\} \\ &\qquad - \int_{|<x,\theta>|\le r} |<x,\bar{\theta}_{i-1}>|^\zeta \mu\{dx\} \\ &= 2^{-\zeta} J_1(r) - J_2(r).\end{aligned} \tag{6.124}$$

Then for $\gamma > 0$ (which we will specify later) we decompose further

$$\begin{aligned} J_1(r) &= \int_{\substack{|<x,\theta>|\le r \\ |<x,\theta_i>|\le \gamma r}} |<x,\theta_i>|^\zeta \mu\{dx\} \\ &\quad + \int_{\substack{|<x,\theta>|\le r \\ |<x,\theta_i>|> \gamma r}} |<x,\theta_i>|^\zeta \mu\{dx\} \\ &\ge \int_{\substack{|<x,\theta>|\le r \\ |<x,\theta_i>|\le \gamma r}} |<x,\theta_i>|^\zeta \mu\{dx\} \\ &= \int_{|<x,\theta_i>|\le \gamma r} |<x,\theta_i>|^\zeta \mu\{dx\} \\ &\quad - \int_{\substack{|<x,\theta>|> r \\ |<x,\theta_i>|\le \gamma r}} |<x,\theta_i>|^\zeta \mu\{dx\} \\ &= U_{i\zeta}(\gamma r, \theta_i) - J_3(r) \end{aligned}$$

in view of (6.114). But $J_3(r) \le (\gamma r)^\zeta V_0(r,\theta)$, and hence for any $\gamma > 0$ and any $r > 0$

$$J_1(r) \ge U_{i\zeta}(\gamma r, \theta_i) - (\gamma r)^\zeta V_0(r,\theta). \tag{6.125}$$

Using $\{|<x,\theta>| > r\} \subset \{|<x,\theta_i>| > r/2\} \cup \{|<x,\bar{\theta}_{i-1}>| > r/2\}$ again, we can bound

$$V_0(r,\theta) \le V_{i0}(r/2,\theta_i) + \bar{V}_{i-1,0}(r/2,\bar{\theta}_{i-1}).$$

Choose $\delta > 0$ such that $2\delta + 1/a_i < 1/a_{i-1}$. Part (b) of Theorem 6.4.6 implies that for some $C_3 > 0$ we have $C_3 r^{-\delta-1/a_i} \le V_{i0}(r,\theta_i)$ for all $\theta_i \in S_i$ and all $r > 0$ sufficiently large. Then for any $\varepsilon > 0$ we have using (6.123) with $\eta = 0$ that

$$\frac{\bar{V}_{i-1,0}(r,\bar{\theta}_{i-1})}{V_{i0}(r,\theta_i)} \le \frac{C_4 r^{\delta-1/a_{i-1}}}{C_3 r^{-\delta-1/a_i}} < \varepsilon$$

for all $\theta \in S$ and all $r > 0$ sufficiently large. Hence by (6.125) we have for any $\gamma > 0$ that

$$J_1(r) \ge U_{i\zeta}(\gamma r,\theta_i) - (\gamma r)^\zeta (1+\varepsilon) V_{i0}(r/2,\theta_i) \tag{6.126}$$

for all $\theta \in S$ and all $r > 0$ sufficiently large.

If $\theta \in W_i$, then Lemmas 6.4.10 and 6.4.13 imply that $U_{i\zeta}(r,\theta_i) = U_\zeta(r,\theta) = \bar{U}_{i\zeta}(r,\bar{\theta}_i)$ so $U_{i\zeta}$ exists. Since $\zeta > 1/a_i$ Theorem 6.4.6 yields that $V_{i0}(r,\theta_i)$ and $U_{i\zeta}(r,\theta_i)$ are both uniformly R–O varying on compact subsets of $\theta_i \in W_i$. Then part (a) of Corollary 6.3.9 implies that for some $M_2 < \infty$ we have

$$\frac{r^\zeta V_{i0}(r,\theta_i)}{U_{i\zeta}(r,\theta_i)} \le M_2$$

for all $\theta_i \in S_i$ and all $r > 0$ sufficiently large. Also Theorem 6.4.6 shows that for some $0 < C_1 < C_2$ we have

$$C_1 \lambda^{\zeta-\delta-1/a_i} \leq \frac{U_{i\zeta}(\lambda r, \theta_i)}{U_{i\zeta}(r, \theta_i)} \leq C_2 \lambda^{\zeta+\delta-1/a_i} \tag{6.127}$$

for all $\lambda \geq 1$, all $\theta_i \in S_i$ and all $r > 0$ sufficiently large. Then

$$\begin{aligned} r^\zeta V_{i0}(r/2, \theta_i) &= \frac{r^\zeta V_{i0}(r/2, \theta_i)}{U_{i\zeta}(r/2, \theta_i)} \cdot \left(\frac{U_{i\zeta}(2(r/2), \theta_i)}{U_{i\zeta}(r/2, \theta_i)} \right)^{-1} \cdot U_{i\zeta}(r, \theta_i) \\ &\leq M_2 \cdot (C_1 2^{\zeta-\delta-1/a_i})^{-1} \cdot U_{i\zeta}(r, \theta_i) = C U_{i\zeta}(r, \theta_i) \end{aligned}$$

for all $\theta_i \in S_i$ and all large $r > 0$, where $C = M_2/(C_1 2^{\zeta-\delta-1/a_i})$. Hence in view of (6.126) we have shown that for any $\gamma > 0$ we have

$$J_1(r) \geq U_{i\zeta}(\gamma r, \theta_i) - \gamma^\zeta C(1+\varepsilon) U_{i\zeta}(r, \theta_i) \tag{6.128}$$

for all $\theta \in S$ and all $r > 0$ sufficiently large.

For any $\delta > 0$ and any fixed $0 < \gamma < 1$, the upper bound in (6.127) implies that

$$\begin{aligned} U_{i\zeta}(\gamma r, \theta_i) &= \frac{U_{i\zeta}(\gamma r, \theta_i)}{U_{i\zeta}(r, \theta_i)} \cdot U_{i\zeta}(r, \theta_i) \\ &= \left(\frac{U_{i\zeta}(\gamma^{-1}(\gamma r), \theta_i)}{U_{i\zeta}(\gamma r, \theta_i)} \right)^{-1} \cdot U_{i\zeta}(r, \theta_i) \\ &\geq C_2^{-1} \gamma^{\zeta+\delta-1/a_i} \cdot U_{i\zeta}(r, \theta_i) \end{aligned}$$

for all $\theta_i \in S_i$ and all large $r > 0$. Then for any fixed $0 < \gamma < 1$ we obtain from (6.128) that

$$J_1(r) \geq \gamma^\zeta \left(C_2^{-1} \gamma^{\delta-1/a_i} - C(1+\varepsilon) \right) U_{i\zeta}(r, \theta_i)$$

for all $\theta \in S$ and all $r > 0$ sufficiently large. Choose $0 < \gamma < 1$ small enough so that $C_2^{-1}\gamma^{\delta-1/a_i} - C(1+\varepsilon) \geq 1$. Then we have shown that

$$J_1(r) \geq D U_{i\zeta}(r, \theta_i) \tag{6.129}$$

for all $\theta \in S$ and all $r > 0$ sufficiently large, where $D = \gamma^\zeta$.

Next we bound J_2 from above. Note that

$$\begin{aligned} J_2(r) &= \int\limits_{\substack{|<x,\theta>| \leq r \\ |<x,\bar{\theta}_{i-1}>| \leq r}} |<x, \bar{\theta}_{i-1}>|^\zeta \mu\{dx\} \\ &\quad + \int\limits_{\substack{|<x,\theta>| \leq r \\ |<x,\bar{\theta}_{i-1}>| > r}} |<x, \bar{\theta}_{i-1}>|^\zeta \mu\{dx\} \\ &\leq \bar{U}_{i-1,\zeta}(r, \bar{\theta}_{i-1}) + \bar{V}_{i-1,\zeta}(r, \bar{\theta}_{i-1}). \end{aligned}$$

Since $\bar{S}_{i-1}$ is compact there exists some $B > 0$ such that $\|\bar{\theta}_{i-1}\| \leq B$ for all $\bar{\theta}_{i-1} \in \bar{S}_{i-1}$. Since $\zeta < 1/a_{i-1}$ it follows from Theorem 6.4.12 that $\bar{V}_{i-1,\zeta}$ exists. Then for any $\bar{\theta}_{i-1} \in \bar{S}_{i-1}$ and any $r > 0$ we have

$$\begin{aligned} \bar{U}_{i-1,\zeta}(r, \bar{\theta}_{i-1}) &\leq \bar{U}_{i-1,\zeta}(r, \bar{\theta}_{i-1}) + \bar{V}_{i-1,\zeta}(r, \bar{\theta}_{i-1}) \\ &= \int_{\mathbb{R}^d} |<x, \bar{\theta}_{i-1}>|^\zeta \mu\{dx\} \\ &\leq B^\zeta \int_{\mathbb{R}^d} \|x\|^\zeta \mu\{dx\} \end{aligned}$$

so that $\bar{U}_{i-1,\zeta}(r, \bar{\theta}_{i-1}) \leq M < \infty$ for all $\bar{\theta}_{i-1} \in \bar{S}_{i-1}$ and all $r > 0$. Choose $\delta > 0$ such that $2\delta + 1/a_i < 1/a_{i-1}$ and, recalling that $U_{i\zeta}$ exists, apply part (d) of Theorem 6.4.6 to obtain $C_3 > 0$ such that $U_{i\zeta}(r, \theta_i) \geq C_3 r^{\zeta-\delta-1/a_i}$ for all $\theta_i \in S_i$ and all $r > 0$ sufficiently large. Since $\zeta < 1/a_{i-1}$ we can apply (6.123) with $\eta = \zeta$, and this together with the bounds on $\bar{U}_{i-1,\zeta}$ and $U_{i\zeta}$ yields

$$\begin{aligned} \frac{J_2(r)}{U_{i\zeta}(r, \theta_i)} &\leq \frac{M}{U_{i\zeta}(r, \theta_i)} + \frac{\bar{V}_{i-1,\zeta}(r, \bar{\theta}_{i-1})}{U_{i\zeta}(r, \theta_i)} \\ &\leq \frac{M}{C_3 r^{\zeta-\delta-1/a_i}} + \frac{C_4 r^{\zeta+\delta-1/a_{i-1}}}{C_3 r^{\zeta-\delta-1/a_i}} \end{aligned}$$

for all $\theta \in S$ and all $r > 0$ sufficiently large. It follows that for any $\varepsilon > 0$ we have

$$J_2(r) \leq \varepsilon U_{i\zeta}(r, \theta_i) \tag{6.130}$$

for all $\theta \in S$ and all $r > 0$ sufficiently large. This together with (6.129) and (6.124) shows that

$$U_\zeta(r, \theta) \geq \left(2^{-\zeta} D - \varepsilon\right) U_{i\zeta}(r, \theta_i)$$

for all $\theta \in S$ and all $r > 0$ sufficiently large, which proves the lower bound in (6.118) for $1/a_i < \zeta < 1/a_{i-1}$.

If $\zeta \geq 1/a_{i-1}$ we have to argue differently. Write $\zeta = \zeta_1 + \rho_1$ for some $1/a_i < \zeta_1 < 1/a_{i-1}$ and some $\rho_1 > 0$. Then for any $0 < \gamma < 1$ we obtain

$$\begin{aligned} U_\zeta(r, \theta) &\geq \int_{\gamma r < |<x,\theta>| \leq r} |<x, \theta>|^\zeta \mu\{dx\} \\ &\geq (\gamma r)^{\rho_1} \int_{\gamma r < |<x,\theta>| \leq r} |<x, \theta>|^{\zeta_1} \mu\{dx\} \\ &= (\gamma r)^{\rho_1} U_{\zeta_1}(r, \theta) \left(1 - \frac{U_{\zeta_1}(\gamma r, \theta)}{U_{\zeta_1}(r, \theta)}\right). \end{aligned} \tag{6.131}$$

Using the fact that the upper bound in (6.117) holds and that (6.118) and (6.127) hold for $\zeta = \zeta_1$ along with part (a) of Corollary 6.3.9, we get for any

$\eta < 1/a_i$ and any fixed $0 < \gamma < 1$ that

$$\begin{aligned}\frac{U_{\zeta_1}(\gamma r,\theta)}{U_{\zeta_1}(r,\theta)} &\le \frac{C_4(\gamma r)^{\zeta_1-\eta}V_{i\eta}(\gamma r,\theta_i)}{C_3 U_{i\zeta_1}(r,\theta_i)} \\ &= \frac{C_4}{C_3}\cdot\frac{(\gamma r)^{\zeta_1-\eta}V_{i\eta}(\gamma r,\theta_i)}{U_{i\zeta_1}(\gamma r,\theta_i)}\cdot\left(\frac{U_{i\zeta_1}(\gamma^{-1}(\gamma r),\theta_i)}{U_{i\zeta_1}(\gamma r,\theta_i)}\right)^{-1} \\ &\le \frac{C_4}{C_3}\cdot M_2\cdot C_1^{-1}\gamma^{\zeta_1-\delta-1/a_i}\end{aligned}$$

for all $\theta \in S$ and all $r > 0$ sufficiently large. Choose $0 < \gamma < 1$ so that $\gamma^{\zeta_1-\delta-1/a_i}(C_4M_2)/(C_3C_1) < 1/2$. Then for all $r > 0$ sufficiently large and all $\theta \in S$ it follows from (6.131) that

$$U_\zeta(r,\theta) \ge \frac{1}{2}(\gamma r)^{\rho_1}U_{\zeta_1}(r,\theta)$$

and using the fact that (6.118) holds for $\zeta = \zeta_1$ again we conclude that

$$U_\zeta(r,\theta) \ge (C_3/2)(\gamma r)^{\rho_1}U_{i\zeta_1}(r,\theta_i) \tag{6.132}$$

for all $\theta \in S$ and all $r > 0$ sufficiently large. Since $\zeta_1 > 1/a_i$ we can apply both parts of Corollary 6.3.9 to see that

$$\begin{aligned}r^{\rho_1}U_{i\zeta_1}(r,\theta_i) &= \frac{U_{i\zeta_1}(r,\theta_i)}{r^{\zeta_1}V_{i0}(r,\theta_i)}\cdot\frac{r^{\zeta_1+\rho_1}V_{i0}(r,\theta_i)}{U_{i\zeta}(r,\theta_i)}\cdot U_{i\zeta}(r,\theta_i) \\ &\ge \frac{1}{M_2}\cdot m_2\cdot U_{i\zeta}(r,\theta_i)\end{aligned}$$

for all $\theta \in S$ and all $r > 0$ sufficiently large. Putting this together with (6.132) proves the lower bound in (6.118) in the general case.

Now we prove the upper bound in (6.118). If $\theta \in L_i$, then Lemma 6.4.13 implies that $V_\eta(r,\theta) = \bar{V}_{i\eta}(r,\bar{\theta}_i)$ and $U_\zeta(r,\theta) = \bar{U}_{i\zeta}(r,\bar{\theta}_i)$. Since $\bar{U}_{i\zeta}$ exists this implies that $U_\zeta(r,\theta)$ exists for $\theta \in L_i$. Since $\eta < 1/a_i$ we know that $\bar{V}_{i\eta}(r,\bar{\theta}_i)$ is uniformly R–O varying on compact subsets of $\bar{\theta}_i \in L_i$. Since S is a compact subset of $L_i \setminus L_{i-1}$, the set $\bar{S}_i = P_i(S)$ is a compact subset of $L_i \setminus \{0\}$. Then we can apply part (b) of Corollary 6.3.9 to obtain some $m_2 > 0$ and $r_0 > 0$ such that

$$\frac{r^{\zeta-\eta}\bar{V}_{i\eta}(r,\bar{\theta}_i)}{\bar{U}_{i\zeta}(r,\bar{\theta}_i)} \ge m_2 \tag{6.133}$$

for all $r \ge r_0$ and all $\bar{\theta}_i \in \bar{S}_i$. Then in view of the upper bound in (6.117) we have

$$U_\zeta(r,\theta) = \bar{U}_{i\zeta}(r,\bar{\theta}_i) \le \frac{r^{\zeta-\eta}\bar{V}_{i\eta}(r,\bar{\theta}_i)}{m_2} = \frac{r^{\zeta-\eta}V_\eta(r,\theta)}{m_2} \le \frac{C_2}{m_2}r^{\zeta-\eta}V_{i\eta}(r,\theta_i)$$

which proves the upper bound in (6.118).

Finally we prove the lower bound in (6.117). Using (6.133) along with the upper bound in (6.117), we obtain

$$\begin{aligned} V_\eta(r,\theta) = \bar{V}_{i\eta}(r,\bar{\theta}_i) &\geq m_2 r^{\eta-\zeta}\bar{U}_{i\zeta}(r,\bar{\theta}_i) \\ &= m_2 r^{\eta-\zeta} U_\zeta(r,\theta) \geq m_2 C_3 r^{\eta-\zeta} U_{i\zeta}(r,\theta_i) \end{aligned}$$

which proves the lower bound in (6.117). □

Recall from Definition 6.4.9 that the spectral index $\alpha(\theta) = 1/a_i$ for all $\theta \in L_i \setminus L_{i-1}$, where $L_i = W_1 \oplus \cdots \oplus W_i$ and $W_1, \ldots W_p$ is the spectral decomposition of Theorem 6.4.8. Our next result shows that the tails of an R–O varying measure fall off about as fast as $r^{-\alpha(\theta)}$ in the direction θ for all $\theta \in \Gamma = \mathbb{R}^d \setminus \{0\}$. It also gives sharp bounds on the truncated moments and tail moments in the direction θ.

Theorem 6.4.15. *Suppose that $\mu \in \mathrm{ROV}(E,c)$ for some $c > 1$ where the real parts of the eigenvalues of E are $0 < a_1 < \cdots < a_p$. Let $\alpha(\theta)$ be the index function in Definition 6.4.9 so that $\alpha(\theta) = 1/a_i$ for all $\theta \in L_i \setminus L_{i-1}$, where $L_i = W_1 \oplus \cdots \oplus W_i$ and $W_1, \ldots, W_p$ is the spectral decomposition of Theorem 6.4.8.*

(a) If $0 \leq \eta < \alpha(\theta)$, then $V_\eta(r,\theta)$ exists and is uniformly R–O varying on compact subsets of $\theta \in L_i \setminus L_{i-1}$. Furthermore, for any $\delta > 0$ and any $S \subset L_i \setminus L_{i-1}$ compact:

(i) there exist constants $C_1, C_2 > 0$ and $r_0 \geq 1$ such that

$$C_1 \lambda^{\eta-\alpha(\theta)-\delta} \leq \frac{V_\eta(\lambda r,\theta)}{V_\eta(r,\theta)} \leq C_2 \lambda^{\eta-\alpha(\theta)+\delta} \tag{6.134}$$

whenever $r \geq r_0$, $\lambda \geq 1$ and $\theta \in S$; and

(ii) there exist constants $B_1, B_2 > 0$ and $r_0 \geq 1$ such that

$$B_1 r^{\eta-\alpha(\theta)-\delta} \leq V_\eta(r,\theta) \leq B_2 r^{\eta-\alpha(\theta)+\delta} \tag{6.135}$$

whenever $r \geq r_0$ and $\theta \in S$.

(b) If $\zeta > \alpha(\theta)$ and U_ζ exists, then $U_\zeta(r,\theta)$ is uniformly R–O varying on compact subsets of $\theta \in L_i \setminus L_{i-1}$. Furthermore, for any $\delta > 0$ and any $S \subset L_i \setminus L_{i-1}$ compact:

(i) there exist constants $C_3, C_4 > 0$ and $r_0 \geq 1$ such that

$$C_3 \lambda^{\zeta-\alpha(\theta)-\delta} \leq \frac{U_\zeta(\lambda r,\theta)}{U_\zeta(r,\theta)} \leq C_4 \lambda^{\zeta-\alpha(\theta)+\delta} \tag{6.136}$$

whenever $r \geq r_0$, $\lambda \geq 1$ and $\theta \in S$; and

(ii) there exist constants $B_3, B_4 > 0$ and $r_0 \geq 1$ such that

$$B_3 r^{\zeta-\alpha(\theta)-\delta} \leq U_\zeta(r,\theta) \leq B_4 r^{\zeta-\alpha(\theta)+\delta} \tag{6.137}$$

whenever $r \geq r_0$ and $\theta \in S$.

Proof. For any nonzero $\theta \in \mathbb{R}^d$, there exists a unique $1 \leq i \leq p$ such that $\theta \in L_i \setminus L_{i-1}$, and then $\alpha(\theta) = 1/a_i$. By Proposition 6.4.14 (a) together with Theorem 6.4.6 (c) and Corollary 6.3.9, we have

$$\begin{aligned}
\frac{V_\eta(\lambda r,\theta)}{V_\eta(r,\theta)} &\leq \frac{C_2 V_{i\eta}(\lambda r,\theta_i)}{C_1 r^{\eta-\zeta} U_{i\zeta}(r,\theta_i)} \\
&= \frac{C_2 \lambda^{\eta-\zeta}}{C_1} \cdot \frac{(\lambda r)^{\zeta-\eta} V_{i\eta}(\lambda r,\theta_i)}{U_{i\zeta}(\lambda r,\theta_i)} \cdot \frac{U_{i\zeta}(\lambda r,\theta_i)}{U_{i\zeta}(r,\theta_i)} \\
&\leq \frac{C_2 \lambda^{\eta-\zeta}}{C_1} \cdot M_2 \cdot \tilde{C}_2 \lambda^{\zeta+\delta-1/a_i} \\
&= \frac{C_2 M_2 \tilde{C}_2}{C_1} \lambda^{\eta-\alpha(\theta)+\delta}
\end{aligned}$$

for all $\lambda \geq 1$ and $\theta \in S$ for $r > 0$ sufficiently large.

To obtain the lower bound in (6.134), we use Proposition 6.4.14 (a) together with Theorem 6.4.6 (a) and Corollary 6.3.9 to obtain

$$\begin{aligned}
\frac{V_\eta(\lambda r,\theta)}{V_\eta(r,\theta)} &\geq \frac{C_1 (\lambda r)^{\eta-\zeta} U_{i\zeta}(\lambda r,\theta_i)}{C_2 V_{i\eta}(r,\theta_i)} \\
&= \frac{C_1}{C_2} \cdot \frac{U_{i\zeta}(\lambda r,\theta_i)}{(\lambda r)^{\zeta-\eta} V_{i\eta}(\lambda r,\theta_i)} \cdot \frac{V_{i\eta}(\lambda r,\theta_i)}{V_{i\eta}(r,\theta_i)} \\
&\geq \frac{C_1}{C_2} \cdot \frac{1}{M_2} \cdot \tilde{C}_1 \lambda^{\eta-\delta-1/a_i} \\
&= \frac{C_1 \tilde{C}_1}{C_2 M_2} \lambda^{\eta-\alpha(\theta)-\delta}
\end{aligned}$$

for all $\lambda \geq 1$ and $\theta \in S$ for $r > 0$ sufficiently large.

Choose any $\delta > 0$ and $\lambda_0 > 1$. Then it follows from the monotonicity of V_η along with the lower bound in (6.134) that

$$C_1 \lambda_0^{\eta-\alpha(\theta)-\delta} \leq \frac{V_\eta(\lambda r,\theta)}{V_\eta(r,\theta)} \leq 1$$

whenever $r \geq r_0$, $1 \leq \lambda \leq \lambda_0$ and $\theta \in S$, so that $V_\eta(r,\theta)$ is uniformly R–O varying on compact subsets of $\theta \in L_i \setminus L_{i-1}$.

Combining Proposition 6.4.14 (a) together with Theorem 6.4.6 (b,d) we have for any compact $S \subset L_i \setminus L_{i-1}$ that

$$V_\eta(r,\theta) \leq C_2 V_{i\eta}(r,\theta_i) \leq C_2 C_4 r^{\eta+\delta-1/a_i} = C_2 C_4 r^{\eta-\alpha(\theta)+\delta}$$

and

$$V_\eta(r,\theta) \ge C_1 r^{\eta-\varsigma} U_{i\varsigma}(r,\theta_i) \ge C_1 r^{\eta-\varsigma} C_3 r^{\varsigma-\delta-1/a_i} = C_1 C_3 r^{\eta-\alpha(\theta)-\delta}$$

for all $\lambda \ge 1$ and $\theta \in S$ for $r > 0$ sufficiently large, which finishes the proof of part (a).

The proof of part(b) is similar. By Proposition 6.4.14 (b) together with Theorem 6.4.6 (c) and Corollary 6.3.9, we have

$$\begin{aligned}\frac{U_\varsigma(\lambda r,\theta)}{U_\varsigma(r,\theta)} &\le \frac{C_4(\lambda r)^{\varsigma-\eta} V_{i\eta}(\lambda r,\theta_i)}{C_3 U_{i\varsigma}(r,\theta_i)} \\ &= \frac{C_4}{C_3}\cdot\frac{(\lambda r)^{\varsigma-\eta} V_{i\eta}(\lambda r,\theta_i)}{U_{i\varsigma}(\lambda r,\theta_i)}\cdot\frac{U_{i\varsigma}(\lambda r,\theta_i)}{U_{i\varsigma}(r,\theta_i)} \\ &\le \frac{C_4}{C_3}\cdot M_2\cdot C_2\lambda^{\varsigma+\delta-1/a_i} \\ &= \frac{C_4 M_2 C_2}{C_3}\lambda^{\varsigma-\alpha(\theta)+\delta}\end{aligned}$$

for all $\lambda \ge 1$ and $\theta \in S$ for $r > 0$ sufficiently large.

To obtain the lower bound in (6.136), we use Proposition 6.4.14 (b) together with Theorem 6.4.6 (a) and Corollary 6.3.9 to obtain

$$\begin{aligned}\frac{U_\varsigma(\lambda r,\theta)}{U_\varsigma(r,\theta)} &\ge \frac{C_3 U_{i\varsigma}(\lambda r,\theta_i)}{C_4 r^{\varsigma-\eta} V_{i\eta}(r,\theta_i)} \\ &= \frac{C_3\lambda^{\varsigma-\eta}}{C_4}\cdot\frac{U_{i\varsigma}(\lambda r,\theta_i)}{(\lambda r)^{\varsigma-\eta} V_{i\eta}(\lambda r,\theta_i)}\cdot\frac{V_{i\eta}(\lambda r,\theta_i)}{V_{i\eta}(r,\theta_i)} \\ &\ge \frac{C_3\lambda^{\varsigma-\eta}}{C_4}\cdot\frac{1}{M_2}\cdot C_1\lambda^{\eta-\delta-1/a_i} \\ &= \frac{C_3 C_1}{C_4 M_2}\lambda^{\varsigma-\alpha(\theta)-\delta}\end{aligned}$$

for all $\lambda \ge 1$ and $\theta \in S$ for $r > 0$ sufficiently large.

Choose and $\delta > 0$ and $\lambda_0 > 1$. Then it follows from the monotonicity of U_ς along with the upper bound in (6.136) that

$$1 \le \frac{U_\varsigma(\lambda r,\theta)}{U_\varsigma(r,\theta)} \le C_4\lambda_0^{\varsigma-\alpha(\theta)+\delta}$$

whenever $r \ge r_0$, $1 \le \lambda \le \lambda_0$, and $\theta \in S$, so that $U_\varsigma(r,\theta)$ is uniformly R–O varying on compact subsets of $\theta \in L_i \setminus L_{i-1}$.

Combining Proposition 6.4.14 (b) together with Theorem 6.4.6 (b,d) we have for any compact $S \subset L_i \setminus L_{i-1}$ that

$$U_\varsigma(r,\theta) \le C_4 r^{\varsigma-\eta} V_{i\eta}(r,\theta_i) \le C_4 r^{\varsigma-\eta}\cdot\tilde{C}_4 r^{\eta+\delta-1/a_i} = C_4\tilde{C}_4 r^{\varsigma-\alpha(\theta)+\delta}$$

and

$$U_\varsigma(r,\theta) \ge C_3 U_{i\varsigma}(r,\theta_i) \ge C_3\tilde{C}_3 r^{\varsigma-\delta-1/a_i} = C_3\tilde{C}_3 r^{\varsigma-\alpha(\theta)-\delta}$$

for all $\lambda \geq 1$ and $\theta \in S$ for $r > 0$ sufficiently large, which finishes the proof of part (b). □

In the situation of Theorem 6.4.15, for easy reference, we include a special case of Corollary 6.3.9.

Corollary 6.4.16. *Suppose that $\mu \in \mathrm{ROV}(E, c)$ for some $c > 1$ where the real parts of the eigenvalues of E are $0 < a_1 < \cdots < a_p$. Let $\alpha(\theta)$ be the index function in Definition 6.4.9 so that $\alpha(\theta) = 1/a_i$ for all $\theta \in L_i \setminus L_{i-1}$, where $L_i = W_1 \oplus \cdots \oplus W_i$ and $W_1, \ldots, W_p$ is the spectral decomposition of Theorem 6.4.8. If $0 \leq \eta < \alpha(\theta) < \zeta$, then there exist positive real constants $m, M > 0$ and a $r_0 > 0$ such that*

$$m \leq \frac{r^{\zeta-\eta} V_\eta(r, \theta)}{U_\zeta(r, \theta)} \leq M$$

for all $r \geq r_0$.

Proof. This is just a combination of Theorem 6.4.15 together with Corollary 6.3.9. □

Corollary 6.4.17. *If μ is R–O varying at infinity, then the spectral decomposition $W_1, \ldots W_p$ of Theorem 6.4.8 is unique, and every exponent E of μ has the same real spectrum.*

Proof. Since (6.135) holds for any nonzero vector $\theta \in \mathbb{R}^d$ and any $\delta > 0$, the index function $\alpha(\theta)$ is uniquely determined by the measure μ. This implies that the sets $L_i \setminus L_{i-1}$ are uniquely determined for all $1 \leq i \leq p$, and then $W_1, \ldots, W_p$ are unique as well. It also implies that the values $0 < a_1 < \cdots < a_p$ are uniquely determined, since $1/\alpha(\theta)$ ranges over these values. □

Remark 6.4.18. If μ is regularly varying at infinity and E is an exponent of μ, then Theorem 6.2.3 implies that μ is also R–O varying at infinity with the same exponent E. Hence all of the results of this section also apply to measures which are regularly varying at infinity.

Remark 6.4.19. If μ is an R–O varying measure and E, (B_n) and $W_1, \cdots, W_p$ are as in Theorem 6.4.8, then (B_n) is spectrally compatible with E so that the conclusions of Theorem 4.3.1 hold with $V_i = W_i$ and $L_i = W_1 \oplus \cdots \oplus W_i$ for each $i = 1, \ldots, p$. This gives sharp bounds on the growth rate of the norming operators as well.

6.5 NOTES AND COMMENTS

Regularly varying measures first appeared in [112] with an application to scalar-normed domains of attraction in $\mathbb{R}^d$. The notion of a regularly varying measure is implicit in a result of Feller [40] (see Corollary 8.2.19). Regular variation of a probability measure μ on $\mathbb{R}^1$ with distribution function $F(x) = \mu(-\infty, x]$ is equivalent to regular variation of the tail function $T(x) = F(x) + (1 - F(-x))$ along with a balanced tails condition

$$\frac{F(x)}{T(x)} \to p \quad \text{and} \quad \frac{1 - F(-x)}{T(x)} \to q$$

as $x \to \infty$ for some $0 \leq p, q \leq 1$ with $p + q = 1$. The special case where the norming operators $f(t)$ in (6.1) are all diagonal with respect to the standard basis for $\mathbb{R}^d$ is called *vector-normed* regular variation. In Resnick and Greenwood [151] and Meerschaert [115] this variety of regular variation is used to characterize vector-normed domains of attraction on $\mathbb{R}^d$. The general case of operator norming originated in [113] but its precursors may be found in Jurek [73] and Meerschaert [111]. The disintegration formula in Theorem 6.1.7 is due to Jurek [74]. The convergence $n(A_n\mu) \to \phi$ of Proposition 6.1.10 is basic to the theory of point processes (see Remark 3.3.30). In this form, the notion of regularly varying measures is also quite useful in extreme value theory. The material in Section 6.1 on convergence of types, and exponents and symmetries, mostly follows the arguments laid out in the excellent book of Jurek and Mason [75]. Original references are given at the end of Chapter 5. Theorem 6.1.29, which links regularly varying functions and measures, and Theorems 6.1.31 and 6.1.32 characterizing regularly varying measures in terms of tail behavior, appeared in [113]. R–O varying measures originated in [166], but the basic ideas were already present in [164]. The disintegration formula of Theorem 6.2.8 and the argument for Lemma 6.2.13 are from Chorny [24]. Some of the results of Section 6.3 appeared in [118] in the special case of regularly varying measures, with an application to nonnormal generalized domains of attraction. The same results in the more general case of R–O varying measures appeared in [121] and were generalized by Scheffler in [172, 173], but Section 6.3 contains additional results and more detailed proofs. Ultimately, most of these results go back to Section VIII.8 in Feller [40] (see also Appendix A in Seneta [177]). The spectral decomposition for regularly varying probability measures appeared in [116] and was extended to R–O varying probability measures in [122]. Sharp tail bounds for regularly varying and R–O varying probability measures were also obtained in [116, 122] using this spectral decomposition. The proof of Theorem 6.4.15 is new, using the two-sided bounds provided by Theorems 5.3.14 and 5.3.18.

Part III

MULTIVARIATE LIMIT THEOREMS

In Chapter 3, we developed the theory of infinitely divisible laws and triangular arrays. Then in Part II, we introduced multivariable regular variation. Now we put this all together to prove the fundamental limit theorems for sums of independent and identically distributed random variables.

7

The Limit Distributions

In this chapter, we investigate the limit distributions that occur in the most important limit theorems for sums of independent random variables. These limit distributions have certain scaling properties that are very useful in applications. The methods used here are a combination of the material on infinitely divisible laws and convergence of triangular arrays from Chapter 3 along with some regular variation results from Part II. These results also set the stage for Chapter 8, where we finally prove the limit theorems that are the main purpose of this book. For reasons of efficiency, we begin with the most general case of operator semistable laws in Section 7.1. Then in Sections 7.2–7.4 we consider the important special cases of stable, semistable, and operator stable laws. Finally, in Section 7.5 we prove some structural results concerning the class of operator semistable laws.

7.1 OPERATOR SEMISTABLE LAWS

Recall from Definition 3.3.20 that a probability distribution ν on $\mathbb{R}^d$ is operator semistable if there exist linear operators $A_n \in \mathrm{L}(\mathbb{R}^d)$, an increasing sequence of positive integers $k_n \to \infty$ with $k_{n+1}/k_n \to c \geq 1$ and nonrandom vectors $b_n \in \mathbb{R}^d$ such that

$$A_n \mu^{k_n} * \varepsilon_{b_n} \Rightarrow \nu \tag{7.1}$$

for some probability measure μ on $\mathbb{R}^d$. Then we say that μ belongs to the generalized domain of semistable attraction of ν, and we write $\mu \in \mathrm{GDOSA}(\nu, c)$.

The purpose of this section is to investigate the structure of these laws, and derive some important basic properties.

The following characterization of operator semistable laws is due to Jajte [69] and will be the key result in our characterizations of other limit distributions in Section 7.2. Recall from Definition 3.1.12 that any infinitely divisible law ν has a unique Lévy representation $[a, Q, \phi]$ that decomposes ν into two independent components, one Gaussian and one Poissonian.

Theorem 7.1.1. *A full probability measure ν is operator semistable if and only if ν is infinitely divisible and there exist a number $c > 1$, a vector $b \in \mathbb{R}^d$ and an operator $B \in \mathrm{GL}(\mathbb{R}^d)$ such that*

$$\nu^c = B\nu * \varepsilon_b \tag{7.2}$$

holds. The spectrum of B is contained in the set $\{z \in \mathbb{C} : |z|^2 \geq c\}$. Eigenvalues of B satisfying $|\lambda|^2 = c$ are simple. If $k_{n+1}/k_n \to r > 1$ in Definition 3.3.20, then we may take $c = r$ in (7.2).

Furthermore, ν can be decomposed uniquely into a convolution product $\nu = \nu_1 * \nu_2$ of two probability measures ν_1 and ν_2, concentrated[1] on B-invariant subspaces V_1 and V_2, respectively, such that $\mathbb{R}^d = V_1 \oplus V_2$. Then ν_1 is a full Poissonian operator semistable measure on V_1 and ν_2 is a full Gaussian measure on V_2. The spectrum of $B|_{V_1}$ is then contained in the set $\{|z|^2 > c\}$ and for the eigenvalues z of $B|_{V_2}$ the equality $|z|^2 = c$ holds.

Definition 7.1.2. If (7.2) holds, we say that ν is (B, c) *operator semistable.* For any operator semistable law ν we will call the probability measures ν_1, ν_2 from Theorem 7.1.1 the *Lévy decomposition* of ν.

The proof of the theorem will be preceded by several lemmas.

Lemma 7.1.3. *Let ν be a full measure for which (7.1) holds, where $A_n \in \mathrm{GL}(\mathbb{R}^d)$, $b_n \in \mathbb{R}^d$ and $k_n \to \infty$ with $k_{n+1}/k_n \to r < \infty$. Then $A_n\mu \Rightarrow \varepsilon_0$. Especially, ν is infinitely divisible.*

Proof. Since $\mu \in \mathrm{GDOPA}(\nu)$ the result follows from Lemma 3.3.3 together with Theorem 3.2.14. □

Lemma 7.1.4. *If ν is full and (7.1) holds for some increasing sequence (k_n) with $k_{n+1}/k_n \to c > 1$, then (7.2) holds for some $B \in \mathrm{GL}(\mathbb{R}^d)$ and $b \in \mathbb{R}^d$.*

Proof. Since (7.1) holds we also have $A_{n-1}\mu^{k_{n-1}} * \varepsilon_{b_{n-1}} \Rightarrow \nu$, where ν is infinitely divisible in view of Lemma 7.1.3. Then Proposition 3.3.7 with $t_n = k_n/k_{n-1}$ yields $A_{n-1}\mu^{k_n} * \varepsilon_{s_n} \Rightarrow \nu^c$, where the shifts $s_n = t_n b_{n-1}$. Then convergence of types, Theorem 2.3.17, implies that (7.2) holds for some $B \in \mathrm{GL}(\mathbb{R}^d)$ and $b \in \mathbb{R}^d$. □

[1] Here we identify a measure on $\mathbb{R}^d$ that is supported on a subspace V_i with its projection onto V_i, in order to avoid needlessly complicated notation.

Lemma 7.1.5. *If ν is full and (7.1) holds for some increasing sequence (k_n) with $k_{n+1}/k_n \to 1$, then ν is operator stable and (7.2) holds for any $c > 1$.*

Proof. Given $a > 0$ define $n(1) = \min\{n \geq 1 : k_n \geq a\}$ and let $n(j+1) = \min\{n > n(j) : k_{n+1}/k_{n(j)} > a\}$ so that $k_{n(j+1)}/k_{n(j)} \to a$ as $j \to \infty$. Then (7.1) implies that $A_{n(j)}\mu^{k_{n(j)}} * \varepsilon_{b_{n(j)}} \Rightarrow \nu$ and $A_{n(j-1)}\mu^{k_{n(j-1)}} * \varepsilon_{b_{n(j-1)}} \Rightarrow \nu$ where ν is infinitely divisible in view of Lemma 7.1.3. Then Proposition 3.3.7 with $t_j = k_{n(j)}/k_{n(j-1)}$ yields $A_{n(j-1)}\mu^{k_{n(j)}} * \varepsilon_{s_j} \Rightarrow \nu^a$ where the shifts $s_j = t_j b_{n(j-1)}$. Then convergence of types, Theorem 2.3.17, implies that $\nu^a = C_a\nu * \varepsilon_{b_a}$ for some $b_a \in \mathbb{R}^d$ and $C_a \in \mathrm{GL}(\mathbb{R}^d)$. But then $\nu = D_n\nu^n * \varepsilon_{d_n}$ for some $D_n \in \mathrm{GL}(\mathbb{R}^d)$ and $d_n \in \mathbb{R}^d$ and for all $n \geq 1$, and hence $\nu \in \mathrm{GDOA}(\nu)$ so ν is operator stable. Then Lemma 3.3.27 implies that $\nu \in \mathrm{GDOSA}(\nu, c)$ and hence Lemma 7.1.4 yields (7.2) for any $c > 1$. □

Lemma 7.1.6. *Let ν be a full infinitely divisible measure with Lévy triple $[a, Q, \phi]$. Given $\beta > 0$ and $C \in \mathrm{GL}(\mathbb{R}^d)$ we have $\nu^\beta = C\nu * \varepsilon_b$ for some $b \in \mathbb{R}^d$ if and only if*

(a) *$Q(C^*t) = \beta \cdot Q(t)$ for all $t \in \mathbb{R}^d$ and*

(b) *$C\phi = \beta \cdot \phi$.*

Proof. Definition 3.1.23 shows that ν^β has Lévy representation $[a\beta, \beta\cdot Q, \beta\cdot\phi]$, and Proposition 3.1.21 shows that $[Ca + b + b', Q', C\phi]$ is the Lévy representation of $C\nu * \varepsilon_b$, where $Q'(t) = Q(C^*t)$ for all $t \in \mathbb{R}^d$ and b' is given by (3.28). If $\nu^\beta = C\nu * \varepsilon_b$, then since the Lévy representation is unique it follows that (a) and (b) hold. Conversely, if (a) and (b) hold, then it follows that $\nu^\beta = C\nu * \varepsilon_b$ with $b = a\beta - Ca - b'$. □

Lemma 7.1.7. *Let $C \in \mathrm{GL}(\mathbb{R}^m)$ for some $m \geq 1$, $0 < \beta < 1$ and let ϕ be an $\mathcal{M}$-full Lévy measure on $\mathbb{R}^m$ such that $C\phi = \beta \cdot \phi$. Then $C^n x \to 0$ for all $x \in \mathbb{R}^m$ and $r(C) < \beta^{1/2}$, where $r(C)$ denotes the spectral radius of C.*

Proof. Let $\varphi(x) = \|x\|^2/(1 + \|x\|^2)$. Then by (3.6) $\int \varphi d\phi < \infty$. Since $0 < \beta < 1$ and $C^n\phi = \beta^n \cdot \phi$ for all integers n we conclude using Monotone Convergence that

$$\begin{aligned}\int_{x\neq 0}\Big(\sum_{n=0}^{\infty}\varphi(C^n x)\Big)\phi\{dx\} &= \sum_{n=0}^{\infty}\int_{x\neq 0}\varphi(C^n x)\phi\{dx\}\\ &= \Big(\sum_{n=0}^{\infty}\beta^n\Big)\int_{x\neq 0}\varphi(x)\phi\{dx\} < \infty.\end{aligned}$$

Hence $\sum_n \varphi(C^n x) < \infty$ so $\lim_{n\to\infty}\varphi(C^n x) = 0$ and thus $\lim_{n\to\infty}\|C^n x\| = 0$ for ϕ-almost all $x \in \mathbb{R}^m \setminus \{0\}$. Proposition 2.1.8 shows that this is equivalent to $\lim_{n\to\infty} C^n x = 0$ for ϕ-almost all $x \in \mathbb{R}^m \setminus \{0\}$. Let

$$K(C) = \{x \in \mathbb{R}^m : C^n x \to 0 \quad \text{as } n \to \infty\}$$

denote the contraction subspace of C. Since ϕ is supported on the linear subspace $K(C)$, and since ϕ is $\mathcal{M}$-full, we must have $K(C) = \mathbb{R}^m$. Choose $\lambda \in \mathbb{C}$ and $x \neq 0$ in $\mathbb{R}^m$ such that $Cx = \lambda x$, and such that the complex absolute value $|\lambda| = r(C)$, the spectral radius. Since $x \in K(C)$ we have $C^n x = \lambda^n x \to 0$ as $n \to \infty$, so we must have $|\lambda| = r(C) < 1$.

Now let

$$L = \{x \in \mathbb{R}^m : ||x|| < 1 \text{ and } ||C^{-n}x|| \geq 1 \text{ for all } n \geq 1\}.$$

Choose $\lambda \in \mathbb{C}$ and $x \in \mathbb{R}^m$ such that $Cx = \lambda x$ and $|\lambda| = r(C) < ||x|| < 1$. Then $||C^{-n}x|| = |\lambda|^{-n}||x|| > 1$ for all $n \geq 1$ so that $L \neq \emptyset$. If $x \in \mathbb{R}^m \setminus \{0\}$ let $n(x) = \inf\{n \in \mathbb{Z} : ||C^n x|| < 1\}$. Since $K(C) = \mathbb{R}^m$ we have $C^n x \to 0$ for all $x \in \mathbb{R}^m$, so that $||C^n x|| \to 0$. Proposition 2.1.3 (d) shows that $||C^{-n}x|| \geq ||x||/||C^n|| \to \infty$ so $n(x)$ is well defined. Since $||C^{n(x)}x|| < 1$ and $||C^{n(x)-k}x|| \geq 1$ for all $k \geq 1$ we have $C^n x \in L$ if and only if $n = n(x)$.

If $y = C^n x$ for some $x \in L$ then $||y|| = ||C^n x|| \leq ||C^n|| \cdot ||x|| < ||C^n||$ since $||x|| < 1$. It follows that, since $||C^n|| \to 0$, there exists an $r > 0$ such that

$$\bigcup_{n=0}^{\infty} C^n(L) \subset B(r) = \{x : ||x|| < r\}$$

and by definition of L this union is disjoint. Consequently we have

$$\begin{aligned}
\int_L \Big(\sum_{n=0}^{\infty} \beta^{-n}||C^n x||^2\Big)\phi\{dx\} &= \sum_{n=0}^{\infty} \int_L ||C^n x||^2 (\beta^{-n} \cdot \phi)\{dx\} \\
&= \sum_{n=0}^{\infty} \int_L ||C^n x||^2 \phi\{C^n dx\} \\
&= \sum_{n=0}^{\infty} \int_{C^n(L)} ||x||^2 \phi\{dx\} \\
&\leq \int_{B(r)\setminus\{0\}} ||x||^2 \phi\{dx\} < \infty.
\end{aligned}$$

Then $\sum_{n=0}^{\infty} \beta^{-n}||C^n x||^2 < \infty$, and hence $\lim_{n\to\infty} \beta^{-n}||C^n x||^2 = 0$, for ϕ-almost all $x \in L$. In other words, $\beta^{-n}||C^n x||^2 \to 0$ for all $x \in L \setminus U$ where $U \subset L$ is a Borel set with $\phi(U) = 0$. If $y = C^k x$ for some $x \in L$ then $\beta^{-n}||C^n y||^2 = \beta^{-n}||C^{n+k}x||^2 = \beta^k \beta^{-(n+k)}||C^{n+k}x||^2 \to 0$ whenever $x \in L \setminus U$, hence this convergence holds whenever $y \in C^k(L) \setminus C^k(U)$. Since $\phi(C^k(U)) = C^{-k}\phi(U) = \beta^{-k} \cdot \phi(U) = 0$ we have $\beta^{-n}||C^n y||^2 \to 0$ ϕ-almost everywhere on $C^k(L)$ for each $k \in \mathbb{Z}$. It follows that $\beta^{-n}||C^n y||^2 \to 0$ ϕ-almost everywhere on $\mathbb{R}^m \setminus \{0\}$. Then ϕ is supported on $K(\beta^{-1/2}C)$, the contraction subspace of $\beta^{-1/2}C$, and hence $K(\beta^{-1/2}C) = \mathbb{R}^m$ since ϕ is $\mathcal{M}$-full. This yields $\beta^{-1/2}r(C) = r(\beta^{-1/2}C) < 1$ and the proof is complete. □

Remark 7.1.8. Following [55] the operator C is called a contraction, $K(C)$ is called the contraction subspace of C. Moreover the set L constructed above is called a cross-section for $(C^n)_{n\in\mathbb{Z}}$. For more information about contractions and cross-sections see Section 1.1.II in [55].

Lemma 7.1.9. *Let $C \in \mathrm{GL}(\mathbb{R}^m)$ for some $m \geq 1$, $0 < \beta < 1$ and let Q be a positive definite quadratic form on $\mathbb{R}^m$ such that $Q(C^*t) = \beta Q(t)$ for all $t \in \mathbb{R}^m$. Then every eigenvalue z of C is simple with $|z|^2 = \beta$.*

Proof. Let $B = \beta^{-1/2}C$ so that $Q(B^*t) = Q(t)$ for all $t \in \mathbb{R}^m$. Then we have $(B^*)^{-1} \in \mathcal{S}(Q)$ in the sense of Definition 5.2.3. Since $\mathcal{S}(Q)$ is compact, Theorem 2.3.11 implies that $(B^*)^{-1} = LUL^{-1}$ for some symmetric $L \in \mathrm{GL}(\mathbb{R}^d)$ and some orthogonal $U \in \mathrm{GL}(\mathbb{R}^d)$. Then $B = L^{-1}UL$. Theorem 2.1.19 implies that every eigenvalue λ of U is simple with complex absolute value $|\lambda| = 1$, and since $B = L^{-1}UL$ the same is true of B. Since $C = \beta^{1/2}B$ the result follows. □

Proof of Theorem 7.1.1. First we will prove sufficiency. Let ν be an infinitely divisible measure satisfying (7.2) for some $c > 1$, $B \in \mathrm{GL}(\mathbb{R}^d)$ and $b \in \mathbb{R}^d$. From (7.2) it follows that there exists a sequence of nonrandom vectors (b_n) such that

$$\nu = (B^{-n}\nu)^{c^n} * \varepsilon_{b_n} \quad \text{for all } n \geq 1$$

holds. Putting $k_n = [c^n] \sim c^n$, an easy argument using characteristic functions shows that

$$B^{-n}\nu^{k_n} * \varepsilon_{b_n} \Rightarrow \nu$$

where $k_n \uparrow \infty$ and $k_{n+1}/k_n \to c$, so that $\nu \in \mathrm{GDOSA}(\nu)$ and hence by Definition 3.3.20 ν is operator semistable.

Now we prove necessity. It follows from Lemma 7.1.3 that ν is infinitely divisible and from Lemmas 7.1.4 and 7.1.5 that (7.2) holds for some $c > 1$ and $B \in \mathrm{GL}(\mathbb{R}^d)$. Note that if $r > 1$ in Definition 3.3.20, then $c = r$ and otherwise $c > 1$ is arbitrary. If $[a, Q, \phi]$ is the Lévy representation of ν, then Lemma 7.1.6 implies that $C\phi = \beta \cdot \phi$ and $Q(C^*t) = \beta \cdot Q(t)$ for all $t \in \mathbb{R}^d$. Note that (7.2) is equivalent to $\nu^\beta = C\nu * \varepsilon_s$, where $\beta = c^{-1} < 1$, $C = B^{-1}$ and s is chosen appropriately. Let μ_1, μ_2 be infinitely divisible with Lévy representations $[0, 0, \phi]$ and $[0, Q, 0]$, respectively, and define $V_1 = \mathrm{span}(\mathrm{supp}(\phi))$ and $V_2 = \mathrm{span}(\mathrm{supp}(\mu_2))$. Since $C\phi = \beta \cdot \phi$ we have $\mathrm{supp}(C\phi) = \mathrm{supp}(\phi)$. Then Lemma 1.1.9 shows that $C\,\mathrm{supp}(\phi) = \mathrm{supp}(\phi)$, and it follows that $CV_1 = V_1$. Since $Q(C^*t) = \beta \cdot Q(t)$ for all $t \in \mathbb{R}^d$ we have $C\mu_2 = \mu_2^\beta$ so that $\mathrm{supp}(C\mu_2) = \mathrm{supp}(\mu_2)$ and another application of Lemma 1.1.9 yields $CV_2 = V_2$. Let C_i denote the restriction of C to V_i. Proposition 3.1.20 shows that ϕ is $\mathcal{M}$-full on V_1, and hence Lemma 7.1.7 implies that every eigenvalue $a + ib$ of C_1 satisfies $a^2 + b^2 < \beta$. Since μ_2 is a full Gaussian law on V_2, its quadratic form is positive definite on V_2, and then Lemma 7.1.9 implies that every eigenvalue $a + ib$ of C_2 is simple with $a^2 + b^2 = \beta$. Since C_1, C_2 have no eigenvalues in common, $V_1 \cap V_2 = \{0\}$. Since ν is full, it follows that $\mathbb{R}^d = V_1 \oplus V_2$.

Write $a = a_1 + a_2$ with respect to this direct sum decomposition, and define $\nu_i = \mu_i * \varepsilon_{a_i}$. Uniqueness follows easily from the uniqueness of the Lévy representation. □

The following result, due to Chorny [24], shows that the operator B in (7.2) can always be embedded into a one-parameter group $(t^E)_{t>0}$ of linear operators for some linear operator $E \in \mathrm{L}(\mathbb{R}^d)$ called an *exponent* of ν. We will see later in Chapter 8 that this exponent captures many important properties of the generalized domain of semistable attraction of the operator semistable law ν.

Theorem 7.1.10. *A full probability measure ν on $\mathbb{R}^d$ is operator semistable, if and only if it is infinitely divisible and there exists a number $c > 1$, a vector $b \in \mathbb{R}^d$ and a linear operator $E \in \mathrm{L}(\mathbb{R}^d)$ such that*

$$\nu^c = c^E \nu * \varepsilon_b. \tag{7.3}$$

The roots of the minimal polynomial of E lie in the half-plane $\{\operatorname{Re} z \geq 1/2\}$ and all roots lying on the line $\{\operatorname{Re} z = 1/2\}$ are simple.

Moreover, ν can be decomposed uniquely into a product $\nu = \nu_1 * \nu_2$ concentrated on E invariant subspaces V_1 and V_2 respectively, and such that $\mathbb{R}^d = V_1 \oplus V_2$. Then ν_1 is a full operator semistable measure on V_1 of Poisson type and ν_2 is a full Gaussian measure on V_2. The spectrum of $E|_{V_1}$ is contained in $\{\operatorname{Re} z > 1/2\}$ and for the eigenvalues z of $E|_{V_2}$ we have $\operatorname{Re} z = 1/2$.

Definition 7.1.11. We call E an *exponent* of ν and say that ν is (c^E, c) *operator semistable* if (7.3) holds.

Proof. Assume ν is a full operator semistable law on $\mathbb{R}^d$. Apply Theorem 7.1.1 to obtain $B \in \mathrm{GL}(\mathbb{R}^d)$ and $\bar{c} > 1$ such that $\nu^{\bar{c}} = B\nu * \varepsilon_b$ for some $b \in \mathbb{R}^d$. Let $\nu = \nu_1 * \nu_2$ be the Lévy decomposition of ν from Theorem 7.1.1, so that $\mathbb{R}^d = V_1 \oplus V_2$, $BV_i = V_i$, ν_1 is a full Poissonian operator semistable measure on V_1 and ν_2 is a full Gaussian measure on V_2. Let $[a, Q, \phi]$ be the Lévy triple of ν given by Theorem 3.1.11. Let $\pi_i : \mathbb{R}^d \to V_i$ be the natural projections associated with the direct sum decomposition, and write $B = B_1 \oplus B_2$ with respect to this direct sum decomposition. Projecting (7.2) onto V_1 we see that ν_1 is a full $(B_1, \bar{c})$ operator semistable law on V_1 with Lévy representation $[a_1, 0, \phi_1]$, where $a_1 = \pi_1 a$ and $\phi_1 = \pi_1 \phi$. Then Lemma 7.1.6 implies that $B_1\phi = \bar{c} \cdot \phi$, and hence $\bar{c}^n \cdot \phi_1 = B_1^n \phi_1$ for all $n \geq 1$. It follows easily that

$$[\bar{c}^n] \cdot (B_1^{-n} \phi_1) \to \phi_1 \quad \text{as } n \to \infty.$$

Then Theorem 6.2.10 shows that $\phi_1 \in \mathrm{ROV}(E_1, c)$, where $c = \bar{c}^2$ and $E_1 \in \mathrm{L}(V_1)$ is some exponent, and Lemma 6.2.5 yields that $c^{E_1}\phi_1 = c \cdot \phi_1$. Definition 3.1.23 shows that ν_1^c has Lévy triple $[ca_1, 0, c \cdot \phi_1]$ and Proposition 3.1.21 implies that $c^{E_1}\nu_1$ has Lévy triple $[a_1', 0, c^{E_1}\phi_1]$. Then it follows from the uniqueness of the Lévy representation that $\nu_1^c = c^{E_1}\nu * \varepsilon_{b_1}$ for some $b_1 \in V_1$.

Then ν_1 is a full operator semistable law on V_1 with exponent E_1. Furthermore, ν_2 has Lévy representation $[a_2, Q_2, 0]$, where $a_2 = \pi_2 a$ and Q_2 is the covariance operator of ν_2. Define $E_2 = (1/2)I_2$ where I_2 is the identity operator on V_2. Then ν_2^c has Lévy triple $[ca_2, cQ_2, 0]$ while $c^{E_2}\nu_2$ has Lévy triple $[a_2', Q', 0]$, where $Q'(t) = Q_2(c^{E_2^*}t) = Q_2(c^{1/2}t) = cQ_2(t)$ for all $t \in V_2$, and it follows from the uniqueness of the Lévy representation that $\nu_2^c = c^{E_2}\nu * \varepsilon_{b_2}$ for some $b_2 \in V_2$. Let $E = E_1 \oplus E_2$ with respect to the direct sum decomposition. By letting $b = b_1 + b_2$, we have

$$\begin{aligned}\nu^c &= \nu_1^c * \nu_2^c \\ &= (c^{E_1}\nu_1 * \varepsilon_{b_1}) * (c^{E_2}\nu_2 * \varepsilon_{b_2}) \\ &= c^{E_1 \oplus E_2}(\nu_1 * \nu_2) * \varepsilon_{b_1+b_2} \\ &= (c^E\nu) * \varepsilon_b,\end{aligned}$$

so ν is (c^E, c) operator semistable. By Theorem 7.1.1, the eigenvalues of the operator c^{E_1} satisfy $|z|^2 > c$ and the eigenvalues of c^{E_2} are simple with $|z|^2 = c$. Proposition 2.2.2 (f) shows that $z = c^\lambda$ is an eigenvalue of c^{E_i} if and only if λ is an eigenvalue of E_i, hence the eigenvalues of E_1 satisfy $|c^\lambda|^2 > c$ and the eigenvalues of c^{E_2} are simple with $|c^\lambda|^2 = c$. Since $|c^\lambda| = c^{\operatorname{Re}\lambda}$ it follows that the eigenvalues of E_1 satisfy $2\operatorname{Re}\lambda > 1$ and the eigenvalues of c^{E_2} are simple with $2\operatorname{Re}\lambda = 1$. This proves necessity, and sufficiency follows immediately from Theorem 7.1.1. □

Corollary 7.1.12. *If ν is a full operator semistable law on $\mathbb{R}^d$ and $\hat{\nu}$ is the corresponding characteristic function, then $|\hat{\nu}(t)| < 1$ for all $t \neq 0$.*

Proof. If $|\hat{\nu}(t)| = 1$ for some $t \neq 0$ in $\mathbb{R}^d$, then (7.3) implies that $|\hat{\nu}(c^{-kE^*}t)| = 1$ for all $k \geq 0$, where the eigenvalues of E^* all have positive real part. But Corollary 4.2.6 with $\alpha = 0$ implies that $\|c^{-kE^*}t\| \to 0$ and in view of Lemma 1.3.15 this contradicts the fullness of ν. □

Recall from Definition 2.3.3 that $\mathcal{S}^1(\nu)$ is the symmetry group of a probability measure ν, so that $A \in \mathcal{S}^1(\nu)$ if $\nu = A\nu * \varepsilon_a$ for some $a \in \mathbb{R}^d$.

Proposition 7.1.13. *Let ν be a full (c^E, c) operator semistable law. Then*

$$c^E\mathcal{S}^1(\nu) = \mathcal{S}^1(\nu)c^E.$$

Proof. It follows from Proposition 3.1.24 that $\mathcal{S}^1(\nu) = \mathcal{S}^1(\nu^t)$ for any $t > 0$. Since ν is (c^E, c) operator semistable, (7.3) holds and hence it follows that $\nu = c^E\nu^{1/c} * \varepsilon_{b/c}$ and $c^{-E}\nu = \nu^{1/c} * \varepsilon_{c^{-E}b/c}$. We show that $c^E\mathcal{S}^1(\nu)c^{-E} = \mathcal{S}^1(\nu)$. Let $D \in c^E\mathcal{S}^1(\nu)c^{-E}$. Then $D = c^EAc^{-E}$ for some $A \in \mathcal{S}^1(\nu)$, and $A\nu = \nu * \varepsilon_a$ for some $a \in \mathbb{R}^d$ and so $(A\nu)^{1/c} = \nu^{1/c} * \varepsilon_{a/c}$. Now for some

$a' \in \mathbb{R}^d$ we have

$$\begin{aligned} D\nu &= (c^E A c^{-E})\nu = (c^E A)(c^{-E}\nu) = (c^E A)(\nu^{1/c} * \varepsilon_{c^{-E}(b/c)}) \\ &= c^E((A\nu)^{1/c} * \varepsilon_{Ac^{-E}(b/c)}) = c^E(\nu^{1/c} * \varepsilon_{a/c+Ac^{-E}(b/c)}) \\ &= \nu * \varepsilon_{-b/c+c^E(a/c)+c^E Ac^{-E}(b/c)} \\ &= \nu * \varepsilon_{a'}, \end{aligned}$$

showing $D \in \mathcal{S}^1(\nu)$. Conversely, if $D \in \mathcal{S}^1(\nu)$, then writing $D = c^E A c^{-E}$ we have $A = c^{-E} D c^E$, and an argument similar to the one above shows that $A \in \mathcal{S}^1(\nu)$. □

Corollary 7.1.14. *Let ν be a full (c^E, c) operator semistable law with Lévy representation $[a, 0, \phi]$. Then there exists a norm $\|\cdot\|_0$ on $\mathbb{R}^d$ such that for all Borel subsets $B \subset \Gamma = \mathbb{R}^d \setminus \{0\}$ we have*

$$\phi(B) = \int_{S_0} \int_0^\infty I(t^E\theta \in B) H_\theta(-dt)\lambda\{d\theta\} \tag{7.4}$$

where

$$\lambda(D) = \phi\{t^E\theta : \theta \in D; t > 1\} \tag{7.5}$$

is a finite Borel measure on the unit sphere $S_0 = \{x \in \mathbb{R}^d : \|x\|_0 = 1\}$ and $H_\theta(t)$ is a monotone decreasing function of $t > 0$ for all $\theta \in S_0$. Furthermore, we can write $H_\theta(s) = K_\theta(\log s)/s$ where $K_\theta(s)$ is a nonnegative $\log c$*-periodic function such that*

$$K_\theta(t+h)e^{-h} \le K_\theta(t-h)e^h \tag{7.6}$$

for all $t \in \mathbb{R}$, $\theta \in S_0$ and $h > 0$.

Proof. Proposition 3.1.20 implies that ϕ is $\mathcal{M}$-full, and in view of Lemma 7.1.6 together with Theorem 7.1.10 we know that ϕ satisfies (6.39) where the real parts of the eigenvalues of E are all greater than $1/2$. Now the assertion follows from Theorem 6.2.8. □

It is well known that full normal laws have a density with respect to Lebesgue measure. Obviously, full normal laws are operator stable and hence operator semistable with exponent $(1/2)I$. It turns out that every full operator semistable law has a density with respect to Lebesgue measure as shown now.

Theorem 7.1.15. *A full operator semistable law ν on $\mathbb{R}^d$ has a density with respect to Lebesgue measure.*

Proof. In view of Theorem 1.3.7 and the fact that $|\hat{\nu}(t)| \le 1$ for all $t \in \mathbb{R}^d$, it suffices to show that $\int_D |\hat{\nu}(t)| dt < \infty$ where $D = \{t : \|t\| \ge 1\}$. Define

$A_0 = \{x : 1 \leq \|x\| \leq b\}$ where $b \geq \|c^{E^*}\|$ and $r = \sup\{|\hat{\nu}(t)| : t \in A_0\}$, so that $r < 1$ in view of Corollary 7.1.12. As in the proof of Lemma 6.3.10 (a) it is easy to check that $D \subseteq \bigcup_{k=0}^{\infty} c^{kE^*} A_0$. Note that c^{E^*} has determinant $c^{\operatorname{trace}(E^*)}$ and apply the multivariable change of variables theorem to obtain

$$\begin{aligned}
\int_D |\hat{\nu}(t)| dt &\leq \sum_{k=0}^{\infty} \int_{c^{kE^*} A_0} |\hat{\nu}(t)| dt \\
&= \sum_{k=0}^{\infty} \int_{A_0} |\hat{\nu}(c^{kE^*} t)| c^{k \operatorname{trace}(E^*)} dt \\
&= \sum_{k=0}^{\infty} \int_{A_0} |\hat{\nu}(t)|^{c^k} c^{k \operatorname{trace}(E^*)} dt \\
&\leq \sum_{k=0}^{\infty} r^{c^k} c^{k \operatorname{trace}(E^*)} \int_{A_0} dt < \infty
\end{aligned} \tag{7.7}$$

since $u^p r^u \to 0$ as $u \to \infty$ for any $p > 0$. □

Theorem 7.1.1 shows that an operator semistable law can be decomposed into two independent components, one normal and the other having no normal component. Now we will decompose the nonnormal part further, using the spectral decomposition. Even though the resulting components are not necessarily independent, this decomposition is of considerable practical value, as we will see in Chapter 8 and the applications to follow. Recall from Definition 2.1.15 that for any linear operator E on $\mathbb{R}^d$ there exists a direct sum decomposition $\mathbb{R}^d = V_1 \oplus \cdots \oplus V_p$ into E-invariant subspaces such that, if we write $E = E_1 \oplus \cdots \oplus E_p$, where $E_i : V_i \to V_i$, then every eigenvalue of E_i has a real part equal to a_i and $a_1 < \cdots < a_p$. This is called the spectral decomposition of $\mathbb{R}^d$ relative to E.

Definition 7.1.16. Suppose Y is a full random vector on $\mathbb{R}^d$ whose distribution ν is (c^E, c) operator semistable. Let $V_1, \ldots, V_p$ denote the spectral decomposition of $\mathbb{R}^d$ relative to E, and $\pi_i : \mathbb{R}^d \to V_i$ the projections associated with this direct sum decomposition. If we let $Y_i = \pi_i Y$ for $i = 1, \ldots, p$ then $Y = Y_1 + \cdots + Y_p$. This is called the *spectral decomposition of* Y *with respect to* E. The probability distributions $\nu_i = \pi_i \nu$ of the random vectors Y_i are called the *spectral decomposition of* ν *with respect to* E. If every eigenvalue of E has the same real part, then there is only one component in the spectral decomposition. In this case, we say that Y (or ν) is *spectrally simple.*

Lemma 7.1.17. *Suppose that ν is (c^E, c) operator semistable. Then every ν_i in the spectral decomposition of ν relative to $E = E_1 \oplus \cdots \oplus E_p$ is a full (c^{E_i}, c) operator semistable law on the space V_i. Furthermore, ν_i is spectrally simple.*

Proof. If some ν_i is not full, then $\nu_i = \pi_i \nu$ is supported on a proper affine subspace $L_i \subset V_i$. Then ν is supported on the preimage $\pi_i^{-1}(L_i)$, which is a proper affine subspace of $\mathbb{R}^d$, and this contradicts the assumption that ν is full. Hence every ν_i is full. Theorem 7.1.10 shows that $\nu^c = c^E\nu * \varepsilon_b$ for some $b \in \mathbb{R}^d$. Since $\pi_i E = E_i \pi_i$ this implies that $\nu_i^c = (\pi_i\nu)^c = \pi_i(\nu^c) = \pi_i(c^E\nu * \varepsilon_b) = c^{E_i}\pi_i\nu * \varepsilon_{b_i} = c^{E_i}\nu_i * \varepsilon_{b_i}$, where $b_i = \pi_i b$, and then Theorem 7.1.10 shows that ν_i is (c^{E_i}, c) operator semistable. Since every eigenvalue of E_i is equal to a_i, ν_i is spectrally simple. □

Theorem 7.1.18. *If ν is a full operator semistable law on $\mathbb{R}^d$, then for every $\theta \neq 0$ there exists some $\alpha(\theta) \in (0, 2]$ such that*

$$\int |\langle x, \theta\rangle|^\rho \nu\{dx\}$$

exists for all $0 \leq \rho < \alpha(\theta)$, and if $\alpha(\theta) < 2$, then the integral diverges for all $\rho > \alpha(\theta)$.

Proof. We defer the proof to Chapter 8 (see Theorem 8.3.10). □

Remark 7.1.19. The function $\alpha(\theta)$ in Theorem 7.1.18 is closely related to the spectral index of Definition 6.4.9. Since $\alpha(\theta)$ governs the moments of ν, it is uniquely determined by ν, and it follows that every exponent of ν has the same real spectrum.

Corollary 7.1.20. *Suppose that Y is a full random vector on $\mathbb{R}^d$ whose distribution is (c^B, c) operator semistable. Let $Y = Y_1 + \cdots Y_p$ denote the spectral decomposition of Y with respect to B and $a_1 < \cdots < a_p$ the real spectrum of B. Then $E\|Y_i\|^\rho$ exists for all $0 \leq \rho < 1/a_i$, and if $a_i > 1/2$ then $E\|Y_i\|^\rho = \infty$ for all $\rho > 1/a_i$.*

Proof. See Corollary 8.3.13. □

Corollary 7.1.21. *Suppose that Y is a full random vector on $\mathbb{R}^d$ whose distribution is (c^B, c) operator semistable, and let $a_1 < \cdots < a_p$ denote the real spectrum of B. Then $E\|Y\|^\rho$ exists for all $0 \leq \rho < 1/a_p$, and if $a_p > 1/2$, then $E\|Y\|^\rho$ is infinite for all $\rho > 1/a_p$.*

Proof. See Corollary 8.3.15. □

Remark 7.1.22. The exponent E of an operator semistable law is not necessarily unique, because of possible symmetries. For example, while $E = (1/2)I$ is an exponent of the standard normal law, if Q is skew symmetric then $E' = E + Q$ is another exponent. The problem of characterizing the possible exponents of an operator semistable law, and their relation to the symmetry group, is still open.

7.2 OPERATOR STABLE LAWS

In this section we investigate some limit distributions that are special cases of operator semistable laws. Our main focus is on the class of operator stable laws. Recall from Definition 3.3.24 that a probability measure ν is operator stable if and only if it possesses a nonempty generalized domain of attraction. Theorem 7.1.10 showed that operator semistable laws possess a certain scaling property along a geometric sequence of scales. Our main result in this section is Theorem 7.2.1, which shows that an operator stable law obeys a similar scaling property at every scale. For operator semistable laws, one of our main technical tools was the theory of R–O varying measures from Section 6.2. In this section, we will rely instead on the theory of regularly varying measures from Section 6.1.

Theorem 7.2.1. *A full probability measure ν on $\mathbb{R}^d$ is operator stable if and only if it is infinitely divisible and there exists a linear operator $E \in \mathrm{L}(\mathbb{R}^d)$ such that for suitable shifts $a_t \in \mathbb{R}^d$ we have*

$$\nu^t = t^E \nu * \varepsilon_{a_t} \quad \textit{for all } t > 0. \tag{7.8}$$

The roots of the minimal polynomial of E lie in the half-plane $\{\mathrm{Re}\, z \geq 1/2\}$ and all roots on the line $\{\mathrm{Re}\, z = 1/2\}$ are simple. Moreover, ν can be decomposed uniquely into a convolution product $\nu = \nu_1 * \nu_2$ of two probability measures concentrated on E-invariant subspaces V_1 and V_2 respectively, and such that $\mathbb{R}^d = V_1 \oplus V_2$. Then ν_1 is a full operator stable law on V_1 of Poisson type and ν_2 is a full Gaussian measure on V_2. The spectrum of $E|_{V_1}$ is contained in $\{\mathrm{Re}\, z > 1/2\}$ and for the eigenvalues z of $E|_{V_2}$ we have $\mathrm{Re}\, z = 1/2$.

Definition 7.2.2. We call E an *exponent* of ν and say that ν is (t^E) *operator stable* if (7.8) holds. For any operator stable law ν, we will call the probability measures ν_1, ν_2 from Theorem 7.2.1 the *Lévy decomposition* of ν.

Proof. Let ν be operator stable. Then there exists a $\mu \in \mathrm{GDOA}(\nu)$ and hence by Lemma 3.3.27 $\mu \in \mathrm{GDOSA}(\nu, c)$ for any $c > 1$. Especially, by Theorem 7.1.1 ν is infinitely divisible and ν is (B_n, n) operator semistable for any $n \geq 2$ and some $B_n \in \mathrm{GL}(\mathbb{R}^d)$. Now let $[a, Q, \phi]$ denote the Lévy triple of ν. Then using Theorem 7.1.1 again, there exist a decomposition $\mathbb{R}^d = V_1 \oplus V_2$ and full probability measures ν_1 and ν_2 on V_1 and V_2, respectively, such that $\nu = \nu_1 * \nu_2$. Moreover, ν_1 has Lévy triple $[a_1, 0, \phi_1]$ and ν_2 has Lévy triple $[a_2, Q_2, 0]$ where ϕ_1 is the restriction of ϕ to V_1, Q_2 is the restriction of Q to V_2, and $a = a_1 + a_2$ with respect to the direct sum decomposition. Theorem 7.1.1 also shows that every B_n leaves this decomposition invariant, i.e., $B_n(V_i) = V_i$ for all $n \geq 2$ and $i = 1, 2$.

Projecting (7.2) onto V_1 yields

$$\nu_1^n = B_{1n} \nu_1 * \varepsilon_{a_{1n}} \qquad \text{for all } n \geq 2$$

where $B_{1n} = B_n|_{V_1}$ and $a_{1n} \in V_1$. By Lemma 7.1.6 (b), this implies

$$n \cdot \phi_1 = B_{1n}\phi_1 \quad \text{for all } n \geq 2 \tag{7.9}$$

and since ν_1 is full on V_1, Proposition 3.1.20 implies that ϕ_1 is $\mathcal{M}$-full on V_1. By rewriting, (7.9) yields

$$n \cdot (B_{1n})^{-1}\phi_1 = \phi_1 \to \phi_1 \quad \text{as } n \to \infty$$

and Theorem 6.1.20 implies that there exists an exponent $E_1 \in \mathcal{E}(\phi)$ such that $t \cdot \phi_1 = t^{E_1}\phi_1$ for all $t > 0$. Since ν_1^t has Lévy representation $[ta_1, 0, t \cdot \phi_1]$ while $t^{E_1}\nu_1$ has Lévy representation $[\bar{a}_{1t}, 0, t^{E_1}\phi_1]$ for some $\bar{a}_{1t} \in V_1$, Lemma 7.1.6 yields $\nu_1^t = t^{E_1}\nu_1 * \varepsilon_{a_{1t}}$ for all $t > 0$ and some $a_{1t} \in V_1$.

Define $E_2 = (1/2)I_2$ where I_2 is the identity operator on V_2. Then ν_2^t has Lévy triple $[ta_2, tQ_2, 0]$ while $t^{E_2}\nu_2$ has Lévy triple $[\bar{a}_{2t}, Q', 0]$ where $Q'(x) = Q_2(t^{E_2^*}x) = Q_2(t^{1/2}x) = tQ_2(x)$ for all $x \in V_2$, and it follows from Lemma 7.1.6 that $\nu_2^t = t^{(1/2)I_2}\nu_2 * \varepsilon_{a_{2t}}$ for all $t > 0$ and some $a_{2t} \in V_2$. Put $E = E_1 \oplus E_2$ with respect to the direct sum decomposition and let $a_t = a_{1t} + a_{2t}$. Then

$$\begin{aligned} \nu^t &= (\nu_1 * \nu_2)^t = \nu_1^t * \nu_2^t \\ &= \left(t^{E_1}\nu_1 * \varepsilon_{a_{1t}}\right) * \left(t^{E_2}\nu_2 * \varepsilon_{a_{2t}}\right) \\ &= t^{E_1 \oplus E_2}(\nu_1 * \nu_2) * \varepsilon_{a_{1t}} * \varepsilon_{a_{2t}} \\ &= t^E \nu * \varepsilon_{a_t} \end{aligned}$$

for all $t > 0$ so (7.8) holds.

On the other hand, if (7.8) holds, then for $n \geq 1$ we have $\nu^n = n^E\nu * \varepsilon_{a_n}$ and hence $n^{-E}\nu^n * \varepsilon_{-n^{-E}a_n} = \nu$ so $\nu \in \text{GDOA}(\nu)$ and therefore ν is operator stable. Since, by (7.8) and Theorem 7.1.1 ν is (B, c) operator semistable for any $c > 1$ the assertion about the eigenvalues of E follow from Theorem 7.1.1, since by Proposition 2.2.2 (f) c^λ is an eigenvalue of c^E if and only if λ is an eigenvalue of E. □

Now we obtain a couple of technical results that we will need later in Chapter 8, and in the applications to follow.

Proposition 7.2.3. *Let ν be a full (t^E) operator stable law on $\mathbb{R}^d$. Then the family of shifts (a_t) in (7.8) depends continuously on $t > 0$.*

Proof. Theorem 7.2.1 implies that ν is infinitely divisible, and then Theorem 3.1.2 yields that $\hat{\nu}(y) \neq 0$ for all $y \in \mathbb{R}^d$. If we rewrite (7.8) in terms of Fourier transforms we get

$$\hat{\nu}(y)^t = \hat{\nu}(t^{E^*}y)e^{i<a_t, y>}$$

for all $t > 0$ and all $y \in \mathbb{R}^d$ and hence

$$e^{i<a_t, y>} = \frac{\hat{\nu}(y)^t}{\hat{\nu}(t^{E^*}y)}. \tag{7.10}$$

Since $\hat{\nu}$ is continuous, the right-hand side of (7.10) depends continuously on $t > 0$. Therefore, as $t_n \to t > 0$ we have

$$e^{i<a_{t_n},y>} \to e^{i<a_t,y>} \quad \text{as } n \to \infty.$$

Then Lemma 3.1.10 implies that $<a_{t_n}, y> \to <a_t, y>$ as $n \to \infty$ for all $y \in \mathbb{R}^d$. This is only possible if $a_{t_n} \to a_t$ showing the continuity of $t \mapsto a_t$. □

Proposition 7.2.4. *Let ν be a full (t^E) operator stable law. Then*

$$t^E \mathcal{S}^1(\nu) = \mathcal{S}^1(\nu) t^E \quad \textit{for all } t > 0.$$

Proof. The proof is almost identical to the proof of Proposition 7.1.13 and is left to the reader. □

Theorem 7.2.5. *Let ν be a full (t^E) operator stable law with Lévy representation $[a, 0, \phi]$. Then there exists a norm $\|\cdot\|_0$ on $\mathbb{R}^d$ such that for all Borel subsets $A \subset \Gamma$ we have*

$$\phi(A) = \int_{S_0} \int_0^\infty I(t^E \theta \in A) \frac{dt}{t^2} \lambda\{d\theta\} \tag{7.11}$$

where

$$\lambda(D) = \phi\{t^E \theta : \theta \in D; t > 1\} \tag{7.12}$$

is a finite Borel measure on the unit sphere $S_0 = \{x : \|x\|_0 = 1\}$.

Proof. If ν is a full (t^E) operator stable law having no Gaussian component and Lévy measure ϕ then it follows from Theorem 7.2.1 and Lemma 7.1.6 that $t \cdot \phi = t^E \phi$ for all $t > 0$ holds. Then the result follows from Theorem 6.1.7. □

Definition 7.2.6. Let ν be a full operator stable law with Lévy representation $[a, 0, \phi]$. Then the measure λ in (7.12) above is called the *spectral measure* of ν.

Since operator stable laws are by Lemma 3.3.27 operator semistable laws it follows in view of Theorem 7.1.15 that they also have a density.

Theorem 7.2.7. *Every full operator stable law has a density with respect to Lebesgue measure on $\mathbb{R}^d$.*

Proof. Apply Theorem 7.2.1 together with Theorems 7.1.1 and 7.1.15. □

Definition 7.2.8. Suppose Y is a full random vector on $\mathbb{R}^d$ whose distribution ν is (t^E) operator stable. Let $V_1, \ldots, V_p$ denote the spectral decomposition of $\mathbb{R}^d$ relative to E, and $\pi_i : \mathbb{R}^d \to V_i$ the projections associated with this direct sum decomposition. If we let $Y_i = \pi_i Y$ for $i = 1, \ldots, p$, then $Y = Y_1 + \cdots + Y_p$. This is called the *spectral decomposition of Y with respect* to E. The probability distributions $\nu_i = \pi_i \nu$ of the random vectors Y_i are called the *spectral decomposition of ν with respect to E.* If every eigenvalue of E has the same real part, then there is only one component in the spectral decomposition. In this case, we say that Y (or ν) is *spectrally simple.*

Lemma 7.2.9. *Suppose that ν is (t^E) operator stable. Then every ν_i in the spectral decomposition of ν relative to $E = E_1 \oplus \cdots \oplus E_p$ is a full (t^{E_i}) operator stable law on the space V_i. Furthermore, ν_i is spectrally simple.*

Proof. The proof is similar to Lemma 7.1.17. Since ν is full, every ν_i must be full. Theorem 7.2.1 shows that $\nu^t = t^E \nu * \varepsilon_{a_t}$ for some $a_t \in \mathbb{R}^d$. Since $\pi_i E = E_i \pi_i$ this implies that $\nu_i^t = \pi_i(t^E \nu * \varepsilon_{a_t}) = t^{E_i} \nu_i * \varepsilon_{\pi_i a_t}$ and then Theorem 7.2.1 shows that ν_i is (t^{E_i}) operator stable. □

Remark 7.2.10. Suppose Y is a full random vector on $\mathbb{R}^d$ whose distribution ν is (t^B) operator stable. Then ν is also (c^B, c) operator semistable for every $c > 1$, hence the moment results from Section 7.1 apply. In particular, for every $\theta \neq 0$ there exists some $\alpha(\theta) \in (0, 2]$ such that

$$\int |\langle x, \theta \rangle|^\rho \nu\{dx\} \tag{7.13}$$

exists for all $0 \leq \rho < \alpha(\theta)$, and if $\alpha(\theta) < 2$ then (7.13) diverges for all $\rho > \alpha(\theta)$. It follows that every exponent of ν has the same real spectrum. If $a_1 < \cdots < a_p$ is the real spectrum of B, then $E\|Y\|^\rho$ exists for $0 \leq \rho < 1/a_p$, and if $a_p > 1/2$, then $E\|Y\|^\rho$ is infinite for all $\rho > 1/a_p$. Furthermore, if $Y = Y_1 + \cdots + Y_p$ is the spectral decomposition of Y with respect to B, then $E\|Y_i\|^\rho$ exists for all $0 \leq \rho < 1/a_i$, and if $a_i > 1/2$, then $E\|Y_i\|^\rho = \infty$ for all $\rho > 1/a_i$. If $\alpha(\theta) < 2$, then (7.13) also diverges for $\rho = \alpha(\theta)$. The proof of this fact requires different methods, see [114]. It follows as in Corollary 8.3.15 that if $a_p > 1/2$, then $E\|Y\|^{1/a_p} = \infty$, and if $a_i > 1/2$, then $E\|Y_i\|^{1/a_i} = \infty$.

Let $\mathcal{E}(\nu)$ denote the set of all exponents of a full operator stable law ν on $\mathbb{R}^d$. Recall from Definition 2.2.9 that $T\mathcal{S}^1(\nu)$ is the tangent space of the symmetry group of ν.

Theorem 7.2.11. *The exponents and symmetries of a full operator stable law ν on $\mathbb{R}^d$ are related by*

$$\mathcal{E}(\nu) = B + T\mathcal{S}^1(\nu) \tag{7.14}$$

where $B \in \mathcal{E}(\nu)$ is arbitrary. One can always select an exponent B_0 that commutes with every symmetry $A \in \mathcal{S}^1(\nu)$.

Proof. The proof is similar to Theorems 5.2.13 and 6.1.21. Given any exponent E of ν define $G_\lambda = \lambda^{-E}\mathcal{S}^1(\nu)$ and $G = \cup_{\lambda>0} G_\lambda$. By using Proposition 7.2.4, it is not hard to check that the function $\zeta : G \to (0, \infty)$ defined by $\zeta(C) = \lambda$ if $C \in G_\lambda$ is a well-defined continuous group homomorphism. Then it follows as in the proof of Theorem 5.2.13 that the continuous group homomorphism $L : TG \to \mathbb{R}$ defined by letting $L(X) = \log \zeta(\exp(X))$ is linear. Now check that $T\mathcal{S}^1(\nu) = \{X \in TG : L(X) = 0\}$ and $\mathcal{E}(\nu) = \{X \in TG : L(X) = -1\}$. The remainder of the proof is identical to Theorem 5.2.13. □

Corollary 7.2.12. *Suppose that ν is a full operator stable law on $\mathbb{R}^d$. Then the nilpotent part of every $B \in \mathcal{E}(\nu)$ is the same, and it commutes with every element of the symmetry group $\mathcal{S}^1(\nu)$. Furthermore, every $B \in \mathcal{E}(\nu)$ has the same real spectrum, and the spectral decomposition of ν is the same for every $B \in \mathcal{E}(\nu)$.*

Proof. The proof is identical to Theorem 5.2.14. □

7.3 STABLE LAWS

In this section we consider the important special case of stable laws. In this case, we can say more about the form of the limit. In particular, we can obtain equivalent forms of the characteristic function that are useful in many applications.

Definition 7.3.1. Suppose that $X, X_1, X_2, X_3, \ldots$ are independent and identically distributed random vectors on $\mathbb{R}^d$ with common distribution μ and that Y is a random vector whose distribution ν is full.

(a) We say that ν (or Y) is *stable* if ν is (t^E) operator stable and $E = aI$ for some real number a. In this case, we call $\alpha = 1/a$ the *index* of ν.

(b) We say that X (or equivalently μ) belongs to the *domain of attraction* of Y (or ν) if there exist nonrandom real numbers $a_n > 0$ and nonrandom vectors $b_n \in \mathbb{R}^d$ such that

$$a_n(X_1 + \cdots + X_n) + b_n \Rightarrow Y. \tag{7.15}$$

In terms of measures, we can rewrite (7.15) in the form

$$a_n \mu^n * \varepsilon_{b_n} \Rightarrow \nu. \tag{7.16}$$

We write $X \in \mathrm{DOA}(Y)$ (or $\mu \in \mathrm{DOA}(\nu)$) in this case.

Remark 7.3.2. In Chapter 8 we will prove that a full probability measure ν on $\mathbb{R}^d$ is stable if and only if $\mathrm{DOA}(\nu)$ is not empty (see Proposition 8.3.26).

Theorem 7.3.3. *If a full probability measure ν on $\mathbb{R}^d$ is stable with index α, then ν is a full infinitely divisible law on $\mathbb{R}^d$ with Lévy representation $[a, Q, \phi]$ and $0 < \alpha \le 2$.*

(a) If $\alpha = 2$, then ν is normal with mean $a \in \mathbb{R}^d$ and covariance matrix B, $Q(t) = \langle t, Bt \rangle$ for all $t \in \mathbb{R}^d$, and ϕ is the zero measure.

(b) *If $\alpha < 2$, then ν is stable with index α, $Q(t) = 0$ for all $t \in \mathbb{R}^d$, and ϕ is of the form*

$$\phi\{dr, d\theta\} = \phi_\alpha\{dr\}\lambda\{d\theta\} \tag{7.17}$$

in polar coordinates $x = r\theta$ for $r > 0$ and $\|\theta\| = 1$, where $\phi_\alpha(r, \infty) = r^{-\alpha}$ for all $r > 0$ and

$$\lambda(D) = \phi\{r\theta : \theta \in D; r > 1\} \tag{7.18}$$

is a finite Borel measure on the unit sphere in $\mathbb{R}^d$.

Proof. Theorem 7.2.1 implies that $a = 1/\alpha \geq 1/2$ so that $0 < \alpha \leq 2$, and since $E = aI$ it also implies that ν is either normal, or has no normal component. Then part (a) follows immediately from the Lévy representation of Theorem 3.1.11. In the nonnormal case, Lemma 7.1.6 (b) implies that $t \cdot \phi = t^a \phi$ for all $t > 0$. In order to establish (7.17), it suffices to show that $\phi(A(r, D)) = r^{-\alpha} \cdot \lambda(D)$ when

$$A(r, D) = \{t\theta : t > r; \theta \in D\}$$

for every $r > 0$ and every Borel subset D of the unit sphere, since sets of this form uniquely determine a measure $\phi \in \mathcal{M}$, see Example 1.2.22. But for such sets we have

$$\begin{aligned} r^\alpha \cdot \phi(A(r, D)) &= (r^\alpha)^a \phi(A(r, D)) \\ &= (r\phi)\{t\theta : t > r; \theta \in D\} \\ &= \phi\{r^{-1}t\theta : t > r; \theta \in D\} \\ &= \phi\{t\theta : t > 1; \theta \in D\} \\ &= \phi(A(1, D)) = \lambda(D) \end{aligned}$$

which completes the proof of part (b). □

Now we write down the corresponding results for the special case of stable random variables. Recall that (the distribution of) a random variable Y on $\mathbb{R}^1$ is *nondegenerate* if Y is not almost surely concentrated at one point. Hence a random variable is nondegenerate if and only if it is full on $\mathbb{R}^1$.

Corollary 7.3.4. *If a nondegenerate probability measure ν on $\mathbb{R}^1$ is stable with index α, then ν is a nondegenerate infinitely divisible law on $\mathbb{R}^1$ with Lévy representation $[a, Q, \phi]$ and $0 < \alpha \leq 2$.*

(a) *If $\alpha = 2$, then ν is normal with mean $a \in \mathbb{R}^1$ and variance $b > 0$, $Q(t) = bt^2$ for all $t \in \mathbb{R}^1$, and ϕ is the zero measure.*

(b) *If $\alpha < 2$, then ν is stable with index α, $Q(t) = 0$ for all $t \in \mathbb{R}^1$, and ϕ is of the form*

$$\begin{aligned} \phi(r, \infty) &= pCr^{-\alpha} \\ \phi(-\infty, -r) &= qCr^{-\alpha} \end{aligned} \tag{7.19}$$

for every $r > 0$, for some $C > 0$ and $p \geq 0$, $q \geq 0$ with $p + q = 1$.

Proof. Specialize Theorem 7.3.3 to the case $d = 1$ so that in part (b) we have that $S = \{-1, +1\}$ is the unit sphere in $\mathbb{R}^1$. Then $C = \lambda(S) > 0$ by Proposition 3.1.20 and we can take $p = \lambda\{+1\}/C$ and $q = \lambda\{-1\}/C$. □

Theorem 7.3.5. *Suppose that ν is a nondegenerate nonnormal stable law on $\mathbb{R}^1$ with index α and Lévy measure ϕ satisfying* (7.19).

(a) *If $\alpha \neq 1$, then ν has characteristic function* $\exp(\psi(t))$ *where*

$$\psi(t) = iat - \sigma^\alpha |t|^\alpha \left(1 - i\beta \operatorname{sign}(t) \tan\left(\frac{\pi\alpha}{2}\right)\right) \tag{7.20}$$

for some $a \in \mathbb{R}^1$ with $\beta = p - q$ and

$$\sigma^\alpha = C \cdot \frac{\Gamma(2-\alpha)}{1-\alpha} \cdot \cos\left(\frac{\pi\alpha}{2}\right) \tag{7.21}$$

(b) *If $\alpha = 1$, then ν has characteristic function* $\exp(\psi(t))$ *where*

$$\psi(t) = iat - \sigma^\alpha |t|^\alpha \left(1 + i\beta \left(\frac{2}{\pi}\right) \operatorname{sign}(t) \ln|t|\right) \tag{7.22}$$

for some $a \in \mathbb{R}^1$ with $\beta = p - q$ and

$$\sigma^\alpha = C \cdot \frac{\pi}{2} \tag{7.23}$$

If ν is a normal with mean a and variance b, then ν has characteristic function $\exp(\psi(t))$ *where ψ is given by* (7.20) *with $\alpha = 2$, $\sigma = \sqrt{b/2}$, and β arbitrary.*

Definition 7.3.6. In the situation of Theorem 7.3.5, we say that ν has distribution $S_\alpha(\sigma, \beta, a)$. The entire class of nondegenerate stable laws on $\mathbb{R}^1$ is given by these formulas with index $\alpha \in (0, 2]$, *scale* $\sigma \in (0, \infty)$, *skewness* $\beta \in [-1, +1]$, and *center* $a \in (-\infty, \infty)$.

The proof of Theorem 7.3.5 follows Section XVII.3 of Feller [40] but with a different parameterization, and quite a few more details. To begin with, we compute the characteristic function of ν in some special cases, which will also be useful in Chapter 8.

Lemma 7.3.7. *Suppose that ν is infinitely divisible with characteristic function* $\exp(\psi(t))$, *where $\psi(t)$ is given by*

$$\psi(t) = \int_{x \neq 0} (e^{itx} - 1)\phi\{dx\} \tag{7.24}$$

and ϕ is given by (7.19) *with $q = 0$ and $0 < \alpha < 1$. Then ν is stable with distribution $S_\alpha(\sigma, 1, 0)$, where $\sigma > 0$ is given by* (7.21).

Proof. If ν is infinitely divisible with characteristic function $\exp(\psi(t))$ where ψ is given by (7.24) and ϕ is given by (7.19) with $q = 0$ and $0 < \alpha < 1$, then Theorem 3.1.14 (a) shows that ν has Lévy representation $[a, 0, \phi]$ for some $a \in \mathbb{R}^1$, and in particular, ϕ is the Lévy measure of ν. Then Corollary 7.3.4 (b) shows that ν is stable with index α. Substituting (7.19) into (7.24) shows that $\psi(t) = I(\alpha)$, where

$$I(\alpha) = C \int_0^\infty \left(e^{itx} - 1\right) \alpha x^{-\alpha-1} dx. \tag{7.25}$$

We will compute $I(\alpha)$ as the limit of integrals of the form

$$I_s(\alpha) = C \int_0^\infty \left(e^{(it-s)x} - 1\right) \alpha x^{-\alpha-1} dx \tag{7.26}$$

where $s > 0$. Integrate by parts to obtain

$$I_s(\alpha) = C(it - s) \int_0^\infty e^{itx} e^{-sx} x^{-\alpha} dx = -C\Gamma(1-\alpha)(s - it)^\alpha \tag{7.27}$$

by the well-known formula for the characteristic function of a Gamma distribution. The real and imaginary parts of the integrand in (7.26) are dominated by $C_1 x^{-\alpha-1}$ and are $O(x^{-\alpha})$ as $x \to 0$ where $0 < \alpha < 1$, hence $I_s(\alpha) \to I(\alpha)$ as $s \to 0$ by dominated convergence. Therefore

$$\begin{aligned} I(\alpha) &= -C\Gamma(1-\alpha)(-it)^\alpha \\ &= -C\Gamma(1-\alpha)\left(-i|t| \operatorname{sign}(t)\right)^\alpha \\ &= -C\Gamma(1-\alpha)|t|^\alpha \left(e^{-i \operatorname{sign}(t)\pi/2}\right)^\alpha \\ &= -C\Gamma(1-\alpha)|t|^\alpha \left(\cos\left(\frac{\pi\alpha}{2}\right) - i \operatorname{sign}(t) \sin\left(\frac{\pi\alpha}{2}\right)\right) \\ &= -C\frac{\Gamma(2-\alpha)}{1-\alpha} \cos\left(\frac{\pi\alpha}{2}\right)|t|^\alpha \left(1 - i \operatorname{sign}(t) \tan\left(\frac{\pi\alpha}{2}\right)\right) \end{aligned} \tag{7.28}$$

using the well-known fact that $\Gamma(s+1) = s\Gamma(s)$ in the last step. Hence $\psi(t) = I(\alpha)$ is of the form (7.20) with $a = 0$, $\beta = 1$, and σ given by (7.21). Note that $\sigma > 0$ since $0 < \alpha < 1$. □

Lemma 7.3.8. *Suppose that ν is infinitely divisible with characteristic function $\exp(\psi(t))$, where*

$$\psi(t) = \int_{x \neq 0} \left(e^{itx} - 1 - itx\right) \phi\{dx\} \tag{7.29}$$

and ϕ is given by (7.19) with $q = 0$ and $1 < \alpha < 2$. Then ν is a stable law with distribution $S_\alpha(\sigma, 1, 0)$, where $\sigma > 0$ satisfies (7.21).

Proof. If ν is infinitely divisible with characteristic function $\exp(\psi(t))$ where ψ is given by (7.29) and ϕ is given by (7.19) with $q = 0$ and $1 < \alpha < 2$, then

Theorem 3.1.14 (b) shows that ν has Lévy representation $[a, 0, \phi]$ for some $a \in \mathbb{R}^1$, and in particular, ϕ is the Lévy measure of ν. Then Corollary 7.3.4 (b) shows that ν is stable with index α. Substituting (7.19) into (7.29) shows that $\psi(t) = J(\alpha)$, where

$$J(\alpha) = C \int_0^\infty \left(e^{itx} - 1 - itx\right)\alpha x^{-\alpha-1}dx. \tag{7.30}$$

We will compute $J(\alpha)$ as the limit of integrals of the form

$$J_s(\alpha) = C \int_0^\infty \left(e^{(it-s)x} - 1 - (it-s)x\right)\alpha x^{-\alpha-1}dx \tag{7.31}$$

where $s > 0$. Integrate by parts to obtain

$$\begin{aligned} J_s(\alpha) &= C(it-s) \int_0^\infty \left(e^{(it-s)x} - 1\right)x^{-\alpha}dx \\ &= \frac{(it-s)}{\alpha-1} \cdot C \int_0^\infty \left(e^{(it-s)x} - 1\right)(\alpha-1)x^{-\alpha}dx \\ &= \frac{(it-s)}{\alpha-1} I_s(\alpha-1) \\ &= \frac{(it-s)}{\alpha-1}\left(-C\Gamma\left((1-(\alpha-1)\right)(s-it)^{\alpha-1}\right) \\ &= -C\frac{\Gamma(2-\alpha)}{1-\alpha}(s-it)^\alpha \end{aligned}$$

by (7.27). The real and imaginary parts of the integrand in (7.31) are dominated by $C_2 x^{-\alpha}$ and are $O(x^{1-\alpha})$ as $x \to 0$ where $1 < \alpha < 2$, hence $J_s(\alpha) \to J(\alpha)$ as $s \to 0$ by dominated convergence. Therefore

$$\begin{aligned} J(\alpha) &= -C\frac{\Gamma(2-\alpha)}{1-\alpha}(-it)^\alpha \\ &= -C\frac{\Gamma(2-\alpha)}{1-\alpha}\cos\left(\frac{\pi\alpha}{2}\right)|t|^\alpha\left(1 - i\,\mathrm{sign}(t)\tan\left(\frac{\pi\alpha}{2}\right)\right) \end{aligned}$$

as in (7.28). Hence $\psi(t) = J(\alpha)$ is of the form (7.20) with $a = 0$, $\beta = 1$ and $\sigma > 0$ satisfying (7.21). Note that in this case both $1 - \alpha$ and $\cos(\pi\alpha/2)$ are negative. □

Lemma 7.3.9. *Suppose that ν is infinitely divisible with characteristic function $\exp(\psi(t))$, where*

$$\psi(t) = it\bar{a} + \int_{x\neq 0}\left(e^{itx} - 1 - it\sin x\right)\phi\{dx\} \tag{7.32}$$

with $\bar{a} = 0$ and ϕ is given by (7.19) with $q = 0$ and $\alpha = 1$. Then ν is a stable law with distribution $S_1(\sigma, 1, 0)$, where $\sigma > 0$ is given by (7.23).

Proof. In the one variable case, if ν is infinitely divisible with Lévy representation $[b, 0, \phi]$, then ν also has characteristic function $\exp(\psi(t))$ where $\psi(t)$ is given by (7.32) with

$$\bar{a} = b + \int_{x \neq 0} \Big(\sin x - \frac{x}{1 + x^2} \Big) \phi\{dx\} \tag{7.33}$$

where $\bar{a}$ exists since the integrand is bounded and $O(x^3)$ as $x \to 0$. Then ϕ is the Lévy measure of ν, and Corollary 7.3.4 (b) shows that ν is stable with index $\alpha = 1$. Substituting (7.19) with $q = 0$ and $\alpha = 1$ into (7.32) with $\bar{a} = 0$ shows that $\psi(t) = A + iB$ where

$$\begin{aligned} A &= C \int_0^\infty (\cos(tx) - 1) x^{-2} dx \\ &= -\frac{1}{2} C t \pi \int_{-\infty}^{\infty} \frac{1 - \cos(tx)}{\pi t x^2} dx \\ &= -\frac{\pi}{2} C t \end{aligned} \tag{7.34}$$

for $t > 0$, since the last integrand is a probability density (e.g., see Feller [40] p. 503), and

$$\begin{aligned} B &= C \int_0^\infty (\sin(tx) - t \sin(x)) x^{-2} dx \\ &= C \lim_{\varepsilon \to 0} \int_\varepsilon^\infty \Big(\frac{\sin(tx)}{x^2} - \frac{t \sin(x)}{x^2} \Big) dx \\ &= C \lim_{\varepsilon \to 0} \Big(\int_{t\varepsilon}^\infty \Big(\frac{t \sin(x)}{x^2} \Big) dx - \int_\varepsilon^\infty \Big(\frac{t \sin(x)}{x^2} \Big) dx \\ &= C \lim_{\varepsilon \to 0} \int_{t\varepsilon}^\varepsilon \Big(\frac{t \sin(x)}{x^2} \Big) dx \\ &= C t \lim_{\varepsilon \to 0} \int_t^1 \Big(\frac{\sin(\varepsilon x)}{\varepsilon x} \Big) \frac{dx}{x} \\ &= -C t \ln t \end{aligned} \tag{7.35}$$

assuming that $0 < t \leq 1$, since $\sin(\varepsilon x)/(\varepsilon x) \to 1$ as $\varepsilon \to 0$ uniformly on $t \leq x \leq 1$. If $t > 1$, a similar calculation yields $B = -Ct \ln t$. If $t < 0$, a simple substitution in (7.34) and (7.35) yields $A = (-\pi/2)C|t|$ and $B = C|t| \ln |t|$. If $t = 0$, then of course $A = B = 0$. Then for all $t \in \mathbb{R}^1$ we can write

$$\begin{aligned} A + iB &= -\frac{\pi}{2} C|t| - i \operatorname{sign}(t) C|t| \ln |t| \\ &= -\frac{\pi}{2} C|t| \Big(1 + i \operatorname{sign}(t) \Big(\frac{2}{\pi} \Big) \ln |t| \Big) \end{aligned} \tag{7.36}$$

where we use the convention that $|t| \ln |t| = 0$ when $t = 0$. Hence $\psi(t) = A + iB$ is of the form (7.22) with $a = 0$, $\beta = 1$ and $\sigma > 0$ satisfying (7.23). □

Lemma 7.3.10. *Suppose that ν is stable with index $0 < \alpha < 1$ and Lévy representation $[a, 0, \phi]$ where ϕ is given by (7.19). Then ν has a $S_\alpha(\sigma, \beta, b)$ distribution with σ given by (7.21), $\beta = p - q$ and*

$$b = a - \int_{x \neq 0} \left(\frac{x}{1 + x^2} \right) \phi\{dx\}. \tag{7.37}$$

Proof. If ν is stable with index $0 < \alpha < 1$ and Lévy representation $[a, 0, \phi]$, where ϕ is given by (7.19), then we can write $\nu = \nu_1 * \nu_2 * \varepsilon_a$, where ν_1 has characteristic function $\exp(\psi_1(t))$ with

$$\psi_1(t) = \int_0^\infty \left(e^{itx} - 1 - \frac{itx}{1 + x^2} \right) \phi\{dx\} \tag{7.38}$$

and ν_2 has characteristic function $\exp(\psi_2(t))$ with

$$\begin{aligned} \psi_2(t) &= \int_{-\infty}^0 \left(e^{itx} - 1 - \frac{itx}{1 + x^2} \right) \phi\{dx\} \\ &= -\int_0^\infty \left(e^{-itx} - 1 + \frac{itx}{1 + x^2} \right) \phi\{-dx\}. \end{aligned} \tag{7.39}$$

The log-characteristic function $\psi_1(t) = ia_1 t + \psi(t)$ with $\psi(t)$ given by (7.24) with $\phi(r, \infty) = pCr^{-\alpha}$ and

$$a_1 = -\int_0^\infty \left(\frac{x}{1 + x^2} \right) \phi\{dx\} \tag{7.40}$$

where the integral in (7.40) converges since $0 < \alpha < 1$. Then Lemma 7.3.7 implies that the probability distribution with log-characteristic function $\psi(t)$ is $S_\alpha(\sigma_1, 1, 0)$, where $\sigma_1^\alpha = p\sigma^\alpha$ and σ is given by (7.21). Hence

$$\psi_1(t) = ia_1 t - p\sigma^\alpha |t|^\alpha \left(1 - i \operatorname{sign}(t) \tan\left(\frac{\pi\alpha}{2} \right) \right). \tag{7.41}$$

Write $\psi_2(t) = ia_2 t - \psi(-t)$, where

$$a_2 = -\int_0^\infty \left(\frac{x}{1 + x^2} \right) \phi\{-dx\} = -\int_{-\infty}^0 \left(\frac{x}{1 + x^2} \right) \phi\{dx\} \tag{7.42}$$

and

$$\psi(t) = \int_0^\infty (e^{itx} - 1) \phi\{-dx\}.$$

Since the Lévy measure $\phi\{-dx\}$ assigns measure $qCr^{-\alpha}$ to the interval (r, ∞), Lemma 7.3.7 implies that the probability distribution with log-characteristic function $\psi(t)$ is $S_\alpha(\sigma_2, 1, 0)$, where $\sigma_2^\alpha = q\sigma^\alpha$ and σ is given by (7.21). Then

$$\psi_2(t) = ia_2 t - \psi(-t) = ia_2 t - q\sigma^\alpha |t|^\alpha \left(1 + i \operatorname{sign}(t) \tan\left(\frac{\pi\alpha}{2} \right) \right). \tag{7.43}$$

Since $\nu = \nu_1 * \nu_2 * \varepsilon_a$ decomposes ν into three independent components, we can add the log-characteristic functions of each component to conclude that ν has characteristic function $\exp(\psi(t))$ where

$$\begin{aligned}\psi(t) &= iat + \psi_1(t) + \psi_2(t)\\ &= iat + ia_1 t - p\sigma^\alpha |t|^\alpha \left(1 - i\,\mathrm{sign}(t)\tan\left(\frac{\pi\alpha}{2}\right)\right)\\ &\qquad + ia_2 t - q\sigma^\alpha |t|^\alpha \left(1 + i\,\mathrm{sign}(t)\tan\left(\frac{\pi\alpha}{2}\right)\right)\\ &= i(a + a_1 + a_2)t\\ &\qquad - \sigma^\alpha |t|^\alpha \left((p+q) - i(p-q)\,\mathrm{sign}(t)\tan\left(\frac{\pi\alpha}{2}\right)\right)\end{aligned}$$

which shows that ν has a $S_\alpha(\sigma, \beta, b)$ distribution with σ given by (7.21), $\beta = p - q$ and $b = a + a_1 + a_2$ satisfies (7.37) □

Lemma 7.3.11. *Suppose that ν is stable with index $1 < \alpha < 2$ and Lévy representation $[a, 0, \phi]$, where ϕ is given by (7.19). Then ν has a $S_\alpha(\sigma, \beta, b)$ distribution with σ given by (7.21), $\beta = p - q$ and*

$$b = a + \int_{x\neq 0} \left(x - \frac{x}{1+x^2}\right)\phi\{dx\}. \tag{7.44}$$

Proof. If ν is stable with index $1 < \alpha < 2$ and Lévy representation $[a, 0, \phi]$, where ϕ is given by (7.19), then we can write $\nu = \nu_1 * \nu_2 * \varepsilon_a$ where ν_i has characteristic function $\exp(\psi_i(t))$ with ψ_i given by (7.38) and (7.39) for $i = 1, 2$ respectively. The log-characteristic function $\psi_1(t) = ia_1 t + \psi(t)$ with $\psi(t)$ given by (7.29) with $\phi(r, \infty) = pCr^{-\alpha}$ and

$$a_1 = \int_0^\infty \left(x - \frac{x}{1+x^2}\right)\phi\{dx\} \tag{7.45}$$

where the integral in (7.45) converges since $1 < \alpha < 2$. Lemma 7.3.8 implies that the probability distribution with log-characteristic function $\psi(t)$ is $S_\alpha(\sigma_1, 1, 0)$, where $\sigma_1^\alpha = p\sigma^\alpha$ and σ is given by (7.21). Hence (7.41) holds once again with a_1 given by (7.45) in this case.

Now write $\psi_2(t) = ia_2 t - \psi(-t)$ where

$$a_2 = \int_0^\infty \left(x - \frac{x}{1+x^2}\right)\phi\{-dx\} = \int_{-\infty}^0 \left(x - \frac{x}{1+x^2}\right)\phi\{dx\} \tag{7.46}$$

and

$$\psi(t) = \int_0^\infty (e^{itx} - 1 - itx)\phi\{-dx\}.$$

Since the Lévy measure $\phi\{-dx\}$ assigns measure $qCr^{-\alpha}$ to the interval (r, ∞), Lemma 7.3.8 implies that the probability distribution with log-characteristic function $\psi(t)$ is $S_\alpha(\sigma_2, 1, 0)$, where $\sigma_2^\alpha = q\sigma^\alpha$ and σ is given by (7.21). Hence (7.43) holds once again with a_2 given by (7.46) in this case.

Now, it follows exactly as in Lemma 7.3.10 that ν has a $S_\alpha(\sigma, \beta, b)$ distribution with σ given by (7.21), $\beta = p - q$ and $b = a + a_1 + a_2$ satisfying (7.44). □

Lemma 7.3.12. *Suppose that ν is stable with index $\alpha = 1$ and Lévy representation $[a, 0, \phi]$, where ϕ is given by (7.19). Then ν has a $S_\alpha(\sigma, \beta, b)$ distribution with σ given by (7.23), $\beta = p - q$ and*

$$b = a + \int_{x \neq 0} \left(\sin x - \frac{x}{1 + x^2} \right) \phi\{dx\}. \tag{7.47}$$

Proof. If ν is stable with index $\alpha = 1$ and Lévy representation $[a, 0, \phi]$ where ϕ is given by (7.19), then we can write $\nu = \nu_1 * \nu_2 * \varepsilon_a$, where ν_i has characteristic function $\exp(\psi_i(t))$ with ψ_i given by (7.38) and (7.39) for $i = 1, 2$ respectively. The log-characteristic function $\psi_1(t) = ia_1 t + \psi(t)$ with $\psi(t)$ given by (7.29) with $\phi(r, \infty) = pCr^{-\alpha}$ and

$$a_1 = \int_0^\infty \left(\sin x - \frac{x}{1 + x^2} \right) \phi\{dx\}. \tag{7.48}$$

Lemma 7.3.9 implies that the probability distribution with log-characteristic function $\psi(t)$ is $S_1(\sigma_1, 1, 0)$, where $\sigma_1^\alpha = p\sigma^\alpha$ and σ is given by (7.23). Hence

$$\psi_1(t) = ia_1 t - p\sigma^\alpha |t|^\alpha \left(1 + i \left(\frac{2}{\pi} \right) \operatorname{sign}(t) \ln |t| \right). \tag{7.49}$$

Now, write $\psi_2(t) = ia_2 t - \psi(-t)$, where

$$a_2 = \int_0^\infty \left(\sin x - \frac{x}{1 + x^2} \right) \phi\{-dx\} = \int_{-\infty}^0 \left(\sin x - \frac{x}{1 + x^2} \right) \phi\{dx\} \tag{7.50}$$

and

$$\psi(t) = \int_0^\infty (e^{itx} - 1 - it \sin x) \phi\{-dx\}.$$

Since the Lévy measure $\phi\{-dx\}$ assigns measure $qCr^{-\alpha}$ to the interval (r, ∞), Lemma 7.3.9 implies that the probability distribution with log-characteristic function $\psi(t)$ is $S_\alpha(\sigma_2, 1, 0)$, where $\sigma_2^\alpha = q\sigma^\alpha$ and σ is given by (7.23). Hence

$$\psi_2(t) = ia_2 t - q\sigma^\alpha |t|^\alpha \left(1 - i \left(\frac{2}{\pi} \right) \operatorname{sign}(t) \ln |t| \right). \tag{7.51}$$

Now it follows exactly as in Lemma 7.3.10 that ν has a $S_\alpha(\sigma, \beta, b)$ distribution with σ given by (7.21), $\beta = p - q$ and $b = a + a_1 + a_2$ satisfying (7.47). □

Proof of Theorem 7.3.5. If $0 < \alpha < 1$, then the result follows from Lemma 7.3.10. If $1 < \alpha < 2$ use Lemma 7.3.11, and if $\alpha = 1$ use Lemma 7.3.12. In the remaining case, when $\alpha = 2$, ν is normal with some mean a and variance b. Corollary 7.3.4 (a) shows that ν has Lévy representation $[a, Q, 0]$, where $Q(t) = bt^2$ for all $t \in \mathbb{R}^1$ and ϕ is the zero measure. Thus the log-characteristic function of ν is

$$\psi(t) = iat - \frac{1}{2}bt^2$$

which is of the form (7.20) with $\alpha = 2$, $\sigma^2 = b/2$, and β is arbitrary since $\tan(\pi\alpha/2) = 0$. □

Remark 7.3.13. In this remark, we indicate the meaning of the four parameters of a stable distribution. More details can be found in Samorodnitsky and Taqqu [161]. Since $\nu^t = t^{1/\alpha}\nu * \varepsilon_{a_t}$ for all $t > 0$, it is clear that in some sense the stable index α governs the tails of the distribution ν. In Chapter 8, we will make this more explicit, and we will see that α also governs the tails of any law $\mu \in \text{DOA}(\nu)$ (see Theorem 8.2.14). The stable index α also governs the moments of ν (see Remark 7.3.21). The skewness governs the deviations of the distribution from symmetry. A stable law with distribution $S_\alpha(\sigma, 0, 0)$ is symmetric since its characteristic function $\exp(-\sigma^\alpha|t|^\alpha)$ is real valued. The scale σ and the center b have the usual meaning that if X has a $S_\alpha(1, \beta, 0)$ distribution then $\sigma X + b$ has a $S_\alpha(\sigma, \beta, b)$ distribution, except that for $\alpha = 1$ and $\beta \neq 0$ multiplication by σ introduces a nonlinear change in the shift.

Remark 7.3.14. Feller [40] and Zolotarev [199] use different parameterizations of the family of stable laws on $\mathbb{R}^1$, but by now the parameterization of Samorodnitsky and Taqqu [161] that we use in Theorem 7.3.5 has become standard. Another useful alternative parameterization is to use the *dispersion* C rather than the scale σ. These two parameters are related by (7.21) when $\alpha \neq 1$ and (7.23) when $\alpha = 1$ so that both range over the positive reals. The dispersion is related directly to the Lévy measure, and as we shall see in Chapter 8, is also convenient in limit theorems.

Remark 7.3.15. In some applications, it may be necessary to explicitly compute the centering constants b given in (7.37), (7.44), and (7.47). In fact, if ϕ is given by (7.19) with $\alpha = 1$, then one can compute using standard integral forms that the centering in (7.47) reduces to

$$b = a + C\beta(1 - \gamma) \tag{7.52}$$

where $\gamma \approx 0.5772$ is Euler's constant. The other cases may also be explicitly computed in terms of the Beta function $B(x, y) = \Gamma(x)\Gamma(x)/\Gamma(x + y)$.

Theorem 7.3.16. *Suppose that ν is a full nonnormal stable law on $\mathbb{R}^d$ with index α and Lévy measure ϕ.*

(a) *If* $\alpha \neq 1$*, then* ν *has characteristic function* $\exp(\psi(t))$ *where*

$$\psi(t) = i\langle a, t\rangle - \int_{\|\theta\|=1} |\langle\theta, t\rangle|^{\alpha} \left(1 - i\,\mathrm{sign}(\langle\theta, t\rangle) \tan\left(\frac{\pi\alpha}{2}\right)\right) \Lambda\{d\theta\} \quad (7.53)$$

for some $a \in \mathbb{R}^d$*, where*

$$\Lambda(D) = \sigma^{\alpha} \cdot \phi\{r\theta : \theta \in D; r > 1\} \quad (7.54)$$

is a finite Borel measure on the unit sphere in $\mathbb{R}^d$*, with* σ *given by* (7.21) *with* $C = 1$.

(b) *If* $\alpha = 1$*, then* ν *has characteristic function* $\exp(\psi(t))$ *where*

$$\psi(t) = i\langle a, t\rangle - \int_{\|\theta\|=1} |\langle\theta, t\rangle| \left(1 + i\left(\frac{2}{\pi}\right) \mathrm{sign}(\langle\theta, t\rangle) \ln|\langle\theta, t\rangle|\right) \Lambda\{d\theta\} \quad (7.55)$$

for some $a \in \mathbb{R}^d$*. where* Λ *is a finite Borel measure on the unit sphere in* $\mathbb{R}^d$ *defined by* (7.54) *with* σ *given by* (7.23) *with* $C = 1$.

If ν *is a full normal law with mean a and covariance matrix B, then* ν *has characteristic function* $\exp(\psi(t))$*, where* ψ *is given by* (7.53) *with* $\alpha = 2$ *and*

$$B = \int_{\|\theta\|=1} \theta\theta' \Lambda\{d\theta\} \quad (7.56)$$

where θ' *is the transpose of the column vector* θ *and* Λ *is a finite Borel measure on the unit sphere of* $\mathbb{R}^d$*. The entire class of full stable laws on* $\mathbb{R}^d$ *is given by these formulas with index* $\alpha \in (0, 2]$*, center* $a \in (-\infty, \infty)$ *and spectral measure* Λ *an arbitrary finite Borel measure on the unit sphere in* $\mathbb{R}^d$*. We say that* ν *has distribution* $S_{\alpha}(\Lambda, a)$ *in this case.*

The proof follows quite easily from the corresponding one variable results, using the polar representation of the Lévy measure given in Theorem 7.3.3. For the sake of clarity, we will begin with a few simple lemmas.

Lemma 7.3.17. *Suppose* ν *is a full stable law on* $\mathbb{R}^d$ *with index* $0 < \alpha < 1$ *and Lévy representation* $[a, 0, \phi]$ *where* $\phi\{dr, d\theta\} = \phi_{\alpha}\{dr\}\lambda\{d\theta\}$ *as in Theorem 7.3.3 (b). Then* ν *has a* $S_{\alpha}(\Lambda, b)$ *distribution where* $\Lambda\{d\theta\} = \sigma^{\alpha} \cdot \lambda\{d\theta\}$ *with* σ *given by* (7.21) *and*

$$b = a - \int_{x\neq 0} \left(\frac{x}{1 + \|x\|^2}\right) \phi\{dx\}. \quad (7.57)$$

Proof. Suppose that ν is a full stable law on $\mathbb{R}^d$ with index $0 < \alpha < 1$ and Lévy representation $[a, 0, \phi]$, where $\phi\{dr, d\theta\} = \phi_\alpha\{dr\}\lambda\{d\theta\}$ with $\phi_\alpha(r, \infty) = r^{-\alpha}$ for all $r > 0$ and λ a finite Borel measure on the unit sphere in $\mathbb{R}^d$ satisfying (7.18). Then ν has characteristic function $\exp(\psi(t))$, where

$$\begin{aligned}\psi(t) &= i{<}a,t{>} + \int_{x\neq 0}\left(e^{i<t,x>} - 1 - \frac{i{<}t,x{>}}{1+\|x\|^2}\right)\phi\{dx\} \\ &= i{<}a,t{>} + \int_{\|\theta\|=1}\int_0^\infty \left(e^{ir<t,\theta>} - 1 - \frac{ir{<}t,\theta{>}}{1+r^2}\right)\phi_\alpha\{dr\}\lambda\{d\theta\} \\ &= i{<}a,t{>} + \int_{\|\theta\|=1} \psi_\alpha({<}t,\theta{>})\lambda\{d\theta\}\end{aligned} \tag{7.58}$$

where

$$\psi_\alpha(s) = \int_0^\infty \left(e^{irs} - 1 - \frac{irs}{1+r^2}\right)\phi_\alpha\{dr\} \tag{7.59}$$

is the log-characteristic function of a stable law on $\mathbb{R}^1$ with Lévy representation $[0, 0, \phi_\alpha]$. Since ϕ_α is of the form (7.19) with $C = 1$ and $q = 0$, it follows from Lemma 7.3.10 with $\beta = 1$ that

$$\psi_\alpha(s) = ibs - \sigma^\alpha |s|^\alpha \left(1 - i\,\mathrm{sign}(s)\tan\left(\frac{\pi\alpha}{2}\right)\right) \tag{7.60}$$

where b satisfies (7.37) and σ given by (7.21). Then

$$\begin{aligned}\psi(t) &= i{<}a,t{>} + \int_{\|\theta\|=1} \psi_\alpha({<}t,\theta{>})\lambda\{d\theta\} \\ &= i{<}a,t{>} + \int_{\|\theta\|=1} ib{<}t,\theta{>}\lambda\{d\theta\} \\ &\quad - \int_{\|\theta\|=1} |{<}t,\theta{>}|^\alpha \left(1 - i\,\mathrm{sign}({<}t,\theta{>})\tan\left(\frac{\pi\alpha}{2}\right)\right)\sigma^\alpha\lambda\{d\theta\}\end{aligned} \tag{7.61}$$

which shows that ν has a $S_\alpha(\Lambda, c)$ distribution with $\Lambda\{d\theta\} = \sigma^\alpha \cdot \lambda\{d\theta\}$ and

$$\begin{aligned}c &= a + \int_{\|\theta\|=1} b\theta\lambda\{d\theta\} \\ &= a - \int_{\|\theta\|=1}\int_0^\infty \left(\frac{r\theta}{1+r^2}\right)\phi_\alpha\{dr\}\lambda\{d\theta\} \\ &= a - \int_{x\neq 0}\left(\frac{x}{1+\|x\|^2}\right)\phi\{dx\}\end{aligned} \tag{7.62}$$

so that (7.57) holds. □

Lemma 7.3.18. *Suppose ν is a full stable law on $\mathbb{R}^d$ with index $1 < \alpha < 2$ and Lévy representation $[a, 0, \phi]$ where $\phi\{dr, d\theta\} = \phi_\alpha\{dr\}\lambda\{d\theta\}$ as in Theorem 7.3.3 (b). Then ν has a $S_\alpha(\Lambda, b)$ distribution where $\Lambda\{d\theta\} = \sigma^\alpha \cdot \lambda\{d\theta\}$ with σ given by* (7.21) *and*

$$b = a + \int_{x \neq 0} \left(x - \frac{x}{1 + \|x\|^2} \right) \phi\{dx\}. \tag{7.63}$$

Proof. The proof is similar to Lemma 7.3.18. Again ν has characteristic function $\exp(\psi(t))$ where $\psi(t)$ is given by (7.58) with $\psi_\alpha(s)$ defined by (7.59). Hence $\psi_\alpha(s)$ is the log-characteristic function of a stable law on $\mathbb{R}^1$ with Lévy representation $[0, 0, \phi_\alpha]$. Since ϕ_α is of the form (7.19) with $C = 1$ and $q = 0$, it follows from Lemma 7.3.10 with $\beta = 1$ that (7.60) holds where b satisfies (7.44) and σ is given by (7.21). Then it follows as before that ν has a $S_\alpha(\Lambda, c)$ distribution with $\Lambda\{d\theta\} = \sigma^\alpha \cdot \lambda\{d\theta\}$ and

$$\begin{aligned} c &= a + \int_{\|\theta\|=1} b\theta\lambda\{d\theta\} \\ &= a + \int_{\|\theta\|=1} \int_0^\infty \left(r\theta - \frac{r\theta}{1 + r^2} \right) \phi_\alpha\{dr\}\lambda\{d\theta\} \end{aligned} \tag{7.64}$$

so that (7.63) holds. □

Lemma 7.3.19. *Suppose ν is a full stable law on $\mathbb{R}^d$ with index $\alpha = 1$ and Lévy representation $[a, 0, \phi]$, where $\phi\{dr, d\theta\} = \phi_\alpha\{dr\}\lambda\{d\theta\}$ as in Theorem 7.3.3 (b). Then ν has a $S_\alpha(\Lambda, b)$ distribution, where $\Lambda\{d\theta\} = \sigma^\alpha \cdot \lambda\{d\theta\}$ with σ given by* (7.23) *and*

$$b = a + \int_{x \neq 0} \left(\frac{x}{\|x\|} \sin(\|x\|) - \frac{x}{1 + \|x\|^2} \right) \phi\{dx\}. \tag{7.65}$$

Proof. The proof is identical to Lemma 7.3.18, except that now b satisfies (7.47) and σ is given by (7.23). Then ν has a $S_\alpha(\Lambda, c)$ distribution with $\Lambda\{d\theta\} = \sigma^\alpha \cdot \lambda\{d\theta\}$ and

$$\begin{aligned} c &= a + \int_{\|\theta\|=1} b\theta\lambda\{d\theta\} \\ &= a + \int_{\|\theta\|=1} \int_0^\infty \left(\theta \sin(r) - \frac{r\theta}{1 + r^2} \right) \phi_\alpha\{dr\}\lambda\{d\theta\} \end{aligned} \tag{7.66}$$

so that (7.65) holds. □

Proof of Theorem 7.3.16. If $0 < \alpha < 1$, then result follows from Lemma 7.3.17. If $1 < \alpha < 2$ use Lemma 7.3.18, and if $\alpha = 1$ use Lemma 7.3.19. In the remaining case when $\alpha = 2$, ν is normal with some mean a and covariance matrix B. In this case, Theorem 7.3.3 (a) shows that ν has Lévy representation $[a, Q, 0]$, where $Q(t) = <t, Bt>$ for all $t \in \mathbb{R}^d$ and ϕ is the zero measure, and (7.53) reduces to

$$\psi(t) = i<a,t> - \int_{\|\theta\|=1} |<\theta, t>|^2 \Lambda\{d\theta\} \tag{7.67}$$

since $\tan(\pi\alpha/2) = 0$. If (7.56) holds, then

$$\begin{aligned} <t, Bt> = t'Bt &= \int_{\|\theta\|=1} t'\theta\theta' t \Lambda\{d\theta\} \\ &= \int_{\|\theta\|=1} |<t, \theta>|^2 \Lambda\{d\theta\} \end{aligned}$$

so that the log-characteristic function of ν is of the form (7.67). For example, if we write $B = (b_{ij})$ with respect to some basis $e_1, \ldots, e_d$ for $\mathbb{R}^d$, then we can take Λ discrete with $\Lambda\{e_i e_j'\} = b_{ij}$. □

Remark 7.3.20. If $\alpha < 2$, then there exists a unique $S_\alpha(\Lambda, b)$ distribution for each pair (Λ, b). If $\alpha = 2$, then many different measures Λ give the same covariance matrix B in (7.56), so the classical parameterization is preferred.

Remark 7.3.21. If Y is a full random vector on $\mathbb{R}^d$ whose probability distribution ν is stable with index α, then Y is also operator stable with exponent $E = \alpha^{-1} I$. Then it follows from Remark 7.2.10 that $E\|Y\|^\rho$ exists for $0 < \rho < \alpha$ and if $\alpha < 2$, then $E\|Y\|^\rho = \infty$ for $\rho \geq \alpha$. If Z is a stable random variable on $\mathbb{R}^1$ with index α then $E|Z|^\rho$ exists for $0 < \rho < \alpha$ and if $\alpha < 2$, then $E|Z|^\rho = \infty$ for $\rho \geq \alpha$. Since projection onto any lower dimensional subspace commutes with the exponent E, it follows immediately from (7.8) and Definition 7.3.1 that $Z = <Y, \theta>$ is stable with index α for any nonzero $\theta \in \mathbb{R}^d$.

7.4 SEMISTABLE LAWS

After considering the special case of stable laws in Section 7.3, we now investigate the case of semistable limit laws. As for stable laws, in this case we can say more about the limit.

Definition 7.4.1. Suppose that $X, X_1, X_2, X_3, \ldots$ are independent and identically distributed random vectors on $\mathbb{R}^d$ with common distribution μ and that Y is a random vector whose distribution ν is full.

(a) We say that ν (or Y) is (b, c) *semistable* if ν is (B, c) operator semistable for some $c > 1$ and $B = bI$ for some real number b. We will also call

$$\alpha = \frac{\log c}{\log b} \tag{7.68}$$

the *index* of ν.

(b) We say that X (or equivalently μ) belongs to the *domain of semistable attraction* of Y (or ν) if there exist nonrandom real numbers $a_n > 0$, an increasing sequence of positive integers $k_n \to \infty$ with

$$k_{n+1}/k_n \to c \geq 1 \tag{7.69}$$

and nonrandom vectors $s_n \in \mathbb{R}^d$ such that

$$a_n(X_1 + \cdots + X_{k_n}) + s_n \Rightarrow Y. \tag{7.70}$$

In terms of measures we can rewrite (7.70) in the form

$$a_n \mu^{k_n} * \varepsilon_{s_n} \Rightarrow \nu. \tag{7.71}$$

We write $X \in \mathrm{DOSA}(Y, c)$ (or $\mu \in \mathrm{DOSA}(\nu, c)$) in this case.

Remark 7.4.2. In Chapter 8, we will prove that a full probability measure ν on $\mathbb{R}^d$ is (b, c) semistable if and only if $\mathrm{DOSA}(\nu, c)$ is not empty (see Proposition 8.3.16).

Theorem 7.4.3. *If a full probability measure ν on $\mathbb{R}^d$ is (b, c) semistable, then ν is a full infinitely divisible law on $\mathbb{R}^d$ with Lévy representation $[m, Q, \phi]$ and $b^2 \geq c$ so that $0 < \alpha \leq 2$. Also ν is (c^E, c) operator semistable with $E = aI$ where $b = c^a$, so that $a = 1/\alpha$.*

(a) If $\alpha = 2$, then ν is normal with mean $m \in \mathbb{R}^d$ and covariance matrix B, $Q(t) = \langle t, Bt\rangle$ for all $t \in \mathbb{R}^d$, and ϕ is the zero measure.

(b) If $\alpha < 2$, then ν is (b, c) semistable with index α given by (7.68), $Q(t) = 0$ for all $t \in \mathbb{R}^d$, and $c \cdot \phi = b\phi$. In this case,

$$\phi\{t\theta : t > r, \theta \in D\} = \int_{\theta \in D} r^{-\alpha} K_\theta(\alpha \log r) \lambda\{d\theta\} \tag{7.72}$$

in polar coordinates $x = r\theta$ for $r > 0$ and $\|\theta\| = 1$, where

$$\lambda(D) = \phi\{t\theta : \theta \in D; t > 1\} \tag{7.73}$$

is a finite Borel measure on the unit sphere $S = \{x \in \mathbb{R}^d : \|x\| = 1\}$ and $K_\theta(s)$ is a nonnegative $\log c$*-periodic function such that*

$$K_\theta(t + h)e^{-h} \leq K_\theta(t - h)e^h \tag{7.74}$$

for all $t \in \mathbb{R}$, $\theta \in S$ and $h > 0$.

Proof. If ν is (b, c) semistable, then Theorem 7.1.1 shows that $b^2 \geq c$ and that ν is normal if $b^2 = c$ and has no normal component if $b^2 > c$. In other words, the index $\alpha \in (0, 2]$, ν is normal if $\alpha = 2$ and if $\alpha < 2$, then ν has no normal component. If $\alpha = 2$, then ν is normal so $E = (1/2)I$ is an exponent of ν and hence ν is (c^E, c) operator semistable. Then part (a) follows immediately from the Lévy representation of Theorem 3.1.11. If $\alpha < 2$, then Lemma 7.1.6 implies that $c \cdot \phi = b\phi$. Since $\nu^c = b\nu * \varepsilon_s$ for some $s \in \mathbb{R}^d$, Lemma 7.1.6 also implies that ν is (c^E, c) operator semistable where $E = aI$ so that $b = c^a$. Then Corollary 7.1.14 shows that

$$\begin{aligned} \phi\{t\theta : t > r, \theta \in D\} &= \int_S \int_0^\infty I(t^a > r, \theta \in D) H_\theta(-dt)\lambda\{d\theta\} \\ &= \int_S H_\theta(r^\alpha)\lambda\{d\theta\} \end{aligned} \tag{7.75}$$

where λ is a finite Borel measure on the unit sphere S defined by (7.73) and $H_\theta(s) = K_\theta(\log s)/s$ for some nonnegative $\log c$-periodic function K_θ such that (7.74) holds. Then $H_\theta(r^\alpha) = r^{-\alpha} K_\theta(\alpha \log r)$, which completes the proof of part (b). □

Corollary 7.4.4. *If a nondegenerate probability measure ν on $\mathbb{R}^1$ is (b, c) semistable with index α, then ν is a nondegenerate infinitely divisible law on $\mathbb{R}^1$ with Lévy representation $[m, Q, \phi]$ and $0 < \alpha \le 2$.*

(a) *If $\alpha = 2$, then ν is normal with mean $m \in \mathbb{R}^1$ and variance $b \ge 0$, $Q(t) = bt^2$ for all $t \in \mathbb{R}^1$, and ϕ is the zero measure.*

(b) *If $\alpha < 2$, then ν is (b, c) semistable with index α, $Q(t) = 0$ for all $t \in \mathbb{R}^1$, and ϕ is of the form*

$$\begin{aligned} \phi(r, \infty) &= r^{-\alpha}\theta_1(\log r) \\ \phi(-\infty, -r) &= r^{-\alpha}\theta_2(\log r) \\ \phi\{t : |t| > r\} &= r^{-\alpha}\theta(\log r) \end{aligned} \tag{7.76}$$

for every $r > 0$, where $\theta = \theta_1 + \theta_2$, $\theta(r)$ is strictly positive and periodic with period $\log b$, and $\theta_i(r)$ is nonnegative and periodic with period $\log b$. Furthermore, we have

$$\begin{aligned} \theta_1(y + \delta) \le e^{\alpha\delta}\theta_1(y) \quad &\text{and} \quad \theta_1(y - \delta) \ge e^{-\alpha\delta}\theta_1(y) \\ \theta_2(y + \delta) \le e^{\alpha\delta}\theta_2(y) \quad &\text{and} \quad \theta_2(y - \delta) \ge e^{-\alpha\delta}\theta_2(y) \\ \theta(y + \delta) \le e^{\alpha\delta}\theta(y) \quad &\text{and} \quad \theta(y - \delta) \ge e^{-\alpha\delta}\theta(y) \end{aligned} \tag{7.77}$$

for all $y > 0$ and all $\delta > 0$.

Proof. Specialize Theorem 7.4.3 to the case $d = 1$ so that in part (b) we have that $S = \{-1, +1\}$ is the unit sphere in $\mathbb{R}^1$. Then we can take $\theta_1(r) = K_1(\alpha r)\lambda\{1\}$ and $\theta_2(r) = K_{-1}(\alpha r)\lambda\{-1\}$. Then θ_i is periodic with period $\log b$ since $\alpha \log b = \log c$, and so θ is periodic with the same period. The inequalities (7.77) follow from (7.74), or directly from the fact that $r^{-\alpha}\theta_i(\log r)$

is monotone decreasing. Proposition 3.1.20 shows that ϕ is $\mathcal{M}$-full, and so we must have $\theta(y) > 0$ for some $y > 0$, and then it follows from (7.77) that $\theta(y) > 0$ for all $y > 0$. □

Remark 7.4.5. The index α governs the moments of a semistable law. Suppose Y is a full random vector on $\mathbb{R}^d$ whose distribution ν is (b, c) semistable. Then ν is also (c^E, c) operator semistable with $E = \alpha^{-1}I$ and $\alpha = \log c / \log b$. Then Corollary 7.1.21 implies that $E\|Y\|^\rho$ exists for all $0 \leq \rho < \alpha$, and if $\alpha < 2$, then $E\|Y\|^\rho$ is infinite for all $\rho > \alpha$. If Z is a nondegenerate (b, c) semistable random variable on $\mathbb{R}^1$ with index α, then $E|Z|^\rho$ exists for $0 < \rho < \alpha$ and if $\alpha < 2$, then $E|Z|^\rho = \infty$ for $\rho > \alpha$. Since projection onto any lower dimensional subspace commutes with the exponent E, it follows immediately from (7.3) and Definition 7.4.1 that $Z = \langle Y, \theta \rangle$ is (b, c) semistable for any nonzero $\theta \in \mathbb{R}^d$.

7.5 STRUCTURE THEOREMS

A natural question arising in the study of operator semistable laws ν is whether the number $c > 1$ in (7.2) is uniquely determined. We will now characterize the numbers $c > 1$ that can appear. If ν is operator stable, then Lemma 3.3.27 shows that ν is operator semistable for any $c > 1$ and hence by Theorem 7.1.1 we have (7.2) for every $c > 1$. On the other hand, if ν is operator semistable and not operator stable, and if (7.2) holds for some $c > 1$, then it also holds for c^n for any natural number $n \geq 1$. We show below that for properly operator semistable laws ν there exists a smallest $c > 1$ such that ν is (B, c) operator semistable.

We start with some preparatory results.

Definition 7.5.1. If ν is infinitely divisible we define

$$\mathcal{D}(\nu) = \{t > 0 : \exists A \in \mathrm{GL}(\mathbb{R}^d), a \in \mathbb{R}^d \text{ such that } \nu^t = A\nu * \varepsilon_a\}.$$

Lemma 7.5.2. *If ν is a full infinitely divisible measure on $\mathbb{R}^d$, then $\mathcal{D}(\nu)$ is a closed multiplicative subgroup of $\{t > 0\}$.*

Proof. If $s, t \in \mathcal{D}(\nu)$, then we have $\nu^t = A\nu * \varepsilon_a$ and $\nu^s = B\nu * \varepsilon_b$ for some $A, B \in \mathrm{GL}(\mathbb{R}^d)$ and $a, b \in \mathbb{R}^d$. A simple calculation using Propositions 1.1.7 and 3.1.24 shows that $\nu^{st} = (AB)\nu * \varepsilon_{Ab+sa}$, so that $st \in \mathcal{D}(\nu)$, and $\nu^{1/t} = A^{-1}\nu * \varepsilon_{-A^{-1}a/t}$, so that $1/t \in \mathcal{D}(\nu)$. Hence $\mathcal{D}(\nu)$ is a multiplicative subgroup of $\{t > 0\}$. If $t_n \in \mathcal{D}(\nu)$ and $t_n \to t > 0$ then $\nu^{t_n} = A_n\nu * \varepsilon_{a_n} \Rightarrow \nu^t$ for some $A_n \in \mathrm{GL}(\mathbb{R}^d)$ and $a_n \in \mathbb{R}^d$. Trivially, $\nu \Rightarrow \nu$ and hence by convergence of types, Theorem 2.3.17, $\nu^t = A\nu * \varepsilon_a$ for some $A \in \mathrm{GL}(\mathbb{R}^d)$ and $a \in \mathbb{R}^d$. This shows that $t \in \mathcal{D}(\nu)$ so that $\mathcal{D}(\nu)$ is closed in $\{t > 0\}$. □

Definition 7.5.3. An operator semistable law ν on $\mathbb{R}^d$ is *properly operator semistable* if it is not operator stable.

Theorem 7.5.4. *Let ν be a full infinitely divisible measure on $\mathbb{R}^d$. Then*

(a) *ν is operator stable if and only if $\mathcal{D}(\nu) = \{t > 0\}$.*

(b) *ν is properly operator semistable if and only if there exists a $c > 1$ such that $\mathcal{D}(\nu) = \{c^n : n \in \mathbb{Z}\}$. Then this is the smallest $c > 1$ such that (7.2) holds.*

Proof. (a) If ν is operator stable, then Theorem 7.2.1 shows that $\nu^t = t^E \nu * \varepsilon_{a_t}$ for $t > 0$ so that $\mathcal{D}(\nu) = \{t > 0\}$. Conversely, if $\mathcal{D}(\nu) = \{t > 0\}$, then $\nu^n = A_n \nu * \varepsilon_{a_n}$, so for some shifts $a'_n \in \mathbb{R}^d$ we have $A_n^{-1}\nu^n * \varepsilon_{a'_n} \Rightarrow \nu$, which shows that ν is operator stable.

(b) If ν is properly operator semistable, then part (a) implies that $\mathcal{D}(\nu) \neq \{t > 0\}$, and Theorem 7.1.1 implies that $\mathcal{D}(\nu) \neq \{1\}$. This implies that there exists some (uniquely determined) $c > 1$ such that $\mathcal{D}(\nu) = \{c^n : n \in \mathbb{Z}\}$. To prove this, consider the closed subgroup $G = \{\log x : x \in \mathcal{D}(\nu)\}$ of the additive group of real numbers. If for some $g > 0$ we have $|h| \geq g$ for all $h \in G$ with $h \neq 0$, then take $c = e^g$. Otherwise for each $\varepsilon > 0$ there exists some $h > 0$ with $h < \varepsilon$. Then for any $x \in \mathbb{R}^1$ and any $\varepsilon > 0$ there exists some integer n such that $|nh - x| < \varepsilon$. Since G is closed this implies that $x \in G$ and then $\mathcal{D}(\nu) = \{t > 0\}$.

Conversely, if $\mathcal{D}(\nu) = \{c^n : n \in \mathbb{Z}\}$ then (7.2) holds for some $c > 1$, so Theorem 7.1.1 implies that ν is operator semistable, and since $\mathcal{D}(\nu) \neq \{t > 0\}$ we know from part (a) that ν is not operator stable. □

Theorem 7.5.5. *If ν is (B, c) operator semistable for some $c > 1$ and some $B \in \mathrm{GL}(\mathbb{R}^d)$, then $\mathrm{GDOSA}(\nu, c) = \mathrm{GDOSA}(\nu, c^\ell)$ for every natural $\ell \geq 1$.*

Proof. If $\mu \in \mathrm{GDOSA}(\nu, c)$, then there exists an increasing sequence (k_n) with $k_{n+1}/k_n \to c$ such that (3.94) holds for some $A_n \in \mathrm{GL}(\mathbb{R}^d)$ and $b_n \in \mathbb{R}^d$. Let $k'_n = k_{\ell n}$, $A'_n = A_{\ell n}$ and $b'_n = b_{\ell n}$. Then $k'_{n+1}/k'_n \to c^\ell$ and it follows from (3.94) that $A'_n \mu^{k'_n} * \varepsilon_{b'_n} \Rightarrow \nu$, so $\mu \in \mathrm{GDOSA}(\nu, c^\ell)$.

On the other hand, if $\mu \in \mathrm{GDOSA}(\nu, c^\ell)$ for some $\ell \geq 2$, then there exists an increasing sequence (k_n) with $k_{n+1}/k_n \to c^\ell$ such that (3.94) holds for some $A_n \in \mathrm{GL}(\mathbb{R}^d)$ and $b_n \in \mathbb{R}^d$. Furthermore, by Theorem 7.1.1, rewriting (7.2) using Proposition 1.1.7 we get $B^{-1}\nu^c * \varepsilon_{-B^{-1}b} = \nu$ and iterating this equation yields a sequence (s_j) of vectors such that

$$B^{-j}\nu^{c^j} * \varepsilon_{s_j} = \nu \quad \text{for } j \geq 1. \tag{7.78}$$

Define for all $n \in \mathbb{N}$ natural numbers $m(n)$ and $r(n)$ by letting $n = \ell m(n) + r(n)$, where $0 \leq r(n) < \ell$. Put $k'_n = [k_{m(n)} c^{r(n)}]$. Then it is easy to see that $k'_{n+1}/k'_n \to c$. Further let $A'_n = B^{-r(n)} A_{m(n)}$ and choose shifts

$$b'_n = B^{-r(n)} \frac{[k_{m(n)} c^{r(n)}]}{k_{m(n)}} b_{m(n)} + s_{r(n)}.$$

Then, along the subsequence where, say $r(n) = j$ for some fixed $j = 0, \ldots, \ell-1$ we get from Proposition 3.3.7 by using (3.71) with $t = c^j$ that

$$\begin{aligned} A'_n \mu^{k'_n} * \varepsilon_{b'_n} &= B^{-j}\Big(A_{m(n)}\mu^{[c^j k_{m(n)}]} * \varepsilon_{\frac{[k_{m(n)}c^j]}{k_{m(n)}} b_{m(n)}}\Big) * \varepsilon_{s_j} \\ &\Rightarrow B^{-j}\nu^{c^j} * \varepsilon_{s_j} = \nu \end{aligned}$$

Since there are only ℓ subsequences with that property it follows that $\mu \in \mathrm{GDOSA}(\nu, c)$. The proof is now complete. □

Theorem 3.3.19 shows that, without some restriction on the sampling sequence (k_n), the convergence in (7.1) to a limit ν tells us nothing about the probability distribution μ. In the proof of Theorem 3.3.19, the sampling sequence (k_n) satisfies $k_{n+1}/k_n \to \infty$. In Definition 3.3.20, we assume that

$$\lim_{n\to\infty} \frac{k_{n+1}}{k_n} = c \tag{7.79}$$

for some increasing sequence (k_n) of positive integers tending to infinity. Now we show that if (7.1) holds with (7.79) replaced by the weaker condition

$$\limsup_{n\to\infty} \frac{k_{n+1}}{k_n} = \bar{c} < \infty \tag{7.80}$$

then ν is still operator semistable and the measure μ still belongs to the generalized domain of semistable attraction of ν. In this sense, generalized domains of semistable attraction represent the most general situation in which convergence of the centered and normalized partial sums of a sequence of i.i.d. random vectors to some full limit allows us to deduce useful information about the underlying distribution of the summands from that of the limit law.

Theorem 7.5.6. *Let μ, ν be probability measures on $\mathbb{R}^d$ and assume that ν is full. Assume further that there exists an increasing sequence (k_n) of natural numbers tending to infinity such that (7.80) holds and for some sequence (A_n) of linear operators and some sequence (a_n) of shifts we have*

$$A_n \mu^{k_n} * \varepsilon_{a_n} \Rightarrow \nu. \tag{7.81}$$

Then,

(a) *there exist $B \in \mathrm{GL}(\mathbb{R}^d)$ and $1 < c \le \bar{c}$ such that ν is (B, c) operator semistable. Furthermore, if ν is properly operator semistable, then $\bar{c} = c^{\bar{p}}$ for some natural $\bar{p} \ge 1$, where c is the smallest $c > 1$ such that ν is (B, c) operator semistable.*

(b) *if ν is properly operator semistable, then there exists an increasing sequence (k'_n) of natural numbers tending to infinity with $k'_{n+1}/k'_n \to c$,* *where c is the smallest $c > 1$ such that ν is (B, c) operator semistable, and linear operators $A_n' \in \mathrm{GL}(\mathbb{R}^d)$ such that*

$$A_n' \mu^{k_n'} * \varepsilon_{a_n'} \Rightarrow \nu$$

for some shifts $a_n' \in \mathbb{R}^d$. In other words $\mu \in \mathrm{GDOSA}(\nu, c)$.

(c) if ν is operator stable, then $\mu \in \mathrm{GDOA}(\nu)$.

The proof will be preceded by several lemmas.

Lemma 7.5.7. *Suppose that (7.81) holds for some increasing sequence (k_n) such that (7.80) holds, and let (n_j) be a subsequence such that $k_{n_j+1}/k_{n_j} \to \tilde{c}$ for some $1 < \tilde{c} \leq \bar{c}$. Then there exists a $\tilde{B} \in \mathrm{GL}(\mathbb{R}^d)$ such that ν is $(\tilde{B}, \tilde{c})$ operator semistable.*

Proof. Let $\rho_j = A_{n_j} \mu^{k_{n_j+1}}$ and $T_j = A_{n_j+1} A_{n_j}^{-1}$, where (A_n) is the sequence in (7.81). Then we have for suitable shifts b_{n_j} that

$$\rho_j * \varepsilon_{b_{n_j}} = A_{n_j} \mu^{k_{n_j}(k_{n_j+1}/k_{n_j})} * \varepsilon_{b_{n_j}} \Rightarrow \nu^{\tilde{c}}$$

and

$$T_j \rho_j * \varepsilon_{a_{n_j+1}} = A_{n_j+1} \mu^{k_{n_j+1}} * \varepsilon_{a_{n_j+1}} \Rightarrow \nu.$$

Since ν is full, $\nu^{\tilde{c}}$ is also full. Then it follows from convergence of types, Theorem 2.3.17, that there exist a $\tilde{B} \in \mathrm{GL}(\mathbb{R}^d)$ and a vector $\tilde{b} \in \mathbb{R}^d$ such that $\nu^{\tilde{c}} = \tilde{B}\nu * \varepsilon_{\tilde{b}}$. Now Theorem 7.1.1 implies that ν is $(\tilde{B}, \tilde{c})$ operator semistable. □

Lemma 7.5.8. *If (7.81) holds for some increasing sequence (k_n) such that (7.80) holds, and if ν is operator stable, then $\mu \in \mathrm{GDOA}(\nu)$, so part (c) of Theorem 7.5.6 follows.*

Proof. Since ν is operator stable, by Theorem 7.2.1 there exists an exponent E such that $\nu^t = t^E \nu * \varepsilon_{b_t}$ for all $t > 0$. It follows that $\nu = t^{-E} \nu^t * \varepsilon_{s(t)}$, where $s(t) = -t^{-E} b_t$ for all $t > 0$. Note that by Proposition 7.2.3 the mapping $t \mapsto s(t)$ is continuous. Define real numbers t_n and natural $\ell(n)$ by writing $n = t_n k_{\ell(n)}$ if $k_{\ell(n)} \leq n < k_{\ell(n)+1}$. Now let $A_n' = t_n^{-E} A_{\ell(n)}$ and

$$a_n' = \frac{[k_{\ell(n)t_n}]}{k_{\ell(n)}} t_n^{-E} a_{\ell(n)} + s(t_n).$$

Since the sequence (t_n) is relatively compact with limit set contained in $[1, \bar{c}]$, every subsequence contains a further subsequence along which $t_n \to t \in [1, \bar{c}]$. Then along that subsequence we have by Proposition 3.3.7 that

$$\begin{aligned} A_n' \mu^n * \varepsilon_{a_n'} &= t_n^{-E} \left[A_{\ell(n)} \mu^{[k_{\ell(n)} t_n]} * \varepsilon_{([k_{\ell(n)} t_n]/k_{\ell(n)}) a_{\ell(n)}} \right] * \varepsilon_{s(t_n)} \\ &\Rightarrow t^{-E} \nu^t * \varepsilon_{s(t)} = \nu. \end{aligned}$$

Since every subsequence contains a further subsequence with that property, we have shown that $\mu \in \mathrm{GDOA}(\nu)$. □

Lemma 7.5.9. *If* (7.81) *holds for some increasing sequence* (k_n) *such that* (7.80) *holds, and if* ν *is full but not operator stable, then the set of limit points of the sequence* (k_{n+1}/k_n) *is of the form*

$$K = \{c_0, \dots, c_m\} = \{c^{p_0}, \dots, c^{p_m}\} \tag{7.82}$$

where $m \geq 1$ *and* $1 \leq p_0 < \cdots < p_m$ *are natural numbers, and* c *is the smallest* $c > 1$ *such that* ν *is* (B, c) *operator semistable.*

Proof. If (7.81) holds for some ν full and some increasing sequence (k_n) such that (7.80) holds, then Lemma 7.5.7 shows that ν is operator semistable. If ν is not operator stable, then by Theorem 7.5.4 we can chose a smallest $c > 1$ such that ν is (B, c) operator semistable. Furthermore, since by Lemma 7.5.7 ν is $(\tilde{B}, \tilde{c})$ operator semistable for every limit point $\tilde{c}$ of (k_{n+1}/k_n) we get from Theorem 7.5.4 that $\tilde{c} = c^{\tilde{q}}$ for some natural $\tilde{q} \geq 1$. Since $1 \leq \tilde{c} \leq \bar{c}$ it follows that there are only a finite number of limit points. □

Lemma 7.5.10. *Suppose that* (k_n) *is increasing and that* (7.80) *holds. If* (7.81) *holds with* ν *properly operator semistable, so that the set of limit points of the sequence* (k_{n+1}/k_n) *is given by* (7.82), *then there exist a natural* $N_0 \geq 1$ *and disjoint subsequences* $\{n_i^{(r)} : i \in \mathbb{N}\}$, $0 \leq r \leq m$ *such that*

$$\{N_0, N_0 + 1, \dots\} = \bigcup_{r=0}^{m} \{n_i^{(r)} : i \in \mathbb{N}\},$$

and

$$\lim_{i \to \infty} \frac{k_{n_i^{(r)}+1}}{k_{n_i^{(r)}}} = c^{p_r} \quad \textit{for } 0 \leq r \leq m.$$

Proof. Let $K_\varepsilon(x) = (x - \varepsilon, x + \varepsilon)$ and chose $\varepsilon > 0$ such that $K_\varepsilon(c^{p_r}) \cap K_\varepsilon(c^{p_\ell}) = \emptyset$ if $r \neq \ell$. Since all limit points of (k_{n+1}/k_n) belong to K there exists a number $N_0 \geq 1$ such that

$$\frac{k_{n+1}}{k_n} \in \bigcup_{r=0}^{m} K_\varepsilon(c^{p_r})$$

for all $n \geq N_0$. Hence, since the sets $K_\varepsilon(c^{p_r})$ are mutually disjoint there exist subsequences $\{n_i^{(r)} : i \in \mathbb{N}\}$, $0 \leq r \leq m$ forming a partition of $\{N_0, N_0 + 1, \dots\}$ such that for all $0 \leq r \leq m$ and $i \geq 1$ we have $k_{n_i^{(r)}+1}/k_{n_i^{(r)}} \in K_\varepsilon(c^{p_r})$. It follows that $k_{n_i^{(r)}+1}/k_{n_i^{(r)}} \to c^{p_r}$ as $i \to \infty$ for all $0 \leq r \leq m$. □

Proof of Theorem 7.5.6. Since $\bar{c}$ is a limit point of (k_{n+1}/k_n) we have from Lemma 7.5.7 that ν is $(\bar{B}, \bar{c})$ operator semistable for some $\bar{B} \in \mathrm{GL}(\mathbb{R}^d)$. If

$\bar{c} = 1$, then Lemma 7.1.5 implies that ν is operator stable, and there is nothing more to prove. Otherwise $\bar{c} > 1$, and then Theorem 7.5.4 implies that $\bar{c} = c^{\bar{p}}$ for some $\bar{p} \geq 1$, where c is the smallest $c > 1$ such that ν is (B, c) operator semistable for some $B \in \mathrm{GL}(\mathbb{R}^d)$. This finishes the proof of part (a).

If ν is properly operator semistable, then we can apply Lemma 7.5.10 to obtain subsequences $\{n_i^{(r)} : i \in \mathbb{N}\}$ for $0 \leq r \leq m$. To complete the proof, we now construct an increasing sequence k'_n of natural numbers with $k'_{n+1}/k'_n \to c$, linear operators A'_n such that for some shifts a'_n

$$A'_n \mu^{k'_n} * \varepsilon_{a'_n} \Rightarrow \nu. \tag{7.83}$$

Set $k'_0 = k_{n_1^{(0)}}, k'_1 = [ck_{n_1^{(0)}}], \ldots, k'_{p_0-1} = [c^{p_0-1}k_{n_1^{(0)}}]$ and $k'_{p_0} = k_{n_1^{(0)}+1}$. Since by Lemma 7.5.10 the union of these subsequences is equal to $\{N_0, N_0+1, \ldots,\}$ there exist $0 \leq r \leq m$ and $i \geq 1$ such that $n_1^{(0)} + 1 = n_i^{(r)}$. Then $k'_{p_0} = k_{n_1^{(0)}+1} = k_{n_i^{(r)}}$. Now let $k'_{p_0+1} = [ck_{n_i^{(r)}}], \ldots, k'_{p_0+p_r-1} = [c^{p_r-1}k_{n_i^{(r)}}]$ and $k'_{p_0+p_r} = k_{n_i^{(r)}+1}$ and continue inductively. Then it follows from the definition of (k'_n) that $k'_{n+1}/k'_n \to c$.

Analogously, let $A'_0 = A_{n_1^{(0)}}, A'_1 = B^{-1}A_{n_1^{(0)}}, \ldots, A'_{p_0-1} = B^{-(p_0-1)}A_{n_1^{(0)}}$ and $A'_{p_0} = A_{n_1^{(0)}+1} = A_{n_j^{(i)}}$ and so on. Then we have $k'_\ell = [c^j k_{n_i^{(r)}}]$ and $A'_\ell = B^{-j}A_{n_i^{(r)}}$ for some $i = i_\ell$, $j = j_\ell$ and $r = r_\ell$ with $0 \leq r \leq m$ and $0 \leq j \leq \max\{p_0, \ldots, p_m\} - 1$. Define shifts $a'_\ell = \tilde{a}_\ell + s_j$ where

$$\tilde{a}_\ell = B^{-j} \frac{[c^j k_{n_i^{(r)}}]}{k_{n_i^{(r)}}} a_{n_i^{(r)}}$$

where s_j is chosen as in (7.78). Then Proposition 3.3.7 implies that

$$\begin{aligned} A'_\ell \mu^{k'_\ell} * \varepsilon_{a'_\ell} &= B^{-j}\left(A_{n_i^{(r)}} \mu^{[k_{n_i^{(r)}} c^j]} * \varepsilon_{\tilde{a}_\ell}\right) * \varepsilon_{s_j} \\ &\Rightarrow B^{-j}\nu^{c^j} * \varepsilon_{s_j} = \nu \end{aligned}$$

along all subsequences (ℓ') of (ℓ) on which $j = j_\ell$ remains constant. Since every subsequence contains another subsequence with this property, we have shown that (7.83) holds. □

It follows from the proof of Theorem 7.5.6 that for operator stable laws ν the opposite inclusion in Lemma 3.3.27 is also true.

Corollary 7.5.11. *Assume ν is a full operator stable law. Then we have for every $c \geq 1$ that* $\mathrm{GDOA}(\nu) = \mathrm{GDOSA}(\nu, c)$.

Proof. In view of Lemma 3.3.27, we only have to show that $\mathrm{GDOSA}(\nu, c) \subset \mathrm{GDOA}(\nu)$. If $\mu \in \mathrm{GDOSA}(\nu, c)$ for some $c \geq 1$, then there exist (k_n), (A_n) and (a_n) such that (7.80) and (7.81) hold. Since ν is operator stable, Lemma 7.5.8 yields the assertion. □

Corollary 7.5.12. *Let ν be a full operator semistable probability measure. Then* $\mathrm{GDOSA}(\nu, c_1) = \mathrm{GDOSA}(\nu, c_2)$ *for any* $c_1, c_2 \in \mathcal{D}(\nu)$*, where* $c_1, c_2 > 1$.

Proof. In view of Theorem 7.5.4, we have either $\mathcal{D}(\nu) = \{t > 0\}$ or $\mathcal{D}(\nu) = \{c^k : k \in \mathbb{Z}\}$ for some $c > 1$. In the first case, ν is actually operator stable and then by Corollary 7.5.11 we have

$$\mathrm{GDOSA}(\nu, c_1) = \mathrm{GDOA}(\nu) = \mathrm{GDOSA}(\nu, c_2).$$

In the second case, ν is properly operator semistable and it follows from Theorem 7.5.4 that $c_1 = c^{p_1}$ and $c_2 = c^{p_2}$ for some natural p_1 and p_2, where c is the smallest $c > 1$ such that ν is (B, c) operator semistable. Then Theorem 7.5.5 implies

$$\mathrm{GDOSA}(\nu, c_1) = \mathrm{GDOSA}(\nu, c) = \mathrm{GDOSA}(\nu, c_2)$$

which concludes the proof. □

Remark 7.5.13. If μ belongs to the generalized domain of semistable attraction of ν full and operator semistable, Theorem 7.1.10 shows that ν is (c^E, c) operator semistable for some exponent E and some $c > 1$. Then $c \in \mathcal{D}(\nu)$ and so Corollary 7.5.12 implies that $\mu \in \mathrm{GDOSA}(\nu, c)$. In other words, we may assume that (3.94) holds for some increasing sequence (k_n) with $k_{n+1}/k_n \to c$.

Remark 7.5.14. It is a simple matter to specialize the results of this section to the case of scalar norming. Then (7.71) with (7.69) replaced by (7.80) is sufficient for $\mu \in \mathrm{DOSA}(\nu)$, $\mathrm{DOA}(\nu) = \mathrm{DOSA}(\nu, c)$ for every $c \geq 1$ when ν is stable, and $\mathrm{DOSA}(\nu, c_1) = \mathrm{DOSA}(\nu, c_2)$ for any $c_1, c_2 \in \mathcal{D}(\nu) = \{t > 0 : \exists a > 0, b \in \mathbb{R}^d \text{ such that } \nu^t = a\nu * \varepsilon_b\}$ where $c_1, c_2 > 1$. In particular, we may always assume without loss of generality that $c > 1$. We can also specialize Lemma 7.5.8 to the case of scalar norming. If $\mu \in \mathrm{DOSA}(\nu, c)$ where ν is stable and $a_n \mu^{k_n} * \varepsilon_{s_n} \Rightarrow \nu$ for some increasing sequence of positive integers (k_n) with $k_{n+1}/k_n \to c$ then we can specialize the construction of Lemma 7.5.8 to obtain a sequence of positive reals (b_n) such that $a_n = b_{k_n}$ and $b_n \mu^n * \varepsilon_{s'_n} \Rightarrow \nu$ for some suitable shifts s'_n.

7.6 NOTES AND COMMENTS

Jajte [69] pioneered the study of operator semistable laws by proving Theorem 7.1.1, which characterizes the spectrum of the linear operator B for a (B, c) operator semistable law. It took almost ten years until Chorny [24] proved the existence of exponents for operator semistable laws. The disintegration formula of Corollary 7.1.14 is also due to Chorny [24]. Łucsak [91] proved Theorem 7.1.15, that operator semistable laws have a density. The spectral decomposition for operator semistable laws is from Meerschaert and Scheffler [122]. Scheffler [165] showed that if Y is (B, c) operator semistable and B

has real spectrum $a_1 < \cdots < a_p$ then $E\|Y\|^\rho$ exists whenever $0 \leq \rho < \log c / \log a_p$ and $E\|Y\|^\rho = \infty$ whenever $\rho > \log c / \log a_1$. This is essentially equivalent to Corollary 7.1.21. Meerschaert and Scheffler [122] proved the sharp moment results of Theorem 7.1.18 and Corollary 7.1.20. There are still some important unsolved problems concerning operator semistable laws, in particular the extension of Theorem 7.2.11. The main technical problem with extending this result to operator semistable laws seems to be the fact that the structure of the set of exponents of a given operator semistable law is unknown. The construction of a commuting exponent relies on the fact that the set of exponents is convex.

Sharpe [181] coined the term *operator stable* to describe those full probability distributions that arise as weak limits of operator normalized partial sums of i.i.d. random vectors. He also proved Theorem 7.2.1 establishing the existence of exponents for operator stable laws. The disintegration formula of Theorem 7.2.5 is due to Jurek [74]. Hudson [64] proved Theorem 7.2.7, that operator stable laws have a density. Hudson, Veeh, and Weiner [67] and Meerschaert [110] independently established the existence and nonexistence of certain radial moments for operator stable laws. They showed that if Y is operator stable with index E having real spectrum $a_1 < \cdots < a_p$ and θ is any nonzero vector in $\mathbb{R}^d$ then $E|\langle Y, \theta \rangle|^\rho$ exists for all $0 < \rho < 1/a_p$, and if $a_1 > 1/2$ then $E|\langle Y, \theta \rangle|^\rho = \infty$ for all $\rho > 1/a_1$. The sharp moment results of Remark 7.2.10 follow immediately, using the spectral decomposition for operator stable laws from Definition 7.2.8. This spectral decomposition is due to Meerschaert [114]. The portion of Theorem 7.2.11 relating exponents and symmetries of operator stable laws is due to Holmes, Hudson, and Mason [62]. The construction of commuting exponents is from Hudson, Jurek, and Veeh [66]. Corollary 7.2.12 showing that every exponent has the same real spectrum is due to Meerschaert and Veeh [117]. That paper also contains additional information on the structure of exponents and symmetries.

Stable and semistable laws are generally credited to Lévy [88]. Samorodnitsky and Taqqu [161] introduced the parameterization of stable laws that is now standard, and their book is the best modern reference for the most important properties of stable laws on $\mathbb{R}^1$ and $\mathbb{R}^d$. The material on stable random variables in Section 7.3 mostly follows Feller [40]. Zolotarev [199] contains many additional results on stable random variables, including explicit formulas for fractional and logarithmic moments. The recent book of Uchaikan and Zolotarev [192] is a more accessible reference for some of these results, and includes quite a wide variety of applications. Kanter [76] developed an analytical formula for the density of a stable random variable. The portion of Section 7.3 dealing with stable random vectors is mainly taken from Samorodnitsky and Taqqu [161]. Theorems 7.3.3 and 7.3.16 on stable random vectors are due to Rvačeva [158], who generalized the arguments of Gnedenko and Kolmogorov [44]. A variety of different parameterizations for stable random variables has caused considerable confusion in the literature, see Hall [52] and McCulloch [108]. The parameterization of Samorodnitsky and Taqqu [161] is now standard, although many authors continue to use the dispersion as a measure of spread (see, e.g., Brockwell and Davis [22]). The dispersion also occurs naturally in certain limit theorems (see Remark 8.2.20).

Pillai [147] showed that semistable laws are those with nonempty domains of semistable attraction. The form of the Lévy measure of a semistable law on $\mathbb{R}^1$ in Corollary 7.4.4 was discovered independently by Kruglov [84], Mejzler [135], and Shimuzu [182]. Moment results for semistable laws on $\mathbb{R}^1$ were also obtained by these authors. Theorem 7.5.4 is due to Łuczak [92] and Chorny [24], who generalized the one variable result of Kruglov [84]. Theorem 7.5.6 is due to Phuc [146]. Together with Theorem 3.3.19, this result shows that in some sense the generalized domain of semistable attraction is the most general situation in which one can infer useful information about a sum of i.i.d. random vectors from the limiting distribution. Meerschaert and Scheffler [123] proved Corollary 7.5.11, that every law in the generalized domain of semistable attraction of an operator stable law is also in its generalized domain of attraction. This extends a one variable result of Mejzler [135]. The definition and characterization of stable and semistable laws on locally compact groups is laid out in the recent book of Hazod and Siebert [55].

8

Central Limit Theorems

In this chapter, we state and prove the limit theorems for sums of independent random vectors that are the central focus of this book. Gnedenko and Kolmogorov [44] treat much of this material for random variables by way of some intricate analytical arguments. Much later, Feller [40] gave an elegant treatment using regular variation for functions of one variable.[1] Our approach mirrors that of Feller. All of the preliminary technical results about infinitely divisible laws, triangular arrays, and regular variation are put to good use here, and they greatly streamline both the statement and proof of our results.

8.1 NORMAL LIMITS

In this section, we characterize the probability distributions that are attracted to some full normal law on $\mathbb{R}^d$. Since normal laws are operator stable, Corollary 7.5.11 implies that the generalized domain of attraction and the generalized domain of semistable attraction of a normal law are equal. Hence we only need to consider generalized domains of attraction. Since every full normal law on $\mathbb{R}^d$ is of the same type, Lemma 3.3.26 implies that every full normal law on $\mathbb{R}^d$ has the same generalized domain of attraction.

[1] See also Sakovich [159], who anticipated Feller to some extent.

Recall from Definition 3.3.24 that $\mu \in \mathrm{GDOA}(\nu)$ if and only if there exist linear operators $A_n \in \mathrm{L}(\mathbb{R}^d)$ and nonrandom $a_n \in \mathbb{R}^d$ such that

$$A_n\mu^n * \varepsilon_{a_n} \Rightarrow \nu \quad \text{as } n \to \infty. \tag{8.1}$$

Lemma 3.3.25 implies that if (8.1) holds, then A_n is automatically invertible for all n large, and hence we may assume without loss of generality that $A_n \in \mathrm{GL}(\mathbb{R}^d)$. If $X, X_1, X_2, \ldots$ are independent and identically distributed random vectors on $\mathbb{R}^d$ with common distribution μ, and Y is a random vector on $\mathbb{R}^d$ with distribution ν, then (8.1) is equivalent to

$$A_n(X_1 + \cdots + X_n) + a_n \Rightarrow Y. \tag{8.2}$$

Since A_n^{-1} exists, this means that the distribution of $S_n = X_1 + \cdots + X_n$ can be usefully approximated by that of $A_n^{-1}(Y - a_n)$ for n large. This approximation is the basis for a large number of commonly used statistical procedures, including confidence intervals and hypothesis testing.

If $E\|X\|^2 < \infty$, then the following facts are well known: The mean $m = EX$ and the covariance matrix $C = E\big((X - m)(X - m)'\big)$ exist, and (8.2) follows from the Central Limit Theorem. In this case, we may take $A_n = n^{-1/2}I$ and $a_n = -n^{1/2}m$ in (8.2), yielding Y mean zero normal with covariance matrix C. If we take $A_n = n^{-1/2}T$ instead, where $T \in \mathrm{GL}(\mathbb{R}^d)$ is arbitrary, then the limit is also mean zero normal with covariance matrix TCT^*. An appropriate choice of T yields a standard normal limit. If we take $a_n = b - n^{1/2}m$, then the limit has mean $b \in \mathbb{R}^d$, so we recover the entire class of multivariable normal laws. In this section, we will recover all of these facts and more, but we will not assume that $E\|X\|^2 < \infty$. The Central Limit Theorem states that $E\|X\|^2 < \infty$ is a sufficient condition for (8.2) to hold for some full normal limit. In fact, it is also possible to obtain a normal limit in (8.2) even when $E\|X\|^2 = \infty$. We will obtain necessary and sufficient conditions in terms of regular variation. The theory of regularly varying real-valued functions on $\mathbb{R}^d$ was laid out in Chapter 5. The main result of this section is Theorem 8.1.7, which shows that regular variation of a certain truncated variance function of X is necessary and sufficient for X to belong to the generalized domain of attraction of a full normal law on $\mathbb{R}^d$. For technical reasons, we begin with the special case $EX = 0$.

Lemma 8.1.1. *Suppose that X is a random vector on $\mathbb{R}^d$ with distribution μ and that $E(X) = 0$. Let ν be a full normal law on $\mathbb{R}^d$ with Lévy representation $[a, Q, 0]$. Then $\mu \in \mathrm{GDOA}(\nu)$ and (8.1) holds for some $a_n \in \mathbb{R}^d$ if and only if*

$$nF(A_n^* t_n) \to Q(t) \tag{8.3}$$

whenever $t_n \to t$ in $\Gamma = \mathbb{R}^d \setminus \{0\}$, where

$$F(t) = E\big(\langle X, t\rangle^2 I(|\langle X, t\rangle| \leq 1)\big). \tag{8.4}$$

In this case, $\|A_n\| \to 0$.

Proof. If $\mu \in \mathrm{GDOA}(\nu)$ and (8.1) holds, then Corollary 3.3.15 with $k_n = n$ implies that

$$\lim_{n\to\infty} Q_n(t_n) = Q(t) \tag{8.5}$$

whenever $t_n \to t$ in $\mathbb{R}^d$, where

$$\begin{aligned} Q_n(t) &= n \int_{|<x,t>|\le 1} <x,t>^2 A_n\mu\{dx\} \\ &= n \int_{|<A_nx,t>|\le 1} <A_nx,t>^2 \mu\{dx\} \\ &= n \int_{|<x,A_n^*t>|\le 1} <x,A_n^*t>^2 \mu\{dx\} \\ &= nE\big(<X,A_n^*t>^2 I(|<X,A_n^*t>| \le 1)\big) \\ &= nF(A_n^*t) \end{aligned} \tag{8.6}$$

so that (8.5) is equivalent to (8.3).

Conversely, suppose that (8.3) holds for some linear operators A_n and some nonnegative definite quadratic form Q. It follows from (8.3) that (8.5) holds with $t_n = t$. Then in view of Corollary 3.3.14 applied to each $t \in \Gamma$, with $\varepsilon = \|t\|^{-1}$, in order to show that $\mu \in \mathrm{GDOA}(\nu)$ and (8.1) holds it suffices to show that $n(A_n\mu) \to 0$. Since Q is a quadratic form we have $Q(\lambda t) = \lambda^2 Q(t)$ for all $\lambda > 0$ and all $t \in \mathbb{R}^d$. By using the notation introduced in Section 5.2, we can rewrite this in the equivalent form $\lambda \cdot Q = \lambda^B Q$ for all $\lambda > 0$, where $B = (-1/2)I$. Then B is an exponent of Q in terms of Definition 5.1.4, and since $Q(t) > 0$ for $t \neq 0$ it follows that Q is admissible in terms of Definition 5.2.1. Hence we can complete the proof by applying the following lemma. □

Lemma 8.1.2. *Suppose that X is a random vector on $\mathbb{R}^d$ with distribution μ and that $E(X) = 0$. If the function F defined by (8.4) satisfies (8.3) whenever $t_n \to t$ in $\Gamma = \mathbb{R}^d \setminus \{0\}$ for some admissible function $Q(t) > 0$ with exponent $B = (-1/2)I$, then*

$$n \cdot (A_n\mu) \to 0 \tag{8.7}$$

and $\|A_n\| \to 0$ *as* $n \to \infty$.

Proof. Since F is monotone on rays and Q is admissible, we can rewrite (8.3) in the equivalent form $n \cdot (A_n^*)^{-1}F \to Q$ in the notation of Section 5.2. Then Theorem 5.2.16 implies that for some $G_n \in \mathcal{S}(Q)$ we have $B_n = G_n(A_n^*)^{-1}$ regularly varying with index $-B = (1/2)I$ and $n \cdot B_nF \to Q$, so that F is regularly varying at zero with exponent B.

Recall from Remark 2.1.5 that $\|G_n^*\| = \|G_n\|$ in the Euclidean operator norm. Then $A_n = (B_n^*)^{-1}G_n^*$ where $\|G_n^*\| = \|G_n\|$ is bounded since

$\mathcal{S}(Q)$ is compact. Since $(B_n^*)^{-1}$ varies regularly with index $B^* = (-1/2)I$, Theorem 4.2.9 implies that $f(t) = (B_{[t]}^*)^{-1}$ varies regularly with index B^* and then Theorem 4.2.4 with $\beta = 0$ implies that $\|(B_n^*)^{-1}\| \to 0$, hence $\|A_n\| \leq \|(B_n^*)^{-1}\| \cdot \|G_n^*\| \to 0$.

Define

$$\begin{aligned} V_0(r,\theta) &= P(|<X,\theta>| > r) \\ U_2(r,\theta) &= E\big(<X,\theta>^2 I(|<X,\theta>| \leq r)\big) \end{aligned} \tag{8.8}$$

so that U_2 and V_0 are the truncated second moment and tail function given by Definition 5.3.10 with $R = V_0$. Since F varies regularly at zero with exponent $B = (-1/\beta)E = (-1/\beta)(\alpha I) = (-1/2)I$, it follows from Theorem 5.3.4 (b) that $(r,\theta) \mapsto F(\theta/r)$ is a regularly varying function of $r > 0$ with index $(-\beta/\alpha) = -2$ uniformly on the unit sphere $S = \{\theta : \|\theta\| = 1\}$. Since

$$\begin{aligned} r^2 F(\theta/r) &= r^2 \int_{|<x,\theta/r>| \leq 1} <x,\theta/r>^2 \mu\{dx\} \\ &= \int_{|<x,\theta>| \leq r} <x,\theta>^2 \mu\{dx\} \\ &= U_2(r,\theta) \end{aligned} \tag{8.9}$$

it follows that $U_2(r,\theta)$ is a slowly varying function of $r > 0$ uniformly on $\theta \in S$. Define $H_t = \{x \in \mathbb{R}^d : |<x,t>| > 1\}$ and choose $r_n > 0$, $\|\theta_n\| = 1$ such that $A_n^* t = r_n\theta_n$. Then $r_n^{-1} \to \infty$ and

$$\begin{aligned} n(A_n\mu)(H_t) &= n\mu(A_n^{-1}H_t) \\ &= n\mu\{A_n^{-1}x : |<x,t>| > 1\} \\ &= n\mu\{x : |<A_nx,t>| > 1\} \\ &= n\mu\{x : |<x,A_n^*t>| > 1\} \\ &= n\mu\{x : |<x,r_n\theta_n>| > 1\} \\ &= n\mu\{x : |<x,\theta_n>| > r_n^{-1}\} \\ &= nV_0(r_n^{-1},\theta_n). \end{aligned}$$

Apply Theorem 5.3.11 (a) with $\zeta = 2$, $\eta = 0$ and $\rho = 0$. Then (5.40) holds with $\alpha = 2$ and so

$$\frac{r_n^{-2}V_0(r_n^{-1},\theta_n)}{U_2(r_n^{-1},\theta_n)} \to 0 \quad \text{as } n \to \infty.$$

Furthermore, by (8.9) and (8.3) with $t_n = t$ we have $nr_n^2U_2(r_n^{-1},\theta_n) = nF(r_n\theta_n) = nF(A_n^*t) \to Q(t)$. Hence

$$n(A_n\mu)(H_t) = nV_0(r_n^{-1},\theta_n) = \frac{r_n^{-2}V_0(r_n^{-1},\theta_n)}{U_2(r_n^{-1},\theta_n)} \cdot nr_n^2U_2(r_n^{-1},\theta_n) \to 0$$

as $n \to \infty$. Since any set that is bounded away from the origin is contained in a finite union of sets of the form H_t, it follows that (8.7) holds. □

Theorem 8.1.3. *Suppose that X is a random vector on $\mathbb{R}^d$ with distribution μ and that $E(X) = 0$. Then X belongs to the generalized domain of attraction of a full normal law on $\mathbb{R}^d$ if and only if the truncated second moment function F defined in* (8.4) *varies regularly at zero with exponent* $(-1/2)I$.

Proof. If F varies regularly at zero with exponent $B = (-1/2)I$ then Corollary 5.1.9 (b) implies that there exists some admissible function $Q(t)$ and some $(B_n) \in \mathrm{RV}(-B)$ such that $nF(B_n^{-1}t_n) \to Q(t) > 0$ whenever $t_n \to t$ in Γ. Define $Q(0) = 0$, and let $A_n^* = B_n^{-1}$ so that (8.3) holds. Then (8.5) holds with $t_n = t$. Since B is an exponent of Q it follows from Lemma 8.1.2 that $nA_n\mu \to 0$. Define $I_n(t) = \langle t, C_n t\rangle$ where

$$C_n = n \int_{\|x\| \leq 1} xx' A_n\mu\{dx\}$$

and xx' is the outer product on $\mathbb{R}^d$. If $\|t\| = 1$, then $0 \leq Q_n(t) - I_n(t) \leq n \cdot A_n\mu\{x : \|x\| > 1\} \to 0$ and hence $I_n(t) \to Q(t)$. Since $I_n(\lambda t) = \lambda^2 I_n(t)$ and $Q(\lambda t) = \lambda^2 Q(t)$ for any $t \in \mathbb{R}^d$ and any $\lambda \geq 0$, it follows that $I_n(t) \to Q(t)$ for all $t \in \mathbb{R}^d$. Then Corollary 2.1.9 implies that $Q(t)$ is a quadratic form on $\mathbb{R}^d$. Since $Q(t) > 0$ for all $t \in \Gamma$, in view of Example 1.3.16, the probability distribution ν with Lévy representation $[a, Q, 0]$ is a full normal law. Then Lemma 8.1.1 implies that $\mu \in \mathrm{GDOA}(\nu)$ and (8.1) holds for some $a_n \in \mathbb{R}^d$.

Conversely, if X belongs to the generalized domain of attraction of some full normal law ν on $\mathbb{R}^d$ with Lévy representation $[a, Q, 0]$, then Lemma 8.1.1 implies that (8.3) holds, and we can rewrite this in the notation of Section 5.2 as $n \cdot (A_n^*)^{-1}F \to Q$. Then Theorem 5.2.16 implies that F is regularly varying. Since Q is a positive definite quadratic form, $B = (-1/2)I$ is an exponent of Q and $F \in \mathrm{RVF}_0(B)$. □

Theorem 8.1.3 characterizes the measures that belong to the generalized domain of attraction of some full normal law. However, if we are given a specific normal law, then it is also interesting to characterize the generalized domain of attraction of that distribution. Our next result gives necessary and sufficient conditions for a probability measure μ to belong to the generalized domain of attraction of a full normal law ν on $\mathbb{R}^d$. It also shows that the norming operators A_n in (8.2) can always be chosen to be regularly varying.

Corollary 8.1.4. *Let ν be a full normal law on $\mathbb{R}^d$. Suppose that X is a random vector on $\mathbb{R}^d$ with distribution μ and $E(X) = 0$, and let F denote the truncated second moment function defined in* (8.4). *Then the following are equivalent:*

(a) $\mu \in \mathrm{GDOSA}(\nu, c)$ *for every* $c \geq 1$;

(b) $\mu \in \mathrm{GDOA}(\nu)$;

(c) F varies regularly at zero with exponent $(-1/2)I$ and limit function Q;

(d) for some regularly varying sequence (A_n) with index $(-1/2)I$ we have

$$nF(A_n^* t_n) \to Q(t) \tag{8.10}$$

whenever $t_n \to t$ in $\Gamma = \mathbb{R}^d \setminus \{0\}$; and

(e) for some regularly varying sequence (A_n) with index $(-1/2)I$ we have

$$A_n \mu^n * \varepsilon_{a_n} \Rightarrow \nu \tag{8.11}$$

as $n \to \infty$ for some $a_n \in \mathbb{R}^d$.

In this case, ν has Lévy representation $[a, Q, 0]$. Furthermore, we can use the same sequence (A_n) in both (d) and (e).

Proof. Corollary 7.5.11 implies that (a) holds if and only if (b) holds. The proof of Theorem 8.1.3 shows that (b) implies (c), and that (c) implies (d) implies (e). Part (b) follows directly from (e) by definition, which completes the chain of implications. □

In the remainder of this section, we show how to deal with the general case, in which do not assume $EX = 0$. Our first result shows that we can always choose the norming operators (A_n) in (8.1) to be regularly varying.

Theorem 8.1.5. *Assume $\mu \in \mathrm{GDOA}(\nu)$ for some full (t^E) operator stable law ν and (8.1) holds. Then there exists a sequence $(B_n) \in \mathrm{RV}(-E)$ such that for a suitable choice of centering constants (b_n) we have*

$$B_n \mu^n * \varepsilon_{b_n} \Rightarrow \nu \quad \text{as } n \to \infty. \tag{8.12}$$

Proof. Since ν is operator stable, by Theorem 7.2.1 there exists an exponent E such that $\nu^\lambda = \lambda^E \nu * \varepsilon_{b_\lambda}$ for all $\lambda > 0$. It follows that $\nu = \lambda^{-E} \nu^\lambda * \varepsilon_{s_\lambda}$ where $s_\lambda = -\lambda^{-E} b_\lambda$ for all $\lambda > 0$. Fix $\lambda > 0$ and let $\ell(n) = [\lambda n]$. It follows immediately from (8.1) that

$$A_{\ell(n)} \mu^{\ell(n)} * \varepsilon_{a_{\ell(n)}} \Rightarrow \nu,$$

and then Proposition 3.3.7 implies that

$$A_{\ell(n)} \mu^{[\ell(n) t_n]} * \varepsilon_{a_{\ell(n)}(t_n)} \Rightarrow \nu^t$$

whenever $t_n \to t > 0$, where $a_{\ell(n)}(t_n) = ([\ell(n)t_n]/\ell(n)) a_{\ell(n)}$. Letting $t_n = n/\ell(n) = n/[\lambda n]$ and $t = 1/\lambda$, it follows that

$$A_{[\lambda n]} \mu^n * \varepsilon_{(n/[\lambda n]) a_n} = A_{\ell(n)} \mu^{[\ell(n) t_n]} * \varepsilon_{a_{\ell(n)}(t_n)} \Rightarrow \nu^t = \lambda^{-E} \nu * \varepsilon_{s_{1/\lambda}}$$

as $n \to \infty$. Then in view of (8.1) we obtain using the Convergence of Types Theorem 2.3.17 that

$$A_{[\lambda n]} A_n^{-1} \rightharpoonup \lambda^{-E} \mathcal{S}^1(\nu) \quad \text{as } n \to \infty$$

for all $\lambda > 0$. By Theorem 2.3.10, we know that $\mathcal{S}^1(\nu)$ is a compact subgroup of $\mathrm{GL}(\mathbb{R}^d)$, and Proposition 7.2.4 implies that $\mathcal{S}^1(\nu) = \lambda^E \mathcal{S}^1(\nu)\lambda^{-E}$ for all $\lambda > 0$. Then Proposition 4.2.12 implies that there exist $G_n \in \mathcal{S}^1(\nu)$ such that $B_n = G_n A_n$ is regularly varying with index $(-E)$, and another application of convergence of types, Theorem 2.3.17, shows that (8.12) holds. □

Our next result shows that, for any X in the generalized domain of attraction of a normal law, EX exists. This will allow us to reduce to the case where $EX = 0$.

Proposition 8.1.6. *Suppose that X is a random vector with distribution $\mu \in$ GDOA(ν) where ν is a full normal law on $\mathbb{R}^d$. Then $E\|X\|^\beta < \infty$ for any $0 \le \beta < 2$. In particular, the mean $E(X)$ exists.*

Proof. We know that ν is $(t^{(1/2)I})$ operator stable and then by Theorem 8.1.5 we can choose $(B_n) \in \mathrm{RV}(-(1/2)I)$ such that (8.12) holds. Corollary 3.3.9 implies that

$$n \cdot (B_n \mu) \to 0 \quad \text{as } n \to \infty. \tag{8.13}$$

It follows from Theorem 4.2.4 together with Theorem 4.2.9 that for any $\varepsilon > 0$ we have $n^{(1/2)+\varepsilon}\|B_n\theta\| \to \infty$ as $n \to \infty$ uniformly in $\|\theta\| = 1$. Hence there exists a $n_0 \ge 1$ such that for all $n \ge n_0$ and all $\|\theta\| = 1$ we have $\|B_n\theta\| \ge n^{-(1/2)-\varepsilon}$. Since $n^{-(1/2)-\varepsilon}\|x\| \le \|x\| \cdot \|B_n(x/\|x\|)\| = \|B_n x\|$ it follows that $\{x : \|x\| \ge n^{(1/2)+\varepsilon}\} \subset \{x : \|B_n x\| \ge 1\}$ for all $n \ge n_0$. Hence, in view of (8.13) we get for all $n \ge n_1$ and some $n_1 \ge n_0$ that

$$\mu\{x : \|x\| \ge n^{(1/2)+\varepsilon}\} \le \frac{1}{n}.$$

If we let $n = [t]$ for some real $t \ge n_1$ we obtain that for some constant $C > 0$

$$\mu\{x : \|x\| \ge t^{(1/2)+\varepsilon}\} \le \frac{C}{t}.$$

Substituting $t = s^{2/(1+2\varepsilon)}$ we get for some $\delta = \delta(\varepsilon)$ with $\delta(\varepsilon) \to 0$ as $\varepsilon \to 0$, that for some $s_0 > 0$, for all $s \ge s_0$ we have

$$\mu\{x : \|x\| \ge s\} \le C s^{-2+\delta}. \tag{8.14}$$

Using the well-known formula

$$E\|X\|^\beta = \beta \int_0^\infty s^{\beta-1}\mu\{x : \|x\| \ge s\} ds$$

it follows from (8.14) that $E\|X\|^\beta$ exists if $\beta + \delta < 2$. Since $\beta < 2$ there exists an $\varepsilon > 0$ with $\beta + \delta(\varepsilon) < 2$, which concludes the proof. □

Now we come to the main result of this section, that $\mu \in$ GDOA(ν) for some full normal law ν on $\mathbb{R}^d$ if and only if the truncated variance of μ is regularly varying with the correct index.

Theorem 8.1.7. *A random vector X on $\mathbb{R}^d$ belongs to the generalized domain of attraction of a full normal law if and only if EX exists and the truncated variance function defined by*

$$M(t) = E\big(\langle X - EX, t\rangle^2 I(|\langle X - EX, t\rangle| \le 1)\big) \tag{8.15}$$

varies regularly at zero with exponent $(-1/2)I$.

Proof. If $X \in \mathrm{GDOA}(Y)$, then (8.2) holds for some linear operators A_n and some shifts $a_n \in \mathbb{R}^d$. Proposition 8.1.6 implies that EX exists, so we can define $X_i' = X_i - E(X_i)$ and $a_n' = a_n + nA_nE(X_1)$. Then it follows from (8.2) that

$$A_n(X_1' + \cdots + X_n') + a_n' \Rightarrow Y \tag{8.16}$$

so that $X' = X - EX \in \mathrm{GDOA}(Y)$ as well. Since $M(t)$ is the truncated second moment function of X', it follows from Theorem 8.1.3 that $M \in \mathrm{RVF}_0((-1/2)I)$.

Conversely, if EX exists and $M \in \mathrm{RVF}_0((-1/2)I)$ then, since $M(t)$ is the truncated second moment function of X', it follows from Theorem 8.1.3 that $X - EX$ is in the generalized domain of attraction of some full normal law Y, so that (8.16) holds for some linear operators A_n and some $a_n' \in \mathbb{R}^d$. It follows that (8.2) holds with $a_n = a_n' - nA_nE(X_1)$, and hence $X \in \mathrm{GDOA}(Y)$ as well. □

Corollary 8.1.8. *Let ν be a full normal law on $\mathbb{R}^d$. Suppose that X is a random vector on $\mathbb{R}^d$ with distribution μ, and let M denote the truncated variance function defined in* (8.15). *Then the following are equivalent:*

(a) $\mu \in \mathrm{GDOSA}(\nu, c)$ *for every* $c \ge 1$*;*

(b) $\mu \in \mathrm{GDOA}(\nu)$*;*

(c) *M varies regularly at zero with exponent $(-1/2)I$ and limit function Q;*

(d) *for some regularly varying sequence (A_n) with index $(-1/2)I$ we have*

$$nM(A_n^* t_n) \to Q(t) \tag{8.17}$$

whenever $t_n \to t$ in $\Gamma = \mathbb{R}^d \setminus \{0\}$; and

(e) *for some regularly varying sequence (A_n) with index $(-1/2)I$ we have*

$$A_n\mu^n * \varepsilon_{b_n} \Rightarrow \nu \tag{8.18}$$

as $n \to \infty$ for some $b_n \in \mathbb{R}^d$.

Then ν has Lévy representation $[a, Q, 0]$. Furthermore, we can use the same sequence (A_n) in both (d) and (e).

Proof. If $\mu \in \text{GDOA}(\nu)$, then Proposition 8.1.6 shows that EX exists. Let ω denote the probability distribution of the random vector $X' = X - EX$. Then $\mu \in \text{GDOA}(\nu)$ and (8.18) holds if and only if $\omega \in \text{GDOA}(\nu)$ and

$$A_n \omega^n * \varepsilon_{a_n} \Rightarrow \nu$$

where $a_n = b_n + nA_n EX$. Now the result follows immediately from Corollary 8.1.4, using the fact that $M(t)$ is also the truncated second moment function of the random vector $X' = X - EX$. □

Now we show that we can always choose the centering constants in (8.2) to zero expectation.

Theorem 8.1.9. *If X belongs to the generalized domain of attraction of some full normal law on $\mathbb{R}^d$ and (8.2) holds, then (8.2) also holds with $a_n = -nA_n EX$ and $EY = 0$.*

Proof. If X belongs to the generalized domain of attraction of some full normal law, then Proposition 8.1.6 shows that EX exists. First, suppose that $EX = 0$. Then Theorem 8.1.3 shows that the truncated second moment function F defined by (8.4) varies regularly at zero with index $B = (-1/2)I$. Since $B = (-1/\beta)E = (-1/\beta)(\alpha I) = (-1/2)I$, it follows from Theorem 5.3.4 (b) that $(r, \theta) \mapsto F(\theta/r)$ is a regularly varying function of $r > 0$ with index $(-\beta/\alpha) = -2$ uniformly on the unit sphere $S = \{\theta : \|\theta\| = 1\}$. Then in view of (8.9) it follows that the function $U_2(r, \theta)$ defined by (8.8) is a slowly varying function of $r > 0$ uniformly on $\theta \in S$. Theorem 3.3.8 implies that we may choose

$$a_n = -n \int_{\|x\|<R} x(A_n\mu)\{dx\} \tag{8.19}$$

in (8.2) for any $R > 0$ (since the Lévy measure of a normal law is the zero measure, any such domain of integration is a ϕ-continuity set). We will choose $R = 1$. Then in order to show that we can also choose $a_n = -nA_n EX = 0$, it suffices to show that

$$d_n = -n \int_{\|x\|>1} x(A_n\mu)\{dx\} \to 0$$

as $n \to \infty$. Given $\|\theta\| = 1$ write $A_n^*\theta = r_n\theta_n$ for some $r_n > 0$ and $\|\theta_n\| = 1$. Lemma 8.1.1 shows that $r_n \to 0$. Then we have

$$\begin{aligned}
|\langle d_n, \theta\rangle| &\le n \int_{\|x\|>1} |\langle x, \theta\rangle| (A_n\mu)\{dx\} \\
&= n \int_{|\langle x,\theta\rangle|>1} |\langle x, \theta\rangle| (A_n\mu)\{dx\} + n \int_{A(\theta)} |\langle x, \theta\rangle| (A_n\mu)\{dx\} \\
&= nr_n V_1(r_n^{-1}, \theta_n) + n \int_{A(\theta)} |\langle x, \theta\rangle| (A_n\mu)\{dx\}
\end{aligned} \tag{8.20}$$

where V_1 is the truncated first tail moment function defined by

$$V_1(r,\theta) = E\big(|\langle X,\theta\rangle| I(|\langle X,\theta\rangle| > r)\big) \tag{8.21}$$

and $A(\theta) = \{x : \|x\| > 1, |\langle x,\theta\rangle| \le 1\}$. The second integral on the right-hand side of (8.20) is bounded above by $n(A_n\mu)\{x : \|x\| > 1\}$, so this term tends to zero in view of Corollary 3.3.9, which implies that $n(A_n\mu) \to 0$.

Since EX exists it follows that V_1 exists, and then since the truncated second moment function U_2 is slowly varying uniformly on $\|\theta\| = 1$, Theorem 5.3.11 (a) with $\zeta = 2$, $\eta = 1$ and $\rho = 0$ implies that (5.40) holds with $\alpha = 2$ and so

$$\frac{r_n^{-1}V_1(r_n^{-1},\theta_n)}{U_2(r_n^{-1},\theta_n)} \to 0 \quad \text{as } n \to \infty.$$

Lemma 8.1.1 along with (8.9) implies that

$$nr_n^2 U_2(r_n^{-1},\theta_n) = nF(A_n^*\theta) \to Q(\theta)$$

as $n \to \infty$ for any $\|\theta\| = 1$. Then

$$\begin{aligned} nr_nV_1(r_n^{-1},\theta_n) &= \frac{r_n^{-1}V_1(r_n^{-1},\theta_n)}{U_2(r_n^{-1},\theta_n)} \cdot nr_n^2U_2(r_n^{-1},\theta_n) \\ &\to 0 \cdot Q(\theta) = 0 \end{aligned}$$

as $n \to \infty$, which finishes the proof that we can choose $a_n = 0$ in (8.2) when $EX = 0$.

In the general case, let $X_i' = X_i - E(X_i)$ and $a_n' = a_n + nA_nE(X_i)$. Then (8.2) holds if and only if

$$A_n(X_1' + \cdots + X_n') + a_n' \Rightarrow Y.$$

Since $EX_i' = 0$ we have already shown that we can take $a_n' = 0$, and then $a_n = -nA_nEX$ as desired.

Theorem 3.3.8 implies that Y has characteristic function $\exp(\psi_R(t))$ where $\psi_R(t)$ is given by (3.15) with $a_R = 0$. Since $\phi = 0$ in this case, (3.15) reduces to $\psi_R(t) = -(1/2)Q(t)$ so that Y is normal with mean zero. □

If $E\|X\|^2 < \infty$, then we recover the usual Central Limit Theorem results.

Corollary 8.1.10. *If $E\|X\|^2 < \infty$, then the mean $m = EX$ and the covariance matrix $C = E\big((X-m)(X-m)'\big)$ exist, (8.2) holds with $A_n = n^{-1/2}I$ and $a_n = -nA_nE(X)$, and Y is normal with mean zero and covariance matrix C.*

Proof. If $E\|X\|^2 < \infty$, then $E\langle t,X\rangle^2 \le \|t\|^2E\|X\|^2 < \infty$, and the Cauchy-Schwarz inequality implies that $(E|\langle t,X\rangle|)^2 \le E\langle t,X\rangle^2$ so that $E\langle t,X\rangle$ also exists. If $t_1,\ldots,t_d$ is the standard basis for $\mathbb{R}^d$, then m is a vector with entries $m_i = E\langle t_i, X\rangle$ and C is a matrix with entries $c_{ij} = E\langle Z, t_i\rangle\langle Z, t_j\rangle$ where $Z = X - m$, and another application of the Cauchy-Schwarz inequality yields $(E|\langle Z, t_i\rangle\langle Z, t_j\rangle|)^2 \le E\langle Z, t_i\rangle^2\langle Z, t_j\rangle^2 \le (E\|Z\|^2)^2 \le (E\|X\|^2)^2$ so that C exists. If $t_n \to t$ in $\Gamma = \mathbb{R}^d \setminus \{0\}$, then for some $B > 0$ we have $\|t_n\| \le B$ for all n, and then

$$\begin{aligned} nM(A_n^* t_n) &= nM(n^{-1/2} t_n) \\ &= nE\big(\langle Z, n^{-1/2} t_n\rangle^2 I(|\langle Z, n^{-1/2} t_n\rangle| \le 1)\big) \\ &= E\big(\langle Z, t_n\rangle^2 I(|\langle Z, t_n\rangle| \le n^{1/2})\big) \\ &\to E\langle Z, t\rangle^2 \end{aligned}$$

by dominated convergence. Then Corollary 8.1.8 implies that (8.2) holds with $A_n = n^{-1/2} I$ and Y normal with Lévy representation $[a, Q, 0]$ where $Q(t) = E\langle Z, t\rangle^2 = \langle t, Ct\rangle$ so that C is the covariance matrix of Y. Theorem 8.1.9 shows that we can take $a_n = -nA_n E(X)$, and then $EY = 0$. □

In the special case of random variables $d = 1$, we recover the result of Feller [40] giving necessary and sufficient conditions for a random variable to belong to the domain of attraction of a nondegenerate normal law.

Theorem 8.1.11. *A random variable X in $\mathbb{R}^1$ belongs to the domain of attraction of some nondegenerate normal law if and only if EX exists and the truncated variance*

$$\sigma_X^2(r) = E\big((X - EX)^2 I(|X - EX| \le r)\big) \tag{8.22}$$

is slowly varying at infinity.

Proof. If X belongs to the domain of attraction of a nondegenerate normal law on $\mathbb{R}^1$, then Theorem 8.1.7 implies that EX exists and the truncated variance function defined by (8.15) varies regularly at zero with exponent $(-1/2)$. In the scalar case, this function reduces to

$$M(t) = E\big((X - EX)^2 t^2 I(|(X - EX)t| \le 1)\big) = t^2 \sigma_X^2(1/t). \tag{8.23}$$

Since M varies regularly at zero with exponent $B = (-\alpha/\beta) = (-1/2)$, it follows from Theorem 5.3.4 (b) that $M(1/r) = r^{-2}\sigma_X^2(r)$ is regularly varying at infinity with index $(-\beta/\alpha) = -2$ in terms of the definition (4.1). Then $\sigma_X^2(r)$ is slowly varying at infinity. Conversely, if EX exists and $\sigma_X^2(r)$ is slowly varying at infinity, then $M(1/r)$ is regularly varying at infinity with index -2, so that $M(t)$ is regularly varying at zero with index 2 in terms of the definition (5.1). Then it follows from Remark 5.1.5 that $M(t)$ varies regularly at zero with exponent $-1/2$, and Theorem 8.1.7 implies that X belongs to some normal domain of attraction. □

Now we present an interesting result of Klosowska [79] concerning the one-dimensional marginals of a random vector X belonging to the generalized domain of attraction of some full normal law.

Corollary 8.1.12. *Let ν be a full normal law, and let X be a random vector with distribution $\mu \in \mathrm{GDOA}(\nu)$. Then for any $\theta \neq 0$ in $\mathbb{R}^d$ the random variable $<X,\theta>$ belongs to the domain of attraction of some nondegenerate normal law on $\mathbb{R}^1$.*

Proof. Let $Y = <X,\theta>$ denote the marginal of X in the direction θ. Proposition 8.1.6 shows that EX exists, and then

$$<X-EX,\theta> = <X,\theta> - E<X,\theta> = Y - EY.$$

Theorem 8.1.7 implies that the truncated variance function of X defined by (8.15) varies regularly at zero with exponent $(-1/2)I$, and it follows from Theorem 5.3.4 (b) that $M(\theta/r)$ is a regularly varying function of $r > 0$ with index $(-\beta/\alpha) = -2$. Let $\sigma_Y^2(r)$ denote the truncated variance of Y defined by (8.22). Since

$$\begin{aligned} M(\theta/r) &= E\big(<X-EX,\theta/r>^2 I(|<X-EX,\theta/r>| \le 1)\big) \\ &= r^{-2}E\big(<X-EX,\theta>^2 I(|<X-EX,\theta>| \le r)\big) \\ &= r^{-2}E\big((Y-EY)^2 I(|Y-EY| \le r)\big) \\ &= r^{-2}\sigma_Y^2(r) \end{aligned}$$

it follows that $\sigma_Y^2(r)$ is slowly varying at infinity, and then Theorem 8.1.11 yields that Y belongs to the domain of attraction of some nondegenerate normal law. □

Remark 8.1.13. In the one variable case, Corollary 8.1.8 shows that X is in the domain of attraction of the nondegenerate normal law with mean a and variance σ^2 if and only if there exists a regularly varying sequence (a_n) with index $-1/2$ such that $nM(a_n t_n) \to Q(t) = \sigma^2 t^2$ whenever $t_n \to t \neq 0$ in $\mathbb{R}^1$. Let $t_n = t = 1$ and apply (8.23) to obtain

$$n a_n^2 \sigma_X^2(a_n^{-1}) \to \sigma^2. \tag{8.24}$$

Feller [40] remarks that in this case we can construct a suitable sequence of norming operators by letting

$$a_n = \sup\{t : nt^2\sigma_X^2(t^{-1}) \le \sigma^2\}.$$

In the multivariable case, Hahn and Klass [49] show that there exists a sequence of orthonormal bases $t_{n1},\ldots,t_{nd}$ for each n such that if we take a_{nj} to be suitable norming constants for the one-dimensional marginal $<X,t_{nj}>$ then a suitable sequence of norming operators for (8.1) can be obtained by extending linearly, i.e., if we write x in terms of the basis $t_{n1},\ldots,t_{nd}$, we define

$$A_n x = A_n\left(\sum_{i=1}^{d} x_j t_{nj}\right) = \sum_{i=1}^{d} a_{nj} x_j t_{nj}.$$

In the finite second moment case, we can take $a_n = n^{1/2}$ in (8.24) and then $A_n = n^{-1/2}I$ for any choice of bases. But Hahn and Klass [49] give an example to show that, in the infinite second moment case, it may be necessary to choose specific rotating bases to obtain convergence in (8.1). It is not known how one can construct these orthonormal bases. However, our next result shows that these rotations must be very slow.

Corollary 8.1.14. *Suppose $\mu \in \mathrm{GDOA}(\nu)$ and ν is a full (t^E) operator stable law on $\mathbb{R}^d$ with exponent $E = aI$ for some $a \geq 1/2$. Then we may always choose the norming operators A_n in (8.1) to be of the form $A_n = n^{-a}L_n$ where L_n is slowly varying.*

Proof. Theorem 8.1.5 shows that we may always choose the norming operators A_n to vary regularly with index $-E = -aI$. If we let $L_n = n^a A_n$, then for any $\lambda > 0$ we have

$$L_{[\lambda n]}L_n^{-1} = (\lambda n)^a A_{[\lambda n]}A_n^{-1}n^{-1} = \lambda^a A_{[\lambda n]}A_n^{-1} \to \lambda^a\lambda^{-a} = I$$

in view of the regular variation of (A_n), so that (L_n) is slowly varying. □

Remark 8.1.15. If ν is normal, then Corollary 8.1.14 applies with $a = 1/2$. If (L_n) is slowly varying, then $L_{[\lambda n]} \sim L_n$ as $n \to \infty$ for all $\lambda > 0$. This is equivalent to

$$\begin{aligned} \frac{L_{[\lambda n]}t}{\|L_{[\lambda n]}t\|} - \frac{L_n t}{\|L_n t\|} &\to 0 \\ \frac{\|L_{[\lambda n]}t\|}{\|L_n t\|} &\to 1 \end{aligned}$$

for all nonzero $t \in \mathbb{R}^d$. Letting $r_n = \|L_n t\|$ and $\theta_n = L_n t/r_n$, we have $r_{[\lambda n]}/r_n \to 1$ and $\theta_{[\lambda n]} - \theta_n \to 0$. Then (r_n) is slowly varying, and (θ_n) is *very slowly varying*, i.e., if we represent θ_n in terms of coordinates (rectangular, polar, or spherical) $\theta_{n1}, \ldots, \theta_{nd}$, then we have that $\exp(\theta_{ni})$ is slowly varying for all i. Then the orthonormal bases in Remark 8.1.13 must rotate very slowly, like the logarithm of a slowly varying function.

In some applications, it is useful to consider the case of scalar norming. Recall from Definition 7.3.1 that if

$$a_n\mu^n * \varepsilon_{b_n} \Rightarrow \nu \quad \text{as } n \to \infty \tag{8.25}$$

for some $a_n > 0$ and $b_n \in \mathbb{R}^d$, we say that μ belongs to the domain of attraction of ν and we write $\mu \in \mathrm{DOA}(\nu)$. If n is replaced by an increasing sequence of positive integers $k_n \to \infty$ with $k_{n+1}/k_n \to c \geq 1$, then we say that μ belongs to the domain of semistable attraction of ν and we write $\mu \in \mathrm{DOSA}(\nu)$. Necessary and sufficient conditions in this case are stated in terms of the scalar-normed regular variation introduced in Definition 5.2.18.

Lemma 8.1.16. *Suppose that X is a random vector on $\mathbb{R}^d$ with distribution μ and $E(X) = 0$, and let F denote the truncated second moment function defined in (8.4). Suppose that ν is a full normal law on $\mathbb{R}^d$. Then the following are equivalent:*

(a) $\mu \in \mathrm{DOSA}(\nu, c)$;

(b) $\mu \in \mathrm{DOA}(\nu)$;

(c) *there exists a regularly varying sequence of positive real numbers (a_n) with index $(-1/2)$ such that*

$$a_n \mu^n * \varepsilon_{b_n} \Rightarrow \nu \tag{8.26}$$

as $n \to \infty$ for some $b_n \in \mathbb{R}^d$;

(d) *there exists a regularly varying sequence of positive real numbers (a_n) with index $(-1/2)$ such that*

$$nF(a_n t_n) \to Q(t) > 0 \tag{8.27}$$

whenever $t_n \to t$ in $\Gamma = \mathbb{R}^d \setminus \{0\}$; and

(e) *F is scalar-normed regularly varying with exponent $(-1/2)$ and limit function Q.*

In this case ν, has Lévy representation $[a, Q, 0]$. Furthermore, we can use the same sequence (a_n) in both (c) and (d), and it follows from either (8.26) or (8.27) that (a_n) varies regularly.

Proof. If (b) holds, then (8.26) holds for some norming constants (a_n). Then Lemma 8.1.1 with $A_n = a_n I$ implies that (8.27) holds, where $Q(t)$ is the quadratic form in the Lévy representation of ν. Then Q is admissible since ν is full. In the notation of Section 5.2, we can rewrite (8.27) in the form $n(a_n^{-1}F) \to Q$. In that notation, we also have $t \cdot Q = t^{-1/2}Q$ for all $t > 0$ since Q is a quadratic form. Hence $b = -1/2$ is an exponent of Q. Then Proposition 5.2.19 (d) implies that (a_n^{-1}) varies regularly with index $(1/2)$ and F is scalar-normed regularly varying with exponent $(-1/2)$. Then (a_n) varies regularly with index $(-1/2)$, and hence (c), (d), and (e) follow from (b). Conversely, if (e) holds, then Proposition 5.2.19 implies that (d) holds. Then Corollary 8.1.4 with $B_n = a_n I$ shows that (c) holds, where ν is a full normal law with Lévy representation $[a, Q, 0]$. Next, (b) follows by definition and finally, we can show that (a) is equivalent to (b) by specializing the proofs of Lemmas 3.3.27 and 7.5.8 to the case of scalar norming. □

Remark 8.1.17. If $\mu \in \mathrm{DOSA}(\nu, c)$ where ν is normal and

$$a_n \mu^{k_n} * \varepsilon_{s_n} \Rightarrow \nu \tag{8.28}$$

for some increasing sequence of positive integers (k_n) with $k_{n+1}/k_n \to c$, then Remark 7.5.14 yields a sequence of positive reals (b_n) such that $a_n = b_{k_n}$ and $b_n\mu^n * \varepsilon_{s'_n} \Rightarrow \nu$ for some suitable shifts s'_n. Then it follows from Lemma 8.1.16 that (b_n) varies regularly with index $-1/2$. Hence when ν is normal the norming constants (8.28) can always be embedded in a regularly varying sequence.

Theorem 8.1.18. *A random vector X on $\mathbb{R}^d$ belongs to the domain of attraction of a full normal law with Lévy representation $[a, Q, 0]$, if and only if EX exists and the truncated variance function defined by* (8.15) *is scalar-normed regularly varying at zero with exponent* $(-1/2)$ *and limit function Q.*

Proof. Suppose that $X, X_1, X_2, X_3, \ldots$ are i.i.d. random vectors on $\mathbb{R}^d$ whose distribution μ belongs to the domain of attraction of some full normal random vector Y on $\mathbb{R}^d$ whose distribution ν has Lévy representation $[a, Q, 0]$. Then

$$a_n(X_1 + \cdots X_n) + b_n \Rightarrow Y \tag{8.29}$$

for some $b_n \in \mathbb{R}^d$, and since this is of the form (8.2) with $A_n = a_n I$, we also have that $X \in \text{GDOA}(Y)$, so EX exists by virtue of Proposition 8.1.6. Define $X'_i = X_i - E(X_i)$ and $b'_n = b_n + na_nE(X_i)$. Then (8.29) holds if and only if

$$a_n(X'_1 + \cdots + X'_n) + b'_n \Rightarrow Y. \tag{8.30}$$

Since $M(t)$ is the truncated second moment function of $X - EX$, it follows from Lemma 8.1.16 that M is scalar-normed regularly varying at zero with exponent $b = (-1/2)$ and limit function Q.

Conversely, if M is scalar-normed regularly varying at zero with exponent $(-1/2)$ and limit function $Q(t)$, then Lemma 8.1.16 implies that $X - EX$ belongs to the domain of attraction of the full normal law ν on $\mathbb{R}^d$ with Lévy representation $[a, Q, 0]$. That is, (8.30) holds for some $b_n \in \mathbb{R}^d$, which implies that (8.29) holds. □

8.2 NONNORMAL LIMITS

After considering the case of a normal limit in Section 8.1, we now turn to the case of a full Poissonian limit ν with no normal component. In Section 8.3, we will combine the results of Section 8.1 with those developed here to characterize the domains of attraction of full limit distributions with both a normal and a Poissonian part. Recall from Definition 3.3.20 that $\mu \in \text{GDOSA}(\nu, c)$ if and only if there exist linear operators $A_n \in \text{L}(\mathbb{R}^d)$ and an increasing sequence of positive integers $k_n \to \infty$ with

$$k_{n+1}/k_n \to c \geq 1 \tag{8.31}$$

such that

$$A_n\mu^{k_n} * \varepsilon_{a_n} \Rightarrow \nu \tag{8.32}$$

for some nonrandom vectors $a_n \in \mathbb{R}^d$. If $X, X_1, X_2, \ldots$ are independent and identically distributed random vectors on $\mathbb{R}^d$ with common distribution μ and Y is a random vector with distribution ν then (8.32) is equivalent to

$$A_n(X_1 + \cdots + X_{k_n}) + a_n \Rightarrow Y \tag{8.33}$$

and we also write $X \in \mathrm{GDOSA}(Y, c)$. If (8.32) holds, then it follows from Lemma 3.3.21 that $A_n \in \mathrm{GL}(\mathbb{R}^d)$ for all large n. Hence the distribution of $S_n = X_1 + \cdots + X_n$ can be usefully approximated by that of $A_n^{-1}(Y - a_n)$ for n large.

In contrast to the normal law, which is automatically operator stable, for nonnormal laws we have to investigate generalized domains of semistable attraction and generalized domains of attraction separately. We begin with the more general case of operator semistable limit distributions.

Theorem 8.2.1. *Suppose X is a random vector on $\mathbb{R}^d$ with distribution μ. Then μ belongs to the generalized domain of semistable attraction of some full operator semistable law with no normal component if and only if $\mu \in \mathrm{ROV}_\infty(E, c)$ for some $c > 1$ and some linear operator E whose eigenvalues all have real part greater than $1/2$.*

Proof. Suppose that μ belongs to the generalized domain of semistable attraction of some full operator semistable law ν having no normal component. Then by Definition 3.3.20 there exist an increasing sequence $(\bar{k}_n)$ of natural numbers tending to infinity with $\bar{k}_{n+1}/\bar{k}_n \to \bar{c} \geq 1$ and $A_n \in \mathrm{L}(\mathbb{R}^d)$ such that for some $b_n \in \mathbb{R}^d$ we have $A_n \mu^{\bar{k}_n} * \varepsilon_{b_n} \Rightarrow \nu$ for some probability measure μ. If $\bar{c} = 1$, then by Lemma 7.1.5 ν is operator stable and in view of Lemma 7.5.8 $\mu \in \mathrm{GDOA}(\nu)$. Therefore, by Corollary 7.5.11 $\mu \in \mathrm{GDOSA}(\nu, c)$ for any $c > 1$ so we can, and hence will assume that $\bar{c} > 1$.

Apply Corollary 3.3.4 to see that ν is infinitely divisible, and hence by Theorem 3.1.11 has Lévy representation $[a, 0, \phi]$. Note that by Proposition 3.1.20 ϕ is $\mathcal{M}$-full. Then Theorem 3.3.8 yields $\bar{k}_n(A_n \mu) \to \phi$ as $n \to \infty$. An application of Theorem 6.2.10 yields $\mu \in \mathrm{ROV}(E, c)$, where $c = \bar{c}^2$ and $E \in \mathrm{L}(\mathbb{R}^d)$. That is, there exist an increasing sequence (k_n) of natural numbers tending to infinity with $k_{n+1}/k_n \to c$ and $(B_n) \in \mathrm{RV}(-E)$ such that

$$k_n(B_{k_n} \mu) \to \phi \quad \text{as } n \to \infty. \tag{8.34}$$

Theorem 6.2.10 also yields $c^E \phi = c \cdot \phi$, and hence by Lemma 7.1.6

$$\nu^c = c^E \nu * \varepsilon_b \tag{8.35}$$

for some $b \in \mathbb{R}^d$, so that ν is (c^E, c) operator semistable. Theorem 7.1.10 implies that the real parts of the eigenvalues of E are greater than $1/2$, and then Definition 6.2.1 yields $\mu \in \mathrm{ROV}_\infty(E, c)$.

Conversely, suppose that $\mu \in \mathrm{ROV}_\infty(E, c)$ for some $c > 1$. Then by Lemma 6.2.2 there exist an increasing sequence of natural numbers k_n tending to infinity with $k_{n+1}/k_n \to c$ and $(B_n) \in \mathrm{RV}(-E)$ such that (8.34) holds where ϕ is $\mathcal{M}$-full. Lemma 6.2.5 implies that $c^E\phi = c \cdot \phi$, and then it follows from Lemma 6.3.11 with $\zeta = 2$ and $a_1 > 1/2$ that ϕ satisfies (3.6), hence ϕ is a Lévy measure. Lemma 7.1.6 implies that the infinitely divisible law ν with Lévy representation $[a, 0, \phi]$ satisfies (8.35), so that by Theorem 7.1.10 ν is (c^E, c) operator semistable.

Now in order to show that $\mu \in \mathrm{GDOSA}(\nu, c)$, in view of (8.34) and Proposition 3.3.11 it suffices to show that

$$\lim_{\varepsilon\to 0} \limsup_{n\to\infty} k_n \int_{|<x,\theta>|\le\varepsilon} <x,\theta>^2 B_{k_n}\mu\{dx\} = 0 \tag{8.36}$$

for any $\|\theta\| = 1$. Then Proposition 3.3.11 implies

$$B_{k_n}\mu^{k_n} * \varepsilon_{b_n} \Rightarrow \nu \quad \text{as } n \to \infty \tag{8.37}$$

for some $b_n \in \mathbb{R}^d$.

In order to show (8.36), write $B^*_{k_n}\theta = r_n\theta_n$ for some $r_n > 0$ and $\|\theta_n\| = 1$. Theorem 4.2.9 implies that the function $f(t) = B_{[t]}$ is regularly varying with index $-E$, and it follows that $(f(t)^*)^{-1}$ is regularly varying with index E^*. Since the real parts of the eigenvalues of E and E^* are the same, it follows from Corollary 4.2.6 with $\alpha = 0$ that $r_n \to 0$ as $n \to \infty$. Let U_2 denote the truncated second moment function defined by (8.8). Then the integral in (8.36) can be written in the form

$$\begin{aligned}
\int_{|<x,\theta>|\le\varepsilon} <x,\theta>^2 B_{k_n}\mu\{dx\} &= \int_{|<B_{k_n}x,\theta>|\le\varepsilon} <B_{k_n}x,\theta>^2 \mu\{dx\} \\
&= \int_{|<x,B^*_{k_n}\theta>|\le\varepsilon} <x,B^*_{k_n}\theta>^2 \mu\{dx\} \\
&= r_n^2 \int_{|<x,\theta_n>|\le r_n^{-1}\varepsilon} <x,\theta_n>^2 \mu\{dx\} \\
&= r_n^2 U_2(r_n^{-1}\varepsilon, \theta_n).
\end{aligned}$$

Let V_0 denote the tail function defined by (8.8). Then Proposition 1.2.19 along with (8.34) implies that

$$\begin{aligned}
\limsup_{n\to\infty} k_n V_0(r_n^{-1}, \theta_n) &= \limsup_{n\to\infty} k_n\mu\{x : |<x,\theta_n>| > r_n^{-1}\} \\
&= \limsup_{n\to\infty} k_n\mu\{x : |<x, B^*_{k_n}\theta>| > 1\} \\
&\le \limsup_{n\to\infty} k_n\mu\{x : |<B_{k_n}x,\theta>| \ge 1\} \\
&= \limsup_{n\to\infty} k_n(B_{k_n}\mu)\{x : |<x,\theta>| \ge 1\} \le M_\theta
\end{aligned} \tag{8.38}$$

where $M_\theta = \phi\{x \in \mathbb{R}^d : |<x,\theta>| \ge 1\} < \infty$. Since $a_1 > 1/2$ we can choose $\delta > 0$ so that $2 - \delta - 1/a_1 > 0$. Then Theorem 6.3.4 implies that V_0 is uniformly R–O varying, and that there exists a constant $C_1 > 0$ such that for any $0 < \varepsilon < 1$ we have

$$\frac{V_0(\varepsilon^{-1}(\varepsilon r_n^{-1}), \theta_n)}{V_0(\varepsilon r_n^{-1}, \theta_n)} \geq C_1 \varepsilon^{\delta + 1/a_1} \tag{8.39}$$

for all n sufficiently large. Also, Corollary 6.3.9 (b) implies that for some $m_2 > 0$ we have for all $0 < \varepsilon < 1$ that

$$\frac{(\varepsilon r_n^{-1})^2 V_0(\varepsilon r_n^{-1}, \theta_n)}{U_2(\varepsilon r_n^{-1}, \theta_n)} \geq m_2 \tag{8.40}$$

for all n sufficiently large. By putting this together with (8.38) we conclude that for any $0 < \varepsilon < 1$ we have

$$\begin{aligned} k_n r_n^2 U_2(\varepsilon r_n^{-1}, \theta_n) &= \varepsilon^2 \frac{U_2(\varepsilon r_n^{-1}, \theta_n)}{(\varepsilon r_n^{-1})^2 V_0(\varepsilon r_n^{-1}, \theta_n)} \cdot \frac{V_0(\varepsilon r_n^{-1}, \theta_n)}{V_0(\varepsilon^{-1}(\varepsilon r_n^{-1}), \theta_n)} k_n V_0(r_n^{-1}, \theta_n) \\ &\leq \varepsilon^2 m_2^{-1} C_1^{-1} \varepsilon^{-\delta - 1/a_1}(M_\theta + 1) \\ &= C \varepsilon^{\delta'} \end{aligned}$$

for all large n, where $C = (M_\theta + 1)/(m_2 C_1)$ and $\delta' = 2 - \delta - 1/a_1 > 0$. Then (8.36) holds and the proof is complete. □

Theorem 8.2.1 characterizes the measures that belong to the generalized domain of semistable attraction of some operator semistable law having no normal component. However, if we are given a specific (c^E, c) operator semistable law, then it is also interesting to characterize the generalized domain of semistable attraction of that distribution. Our next result gives necessary and sufficient conditions for a probability measure μ to belong to the generalized domain of semistable attraction of a full operator semistable law ν with no normal component. It also shows that the norming operators A_n in (8.32) can always be chosen from a regularly varying sequence.

Corollary 8.2.2. *Let μ be a probability measure on $\mathbb{R}^d$ and let ν be a full (c^E, c) operator semistable law on $\mathbb{R}^d$ having no normal component, where $c > 1$. Then the following are equivalent:*

(a) $\mu \in \mathrm{GDOSA}(\nu, c)$;

(b) $\mu \in \mathrm{ROV}_\infty(E, c)$ with limit measure ϕ;

(c) there exists a sequence $(B_n) \in \mathrm{RV}(-E)$ and an increasing sequence of positive integers (k_n) with $k_{n+1}/k_n \to c$ such that

$$k_n(B_{k_n}\mu) \to \phi \tag{8.41}$$

as $n \to \infty$, where ϕ is $\mathcal{M}$-full; and

(d) *there exists a sequence* $(B_n) \in \mathrm{RV}(-E)$ *and an increasing sequence of positive integers* (k_n) *with* $k_{n+1}/k_n \to c$ *such that*

$$B_{k_n}\mu^{k_n} * \varepsilon_{b_n} \Rightarrow \nu \tag{8.42}$$

as $n \to \infty$ *for some* $b_n \in \mathbb{R}^d$.

In this case, ν *has Lévy representation* $[a, 0, \phi]$. *Furthermore, we can use the same sampling sequence* (k_n) *and the same regularly varying sequence* (B_n) *in both (c) and (d).*

Proof. The proof of Theorem 8.2.1 shows that (a) implies (b) and that (b) implies (c) implies (d). In this case, we already know that ν is (c^E, c) operator semistable, so Lemma 7.1.6 implies that $c \cdot \phi = c^E \phi$, and then Corollary 6.2.15 shows that we can use the same $c > 1$ in (a) and (b). Finally, (a) follows from (d) by definition. □

Our next result establishes a weaker necessary and sufficient condition for a probability measure μ to belong to the generalized domain of semistable attraction of a given full operator semistable law ν having no normal component. In this condition, we drop the requirement that the norming operators are part of a regularly varying sequence.

Corollary 8.2.3. *Let* ν *be a full operator semistable law having no normal component and Lévy measure* ϕ. *Then* μ *belongs to the generalized domain of semistable attraction of* ν *if and only*

$$k_n(A_n\mu) \to \phi \quad \text{as } n \to \infty \tag{8.43}$$

for some linear operators $A_n \in \mathrm{L}(\mathbb{R}^d)$ *and some increasing sequence of positive integers* $k_n \to \infty$ *such that* $k_{n+1}/k_n \to \bar{c} > 1$.

Proof. If μ belongs to the generalized domain of semistable attraction of ν and (8.32) holds, then an application of Theorem 3.3.8 yields (8.43). On the other hand, if (8.43) holds, then Theorem 6.2.10 implies that $\mu \in \mathrm{ROV}(E, c)$ and $c \cdot \phi = c^E \phi$ for some $E \in \mathrm{L}(\mathbb{R}^d)$ and $c = \bar{c}^2$. Then it follows from Lemma 6.3.11 with $\zeta = 2$ and $a_1 > 1/2$ that ϕ satisfies (3.6), hence ϕ is a Lévy measure. Lemma 7.1.6 implies that the infinitely divisible law ν with Lévy representation $[a, 0, \phi]$ satisfies (8.35), and then Theorem 7.1.10 implies that ν is (c^E, c) operator semistable. But then Theorem 7.1.10 implies that the real parts of the eigenvalues of E are greater than $1/2$, so that $\mu \in \mathrm{ROV}_\infty(E, c)$. Then Corollary 8.2.2 shows that $\mu \in \mathrm{GDOSA}(\nu, c)$. □

Remark 8.2.4. Although (8.43) does imply that μ belongs to the generalized domain of semistable attraction of some full operator semistable law ν, the norming operators (A_n) and the sampling sequence (k_n) in (8.43) cannot necessarily be used in (8.32). In the proof of Theorem 6.2.10, we show (in the proof of Lemma 6.2.12) that (8.43) implies

$$k'_n(A'_n\mu) \to \phi \tag{8.44}$$

for some linear operators $A'_n \in \mathrm{L}(\mathbb{R}^d)$ and some increasing sequence of positive integers (k'_n) such that $k'_{n+1}/k'_n \to c = \bar{c}^2$. Then we construct $B_{k_n} = G_n A'_n$ regularly varying for some $G_n \in \mathcal{S}(\phi)$. Note that by Proposition 3.1.22 $\mathcal{S}(\phi) = \mathcal{S}^1(\nu)$. It then follows from Theorem 2.3.17 that we may replace B_{k_n} by A'_n in (8.42). However, the same does not apply to the original sequence (A_n). On the other hand, if ν is (c^E, c) operator semistable and (8.43) holds for some $k_{n+1}/k_n \to c > 1$ then Corollary 6.2.15 applies, and we can use the same (k_n) and (A_n) in (8.32).

As shown in Proposition 8.1.6, for measures attracted to a full normal law all moments less than 2 exist. The following two results investigate the moments of measures attracted to operator semistable laws having no normal component.

Theorem 8.2.5. *Suppose $c > 1$ and ν is a full (c^E, c) operator semistable law having no normal component. Let $a_1 < \cdots < a_p$ denote the real parts of the eigenvalues of E. Then for any $\mu \in \mathrm{GDOSA}(\nu, c)$ there exists a function $\rho : \Gamma \to \{a_p^{-1}, \ldots, a_1^{-1}\}$ such that for all $\theta \in \Gamma = \mathbb{R}^d \setminus \{0\}$ we have:*

$$\int |\langle x, \theta\rangle|^\rho \mu\{dx\} \tag{8.45}$$

exists for all $0 \le \rho < \rho(\theta)$ and the integral diverges for all $\rho > \rho(\theta)$.

Proof. If $\mu \in \mathrm{GDOSA}(\nu, c)$, then Corollary 8.2.2 implies $\mu \in \mathrm{ROV}_\infty(E, c)$. Let X be a random vector with distribution μ and fix $\theta \in \Gamma$. By using the well-known formula

$$E|\langle X, \theta\rangle|^\rho = \rho \int_0^\infty t^{\rho-1} P\{|\langle X, \theta\rangle| > t\} dt$$

it follows that the integral in (8.45) exists if and only if

$$\int_{t_0}^\infty t^{\rho-1} V_0(t, \theta) dt < \infty \tag{8.46}$$

for some $t_0 > 0$. In view of Theorem 6.4.15 (a) (ii), there exists a function $\alpha : \Gamma \to \{-1/a_1, \ldots, -1/a_p\}$ such that for any $\delta > 0$ there exist constants $B_1, B_2 > 0$ and $t_0 > 0$ such that for $t \ge t_0$

$$B_1 t^{\alpha(\theta)-\delta} \le V_0(t, \theta) \le B_2 t^{\alpha(\theta)+\delta}. \tag{8.47}$$

Let $\rho(\theta) = -\alpha(\theta)$. Then (8.46) follows immediately from (8.47). □

Corollary 8.2.6. *Suppose $c > 1$ and ν is a full (c^E, c) operator semistable law having no normal component, and let $a_1 < \cdots < a_p$ denote the real parts of the eigenvalues of E. Then for a random vector X with distribution $\mu \in \mathrm{GDOSA}(\nu, c)$ we have that $E\|X\|^\rho$ exists if $\rho < 1/a_p$ and is infinite if $\rho > 1/a_p$.*

Proof. Since all norms on $\mathbb{R}^d$ are equivalent it suffices to consider the Euclidean norm $\|\cdot\|_2$. If $X = (X_1, \ldots, X_d)$, then $\|X\|^\rho \leq d^{\rho/2}(|X_1|^\rho + \cdots + |X_d|^\rho)$. If $\rho < 1/a_p = \min\{\rho(\theta) : \|\theta\| = 1\}$, then by Theorem 8.2.5 we know $E|X_i|^\rho = E|\langle X, e_i\rangle|^\rho < \infty$ for $i = 1, \ldots, d$ where $e_1, \ldots, e_d$ denotes the standard basis of $\mathbb{R}^d$. Now suppose that $\rho > 1/a_p$. Apply Corollary 8.2.2 to see that $\mu \in \mathrm{ROV}_\infty(E, c)$. Then choose an orthonormal basis $b_1, \ldots, b_d$ for $\mathbb{R}^d$ such that $b_1 \in W_p$, where $W_1, \ldots, W_p$ is the spectral decomposition of Theorem 6.4.8. Write $X = (X_1, \ldots, X_d)$ with respect to these coordinates, so that $X_i = \langle X, b_i\rangle$. Since $b_1 \in W_p$ we have by Definition 6.4.9 that $\rho(b_1) = -\alpha(b_1) = 1/a_p$, and then $E\|X\|^\rho = E(X_1^2 + \cdots + X_d^2)^{\rho/2} \geq E|X_1|^\rho = \infty$ by Theorem 8.2.5. □

Theorem 8.2.7. *Suppose X is a random vector whose distribution μ belongs to the generalized domain of semistable attraction of a full (c^E, c) operator semistable law ν with Lévy measure ϕ and no normal component, and let $a_1 < \cdots < a_p$ be the real parts of the eigenvalues of E.*

(a) *If $a_p < 1$, then we may choose $a_n = -k_n E(A_n X)$ in (8.32), so we center to zero expectation. In this case, the characteristic function of ν can also be written as $\exp(\psi_2(t))$ with $a_2 = 0$, where $\psi_2(t)$ is given by (3.19).*

(b) *If $a_1 > 1$, then we may choose $a_n = 0$ in (8.32), so no centering is required. In this case, the characteristic function of ν can be written in the form $\exp(\psi_1(t))$ with $a_1 = 0$, where $\psi_1(t)$ is given by (3.17).*

Proof. If $a_p < 1$, then Corollary 8.2.6 implies that $E(X)$ exists. Furthermore, Corollary 8.2.2 shows that for some $(B_n) \in \mathrm{RV}(-E)$ we can take $A_n = B_{k_n}$ in (8.32). From Theorem 3.3.8 it follows that we may choose

$$b_n = -k_n \int_{\|x\|<R} x(A_n\mu)\{dx\} \tag{8.48}$$

for any $R > 0$ such that the domain of integration is a ϕ-continuity set. Part (a) states that in (8.32) we can take $b_n = -k_n E(A_n X_1)$. Hence we want to show that

$$d_n = -k_n \int_{\|x\|>R} x(A_n\mu)\{dx\}$$

can be made arbitrary close to zero in norm for all large n by taking $R > 1$ sufficiently large. Given $\|\theta\| = 1$ write $A_n^*\theta = r_n\theta_n$ for some $r_n > 0$ and

$\|\theta_n\| = 1$. Then we have

$$\begin{aligned}
|<d_n,\theta>| &\le k_n \int_{\|x\|>R} |<x,\theta>|(A_n\mu)\{dx\} \\
&= k_n \int_{|<x,\theta>|>R} |<x,\theta>|(A_n\mu)\{dx\} \\
&\qquad + k_n \int_{A(R,\theta)} |<x,\theta>|(A_n\mu)\{dx\} \\
&= k_n r_n V_1(R r_n^{-1}, \theta_n) + k_n \int_{A(R,\theta)} |<x,\theta>|(A_n\mu)\{dx\}
\end{aligned} \tag{8.49}$$

where V_1 is the truncated first tail moment function defined by

$$V_1(r,\theta) = E\big(|<X,\theta>| I(|<X,\theta>| > r)\big) \tag{8.50}$$

and $A(R,\theta) = \{x : \|x\| > R, |<x,\theta>| \le R\}$. The second integral on the right-hand side of (8.49) is bounded above by $Rk_n(A_n\mu)\{x : \|x\| > R\}$. Choose $\{\theta_k\}$ an orthonormal basis for $\mathbb{R}^d$ and note that for some constant $M > 0$ (we can take $M = d^{-1/2}$) we have

$$\{x \in \mathbb{R}^d : \|x\| > R\} \subset \bigcup_{k=1}^{d} \{x \in \mathbb{R}^d : |<x,\theta_k>| > MR\}.$$

Write $A_n^*\theta_k = B_{k_n}^*\theta_k = r_{nk}\theta_{nk}$ for some $r_{nk} > 0$ and $\|\theta_{nk}\| = 1$. Theorem 4.2.9 implies that the function $f(t) = B_{[t]}$ is regularly varying with index $-E$, and it follows that $(f(t)^*)^{-1}$ is regularly varying with index E^*. Since the real parts of the eigenvalues of E and E^* are the same, and since by Theorem 7.1.10 the real parts of the eigenvalues of E are all greater than $1/2$, it follows from Corollary 4.2.6 with $\alpha = 0$ that $r_{nk} \to 0$ as $k \to \infty$. Then

$$\begin{aligned}
k_n \int_{A(R,\theta)} |<x,\theta>|(A_n\mu)\{dx\} &\le Rk_n(A_n\mu)\{x : \|x\| > R\} \\
&\le Rk_n \sum_{k=1}^{d} (A_n\mu)\{x : |<x,\theta_k>| > MR\} \\
&= Rk_n \sum_{k=1}^{d} V_0(r_{nk}^{-1} MR, \theta_{nk}).
\end{aligned}$$

Apply Corollary 8.2.2 to see that $\mu \in \mathrm{ROV}_\infty(E,c)$. Since $a_p < 1$ we can choose $\delta > 0$ such that $1/a_p > 1 + \delta$. Then Theorem 6.3.4 implies that for some positive constants C_1 and C_2 we have

$$C_1\lambda^{-\delta-1/a_1} \le \frac{V_0(r\lambda,\theta)}{V_0(r,\theta)} \le C_2\lambda^{\delta-1/a_p} \tag{8.51}$$

for all large r, all $\lambda \geq 1$ and all $||\theta|| = 1$. Then in view of (8.38) we see that we can bound each individual summand in the sum above for all large n by

$$\begin{aligned} Rk_n V_0(r_{nk}^{-1} MR, \theta_{nk}) &= R \cdot \frac{V_0(r_{nk}^{-1} MR, \theta_{nk})}{V_0(r_{nk}^{-1}, \theta_{nk})} \cdot k_n V_0(r_{nk}^{-1}, \theta_{nk}) \\ &\leq R \cdot C_2 (MR)^{\delta - 1/a_p} \cdot (M_{\theta_k} + 1) = KR^{-\delta'} \end{aligned}$$

where $K = C_2(M_{\theta_k} + 1) M^{\delta - 1/a_p}$ and $\delta' = -1 - \delta + 1/a_p > 0$. Hence the second integral in (8.49) can be made arbitrary small for all large n by taking $R > 1$ large enough.

Theorem 6.3.16 implies that the truncated second moment function U_2 defined by (8.8) is uniformly R–O varying on $||\theta|| = 1$, and Theorem 6.3.23 implies that the first tail moment function V_1 defined by (8.50) exists. Then Corollary 6.3.9 (a) implies that for some positive constant M_2 we have

$$\frac{rV_1(r, \theta)}{U_2(r, \theta)} \leq M_2 \tag{8.52}$$

for all $||\theta|| = 1$ and all r sufficiently large. Since $a_p < 1$ we can choose $\delta > 0$ such that $1/a_p > 1 + \delta$. Then Theorem 6.3.16 also implies that for some positive constant C_2 we have

$$\frac{U_2(\lambda r, \theta)}{U_2(r, \theta)} \leq C_2 \lambda^{2 + \delta - 1/a_p} \tag{8.53}$$

for all $||\theta|| = 1$ and all r sufficiently large. We have already observed that V_0 is uniformly R–O varying on $||\theta|| = 1$, and so Corollary 6.3.9 (b) implies that for some $m_2 > 0$ we have

$$\frac{r^2 V_0(r, \theta)}{U_2(r, \theta)} \geq m_2 \tag{8.54}$$

for all $||\theta|| = 1$ and all r sufficiently large. Then in view of (8.38) we see that for all large n we have

$$\begin{aligned} k_n r_n V_1(Rr_n^{-1}, \theta_n) &= R^{-1} \cdot \frac{Rr_n^{-1} V_1(Rr_n^{-1}, \theta_n)}{U_2(Rr_n^{-1}, \theta_n)} \cdot \frac{U_2(Rr_n^{-1}, \theta_n)}{U_2(r_n^{-1}, \theta_n)} \\ &\quad \cdot \frac{U_2(r_n^{-1}, \theta_n)}{r_n^{-2} V_0(r_n^{-1}, \theta_n)} \cdot k_n V_0(r_n^{-1}, \theta_n) \\ &\leq R^{-1} \cdot M_2 \cdot C_2 R^{2 + \delta - 1/a_p} \cdot \frac{1}{m_2} \cdot (M_\theta + 1) = KR^{-\delta'} \end{aligned}$$

where $K = M_2 C_2 (M_\theta + 1)/m_2$ and $\delta' = -1 - \delta + 1/a_p > 0$. Hence the first term in (8.49) can be made arbitrary small by taking $R > 1$ sufficiently large, too. This shows that we can center to zero expectation in this case.

Theorem 3.3.8 implies that the limit ν has characteristic function $e^{\psi_R(t)}$ where $\psi_R(t)$ is given by (3.15) with $a_R = 0$. Since $a_p < 1$ Lemma 6.3.17 implies that (3.18) holds, so that the characteristic function of ν can also be written in the form $\exp(\psi_2(t))$ where $\psi_2(t)$ is given by (3.19) with

$$a_2 = a_R + \int_{\|x\|\geq R} x\phi\{dx\}.$$

Hence a_2 vanishes as $R \to \infty$.

The proof of (b) is similar: We want to show that b_n in (8.48) can be made arbitrary close to zero in norm for all large n by taking $R > 0$ sufficiently small. Fix $\|\theta\| = 1$ and write $A_n^*\theta = r_n\theta_n$ for $r_n > 0$ and $\|\theta_n\| = 1$. Let

$$U_1(r,\theta) = E\big(|\langle X,\theta\rangle| I(|\langle X,\theta\rangle| \leq r)\big) \tag{8.55}$$

denote the truncated first moment function. Since V_0 is uniformly R–O varying on $\|\theta\| = 1$ Corollary 6.3.9 (b) implies that for some $m_2 > 0$ we have

$$\frac{rV_0(r,\theta)}{U_1(r,\theta)} \geq m_2 \tag{8.56}$$

for all $\|\theta\| = 1$, for all r sufficiently large. Since $a_1 > 1$ we can choose $\delta > 0$ such that $1 - \delta > 1/a_1$. Then Theorem 6.3.4 implies that for some positive constants C_1 and C_2 (8.51) holds for all large r, all $\lambda \geq 1$ and all $\|\theta\| = 1$. Hence, for any fixed $0 < R < 1$ we have using (8.38) that for all large n

$$\begin{aligned}
|\langle b_n,\theta\rangle| &\leq k_n \int_{|\langle x,\theta\rangle|\leq R} |\langle x,\theta\rangle|(A_n\mu)\{dx\} \\
&= k_n r_n U_1(Rr_n^{-1},\theta_n) \\
&= R\cdot\frac{U_1(Rr_n^{-1},\theta_n)}{Rr_n^{-1}V_0(Rr_n^{-1},\theta_n)}\cdot\frac{V_0(Rr_n^{-1},\theta_n)}{V_0(R^{-1}(Rr_n^{-1}),\theta_n)}\cdot k_nV_0(r_n^{-1},\theta_n) \\
&\leq R\cdot\frac{1}{m_2}\cdot\frac{1}{C_1(R^{-1})^{-\delta-1/a_1}}\cdot(M_\theta+1) = KR^{\delta'}
\end{aligned}$$

where $K = (M_\theta + 1)/(m_2C_1)$ and $\delta' = 1 - \delta - 1/a_1 > 0$. Hence the centering term b_n in (8.48) can be made arbitrary close to zero in norm for all large n by taking $R > 0$ sufficiently small.

Again Theorem 3.3.8 implies that the limit ν has characteristic function $\exp(\psi_R(t))$ where $\psi_R(t)$ is given by (3.15) with $a_R = 0$. Since $a_1 > 1$ Lemma 6.3.11 implies that (3.16) holds, so that the characteristic function of ν can also be written in the form $\exp(\psi_1(t))$ where $\psi_1(t)$ is given by (3.17) with

$$a_1 = a_R - \int_{0<\|x\|<R} x\phi\{dx\},$$

hence a_1 vanishes as $R \to 0$. □

Remark 8.2.8. The limit ν in Theorem 8.2.7 (a) has mean $m = 0$. The proof of Theorem 8.2.7 shows that ν has characteristic function

$$\hat{\nu}(t) = \exp\left(\int_{x \neq 0} \left(e^{i<t,x>} - 1 - i<t,x> \right) \phi\{dx\} \right), \tag{8.57}$$

and then it is easy to check that $\nu^c = c^E \nu$. By taking $k_n = [c^n]$ and $A_n = c^{-nE}$, the proof of Theorem 7.1.1 shows that $A_n \nu^{k_n} \Rightarrow \nu$. But Theorem 8.2.7 (a) shows that $A_n \nu^{k_n} * \varepsilon_{a_n} \Rightarrow \nu$, where $a_n = -k_n A_n m$, hence $a_n \to 0$. Since $a_p < 1$ it follows from Corollary 4.2.6 that $k_n \|A_n x\| \to \infty$ for all $x \neq 0$, so $m = 0$. This fact also follows from Remark 3.1.15, using Theorem 1 in [183].

Now, we consider the special case of semistable limits. Recall from Definition 7.4.1 that Y is (b, c) semistable if it is (B, c) operator semistable where $B = bI$, and that X belongs to the domain of semistable attraction of Y if (8.33) holds with $A_n = a_n I$ for some $a_n > 0$. Recall from Theorem 7.4.3 that for Y nonnormal the index $\alpha = \log c / \log b$ satisfies $0 < \alpha < 2$. The exponent ρ in the next result is related to the semistable index α by $\rho = 1/\alpha$.

Theorem 8.2.9. *Let μ be a probability measure on $\mathbb{R}^d$ and let ν be a full nonnormal (b, c) semistable law on $\mathbb{R}^d$ with index $\alpha = \log c / \log b$. Then the following are equivalent:*

(a) $\mu \in \mathrm{DOSA}(\nu, c)$.

(b) *μ is scalar normed R–O varying with exponent $\rho > 1/2$ and limit measure ϕ in* (6.52).

(c) *There exist an increasing sequence of natural numbers (k_n) tending to infinity with $k_{n+1}/k_n \to c$ and a sequence of positive reals (a_n) such that*

$$k_n(a_n \mu) \to \phi \quad \textit{as } n \to \infty \tag{8.58}$$

where ϕ is $\mathcal{M}$-full and $c \cdot \phi = c^\rho \phi$ for some $\rho > 1/2$.

(d) *There exist an increasing sequence of natural numbers (k_n) tending to infinity with $k_{n+1}/k_n \to c$ and a sequence of positive reals (a_n) such that*

$$a_n \mu^{k_n} * \varepsilon_{b_n} \Longrightarrow \nu \quad \textit{as } n \to \infty \tag{8.59}$$

for some $b_n \in \mathbb{R}^d$.

In this case, ν has Lévy triple $[a, 0, \phi]$ for some $a \in \mathbb{R}^d$, and there exists a regularly varying sequence of positive reals (b_n) with index $-\rho$ such that $a_n = b_{k_n}$. Furthermore, we can use the same (a_n) in both (c) and (d).

Proof. (a) and (d) are equivalent by Definition 7.4.1. If (d) holds, then Theorem 3.3.8 with $A_n = a_n I$ implies that (8.58) holds, where the Lévy measure ϕ of ν is $\mathcal{M}$-full in view of Proposition 3.1.20 and $c \cdot \phi = c^\rho \phi$ for some $\rho > 1/2$

holds true in view of Theorem 7.4.3. If (c) holds, then Proposition 6.2.17 shows that μ is scalar normed R–O varying with exponent $\rho > 1/2$ and limit measure ϕ, so (b) holds. If (b) holds, then (c) follows by parts (a) and (c) of Proposition 6.2.17. Finally, if (c) holds, then Corollary 8.2.2 with $B_{k_n} = a_n I$ shows that (d) holds, and Proposition 6.2.17 (d) shows that $a_n = b_{k_n}$ for some regularly varying sequence of positive reals (b_n) with index $-\rho$. □

Now we consider the special case of operator stable limits. Every operator stable law is also operator semistable, and the generalized domain of attraction of a given operator stable law coincides with its generalized domain of semistable attraction. Hence many of the properties listed here follow immediately from the more general case. However, the necessary and sufficient conditions for a probability measure to belong to some generalized domain of attraction are understandably stronger. They are stated in terms of regular variation of the measure, rather than the weaker assumption of R–O variation.

Theorem 8.2.10. *Suppose X is a random vector on $\mathbb{R}^d$ with distribution μ. Then μ belongs to the generalized domain of attraction of a full operator stable law having no normal component if and only if $\mu \in \mathrm{RVM}_\infty(E)$ for some linear operator E whose eigenvalues all have real part greater than $1/2$.*

Proof. Suppose that $A_n\mu^n * \varepsilon_{b_n} \Rightarrow \nu$ as $n \to \infty$ for some linear operators $A_n \in \mathrm{L}(\mathbb{R}^d)$, constants $b_n \in \mathbb{R}^d$, and some full Poissonian probability measure ν on $\mathbb{R}^d$. Corollary 3.3.4 implies that ν is infinitely divisible and hence, by Theorem 3.1.11 has Lévy representation $[a, 0, \phi]$. Proposition 3.1.20 implies that ϕ is $\mathcal{M}$-full, and Theorem 3.3.8 yields

$$n(A_n\mu) \to \phi \quad \text{as } n \to \infty \tag{8.60}$$

and hence by Theorem 6.1.24 there exist an exponent $E \in \mathcal{E}(\phi)$ and a sequence $(B_n) \in \mathrm{RV}(-E)$ such that μ varies regularly with exponent E and

$$n(B_n\mu) \to \phi \quad \text{as } n \to \infty \tag{8.61}$$

where by Proposition 6.1.2 we have $t^E\phi = t \cdot \phi$ for any $t > 0$. Then in view of Lemma 7.1.6 we have $\nu^t = t^E\nu * \varepsilon_{a_t}$ for all $t > 0$ and suitable $a_t \in \mathbb{R}^d$. Hence ν is (t^E) operator stable, and so it follows from Theorem 7.2.1 that the real parts of the eigenvalues of E are greater than $1/2$, which by Definition 6.1.1 shows $\mu \in \mathrm{RVM}_\infty(E)$ with limit measure ϕ.

On the other hand, if $\mu \in \mathrm{RVM}_\infty(E)$ with limit measure ϕ, then Proposition 6.1.10 (b) implies that (8.61) holds for some sequence $(B_n) \in \mathrm{RV}(-E)$ and some ϕ $\mathcal{M}$-full. Theorem 6.2.3 implies that $\mu \in \mathrm{ROV}_\infty(E, c)$ for any $c > 1$, and now the rest of the proof is similar to Theorem 8.2.1. It follows as before that the limit measure ϕ in (8.61) is a Lévy measure. Then in view of (8.61) and Proposition 3.3.11, in order to show that $\mu \in \mathrm{GDOA}(\nu)$ it suffices to show that for any $\|\theta\| = 1$ we have

$$\lim_{\varepsilon\to 0}\limsup_{n\to\infty} n\int_{|<x,\theta>|\le\varepsilon} <x,\theta>^2 (B_n\mu)\{dx\} = 0. \tag{8.62}$$

Let U_2, V_0 be the truncated second moment and tail functions defined in (8.8). Set $A_n^*\theta = r_n\theta_n$ for some $r_n > 0$ and $\|\theta_n\| = 1$, and write the integral in (8.62) in the form $r_n^2 U_2(r_n^{-1}, \theta_n)$. It follows from Proposition 1.2.19 along with (8.61) that

$$nV_0(r_n^{-1}, \theta_n) = n(B_n\mu)\{x : |\langle x, \theta\rangle| > 1\} \le M_\theta + 1 \tag{8.63}$$

for all large n, where $M_\theta = \phi\{x : |\langle x, \theta\rangle| \ge 1\}$. Since $\mu \in \mathrm{ROV}_\infty(E, c)$ for $c > 1$ Theorem 6.3.4 implies that $V_0(r, \theta)$ is R–O varying uniformly on $\|\theta\| = 1$. Then for any $0 < \varepsilon < 1$ we have as in the proof of Theorem 8.2.1 that (8.39) and (8.40) hold for all large n. Hence for all large n we have

$$\begin{aligned} nr_n^2 U_2(\varepsilon r_n^{-1}, \theta_n) &= \varepsilon^2 \frac{U_2(\varepsilon r_n^{-1}, \theta_n)}{(\varepsilon r_n^{-1})^2 V_0(\varepsilon r_n^{-1}, \theta_n)} \cdot \frac{V_0(\varepsilon r_n^{-1}, \theta_n)}{V_0(\varepsilon^{-1}(\varepsilon r_n^{-1}), \theta_n)} \cdot nV_0(r_n^{-1}, \theta_n) \\ &\le \varepsilon^2 m_2^{-1} C_1^{-1} \varepsilon^{-\delta - 1/a_1}(M_\theta + 1) \\ &= C\varepsilon^{\delta'} \end{aligned}$$

where $C = (M_\theta + 1)/(m_2 C_1)$ and $\delta' = 2 - \delta - 1/a_1 > 0$, which implies that (8.62) holds. □

Our next result gives necessary and sufficient conditions for a probability measure μ to belong to the generalized domain of attraction of a given full operator stable law ν with no normal component. It also shows that the norming operators A_n in (3.96) can always be chosen to be regularly varying.

Corollary 8.2.11. *Let μ be a probability measure on $\mathbb{R}^d$ and let ν be a full (t^E) operator stable law on $\mathbb{R}^d$ having no normal component. Then the following are equivalent:*

(a) $\mu \in \mathrm{GDOA}(\nu)$;

(b) $\mu \in \mathrm{RVM}_\infty(E)$ with limit measure ϕ;

(c) there exists a sequence $(B_n) \in \mathrm{RV}(-E)$ such that

$$n(B_n\mu) \to \phi \tag{8.64}$$

as $n \to \infty$ where ϕ is $\mathcal{M}$-full; and

(d) there exists a sequence $(B_n) \in \mathrm{RV}(-E)$ such that

$$B_n\mu^n * \varepsilon_{b_n} \Rightarrow \nu \quad \text{as } n \to \infty \tag{8.65}$$

for some $b_n \in \mathbb{R}^d$.

In this case, ν has Lévy representation $[a, 0, \phi]$. Furthermore, we can use the same sequence (B_n) in both (c) and (d).

Proof. The proof of Theorem 8.2.10 shows that (a) implies (b), and that (b) implies (c) implies (d). Finally, (d) implies (a) by definition. □

Our next result establishes a weaker necessary and sufficient condition for a probability measure μ to belong to the generalized domain of attraction of a given full operator stable law ν having no normal component. In this condition, we drop the requirement that the norming operators are part of a regularly varying sequence. Notice that this result is somewhat stronger than the analogous result, Corollary 8.2.3, for operator semistable limits, since in the case of operator stable limits we can use the same norming operators in both (8.1) and (8.66).

Corollary 8.2.12. *Let ν be a full operator stable law having no normal component and Lévy measure ϕ. Then $\mu \in \mathrm{GDOA}(\nu)$ and* (8.1) *holds if and only if*

$$n(A_n\mu) \to \phi \quad \text{as } n \to \infty \tag{8.66}$$

for some linear operators $A_n \in \mathrm{L}(\mathbb{R}^d)$. In this case, $\|A_n\| \to 0$.

Proof. If $\mu \in \mathrm{GDOA}(\nu)$ and (8.1) holds, then Theorem 3.3.8 yields (8.66). If (8.66) holds, then Theorem 6.1.24 implies that there exist $E \in \mathrm{L}(\mathbb{R}^d)$ and $(B_n) \in \mathrm{RV}(-E)$ such that (8.64) holds. Then Corollary 8.2.11 shows that $\mu \in \mathrm{GDOA}(\nu)$ and (8.65) holds. Theorem 6.1.24 also implies that $B_n = G_n A_n$ for some $G_n \in \mathcal{S}(\phi)$. Since by Proposition 3.1.22 $\mathcal{S}(\phi) = \mathcal{S}(\nu)$ it follows from the Convergence of Types Theorem for probability measures, Theorem 2.3.17 along with (8.65) that (8.1) holds. Theorem 4.2.9 implies that the function $f(t) = B_{[t]}$ is regularly varying with index $-E$. Since the eigenvalues of E all have positive real part, and since $\mathcal{S}(\phi)$ is a compact group, it follows from Theorem 4.2.4 with $\beta = 0$ that $\|A_n\| \le \|G_n^{-1}\| \cdot \|B_n\| \to 0$ as $n \to \infty$. □

Remark 8.2.13. In Remark 3.3.30, we noted that (8.66) is also necessary and sufficient for the weak convergence of certain point processes. Hence $\mu \in \mathrm{GDOA}(\nu)$ is also equivalent to the convergence of these point processes when the limit ν is Poissonian. An alternative proof of Corollary 8.2.12 can be obtained using point process theory in place of the theory of infinitely divisible laws and triangular arrays, see Resnick [152].

Theorem 8.2.14. *Let ν be a full (t^E) operator stable law having no normal component and let $a_1 < \cdots < a_p$ denote the real parts of the eigenvalues of E. Then for any $\mu \in \mathrm{GDOA}(\nu)$ there exists a function $\rho : \Gamma \to \{a_p^{-1}, \ldots, a_1^{-1}\}$ such that for all $\theta \in \Gamma = \mathbb{R}^d \setminus \{0\}$ we have:*

$$\int |\langle x, \theta\rangle|^\rho \mu\{dx\} \tag{8.67}$$

exists for all $0 \le \rho \le \rho(\theta)$ and the integral diverges for all $\rho > \rho(\theta)$.

Proof. If ν is a full (t^E) operator stable law having no normal component then it follows from Theorems 7.2.1 and 7.1.10 that ν is (c^E, c) operator semistable for every $c > 1$. If $\mu \in \text{GDOA}(\nu)$, then Corollary 7.5.11 implies that $\mu \in \text{GDOSA}(\nu, c)$. Now apply Theorem 8.2.5. □

Corollary 8.2.15. *Let ν be a full (t^E) operator stable law having no normal component and let $a_1 < \cdots < a_p$ denote the real parts of the eigenvalues of E. Then for a random vector X with distribution $\mu \in \text{GDOA}(\nu)$ we have that $E\|X\|^\rho$ exists if $\rho < 1/a_p$ and is infinite if $\rho > 1/a_p$.*

Proof. The proof is identical to Corollary 8.2.6, using Theorem 8.2.14 in place of Theorem 8.2.5. □

The next result is analogous to Theorem 8.2.7 for operator semistable limits, but it is somewhat stronger. Here, in the case of an operator stable limit, we show that the centering constants can be replaced by a more convenient form without altering the sequence of norming operators (A_n).

Theorem 8.2.16. *Suppose X belongs to the generalized domain of attraction of a full (t^E) operator stable law Y without normal component and (8.2) holds. Let $a_1 < \cdots < a_p$ denote the real parts of the eigenvalues of E.*

(a) If $a_p < 1$, then (8.2) also holds with $k_n = n$ and $a_n = -nE(A_n X)$, so we center to zero expectation. In this case, the characteristic function of ν can also be written as $\exp(\psi_2(t))$ with $a_2 = 0$, where $\psi_2(t)$ is given by (3.19).

(b) If $a_1 > 1$, then (8.2) also holds with $a_n = 0$, so no centering is required. In this case, the characteristic function of ν can be written in the form $\exp(\psi_1(t))$ with $a_1 = 0$, where $\psi_1(t)$ is given by (3.17).

Proof. If $a_p < 1$, then $E(X)$ exists by virtue of Corollary 8.2.15. Corollary 8.2.11 implies that $\mu \in \text{RVM}_\infty(E)$ and then Theorem 6.2.3 implies that $\mu \in \text{ROV}_\infty(E, c)$ for every $c > 1$. Corollary 8.2.12 shows that (8.66) holds, and this convergence along with Proposition 1.2.19 shows that for any unit vector $\theta \in \mathbb{R}^d$ we have

$$n(A_n\mu)\{x : |\langle x, \theta\rangle| > 1\} \le M_\theta + 1 \tag{8.68}$$

for all large n, where $M_\theta = \phi\{x : |\langle x, \theta\rangle| \ge 1\}$. The rest of the proof is identical to Theorem 8.2.7, if we substitute $k_n = n$ and use (8.68) in place of (8.38). □

Now we consider the special case of multivariable stable limits. Recall from Definition 7.3.1 that Y is stable with index α if it is operator stable with exponent $E = aI$ for $\alpha = 1/a$, and that X belongs to the domain of attraction of Y if (8.33) holds with $k_n = n$ and $A_n = b_n I$ for some $b_n > 0$. Recall from Definition 6.1.36 that a probability measure μ is scalar-normed regularly varying with exponent $a > 0$ if μ is regularly varying according to Definition 6.1.1 with scalar norming operators $f(t) = a(t)I$ for some $a(t) > 0$ and limit measure ϕ satisfying $\lambda \cdot \phi = \lambda^a \phi$ for all $\lambda > 0$.

Theorem 8.2.17. *Let μ be a probability measure on $\mathbb{R}^d$ and let ν be a full nonnormal stable law on $\mathbb{R}^d$. Then the following are equivalent:*

(a) μ is scalar-normed regularly varying with exponent $a > 1/2$ and limit measure ϕ;

(b) there exists a sequence of positive reals (a_n) such that

$$n(a_n \mu) \to \phi \tag{8.69}$$

as $n \to \infty$ where ϕ is $\mathcal{M}$-full with exponent $a > 1/2$;

(c) there exists a sequence of positive reals (a_n) such that

$$a_n \mu^n * \varepsilon_{b_n} \Rightarrow \nu \quad \text{as } n \to \infty \tag{8.70}$$

for some $b_n \in \mathbb{R}^d$; and

(d) $\mu \in \mathrm{DOA}(\nu)$.

In this case, ν is stable with index $0 < \alpha < 2$ and Lévy representation $[a, 0, \phi]$, and (a_n) is regularly varying with index $(-a)$, where $a = 1/\alpha$. Furthermore, we can use the same sequence (a_n) in both (b) and (c).

Proof. If (d) holds, then (8.70) holds for some norming constants (a_n). Then Theorem 3.3.8 with $k_n = n$ and $A_n = a_n I$ implies that (8.69) holds, where the Lévy measure ϕ of ν is $\mathcal{M}$-full in view of Proposition 3.1.20. Then it follows from Proposition 6.1.37 that (a_n) varies regularly with some index $-a$, μ is scalar-normed regularly varying with exponent a, and the limit measure ϕ has exponent a as well. Since ν is a nonnormal stable law it follows from Theorem 7.3.3 that the stable index $\alpha \in (0, 2)$, and Definition 7.3.1 shows that ν is also operator stable with exponent $E = aI$, where $a = 1/\alpha$ so that $a > 1/2$. Hence (c), (b), and (a) follow from (d).

Conversely, if (a) holds then Proposition 6.1.37 shows that (b) holds, where (a_n) is regularly varying with index $(-a)$. Then it follows from Lemma 6.3.11 with $\zeta = 2$ and $a_1 = a > 1/2$ that ϕ satisfies (3.6), hence ϕ is a Lévy measure. Lemma 7.1.6 together with Theorem 7.2.1 imply that the infinitely divisible law ν with Lévy representation $[a, 0, \phi]$ is (t^E) operator stable with $E = aI$, and then Definition 7.3.1 shows that ν is stable with index α. Then (c) follows from Corollary 8.2.11 with $B_n = a_n I$, and (d) follows by definition. □

Recall from Theorem 7.3.16 that a full nonnormal stable law on $\mathbb{R}^d$ has distribution $S_\alpha(\Lambda, b)$ parameterized by the stable index $0 < \alpha < 2$, center $b \in \mathbb{R}^d$ and spectral measure Λ, a finite Borel measure on the unit sphere S in $\mathbb{R}^d$. Rvačeva [158] gave the following necessary and sufficient conditions for a random vector to belong to the domain of attraction of a nondegenerate nonnormal stable law.

Theorem 8.2.18. *A random vector X on $\mathbb{R}^d$ belongs to the domain of attraction of some full nonnormal stable law with distribution $S_\alpha(\Lambda, b)$ if and only if $V(r) = P(\|X\| > r)$ varies regularly with index $-\alpha$ and*

$$P\left(\tfrac{X}{\|X\|} \in D \big| \|X\| > r\right) = \frac{P(\|X\| > r, \frac{X}{\|X\|} \in D)}{V(r)} \to \frac{\Lambda(D)}{\Lambda(S)} \tag{8.71}$$

as $r \to \infty$ for all Borel subsets D of the unit sphere $S = \{\theta \in \mathbb{R}^d : \|\theta\| = 1\}$ such that $\Lambda(\partial D) = 0$, for some $\mathcal{M}$-full finite Borel measure[2] *Λ on the unit sphere.*

Proof. If $\mu \in \mathrm{DOA}(\nu)$, where ν is a full nonnormal stable law with distribution $S_\alpha(\Lambda, b)$ and Lévy measure ϕ, then Theorem 8.2.17 implies that μ is scalar-normed regularly varying with exponent $a = 1/\alpha$ and limit measure ϕ, and that (8.69) holds with (a_n) regularly varying with index $(-a)$. Then Lemma 4.2.16 (a) with $k_n = n$, $c = 1$ and $x_r = x = 1$ implies that

$$n(r) = \sup\{n : |a_n r| \geq 1\}$$

is well defined for all r large, and $n = n(r)$ tends to infinity and $a_n r \to 1$ as $r \to \infty$. Then $r = a_n^{-1}(1 + \varepsilon_n)$, where $\varepsilon_n \to 0$ as $n = n(r) \to \infty$.

It follows from Proposition 6.1.37 (a) that $\lambda \cdot \phi = \lambda^a \phi$ for all $\lambda > 0$. Then for all $\lambda > 0$ we have

$$\lambda \cdot \phi\{x : \|x\| = 1\} = \lambda^a \phi\{x : \|x\| = 1\} = \phi\{x : \|x\| = \lambda^{-a}\}.$$

If one of these sets has positive measure, then they all do, which contradicts the fact that ϕ is σ-finite. Hence every sphere $\{x : \|x\| = \lambda\}$ is ϕ-continuity set. Similarly, for all $\lambda > 0$ we have $\lambda \cdot \phi\{x : \|x\| > 1\} = \phi\{x : \|x\| > \lambda^{-a}\}$, and a simple substitution yields that $\phi\{x : \|x\| > \lambda\} = \lambda^{-\alpha}\phi\{x : \|x\| > 1\}$ for all $\lambda > 0$.

Hence for all $\lambda > 0$ we have

$$\begin{aligned}
\frac{P(\|X\| > \lambda r)}{P(\|X\| > r)} &= \frac{n\mu\{x : \|x\| > \lambda r\}}{n\mu\{x : \|x\| > r\}} \\
&= \frac{n\mu\{x : \|x\| > \lambda a_n^{-1}(1 + \varepsilon_n)\}}{n\mu\{x : \|x\| > a_n^{-1}(1 + \varepsilon_n)\}} \\
&= \frac{n a_n \mu\{x : \|x\| > \lambda(1 + \varepsilon_n)\}}{n a_n \mu\{x : \|x\| > (1 + \varepsilon_n)\}} \\
&\to \frac{\phi\{x : \|x\| > \lambda\}}{\phi\{x : \|x\| > 1\}} = \lambda^{-\alpha}
\end{aligned}$$

[2] Here, we identify the measure Λ with a measure on $\mathbb{R}^d \setminus \{0\}$, which is supported on S.

so that $V(r) = P(\|X\| > r)$ varies regularly with index $-\alpha$. Furthermore, if D is a Borel subset of S such that $\Lambda(\partial D) = 0$, then

$$\begin{aligned}\frac{P(\|X\| > r, \frac{X}{\|X\|} \in D)}{V(r)} &= \frac{n\mu\{x : \|x\| > r, \frac{x}{\|x\|} \in D\}}{n\mu\{x : \|x\| > r\}} \\ &= \frac{n\mu\{x : \|x\| > a_n^{-1}(1+\varepsilon_n), \frac{x}{\|x\|} \in D\}}{n\mu\{x : \|x\| > a_n^{-1}(1+\varepsilon_n)\}} \\ &= \frac{na_n\mu\{x : \|x\| > (1+\varepsilon_n), \frac{x}{\|x\|} \in D\}}{na_n\mu\{x : \|x\| > (1+\varepsilon_n)\}} \\ &\to \frac{\phi\{x : \|x\| > 1, \frac{x}{\|x\|} \in D\}}{\phi\{x : \|x\| > 1\}} = \frac{\Lambda(D)}{\Lambda(S)}.\end{aligned}$$

Conversely, if $V(t) = P(|X| > t)$ varies regularly with index $-\alpha$ for some $0 < \alpha < 2$ and (8.71) holds, define $R(t) = \sigma^\alpha V(t)/\Lambda(S)$ where σ is given by (7.21) with $C = 1$ if $\alpha \neq 1$ and by (7.23) with $C = 1$ if $\alpha = 1$, so that $R(t)$ varies regularly with index $\beta = -\alpha$. Let $a(t) = 1/t$ so that $a(t)$ varies regularly with index $\rho = -1$. Then for any $\lambda > 0$ and any Borel subset D of S such that $\Lambda(\partial D) = 0$ we have

$$\begin{aligned}\frac{a(t)\mu\{x : \|x\| > \lambda, \frac{x}{\|x\|} \in D\}}{R(t)} &= \frac{\mu\{x : \|a(t)x\| > \lambda, \frac{x}{\|x\|} \in D\}}{R(t)} \\ &= \frac{\mu\{x : \|x\| > \lambda t, \frac{x}{\|x\|} \in D\}}{R(t)} \\ &= \frac{P(X > \lambda t, \frac{X}{\|X\|} \in D\}}{R(\lambda t)} \cdot \frac{R(\lambda t)}{R(t)} \\ &\to \frac{\Lambda(D)}{\sigma^\alpha} \cdot \lambda^{-\alpha}\end{aligned}$$

as $t \to \infty$. Define a σ-finite Borel measure ϕ on $\Gamma = \mathbb{R}^d \setminus \{0\}$ by letting

$$\phi\{x : \|x\| > \lambda, \tfrac{x}{\|x\|} \in D\} = \frac{\Lambda(D)}{\sigma^\alpha} \cdot \lambda^{-\alpha}.$$

It follows from Example 1.2.22 that sets of this form determine convergence in $\mathcal{M}$, and hence μ is scalar-normed regularly varying according to Definition 6.1.36 with exponent $a = (\rho/\beta) > 1/2$ and limit measure ϕ. Then it follows from Theorem 8.2.17 that μ belongs to the domain of attraction of a full nonnormal stable law ν with Lévy measure ϕ and index $0 < \alpha < 2$, and Theorem 7.3.16 implies that Λ is the spectral measure of ν. □

In the special case of random variables $d = 1$, we recover the result of Feller [40] giving necessary and sufficient conditions for a random variable to belong to the domain of attraction of a nondegenerate nonnormal stable law.

Corollary 8.2.19. *A random variable X on $\mathbb{R}^1$ belongs to the domain of attraction of some nondegenerate nonnormal stable law with index $0 < \alpha < 2$ and distribution $S_\alpha(\sigma, \beta, b)$ if and only if $V(r) = P(|X| > r)$ varies regularly with index $-\alpha$ and*

$$\frac{P(X > r)}{V(r)} \to p \quad \text{and} \quad \frac{P(X < -r)}{V(r)} \to q \tag{8.72}$$

as $r \to \infty$ for some $p \geq 0$, $q \geq 0$ with $p + q = 1$. In this case, $\beta = p - q$, and if (8.70) holds for some $b_n \in \mathbb{R}^1$, then the Lévy measure ϕ of ν has the canonical representation (7.19) with this p, q for some $C > 0$ depending on the norming sequence (a_n). Then σ is given by (7.21) if $\alpha \neq 1$ or (7.23) if $\alpha = 1$.

Proof. If the distribution μ of X belongs to the domain of attraction of some nondegenerate nonnormal stable law ν with index $0 < \alpha < 2$ and Lévy measure ϕ, then Corollary 7.3.4 implies that

$$\begin{aligned} \phi(r, \infty) &= pCr^{-\alpha} \\ \phi(-\infty, -r) &= qCr^{-\alpha} \end{aligned} \tag{8.73}$$

for every $r > 0$, for some $C > 0$ and $p \geq 0$, $q \geq 0$ with $p + q = 1$, i.e., (7.19) holds. In this case, the unit sphere is the discrete set $S = \{-1, +1\}$ and $\Lambda\{+1\} = \phi(r, \infty) = Cp\sigma^\alpha$, $\Lambda\{-1\} = \phi(-\infty, -r) = Cq\sigma^\alpha$ where σ is given by (7.21) if $\alpha \neq 1$ or (7.23) if $\alpha = 1$. Now the result follows from Theorem 8.2.18 with $p = \Lambda\{+1\}/\Lambda(S)$ and $q = \Lambda\{-1\}/\Lambda(S)$. □

Remark 8.2.20. Suppose that X is a random variable on $\mathbb{R}^1$ whose distribution $\mu \in \mathrm{DOA}(\nu)$ and (8.70) holds, where ν is stable with index $0 < \alpha < 2$ and Lévy measure (8.73). Then Theorem 8.2.17 shows that (8.69) holds. If we let $V(r) = P(|X| > r)$ then this implies that

$$nV(a_n^{-1}) = na_n\mu\{x : |x| > 1\} \to \phi\{x : |x| > 1\} = C \tag{8.74}$$

In this case, we also have $nV(a_n^{-1}r) \to Cr^{-\alpha}$ since V varies regularly with index $-\alpha$. The convergence (8.74) can be used to construct suitable norming operators. In fact, we can take

$$a_n = \sup\{t > 0 : nV(t) \geq C\}$$

for any $C > 0$. If (8.70) holds for one sequence (a_n), which results in a limit with Lévy measure (8.73), then we can always arrange that $C = 1$ by substituting the norming operators $C^{-1/\alpha}a_n$.

8.3 GENERAL LIMITS

In this section, we establish the general limit theorems that are the main purpose of this book. Our strategy is to combine the results of Sections 8.1 and 8.2

using the spectral decomposition of Section 4.3. Recall that $\mu \in \text{GDOSA}(\nu, c)$ if there exist linear operators $A_n \in \text{L}(\mathbb{R}^d)$, an increasing sequence of positive integers $k_n \to \infty$ with $k_{n+1}/k_n \to c \geq 1$ and nonrandom vectors $b_n \in \mathbb{R}^d$ such that

$$A_n \mu^{k_n} * \varepsilon_{a_n} \Rightarrow \nu. \tag{8.75}$$

In view of Remark 7.5.13, we may assume that ν is a full (c^E, c) operator semistable law and $c > 1$. We begin by showing that we can always choose norming operators A_n in (8.75) from a regularly varying sequence.

Theorem 8.3.1. *Assume $\mu \in \text{GDOSA}(\nu, c)$ for some full (c^E, c) operator semistable law ν and (8.75) holds. Then there exists a sequence $(B_n) \in \text{RV}(-E)$ such that for a suitable choice of centering constants (b_n) we have*

$$B_{k_n} \mu^{k_n} * \varepsilon_{b_n} \Rightarrow \nu \quad \text{as } n \to \infty. \tag{8.76}$$

Furthermore, there exist symmetries $G_n \in \mathcal{S}^1(\nu)$ such that $B_{k_n} = G_n A_n$.

Proof. The proof is similar to Theorem 8.1.5. Since ν is (c^E, c) operator semistable, Theorem 7.1.10 shows that $\nu^c = c^E \nu * \varepsilon_b$ for some $b \in \mathbb{R}^d$. By using the algebra introduced in Section 3.1, this implies that $\nu^{1/c} = c^{-E} \nu * \varepsilon_s$ where $s = -c^{-E} b/c$. It follows immediately from (8.75) that

$$A_{n+1} \mu^{k_{n+1}} * \varepsilon_{a_{n+1}} \Rightarrow \nu,$$

and then Proposition 3.3.7 implies that

$$A_{n+1} \mu^{[k_{n+1} t_n]} * \varepsilon_{a_{n+1}(t_n)} \Rightarrow \nu^t$$

whenever $t_n \to t > 0$, where $a_{n+1}(t_n) = [k_{n+1} t_n] a_{n+1}/k_{n+1}$. By letting $t_n = k_n/k_{n+1}$ and $t = 1/c$, it follows that

$$A_{n+1} \mu^{k_n} * \varepsilon_{a_{n+1}(t_n)} \Rightarrow \nu^{1/c} = c^{-E} \nu * \varepsilon_s$$

as $n \to \infty$. Then in view of (8.75) we obtain using the Convergence of Types Theorem 2.3.17 that

$$A_{n+1} A_n^{-1} \rightharpoonup c^{-E} \mathcal{S}^1(\nu) \quad \text{as } n \to \infty.$$

By Theorem 2.3.10, we have that $\mathcal{S}^1(\nu)$ is a compact subgroup of $\text{GL}(\mathbb{R}^d)$, and Proposition 7.1.13 implies that $c^E \mathcal{S}^1(\nu) = \mathcal{S}^1(\nu) c^E$. Then Theorem 4.2.13 implies that there exists a sequence (B_n) varying regularly with index $-E$ such that $B_{k_n} = G_n A_n$ for some $G_n \in \mathcal{S}^1(\nu)$, and another application of Theorem 2.3.17 shows that (8.76) holds for some $b_n \in \mathbb{R}^d$. □

Remark 8.3.2. In the presence of a large degree of symmetry in the limit, the norming operators in (8.75) may exhibit a wild behavior. For example, if μ has covariance matrix I, the classical Central Limit Theorem, Corollary

8.1.10, implies that (8.75) holds with $k_n = n$, $A_n = n^{-1/2}I$ and ν standard normal. But since the symmetry group of ν is the entire orthogonal group on $\mathbb{R}^d$, it follows from the Convergence of Types Theorem 2.3.17 that (8.75) also holds for $A_n = n^{-1/2}U_n$ for any sequence (U_n) of orthogonal transformations. Theorem 8.3.1 shows that we can always *untwist* the sequence of norming operators A_n in (8.75) by selecting appropriate symmetries G_n, so that the resulting sequence $G_n A_n$ belongs to a regularly varying sequence. This rules out most of the spurious rotations and yields a reasonably well-behaved sequence of norming operators.

Recall from Definition 2.1.15 that for any linear operator E on $\mathbb{R}^d$ there exists a direct sum decomposition $\mathbb{R}^d = V_1 \oplus \cdots \oplus V_p$ into E-invariant subspaces such that, if we write $E = E_1 \oplus \cdots \oplus E_p$, where $E_i : V_i \to V_i$, then every eigenvalue of E_i has real part equal to a_i and $a_1 < \cdots < a_p$. This is called the spectral decomposition of $\mathbb{R}^d$ relative to E. Suppose ν is a full (c^E, c) operator semistable law on $\mathbb{R}^d$. If we let $\pi_i : \mathbb{R}^d \to V_i$ denote the projections associated with this direct sum decomposition, then Lemma 7.1.17 shows that each ν_i is a full (c^{E_i}, c) operator semistable law on the space V_i. If $a_1 = 1/2$, then ν_1 is normal, otherwise $a_i > 1/2$ and ν_i is a full nonnormal operator semistable law, which is spectrally simple, since every eigenvalue of E_i has real part equal to a_i. This is called the spectral decomposition of ν with respect to E.

Definition 8.3.3. Suppose that $\mu \in \mathrm{GDOSA}(\nu, c)$ and (8.75) holds, where ν is a full (c^E, c) operator semistable law on $\mathbb{R}^d$. Let $V_1, \ldots, V_p$ denote the spectral decomposition of $\mathbb{R}^d$ relative to E, and $\pi_i : \mathbb{R}^d \to V_i$ the projections associated with this direct sum decomposition. If every V_i is an A_n-invariant subspace for each n, then we say that μ is *spectrally compatible* with ν, and we call the measures $\mu_i = \pi_i \mu$ the *spectral decomposition* of μ with respect to E.

Lemma 8.3.4. *Suppose that $\mu \in \mathrm{GDOSA}(\nu, c)$ and (8.75) holds, where ν is a full (c^E, c) operator semistable law on $\mathbb{R}^d$, and let $\nu_1, \ldots, \nu_p$ denote the spectral decomposition of ν with respect to E. If μ is spectrally compatible with ν, then the measures $\mu_i = \pi_i \mu \in \mathrm{GDOSA}(\nu_i, c)$ for each $i = 1, \ldots, p$. Furthermore, we can write $A_n = A_{1n} \oplus \cdots \oplus A_{pn}$ with respect to this direct sum decomposition, and then*

$$A_{in} \mu_i^{k_n} * \varepsilon_{a_{in}} \Rightarrow \nu_i \tag{8.77}$$

for each $i = 1, \ldots, p$, where $a_{in} = \pi_i a_n$.

Proof. Since each of the subspaces $V_1, \ldots, V_p$ are A_n-invariant we may write $A_n = A_{1n} \oplus \cdots \oplus A_{pn}$ with respect to this direct sum decomposition. Since $\pi_i A_n = A_{in} \pi_i$ we also have $\pi_i(A_n \mu^{k_n} * \varepsilon_{a_n}) \Rightarrow \pi_i \nu = \nu_i$, where

$$\pi_i(A_n \mu^{k_n} * \varepsilon_{a_n}) = A_{in} \pi_i \mu^{k_n} * \varepsilon_{\pi_i a_n} = A_{in} \mu_i^{k_n} * \varepsilon_{a_{in}}$$

so that (8.77) holds, and hence $\mu_i \in \mathrm{GDOSA}(\nu_i, c)$ for each $i = 1, \ldots, p$. □

Recall that two full probability measures ν, ν' are the same type if $\nu' = T\nu * \varepsilon_s$ for some $T \in \mathrm{GL}(\mathbb{R}^d)$ and $s \in \mathbb{R}^d$. Lemma 3.3.22 shows that if ν, ν' are of the same type then $\mathrm{GDOSA}(\nu) = \mathrm{GDOSA}(\nu')$.

Lemma 8.3.5. *Suppose ν is a full (c^E, c) operator semistable law on $\mathbb{R}^d$ and $V_1, \ldots, V_p$ is the spectral decomposition of $\mathbb{R}^d$ with respect to E. Then for any $T \in \mathrm{GL}(\mathbb{R}^d)$ and $s \in \mathbb{R}^d$ the probability measure $\nu' = T\nu * \varepsilon_s$ is a full $(c^{E'}, c)$ operator semistable law on $\mathbb{R}^d$ with exponent $E' = TET^{-1}$, and the spectral decomposition of $\mathbb{R}^d$ with respect to E' is $W_1, \ldots, W_p$, where $W_i = T(V_i)$ for each $i = 1, \ldots, p$.*

Proof. Theorem 7.1.10 shows that $\nu^c = c^E \nu * \varepsilon_b$ for some $b \in \mathbb{R}^d$, and since $Tc^E T^{-1} = c^{E'}$ this implies that

$$\begin{aligned}(\nu')^c &= (T\nu * \varepsilon_s)^c = T\nu^c * \varepsilon_{cs} = T(c^E \nu * \varepsilon_b) * \varepsilon_{cs} \\ &= Tc^E \nu * \varepsilon_{Tb} * \varepsilon_{cs} = c^{E'}(T\nu) * \varepsilon_{Tb+cs} = c^{E'} \nu' * \varepsilon_{s'}\end{aligned}$$

where $s' = Tb + cs - s$, so that ν' is a full $(c^{E'}, c)$ operator semistable law. If $p(x)$ is any polynomial, then $p(E') = Tp(E)T^{-1}$, hence both E, E' have the same minimal polynomial. Recall that if we factor the minimal polynomial of E into $f_1(x) \cdots f_p(x)$ where all roots of f_i have real part a_i and $a_i < a_j$ for $i < j$, then $V_i = \mathrm{Ker}(f_i(E))$. Then $W_i = \mathrm{Ker}(f_i(E'))$ defines the spectral decomposition of $\mathbb{R}^d$ with respect to E'. If $y \in W_i$, then $f_i(TET^{-1})y = Tf_i(E)T^{-1}y = 0$ which implies that $x = T^{-1}y \in V_i$, and conversely if $x \in V_i$ and $y = Tx$, then $f_i(TET^{-1})y = 0$, hence $W_i = T(V_i)$. □

If μ belongs to the generalized domain of semistable attraction of some full (c^E, c) operator semistable law ν on $\mathbb{R}^d$, then Lemma 3.3.22 shows that μ also belongs to the generalized domain of semistable attraction of any other ν' of the same type. Indeed, if (8.75) holds and $\nu' = T\nu * \varepsilon_s$ then it follows from Convergence of Types, Theorem 2.3.17, that

$$A'_n \mu^{k_n} * \varepsilon_{a'_n} \Rightarrow \nu'$$

where $A'_n = TA_n$ and $a'_n = a_n + s$. If μ is spectrally compatible with ν and $V_1, \ldots, V_p$ is the spectral decomposition of $\mathbb{R}^d$ with respect to E, then $A'_n(V_i) = TA_n(V_i) = T(V_i) = W_i$, and hence μ is not spectrally compatible with ν' unless $W_i = V_i$ for each $i = 1, \ldots, p$. Hence we cannot decompose the sequence (A'_n) as we did before. On the other hand, if we are given a probability distribution μ that belongs to the generalized domain of attraction of some operator semistable law ν, then Theorem 8.3.6 shows that we can always find another operator semistable law ν' of the same type such that μ is spectrally compatible with ν'.

Theorem 8.3.6. *For any $\mu \in \mathrm{GDOSA}(\nu, c)$, there exists $T \in \mathrm{GL}(\mathbb{R}^d)$ such that $\mu' = T\mu$ is spectrally compatible with ν. Equivalently, μ is spectrally compatible with $\nu' = T^{-1}\nu$.*

Proof. Suppose $\mu \in \text{GDOSA}(\nu, c)$, where ν is a full (c^E, c) operator semistable law on $\mathbb{R}^d$. Apply Theorem 8.3.1 to see that there exists a sequence $(B_n) \in \text{RV}(-E)$ such that (8.76) holds for a suitable choice of centering constants (b_n). By Theorem 4.3.10 there exist $T \in \text{GL}(\mathbb{R}^d)$ and $(D_n) \in \text{RV}(-E)$ such that $B_n \sim D_n T$ and every V_i in the spectral decomposition of $\mathbb{R}^d$ with respect to E is D_n invariant for all n. It follows from (8.76) by Convergence of Types, Theorem 2.3.17, that

$$D_n T \mu^{k_n} * \varepsilon_{b_n} \Rightarrow \nu$$

and hence $\mu' = T\mu$ is spectrally compatible with ν. On the other hand, we also have that

$$T^{-1} D_n T \mu^{k_n} * \varepsilon_{b_n} \Rightarrow T^{-1}\nu$$

where $\nu' = T^{-1}\nu$ is a $(c^{E'}, c)$ operator semistable law with $E' = T^{-1}ET$ in view of Lemma 8.3.5, and the spectral decomposition of $\mathbb{R}^d$ with respect to E' is $W_1, \ldots, W_p$ where $W_i = T^{-1}(V_i)$. Since $D_n(V_i) = V_i$ we also have $T^{-1}D_nT(W_i) = T^{-1}D_n(V_i) = T^{-1}(V_i) = W_i$, proving that μ is spectrally compatible with ν'. □

In some applications, it is useful to choose the linear operator T in Theorem 8.3.6 to have a special form, so that the subspaces in the resulting spectral decomposition are mutually orthogonal. We take this opportunity to record some of the important properties of this construction, for later reference.

Theorem 8.3.7. *Suppose μ belongs to the generalized domain of semistable attraction of some full operator semistable law ν' on $\mathbb{R}^d$. Then (8.76) holds for some suitable choice of centering constants (b_n), where the limit ν and the norming operators (B_n) are chosen so that:*

(a) ν is a full (c^E, c) operator semistable law on $\mathbb{R}^d$ of the same type as ν' and (B_n) is regularly varying with index $-E$;

(b) the subspaces W_i in the spectral decomposition of $\mathbb{R}^d$ with respect to E are mutually orthogonal and B_n invariant, so that μ is spectrally compatible with ν;

(c) if we write $E = E_1 \oplus \cdots \oplus E_p$ and $B_n = B_{1n} \oplus \cdots \oplus B_{pn}$ with respect to this direct sum decomposition then each $B_{in} : W_i \to W_i$ is regularly varying with index E_i and every eigenvalue of E_i has real part a_i;

(d) if $\pi_i : \mathbb{R}^d \to W_i$ denotes orthogonal projection and $\mu_i = \pi_i\mu$, $\nu_i = \pi_i\nu$, then ν_i is a full (c^{E_i}, c) operator semistable law on W_i, $\mu_i \in \text{GDOSA}(\nu_i, c)$ and

$$B_{ik_n} \mu_i^{k_n} * \varepsilon_{b_{in}} \Rightarrow \nu_i \tag{8.78}$$

for each $i = 1, \ldots, p$ with $b_{in} = \pi_i b_n$.

Proof. If μ belongs to some generalized domain of semistable attraction, then Remark 7.5.13 show that for some full $(c^{E'}, c)$ operator semistable law ν' and some sequence of linear operators (B'_n) we have

$$B'_{k_n}\mu^{k_n} * \varepsilon_{b'_n} \Rightarrow \nu' \quad \text{as } n \to \infty \tag{8.79}$$

for a suitable choice of centering constants (b'_n) and a sampling sequence (k_n) with $k_{n+1}/k_n \to c > 1$. In view of Theorem 8.3.1, we may also assume that (B'_n) varies regularly with index $-E'$. Now Corollary 4.3.12 implies that $TB'_n \sim B_n$ for some $T \in \mathrm{GL}(\mathbb{R}^d)$ and some (B_n) regularly varying with index $E = TE'T^{-1}$ such that (b) and (c) hold. Convergence of Types along with (8.79) yields that

$$T(B'_{k_n}\mu^{k_n} * \varepsilon_{b'_n}) \Rightarrow T\nu'$$

and hence (8.76) holds where $b_n = Tb'_n$ and $\nu = T\nu'$. Lemma 8.3.5 shows that $\nu = T\nu'$ is a full (c^E, c) operator semistable on $\mathbb{R}^d$ with $E = TE'T^{-1}$ and that the subspaces $W_i = T(V_i)$ comprise the spectral decomposition of $\mathbb{R}^d$ with respect to E. Then (a) holds, and Theorem 7.1.17 shows that each ν_i is a full (c^{E_i}, c) operator semistable law on W_i. Finally, it follows from Lemma 8.3.4 that $\mu_i \in \mathrm{GDOSA}(\nu_i, c)$ and (8.78) holds for each $i = 1, \ldots, p$ with $b_{in} = \pi_i b_n$, so that (d) holds. □

Theorem 8.3.7 shows that for any collection $X, X_1, X_2, X_3, \ldots$ of i.i.d. random vectors that belong to some generalized domain of semistable attraction, there exists a full limit Y and an orthonormal basis $v_1, \ldots, v_d$ for $\mathbb{R}^d$ such that

$$B_n(X_1 + \cdots + X_{k_n}) + b_n \Rightarrow Y \tag{8.80}$$

where B_n has a block-diagonal form

$$B_n = \begin{pmatrix} B_{1n} & & \\ & \ddots & \\ & & B_{pn} \end{pmatrix} \tag{8.81}$$

in this basis. If $\dim(W_i) = d_i$, then B_{in} is a square $d_i \times d_i$ matrix whose eigenvalues all have real part equal to a_i. In the special case where E has d distinct real eigenvalues, we can take B_n to be of the form

$$B_n = \begin{pmatrix} g_{1n} & & \\ 0 & \ddots & \\ & & g_{dn} \end{pmatrix} \tag{8.82}$$

where $g_{in} \neq 0$ are real numbers. This leads to the following interesting and useful corollary.

Corollary 8.3.8. *If X belongs to the generalized domain of semistable attraction of some full (c^E, c) operator semistable random vector Y' on $\mathbb{R}^d$ and E has d distinct real eigenvalues $a_1, \ldots, a_d$, then there exists an orthonormal basis $v_1, \cdots, v_d$ for $\mathbb{R}^d$ and a full (c^E, c) operator semistable random vector Y of the same type as Y' such that each of the marginals $<X, v_i>$ belongs to the domain of semistable attraction of $<Y, v_i>$, which is a nondegenerate (c^{a_i}, c) semistable law on $\mathbb{R}^1$. Furthermore, for $X_1, X_2, X_3, \ldots$ i.i.d. with X if we let $X_{in} = <X_n, v_i>$ and $Y_i = <Y, v_i>$, then there exist regularly varying sequences of positive real numbers (g_{in}) with index $-a_i$ such that*

$$g_{ik_n}(X_{i1} + \cdots X_{ik_n}) + b_{in} \Rightarrow Y_i \tag{8.83}$$

for each i, for some suitable sequence of real numbers (b_{in}).

Proof. Apply Theorem 8.3.7 to see that

$$B'_{k_n}(X_1 + \cdots X_{k_n}) + b'_n \Rightarrow Y'$$

for some full (c^E, c) operator semistable random vector Y' on $\mathbb{R}^d$, where B'_n is of the form

$$B'_n = \begin{pmatrix} g'_{1n} & & \\ & \ddots & \\ & & g'_{dn} \end{pmatrix}$$

with respect to some fixed orthonormal basis $v_1, \ldots, v_d$ for $\mathbb{R}^d$ and each (g'_{in}) is a regularly varying sequence of invertible linear operators (i.e., nonzero real numbers) on $\mathbb{R}^1$ with index a_i. Then it follows from Lemma 4.2.10 that $\text{sign}(g_{in}) \to \sigma_i = \pm 1$ for each i, i.e., $\text{sign}(g_{in})$ is eventually constant. Define a linear operator $T \in \text{GL}(\mathbb{R}^d)$ by setting $T(v_i) = \sigma_i v_i$ and extending by linearity. Then

$$TB'_{k_n}(X_1 + \cdots X_{k_n}) + Tb'_n \Rightarrow TY' \tag{8.84}$$

and TB'_n is diagonal with entries $\sigma_i g'_{in}$, which are eventually all positive. Since Y' is (c^E, c) operator semistable and $TET^{-1} = E$, it follows as in the proof of Theorem 8.3.7 that $Y = TY'$ is also (c^E, c) operator semistable. Define $B_n v_i = |g_{in}| v_i$ and extend by linearity so that B_n is of the form (8.82) with respect to $v_1, \ldots, v_d$. Since $B_n = TB'_n$ for all large n, the diagonal elements g_{in} form regularly varying sequences (with i fixed and letting n vary) of positive reals with index a_i, and it follows from (8.84) that (8.80) holds. Orthogonal projection onto the one-dimensional subspace spanned by v_i (i.e., the ith coordinate axis in this coordinate system) shows that (8.83) holds. Then Definition 7.4.1 shows that $<X, v_i> \in \text{DOSA}(Y_i, c)$ for each $i = 1, \ldots, d$. Furthermore, projecting the equation $\nu^c = c^E \nu * \varepsilon_s$ onto the subspace generated by v_i yields that each Y_i is (c^{a_i}, c) semistable. □

The following decomposition result will be used to reduce the study of GDOSA to the two cases, normal and nonnormal, which have already been investigated in Sections 8.1 and 8.2. In view of Theorem 8.3.6, this result is the best possible.

Theorem 8.3.9. *Suppose that ν is a full operator semistable law on $\mathbb{R}^d$ and let $\nu = \nu_1 * \nu_2$ be the Lévy decomposition of Definition 7.1.2, so that ν_1 is a full operator semistable law of Poisson type on V_1, ν_2 is a full Gaussian measure on V_2, and $\mathbb{R}^d = V_1 \oplus V_2$. Let $\pi_i : \mathbb{R}^d \to V_i$ be the natural projections associated with this direct sum decomposition, so that $\nu_i = \pi_i \nu$, and define $\mu_i = \pi_i \mu$.*[3] *If $\mu_i \in \mathrm{GDOSA}(\nu_i, c)$ for $i = 1, 2$, then $\mu \in \mathrm{GDOSA}(\nu, c)$, and conversely if $\mu \in \mathrm{GDOSA}(\nu', c)$, then there exists another full operator semistable law ν of the same type such that $\mu_i \in \mathrm{GDOSA}(\nu_i, c)$ for $i = 1, 2$. The limit ν can be chosen so that V_1 and V_2 are mutually orthogonal.*

Proof. In view of Remark 7.5.13, we can assume that $c > 1$. Suppose that $\mu_i \in \mathrm{GDOSA}(\nu_i, c)$ and

$$A_{in}\mu_i^{k_{in}} * \varepsilon_{a_{in}} \Rightarrow \nu_i \tag{8.85}$$

for $i = 1, 2$ for some sampling sequences (k_{in}) with $k_{i,n+1}/k_{in} \to c$, some linear operators A_{in} on V_i and some nonrandom centering vectors $a_{in} \in V_i$. Corollary 8.1.4 shows that $\mu_2 \in \mathrm{GDOA}(\nu_2)$, hence (8.77) holds for $i = 1, 2$ with $k_n = k_{1n}$, i.e., we can use the same sampling sequence for both components. Define $A_n = A_{1n} \oplus A_{2n}$ and $a_n = a_{1n} + a_{2n}$ so that $\pi_i A_n = A_{in}\pi_i$ and $\pi_i a_n = a_{in}$.

Let $[a, Q, \phi]$ denote the Lévy representation of ν, and recall from the proof of Theorem 7.1.1 that ν_1 has Lévy representation $[a_1, 0, \phi]$ and ν_2 has Lévy representation $[a_2, Q, 0]$, where we have written $a = a_1 + a_2$ in terms of the direct sum decomposition $\mathbb{R}^d = V_1 \oplus V_2$. Then Corollary 3.3.10 implies that

$$k_n(A_{1n}\mu_1) \to \phi \tag{8.86}$$

and the proof of Theorem 8.2.1 shows that

$$\lim_{\varepsilon \to 0} \limsup_{n \to \infty} k_n \int_{|<x,t>|<\varepsilon\|t\|} <x,t>^2 A_{1n}\mu_1\{dx\} = 0 \tag{8.87}$$

for all $t \in \mathbb{R}^d$, while Corollary 3.3.9 implies that

$$k_n(A_{2n}\mu_2) \to 0 \tag{8.88}$$

and

$$Q(t) = \lim_{n \to \infty} k_n \left[\int_{\|x\|<\varepsilon} <t,x>^2 A_{2n}\mu_2\{dx\} - \left(\int_{\|x\|<\varepsilon} <t,x> A_{2n}\mu_2\{dx\} \right)^2 \right] \tag{8.89}$$

[3] Here, we identify a measure on $\mathbb{R}^d$ that is supported on a subspace V_i with its projection onto V_i, in order to avoid needlessly complicated notation.

for all $\varepsilon > 0$ and all $t \in \mathbb{R}^d$. Note that in (8.86)–(8.89) we consider $A_{in}\mu_i$ as measures on $\mathbb{R}^d$.

In order to show that $\mu \in \text{GDOSA}(\nu, c)$ and (8.75) holds, in view of Theorem 3.3.8 it suffices to show that

$$k_n \cdot (A_n\mu) \to \phi \quad \text{as } n \to \infty \tag{8.90}$$

and

$$\begin{aligned} Q(t) &= \lim_{\varepsilon \to 0} \limsup_{n \to \infty} k_n \Big[\int_{\|x\|<\varepsilon} \langle t,x\rangle^2 A_n\mu\{dx\} - \Big(\int_{\|x\|<\varepsilon} \langle t,x\rangle A_n\mu\{dx\} \Big)^2 \Big] \\ &= \lim_{\varepsilon \to 0} \liminf_{n \to \infty} k_n \Big[\int_{\|x\|<\varepsilon} \langle t,x\rangle^2 A_n\mu\{dx\} - \Big(\int_{\|x\|<\varepsilon} \langle t,x\rangle A_n\mu\{dx\} \Big)^2 \Big] \end{aligned} \tag{8.91}$$

for all $t \in \mathbb{R}^d$.

First, we prove that (8.90) holds. Let

$$B = B_1 \oplus B_2 = \{x_1 + x_2 : x_1 \in B_1, x_2 \in B_2\}$$

where $B_i \subset V_i$ are Borel subsets. If B_2 is bounded away from zero in V_2, then in view of (8.88) we have

$$k_n A_n\mu(B) \le k_n A_n\mu(V_1 \oplus B_2) = k_n A_{2n}\mu_2(B_2) \to 0 = \phi(B)$$

using the fact that we define $V_1 = \text{span}(\text{supp}(\phi))$ in the proof of Theorem 7.1.1. If B_1 is bounded away from zero in V_1 and $B_2 = V_2$, then a similar argument using (8.86) shows that

$$k_n A_n\mu(B) = k_n A_n\mu(B_1 \oplus V_2) = k_n A_{1n}\mu_1(B_1 \oplus V_2) \to \phi(B_1 \oplus V_2).$$

By using Theorem 1.2.21, it is not hard to show that sets of these two types determine convergence, hence (8.90) holds.

Now we prove that (8.91) holds. Define

$$V_1^* = V_2^{\perp} = \{y \in \mathbb{R}^d : \langle x,y\rangle = 0 \ \forall x \in V_2\}$$

and $V_2^* = V_1^{\perp}$. Then $\mathbb{R}^d = V_1^* \oplus V_2^*$ and $\pi_i^* : \mathbb{R}^d \to \mathbb{V}_i^*$ are the natural projections associated with this direct sum decomposition. If $t \in V_i^*$, then for integers $i = 1, 2$ we have

$$\begin{aligned} \int_{\|x\|<\varepsilon} \langle t,x\rangle^p A_n\mu\{dx\} &= \int_{\|x\|<\varepsilon} \langle \pi_i^* t,x\rangle^p A_n\mu\{dx\} \\ &= \int_{\|x\|<\varepsilon} \langle t,\pi_i x\rangle^p A_n\mu\{dx\} \\ &= \int_{\|x\|<\varepsilon} \langle t,x\rangle^p \pi_i A_n\mu\{dx\} \\ &= \int_{\|x\|<\varepsilon} \langle t,x\rangle^p A_{in}\mu_i\{dx\} \end{aligned} \tag{8.92}$$

so if $t \in V_2^*$, then (8.91) follows directly from (8.89). Suppose $t \in V_1^*$. An application of the Cauchy-Schwarz Inequality shows that the term in square brackets in (8.91) is positive, and hence bounded above by

$$\int_{\|x\|<\varepsilon} <t,x>^2 A_n\mu\{dx\}$$

which in turn is bounded above by the integral in (8.87), using (8.92) along with the fact that $\{x : \|x\| < \varepsilon\} \subset \{x : |<x,t>| < \varepsilon\|t\|\}$ for all nonzero $t \in \mathbb{R}^d$. Then if $t \in V_1^*$ it follows from (8.87) that (8.91) holds.

In the remaining case, we can use the direct sum decomposition $\mathbb{R}^d = V_1^* \oplus V_2^*$ to write $t = t_1 + t_2$, so that $<t,x> = <t_1,x> + <t_2,x>$ and

$$<t,x>^2 = <t_1,x>^2 + <t_2,x>^2 + 2<t_1,x><t_2,x>$$

so that the term in square brackets in (8.91) equals $I_1 + I_2 + 2I_3 - 2I_4$, where

$$\begin{aligned}
I_1 &= \int_{\|x\|<\varepsilon} <t_1,x>^2 A_n\mu\{dx\} - \left(\int_{\|x\|<\varepsilon} <t_1,x> A_n\mu\{dx\}\right)^2 \\
I_2 &= \int_{\|x\|<\varepsilon} <t_2,x>^2 A_n\mu\{dx\} - \left(\int_{\|x\|<\varepsilon} <t_2,x> A_n\mu\{dx\}\right)^2 \\
I_3 &= \int_{\|x\|<\varepsilon} <t_1,x><t_2,x> A_n\mu\{dx\} \\
I_4 &= \left(\int_{\|x\|<\varepsilon} <t_1,x> A_n\mu\{dx\}\right)\left(\int_{\|x\|<\varepsilon} <t_2,x> A_n\mu\{dx\}\right).
\end{aligned}$$

We have already shown that (8.91) holds for $t \in V_i^*$, and since $t_i = \pi_i^* t \in V_i^*$ it follows that

$$\begin{aligned}
\lim_{\varepsilon\to 0} \limsup_{n\to\infty} k_n I_1 &= 0 \\
\lim_{\varepsilon\to 0} \limsup_{n\to\infty} k_n I_2 &= Q(t)
\end{aligned}$$

for all $t \in \mathbb{R}^d$. An application of the Cauchy-Schwarz Inequality shows that both $(I_3)^2$ and $(I_4)^2$ are bounded above by $J_1 \cdot J_2$ where

$$\begin{aligned}
J_1 &= \int_{\|x\|<\varepsilon} <t_1,x>^2 A_n\mu\{dx\} \\
J_2 &= \int_{\|x\|<\varepsilon} <t_2,x>^2 A_n\mu\{dx\}
\end{aligned}$$

where

$$\lim_{\varepsilon\to 0} \limsup_{n\to\infty} k_n J_1 = 0$$

in view of (8.87) and (8.92). If μ_2 has mean zero, then Corollary 3.3.14 implies that

$$\lim_{n\to\infty} k_n J_2 = Q(t_2) = Q(t)$$

and this can always be arranged by a change of centering. It follows that

$$\lim_{\varepsilon\to 0} \limsup_{n\to\infty} k_n I_3 = 0$$
$$\lim_{\varepsilon\to 0} \limsup_{n\to\infty} k_n I_4 = 0$$

and hence (8.91) holds, proving that $\mu \in \mathrm{GDOSA}(\nu, c)$.

Conversely, suppose that $\mu \in \mathrm{GDOSA}(\nu', c)$. In view of Remark 7.5.13, we can assume that $c > 1$ and ν is $(c^{E'}, c)$ operator semistable. Apply Theorem 8.3.6 to obtain a linear operator $T \in \mathrm{GL}(\mathbb{R}^d)$ such that μ is spectrally compatible with $\nu = T\nu'$. Lemma 8.3.5 implies that ν is (c^E, c) operator semistable where $E = TE'T^{-1}$. Then

$$A_n \mu^{k_n} * \varepsilon_{a_n} \Rightarrow \nu \tag{8.93}$$

for some $A_n \in \mathrm{L}(\mathbb{R}^d)$ and $a_n \in \mathbb{R}^d$ such that every A_n leaves the subspaces in the spectral decomposition $W_1, \ldots, W_p$ of $\mathbb{R}^d$ with respect to E' invariant. If $a_1 > 1/2$ there is no normal component, and hence there is nothing to prove. Otherwise, define $V_1 = W_1$ and $V_2 = W_2 \oplus \cdots \oplus W_p$. Then V_1 and V_2 are the subspaces in the Lévy decomposition, and we can write $A_n = A_{1n} \oplus A_{2n}$ and $a_n = a_{1n} + a_{2n}$ with respect to this direct sum decomposition. Apply the natural projections to (8.93) to see that (8.85) holds, so that $\mu_i \in \mathrm{GDOSA}(\nu_i)$ for $i = 1, 2$. If we apply Theorem 8.3.7 instead of Theorem 8.3.6, we can also obtain V_1, V_2 mutually orthogonal. □

Now we present our main results concerning the moments of random vectors that belong to some generalized domain of semistable attraction.

Theorem 8.3.10. *If μ belongs to some generalized domain of semistable attraction, then for every $\theta \neq 0$ there exists some $\alpha(\theta) \in (0, 2]$ such that*

$$\int |\langle x, \theta\rangle|^\rho \mu\{dx\} \tag{8.94}$$

exists for all $0 \leq \rho < \alpha(\theta)$, and if $\alpha(\theta) < 2$, then the integral diverges for all $\rho > \alpha(\theta)$.

Proof. Remark 7.5.13 implies that $\mu \in \mathrm{GDOSA}(\nu', c)$ for some full $(c^{E'}, c)$ operator semistable law ν' with $c > 1$. Apply Theorem 8.3.7 to obtain a (c^E, c) operator semistable law ν of the same type as ν' such that μ is spectrally compatible with ν and the subspaces $W_1, \ldots, W_p$ in the spectral decomposition of $\mathbb{R}^d$ with respect to E are mutually orthogonal. Let $a_1 < \cdots < a_p$ denote the real spectrum of E and write $\theta = \theta_1 + \cdots + \theta_p$ with respect to the direct sum decomposition $\mathbb{R}^d = W_1 \oplus \cdots \oplus W_p$. Define

$$\alpha(\theta) = \min\{a_i^{-1} : \theta_i \neq 0\} \tag{8.95}$$

for all $\theta \neq 0$. If ν is normal, then Theorem 7.1.10 implies that the real spectrum of E consists of the single value $a_1 = 1/2$, so that $\alpha(\theta) = 2$ for all $\theta \neq 0$. In this case, Proposition 8.1.6 yields that $\int \|x\|^\rho \mu\{dx\}$ exists for all $\rho < 2$, which implies for all $\theta \neq 0$ that (8.94) exists for all $\rho < 2$. If ν has no normal component, then Theorem 7.1.10 implies that every $a_i > 1/2$, hence $0 < \alpha(\theta) < 2$. In this case, Theorem 8.2.5 shows that (8.94) exists for all $0 \leq \rho < \alpha(\theta)$ and diverges for all $\rho > \alpha(\theta)$.

In the remaining case, ν has both a normal and a nonnormal component. Let $L_1 = W_1$ and $L_2 = W_2 \oplus \cdots \oplus W_p$, let P_i denote orthogonal projection onto the space L_i, and define $\mu_i = P_i \mu$ and $\nu_i = P_i \nu$. Then $\mu_i \in \mathrm{GDOSA}(\nu_i, c)$ where ν_1 is a full normal law on L_1 and ν_2 is a full operator semistable law without normal component on L_2. Proposition 8.1.6 implies that

$$\int |{<}x, P_1\theta{>}|^\rho \mu\{dx\}$$

is finite for all $\rho < 2$, and Theorem 8.2.5 shows that

$$\int |{<}x, P_2\theta{>}|^\rho \mu\{dx\}$$

exists for $0 \leq \rho < \alpha(P_2\theta)$ and diverges for $\rho > \alpha(P_2\theta)$, where $\alpha(P_2\theta) < 2$. Then if $0 < \rho < \alpha(\theta)$ we get, using the well-known inequality $(a+b)^\rho \leq 2^\rho(a^\rho + b^\rho)$ for $a, b \geq 0$, that

$$\int |{<}x, \theta{>}|^\rho \mu\{dx\} \leq 2^\rho \left(\int |{<}x, P_1\theta{>}|^\rho \mu\{dx\} + \int |{<}x, P_2\theta{>}|^\rho \mu\{dx\} \right) < \infty$$

On the other hand, if $\alpha(\theta) < 2$, then necessarily $P_2\theta \neq 0$. For any positive random variable Z, it is easy to check that if $EZ^\rho = \infty$ then $EZ^\beta = \infty$ for all $\beta > \rho$.

Hence in order to complete the proof it suffices to show that (8.94) diverges for $\alpha(\theta) < \rho < 2$. It follows from the inequality $|a+b|^\rho \leq 2^\rho(|a|^\rho + |b|^\rho)$ that $|a+b|^\rho \geq 2^{-\rho}|a|^\rho - |b|^\rho$ and hence

$$\begin{aligned}
\int |{<}x, \theta{>}|^\rho \mu\{dx\} &= \int |{<}x, P_2\theta{>} + {<}x, P_1\theta{>}|^\rho \mu\{dx\} \\
&\geq K \int |{<}x, P_2\theta{>}|^\rho \mu\{dx\} \\
&\qquad - \int |{<}x, P_1\theta{>}|^\rho \mu\{dx\} \\
&= \infty
\end{aligned}$$

and this completes the proof. □

Definition 8.3.11. Let $W_1, \ldots, W_p$ and $a_1 < \cdots < a_p$ denote the spectral decomposition of $\mathbb{R}^d$ with respect to the exponent E of the limit law ν in the proof of Theorem 8.3.10. Then the function $\alpha(\theta)$ defined by (8.95) satisfies $\alpha(\theta) = 1/a_i$ if and only if $\theta \in W_1 \oplus \cdots \oplus W_i \setminus W_1 \oplus \cdots \oplus W_{i-1}$ so that $1/\alpha(\theta)$ ranges over the real spectrum of E. In view of the sharp moment results of Theorem 8.3.10, this index function $\alpha(\theta)$ is uniquely defined, and hence the subspaces W_i and the constants a_i in Theorem 8.3.7 are uniquely determined by the probability measure μ. If X is a random vector on $\mathbb{R}^d$ with distribution μ, then we will call $\alpha(\theta)$ the *spectral index* of μ (or X). Write $X = X_1 + \cdots + X_p$ with respect to the direct sum decomposition $\mathbb{R}^d = W_1 \oplus \cdots \oplus W_p$ so that the probability measure μ_i obtained by orthogonal projection of μ onto W_i is the distribution of X_i. We will call $\mu_1, \ldots, \mu_p$ the *spectral decomposition* of μ and $X = X_1 + \cdots + X_p$ the *spectral decomposition* of X. We will also call $W_1, \ldots, W_p$ the *spectral decomposition of* $\mathbb{R}^d$ *with respect to* μ *(or* X*)* and $a_1 < \cdots < a_p$ the *real spectrum* of μ (or X). The spectral index of μ was constructed using the spectral decomposition from Theorem 8.3.7. Hence all of the results listed there still pertain. If ν is a limit law that satisfies all of these conditions, and Y has distribution ν, then we will say that ν (or Y) is *spectrally compatible* with μ (or X). Although such a limit law always exists in view of Theorem 8.3.7, it is not necessarily unique. If (B_n) fulfills the conclusions of Theorem 8.3.7, then we also say that this sequence of norming operators is *spectrally compatible* with μ (or X). In this case, the conclusions of Theorem 4.3.1 hold with $V_i = W_i$ and $L_i = W_1 \oplus \cdots \oplus W_i$ for each $i = 1, \ldots, p$.

Remark 8.3.12. Since every full operator semistable law ν belongs to its own generalized domain of semistable attraction, the moment results above apply to ν as well. Since the spectral index $\alpha(\theta)$ of ν is unique, it follows that every exponent of ν has the same real spectrum.

Corollary 8.3.13. *Suppose that* X *is a random vector on* $\mathbb{R}^d$ *that belongs to some generalized domain of semistable attraction, and let* $X = X_1 + \cdots + X_p$ *denote the spectral decomposition of* X *and* $a_1 < \cdots < a_p$ *the real spectrum of* X*. Then* $E\|X_i\|^\rho$ *exists for all* $0 \le \rho < 1/a_i$*, and if* $a_i > 1/2$*, then* $E\|X_i\|^\rho = \infty$ *for all* $\rho > 1/a_i$*.*

Proof. Since all norms on $\mathbb{R}^d$ are equivalent it suffices to consider the Euclidean norm. Theorem 8.3.7 shows that the distribution μ_i of X_i belongs to the domain of semistable attraction of some spectrally simple operator semistable law ν_i. If ν_i is normal, then the result follows immediately from Proposition 8.1.6 and if ν_i is not normal, then the result follows immediately from Corollary 8.2.6. □

Remark 8.3.14. Suppose that X is a random vector on $\mathbb{R}^d$ that belongs to some generalized domain of semistable attraction, and let $X = X_1 + \cdots + X_p$

denote the spectral decomposition of X and $a_1 < \cdots < a_p$ the real spectrum of X. The spectral decomposition of X is the unique decomposition of X into mutually orthogonal components such that $E\|X_i\|^\rho$ exists for $0 \le \rho < 1/a_i$ and $E\|X_i\|^\rho = \infty$ for $\rho > 1/a_i$ if $a_i > 1/2$.

Corollary 8.3.15. *Suppose that X is a random vector on $\mathbb{R}^d$ that belongs to some generalized domain of semistable attraction, and let $a_1 < \cdots < a_p$ denote the real spectrum of X. Then $E\|X\|^\rho$ exists for all $0 \le \rho < 1/a_p$, and if $a_p > 1/2$ then $E\|X\|^\rho$ is infinite for all $\rho > 1/a_p$.*

Proof. Since all norms on $\mathbb{R}^d$ are equivalent it suffices to consider the Euclidean norm $\|x\| = \sqrt{<x,x>}$. Let $X = X_1 + \cdots + X_p$ denote the spectral decomposition of X. Since these random vectors are supported on orthogonal subspaces we have $\|X\|^2 = \|X_1\|^2 + \cdots + \|X_p\|^2$, and hence $E\|X\|^\rho = E(\|X_1\|^2 + \cdots + \|X_p\|^2)^{\rho/2} \le p^{\rho/2}(E\|X_1\|^\rho + \cdots + E\|X_p\|^\rho)$. If $0 \le \rho < 1/a_p$, then by Corollary 8.3.13 we know that $E\|X_i\|^\rho < \infty$ for $i = 1, \ldots, p$, and hence $E\|X\|^\rho < \infty$. If $\rho > 1/a_p$, then Corollary 8.3.13 implies that $E\|X_p\|^\rho = \infty$, and then $E\|X\|^\rho = E(\|X_1\|^2 + \cdots + \|X_p\|^2)^{\rho/2} \ge E\|X_p\|^\rho = \infty$. □

As a special case we now consider semistable laws on $\mathbb{R}^d$. Recall from Section 7.4 that a law ν on $\mathbb{R}^d$ is said to be (b, c) semistable if it is (c^{aI}, c) operator semistable for some $c > 1$ and $a \ge 1/2$.

Proposition 8.3.16. *A full probability measure ν on $\mathbb{R}^d$ is (b, c) semistable if and only if* $\mathrm{DOSA}(\nu, c)$ *is nonempty.*

Proof. If ν is (b, c) semistable, then it is (B, c) operator semistable, where $B = bI$ for some $b > 0$ and $c > 1$. Then Theorem 7.1.1 implies that $\nu^{c^n} = b^n \nu * \varepsilon_{s_n}$ for all $n \ge 1$ and some $s_n \in \mathbb{R}^d$. Hence $\nu = b^{-n} \nu^{c^n} * \varepsilon_{-b^{-n} s_n}$. If we set $k_n = [c^n]$ and $a_n = b^{-n}$, then Proposition 3.3.7 implies for suitable $\bar{s}_n \in \mathbb{R}^d$ that $a_n \nu^{k_n} * \varepsilon_{\bar{s}_n} \Rightarrow \nu$ as $n \to \infty$ so $\mathrm{DOSA}(\nu, c) \ne \emptyset$.

If $\mu \in \mathrm{DOSA}(\nu, c)$, then $\mu \in \mathrm{GDOSA}(\nu, c)$, so ν is operator semistable. Apply Theorem 7.1.1 and write $\nu = \nu_1 * \nu_2$, where ν_2 is a full normal law on V_2 and ν_1 is a full Poissonian law on V_1 where $\mathbb{R}^d = V_1 \oplus V_2$. An application of Theorem 8.3.9 yields

$$a_n \mu_i^{k_n} * \varepsilon_{b_{in}} \Rightarrow \nu_i \quad \text{as } n \to \infty$$

for $i = 1, 2$. If ν has a Poissonian component, then Theorem 8.2.9 shows that there exists a regularly varying sequence of positive reals (b_n) with index $-\rho$ such that $a_n = b_{k_n}$. If ν has a normal component, then Remark 8.1.17 shows that there exists a regularly varying sequence of positive reals (b_n) with index $-1/2$ such that $a_n = b_{k_n}$. It follows that ν can only have one Lévy component, i.e., ν is either normal (and hence semistable) or it has no normal component. In the latter case, Theorem 8.2.9 implies that $c \cdot \phi = c^\rho \phi$ for some exponent $\rho > 1/2$. Then Lemma 7.1.6 together with Theorem 7.1.10 implies that ν is (c^E, c) operator semistable where $E = \rho I$ and hence is semistable. □

8.3.1 Stochastic Compactness

Suppose $X \in \mathrm{GDOSA}(Y, c)$ and let $S_n = X_1 + \cdots + X_n$ denote the sequence of partial sums of $X_1, X_2, X_3, \ldots$ i.i.d. with X. Then

$$A_n S_{k_n} + a_n \Rightarrow Y \tag{8.96}$$

for some $A_n \in \mathrm{GL}(\mathbb{R}^d)$, $a_n \in \mathbb{R}^d$ and some sampling sequence (k_n), so that the distribution of the partial sums S_{k_n} can be usefully approximated by that of $A_n^{-1}(Y - a_n)$ for n large. In this section, we obtain a *stochastic compactness* result that allows us to approximate any partial sum S_n. The main tools we use in the proofs are the embedding property of the sequence of norming operators proven in Theorem 8.3.1 along with the spectral decomposition of Theorem 8.3.7. We will consider both the sequence of random vectors (S_n) as well as the projection of S_n along any radial direction.

Definition 8.3.17. Let (X_n) be a sequence of random vectors on $\mathbb{R}^d$ and let μ_n denote the distribution of X_n. We say that (X_n) is *stochastically compact* if μ_n is weakly relatively compact and all limit points are full. In other words, for any subsequence of (X_n) there exists a further subsequence $(X_{n'})$ such that $X_{n'} \Rightarrow Y$ where Y is full. The set of all such Y is called the *limit set* of (X_n).

Theorem 8.3.18. *Suppose $X, X_1, X_2, X_3, \ldots$ are i.i.d. with common distribution μ, where $\mu \in \mathrm{GDOSA}(\nu, c)$ with ν a full (c^E, c) operator semistable law on $\mathbb{R}^d$ and $c > 1$. If (8.96) holds with $S_n = X_1 + \cdots + X_n$, then there exists a regularly varying sequence (B_n) with index $(-E)$ and nonrandom vectors (b_n) such that $(B_n S_n + b_n)$ is stochastically compact with limit set contained in*

$$\{\lambda^{-E} Y_\lambda : \lambda \in [1, c]\} \tag{8.97}$$

where Y_λ has distribution ν^λ. Furthermore, for any $\|\theta\| = 1$ the sequence $(\langle B_n S_n, \theta\rangle + \langle b_n, \theta\rangle)$ is stochastically compact with limit set contained in

$$\{\langle Y_\lambda, \theta\rangle : \lambda \in [1, c]\}. \tag{8.98}$$

Proof. Apply Theorem 8.3.1 to obtain a regularly varying sequence (B_n) with index $(-E)$ such that (8.76) holds. Note that by the construction of B_n in Theorem 4.2.13 we have $B_n = \lambda_n^{-E} A_{p_n}$, where we write $n = \lambda_n k_{p_n}$ and $k_{p_n} \leq n < k_{p_n+1}$. Then $k_{n+1}/k_n \to c$ implies that (λ_n) is relatively compact with limit set $[1, c]$. It follows immediately from (8.75) that

$$A_{p_n} \mu^{k_{p_n}} * \varepsilon_{a_{p_n}} \Rightarrow \nu. \tag{8.99}$$

If $\lambda_n \to \lambda$ along a subsequence then Proposition 3.3.7 implies that

$$A_{p_n} \mu^n * \varepsilon_{\lambda_n a_{p_n}} \Rightarrow \nu^\lambda \tag{8.100}$$

along that subsequence, and then it follows from Convergence of Types, Theorem 2.3.17, that

$$B_n\mu^n * \varepsilon_{b_n} \Rightarrow \lambda^{-E}\nu^\lambda \tag{8.101}$$

along that subsequence, where $b_n = \lambda_n\lambda_n^{-E}a_{p_n}$. Since every subsequence has a further subsequence with this property, it follows that $(B_nS_n + b_n)$ is stochastically compact with limit set (8.97). If $\|\theta\| = 1$ define $\pi_\theta x = <x, \theta>$. Apply the Continuous Mapping Theorem 1.2.7 with $T = \pi_\theta$ to see that $(<B_nS_n, \theta> + <b_n, \theta>)$ is stochastically compact with limit set (8.98), which finishes the proof. □

Remark 8.3.19. We pause to list some facts from the proof of Theorem 8.3.18, which may be useful in applications.

(a) If the centering constants $a_n = 0$ in (8.96), then $b_n = 0$ for all n.

(b) Every law $\nu_{(\lambda)} = \lambda^{-E}\nu^\lambda$ in the limit set of Theorem 8.3.18 is also a full (c^E, c) operator semistable law. The Lévy measure of $\nu_{(\lambda)}$ is given by $\phi_{(\lambda)} = \lambda(\lambda^{-E}\phi)$, where ϕ is the Lévy measure of ν.

(c) If ν is operator stable, then E is an exponent of ν and it follows from Theorem 7.2.1 along with Lemma 7.1.6 that $\phi_{(\lambda)} = \phi$ for all $1 \le \lambda \le c$. In this case, we can adjust the centering constants b_n so that the limit set in Theorem 8.3.18 is just the one point set $\{\nu\}$.

(d) In the proof of Theorem 8.3.18, we also showed that $(A_{p_n}S_n + a_{p_n})$ is stochastically compact with limit set $\{\nu^\lambda : \lambda \in [1, c]\}$ and the same remarks apply to this limit set, e.g., the convolutions powers ν^λ are all full (c^E, c) operator semistable laws.

In Theorem 8.3.18, we obtained a stochastic compactness result for the radial projections of the normalized partial sums. Now we will obtain a similar, but more useful, result for the radial projections of the partial sums themselves. In this result, it will be convenient to select a limit Y in (8.96) that is spectrally compatible with X.

Theorem 8.3.20. *Suppose $X, X_1, X_2, X_3, \ldots$ are i.i.d. random vectors on $\mathbb{R}^d$ that belong to some generalized domain of semistable attraction. If (8.96) holds with $S_n = X_1 + \cdots + X_n$ and Y is spectrally compatible with X, then for any unit vector $\theta \in \mathbb{R}^d$ there exists a sequence (r_n) of positive real numbers tending to zero and a sequence (d_n) of shifts such that $(r_n<S_n, \theta> + d_n)$ is stochastically compact. Moreover, if $\alpha(\theta) = 1/a_i$ then the limit set of $(r_n<S_n, \theta> + d_n)$ is contained in the set*

$$\{<Y_\lambda, \theta_0> : \lambda \in [1, c], \theta_0 \in W_i, \|\theta_0\| = 1\} \tag{8.102}$$

where Y_λ has distribution $\lambda^{-E}\nu^\lambda$, ν is the distribution of Y, and $W_1, \ldots, W_p$ is the spectral decomposition of $\mathbb{R}^d$ with respect to X.

Proof. In the proof of Theorem 8.3.18, we apply Theorem 8.3.1 to obtain a regularly varying sequence (B_n) with index $(-E)$ and shifts (b_n) such that $(B_nS_n + b_n)$ is stochastically compact with limit set (8.97). If we apply Theorem 8.3.7 instead, then the same is true, along with the conclusions of Theorem 8.3.7. In particular, the subspaces in the spectral decomposition $\mathbb{R}^d = W_1 \oplus \cdots \oplus W_p$ with respect to X are mutually orthogonal and B_n invariant, and if we write $B_n = B_{1n} \oplus \cdots \oplus B_{pn}$ with respect to this direct sum decomposition then each $B_{in} : W_i \to W_i$ is regularly varying with an index E_i such that every eigenvalue of E_i has real part a_i. Fix any $\|\theta\| = 1$ and let

$$r_n = \left\|\left(B_n^*\right)^{-1}\theta\right\|^{-1}. \tag{8.103}$$

Write $\theta = \theta_1 + \cdots + \theta_p$, where $\theta_i \in W_i$ and $(B_n^*)^{-1}\theta_i = r_{in}\theta_{in}$, θ_{in} is a unit vector in W_i and $r_{in} \geq 0$. If $\alpha(\theta) = 1/a_i$, then $\theta_i \neq 0$ and $\theta_j = 0$ for all $j > i$, so that $r_{in} = 0$ for $j > i$. Note that $(B_{in}^*)^{-1}$ varies regularly with index E_i^*, and that $r_{in} = \|(B_n^*)^{-1}\theta_i\| = \|(B_{in}^*)^{-1}\theta_i\|$. Then it follows from Theorem 4.2.4 together with Theorem 4.2.9 that for any $a_{i-1} < \rho < a_i$ we have $n^{-\rho}r_{in} \to \infty$ as $n \to \infty$ while $n^{-\rho}r_{jn} \to 0$ as for all $j < i$. Then $r_{jn}/r_{in} \to 0$ as $n \to \infty$ for all $j \neq i$, and hence

$$\theta_n = \frac{\left(B_n^*\right)^{-1}\theta}{\left\|\left(B_n^*\right)^{-1}\theta\right\|} = \frac{r_{1n}\theta_{1n} + \cdots + r_{pn}\theta_{pn}}{\sqrt{r_{1n}^2 + \cdots + r_{pn}^2}} \rightharpoonup W_i \tag{8.104}$$

as $n \to \infty$.

Since

$$\langle S_n, \theta\rangle = \langle B_n^{-1}B_nS_n, \theta\rangle = \langle B_nS_n, (B_n^*)^{-1}\theta\rangle = r_n^{-1}\langle B_nS_n, \theta_n\rangle,$$

if we let $d_n = \langle b_n, \theta_n\rangle$, then

$$r_n\langle S_n, \theta\rangle + d_n = \langle B_nS_n, \theta_n\rangle + \langle b_n, \theta_n\rangle = \langle B_nS_n + b_n, \theta_n\rangle.$$

Now every subsequence (n') has a further subsequence $(n'') \subset (n')$ such that $\theta_n \to \theta_0 \in W_i$ along (n''). Hence $h_n(x_n) = \langle x_n, \theta_n\rangle \to h_0(x) = \langle x, \theta_0\rangle$ whenever $x_n \to x$ in $\mathbb{R}^d$ along the same subsequence (n''). Now choose a subsequence $(n''') \subset (n'')$ such that $B_nS_n + b_n \Rightarrow Y_\lambda$ along (n''') for some $\lambda \in [1, c]$. Then Theorem 1.2.8, a variant of the Continuous Mapping Theorem, implies that

$$r_n\langle S_n, \theta\rangle + d_n = h_n\Big(B_nS_n + b_n\Big) \longrightarrow h_0(Y_\lambda) = \langle Y_\lambda, \theta_0\rangle$$

along the subsequence (n'''). This shows that every subsequence (n') has a further subsequence (n''') along which we get convergence to some random variable in the set (8.102), hence $(r_n\langle S_n, \theta\rangle + d_n)$ is stochastically compact with limit set (8.102). □

Remark 8.3.21. The following useful facts follow easily from the proof of Theorem 8.3.20.

(a) If the centering constants $a_n = 0$ in (8.96) then $d_n = 0$, see also Remark 8.3.19 (a).

(b) It follows from (8.104) that $r_n^{-1} \sim r_{in}$ when $\alpha(\theta) = 1/a_i$, hence for every $\delta > 0$ there exists an $n_0 \geq 1$ such that

$$n^{-a_i-\delta} \leq r_n \leq n^{-a_i+\delta}$$

whenever $n \geq n_0$, showing sharp bounds on the rate at which r_n tends to zero. This rate depends on the direction θ.

8.3.2 Operator Stable Limits

Now, we consider the special case of operator stable limits. Every operator stable law is also operator semistable, and the generalized domain of attraction of a given operator stable law coincides with its generalized domain of semistable attraction. Hence some of these results follow immediately from the more general case. We state the most important and useful of these results here without proof, for future reference. However, the present assumptions are stronger, so we also need to prove some additional properties. Recall that $\mu \in \mathrm{GDOA}(\nu)$ if there exist linear operators $A_n \in \mathrm{L}(\mathbb{R}^d)$ and nonrandom vectors $b_n \in \mathbb{R}^d$ such that

$$A_n\mu^n * \varepsilon_{a_n} \Rightarrow \nu. \tag{8.105}$$

Theorem 8.1.5 implies that there exists a sequence $(B_n) \in \mathrm{RV}(-E)$ such that for a suitable choice of centering constants (b_n) we have

$$B_n\mu^n * \varepsilon_{b_n} \Rightarrow \nu \quad \text{as } n \to \infty. \tag{8.106}$$

Lemma 8.3.22. *Suppose ν is a full (t^E) operator stable law on $\mathbb{R}^d$ and $V_1, \ldots, V_p$ is the spectral decomposition of $\mathbb{R}^d$ with respect to E. Then for any $T \in \mathrm{GL}(\mathbb{R}^d)$ and $s \in \mathbb{R}^d$ the probability measure $\nu' = T\nu * \varepsilon_s$ is a full $(t^{E'})$ operator stable law on $\mathbb{R}^d$ with exponent $E' = TET^{-1}$, and the spectral decomposition of $\mathbb{R}^d$ with respect to E' is $W_1, \ldots, W_p$ where $W_i = T(V_i)$ for each $i = 1, \ldots, p$.*

Proof. The proof is similar to Lemma 8.3.5. Theorem 7.2.1 shows that for suitable shifts $a_t \in \mathbb{R}^d$ we have $\nu^t = t^E\nu*\varepsilon_{a_t}$ for all $t > 0$. Since $Tt^ET^{-1} = t^{E'}$ this implies that $(\nu')^t = T\nu^t * \varepsilon_{ts} = Tt^E\nu * \varepsilon_{Ta_t+ts} = t^{E'}\nu' * \varepsilon_{s'}$, where $s' = Ta_t + ts - s$, so that ν' is a full $(t^{E'})$ operator stable law. Since ν is also (c^E, c) operator semistable and ν' is of the same type, it follows from Lemma 8.3.5 that $W_i = T(V_i)$. □

Theorem 8.3.23. *For any $\mu \in \mathrm{GDOA}(\nu)$, there exists $T \in \mathrm{GL}(\mathbb{R}^d)$ such that $\mu' = T\mu$ is spectrally compatible with ν. Equivalently, μ is spectrally compatible with $\nu' = T^{-1}\nu$.*

Proof. If $\mu \in \mathrm{GDOA}(\nu)$, then $\mu \in \mathrm{GDOSA}(\nu, c)$ and Theorem 8.3.6 implies that there exists $T \in \mathrm{GL}(\mathbb{R}^d)$ such that $\mu' = T\mu$ is spectrally compatible with ν and μ is spectrally compatible with $\nu' = T^{-1}\nu$. Since ν' is of the same type as ν, it follows from Lemma 8.3.22 that ν' is operator stable. □

Theorem 8.3.24. *Suppose μ belongs to the generalized domain of attraction of some full operator stable law ν' on $\mathbb{R}^d$. Then (8.106) holds for some suitable choice of centering constants (b_n), where the limit ν and the norming operators (B_n) are chosen so that:*

(a) ν is a full (t^E) operator stable law on $\mathbb{R}^d$ of the same type as ν' and (B_n) is regularly varying with index $-E$;

(b) the subspaces W_i in the spectral decomposition of $\mathbb{R}^d$ with respect to E are mutually orthogonal and B_n invariant, so that μ is spectrally compatible with ν;

(c) if we write $E = E_1 \oplus \cdots \oplus E_p$ and $B_n = B_{1n} \oplus \cdots \oplus B_{pn}$ with respect to this direct sum decomposition, then each $B_{in} : W_i \to W_i$ is regularly varying with index E_i and every eigenvalue of E_i has real part a_i;

(d) if $\pi_i : \mathbb{R}^d \to W_i$ denotes orthogonal projection and $\mu_i = \pi_i\mu$, $\nu_i = \pi_i\nu$, then ν_i is a full (t^{E_i}) operator stable law on W_i, $\mu_i \in \mathrm{GDOA}(\nu_i)$ and

$$B_{in}\mu_i^n * \varepsilon_{b_{in}} \Rightarrow \nu_i \tag{8.107}$$

for each $i = 1, \ldots p$ with $b_{in} = \pi_i b_n$.

We call $\mu_1, \ldots, \mu_p$ the spectral decomposition of μ, $W_1, \ldots, W_p$ the spectral decomposition of $\mathbb{R}^d$ with respect to μ, and $a_1 < \cdots < a_p$ the real spectrum of μ. We also say that ν is spectrally compatible with μ.

Proof. Apply Theorem 8.3.7 together with Lemma 8.3.22. □

Theorem 8.3.25. *Suppose that ν is a full operator stable law on $\mathbb{R}^d$ and let $\nu = \nu_1 * \nu_2$ be the Lévy decomposition of Definition 7.1.2. Let $\pi_i : \mathbb{R}^d \to V_i$ be the natural projections associated with this direct sum decomposition, so that $\nu_i = \pi_i\nu$, and define $\mu_i = \pi_i\mu$. If $\mu_i \in \mathrm{GDOA}(\nu_i)$ for $i = 1, 2$, then $\mu \in \mathrm{GDOA}(\nu)$, and conversely if $\mu \in \mathrm{GDOA}(\nu')$, then there exists another full operator stable law ν of the same type such that $\mu_i \in \mathrm{GDOA}(\nu_i)$ for $i = 1, 2$ and V_1, V_2 are mutually orthogonal.*

Proof. The proof of the direct half is identical to Theorem 8.3.9, using the sampling sequence $k_n = n$. To prove the converse, apply Theorem 8.3.24 and then project (8.106) onto the normal and nonnormal components. □

In Section 7.3, we defined a stable law with index α on $\mathbb{R}^d$ as a (t^E) operator stable law with exponent $E = aI$, where $\alpha = 1/a$. By using the Lévy decomposition of Theorem 8.3.25, we can also show that stable laws are exactly those with a nonempty domain of attraction.

Proposition 8.3.26. *A full probability measure ν on $\mathbb{R}^d$ is stable if and only if* DOA(ν) *is not empty.*

Proof. If ν is stable with index α, then it is also operator stable with exponent $E = aI$ for $a = 1/\alpha$. Then Theorem 7.2.1 implies that $\nu^n = n^a \nu * \varepsilon_{a_n}$ for all $n \geq 1$, so that $n^{-a}\nu^n * \varepsilon_{s_n} \Rightarrow \nu$ with $s_n = -n^{-a}a_n$, and hence $\nu \in \text{DOA}(\nu)$. Conversely, if $\mu \in \text{DOA}(\nu)$ and (7.16) holds, then Theorem 8.3.25 shows that $\mu_i \in \text{DOA}(\nu_i)$ and the proof of that theorem shows that

$$a_n \mu_i^n * \varepsilon_{b_{in}} \Rightarrow \nu_i \tag{8.108}$$

where ν_2 is normal and ν_1 has no normal component. In view of Proposition 8.1.6, we may assume that μ_2 has mean zero. Then Lemma 8.1.16 shows that (a_n) varies regularly with index $-1/2$. But Lemma 8.2.17 shows that (a_n) varies regularly with index $-a$, where $a > 1/2$, hence ν can only have one Lévy component, i.e., ν is either normal (and hence stable) or it has no normal component. In the latter case, Theorem 3.3.8 yields $n(a_n\mu) \to \phi$ as $n \to \infty$, and then Proposition 6.1.37 implies that $t \cdot \phi = t^a \phi$ for all $t > 0$. Then Lemma 7.1.6 along with Theorem 7.2.1 imply that ν is operator stable with exponent $E = aI$, and hence α stable with index $\alpha = 1/a \in (0, 2)$. □

Corollary 8.3.27. *Suppose that $X, X_1, X_2, X_3, \ldots$ are i.i.d. random vectors on $\mathbb{R}^d$, that belong to some generalized domain of attraction, and the spectral index $\alpha(\theta)$ takes on values $a_1 < \cdots < a_d$. Take Y spectrally compatible with X and write $X_n = X_{1n} + \cdots + X_{dn}$ and $Y = Y_1 + \cdots + Y_d$ using the spectral decomposition $\mathbb{R}^d = W_1 \oplus \cdots \oplus W_d$ with respect to X. Then Y_i is stable with index $\alpha_i = 1/a_i$, $X_{1i} \in \text{DOA}(Y_i)$ and for some regularly varying sequence (g_{in}) with index $-a_i$ we have*

$$g_{in}(X_{i1} + \cdots X_{ik_n}) + b_{in} \Rightarrow Y_i \tag{8.109}$$

for each i, for some suitable sequence of real numbers (b_{in}).

Proof. If the spectral index takes on values $a_1 < \cdots < a_d$, then any exponent of any Y spectrally compatible with X has eigenvalues $a_1, \ldots, a_d$ with associated unit eigenvectors $v_1, \ldots, v_d$, and V_i is the one-dimensional subspace spanned by v_i, so that the eigenvectors are mutually orthogonal. Apply Corollary 8.3.27 to see that (8.109) holds. Then Proposition 8.3.26 shows that Y_i is stable, and since (g_{in}) varies regularly with index $-a_i$ it follows from Lemma 8.2.17 that Y_i is stable with index α_i. □

Since GDOA(ν) = GDOSA(ν, c) for ν operator stable, the results on moments apply directly to any $\mu \in \text{GDOA}(\nu)$. The most important results are these.

Theorem 8.3.28. *If μ belongs to some generalized domain of attraction, then*

$$\int |\langle x, \theta\rangle|^\rho \mu\{dx\} \tag{8.110}$$

exists for all $0 \leq \rho < \alpha(\theta)$, and if $\alpha(\theta) < 2$, then the integral diverges for all $\rho > \alpha(\theta)$, where $\alpha(\theta)$ is the spectral index of μ. If $W_1, \ldots, W_p$ is the spectral decomposition of $\mathbb{R}^d$ with respect to μ and $a_1 < \cdots < a_p$ is the real spectrum of μ, then $\alpha(\theta) = 1/a_i$ if and only if $\theta \in W_1 \oplus \cdots \oplus W_i \setminus W_1 \oplus \cdots \oplus W_{i-1}$.

Remark 8.3.29. Since every full (t^E) operator stable law is also (c^E, c) operator semistable, the uniqueness of the spectral index $\alpha(\theta)$ also implies that every exponent of ν has the same real spectrum.

Suppose $X \in \text{GDOA}(Y)$ and let $S_n = X_1 + \cdots + X_n$ denote the sequence of partial sums of $X_1, X_2, X_3, \ldots$ i.i.d. with X. Then

$$A_n S_n + a_n \Rightarrow Y \tag{8.111}$$

for some $A_n \in \text{GL}(\mathbb{R}^d)$ and $a_n \in \mathbb{R}^d$ so that the distribution of the partial sums S_n can be usefully approximated by that of $A_n^{-1}(Y - a_n)$ for n large.

Theorem 8.3.30. *Suppose $X \in \text{GDOA}(Y)$ and (8.111) holds, where Y spectrally compatible with X. Then for any unit vector $\theta \in \mathbb{R}^d$ there exists a sequence (r_n) of positive real numbers tending to zero and a sequence (d_n) of shifts such that $(r_n\langle S_n, \theta\rangle + d_n)$ is stochastically compact. Moreover, if $\alpha(\theta) = 1/a_i$, then the limit set of $(r_n\langle S_n, \theta\rangle + d_n)$ is contained in the set*

$$\{\langle Y, \theta_0\rangle :, \theta_0 \in W_i, \|\theta_0\| = 1\} \tag{8.112}$$

where $W_1, \ldots, W_p$ is the spectral decomposition of $\mathbb{R}^d$ with respect to X. If $a_n = 0$ in (8.111), then we can take $d_n = 0$.

Remark 8.3.31. Samorodnitsky and Taqqu [161] show that if Y is a stable random vector with index α, then the distribution of every one-dimensional marginal $\langle Y, \theta\rangle$ is stable with the same index α. However, Marcus [104] provides an example showing that the converse is not true in general. In fact, there exists a probability distribution μ on $\mathbb{R}^2$ whose one-dimensional marginals are all stable with index $0 < \alpha < 1$ but μ is not multivariate α-stable. It follows from Theorem 1 of Giné and Hahn [43] that μ is not even infinitely divisible even though all one-dimensional marginals are as stable laws of course infinitely divisible. An example in Meerschaert [114] shows that the one-dimensional marginals of an operator stable law are not necessarily stable, or even in the domain of attraction of a stable law. However, Meerschaert and Scheffler [131] show that for any operator stable random vector Y there is a basis $b_1, \ldots, b_d$ such that every marginal $\langle Y, b_i\rangle$ belongs to the domain of semistable attraction of some semistable law on $\mathbb{R}^1$. If the exponent of Y is semisimple, then the marginals are actually semistable.

8.4 NOTES AND COMMENTS

The generalized domain of attraction of a full normal law was first characterized by Hahn and Klass [49], see Remark 8.1.13. Their proof provides useful information about the norming operators A_n in (8.1). Lemma 8.1.1 and Theorem 8.1.9 are due to Meerschaert [110, 118]. Theorem 8.1.5 is from Meerschaert [120]. Regular variation of the norming operators was an open problem for over ten years (see, e.g., Hudson and Mason [65]). Theorem 8.1.3 was originally proved by combining Lemma 8.1.1 and Theorem 8.1.5. Corollary 8.1.12 is due to Klosowska [79]. Using this result, moment results can be obtained by an application of the one variable theory. This was the first proof of Proposition 8.1.6. Rvačeva [158] characterized scalar-normed domains of attraction in $\mathbb{R}^d$ by generalizing the arguments in Gnedenko and Kolmogorov [44]. Meerschaert [112] applied scalar-normed regular variation to the problem, and Meerschaert [115] considered vector-normed domains of attraction using vector-normed regular variation. All of these are special cases of the general theory with operator norming. Examples in Hahn and Klass [49] show that scalar-normed, vector-normed, and generalized (operator-normed) domains of attraction are strictly nested, even in the case of a normal limit.

Around 1950, Rvačeva [158] pioneered the study of domains of attraction in $\mathbb{R}^d$, and was the first to pose the question of characterizing generalized domains of attraction, as well as the class of possible limit laws, which are now called operator stable. Rvačeva [158] proved Theorem 8.2.18, which was the first useful characterization of (scalar-normed) domains of attraction in $\mathbb{R}^d$. About thirty years later, Resnick and Greenwood [151] used point process methods to prove a special case of Corollary 8.2.12 where the norming operators A_n are all diagonal with respect to the standard basis for $\mathbb{R}^d$. Around the same time, Jurek [73] handled the case where the norming operators $A_n = n^{-E}$. This is called the domain of normal attraction of a nonnormal operator stable law, an unfortunate terminology. Over the next decade, generalized domains of attraction for nonnormal limits were characterized by Hahn and Klass [50, 51] and Meerschaert [110, 111]. The Hahn and Klass approach applies the one variable methods of Feller [40] uniformly in each radial direction. The proof of Corollary 8.2.12, due to Meerschaert [111], was the first application of multivariable regular variation to central limit theory. Theorem 8.2.10 was originally proven by combining Corollary 8.2.12 with the result in Theorem 8.1.5 that norming operators can always be chosen to vary regularly. The sharp moment results of Theorem 8.2.14 are from Meerschaert [116]. The centering results of Theorem 8.2.16 are from Meerschaert [110, 118], extending the results of Feller [40] in the one variable case. The connection between scalar-normed domains of attraction and regular variation in Theorem 8.2.17 is from Meerschaert [112], who generalized the one variable result of Corollary 8.2.19, obtained by Feller [40] about twenty years after Rvačeva [158] proved Theorem 8.2.18. There are also some results on domains of attraction in more general settings. See [55] for a discussion of domains of attraction on locally compact groups and [101] for a survey of domains of attraction on Banach spaces.

Generalized domains of semistable attraction for nonnormal limits were treated in Meerschaert and Scheffler [121], where the result in Corollary 8.2.3 appears. The scalar-normed version of this result, Theorem 8.2.9, is due to Scheffler [164, 166]. A vector-normed version can also be found in Scheffler [164]. The sharp moment results of Theorem 8.2.5 for generalized domains of semistable attraction are due to Meerschaert and Scheffler [122]. Theorem 8.2.7 characterizing the centering vectors for generalized domains of semistable attraction is from Meerschaert and Scheffler [121]. Theorem 8.2.9 characterizing scalar-normed domains of semistable attraction is from Scheffler [164], which also considers the vector-normed case. Theorem 8.3.1, which shows that the norming operators for generalized domains of semistable attraction can always be chosen from a regularly varying sequence, is from Scheffler [171]. The spectral decomposition for generalized domains of semistable attraction in Theorem 8.3.6 and the sharp moment results of Theorem 8.3.10 are also from Meerschaert and Scheffler [122]. The portion of Theorem 8.3.9 showing that convergence of the normal and Poissonian components is sufficient for convergence of the entire sum is from Meerschaert and Scheffler [123]. The stochastic compactness results in Theorems 8.3.18 and 8.3.20 are due to Scheffler [166], [171]. The spectral decomposition for generalized domains of attraction in Theorem 8.3.23, and the sharp moment results of Theorem 8.3.28, are from Meerschaert [116]. The portion of Theorem 8.3.25 showing that convergence of the normal and Poissonian components is sufficient for convergence of the entire sum is from Meerschaert [118].

The most important open problem in the theory of generalized domains of attraction is the construction of suitable norming operators. Hahn and Klass [49, 49, 51] show that the construction of Feller [40] can be used to obtain suitable norming operators once a suitable coordinate system is identified, see Remark 8.1.13. This coordinate system varies with the sample size n, and construction of the coordinate vectors from a known distribution requires solving a nonlinear optimization problem. A practical implementation of this algorithm has yet to be accomplished. For generalized domains of semistable attraction, results similar to those of Hahn and Klass are not yet available. Perhaps the formulas (8.3) and (8.43) can be used to construct suitable norming operators, once a suitable sampling sequence has been chosen. However, the problem of constructing a suitable sampling sequence also remains open, even for domains of semistable attraction in $\mathbb{R}^1$.

Generalized domains of semistable attraction provide a flexible and mathematically sound foundation for statistical applications involving sums of i.i.d. random vectors. The results of this chapter extend the central limit theorem to allow the consideration of probability distributions with heavy tails. Heavy tails have been observed in a wide variety of applications, including electrical engineering [140], hydrology [4, 63], and finance [38, 39, 99]. Several additional applications to economics and computer science appear in Adler,

Feldman, and Taqqu [2]. There is extensive empirical evidence of heavy tail price fluctuations in stock markets, futures markets, and currency exchange rates [72, 90, 107]. The most popular estimator for the index α of a stable law is due to Hill [59], see also Hall [53]. Maximum likelihood estimation for stable laws on $\mathbb{R}^1$ is also practical, using the efficient method of Nolan [144] for computing stable densities.

Mittnik and Rachev [138] and Nolan, et al. [142, 143] discuss multivariable heavy tail models in finance. These models are used for portfolio analysis involving several different stock issues or mutual funds. If different stocks have different probability tails with different values of α, there is a need for multivariable techniques which allow the tail parameter α to vary. For such problems, it is typical to restrict attention to domains of normal attraction $A_n = n^{-E}$ where E is diagonal in some basis. Meerschaert and Scheffler [127] discuss methods for estimating this exponent E for a multivariable data set with heavy tails (see Theorem 10.4.10). In short, one computes the eigenvectors of the sample covariance matrix and applies one variable methods in those directions. Scheffler [169] provides a method for estimating the spectral measure (7.12) of the limit.

9

Related Limit Theorems

In this chapter, we present some basic limit theorems that complement the central limit theorems of Chapter 8. Large deviation results give additional information about the distribution of the sum of i.i.d. random vectors. Laws of the iterated logarithm provide an envelope in which the random walk defined by this sequence of partial sums must lie, with probability one. Finally, we discuss convergence of semitypes, a useful generalization of convergence of types for random variables.

9.1 LARGE DEVIATIONS

In this section, we will establish two large deviation results, using the stochastic compactness results of Chapter 8 along with the theory of R–O varying measures from Chapter 6. Then in Section 9.2 we will prove, using these results, two different versions of laws of the iterated logarithm. Here, as in Chapter 8, we will consider the behavior of the norm and of the projection of S_n on any radial direction separately.

The following result is due to Heyde [57]. It allows us to compare the probability of large deviations for the sum S_n with the probability of large deviations for an individual summand. This reflects the fact that, when the sum is unusually large, it is primarily due to the fact that one of the summands is unusually large.

Theorem 9.1.1. *Suppose that $X, X_1, X_2, X_3, \ldots$ are i.i.d. random variables with distribution μ and let $S_n = X_1 + \cdots + X_n$. Suppose that (x_n) is a* *monotone increasing sequence of positive real numbers such that* $x_n \to \infty$ *and* $S_n/x_n \to 0$ *in probability. If*

$$\liminf_{t\to\infty} \frac{t^2 P(|X| > t)}{\int_{|x|\le t} x^2 \mu\{dx\}} > 0 \tag{9.1}$$

and

$$\liminf_{t\to\infty} \frac{P(|X| > \lambda x_n)}{P(|X| > x_n)} > 0 \tag{9.2}$$

for all $\lambda > 1$ *then*

$$0 < \liminf_{n\to\infty} \frac{P(|S_n| > x_n)}{nP(|X| > x_n)} \le \limsup_{n\to\infty} \frac{P(|S_n| > x_n)}{nP(|X| > x_n)} < \infty$$

or equivalently

$$0 < \liminf_{n\to\infty} \frac{P(|S_n| > x_n)}{P(\max_{k\le n} |X_k| > x_n)} \le \limsup_{n\to\infty} \frac{P(|S_n| > x_n)}{P(\max_{k\le n} |X_k| > x_n)} < \infty.$$

Now we want to apply Theorem 9.1.1 to random vectors. We begin by considering the radial projections, normalized according to Theorem 8.3.20 to be stochastically compact.

Definition 9.1.2. Suppose that $\mu \in \mathrm{GDOSA}(\nu, c)$ for some $c > 1$, where ν is a full operator semistable law, and that for $X, X_1, X_2, \ldots$ independent and identically distributed according to μ we have

$$A_n S_{k_n} \Rightarrow Y \tag{9.3}$$

for some sequence of norming operators (A_n) and some sequence (k_n) tending to infinity with $k_{n+1}/k_n \to c$, where $S_n = X_1 + \cdots + X_n$. Then we say that μ is in the *strict generalized domain of semistable attraction* of ν.

Theorem 9.1.3. *Suppose* X *is a random vector with distribution* μ *in the strict* GDOSA *of some full operator semistable law* ν *without normal component. Then for all unit vectors* $\theta \in \mathbb{R}^d$ *and for all* $x_n \to \infty$ *we have that*

$$0 < \liminf_{n\to\infty} \frac{P\{|\langle S_n, \theta\rangle| > r_n^{-1} x_n\}}{nP\{|\langle X, \theta\rangle| > r_n^{-1} x_n\}} \le \limsup_{n\to\infty} \frac{P\{|\langle S_n, \theta\rangle| > r_n^{-1} x_n\}}{nP\{|\langle X, \theta\rangle| > r_n^{-1} x_n\}} < \infty,$$

where (r_n) *is the sequence of real numbers obtained in Theorem 8.3.20.*

Proof. We want to apply the Theorem 9.1.1. Since by Theorem 8.3.20 the sequence $(r_n\langle S_n, \theta\rangle)$ is uniformly tight and $x_n \to \infty$ we conclude that $\frac{1}{x_n} r_n\langle S_n, \theta\rangle \to 0$ in probability. Note that $\langle S_n, \theta\rangle = \langle X_1, \theta\rangle + \cdots +$

$<X_n, \theta>$ is a sum of i.i.d. random variables. Hence all we have to check is that

$$\liminf_{t \to \infty} \frac{t^2 V_0(t, \theta)}{U_2(t, \theta)} > 0 \tag{9.4}$$

and for all $\lambda > 1$

$$\liminf_{t \to \infty} \frac{P\{|<X, \theta>| > \lambda r_n^{-1} x_n\}}{P\{|<X, \theta>| > r_n^{-1} x_n\}} > 0. \tag{9.5}$$

Apply Theorem 8.2.1 to see that μ is $\mathrm{ROV}_\infty(F, c)$, where every eigenvalue of F has real part equal to or exceeding $a_1 > 1/2$. Then (9.4) follows immediately from Theorem 6.3.4 and Corollary 6.3.9 (b). We also have by Theorem 6.3.4 that for all large n

$$\frac{P\{|<X, \theta>| > \lambda r_n^{-1} x_n\}}{P\{|<X, \theta>| > r_n^{-1} x_n\}} = \frac{V_0(\lambda r_n^{-1}, \theta)}{V_0(r_n^{-1} x_n, \theta)} \geq C \lambda^{-\delta - 1/a_1}$$

for some $\delta > 0$ and some constant $C > 0$, proving (9.5). Now the result follows from Theorem 9.1.1. □

We now consider the case where we first normalize and then project. We will restrict our attention to the case of a spectrally simple limit, where all the real parts of the eigenvalues of the exponent have the same real part.

Theorem 9.1.4. *Suppose X is a symmetric random vector with distribution μ in the strict* GDOSA *of some full spectrally simple (c^E, c) operator semistable law ν without normal component, where $c > 1$. Assume further that one is not the real part of the eigenvalues of E. Then there exists a sequence (B_n) regularly varying with index $(-E)$ such that for every compact subset $K \subset \Gamma$ and every increasing sequence $x_n \to \infty$ there exist positive real constants C_1 and C_2 with*

$$C_1 \leq \liminf_{n \to \infty} \frac{P\{|<B_n S_n, \theta>| > x_n\}}{nP\{|<B_n X, \theta>| > x_n\}} \leq \limsup_{n \to \infty} \frac{P\{|<B_n S_n, \theta>| > x_n\}}{nP\{|<B_n X, \theta>| > x_n\}} \leq C_2$$

whenever $\theta \in K$.

Proof. Apply Theorem 8.3.18 to obtain a sequence (B_n) regularly varying with index $(-E)$ such that $(B_n S_n)_n$ is stochastically compact with limit set contained in $\{\lambda^{-E} \nu^\lambda : \lambda \in [1, c]\}$. It suffices to assume that $K = \{\theta \in \mathbb{R}^d : A \leq \|\theta\| \leq B\}$ for some $0 < A < B < \infty$.

First, we prove the lower bound. For $\varepsilon > 0$ and $1 \leq i \leq n$ let

$$C_i^{(n)} = \left\{ \left| \langle B_n \sum_{j=1, j \neq i}^{n} X_j, \theta \rangle \right| < \varepsilon x_n \right\}$$

and

$$D_i^{(n)} = \{|\langle B_n X_i, \theta\rangle| > (1+\varepsilon)x_n\}.$$

Then we have $\{|\langle B_n S_n, \theta\rangle| > x_n\} \supset \bigcup_{i=1}^{n} (D_i^{(n)} \cap C_i^{(n)})$ and hence by the i.i.d. assumption of the X_i we get

$$\begin{aligned} P\{|\langle B_n S_n, \theta\rangle| > x_n\} &\geq P\Big(\bigcup_{i=1}^{n}(D_i^{(n)} \cap C_i^{(n)})\Big) \\ &= \sum_{i=1}^{n} P\Big((D_i^{(n)} \cap C_i^{(n)}) \cap \bigcap_{j=1}^{i-1}(D_j^{(n)} \cap C_j^{(n)})^c\Big) \\ &\geq \sum_{i=1}^{n} P\Big((D_i^{(n)} \cap C_i^{(n)}) \setminus \bigcup_{j=1}^{i-1}(D_j^{(n)} \cap D_i^{(n)})\Big) \\ &\geq \sum_{i=1}^{n} \Big[P(D_i^{(n)})P(C_i^{(n)}) - \sum_{j=1}^{i-1} P(D_j^{(n)})P(D_i^{(n)})\Big] \\ &\geq nP(D_1^{(n)})[P(C_1^{(n)}) - nP(D_1^{(n)})]. \end{aligned} \tag{9.6}$$

From the regular variation of (B_n) and the compactness of K, we conclude that

$$\{(B_n B_{n-1}^{-1})^*\theta : n \geq 1, \theta \in K\} \subset \{x \in \mathbb{R}^d : a \leq \|x\| \leq b\}$$

for some $0 < a < b < \infty$. Therefore the stochastic compactness of $(B_n S_n)$ implies that

$$\begin{aligned} P(C_1^{(n)}) &= P\{|\langle B_n S_{n-1}, \theta\rangle| < \varepsilon x_n\} \\ &= P\{|\langle B_{n-1} S_{n-1}, (B_n B_{n-1}^{-1})^*\theta\rangle| < \varepsilon x_n\} \\ &\geq P\{\|B_{n-1} S_{n-1}\| < \varepsilon x_n / b\} \to 1 \end{aligned}$$

as $n \to \infty$ uniformly in $\theta \in K$. Hence for any $0 < \delta < 1$ there exists a number N_1 such that

$$P(C_1^{(n)}) > 1 - \delta \tag{9.7}$$

for all $n \geq N_1$ uniformly in $\theta \in K$.

On the other hand, it follows from Theorem 3.2.2 that every subsequence contains a further subsequence (n') such that $n(B_n\mu) \to \phi_{(\lambda)} = \lambda(\lambda^{-E}\phi)$ along that subsequence, where $\phi_{(\lambda)}$ is the Lévy measure of $\lambda^{-E}\nu^{\lambda}$ for some $\lambda \in [1, c]$ and ϕ is the Lévy measure of ν (see Remark 8.3.19 (b)). Given $\varepsilon_1 > 0$ there exists a $n_0 \geq 1$ such that $\phi_{(\lambda)}\{x : \|x\| \geq (1+\varepsilon)x_{n_0}/B\} < \varepsilon_1$. On the other hand, an application of Theorem 1.2.19 yields for all $n = n' \geq n_0$ and all $\theta \in K$ that

$$\begin{aligned} nP\{|<B_nX, \theta>| > (1+\varepsilon)x_n\} &\leq nP\{\|B_nX\| \geq (1+\varepsilon)x_{n_0}/B\} \\ &\leq 2\phi_{(\lambda)}\{x : \|x\| \geq (1+\varepsilon)x_{n_0}/B\} < 2\varepsilon_1. \end{aligned}$$

Since every subsequence contains a further subsequence with this property we have shown that $nP\{|<B_nX, \theta>| > (1+\varepsilon)x_n\} \to 0$ as $n \to \infty$ uniformly in $\theta \in K$. Therefore there exists a number N_2 such that

$$nP(D_1^{(n)}) < \delta \tag{9.8}$$

for all $n \geq N_2$ and all $\theta \in K$. By using (9.6)–(9.8), we then get for all $n \geq \max(N_1, N_2)$ and all $\theta \in K$ that

$$P\{|<B_nS_n, \theta>| > x_n\} \geq (1-2\delta)nP\{|<B_nX, \theta>| > (1+\varepsilon)x_n\}. \tag{9.9}$$

Writing $B_n^*\theta = r_n\theta_n$ with $\|\theta_n\| = 1$ and $r_n = r_n(\theta) > 0$, we get from the regular variation of (B_n) that $\|B_n\| \to 0$ and hence $r_n \to 0$ as $n \to \infty$ uniformly in $\theta \in K$. Apply Theorem 8.2.1 to see that μ is $\mathrm{ROV}_\infty(E, c)$. Then by Theorem 6.3.4 we conclude that for large n, some $\beta > 0$ and some positive constant C we have

$$\begin{aligned} \frac{P\{|<B_nS_n, \theta>| > x_n\}}{nP\{|<B_nX, \theta>| > x_n\}} &\geq (1-2\delta)\frac{V_0(r_n^{-1}(1+\varepsilon)x_n, \theta_n)}{V_0(r_n^{-1}x_n, \theta_n)} \\ &\geq C(1-2\delta)(1+\varepsilon)^{-\frac{1}{a}-\beta} \end{aligned}$$

uniformly in $\theta \in K$, where $a > 1/2$ is the real part on the eigenvalues of E. This concludes the proof of the lower bound.

Now we will prove the upper bound. For $1 \leq k \leq n$ and $\theta \in K$, let $X_{k,n}^\theta = X_k I(|<B_nX_k, \theta>| \leq x_n)$ and write $S_{n,n}^\theta = \sum_{k=1}^{n} X_{k,n}^\theta$. Define

$$E_n = \{|<B_nX_k, \theta>| > x_n \text{ for at least one } k \leq n\}$$

and

$$G_n = \{|<B_nS_{n,n}^\theta, \theta>| > x_n\}.$$

Then a simple calculation shows that $\{|<B_nS_n, \theta>| > x_n\} \subset E_n \cup G_n$. Therefore we get for every $\theta \in K$ that

$$P\{|<B_nS_n, \theta>| > x_n\} \leq nP\{|<B_nX_1, \theta>| > x_n\} + P(G_n). \tag{9.10}$$

By using Markov's inequality, we get

$$P(G_n) \leq \frac{1}{x_n^2} E\big(\langle B_nS_{n,n}^\theta, \theta\rangle^2\big) \tag{9.11}$$

and the i.i.d. assumption of the X_i gives

$$E\big(\langle B_nS^\theta_{n,n},\theta\rangle^2\big) = nE\big(\langle B_nX^\theta_{1,n},\theta\rangle^2\big) + n(n-1)\big(E\langle B_nX^\theta_{1,n},\theta\rangle\big)^2.$$

Therefore from (9.10) and (9.11) we have for all $\theta \in K$ that

$$\begin{aligned}\frac{P\{|<B_nS_n,\theta>| > x_n\}}{nP\{|<B_nX,\theta>| > x_n\}} \le 1 + &\frac{E\big(\langle B_nX^\theta_{1,n},\theta\rangle^2\big)}{x_n^2P\{|<B_nX,\theta>| > x_n\}}\\ &+ \frac{n\big(E\langle B_nX^\theta_{1,n},\theta\rangle\big)^2}{x_n^2P\{|<B_nX,\theta>| > x_n\}}.\end{aligned} \tag{9.12}$$

Since by Theorem 8.2.1 μ is $\mathrm{ROV}_\infty(E,c)$ we can apply Theorem 6.3.4 and Corollary 6.3.9 (b) to get

$$\begin{aligned}E\big(\langle B_nX^\theta_{1,n},\theta\rangle^2\big) &= r_n^2U_2(x_nr_n^{-1},\theta_n)\\ &\le M_2x_n^2V_0(x_nr_n^{-1},\theta_n)\\ &= M_2x_n^2P\{|<B_nX,\theta>| > x_n\},\end{aligned}$$

where again $B_n^*\theta = r_n\theta_n$ and M_2 is a constant that is independent of $\theta \in K$ and n. Hence the first fraction on the right-hand side of (9.12) is bounded by M_2.

Now let $a > 1/2$ be the real part of the eigenvalues of E. Then either $1/2 < a < 1$ or $a > 1$. We will consider these two cases separately. First, if $1/2 < a < 1$ then we know by Theorem 8.3.10 that $E|<X,\theta>|$ is finite for all $\theta \in \Gamma$. Then V_1 exists, and by the symmetry of X we get

$$\begin{aligned}\big|E\langle B_nX^\theta_{1,n}\rangle\big| &= \Big|\int_{|<B_nx,\theta>|\le x_n} <B_nx,\theta>d\mu(x)\Big|\\ &= \Big|\int_{|<B_nx,\theta>|>x_n} <B_nx,\theta>d\mu(x)\Big|\\ &\le r_nV_1(r_n^{-1}x_n,\theta_n).\end{aligned}$$

Theorem 6.3.4 implies that V_0 is uniformly R–O varying, and Theorem 6.3.16 implies that U_2 is R–O varying. Then we conclude from Corollary 6.3.9 that

$$\begin{aligned}&r_nV_1(r_n^{-1}x_n,\theta_n)\\ &\quad= \frac{r_n^{-1}x_nV_1(r_n^{-1}x_n,\theta_n)}{U_2(r_n^{-1}x_n,\theta_n)}\,\frac{U_2(r_n^{-1}x_n,\theta_n)}{(r_n^{-1}x_n)^2V_0(r_n^{-1}x_n,\theta_n)}x_nV_0(r_n^{-1}x_n,\theta_n)\\ &\quad\le M_1x_nP\{|<B_nX,\theta>| > x_n\},\end{aligned}$$

for all n and all $\theta \in K$, where M_1 is a constant independent of n and θ. Hence we have shown

$$\big|E\langle B_nX^\theta_{1,n},\theta\rangle\big|^2 \le M_1^2x_n^2\big(P\{|<B_nX,\theta>| > x_n\}\big)^2 \tag{9.13}$$

for all n uniformly in $\theta \in K$.

On the other hand, if $a > 1$, we get from Theorem 6.3.4 and Corollary 6.3.9 (b) that

$$\begin{aligned}
\left|E\langle B_n X^{\theta}_{1,n}, \theta\rangle\right| &\leq E\left|\langle B_n X^{\theta}_{1,n}, \theta\rangle\right| = r_n U_1(r_n^{-1} x_n, \theta_n) \\
&= \frac{U_1(r_n^{-1} x_n, \theta_n)}{r_n^{-1} x_n V_0(r_n^{-1} x_n, \theta_n)} x_n V_0(r_n^{-1} x_n, \theta_n) \\
&\leq M_1 x_n P\{|<B_n X, \theta>| > x_n\},
\end{aligned}$$

whenever $\theta \in K$, where again M_1 is a constant. Hence we get (9.13) in either case.

Finally, this gives for all $\theta \in K$ that

$$\frac{n\left(E\langle B_n X^{\theta}_{1,n}, \theta\rangle\right)^2}{x_n^2 P\{|<B_n X, \theta>| > x_n\}} \leq M_1^2 n P\{|<B_n X, \theta>| > x_n\},$$

where $nP\{|<B_n X, \theta>| > x_n\} \to 0$ as $n \to \infty$ uniformly in $\theta \in K$ as in the proof of the lower bound. The proof is now complete. □

9.2 LAW OF THE ITERATED LOGARITHM

In this section, we prove two laws of the iterated logarithm, using the large deviation results developed in Section 9.1. Assume that ν is a full operator semistable law without normal component and that μ belongs to the strict generalized domain of semistable attraction of ν, so that (9.3) holds for $X, X_1, X_2, \ldots$ independent and identically distributed according to $\mu \in \mathrm{GDOSA}(\nu, c)$ for some $c > 1$, for some sequence of norming operators (A_n) and some sequence (k_n) tending to infinity with $k_{n+1}/k_n \to c$, where $S_n = X_1 + \cdots + X_n$. Furthermore, assume that the (c^E, c) operator semistable law ν is spectrally compatible with μ and that $\alpha(\theta)$ is the spectral index of μ, in the sense of Definition 8.3.11. Let $W_1, \ldots, W_p$ and $a_1 < \cdots < a_p$ denote the spectral decomposition of $\mathbb{R}^d$ with respect to E, and let $L_i = W_1 \oplus \cdots \oplus W_i$. Then $\alpha(\theta) = 1/a_i$ if and only if $\theta \in L_i \setminus L_{i-1}$ so that $1/\alpha(\theta)$ ranges over the real spectrum of E. In view of the results of Section 8.3, these assumptions entail no loss of generality.

Theorem 9.2.1. *Suppose X is a random vector with distribution μ in the strict* GDOSA *of some full (c^E, c) operator semistable law ν without normal component. Let $\alpha(\theta)$ be the spectral index function from Definition 8.3.11. Then we have for all $\theta \in \Gamma$ that*

$$\limsup_{n\to\infty} \left| r_n \sum_{i=1}^{n} <X_i, \theta> \right|^{\frac{1}{\log\log n}} = e^{1/\alpha(\theta)} \quad \textit{almost surely,} \tag{9.14}$$

where $r_n = r_n(\theta)$ is the sequence obtained in Theorem 8.3.20.

Before we prove Theorem 9.2.1, we first need some information on the norming sequence (r_n).

Lemma 9.2.2. *Let* $n_k = 2^k$. *Then*

(a) $\sup_{k\geq 1} \sup_{n_k \leq n < n_{k+1}} r_n r_{n_k}^{-1} < \infty$

(b) $\sup_{k\geq 1} \sup_{n_k \leq n < n_{k+1}} r_{n_k} r_{n_{k+1}-n}^{-1} < \infty$

(c) $\sup_{k\geq 1} r_{n_k} r_{n_{k+1}}^{-1} < \infty$

(d) $\sup_{n\geq 2} r_n r_{n-1}^{-1} < \infty$

Now, if $n_k = [2^{k^\delta}]$ *for some* $\delta \geq 1$, *then*

(e) $\sup_{k\geq 1} r_{n_k - n_{k-1}} r_{n_k}^{-1} < \infty$

(f) $\sup_{k\geq 1} r_{n_k} r_{n_{k-1}}^{-1} < \infty$

Proof. Apply Theorem 8.3.7 to obtain a regularly varying sequence (B_n) that respects the spectral decomposition of $\mathbb{R}^d$ with respect to μ from Definition 8.3.11. Then $((B_n^*)^{-1})$ varies regularly with index E^*, and since $W_1, \ldots, W_p$ are mutually orthogonal, this sequence also respects the spectral decomposition of the $\mathbb{R}^d$ with respect to μ. Note that $\alpha(\theta) = 1/a_i$ if and only if $\theta \in L_i \setminus L_{i-1}$. From Theorem 8.3.20, we get a sequence $(r_n = r_n(\theta))$ of positive real numbers such that $(r_n \langle S_n, \theta \rangle)$ is stochastically compact. Note that by (8.103) $r_n = \|(B_n^*)^{-1}\theta\|^{-1}$. The regular variation of (B_n) implies

$$B_n B_{[\lambda n]}^{-1} \to \lambda^E \quad \text{as } n \to \infty \tag{9.15}$$

uniformly on compact subsets of $(0, \infty)$. For the proof of (a), we write $n_k = \lambda n$. Then $n_k \leq n < n_{k+1}$ implies $1/2 \leq \lambda \leq 1$. Now

$$\begin{aligned} r_n r_{n_k}^{-1} &= \frac{\left\|\left(B_{n_k}^*\right)^{-1}\theta\right\|}{\left\|\left(B_n^*\right)^{-1}\theta\right\|} \\ &= \frac{\left\|\left(B_{[\lambda n]}^*\right)^{-1} B_n^* \left(B_n^*\right)^{-1}\theta\right\|}{\left\|\left(B_n^*\right)^{-1}\theta\right\|} \\ &\leq \left\|B_n B_{[\lambda n]}^{-1}\right\| \leq C \sup_{\frac{1}{2}\leq\lambda\leq 1} \left\|\lambda^E\right\| < \infty, \end{aligned}$$

for some constant $C > 0$, using Proposition 2.1.3 (a) and Remark 2.1.5 along with the uniform convergence in (9.15).

For the proof of (b), we first show an auxiliary result that is also of independent interest. We claim that (9.15) implies

$$\sup_{n\geq 1}\sup_{\frac{1}{n}\leq\lambda\leq 1}\left\|B_n B_{[\lambda n]}^{-1}\right\| < \infty. \tag{9.16}$$

In fact, for all $n \geq 1$ there exists a $1/n \leq \lambda_n \leq 1$ such that

$$M_n = \|B_n B_{[\lambda_n n]}^{-1}\| = \sup_{1/n\leq\lambda\leq 1}\|B_n B_{[\lambda n]}^{-1}\|.$$

If (9.16) does not hold there exists a subsequence (n_k) such that $M_{n_k} \to \infty$ as $k \to \infty$. Since $0 < \lambda_n \leq 1$ every subsequence of (n_k) contains a further subsequence (n') along which $\lambda_n \to \lambda \geq 0$. Now we have to consider several cases separately.

First, assume that $\lambda > 0$. Then (9.15) implies $M_n \to \|\lambda^E\| < \infty$ along (n'), which is a contradiction.

If $\lambda = 0$, then every subsequence of (n') contains a further subsequence (n'') such that either $\lambda_n n \leq M$ along (n'') for some $M > 0$ or $\lambda_n n \to \infty$ along (n''). If $\lambda_n n \leq M$ along (n''), then

$$M_n \leq \|B_n\| \max_{1\leq k\leq M}\|B_k^{-1}\|$$

along (n''). But Theorem 4.2.4 implies $\|B_n\| \to 0$, which is again a contradiction.

Finally, assume that $\lambda_n n \to \infty$ along (n''). Given $0 < \varepsilon < 1$, in view of Theorem 2.2.4 there exists a $\lambda_0 > 1$ such that $\|\lambda_0^{-E}\| + \varepsilon \leq 1$. By uniform convergence there exists a $m_0 \geq 1$ such that

$$\|B_{[\lambda n]}B_n^{-1} - \lambda^{-E}\| < \varepsilon$$

for all $n \geq m_0$ and $1 \leq \lambda \leq \lambda_0$. Hence, for all $n \geq m_0$

$$\|B_{[\lambda_0 n]}B_n^{-1}\| \leq 1.$$

Choose $n_0 \geq 1$ such that $\lambda_n n \geq m_0$ for all $n'' \geq n_0$. For $n'' \geq n_0$ let $m = [n''\lambda_{n''}] \geq m_0$ and write $n'' = m\lambda(m)$ so that $\lambda(m) = n''/[n''\lambda_{n''}]$. Then, if we write $\lambda(m) = \lambda_0^k\alpha$ for some nonnegative integer k and some $1 \leq \alpha < \lambda_0$, we get for all $n'' \geq n_0$ that

$$\begin{aligned}\|B_{n''}B_{[\lambda_{n''}n'']}^{-1}\| &= \|B_{m\lambda(m)}B_m^{-1}\| \\ &= \|B_{[\lambda_0^k\alpha m]}B_m^{-1}\| \\ &\leq \|B_{[\lambda_0^k m\alpha]}B_{[\lambda_0^k m]}^{-1}\|\cdot\|B_{[\lambda_0^k m]}B_{[\lambda_0^{k-1}m]}^{-1}\|\cdots\|B_{[\lambda_0 m]}B_m^{-1}\| \\ &\leq D\cdot 1^k = D\end{aligned}$$

where

$$D = \sup_{n \geq 1} \sup_{1 \leq \alpha \leq \lambda_0} \|B_{[\alpha n]} B_n^{-1}\| < \infty$$

in view of uniform convergence. Hence M_{n_k} remains bounded along (n''), which is again a contradiction. Since every subsequence of (n_k) contains a further subsequence for which either of the above cases applies (9.16) follows.

For the proof of (b), note that if $n_k \leq n < n_{k+1}$ then $(n_{k+1} - n)/n_k \in [\frac{1}{n_k}, 1]$. Since

$$\begin{aligned} r_{n_k} r_{n_{k+1}-n}^{-1} &= \frac{\left\|\left(B_{n_{k+1}-n}^*\right)^{-1}\theta\right\|}{\left\|\left(B_{n_k}^*\right)^{-1}\theta\right\|} \\ &\leq \left\|B_{n_k} B_{n_{k+1}-n}^{-1}\right\| \\ &= \left\|B_{n_k} B_{n_k \frac{n_{k+1}-n}{n_k}}\right\| \\ &\leq \sup_{n \geq 1} \sup_{\frac{1}{n} \leq \lambda \leq 1} \left\|B_n B_{[\lambda n]}^{-1}\right\|, \end{aligned}$$

the assertion follows from (9.16).

(c) and (d) follow directly from (9.15). For the proof of (e), first note that

$$r_{n_k - n_{k-1}} r_{n_k}^{-1} \leq \|B_{n_k - n_{k-1}} B_{n_k}^{-1}\|. \tag{9.17}$$

Write $n_k = (n_k - n_{k-1})\lambda_k$, where $\lambda_k = n_k/(n_k - n_{k-1})$. Then $\delta \geq 1$ implies that $n_{k-1}/n_k \to 0$ as $k \to \infty$ and hence $\lambda_k = (1 - \frac{n_{k-1}}{n_k})^{-1} \to 1$ as $k \to \infty$. Now (e) follows from the uniform convergence in (9.15).

Finally, (f) follows from the bound

$$\begin{aligned} r_{n_k} r_{n_{k-1}}^{-1} &\leq \|B_{n_k} B_{n_{k-1}}^{-1}\| \\ &= \|B_{n_k} B_{n_k \frac{n_{k-1}}{n_k}}\| \\ &\leq \sup_{\frac{1}{n_k} \leq \lambda \leq 1} \|B_{n_k} B_{[n_k \lambda]}^{-1}\| \end{aligned}$$

together with (9.16). This concludes the proof of the lemma. □

Proof of Theorem 9.2.1. Fix any $\theta \in \Gamma$, and suppose that $\theta \in L_i \backslash L_{i-1}$ so that $1/\alpha(\theta) = a_i$. Due to the nature of the power in (9.14) it suffices to consider $\|\theta\| = 1$. Then we have to show that for all $\varepsilon > 0$ we have almost surely that

$$\left|r_n \sum_{i=1}^{n} \langle X_i, \theta \rangle\right| > (\log n)^{(1+\varepsilon)a_i} \text{ for at most finitely many } n \tag{9.18}$$

and

$$\left|r_n \sum_{i=1}^{n} \langle X_i, \theta \rangle\right| > (\log n)^{(1-\varepsilon)a_i} \text{ for infinitely many } n. \tag{9.19}$$

Let

$$D_n = \{|r_n{<}S_n,\theta{>}| > (\log n)^{(1+\varepsilon)a_i}\}.$$

In order to prove (9.18), we have to show $P(\limsup D_n) = 0$. For $n_k = 2^k$, let

$$C^{-1} = \sup_{k\geq 1} \sup_{n_k \leq n < n_{k+1}} r_n r_{n_k}^{-1}$$

which is by Lemma 9.2.2 (a) a positive real constant. Furthermore, let

$$B_k = \Big\{ \max_{n_k \leq n < n_{k+1}} |r_{n_k}{<}S_n,\theta{>}| > C(\log n_k)^{(1+\varepsilon)a_i} \Big\}.$$

Since for $n_k \leq n < n_{k+1}$ we have

$$|r_n{<}S_n,\theta{>}| = r_n r_{n_k}^{-1} |r_{n_k}{<}S_n,\theta{>}| \leq C^{-1} |r_{n_k}{<}S_n,\theta{>}|$$

it follows that $D_n \subset B_k$ if $n_k \leq n < n_{k+1}$. Hence $\limsup D_n \subset \limsup B_k$.

In view of Ottaviani's inequality (see, e.g., Billingsley [18] p. 296) we have to estimate

$$D_k = \max_{n_k \leq n < n_{k+1}} P\left\{ \Big|r_{n_k} \sum_{i=n+1}^{n_{k+1}} {<}X_i,\theta{>}\Big| > \frac{C}{2}(\log n_k)^{(1+\varepsilon)a_i} \right\}.$$

Let

$$M = \sup_{k\geq 1} \sup_{n_k \leq n < n_{k+1}} r_{n_k} r_{n_{k+1}-n}^{-1},$$

which is by Lemma 9.2.2 (b) a positive real constant. Prohorov's Theorem 1.2.10 implies that $(r_n{<}S_n,\theta{>})_n$ is tight, so for $0 < D < 1$ there exists some $k_0 \geq 1$ such that $P\{|r_n{<}S_n,\theta{>}| > \frac{C}{2M}(\log n_{k_0})^{(1+\varepsilon)a_i}\} < D$ for all $n \geq 1$. Hence for all $k \geq k_0$ and $n_k \leq n < n_{k+1}$ we get

$$\begin{aligned}
&P\left\{ \Big|r_{n_k} \sum_{i=n+1}^{n_{k+1}} {<}X_i,\theta{>}\Big| > \frac{C}{2}(\log n_k)^{(1+\varepsilon)a_i} \right\} \\
&= P\Big\{ r_{n_k} r_{n_{k+1}-n}^{-1} \big|r_{n_{k+1}-n}{<}S_{n_{k+1}} - S_n,\theta{>}\big| > \frac{C}{2}(\log n_k)^{(1+\varepsilon)a_i} \Big\} \\
&\leq P\Big\{ \big|r_{n_{k+1}-n}{<}S_{n_{k+1}-n},\theta{>}\big| > \frac{C}{2M}(\log n_k)^{(1+\varepsilon)a_i} \Big\} \\
&\leq P\Big\{ \big|r_{n_{k+1}-n}{<}S_{n_{k+1}-n},\theta{>}\big| > \frac{C}{2M}(\log n_{k_0})^{(1+\varepsilon)a_i} \Big\} < D,
\end{aligned}$$

proving that $D_k \leq D < 1$ for all $k \geq k_0$. On the other hand, let

$$K = \sup_{k\geq 1} r_{n_k} r_{n_{k+1}}^{-1}$$

and apply Lemma 9.2.2 (c) to see that K is finite. Then by an application of Ottaviani's inequality we conclude that for some constant $E > 0$ and all $k \geq k_0$

$$
\begin{aligned}
P(B_k) &\leq \frac{1}{1-D} P\Big\{\big|r_{n_k} <S_{n_{k+1}}, \theta>\big| > \frac{C}{2}(\log n_k)^{(1+\varepsilon)a_i}\Big\} \\
&\leq \frac{1}{1-D} P\Big\{\big|r_{n_{k+1}} <S_{n_{k+1}}, \theta>\big| > \frac{C}{2K}(\log n_k)^{(1+\varepsilon)a_i}\Big\} \\
&\leq \frac{1}{1-D} P\Big\{\big|r_{n_{k+1}} <S_{n_{k+1}}, \theta>\big| > E k^{(1+\varepsilon)a_i}\Big\}.
\end{aligned}
$$

But by Theorem 9.1.3 and Theorems 6.4.15 and 8.2.1 for every $\delta > 0$ there exist positive real constants C_1, C_2 and C_3 such that for large k

$$
\begin{aligned}
&P\{|r_{n_{k+1}} <S_{n_{k+1}}, \theta>| > E k^{(1+\varepsilon)a_i}\} \\
&\leq C_1 n_{k+1} P\{|r_{n_{k+1}} <X, \theta>| > E k^{(1+\varepsilon)a_i}\} \\
&= C_1 \frac{V_0(r_{n_{k+1}}^{-1} E k^{(1+\varepsilon)a_i}, \theta)}{V_0(r_{n_{k+1}}^{-1}, \theta)} n_{k+1} P\{|r_{n_{k+1}}^{-1} <X, \theta>| > 1\} \qquad (9.20) \\
&\leq C_2 \left(k^{(1+\varepsilon)a_i}\right)^{-\frac{1}{a_i}+\delta} n_{k+1} P\{|r_{n_{k+1}} <X, \theta>| > 1\} \\
&\leq C_3 k^{-(1+\varepsilon_1)}
\end{aligned}
$$

where the last inequality holds with $\varepsilon_1 = \varepsilon_1(\varepsilon, \delta)$ positive for $\delta > 0$ is small enough by the following argument.

In order to justify the last inequality in (9.20) above, we have to show that

$$\sup_{n \geq 1} n P\{|r_n <X, \theta>| > 1\} < \infty. \qquad (9.21)$$

By Theorem 8.3.20, the sequence $(r_n <S_n, \theta>)$ is stochastically compact with all limit points in the set $\{<Y_\lambda, \theta_0> : \lambda \in [1, c], \|\theta_0\| = 1\}$, where $\mathcal{L}(Y_\lambda) = \lambda^{-E} \nu^\lambda$. Let ϕ denote the Lévy measure of ν and let $\phi_\lambda(\theta_0)$ be the Lévy measure of $<Y_\lambda, \theta_0>$. Then an easy calculation shows that

$$\phi_\lambda(\theta_0)([1, \infty)) = \lambda \cdot (\lambda^{-E} \phi)\{x \in \mathbb{R}^d : |<x, \theta_0>| \geq 1\} \qquad (9.22)$$

where

$$
\begin{aligned}
\lambda \cdot (\lambda^{-E} \phi)\{x \in \mathbb{R}^d : |<x, \theta_0>| \geq 1\} &\leq c\phi\{x \in \mathbb{R}^d : \|\lambda^{-E} x\| \geq 1\} \\
&\leq c\phi\{x \in \mathbb{R}^d : \|x\| \geq \rho\}
\end{aligned}
$$

for some constant $\rho > 0$. Hence

$$E_1 = \sup_{\lambda \in [1,c]} \sup_{\|\theta_0\|=1} \phi_\lambda(\theta_0)([1, \infty)) < \infty.$$

By using the relative compactness of $(r_n <S_n, \theta>)$, every subsequence contains a further subsequence (n') such that for some $\|\theta_0\| = 1$ and some $\lambda \in [1, c]$ we have $r_n \langle S_n, \theta \rangle \Rightarrow \langle Y_\lambda, \theta_0 \rangle$ along (n'). Then Theorem 3.3.8 together with Theorem 1.2.19 implies that

$$\limsup_{(n')} nP\{|r_n \langle X, \theta \rangle| > 1\} \leq \phi_\lambda(\theta_0)([1, \infty)) \leq E_1$$

along (n') and hence (9.21) is immediate.

Hence we have shown $P(B_k) \leq C_4 k^{-(1+\varepsilon_1)}$ for all $k \geq k_0$, where C_4 is a positive constant and $\varepsilon_1 > 0$. Now an application of the first Borel-Cantelli Lemma (see, e.g., Theorem 4.3 in [18]) yields $P(\limsup D_n) \leq P(\limsup B_k) = 0$ which proves (9.18).

For the proof of (9.19), let

$$K = \sup_{n \geq 2} r_n r_{n-1}^{-1}$$

which is by Lemma 9.2.2 (d) finite. Enlarge K if necessary to ensure $K \geq 1$. Let i.o. mean infinitely often. Then, using the inequality $|r_n \langle X_n, \theta \rangle| \leq |r_n \langle S_n, \theta \rangle| + |r_n \langle S_{n-1}, \theta \rangle|$ we get

$$\begin{aligned} &\{|r_n \langle X_n, \theta \rangle| > 2K(\log n)^{(1-\varepsilon)a_i} \text{ i.o.}\} \\ &\qquad \subset \Big(\{|r_n \langle S_n, \theta \rangle| > K(\log n)^{(1-\varepsilon)a_i} \text{ i.o.}\} \\ &\qquad\qquad \cup \{|r_n \langle S_{n-1}, \theta \rangle| > K(\log n)^{(1-\varepsilon)a_i} \text{ i.o.}\}\Big). \end{aligned}$$

But by the definition of K

$$\{|r_n \langle S_{n-1}, \theta \rangle| > K(\log n)^{(1-\varepsilon)a_i} \text{ i.o.}\} \subset \{|r_n \langle S_n, \theta \rangle| > (\log n)^{(1-\varepsilon)a_i} \text{ i.o.}\}$$

and since $K \geq 1$ we conclude that

$$\{|r_n \langle X_n, \theta \rangle| > 2K(\log n)^{(1-\varepsilon)a_i} \text{ i.o.}\} \subset \{|r_n \langle S_n, \theta \rangle| > (\log n)^{(1-\varepsilon)a_i} \text{ i.o.}\}.$$

Hence, in view of the second Borel-Cantelli Lemma (see, e.g., Theorem 4.4 in [18]), it suffices to show that

$$\sum_{n=1}^{\infty} P\{|r_n \langle X_n, \theta \rangle| > 2K(\log n)^{(1-\varepsilon)a_i}\} = \infty. \tag{9.23}$$

We first show that complementary to (9.21) we also have

$$\inf_{n \geq 1} nP\{|r_n \langle X, \theta \rangle| > 1\} > 0. \tag{9.24}$$

In fact, let us assume that there exists a subsequence (n_k) such that

$$n_k P\{|r_{n_k} \langle X, \theta \rangle| > 1\} \to 0 \quad \text{as } k \to \infty. \tag{9.25}$$

Since by Theorem 8.3.20 the sequence $(r_n\langle S_n,\theta\rangle)$ is stochastically compact there exists a subsequence $(n')\subset(n_k)$, a $\lambda\in[1,c]$ and a unit vector θ_0 such that $r_n\langle S_n,\theta\rangle\Rightarrow\langle Y_\lambda,\theta_0\rangle$ along (n') where $\mathcal{L}(Y_\lambda)=\lambda^{-E}\nu^\lambda$. Then $\phi_\lambda(\theta_0)$ is the Lévy measure of $\langle Y_\lambda,\theta_0\rangle$ and (9.22) holds. Note that by Lemma 6.1.28

$$\phi_\lambda(\theta_0)((1,\infty))=\lambda\phi\{x\in\mathbb{R}^d:|\langle x,\lambda^{-E^*}\theta_0\rangle|>1\}>0.$$

On the other hand, Theorem 3.3.8 together with Theorem 1.2.19 shows

$$\liminf_{(n')} nP\{|r_n\langle X,\theta\rangle|>1\}\geq\phi_\lambda(\theta_0)((1,\infty))>0$$

in contrast to (9.25). This proves (9.24).

Then Theorem 6.4.15 gives for an appropriate choice of $\delta>0$ that for some constants $C_5,C_6>0$ and large n we have

$$\begin{aligned}&P\{|r_n\langle X_n,\theta\rangle|>2K(\log n)^{(1-\varepsilon)a_i}\}\\&=V_0(r_n^{-1}2K(\log n)^{(1-\varepsilon)a_i},\theta)\\&=\frac{1}{n}\frac{V_0(r_n^{-1}2K(\log n)^{(1-\varepsilon)a_i},\theta)}{V_0(r_n^{-1},\theta)}nP\{|r_n\langle X,\theta\rangle|>1\}\\&\geq C_5\frac{1}{n}\left((\log n)^{(1-\varepsilon)a_i}\right)^{-\frac{1}{a_i}-\delta}\geq C_6\frac{1}{n\log n}.\end{aligned}$$

But $\sum_{n=2}^{\infty}\frac{1}{n\log n}=\infty$, proving (9.23), which is sufficient for (9.19). □

Inherent in the assertion of Theorem 9.2.1 is information about the maximal growth rate of the random walk (S_n), which varies along the direction θ in a very specific way. In fact, it follows from Corollary 9.2.3 below that the maximal growth rate of S_n projected onto the direction θ is of the order $(n\log n)^{a_i}$ when $\theta\in L_i\setminus L_{i-1}$.

Corollary 9.2.3. *Under the assumptions of Theorem 9.2.1 we have for every $\|\theta\|=1$ and every $\varepsilon>0$ that almost surely*

$$|\langle S_n,\theta\rangle|>(n\log n)^{\varepsilon+1/\alpha(\theta)}\quad\textit{for at most finitely many } n$$

and

$$|\langle S_n,\theta\rangle|>(n\log n)^{-\varepsilon+1/\alpha(\theta)}\quad\textit{for infinitely many } n.$$

Here $\alpha(\theta)$ is the spectral index function of μ.

Proof. Note that by Remark 8.3.21 we know that for every $\varepsilon>0$

$$n^{-a_i-\varepsilon}\leq r_n\leq n^{-a_i+\varepsilon}$$

for all large n, where $\alpha(\theta)=1/a_i$. Then the assertion follows from (9.18) and (9.19). □

Complementary to Theorem 9.2.1 we can give the following clustering statement, which gives additional information about the path behavior of the random walk (S_n).

Corollary 9.2.4. *Under the assumptions of Theorem 9.2.1, for any $\theta \in \Gamma$, with probability one every point in the interval $(1, e^{1/\alpha(\theta)}]$ is a cluster point of the sequence*

$$\left\{ |r_n \langle S_n, \theta \rangle|^{\frac{1}{\log \log n}} : n \geq 1 \right\}.$$

Here $\alpha(\theta)$ is the spectral index function of μ.

Proof. Again it suffices to consider $\|\theta\| = 1$. Then $1/\alpha(\theta) = a_i$ for some $1 \leq i \leq p$, where $a_1 < \cdots < a_p$ are the real parts of the eigenvalues of E. Fix $0 < \lambda \leq a_i$, let $\delta = a_i/\lambda$ and let $n_k = [2^{k^\delta}]$. We will show that

$$\limsup_{k \to \infty} |r_{n_k} \langle S_{n_k}, \theta \rangle|^{\frac{1}{\log \log n_k}} = e^{\lambda} \quad \text{almost surely.} \tag{9.26}$$

For that, it is sufficient to argue that for any small $\varepsilon > 0$ and for almost all sample points we have

$$|r_{n_k} \langle S_{n_k}, \theta \rangle| > (\log n_k)^{(1+\varepsilon)\lambda} \quad \text{for at most finitely many } k \tag{9.27}$$

and

$$|r_{n_k} \langle S_{n_k}, \theta \rangle| > (\log n_k)^{(1-\varepsilon)\lambda} \quad \text{for infinitely many } k. \tag{9.28}$$

But $E_1 k^{(1+\varepsilon)a_i} \leq (\log n_k)^{(1+\varepsilon)\lambda} \leq E_2 k^{(1+\varepsilon)a_i}$ for some positive constants E_1 and E_2. Then (9.27) follows by the same lines as (9.18) without using Ottaviani's maximal inequality. See also (9.20).

The proof of (9.28) is more involved than the proof of (9.19) above. First, we show that with probability one

$$\left| r_{n_k} \left\langle \sum_{j=n_{k-1}+1}^{n_k} X_j, \theta \right\rangle \right| > (\log n_k)^{(1-\varepsilon/2)\lambda} \quad \text{for infinitely many } k. \tag{9.29}$$

Note that these are independent events.

Let

$$M = \sup_{k \geq 1} r_{n_k - n_{k-1}} r_{n_k}^{-1}$$

and apply Lemma 9.2.2 (e) to see that M is finite. Note that here $0 < \lambda \leq a_i$ is crucial. Then we get from Theorem 9.1.3, for some constants $E, K > 0$, that

$$
\begin{aligned}
&P\Big\{\Big|r_{n_k}\Big\langle \sum_{j=n_{k-1}+1}^{n_k} X_j, \theta\Big\rangle\Big| > (\log n_k)^{(1-\varepsilon/2)\lambda}\Big\}\\
\geq & P\big\{\big|r_{n_k}{<}S_{n_k-n_{k-1}}, \theta{>}\big| > E k^{(1-\varepsilon/2)a_i}\big\}\\
\geq & P\big\{\big|r_{n_k-n_{k-1}}{<}S_{n_k-n_{k-1}}, \theta{>}\big| > M E k^{(1-\varepsilon/2)a_i}\big\}\\
\geq & K(n_k-n_{k-1})P\big\{|r_{n_k-n_{k-1}}{<}X, \theta{>}| > M E k^{(1-\varepsilon/2)a_i}\big\}\\
= & \frac{V_0(r^{-1}_{n_k-n_{k-1}} M E k^{(1-\varepsilon/2)a_i}, \theta)}{V_0(r^{-1}_{n_k-n_{k-1}}, \theta)} \cdot K(n_k-n_{k-1})P\{|r_{n_k-n_{k-1}}{<}X, \theta{>}| > 1\}.
\end{aligned}
$$

But by (9.24) we know that

$$
\inf_{k\geq 1}(n_k-n_{k-1})P\{|r_{n_k-n_{k-1}}{<}X, \theta{>}| > 1\} > 0
$$

and in view of Theorem 6.4.15 we conclude that for all large k

$$
\frac{V_0(r^{-1}_{n_k-n_{k-1}} M E k^{(1-\varepsilon/2)a_i}, \theta)}{V_0(r^{-1}_{n_k-n_{k-1}}, \theta)} \geq C k^{-1+\varepsilon_1},
$$

for some constant $C > 0$, where $\varepsilon_1 = \varepsilon_1(\varepsilon) > 0$. Hence

$$
P\Big\{\Big|r_{n_k}\Big\langle \sum_{j=n_{k-1}+1}^{n_k} X_j, \theta\Big\rangle\Big| > (\log n_k)^{(1-\varepsilon/2)\lambda}\Big\} \geq C k^{-1+\varepsilon_1}
$$

for all large k and some constants $C, \varepsilon_1 > 0$. Then (9.29) follows from the second Borel-Cantelli Lemma.

Finally, suppose that there is a set with positive probability on which (9.28) does not hold. Then for almost all points in this set we have

$$
\begin{aligned}
(\log n_k)^{(1-\varepsilon)\lambda} &\geq |r_{n_k}{<}S_{n_k}, \theta{>}|\\
&\geq \Big|r_{n_k}\Big\langle \sum_{j=n_{k-1}+1}^{n_k} X_j, \theta\Big\rangle\Big| - |r_{n_k}{<}S_{n_{k-1}}, \theta{>}|\\
&> (\log n_k)^{(1-\varepsilon/2)\lambda} - r_{n_k} r^{-1}_{n_{k-1}} (\log n_{k-1})^{(1-\varepsilon)\lambda}
\end{aligned}
$$

for infinitely many k using (9.29) and the assumption. But by Lemma 9.2.2 (f) $r_{n_k} r^{-1}_{n_{k-1}} \leq C$ for all k and some constant $C > 0$. Hence the last difference above is greater than $(\log n_k)^{(1-\varepsilon)\lambda}$ for all large k, which is a contradiction. Therefore (9.28) holds and the proof is complete. □

After considering the sample path behavior of the projection of the random walk (S_n) along any radial direction θ in Theorem 9.2.1 and Corollary 9.2.4

above, we now investigate the behavior of the norm of S_n. Here, as in the large deviation result (Theorem 9.1.4), which we will use in the proof, we first have to normalize S_n by some regularly varying sequence (B_n) of linear operators and then take the norm. This will give complementary information on the almost sure behavior of (S_n).

Let $\mu \in \mathrm{GDOSA}(\nu, c)$ be symmetric such that (9.3) holds, where ν is a full symmetric (c^E, c) operator semistable law without normal component that is spectrally compatible with μ. Assume further that one is not the real part of the eigenvalues of E. The formulation as well as the proof of Theorem 9.2.5 rely heavily on the spectral decomposition of $\mathbb{R}^d$ with respect to μ from Definition 8.3.11. Let $\mathbb{R}^d = W_1 \oplus \cdots \oplus W_p$ be the spectral decomposition of $\mathbb{R}^d$ with respect to μ. Recall that $E = E_1 \oplus \cdots \oplus E_p$ and that every eigenvalue of E_i has real part a_i for $1 \leq i \leq p$. Theorem 7.1.10 implies that $1/2 < a_1 < \cdots < a_p$. Apply Theorem 8.3.7 to obtain a sequence (B_n) of linear operators varying regularly with index $(-E)$ such that $A_n = B_{k_n}$ in (9.3), and the spaces W_i are B_n invariant for all n and all $1 \leq i \leq p$. Given any random vector X we write $X = X^{(1)} + \cdots + X^{(p)}$ with respect to the spectral decomposition of $\mathbb{R}^d$ obtained above and for $1 \leq i \leq p$ set $X^{(1,\ldots,i)} = X^{(1)} + \cdots + X^{(i)}$.

If we use the assumptions of the paragraph above, the following law of the iterated logarithm holds:

Theorem 9.2.5. *For any* $1 \leq i \leq p$

$$\limsup_{n\to\infty} \left\| B_n S_n^{(1,\ldots,i)} \right\|^{\frac{1}{\log\log n}} = e^{a_i} \quad \textit{almost surely.} \tag{9.30}$$

Before going into the details of the proof let us describe its structure. First, we will show that (9.30) is true if ν is spectrally simple, i.e., $\mathrm{Re}(\lambda) = a > 1/2$ for all eigenvalues λ of E. This will be done in Proposition 9.2.6. Then we show that this special result already implies the general case.

In the following, let ν be a full symmetric (c^E, c) operator semistable law without normal component on a finite dimensional vector space V such that $\mathrm{Re}(\lambda) = a > 1/2$ for all eigenvalues λ of E and $a \neq 1$. Let ϕ denote the Lévy measure of ν and let $\mu \in \mathrm{GDOSA}(\nu, c)$ be symmetric. Then an application of Theorem 8.3.18 yields a sequence (B_n) regularly varying with index $(-E)$ such that if $X_1, X_2, \ldots$ are i.i.d. as μ, we have for $S_n = \sum_{i=1}^n X_i$ that $(B_n S_n)$ is stochastically compact.

In this case, the following law of the iterated logarithm holds:

Proposition 9.2.6.

$$\limsup_{n\to\infty} \left\| B_n S_n \right\|^{\frac{1}{\log\log n}} = e^{a} \quad \textit{almost surely.} \tag{9.31}$$

Proof. As in the proof of Theorem 9.2.1 it suffices to show that for any $0 < \varepsilon < 1$ with probability one we have

$$\|B_n S_n\| > (\log n)^{(1+\varepsilon)a} \quad \text{for at most finitely many } n \tag{9.32}$$

and

$$\|B_n S_n\| > (\log n)^{(1-\varepsilon)a} \quad \text{for infinitely many } n. \tag{9.33}$$

For the proof of (9.32), define the events $D_n = \{\|B_n S_n\| > (\log n)^{(1+\varepsilon)a}\}$. Let $n_k = 2^k$ and let

$$C^{-1} = \sup_{k \geq 1} \sup_{n_k \leq n < n_{k+1}} \|B_n B_{n_k}^{-1}\|.$$

Then it follows from the proof of Lemma 9.2.2 that C is finite. Furthermore, let

$$F_k = \Big\{ \max_{n_k \leq n < n_{k+1}} \|B_{n_k} S_n\| > C(\log n_k)^{(1+\varepsilon)a} \Big\}.$$

If $n_k \leq n < n_{k+1}$, then

$$\|B_n S_n\| \leq \|B_n B_{n_k}^{-1}\| \|B_{n_k} S_n\| \leq C^{-1} \|B_{n_k} S_n\|$$

and hence the monotonicity of the function $\log t$ implies that $D_n \subset F_k$ in this case. Hence limsup $D_n \subset$ limsup F_k.

Now let $\{\theta^{(1)}, \ldots, \theta^{(m)}\}$ be an orthonormal basis of V. Then it is easy to see that for any random vector Z with values in V we have for some constant $C_1 > 0$ that

$$P\{\|Z\| > t\} \leq \sum_{j=1}^{m} P\{|\langle Z, \theta^{(j)} \rangle| > C_1 t\} \tag{9.34}$$

for all $t > 0$. Let $\bar{C} = \sup_{k \geq 1} \|B_{n_k} B_{n_{k+1}}^{-1}\|$, which is again a positive real constant. Then an application of Lévy's maximal inequality (e.g., see [6] p. 102) along with (9.34) implies that for some constant $E > 0$

$$\begin{aligned} P(F_k) &\leq 2P\big\{\big\|B_{n_k} S_{n_{k+1}}\big\| > C(\log n_k)^{(1+\varepsilon)a}\big\} \\ &\leq 2P\big\{\big\|B_{n_{k+1}} S_{n_{k+1}}\big\| > C\bar{C}^{-1}(\log n_k)^{(1+\varepsilon)a}\big\} \\ &\leq 2\sum_{j=1}^{m} P\big\{|\langle B_{n_{k+1}} S_{n_{k+1}}, \theta^{(j)} \rangle| > E k^{(1+\varepsilon)a}\big\}. \end{aligned}$$

But in view of Theorem 9.1.4 we conclude that for some constant $D_1 > 0$

$$\begin{aligned} &P\big\{|\langle B_{n_{k+1}} S_{n_{k+1}}, \theta^{(j)} \rangle| > E k^{(1+\varepsilon)a}\big\} \\ &\qquad \leq D_1 n_{k+1} P\big\{|\langle B_{n_{k+1}} X_1, \theta^{(j)} \rangle| > E k^{(1+\varepsilon)a}\big\}. \end{aligned}$$

Write $B_{n_{k+1}}^{*} \theta^{(j)} = r_k \theta_k$ with $\|\theta_k\| = 1$ and $r_k \geq 0$. If $r_k = 0$, then the right-hand side of the last inequality above vanishes, and otherwise this term is equal to

$$D_1 \frac{V_0(r_k^{-1} E k^{(1+\varepsilon)a}, \theta_k)}{V_0(r_k^{-1}, \theta_k)} n_{k+1} P\big\{|\langle B_{n_{k+1}} X_1, \theta^{(j)} \rangle| > 1\big\}. \tag{9.35}$$

Apply Theorem 8.2.1 to see that μ is $\mathrm{ROV}_\infty(E,c)$. Then by Theorem 6.4.15 the fraction in (9.35) is bounded above for all large k by some constant times $k^{-(1+\varepsilon_1)}$ for some $\varepsilon_1 > 0$ depending on ε.

To bound $n_{k+1}P\{|<B_{n_{k+1}}X_1, \theta^{(j)}>| > 1\}$ we argue as in the proof of (9.21): By Theorem 8.3.18, the sequence (B_nS_n) is stochastically compact with limit set $\{\lambda^{-E}\nu^\lambda : \lambda \in [1,c]\}$ and by Remark 8.3.19 (a) we know that $\phi_{(\lambda)} = \lambda(\lambda^{-E}\phi)$ is the Lévy measure of $\lambda^{-E}\nu^\lambda$. Then

$$\begin{aligned}\phi_{(\lambda)}\{x \in \mathbb{R}^d : |<x, \theta^{(j)}>| \geq 1\} &= \lambda\phi\{x \in \mathbb{R}^d : |<x, \lambda^{-E^*}\theta^{(j)}>| \geq 1\} \\ &\leq c\phi\{x \in \mathbb{R}^d : \|x\| \geq \|\lambda^{-E^*}\theta^{(j)}\|^{-1}\} \\ &\leq c\phi\{x \in \mathbb{R}^d : \|x\| \geq \rho\}\end{aligned}$$

for some constant $\rho > 0$ and hence

$$E_1 = \sup_{1\leq\lambda\leq c} \phi_{(\lambda)}\{x : |<x, \theta^{(j)}>| \geq 1\} < \infty.$$

By using the relative compactness of (B_nS_n), every subsequence contains a further subsequence (n') such that for some $1 \leq \lambda \leq c$ we have $B_nS_n \Rightarrow \lambda^{-E}\nu^\lambda$ along (n'). Then Theorem 3.3.8 together with Theorem 1.2.19 implies that

$$\limsup_{(n')} nP\{|<B_nX_1, \theta^{(j)}>| > 1\} \leq \phi_{(\lambda)}\{x : |<x, \theta^{(j)}>| \geq 1\} \leq E_1$$

along (n') and hence it follows easily that

$$\sup_{n\geq 1} nP\{|<B_nX_1, \theta^{(j)}>| > 1\} < \infty. \tag{9.36}$$

Hence (9.35) is bounded above by some positive constant times $k^{-(1+\varepsilon_1)}$ and therefore $P(F_k) \leq C'k^{-(1+\varepsilon_1)}$ for some constant $C' > 0$ and all large k. Finally, an application of the first Borel-Cantelli Lemma shows that we have $P(\limsup D_n) \leq P(\limsup F_k) = 0$, proving (9.32).

For the proof of (9.33), let $K = \sup_{n\geq 2} \|B_nB_{n-1}^{-1}\|$ which is, using the regular variation of (B_n), a finite number. Enlarge K if necessary to ensure $K \geq 1$. Then, using the inequality $\|B_nX_n\| \leq \|B_nS_n\| + \|B_nS_{n-1}\|$ we get

$$\begin{aligned}&\{\|B_nX_n\| > 2K(\log n)^{(1-\varepsilon)a} \text{ i.o.}\} \\ &\quad\subset \{\|B_nS_n\| > K(\log n)^{(1-\varepsilon)a} \text{ i.o.}\} \cup \{\|B_nS_{n-1}\| > (\log n)^{(1-\varepsilon)a} \text{ i.o.}\}.\end{aligned}$$

But by the definition of K

$$\{\|B_nS_{n-1}\| > K(\log n)^{(1-\varepsilon)a} \text{ i.o.}\} \subset \{\|B_nS_n\| > (\log n)^{(1-\varepsilon)a} \text{ i.o.}\}$$

and since $K \geq 1$ we conclude that

$$\{\|B_nX_n\| > 2K(\log n)^{(1-\varepsilon)a} \text{ i.o.}\} \subset \{\|B_nS_n\| > (\log n)^{(1-\varepsilon)a} \text{ i.o.}\}.$$

Hence in view of the second Borel-Cantelli Lemma, it suffices to show that

$$\sum_{n=1}^{\infty} P\{\|B_n X_n\| > 2K(\log n)^{(1-\varepsilon)a}\} = \infty. \tag{9.37}$$

First, note that by an argument similar to those proving (9.36) we also have for any $\|\theta\| = 1$,

$$\inf_{n\geq 1} nP\{|<B_n X_1, \theta>| > 1\} > 0. \tag{9.38}$$

In fact, assume that there exists a subsequence (n_k) such that

$$\lim_{k\to\infty} n_k P\{|<B_{n_k} X_1, \theta>| > 1\} = 0. \tag{9.39}$$

Since by Theorem 8.3.18 the sequence $(B_n S_n)$ is stochastically compact there exists a subsequence $(n') \subset (n_k)$ and a $1 \leq \lambda \leq c$ such that $B_n S_n \Rightarrow Y_\lambda$ along (n'), where $\mathcal{L}(Y_\lambda) = \lambda^{-E}\nu^\lambda$ has Lévy measure $\phi_{(\lambda)} = \lambda \cdot (\lambda^{-E}\phi)$ (see Remark 8.3.19). Note that by Lemma 6.1.28

$$\phi_{(\lambda)}\{x \in \mathbb{R}^d : |<x, \theta>| > 1\} = \lambda\phi\{x \in \mathbb{R}^d : |<x, \lambda^{-E^*}\theta>| > 1\} > 0.$$

On the other hand, Theorem 3.3.8 together with Theorem 1.2.19 imply that

$$\liminf_{(n')} nP\{|<B_n X_1, \theta>| > 1\} \geq \phi_{(\lambda)}\{x \in \mathbb{R}^d : |<x, \theta>| > 1\} > 0$$

in contrast to (9.39). Hence (9.38) holds true.

Fix any $\|\theta\| = 1$ and write $B_n^*\theta = r_n\theta_n$ with $r_n > 0$ and $\|\theta_n\| = 1$ as before. Then

$$\begin{aligned} &P\{\|B_n X_n\| > 2K(\log n)^{(1-\varepsilon)a}\} \\ \geq &P\{|<B_n X_n, \theta>| > 2K(\log n)^{(1-\varepsilon)a}\} \\ = &\frac{1}{n}\frac{V_0(r_n^{-1}2K(\log n)^{(1-\varepsilon)a}, \theta_n)}{V_0(r_n^{-1}, \theta_n)} nP\{|<B_n X_1, \theta>| > 1\}. \end{aligned}$$

Apply Theorem 6.4.15 to get that for some $\delta > 0$ there exists a constant $C > 0$ such that for all large n

$$\begin{aligned} \frac{V_0(r_n^{-1}2K(\log n)^{(1-\varepsilon)a}, \theta_n)}{V_0(r_n^{-1}, \theta_n)} &\geq C\big(2K(\log n)^{(1-\varepsilon)a}\big)^{-1/a-\delta} \\ &\geq C_1\frac{1}{\log n}, \end{aligned}$$

where $C_1 > 0$ is some positive constant. Hence, by (9.38) there exists a constant $C_2 > 0$ such that

$$P\{\|B_n X_n\| > 2K(\log n)^{(1-\varepsilon)a}\} \geq C_2\frac{1}{n\log n}$$

for all large n. This yields (9.37) and concludes the proof. $\square$

Proof of Theorem 9.2.5. Fix any $1 \leq i \leq p$. As in the proof of Proposition 9.2.6 it suffices to show that for any sufficiently small $\varepsilon > 0$ and some constant $C = C(\varepsilon) > 0$ we have with probability one that

$$\left\| B_n S_n^{(1,\ldots,i)} \right\| \leq C(\log n)^{(1+\varepsilon)a_i} \quad \text{for almost all } n \tag{9.40}$$

and

$$\left\| B_n S_n^{(1,\ldots,i)} \right\| > C(\log n)^{(1-\varepsilon)a_i} \quad \text{for infinitely many } n. \tag{9.41}$$

By Proposition 9.2.6, we know that for every $1 \leq j \leq i$, almost all sample points and almost all n, $\|B_n S_n^{(j)}\| \leq (\log n)^{(1+\varepsilon)a_j}$ holds. Since $a_j \leq a_i$ for $1 \leq j \leq i$, we then get

$$\left\| B_n S_n^{(1,\ldots,i)} \right\| \leq \sum_{j=1}^{i} \|B_n S_n^{(j)}\| \leq i(\log n)^{(1+\varepsilon)a_i}$$

for almost all n almost surely, proving (9.40).

For the proof of (9.41), let $\varepsilon < (a_i - a_{i-1})/(a_{i-1} + a_i)$. From Proposition 9.2.6 we get that with probability one $\|B_n S_n^{(i)}\| > (\log n)^{(1-\varepsilon/2)a_i}$ for infinitely many n. Furthermore, for $1 \leq j \leq i-1$, $\|B_n S_n^{(j)}\| \leq (\log n)^{(1+\varepsilon)a}$ for almost all n almost surely. Hence, with probability one, for infinitely many n

$$\begin{aligned}
\left\| B_n S_n^{(1,\ldots,i)} \right\| &\geq \|B_n S_n^{(i)}\| - \sum_{j=1}^{i-1} \|B_n S_n^{(j)}\| \\
&> (\log n)^{(1-\varepsilon/2)a_i} - \sum_{j=1}^{i-1} (\log n)^{(1+\varepsilon)a_j} \\
&\geq (\log n)^{(1-\varepsilon/2)a_i} - (i-1)(\log n)^{(1+\varepsilon)a_{i-1}}.
\end{aligned}$$

But by the choice of $\varepsilon > 0$ the last difference is greater than $(\log n)^{(1-\varepsilon)a_i}$ for all large n, proving (9.41). The proof is now complete. $\square$

Similar to Corollary 9.2.4, where we considered the cluster set of $<S_n, \theta>$, we will now give the following corollary to Theorem 9.2.5 describing the cluster set of the norm of $\|B_n S_n\|$.

Corollary 9.2.7. *Under the assumptions of Theorem 9.2.5 for any $1 \leq i \leq p$, with probability one every point in the interval $(1, e^{a_i}]$ is a cluster point of the sequence*

$$\left\{ \left\| B_n S_n^{(1,\ldots,i)} \right\|^{\frac{1}{\log \log n}} : n \geq 1 \right\}.$$

Proof. Fix any $1 \le i \le p$. Then, as in the proof of Corollary 9.2.4 let $0 < \lambda \le a_i$, $\delta = a_i/\lambda$ and let $n_k = [2^{k^\delta}]$. We will show that

$$\limsup_{k\to\infty} \big\| B_{n_k} S_{n_k}^{(1,\ldots,i)} \big\|^{\frac{1}{\log\log n_k}} = e^\lambda \quad \text{almost surely.} \tag{9.42}$$

Again for this it suffices to show that for any small $\varepsilon > 0$ and for almost all sample points we have

$$\big\| B_{n_k} S_{n_k}^{(1,\ldots,i)} \big\| > (\log n_k)^{(1+\varepsilon)\lambda} \quad \text{for at most finitely many } k \tag{9.43}$$

and

$$\big\| B_{n_k} S_{n_k}^{(1,\ldots,i)} \big\| > (\log n_k)^{(1-\varepsilon)\lambda} \quad \text{for infinitely many } k. \tag{9.44}$$

For $1 \le l \le p$ let $\{\theta^{(l,1)}, \ldots, \theta^{(l,m_l)}\}$, $m_l = \dim W_l$ be an orthonormal basis of W_l. Then, by (9.34), we have for some constant $C > 0$ that

$$\begin{aligned} &P\{\big\| B_{n_k} S_{n_k}^{(1,\ldots,i)} \big\| > (\log n_k)^{(1+\varepsilon)\lambda}\} \\ &\qquad\le \sum_{l=1}^{i}\sum_{j=1}^{m_l} P\{|<B_{n_k} S_{n_k}^{(1,\ldots,i)}, \theta^{(l,j)}>| > C(\log n_k)^{(1+\varepsilon)\lambda}\}. \end{aligned}$$

One gets, as in the proof of Theorem 9.2.5 even without using Lévy's maximal inequality, that the right-hand side of the inequality above is bounded by some constant times $k^{-(1+\varepsilon_1)}$ for all large k and some $\varepsilon_1 > 0$ depending on $\varepsilon > 0$. Hence (9.43) follows.

The proof of (9.44) is similar to those of (9.28). We will include it here for sake of completeness. Hence we first have to show that with probability one

$$\Big\| B_{n_k} \sum_{j=n_{k-1}+1}^{n_k} X_j^{(1,\ldots,i)} \Big\| > (\log n_k)^{(1-\varepsilon/2)\lambda} \quad \text{for infinitely many } k. \tag{9.45}$$

These are independent events. Let $M = \sup_{k\ge 1} \|B_{n_k - n_{k-1}} B_{n_k}^{-1}\|$. Then, since $\delta \ge 1$, we see as in (9.17) that M is finite. Fix any unit vector $\theta \in W_i$ and set $B^*_{n_k - n_{k-1}}\theta = r_k\theta_k$, where $r_k > 0$, $\|\theta_k\| = 1$ and since the (B_n) are block diagonal with respect to $W_1, \ldots, W_p$ we have $\theta_k \in W_i$ for all k. Let $\tilde V_0(t,\theta) = P\{|<X^{(i)}, \theta>| > t\}$. Then, using the fact that the spaces W_i are mutually orthogonal we get

$$\begin{aligned} &P\Big\{\big\| B_{n_k} \sum_{j=n_{k-1}+1}^{n_k} X_j^{(1,\ldots,i)} \big\| > (\log n_k)^{(1-\varepsilon/2)\lambda}\Big\} \\ =&P\Big\{\big\| B_{n_k} S_{n_k-n_{k-1}}^{(1,\ldots,i)} \big\| > (\log n_k)^{(1-\varepsilon/2)\lambda)}\Big\} \\ \ge&P\Big\{\big\| B_{n_k-n_{k-1}} S_{n_k-n_{k-1}}^{(1,\ldots,i)} \big\| > M(\log n_k)^{(1-\varepsilon/2)\lambda}\Big\} \\ \ge&P\Big\{|<B_{n_k-n_{k-1}} S_{n_k-n_{k-1}}^{(1,\ldots,i)}, \theta>| > M(\log n_k)^{(1-\varepsilon/2)\lambda}\Big\} \\ =&P\Big\{|<B_{n_k-n_{k-1}} S_{n_k-n_{k-1}}^{(i)}, \theta>| > M(\log n_k)^{(1-\varepsilon/2)\lambda}\Big\} \end{aligned}$$

If we apply Theorem 9.1.4 to the spectrally simple component $X_1^{(i)}$, the last expression above is bounded below by some constant times

$$\begin{aligned}
&(n_k - n_{k-1})P\{|\langle B_{n_k - n_{k-1}} X_1^{(i)}, \theta\rangle| > M(\log n_k)^{(1-\varepsilon/2)\lambda}\} \\
&= \frac{\tilde{V}_0(r_k^{-1} M(\log n_k)^{(1-\varepsilon/2)\lambda}, \theta_k)}{\tilde{V}_0(r_k^{-1}, \theta_k)} (n_k - n_{k-1})P\{|\langle B_{n_k - n_{k-1}} X_1^{(i)}, \theta\rangle| > 1\}.
\end{aligned}$$

Then, by (9.38) and an application of Theorem 6.4.15 to $\tilde{V}_0$ we can bound the right-hand side of the equality above from below by some constant times $k^{-(1-\varepsilon_1)}$ for some $\varepsilon_1 > 0$ depending on $\varepsilon > 0$. Now the second Borel-Cantelli Lemma yields (9.45).

Finally, suppose that (9.44) does not hold on a set of positive probability. Then, for almost all sample points in that set

$$\begin{aligned}
(\log n_k)^{(1-\varepsilon)\lambda} &\geq \left\| B_{n_k} S_{n_k}^{(1,\ldots,i)} \right\| \\
&\geq \left\| B_{n_k} \sum_{j=n_{k-1}+1}^{n_k} X_j^{(1,\ldots,i)} \right\| - \left\| B_{n_k} S_{n_{k-1}}^{(1,\ldots,i)} \right\| \\
&> (\log n_k)^{(1-\varepsilon/2)\lambda} - \left\| B_{n_k} B_{n_{k-1}}^{-1} \right\| (\log n_{k-1})^{(1-\varepsilon)\lambda}
\end{aligned}$$

for infinitely many k, using (9.45) and the assumption. The regular variation of (B_n) implies that $\|B_{n_k} B_{n_{k-1}}^{-1}\| \leq C$ for all k and some constant $C > 0$. Hence the last difference above is greater than $(\log n_k)^{(1-\varepsilon)\lambda}$ for all large k. This is a contradiction and therefore (9.44) does hold. The proof is now complete. □

9.3 CONVERGENCE OF SEMITYPES

Convergence of types is a basic tool in probability theory (see Section 2.3). In this section, we present the new concept of convergence of semitypes, which is the appropriate generalization for problems concerning domains of semistable attraction on $\mathbb{R}^1$. Recall from Definition 7.4.1 that $\mu \in \mathrm{DOSA}(\nu, c)$ if

$$b_n^{-1} \mu^{k_n} * \varepsilon_{s_n} \Rightarrow \nu \tag{9.46}$$

for some $b_n > 0$ and some $s_n \in \mathbb{R}^1$, where the sampling sequence (k_n) satisfies $k_{n+1}/k_n \to c \geq 1$ as $n \to \infty$. Remark 7.5.14 shows there is no loss of generality in assuming $c > 1$. Now suppose that we also have $\mu \in \mathrm{DOSA}(\nu_1, c)$ so that

$$a_n^{-1} \mu^{k'_n} * \varepsilon_{s'_n} \Rightarrow \nu_1 \tag{9.47}$$

for some $a_n > 0$, $s'_n \in \mathbb{R}^1$ and another sampling sequence (k'_n) with $k'_{n+1}/k'_n \to c$. Convergence of semitypes describes the behavior of b_n/a_n, k'_n/k_n, and the relationship between ν, ν_1 that follows from (9.46) together with (9.47). In particular, it turns out that the growth rate of the sampling sequence (k_n) is uniquely determined by μ. In the course of the proof, we also derive a useful representation for the tail function

$$V_0(x) = \mu\{y : |y| > x\}. \tag{9.48}$$

Definition 9.3.1. We say that two different infinitely divisible laws ν, ν_1 are of the same *semitype* if

$$\nu_1 = a\nu^\lambda * \varepsilon_s \tag{9.49}$$

for some $a > 0$, $s \in \mathbb{R}^1$, and $\lambda > 0$.

Theorem 9.3.2 (Convergence of Semitypes). *Suppose that* (9.46) *holds for some* $b_n > 0$, $s_n \in \mathbb{R}^1$, *and some sequence of positive integers* $k_n \to \infty$ *with* $k_{n+1}/k_n \to c > 1$, *where* ν *is a nondegenerate semistable law that is not stable. Suppose that* (9.47) *also holds for some* $a_n > 0$, $s'_n \in \mathbb{R}^1$, *some sequence of positive integers satisfying* $k'_{n+1}/k'_n \to c$ *and some nondegenerate limit* ν_1. *Then* $b_n/a_n \to a > 0$, $k'_n/k_n \to \lambda > 0$, *and* $\nu_1 = a\nu^\lambda * \varepsilon_s$ *for some* $s \in \mathbb{R}^1$, *so that* ν, ν_1 *are of the same semitype.*

Before we prove Theorem 9.3.2, we recall a few facts about semistable laws and their domains of semistable attraction. Corollary 7.4.4 implies that the Lévy measure ϕ of the nonnormal (b, c) semistable law ν satisfies

$$\phi\{t : |t| > x\} = x^{-\alpha}\theta(\log x) \tag{9.50}$$

where the index $\alpha = \log c/\log b$ is necessarily in the interval $(0, 2)$, and θ is periodic with period $\log b$. If ν is stable, then Corollary 7.3.4 shows that θ is a constant function, but we have assumed that this is not the case. Corollary 7.4.4 also yields

$$\begin{aligned} \theta(y + \delta) &\le e^{\alpha\delta}\theta(y) \\ \theta(y - \delta) &\ge e^{-\alpha\delta}\theta(y) \end{aligned} \tag{9.51}$$

for all $y, \delta > 0$. Theorem 3.3.8 shows that (9.46) implies $k_n(b_n^{-1}\mu) \to \phi$, and it follows immediately that

$$k_n V_0(b_n x) \to x^{-\alpha}\theta(\log x) \tag{9.52}$$

for all $x > 0$ such that x is a continuity point of the limit, where $V_0(x)$ is the tail function (9.48). Since V_0 is monotone, it follows as in Section 0.1 of [153] that if $x_n \to x > 0$ and x is a continuity point of the limit in (9.52), then $k_n V_0(b_n x_n) \to x^{-\alpha}\theta(\log x)$.

The following construction will be useful in the proof of Theorem 9.3.2. For all $x > b_1$, define $n(x) = \sup\{n : b_n \le x\}$, and let $b(x) = b_{n(x)}$, $g(x) = x/b(x)$ so that $x = g(x)b(x)$. Convergence of Types, Theorem 2.3.2, along with the

(b, c) semistability of ν yields $b_{n+1}/b_n \to b$, and it follows that $\{g(x) : x > b_1\}$ is relatively compact in $(0, \infty)$ with every limit point lying in the interval $[1, b]$.

Lemma 9.3.3. *Let $1 \leq \lambda < b$. Then for some $x_0 > b_1$, for all $x \geq x_0$ we have*

$$\begin{aligned} g(x/\lambda) &= g(x)/\lambda && \text{if } g(x) \geq \lambda \\ g(x/\lambda) &= g(x)/\lambda \cdot \frac{b_{n(x)}}{b_{n(x)-1}} && \text{if } g(x) < \lambda \\ g(\lambda x) &= \lambda g(x) && \text{if } \lambda g(x) \leq b \\ g(\lambda x) &= \lambda g(x) \cdot \frac{b_{n(x)}}{b_{n(x)+1}} && \text{if } \lambda g(x) > b \end{aligned} \tag{9.53}$$

Proof. Note that $x = g(x)b_n$ for all $x > b_1$, where $n = n(x)$. If $\lambda \leq g(x)$, then $n(x/\lambda) = n(x)$ and so $g(x/\lambda) = (x/\lambda)/b_n = (x/b_n)/\lambda = g(x)/\lambda$. Similarly, if $\lambda g(x) \leq b$, then $n(\lambda x) = n(x)$ and so $g(\lambda x) = \lambda x/b_n = \lambda g(x)$. Choose n_0 so that $b_n/b_{n-1} > \lambda$ for all $n \geq n_0$. Since $n(x) \to \infty$ as $x \to \infty$ we may choose x_0 so that $n(x) \geq n_0$ for all $x \geq x_0$. If $\lambda > g(x)$, then $n(x/\lambda) < n(x)$. For all such x, setting $n = n(x)$ as before, we have $x/\lambda \geq b_n/\lambda > b_{n-1}$ and so $n(x/\lambda) = n(x) - 1$. Then $g(x/\lambda) = (x/\lambda)/b_{n-1} = (g(x)b_n)/\lambda b_{n-1} = (g(x)/\lambda) \cdot (b_n/b_{n-1})$. Similarly, if $\lambda g(x) > b$, then $n(\lambda x) > n(x)$. For all such x, we have $\lambda x < \lambda b_{n+1} < b_{n+2}$ and so $n(\lambda x) = n(x) + 1$. Then $g(\lambda x) = (\lambda x)/b_{n+1} = \lambda(g(x)b_n)/b_{n+1} = (\lambda g(x)) \cdot (b_n/b_{n+1})$. □

Since $\mu \in \text{DOSA}(\nu, c)$ and ν is a nonnormal (b, c) semistable law, ν is also (c^E, c) semistable with $E = aI$, where $a = 1/\alpha$. Then Theorem 8.2.5 shows that $\int |x|^\rho d\mu(x)$ is finite if $0 < \rho < \alpha$ and infinite if $\rho > \alpha$. Since by (9.47) we also have $\mu \in \text{DOSA}(\nu_1, c)$, where ν_1 is (b', c) semistable for some $b' \geq \sqrt{c}$, in view of this moment result we see that $\log c/\log b' = \alpha$. Then $b' = b$ and hence $\nu_1^c = b\nu_1 * \varepsilon_{s_1}$ for some $s_1 \in \mathbb{R}$. If ν_1 is stable, then Remark 7.5.14 implies that $\mu \in \text{DOA}(\nu_1)$ and hence we also have $d_n^{-1}\mu^{k_n} * \varepsilon_{s'_n} \Rightarrow \nu_1$ for some $d_n > 0$ and some shifts $s'_n \in \mathbb{R}$. Then Convergence of Types, Theorem 2.3.2, together with (9.46) implies that ν is also stable, which is a contradiction. Hence ν_1 is also a (b, c) semistable law that is not stable. Let ϕ_1 be the Lévy measure of ν_1. Another application of Corollary 7.4.4 yields $\phi_1\{t : |t| > x\} = x^{-\alpha}\theta_1(\log x)$ where θ_1 is $\log b$ periodic. As in (9.52) we obtain

$$k'_n V_0(a_n x) \to x^{-\alpha}\theta_1(\log x) \tag{9.54}$$

at continuity points x of the limit. By a simple change of scale, we may assume without loss of generality that $x = 1$ is a continuity point of both limits in (9.52) and in (9.54), and that $\theta(0) = 1$. Now write $V_0(x) = x^{-\alpha}L(x)f(x)$ where

$$\begin{aligned} L(x) &= x^\alpha g(x)^{-\alpha} V_0(b_{n(x)}) \\ f(x) &= g(x)^\alpha V_0(x)/V_0(b_{n(x)}). \end{aligned} \tag{9.55}$$

Lemma 9.3.4. *The function $L(x)$ defined in (9.55) is slowly varying.*

Proof. Apply Lemma 9.3.3. It suffices to show that $L(\lambda x)/L(x) \to 1$ for all $1 \leq \lambda < b$. When $\lambda g(x) \leq b$ we have $L(\lambda x)/L(x) = 1$ and in the remaining case when $\lambda g(x) > b$, setting $n = n(x)$ as before, we have

$$\frac{L(\lambda x)}{L(x)} = (b_n/b_{n+1})^{-\alpha} \cdot \frac{k_{n+1}V_0(b_{n+1})}{k_n V_0(b_n)} \cdot \frac{k_n}{k_{n+1}}$$

where $b_n/b_{n+1} \to b^{-1}$ and $k_n/k_{n+1} \to c^{-1} = b^{-\alpha}$. By using this along with (9.52), the result follows easily. □

Lemma 9.3.5. *The function $f(x)$ defined in (9.55) above satisfies:*

(a) $f(\lambda x) \leq \lambda^{\alpha} f(x)E(x)$ with $E(x) \to 1$ as $x \to \infty$; *and*

(b) $f(x/\lambda) \geq \lambda^{-\alpha} f(x)\tilde{E}(x)$ with $\tilde{E}(x) \to 1$ as $x \to \infty$.

Proof. The proof of (a) and (b) is similar, so we will only prove part (b). Apply Lemma 9.3.3. In the case where $g(x) \geq \lambda$, we have that $f(x/\lambda)/f(x) = \lambda^{-\alpha}V_0(x/\lambda)/V_0(x) \geq \lambda^{-\alpha}$ because V_0 is monotone. In the case $g(x) < \lambda$, we have $f(x/\lambda)/f(x) = \lambda^{-\alpha}\tilde{E}(x)V_0(x/\lambda)/V_0(x) \geq \lambda^{-\alpha}\tilde{E}(x)$ by monotonicity of V_0 where, setting $n = n(x)$ as before, we have

$$\tilde{E}(x) = \frac{b_n^{\alpha}V_0(b_n)}{b_{n-1}^{\alpha}V_0(b_{n-1})} \to 1$$

since $b_n/b_{n-1} \to b$ and $V_0(b_n)/V_0(b_{n-1}) \to b^{-\alpha}$ as $x \to \infty$. □

Lemma 9.3.6. *For all $x > 0$, every limit point of $\{f(a_n x)\}$ lies between m, M where*

$$\begin{aligned} m &= \inf\{\theta(\log y) : 1 \leq y \leq b\} \\ M &= \sup\{\theta(\log y) : 1 \leq y \leq b\} \end{aligned} \tag{9.56}$$

Proof. If $f(a_\ell x) \to t$ along a subsequence, then we can choose a further subsequence along which $g(a_\ell x) \to y$ as well. If y is a continuity point of the limit in (9.52), then along this subsequence, writing $n_\ell = n(a_\ell x)$, we have

$$f(a_\ell x) = g(a_\ell x)^{\alpha}\frac{V_0(g(a_\ell x)b_{n_\ell})}{V_0(b_{n_\ell})} \to y^{\alpha} \cdot y^{-\alpha}\theta(\log y) \tag{9.57}$$

and so $t = \theta(\log y)$. This certainly gives $m \leq t \leq M$. On the other hand, if y is not a continuity point, then we will apply Lemma 9.3.5 to show that $\theta(\log y+) \leq t \leq \theta(\log y-)$. Note that in view of the definition (9.50) we will always have $\theta(\log y-) \geq \theta(\log y+)$. Choose $\lambda > 1$ arbitrarily close to 1 such that y/λ is a continuity point, with $g(a_\ell x/\lambda) = g(a_\ell x)/\lambda$ for all large ℓ. Then $g(a_\ell x/\lambda) \to y/\lambda$ and it follows as above that $f(a_\ell x/\lambda) \to \theta(\log y/\lambda)$ along the chosen subsequence. Then Lemma 9.3.5 implies that $f(a_\ell x) \leq$

$\lambda^\alpha f(a_\ell x/\lambda)/\tilde{E}(a_\ell x)$ and so $t \le \lambda^\alpha \theta(\log y/\lambda)$ for $\lambda > 1$ arbitrarily close to 1, and it follows that $t \le \theta(\log y-)$. Next, choose $\lambda > 1$ arbitrarily close to 1 such that λy is a continuity point, with $g(\lambda a_\ell x) = \lambda g(a_\ell x)$ for all large ℓ. Essentially the same argument as before yields $t \ge \theta(\log y+)$. □

Lemma 9.3.7. *Suppose that for all $x > 0$, either 1 or b is a limit point of $\{g(a_n x)\}$. Then for every $x > 0$, every $\lambda \in [1, b]$ is a limit point of $\{g(a_n x)\}$.*

Proof. Fix any $x > 0$ and $1 < \lambda < b$ and let $y = x/\lambda$. Then we have to consider two cases separately.

Suppose that $g(a_\ell y) \to 1$ along a subsequence. Then along this same subsequence we have for all large ℓ that $\lambda g(a_\ell y) \le b$ and hence by Lemma 9.3.3 $g(a_\ell x) = g(a_\ell \lambda y) = \lambda g(a_\ell y) \to \lambda$. On the other hand, suppose that $g(a_\ell y) \to b$ along a subsequence. Then along this same subsequence we have for all large ℓ that $\lambda g(a_\ell y) > b$ and hence by Lemma 9.3.3 we obtain $g(a_\ell x) = g(a_\ell y \lambda) = \lambda g(a_\ell y) b_{n(a_\ell y)}/b_{n(a_\ell y)+1} \to \lambda b b^{-1} = \lambda$. Then every $\lambda \in (1, b)$ is a limit point of $\{g(a_n x)\}$, and it follows easily that both $1, b$ are also limit points. □

Lemma 9.3.8. *Both $\{b_n/a_n\}$ and $\{k_n/k'_n\}$ are relatively compact in $(0, \infty)$.*

Proof. First, suppose that for some $x > 0$ we have $1 < a \le g(a_n x) \le d < b$ for all $n \ge n_0$. Choose $0 < \varepsilon < \min\{ba/d - 1, 1 - d/ba\}$ and enlarge n_0 if necessary to ensure that $(1-\varepsilon)b \le a_{n+1}/a_n \le (1+\varepsilon)b$ for all $n \ge n_0$. Let $n_\ell = n(a_\ell x)$ so that $a_\ell x = g(a_\ell x) b_{n_\ell}$ and observe that for all ℓ sufficiently large to make $n_\ell \ge n_0$ we have

$$\frac{b_{n_{l+1}}}{b_{n_\ell}} = \frac{a_{l+1}}{a_\ell}\frac{g(a_\ell x)}{g(a_{l+1} x)} \le (1+\varepsilon) b \frac{d}{a} < b^2$$

as well as $b_{n_{l+1}}/b_{n_\ell} \ge (1-\varepsilon)ba/d > 1$. This second inequality implies that $n_{l+1} \ge n_\ell + 1$ for all large ℓ. But if $n_{l+1} \ge n_\ell + 2$ infinitely often, then we contradict the first inequality, since then $b_{n_{l+1}}/b_{n_\ell} \ge b_{n_\ell+2}/b_{n_\ell}$ infinitely often, and this last term tends to b^2 in the limit. Hence for all large ℓ we have $n_{l+1} = n_\ell + 1$, and this implies that for some integer k we have $n_\ell = l + k$ for all large ℓ. Then we may write $a_\ell = b_{l+k} g(a_\ell x)/x$ for all large ℓ, and since g is bounded and $b_{l+k}/b_\ell \to b^k$ we see that $\{b_n/a_n\}$ is relatively compact.

Otherwise Lemma 9.3.7 holds, and in this case we will derive a contradiction. If $x > 0$ is a continuity point of the limit in (9.54), then

$$k'_n a_n^{-\alpha} L(a_n) f(a_n x) \to \theta_1(\log x)$$

in view of Lemma 9.3.4. Then every limit point of $k'_n a_n^{-\alpha} L(a_n)$ is of the form $\theta_1(\log x)/t$, where t is a limit point of $\{f(a_n x)\}$. Given $\varepsilon > 0$ we can choose $y_1, y_2 \in [1, b]$ such that both are continuity points of $\theta(\log y)$ and $t_1 = \theta(\log y_1) < m + \varepsilon$ while $t_2 = \theta(\log y_2) > M - \varepsilon$. Then (9.57) implies that

$f(a_n x)$ has both t_1, t_2 as limit points, and so in view of Lemma 9.3.6 and the fact that $\varepsilon > 0$ is arbitrarily small we have

$$\limsup_{n\to\infty} k'_n a_n^{-\alpha} L(a_n) = \frac{\theta_1(\log x)}{m}$$
$$\liminf_{n\to\infty} k'_n a_n^{-\alpha} L(a_n) = \frac{\theta_1(\log x)}{M}$$

for any continuity point x. But the left-hand side of the above expressions do not depend on x, hence $\theta_1(\log x)$ is a constant. This is a contradiction, since we have assumed that ν_1 is not stable.

Then $\{b_n/a_n\}$ is relatively compact. Suppose $b_n/a_n \to a$ along a subsequence, and choose $x > 0$ such that x is a continuity point of $\theta(\log x)$ and ax is a continuity point of $\theta_1(\log x)$. From (9.52) and (9.54), we have along this same subsequence that

$$\frac{k_n}{k'_n} k'_n V_0\left(a_n \frac{b_n}{a_n} x\right) \to x^{-\alpha}\theta(\log x)$$
$$k'_n V_0\left(a_n \frac{b_n}{a_n} x\right) \to (ax)^{-\alpha}\theta_1(\log ax)$$

which implies that

$$\frac{k_n}{k'_n} \to \frac{x^{-\alpha}\theta(\log x)}{(ax)^{-\alpha}\theta_1(\log ax)} \tag{9.58}$$

along this subsequence. It follows that $\{k_n/k'_n\}$ is relatively compact. □

Proof of Theorem 9.3.2. Apply Lemma 9.3.8. If along a subsequence we have $(b_n/a_n, k'_n/k_n) \to (a, \lambda)$, then we have both

$$\begin{aligned} a_n^{-1}\mu^{k'_n} * \varepsilon_{s'_n} &\Rightarrow \nu_1 \\ a_n^{-1} b_n b_n^{-1} \mu^{k_n k'_n/k_n} * \varepsilon_{\bar{s}_n} &\Rightarrow a\nu^\lambda \end{aligned} \tag{9.59}$$

for some sequence of shifts $\bar{s}_n$, so that $\nu_1 = a\nu^\lambda * \varepsilon_s$ for some $s \in \mathbb{R}^1$. We know that ν is (b, c) semistable, and although (b, c) are not unique, Theorem 7.5.4 implies that we can always choose the unique smallest $c > 1$. If (a_0, λ_0) is another limit point of the sequence $(b_n/a_n, k_n/k'_n)$ then $a\nu^\lambda = a_0\nu^{\lambda_0}$, and then it follows from (9.50) that $a_0 = ab^k$ for some integer k. Then all limit points of b_n/a_n are of the form ab^k, and by relative compactness there are only a finite number of these. To show that in fact there is only one limit point, we will argue by contradiction.

Assume that there are more than one limit points and let $\{a, ab^{k_1}, \ldots, ab^{k_\ell}\}$ denote that set of all limit points of $\{b_n/a_n\}$, where $k_1 \ldots, k_\ell \in \mathbb{Z}\backslash\{0\}$. Choose subsequences $N_j = \{(n_k^{(j)})\}$ with $N_j \cap N_i = \emptyset$ if $j \neq i$ such that $\mathbb{N} = \bigcup_{j=0}^{\ell} N_j$ and $\lim_{n\in N_j} b_n/a_n = ab^{k_j}$ for $j = 0, \ldots, \ell$. If $n_k^{(0)} + 1 \in N_0$ for all but finitely many k it then follows that $N_1, \ldots, N_\ell$ are just finite sets and hence we only have one limit point. That is a contradiction.

On the other hand, if $n_k^{(0)} + 1 \notin N_0$ for infinitely many k then there exist a subsequence (k_j) of k and a $1 \leq m \leq \ell$ such that $n_{k_j}^{(0)} + 1 \in N_m$ for all j. If we put $\bar{n}_j = n_{k_j}^{(0)}$ we have shown that $b_{\bar{n}_j}/a_{\bar{n}_j} \to a$ and $b_{\bar{n}_j+1}/a_{\bar{n}_j+1} \to ab^{k_m}$ as $j \to \infty$. But since $a_{n+1}/a_n \to b$ and $b_{n+1}/b_n \to b$ we must have $b_{n+1}/a_{n+1} \sim b_n/a_n$, which is a contradiction. So $b_n/a_n \to a$, and then it follows by the same argument as for (9.58) that k_n'/k_n converges, and so we must have $k_n'/k_n \to \lambda$. This proves the direct half of the theorem, and the converse follows immediately from (9.59). □

In the theorem above, the assumption that $b_n^{-1}\mu^{k_n} * \varepsilon_{s_n} \Rightarrow \nu$ for some $k_{n+1}/k_n \to c > 1$ already implies that ν is either stable or semistable. If ν is not stable, then neither is ν_1. If ν is stable, then so is ν_1, and in this case we can say nothing about the behavior of the sequence $\{k_n/k_n'\}$ or the sequence $\{b_n/a_n\}$. But if ν is semistable and not stable we have shown:

Corollary 9.3.9. *Suppose $\mu \in \mathrm{DOSA}(\nu, c)$ for some nondegenerate semistable law ν that is not stable. Then the growth rate of the sampling sequence (k_n) is uniquely determined by μ, meaning that if*

$$a_n^{-1}\mu^{k_n} * \varepsilon_{s_n} \Rightarrow \nu \quad \text{for some } \{k_n\} \text{ with } k_{n+1}/k_n \to c$$

and

$$b_n^{-1}\mu^{k_n'} * \varepsilon_{s_n'} \Rightarrow \nu \quad \text{for some } \{k_n'\} \text{ with } k_{n+1}'/k_n' \to c$$

then there exists an integer j such that $k_{n+j}/k_n' \to 1$ as $n \to \infty$.

Proof. Convergence of semitypes yields that $k_n'/k_n \to \lambda$ where $\nu = a\nu^\lambda * \varepsilon_s$ for some $\lambda, a > 0$ and some $s \in \mathbb{R}$. Since ν is (b, c) semistable and not stable we must have $\lambda = c^j$ for some integer j. The result follows easily. □

Remark 9.3.10. Remark 8.2.20 shows how to construct suitable norming constants for $\mu \in \mathrm{DOA}(\nu)$ when ν is stable with index $\alpha < 2$. The results of this section illuminate the tail behavior of a probability measure μ that belongs to some domain of semistable attraction, as well as the relation between the tails and the norming constants. Suppose that (9.46) holds. Then arguing as in the proof of Lemma 9.3.8 it is easy to see that $k_n b_n^{-\alpha} L(b_n) f(b_n x) \to \theta(\log x)$ for continuity points $x > 0$ of the limit. In particular, since $f(b_n) = 1$ by definition, we have $k_n b_n^{-\alpha} L(b_n) \to 1$. Define $R(t) = t^{-\alpha} L(t)$ regularly varying. Then we can always take $b_n = \sup\{t : k_n R(t) \geq 1\}$.

Remark 9.3.11. The representation

$$V_0(x) = x^{-\alpha} L(x) f(x) \tag{9.60}$$

expresses the tail of μ as the product of a regularly varying function and a bounded, asymptotically log periodic function f. In fact, we have $f(b_n x) \to$

$\theta(\log x)$ at continuity points, together with the inequalities of Lemma 9.3.5 at discontinuity points. Since continuity points are dense, this gives sharp bounds on the behavior of f. If the Lévy measure ϕ of the limit ν is continuous, then we can write

$$V_0(x) = x^{-\alpha} L(x) \left(\theta(\log g(x)) + h(x)\right) \tag{9.61}$$

where $L(x)$ is slowly varying and $h(x) \to 0$ as $x \to \infty$. In the case where ϕ has jumps, it seems that no such representation is possible.

Remark 9.3.12. A stable measure ν satisfies $\nu\{x : |x| > t\} \sim Ct^{-\alpha}$ as $t \to \infty$ for some $C > 0$. For a (b, c) semistable measure, we obtain from (9.55) that $\nu\{x : |x| > t\} = x^{-\alpha} f(t)$, where f is a bounded, asymptotically log periodic function. In this case, we may take $b_n = b^n$ and $k_n = c^n$, so that $g(x) = x/b^{n(x)}$. If the Lévy measure ϕ of ν is continuous, then (9.61) holds with $V_0(x) = \nu\{y : |y| > x\}$. If $\nu_1 = a\nu^\lambda * \varepsilon_s$, so that both measures are of the same semitype, then it follows from (9.52) and Theorem 9.3.2 that $\theta_1(\log ax) = \lambda a^\alpha \theta(\log x)$. Alternatively, this follows from (9.50) and the Lévy representation. Thus two different semistable laws that are of the same semitype, but not the same type, are related by a phase shift in the log periodic portion of the tail.

9.4 NOTES AND COMMENTS

Sections 9.1 and 9.2 are from Scheffler [166], see also [170]. This law of the iterated logarithm was discovered by Chover [25] for stable random variables. This result was then generalized in Vasudeva [193] to domains of attraction of stable laws and in Scheffler [168] to semistable laws. In the multivariable case, a first version of Chover's law of the iterated logarithm was proved for operator stable summands by Weiner [195]. Later these results were generalized in Scheffler [168] to operator semistable summands and in Scheffler [167] to the strict generalized domain of attraction of a nonnormal operator stable law. The assumption of symmetry in Theorems 9.1.4 and 9.2.5 can probably be relaxed, but this is an open problem.

Section 9.3 follows Meerschaert and Scheffler [126]. The Convergence of Semitypes Theorem 9.3.2 implies that if (9.46) and (9.47) hold, then ν and ν_1 are of the same semitype. This result can be used to show that domains of semistable attraction are either equal or disjoint. This result was first obtained by Mejzler [135], who also proved the relation between ν and ν_1 in Theorem 9.3.2. That result was extended to simply connected nilpotent Lie groups by Nobel [141]. Convergence of Semitypes for generalized domains of semistable attraction is an open problem.

Part IV

APPLICATIONS

In this part of the book, we describe a number of applications of the limit theorems from Part III to various problems in probability and statistics. It would be impossible to include all, or even a majority, of the applications that have appeared in the literature. Our selection is intended to represent a sample of these results, in order to illustrate the application of the limit theorems in this book.

10

Applications to Statistics

Central limit theory is the foundation of statistics, so it is no surprise that the limit theorems from Part III have diverse applications in statistics. Some of these applications involve random vectors, and others use vector methods to compare the statistics of random variables.

10.1 SAMPLE MOMENTS

The sample moments of a random sample $X_1, \ldots, X_n$ are useful in many statistics applications, including hypothesis testing, confidence intervals and linear regression. In these applications, we typically need to establish the joint asymptotics of the first few sample moments. In this section, we use the theory of generalized domains of attraction developed in Chapter 8 to obtain necessary and sufficient conditions for the joint convergence of any number of sample moments, and we compute their asymptotic distribution, which turns out to be operator stable.

Definition 10.1.1. Given a collection of random variables $(X_1, \ldots, X_n)$, for any positive integer k we define the kth *sample moment*

$$\mu_n^k = \frac{1}{n} \sum_{i=1}^{n} X_i^k. \tag{10.1}$$

Recall from Definition 3.3.24 that a random vector Z on $\mathbb{R}^d$ belongs to the generalized domain of attraction of a full random vector U on $\mathbb{R}^d$ if there exist linear operators $A_n \in \mathrm{GL}(\mathbb{R}^d)$ and nonrandom vectors $s_n \in \mathbb{R}^d$ such that

$$A_n(Z_1 + \cdots + Z_n) + s_n \Rightarrow U \tag{10.2}$$

where (Z_n) are i.i.d. with Z. In this case, the distribution of U is operator stable with some exponent $B \in \mathrm{L}(\mathbb{R}^d)$, in view of Theorem 7.2.1. Also recall from Definition 7.3.1 that a random variable X belongs to the domain of attraction of a nondegenerate random variable Y on $\mathbb{R}$ if there exist $a_n > 0$ and $b_n \in \mathbb{R}$ such that

$$a_n(X_1 + \cdots + X_n) + b_n \Rightarrow Y \tag{10.3}$$

where (X_n) are i.i.d. with X. In this case, Proposition 8.3.26 shows that the distribution of Y is stable with some index $\alpha \in (0, 2]$. Recall from Definition 1.1.19 that a symmetric linear operator $M \in \mathrm{L}(\mathbb{R}^d)$ is *positive definite* if $<t, Mt> > 0$ for all nonzero $t \in \mathbb{R}^d$.

Theorem 10.1.2. *Let $Z = (X, X^2, \ldots, X^d)$, where X is a random variable on $\mathbb{R}$, and assume that the covariance matrix of $(X, \ldots, X^i)$ is positive definite whenever it exists. Then Z belongs to the generalized domain of attraction of some operator stable law on $\mathbb{R}^d$ if and only if each component X^i belongs to the domain of attraction of some stable law on $\mathbb{R}$.*

Proof. Suppose that Z is in the generalized domain of attraction of some full random vector U and (10.2) holds. Theorem 7.2.1 shows that the distribution ν of U is infinitely divisible with Lévy representation $[a, Q, \phi]$. If U is normal (i.e., if ϕ is the zero measure) then Corollary 8.1.12 shows that $<Z, \theta>$ belongs to the domain of attraction of a nondegenerate normal law for every nonzero $\theta \in \Gamma = \mathbb{R}^d \setminus \{0\}$. Then each component X^i belongs to the domain of attraction of some nondegenerate normal law. If U has no normal component (i.e., if $Q(t) = 0$ for all $t \in \mathbb{R}^d$) let $\alpha(\theta)$ denote the spectral index of Z from Definition 8.3.11 so that $E|<Z, \theta>|^\rho$ is finite for all $0 \le \rho < \alpha(\theta)$ and infinite for all $\rho > \alpha(\theta)$. Let $e_1, \ldots, e_d$ denote the standard basis for $\mathbb{R}^d$ and define $\alpha = \alpha(e_1)$, so that $E|X|^\rho$ exists for $0 \le \rho < \alpha$ and diverges for $\rho > \alpha$. Since $Z = (X, X^2, \ldots X^d)$ we also have $\rho(e_i) = \alpha/i$ for all $i = 1, \ldots, d$. Let $W_1, \ldots, W_d$ denote the spectral decomposition of $\mathbb{R}^d$ with respect to Z from Definition 8.3.11 and suppose that U is spectrally compatible with Z. Since $\alpha(\theta) = 1/a_i$ if and only if $\theta \in W_1 \oplus \cdots \oplus W_i \setminus W_1 \oplus \cdots \oplus W_{i-1}$ it follows that $e_1 \in W_1$, which is a one-dimensional subspace and hence W_1 is the span of e_1. Then since $e_2 \in W_1 \oplus W_2$ and W_1, W_2 are mutually orthogonal, W_2 is the span of e_2. Continue in this way to show that each W_i is the span of e_i, so that $X, X^2, \ldots, X^d$ are the marginals in the spectral decomposition of Z. Then it follows from Corollary 8.3.27 that X^i belongs to the domain of attraction of some nondegenerate stable law with index α/i. Finally, suppose that U has both a normal and a nonnormal component. Use Theorem 8.3.25 to reduce to the two cases already considered, where the normal and nonnormal components are supported on mutually orthogonal subspaces, and apply the previous argument.

Conversely, suppose that X^i belongs to the domain of attraction of some stable law for each $i = 1, \ldots, d$. The proof that Z belongs to some generalized domain of attraction will be divided into three cases, depending on the form of the limit. In each case, we will obtain an explicit formula for the Lévy representation of U.

Case I: Strictly Nonnormal Limits. Suppose that (10.3) holds and Y is stable with index $\alpha < 2$. Then Theorem 8.2.17 implies that (a_n) varies regularly with index $-1/\alpha$ and

$$n(a_n \mu) \to \phi \tag{10.4}$$

where μ is the probability distribution of X and ϕ is the Lévy measure of Y. Corollary 7.3.4 shows that ϕ is of the form

$$\begin{aligned} \phi(r, \infty) &= pCr^{-\alpha} \\ \phi(-\infty, -r) &= qCr^{-\alpha} \end{aligned} \tag{10.5}$$

for every $r > 0$, for some $C > 0$ and $p \geq 0$, $q \geq 0$ with $p + q = 1$. Since $Z = f(X)$, where $f : \mathbb{R} \to \mathbb{R}^d$ is defined by $f(x) = (x, x^2, \ldots, x^d)$, the random vector Z has probability distribution $f\mu\{dx\} = \mu\{f^{-1}dx\}$, where $f^{-1}(S) = \{x \in \mathbb{R} : f(x) \in S\}$ is the preimage. Define $A_n = \operatorname{diag}(a_n, \ldots, a_n^d)$, by which we mean that

$$A_n = \begin{pmatrix} a_n & & & \\ & a_n^2 & & \\ & & \ddots & \\ & & & a_n^d \end{pmatrix} \tag{10.6}$$

with respect to the standard basis for $\mathbb{R}^d$. Then $f(a_n x) = A_n f(x)$ for all $x \in \mathbb{R}$, and

$$\begin{aligned} n \cdot A_n f\mu\{dz\} = n \cdot f a_n \mu\{dz\} &= n \cdot a_n \mu\{f^{-1}dz\} \\ &\to \phi\{f^{-1}dz\} = f\phi\{dz\} = \Phi\{dz\} \end{aligned} \tag{10.7}$$

where $\Phi = f\phi$ is a σ-finite measure on $\mathbb{R}^d \setminus \{0\}$. The sequence of linear operators (A_n) varies regularly with index $-B$ where $B = \operatorname{diag}(1/\alpha, \ldots, d/\alpha)$.

Now we want to show that Φ is $\mathcal{M}$-full. Suppose not. Then for some $w \neq 0$ in $\mathbb{R}^d$ we have

$$0 = \Phi\{z : <z, w> \neq 0\} = f\phi\{z : <z, w> \neq 0\} = \phi\{x : <f(x), w> \neq 0\}.$$

But the polynomial $<f(x), w> = w_1 x + \cdots + w_d x^d$ has only a finite number of roots, and then ϕ is concentrated on this set, which contradicts (10.5). Then Φ is $\mathcal{M}$-full, and Proposition 6.1.10 (b) implies that $f\mu \in \mathrm{RVM}_\infty(B)$

with limit measure Φ. The eigenvalues of B are the diagonal elements i/α, and since $\alpha < 2$ they are all greater than $1/2$. Then it follows from Theorem 8.2.10 that Z belongs to the generalized domain of attraction of a full operator stable law with exponent B and Lévy representation $[a, 0, \Phi]$.

Case II: Strictly Normal Limits. If $E(X^{2d}) < \infty$, then we have $E\|Z\|^2 = E(X^2) + \cdots + E(X^{2d}) < \infty$ and it follows from the central limit theorem that Z belongs to the domain of attraction of a multivariable normal law. In fact, Corollary 8.1.10 shows that (10.2) holds with $A_n = n^{-1/2}I$ and $s_n = -nA_nEZ$, where Y is normal with Lévy representation $[0, Q, 0]$ where

$$Q(t) = \langle t, Ct\rangle = \sum_{i=1}^{d}\sum_{j=1}^{d} t_i t_j c_{ij} \tag{10.8}$$

and C is the covariance matrix of Z, whose entries are of the form $c_{ij} = E(X_{i0}X_{j0})$ where $X_{i0} = X^i - E(X^i)$. Since C is positive definite, the normal limit is full.

Suppose that $E(X^{2d}) = \infty$ and that X^d belongs to the domain of attraction of a normal law. Then we have

$$a_n(X_1^d + \cdots + X_n^d) + b_n \Rightarrow Y$$

for some $a_n > 0$ and $b_n \in \mathbb{R}$ with Y normal. Remark 8.1.13 shows that $na_n^2 U_2(a_n^{-1}) \to \sigma^2$, where $U_2(r) = EX_{d0}^2 I(|X_{d0}| \le r)$ is the truncated variance of X^d and $\sigma^2 > 0$ is the variance of Y.

The truncated variance function of Z from Theorem 8.1.7 is defined by $M(t) = E\langle Z_0, t\rangle^2 I(|\langle Z_0, t\rangle| \le 1)$ where $Z_0 = Z - EZ = (X_{10}, \ldots, X_{d0})$. Define $A_n = \operatorname{diag}(n^{-1/2}, \ldots, n^{-1/2}, a_n)$ and compute for $t_n \to t \ne 0$ in $\mathbb{R}^d$ that

$$\begin{aligned}
nM(A_n^* t_n) &= nE\langle A_n Z_0, t_n\rangle^2 I\big(|\langle A_n Z_0, t_n\rangle| \le 1\big) \\
&= nE\left(\sum_{i=1}^{d-1} n^{-1/2} X_{i0} t_{in} + a_n X_{d0} t_{dn}\right)^2 I\big(|\langle A_n Z_0, t_n\rangle| \le 1\big) \\
&= I_1 + I_2 + I_3
\end{aligned}$$

where

$$\begin{aligned}
I_1 &= nE\left(\sum_{i=1}^{d-1} n^{-1/2} X_{i0} t_{in}\right)^2 I\big(|\langle A_n Z_0, t_n\rangle| \le 1\big) \\
I_2 &= 2nE\left((a_n X_{d0} t_{dn}) \sum_{i=1}^{d-1} n^{-1/2} X_{i0} t_{in}\right) I\big(|\langle A_n Z_0, t_n\rangle| \le 1\big) \\
I_3 &= nE\big(a_n X_{d0} t_{dn}\big)^2 I\big(|\langle A_n Z_0, t_n\rangle| \le 1\big)
\end{aligned}$$

where we have written $t_n = t_{1n} + \cdots + t_{dn}$ in the standard coordinates for $\mathbb{R}^d$. Note that the indicator set tends to $\mathbb{R}^d$ as $n \to \infty$. Proposition 8.1.6 shows that $E|X|^\beta < \infty$ for all $0 \le \beta < 2d$, and it follows that $c_{ij} = EX_{i0}X_{j0} = \mathrm{Cov}(X^i, X^j)$ exists for $i + j < 2d$. Then

$$I_1 \to \sum_{i=1}^{d-1}\sum_{j=1}^{d-1} c_{ij}t_it_j \quad \text{as } n \to \infty.$$

Since $U_2(r) \to \infty$ as $r \to \infty$ and $na_nU_2(a_n^{-1}) \to \sigma^2 > 0$, it follows that $na_n^2 \to 0$. Then $n^{1/2}a_n \to 0$, and it follows that $I_2 \to 0$ as $n \to \infty$. The remaining term I_3 can be written in the form

$$I_3 = \int_{|<z,t_n>|\le 1} t_{dn}^2 z_d^2 nP\{A_nZ_0 \in dz\}.$$

Next, we wish to show that

$$nP\{A_nZ_0 \in dx\} \to 0.$$

Since X_{d0} is also in the domain of attraction of a normal law with norming constants a_n it follows from Corollary 3.3.9 that

$$nP(|<A_nZ_0, e_d>| > r) = nP(a_n|X_{d0}| > r) \to 0$$

where $e_1, \ldots, e_d$ is the standard basis for $\mathbb{R}^d$. Since X_{i0} is in the domain of attraction of a normal law with norming constants $n^{-1/2}$ for $i < d$ it follows from Corollary 3.3.9 that

$$nP(|<A_nZ_0, e_i>| > r) = nP(n^{-1/2}|X_{i0}| > r) \to 0.$$

This suffices to show that $nP\{A_nZ_0 \in dx\} \to 0$. But then we have

$$\begin{aligned} I_3 &\sim \int_{|z_d|<1} t_{dn}^2 z_d^2 nP\{A_nZ_0 \in dz\} \\ &= t_{dn}^2 \cdot na_n^2U_2(a_n^{-1}) \to t_d^2\sigma^2 \end{aligned}$$

as $n \to \infty$. By putting all this together, we conclude that

$$nM(A_n^*t_n) \to Q(t)$$

whenever $t_n \to t \ne 0$ in $\mathbb{R}^d$, where

$$Q(t) = \sum_{1\le i,j<d} c_{ij}t_it_j + t_d^2\sigma^2 \tag{10.9}$$

is a positive definite quadratic form on $\mathbb{R}^d$. Then it follows from Corollary 8.1.8 that Z belongs to the generalized domain of attraction of a full normal law on $\mathbb{R}^d$ with Lévy representation $[a, Q, 0]$.

Case III: Mixed Limits. Suppose that X^d belongs to the domain of attraction of some nonnormal stable law with index α/d and that $\alpha \geq 2$. Choose j so that $2(j-1) \leq \alpha < 2j$. Let $e_1, \ldots, e_d$ denote the standard basis for $\mathbb{R}^d$ and define $V_1 = \text{span}\{e_1, \ldots, e_{j-1}\}$ and $V_2 = \text{span}\{e_j, \ldots, e_d\}$. Case II shows that $Z_1 = (X, X^2, \ldots, X^{j-1})$ belongs to the generalized domain of attraction of a full normal law on V_1 with Lévy representation $[a_1, Q, 0]$. The quadratic form Q is given by (10.8) if $EX^{2j-2} < \infty$, where (c_{ij}) is the covariance matrix of Z_1, and by (10.9) if $EX^{2j-2} = \infty$, where (c_{ij}) is the covariance matrix of $(X, X^2, \ldots, X^{j-2})$ and σ^2 depends on the choice of norming constants (a_n) in the convergence

$$a_n(X_1^{j-1} + \cdots + X_n^{j-1}) + b_n \Rightarrow Y$$

as before in Case II.

Now we want to show that $Z_2 = (X^j, \ldots, X^d)$ belongs to some generalized domain of attraction. If $j = d$, then this is immediate, otherwise let k be the smallest odd integer satisfying $k \geq j$. Then X^k belongs to the domain of attraction of some nonnormal stable law with Lévy measure ϕ. Note that $Z_2 = f(X^k)$, where $f(x) = (x^{j/k}, \ldots, x^{d/k})$ and argue as before in Case I that Z_2 belongs to the generalized domain of attraction of some full nonnormal operator stable law with Lévy representation $[a_2, 0, \Phi]$ where $\Phi = f\phi$. In this case, the norming operators are of the form $A_n = \text{diag}(a_n^j, \ldots, a_n^d)$ and they vary regularly with index $B = \text{diag}(j/\alpha, \ldots, d/\alpha)$.

If we extend Q to a quadratic form on $\mathbb{R}^d$ and Φ to a Lévy measure on $\mathbb{R}^d$, then Theorem 8.3.9 implies that Z belongs to the generalized domain of attraction of some full operator stable law with Lévy representation $[a, Q, \Phi]$, where $a = a_1 + a_2$. □

Remark 10.1.3. Theorem 10.1.2 remains true without the covariance assumption if we allow that the normal part of the limit in (10.2) is not full.

Corollary 10.1.4. *The random vector $Z = (X, X^2, \ldots, X^d)$ belongs to the generalized domain of attraction of some operator stable law on $\mathbb{R}^d$ with no normal component if and only if X belongs to the domain of attraction of some nonnormal stable random variable Y on $\mathbb{R}$. In this case, Y has Lévy representation $[a, 0, \phi]$, where ϕ satisfies (10.5), and U has characteristic function $e^{\psi_1(t)+\psi_2(t)}$ where*

$$\begin{aligned}
\psi_1(t) &= Cp \int_0^\infty \left(e^{i\sum_{j=1}^d t_j x^{j/\alpha}} - 1 - \frac{i\sum_{j=1}^d t_j x^{j/\alpha}}{1 + \sum_{j=1}^d x^{2j/\alpha}} \right) \frac{dx}{x^2} \\
\psi_2(t) &= Cq \int_0^\infty \left(e^{i\sum_{j=1}^d t_j (-1)^j x^{j/\alpha}} - 1 - \frac{i\sum_{j=1}^d t_j (-1)^j x^{j/\alpha}}{1 + \sum_{j=1}^d x^{2j/\alpha}} \right) \frac{dx}{x^2}
\end{aligned} \tag{10.10}$$

Proof. The proof of Theorem 10.1.2 (Case I) shows that if $X \in \text{DOA}(Y)$ and (10.3) holds, then $Z \in \text{GDOA}(U)$ and (10.2) holds, where U is operator stable with exponent $B = \text{diag}(1/\alpha, \ldots, d/\alpha)$ and Lévy representation $[0, 0, \Phi]$ with $\Phi = f\phi$, $f(x) = (x, \ldots, x^d)$, and ϕ given by (10.5). Theorem 7.2.5 shows that Φ satisfies the desintegration formula (7.11), where the Lévy spectral measure λ is given by (7.12). Since $\Phi = f\phi$ and (10.5) holds, the Lévy measure Φ is concentrated on the set $f(R^1)$ with $\Phi\{f(x) : x > r\} = Cpr^{-\alpha}$ and $\Phi\{f(x) : x < -r\} = Cqr^{-\alpha}$. Note that $r^B f(1) = (r^{1/\alpha}, \ldots, r^{d/\alpha}) = f(r^{1/\alpha})$ and $r^B f(-1) = (-r^{1/\alpha}, \ldots, (-1)^d r^{d/\alpha}) = f((-r)^{1/\alpha})$ for all $r > 0$, where we interpret $(-r)^{j/\alpha} = (-1)^j r^{j/\alpha}$. Then $f^{-1}(r^B f(1)) = r^{1/\alpha}$ and $f^{-1}(r^B f(-1)) = -r^{1/\alpha}$. Hence the spectral measure is concentrated on the two point set $\{f(1), f(-1)\}$ with

$$\begin{aligned}\lambda\{f(1)\} &= \Phi\{r^B\theta : r > 1; \theta = f(1)\} = f\phi\{r^B\theta : r > 1; \theta = f(1)\} \\ &= \phi\{f^{-1}(r^B\theta) : r > 1; \theta = f(1)\} = \phi\{r^{1/\alpha} : r > 1; \theta = f(1)\} = Cp\end{aligned}$$

and similarly $\lambda\{f(-1)\} = Cq$. Then it follows from Theorem 3.1.11 that U has log-characteristic function

$$\begin{aligned}&\int_{z\neq 0}\left(e^{i<t,z>} - 1 - \frac{i<t,z>}{1+\|z\|^2}\right)\Phi\{dz\} \\ &= \int_{S_0}\int_0^\infty\left(e^{i<t,r^B\theta>} - 1 - \frac{i<t,r^B\theta>}{1+\|r^B\theta\|^2}\right)\frac{dr}{r^2}\lambda\{d\theta\} \\ &= Cp\int_0^\infty\left(e^{i<t,f(r^{1/\alpha})>} - 1 - \frac{i<t,f(r^{1/\alpha})>}{1+\|f(r^{1/\alpha})\|^2}\right)\frac{dr}{r^2} \\ &+ Cq\int_0^\infty\left(e^{i<t,f(-r^{1/\alpha})>} - 1 - \frac{i<t,f(-r^{1/\alpha})>}{1+\|f(-r^{1/\alpha})\|^2}\right)\frac{dr}{r^2}\end{aligned}$$

which is of the form (10.10) with $t = (t_1, \ldots, t_d)$. In this computation we have used the norm $\|x\|_0$ given by (6.5), where $\|x\|$ is chosen so that $f(1), f(-1)$ are contained in the unit sphere $S_0 = \{x : \|x\|_0 = 1\}$ in this norm. This can be accomplished by taking $\|x\| = c\|x\|_*$, where

$$c^{-1} = \int_0^1\left(\sum_{j=1}^d x^{2j/\alpha}\right)^{1/2}\frac{dx}{x}$$

and $\|x\|_*$ is the usual Euclidean norm. □

Corollary 10.1.5. *Suppose that X is a random variable on $\mathbb{R}$, and the covariance matrix of $(X, \ldots, X^i)$ is positive definite whenever it exists. Then the vector of sample moments $(\mu_n^1, \ldots, \mu_n^d)$ is asymptotically operator stable if and only if X^i belongs to the domain of attraction of some stable law on $\mathbb{R}$ for each $i = 1, \ldots d$. If X belongs to the domain of attraction of some nonnormal stable law on $\mathbb{R}$ and (10.3) holds, then*

$$nA_n(\mu_n^1, \ldots, \mu_n^d) + s_n \Rightarrow U \tag{10.11}$$

for some sequence of shifts $s_n \in \mathbb{R}^d$, *where* A_n *is given by* (10.6) *and* U *is operator stable with characteristic function* $e^{\psi_1(t)+\psi_2(t)}$, *where* ψ_i *are defined by* (10.10).

Proof. The proof follows immediately from Theorem 10.1.2 and Corollary 10.1.4. □

Corollary 10.1.6. *Suppose* Y *is a nonnormal stable random variable on* $\mathbb{R}$ *with index* α *and Lévy measure* ϕ *given by* (10.5). *If* $X \in \mathrm{DOA}(Y)$ *and* (10.3) *holds, then* $X^i \in \mathrm{DOA}(Y_i)$ *and*

$$a_n^i(X_1^i + \cdots + X_n^i) + b_{in} \Rightarrow Y_i \tag{10.12}$$

for some sequence of shifts $b_{in} \in \mathbb{R}$, *where* Y_i *is nonnormal stable with index* α/i *and Lévy measure* ϕ_i. *If* i *is odd, then* $\phi_i(r,\infty) = Cpr^{-\alpha/i}$ *and* $\phi_i(-\infty,-r) = Cqr^{-\alpha/i}$ *for all* $r > 0$. *If* i *is even, then* $\phi_i(r,\infty) = Cr^{-\alpha/i}$ *and* $\phi_i(-\infty,-r) = 0$ *for all* $r > 0$.

Proof. The converse part of the proof of Theorem 10.1.2 (Case I) shows that (10.2) holds where A_n is given by (10.6) and U is operator stable with exponent $B = \mathrm{diag}(1/\alpha,\ldots,d/\alpha)$. Then the direct part of the proof shows that (10.12) holds where Y_i is the ith component in the spectral decomposition of U. Then Y_i is operator stable with exponent $B_i = (i/\alpha)$ on the one-dimensional subspace $W_i \subset \mathbb{R}^d$ spanned by the ith element e_i in the standard basis. Hence Y_i is stable with index α/i. If we let $\pi_i : \mathbb{R}^d \to W_i$ denote orthogonal projection, then Y_i has Lévy representation $\phi_i = \pi_i\Phi = \pi_i f\phi$, where $\pi_i f(x) = x^i$. If i is odd then $\pi_i f : \mathbb{R} \to \mathbb{R}$ is a bijection and $(\pi_i f)^{-1}(x^i) = x$ for all $x \in \mathbb{R}$. Then

$$\begin{aligned}\phi_i(r,\infty) = \pi_i f\phi\{x : x > r\} &= \phi\{(\pi_i f)^{-1}x : x > r\} = \phi\{x : \pi_i f(x) > r\}\\ &= \phi\{x : x^i > r\} = \phi\{x : x > r^{1/i}\} = Cpr^{-\alpha/i}\end{aligned}$$

and similarly $\phi_i(-\infty,-r) = Cqr^{-\alpha/i}$. If i is even, then $\pi_i f$ maps $\mathbb{R}$ onto $[0,\infty)$ and the preimage $(\pi_i f)^{-1}(x^i) = \{x,-x\}$ for all $x > 0$. Then

$$\begin{aligned}\phi_i(r,\infty) = \phi\{(\pi_i f)^{-1}x : x > r\} &= \phi\{x : x > r^{1/i}\} + \phi\{x : x < -r^{1/i}\}\\ &= Cpr^{-\alpha/i} + Cqr^{-\alpha/i} = Cr^{-\alpha/i}\end{aligned}$$

and $\phi_i(-\infty,-r) = 0$. □

Corollary 10.1.7. *Suppose that* $X \in \mathrm{DOA}(Y)$ *on* $\mathbb{R}$ *where* Y *is stable with index* α, *and that either* $0 < \alpha < 1$, *or* $1 < \alpha < 2$ *and* $EX = 0$. *Then we can take* $s_n = 0$ *in* (10.2) *and* (10.11), $b_n = 0$ *in* (10.3), *and* $b_{in} = 0$ *in* (10.12).

Proof. Corollary 10.1.4 shows that $Z \in \mathrm{GDOA}(U)$ and (10.2) holds with $Z = (X,\ldots,X^k)$, A_n is given by (10.6), and the limit U is operator stable with exponent $B = \mathrm{diag}(1/\alpha,\ldots,k/\alpha)$. Corollary 10.1.6 shows that each

$X^i \in \mathrm{DOA}(Y_i)$ and (10.12) holds, which generalizes (10.3). Then we can rewrite either (10.2) or (10.11) in the form

$$\sum_{i=1}^{n}(a_n X_i^1, \ldots, a_n^k X_i^k) + (b_{1n}, \ldots, b_{kn}) \Rightarrow (Y_1, \ldots, Y_k) \tag{10.13}$$

so that $s_n = (b_{1n}, \ldots, b_{dn})$ in (10.2) and (10.11) and $b_n = b_{1n}$ in (10.3). If $1 < \alpha < 2$ and $EX = 0$, then Theorem 8.2.7 (a) shows that we can take $b_{1n} = 0$ in (10.13). Otherwise $\alpha/i < 1$, and Theorem 8.2.7 (b) shows that we can take $b_{in} = 0$ in (10.13). □

Now the following useful result of Logan, Mallows, Rice, and Shepp [89] is recovered as an easy corollary of Theorem 10.1.2.

Corollary 10.1.8. *Suppose that $X \in \mathrm{DOA}(Y)$ on $\mathbb{R}$ and (10.3) holds, where Y is stable with index α, and that either $0 < \alpha < 1$, or $1 < \alpha < 2$ and $EX = 0$. Then the self-normalized sum*

$$\frac{\sum_{i=1}^n X_i}{\sqrt{\sum_{i=1}^n X_i^2}} \Rightarrow \frac{Y_1}{Y_2} \tag{10.14}$$

where Y_i are dependent stable laws with index α/i given by Corollary 10.1.6, so that $U = (Y_1, Y_2)$ has the joint distribution specified in Corollary 10.1.4.

Proof. Corollary 10.1.7 shows that (10.2) holds with $s_n = 0$. Rewrite this convergence in the form

$$\left(a_n \sum_{i=1}^n X_i, a_n^2 \sum_{i=1}^n X_i^2\right) \Rightarrow (Y_1, Y_2).$$

Define $h(x_1, x_2) = x_1/\sqrt{x_2}$. Since Y_2 is stable it is also operator stable, and then Theorem 7.2.7 shows that Y_2 has a density, so that $P(Y_2 = 0) = 0$. Then it follows from the Continuous Mapping Theorem 1.2.7 that

$$\frac{a_n \sum_{i=1}^n X_i}{\sqrt{a_n^2 \sum_{i=1}^n X_i^2}} = h\left(a_n \sum_{i=1}^n X_i, a_n^2 \sum_{i=1}^n X_i^2\right) \Rightarrow h(Y_1, Y_2) = \frac{Y_1}{Y_2}$$

which is equivalent to (10.14). □

Remark 10.1.9. The components Y_i in Corollary 10.1.6 are dependent stable laws. Note that Y_i is not the same as Y^i. In fact, Y^i is not even stable for $i > 1$, although it does belong to the domain of attraction of Y_i. If i is even, then we may have $X^i \in \mathrm{DOA}(Y_i)$ even if X does not belong to any domain of attraction, since in this case the balancing condition (8.72) is automatically satisfied. The random variable $|X|^i$ also belongs to the domain of attraction of a stable law with index α/i, and once again this only requires that $P(|X| > t)$ varies regularly with index $-\alpha$. In this case, the Lévy measure of the limit is $\bar{\phi}_i(r, \infty) = Cr^{-\alpha/i}$ and $\bar{\phi}_i(-\infty, -r) = 0$ for all $r > 0$. We also have $(|X|, \ldots, |X|^d)$ in the generalized domain of attraction of an operator stable limit with Lévy representation $[a, 0, \bar{\Phi}]$ in this case, where $\bar{\Phi} = g\phi$ with $g : \mathbb{R} \to \mathbb{R}^d$ defined by $g(x) = (|x|, \ldots, |x|^d)$. This is useful in the study of absolute sample moments. Finally, note that all of the above is also true for fractional powers $|X|^\rho$ and $X^\rho = \text{sign}(X)|X|^\rho$ for any $\rho > 0$ such that $\alpha/\rho < 2$, with the obvious change in the formulas for the Lévy measure.

Remark 10.1.10. A random vector U on $\mathbb{R}^\infty$ is called operator stable if U is full and (10.2) holds for some i.i.d. sequence of random vectors (Z_n) for some linear operators A_n and nonrandom centering constants s_n. In this case, we say that Z_1 belongs to the generalized domain of attraction of U. When we say that U is full, we mean that the finite dimensional distributions of U are all full, and we interpret weak convergence in (10.2) to mean weak convergence of finite dimensional distributions. Then it follows immediately from Theorem 10.1.2 that the random vector $Z = (X, X^2, X^3, \ldots)$ belongs to the generalized domain of attraction of some operator stable law on $\mathbb{R}^\infty$ if and only if X^j belongs to the domain of attraction of a stable law for all positive integers j, assuming that the covariance of $(X, \ldots, X^j)$ is positive definite whenever it exists. If X has moments of all orders, then U is Gaussian. If X is attracted to a nonnormal stable law with index α, then U is operator stable with exponent $E = \text{diag}(1/\alpha, 2/\alpha, \ldots)$. Otherwise U is operator stable with a Poissonian component of this nature along with a finite dimensional normal component. When we say that U is operator stable with exponent E we mean that $(U_1, \ldots, U_d)$ is operator stable with exponent $\text{diag}(1/\alpha, \ldots, d/\alpha)$ for every positive integer d. All of our fundamental results on infinitely divisible laws, operator stable laws, and weak convergence generalize immediately to $\mathbb{R}^\infty$ by considering the finite dimensional distributions.

Definition 10.1.11. Given a random sample $(X_1, \ldots, X_n)$ a *symmetric statistic* is a real-valued random variable $f(X_1, \ldots, X_n)$ with the property that $f(x_1, \ldots, x_n) = f(x_{\pi(1)}, \ldots, x_{\pi(n)})$ for any permutation π of the integers $1, \ldots, n$.

While the sample moments are the most familiar symmetric statistics, many other kinds of symmetric statistics are useful in some applications, including U and V statistics (see, e.g., [87, 180]).

Definition 10.1.12. Given a collection of random variables $(X_1, \ldots, X_n)$ we define for each $k = 1, \ldots, n$ the *elementary symmetric statistic of order* k by the formula

$$\sigma_n^k = \sum_{i \in I_k} (X_{i_1} \cdot X_{i_2} \cdots X_{i_k}) \tag{10.15}$$

where I_k denotes the collection of all k-tuples of integers $i = (i_1, \ldots, i_k)$ for which $1 \leq i_1 < i_2 < \cdots < i_k \leq n$. If $k = 0$ then we define $\sigma_n^k = 1$.

Lemma 10.1.13. *The elementary symmetric statistic* σ_n^k *is related to the sample moments* μ_n^k *defined by* (10.1) *according to Newton's identity*

$$\sigma_n^k = P_k(n\mu_n^1, \ldots, n\mu_n^k) \tag{10.16}$$

where

$$P_k(y_1, \ldots, y_k) = (-1)^k \sum_{m \in S_k} \prod_{i=1}^{k} \left(\frac{(-1)^{m_i} y_i^{m_i}}{i^{m_i} m_i!} \right) \tag{10.17}$$

and S_k *is the set of all* k*-tuples of nonnegative integers* $m = (m_1, \ldots, m_k)$ *such that* $m_1 + 2m_2 + \cdots + km_k = k$.

Proof. This is a classical result (see, e.g., [7]). □

Now, we compute the asymptotic distribution of the elementary symmetric statistics. Recall that $X \in \mathrm{DOA}(Y)$ on $\mathbb{R}$ if there exist $a_n > 0$ and $b_n \in \mathbb{R}$ such that

$$\sum_{i=1}^{n} (a_n X_i + b_n) \Rightarrow Y \tag{10.18}$$

where (X_n) are i.i.d. with X. In this case, Y is stable, and if the stable index $\alpha < 2$, then the Lévy measure of Y satisfies (10.5). The centering constants can always be chosen to satisfy

$$b_n = EXI(|a_n X| \leq 1). \tag{10.19}$$

If $\alpha < 1$, or if $EX = 0$ (which requires $\alpha \geq 1$), then (10.18) also holds with the more convenient centering $b_n = 0$. Furthermore, we can always choose the norming sequence (a_n) so that $C = 1$ in (10.5).

Proposition 10.1.14. *Suppose that* $X \in \mathrm{DOA}(Y)$ *on* $\mathbb{R}$ *where* Y *is stable with index* $\alpha < 2$. *Let* μ_n^k *denote the* k*th sample moment of the random sample* $(a_n X_1 + b_n, \ldots, a_n X_n + b_n)$ *and* σ_n^k *the* k*th order elementary symmetric statistic of this random sample. Then*

$$\sigma_n^k \Rightarrow P_k(Y_1, \ldots, Y_k) \tag{10.20}$$

and

$$\frac{\sigma_n^k}{n\mu_n^k} \Rightarrow \frac{P_k(Y_1, \ldots, Y_k)}{Y_k} \tag{10.21}$$

where Y_i *is the stable law with index* α/i *given by Corollary 10.1.6, so that* $U = (Y_1, \ldots, Y_d)$ *has the joint distribution specified in Corollary 10.1.4.*

Proof. The proof is similar to Corollary 10.1.8. Apply Corollary 10.1.5 together with Corollary 10.1.7 to show that

$$(n\mu_n^1, \ldots, n\mu_n^k) \Rightarrow (Y_1, \ldots, Y_k) \tag{10.22}$$

and note that the norming and centering constants have been absorbed into the sample moments. Now it follows from Lemma 10.1.13 along with the Continuous Mapping Theorem 1.2.7 that

$$\sigma_n^k = P_k(n\mu_n^1, \ldots, n\mu_n^k) \Rightarrow P_k(Y_1, \ldots, Y_k)$$

so that (10.20) holds, and since Y_k has a density, it also follows that

$$\frac{\sigma_n^k}{n\mu_n^k} = \frac{P_k(n\mu_n^1, \ldots, n\mu_n^k)}{n\mu_n^k} \Rightarrow \frac{P_k(Y_1, \ldots, Y_k)}{Y_k}$$

so that (10.21) holds. □

Corollary 10.1.15. *Suppose that $X \in \mathrm{DOA}(Y)$ on $\mathbb{R}$ where Y is stable with index $\alpha < 2$. Let μ_n^k denote the kth sample moment of the random sample $(X_1, \ldots X_n)$ and σ_n^k the kth order elementary symmetric statistic of this random sample. If $\alpha < 1$ or if $EX = 0$, then*

$$a_n^k \sigma_n^k \Rightarrow P_k(Y_1, \ldots, Y_k) \tag{10.23}$$

and

$$\frac{\sigma_n^k}{n\mu_n^k} \Rightarrow \frac{P_k(Y_1, \ldots, Y_k)}{Y_k} \tag{10.24}$$

where Y_i is the stable law with index α/i given by Corollary 10.1.6, so that $U = (Y_1, \ldots, Y_d)$ has the joint distribution specified in Corollary 10.1.4.

Proof. Note that $b_n = 0$ in (10.18) in this case. Apply Proposition 10.1.14 to see that

$$\bar{\sigma}_n^k \Rightarrow P_k(Y_1, \ldots, Y_k)$$

where $\bar{\sigma}_n^k$ is the kth order elementary symmetric statistic of the random sample $(a_n X_1, \ldots, a_n X_n)$. Since $\bar{\sigma}_n^k = a_n^k \sigma_n^k$, this is the same as (10.23). Proposition 10.1.14 also yields

$$\frac{\bar{\sigma}_n^k}{n\bar{\mu}_n^k} \Rightarrow \frac{P_k(Y_1, \ldots, Y_k)}{Y_k}$$

where $\bar{\mu}_n^k$ is the kth sample moment of the random sample $(a_n X_1, \ldots, a_n X_n)$. Since $\bar{\mu}_n^k = a_n^k \mu_n^k$, this reduces immediately to (10.24). □

10.2 SAMPLE COVARIANCE MATRIX

In this section, we compute the asymptotic distribution of the sample covariance matrix of an i.i.d. random sample that belongs to some generalized domain of semistable attraction.

Definition 10.2.1. Given i.i.d. random vectors $X, X_1, X_2, \ldots$ on $\mathbb{R}^d$ we define the *sample covariance matrix*

$$M_n = \frac{1}{n}\sum_{i=1}^{n} X_i X_i' \tag{10.25}$$

where X' denotes the transpose of the vector X and hence XX' is a symmetric $d \times d$ matrix.

Let $\mathcal{M}_s^d$ denote the real vector space of $d \times d$ symmetric matrices together with the inner product

$$<A, B> = \sum_{i,j=1}^{d} A_{ij} B_{ij}$$

where A_{ij} is the ij element of A. Then $(\mathcal{M}_s^d, <\cdot,\cdot>)$ is an Euclidean vector space of dimension $d(d+1)/2$. Define $T : \mathbb{R}^d \to \mathcal{M}_s^d$ by $T(x) = xx'$ and for any linear operator A on $\mathbb{R}^d$ let $L_A(B) = ABA^*$ where A^* is the transposed matrix of A. Then the following properties hold.

Lemma 10.2.2. *For any $x, y \in \mathbb{R}^d$, $A, B \in \mathrm{GL}(\mathbb{R}^d)$ and $M, N \in \mathcal{M}_s^d$ we have*

(a) $<T(x), T(y)> = <x, y>^2$;

(b) $L_A^{-1} = L_{A^{-1}}$;

(c) $L_{AB}(M) = L_A(L_B(M))$;

(d) $L_A(T(x)) = T(Ax)$;

(e) $<L_A(M), N> = <M, L_{A^*}(N)>$;

(f) $T : \mathbb{R}^d \to \mathcal{M}_s^d$ *is open and continuous;*

(g) The image of T *spans* $\mathcal{M}_s^d$.

Proof. (a)–(e) and (g) follow from a straight forward calculation, whereas (f) follows from the fact that T is a polynomial mapping. □

Recall that a probability measure $\mu \in \mathrm{ROV}_\infty(E, c)$ for some $c > 1$ if there exists a sequence $(B_n) \subset \mathrm{GL}(\mathbb{R}^d)$ regularly varying with index $(-E)$ and a sequence (k_n) of natural numbers tending to infinity with $k_{n+1}/k_n \to c$ such that

$$k_n(B_{k_n}\mu) \to \phi \tag{10.26}$$

where ϕ is $\mathcal{M}$-full. If every eigenvalue of E has real part less than 1/4 then $E\langle T(X), T(X)\rangle = E\langle X, X\rangle^2 = E\|X\|^4 < \infty$ by Theorem 6.3.23, where X has distribution μ. Notice that nM_n is the sum of i.i.d. random matrices, which are random vectors in the real vector space $\mathcal{M}_s^d$. Then it follows immediately from the central limit theorem, Corollary 8.1.10, that

$$\sqrt{n}(M_n - C_n) \Rightarrow W$$

where $C_n = EXX'$ and W is a random matrix in $\mathcal{M}_s^d$ with a normal distribution. Our next result establishes the asymptotics of the sample covariance matrix in the case where the tails of the measure μ are too heavy to be attracted to a normal law. Recall from Definition 1.1.5 that $T\phi\{dA\} = \phi\{T^{-1}dA\}$.

Theorem 10.2.3. *Suppose that the probability measure μ is* ROV(E, c) *for some $c > 1$ and* (10.26) *holds. If every eigenvalue of E has real part exceeding 1/4 then*

(a) There exist nonrandom $C_{k_n} \in \mathcal{M}_s^d$ such that

$$k_n B_{k_n} M_{k_n} B_{k_n}^* - C_{k_n} \Rightarrow W \tag{10.27}$$

where the distribution Υ of W is infinitely divisible on $\mathcal{M}_s^d$ with Lévy representation $[C, 0, T\phi]$ for some $C \in \mathcal{M}_s^d$.

(b) There exists a sequence (C_n) of nonrandom matrices in $\mathcal{M}_s^d$ such that the sequence

$$\left(nB_n M_n B_n^* - C_n\right)_{n\geq 1} \tag{10.28}$$

is weakly relatively compact with limit set contained in

$$\left\{L_{\lambda^{-E}}\Upsilon^{\lambda} : \lambda \in [1, c]\right\}. \tag{10.29}$$

Proof. To prove (a), we use Theorem 3.2.2, the standard convergence criteria for triangular arrays of random vectors. Since $\|T(x)\|^2 = \langle T(x), T(x)\rangle = \langle x, x\rangle^2 = \|x\|^4$ we have

$$\begin{aligned}\int_{A\neq 0} \min\{\|A\|^2, 1\} T\phi\{dA\} &= \int_{0<\|A\|\leq 1} \|A\|^2 T\phi\{dA\} \\ &\quad + (T\phi)\{A : \|A\| > 1\} \\ &= \int_{0<\|x\|\leq 1} \|x\|^4 \phi\{dx\} + \phi\{x : \|x\| > 1\}.\end{aligned}$$

Now, since all real parts of the eigenvalues of E exceed $1/4$ it follows Lemma 6.3.11 that $\int_{0<\|x\|\le 1}\|x\|^4\phi\{dx\}<\infty$ and hence $T\phi$ is a Lévy measure on $\mathcal{M}_s^d$. Since T is continuous, if S is a $T\phi$-continuity set then $T^{-1}(S)$ is a ϕ-continuity set. Then for all such sets with $\operatorname{dist}(\{0\},S)>0$ we have $\operatorname{dist}(\{0\},T^{-1}(S))>0$ and hence we get from (10.26)

$$\begin{aligned}k_n L_{B_{k_n}}(T\mu)(S) &= k_n T(B_{k_n}\mu)(S)\\ &= k_n(B_{k_n}\mu)(T^{-1}(S)) \to \phi(T^{-1}(S)) = (T\phi)(S).\end{aligned}$$

This proves the first condition of Theorem 3.2.2.

In view of Proposition 3.3.11, it is now sufficient to show that

$$\lim_{\varepsilon\downarrow 0}\limsup_{n\to\infty} k_n\int_{\|A\|\le\varepsilon} <A,B>^2 L_{B_{k_n}}T\mu\{dA\}=0 \tag{10.30}$$

for all unit vectors $B\in\mathcal{M}_s^d$. First, suppose that $B=T(b)$ for some $b\in\mathbb{R}^d$. Then

$$\begin{aligned}&k_n\int_{\|A\|\le\varepsilon} <A,B>^2 L_{B_{k_n}}T\mu\{dA\}\\ \le& k_n\int_{|<A,B>|\le\varepsilon} <A,B>^2 L_{B_{k_n}}T\mu\{dA\}\\ =& k_n\int_{|<A,L_{B^*_{k_n}}(B)>|\le\varepsilon} <A,L_{B^*_{k_n}}(B)>^2 T\mu\{dA\}\\ =& k_n\int_{|<Tx,L_{B^*_{k_n}}(B)>|\le\varepsilon} <Tx,L_{B^*_{k_n}}(B)>^2 \mu\{dx\}\\ =& k_n\int_{|<Tx,T(B^*_{k_n}b)>|\le\varepsilon} <Tx,T(B^*_{k_n}b)>^2 \mu\{dx\}\\ =& k_n\int_{|<x,B^*_{k_n}b>|\le\varepsilon^{1/2}} <x,B^*_{k_n}b>^4 \mu\{dx\}.\end{aligned}$$

If we write $B^*_{k_n}b=r_n\theta_n$ with $r_n>0$ and $\|\theta_n\|=1$, we have shown that

$$k_n\int_{\|A\|\le\varepsilon} <A,B>^2 L_{B_{k_n}}T\mu\{dA\}\le k_n r_n^4 U_4(\varepsilon^{1/2}r_n^{-1},\theta_n)$$

whenever $B=T(b)$. Let $1/4<a_1<\cdots<a_p$ be the real parts of the eigenvalues of E. Theorem 6.3.4 shows that V_0 is uniformly R–O varying, and Theorem 6.3.16 shows that U_4 is also uniformly R–O varying. Theorem 6.3.4 also implies that, for any $\delta>0$, there exists a constant $K>0$ such that for any $0<\varepsilon<1$ we have

$$\liminf_{n\to\infty}\frac{V_0(\varepsilon^{-1/2}(\varepsilon^{1/2}r_n^{-1}),\theta_n)}{V_0(\varepsilon^{1/2}r_n^{-1},\theta_n)}\ge K\varepsilon^{\frac{1}{2a_1}+\frac{\delta}{2}}.$$

Furthermore, it follows from (10.26) that

$$\limsup_{n\to\infty} k_n V_0(r_n^{-1},\theta_n) = \limsup_{n\to\infty} k_n P\{|<B_{k_n}X, b>| > 1\} < \infty. \tag{10.31}$$

Hence, by Corollary 6.3.9 (b), we obtain for any $\delta > 0$ all large n and some constant $C > 0$,

$$\begin{aligned} k_n r_n^4 U_4(\varepsilon^{1/2} r_n^{-1},\theta_n) &= \varepsilon^2 \frac{U_4(\varepsilon^{1/2} r_n^{-1},\theta_n)}{(\varepsilon^{1/2} r_n^{-1})^4 V_0(\varepsilon^{1/2} r_n^{-1},\theta_n)} \cdot \frac{V_0(\varepsilon^{1/2} r_n^{-1},\theta_n)}{V_0(\varepsilon^{-1/2}(\varepsilon^{1/2} r_n^{-1}),\theta_n)} \\ &\quad \times k_n V_0(r_n^{-1},\theta_n) \\ &\le C\varepsilon^{2-\frac{1}{2a_1}-\frac{\delta}{2}}. \end{aligned}$$

Since $a_1 > 1/4$, we can choose $\delta > 0$ such that $2 - \frac{1}{2a_1} - \frac{\delta}{2} > 0$ and hence (10.30) follows whenever $B = T(b)$.

For the general case, choose $\{b_k\}$ unit vectors such that $\{T(b_k)\}$ spans $\mathcal{M}_s^d$. Note that $\|T(b)\| = 1$ if and only if $\|b\| = 1$. Now for any $B \in \mathcal{M}_s^d$ we can write $B = \sum_{k=1}^m c_k T(b_k)$ for some $c_k \in \mathbb{R}$ and some finite natural number m. Then by the Cauchy-Schwarz inequality, Theorem 1.1.23, we get

$$\begin{aligned} & k_n \int_{\|A\|\le\varepsilon} <A, B>^2 L_{B_{k_n}} T\mu\{dA\} \\ =& k_n \int_{\|A\|\le\varepsilon} \langle A, \sum_{k=1}^m c_k T(b_k)\rangle^2 L_{B_{k_n}} T\mu\{dA\} \\ =& \sum_{j=1}^m \sum_{k=1}^m c_j c_k k_n \int_{\|A\|\le\varepsilon} <A, T(b_j)><A, T(b_k)> L_{B_{k_n}} T\mu\{dA\} \\ \le& \sum_{j=1}^m \sum_{k=1}^m |c_j c_k| \Big(k_n \int_{\|A\|\le\varepsilon} <A, T(b_j)>^2 L_{B_{k_n}} T\mu\{dA\}\Big)^{1/2} \\ & \qquad \times \Big(k_n \int_{\|A\|\le\varepsilon} <A, T(b_k)>^2 L_{B_{k_n}} T\mu\{dA\}\Big)^{1/2}. \end{aligned}$$

Since we already know that for all $b \in \mathbb{R}^d$

$$\lim_{\varepsilon\downarrow 0} \limsup_{n\to\infty} k_n \int_{\|A\|\le\varepsilon} <A, T(b)>^2 L_{B_{k_n}} T\mu\{dA\} = 0,$$

(10.30) follows in the general case. Hence (a) is proved.

For the proof of (b), we argue as in the proof of Theorem 8.3.18. Write $n = \lambda(n) k_{p(n)}$ whenever $k_{p(n)} \le n < k_{p(n)+1}$. Then $k_{n+1}/k_n \to c$ implies that $(\lambda(n))$ is relatively compact with limit set contained in $[1, c]$. Note that $p(k_n) = n$ and $\lambda(k_n) = 1$. If we define $\tilde{B}_n = \lambda(n)^{-E} B_{k_{p(n)}}$, then $\tilde{B}_{k_n} = B_{k_n}$.

Furthermore, using the regular variation of (B_n), it follows that $B_n B_{[\lambda n]}^{-1} \to \lambda^E$ uniformly on compact subsets of $\{\lambda > 0\}$. Therefore

$$\tilde{B}_n B_n^{-1} = \lambda(n)^{-E} B_{k_{p(n)}} B_{[\lambda(n)k_{p(n)}]}^{-1} \to \lambda^{-E}\lambda^E = I$$

whenever $\lambda(n) \to \lambda \in [1, c]$ along a subsequence. Similarly, $B_n \tilde{B}_n^{-1} \to I$ as $n \to \infty$. Hence $(\tilde{B}_n)$ varies regularly with index $(-E)$ too and the norming sequences (B_n) and $(\tilde{B}_n)$ are equivalent, meaning that $L_{B_n}(M_n)$ and $L_{\tilde{B}_n}(M_n)$ have the same limiting behavior. In fact, if $nB_n M_n B_n^* - C_n$ is weakly converging along a subsequence to Z, using $L_{\tilde{B}_n B_n^{-1}} \to I$ uniformly on compact subsets of $\mathcal{M}_s^d$, Theorem 1.2.8 implies that

$$L_{\tilde{B}_n B_n^{-1}}\big(nB_n M_n B_n^* - C_n\big) = n\tilde{B}_n M_n \tilde{B}_n^* - \tilde{B}_n B_n^{-1} C_n (B_n^{-1})^* \tilde{B}_n^* \Rightarrow Z.$$

It therefore suffices to show the assertion for the norming sequence $(\tilde{B}_n)$, which is again regular varying with index $(-E)$. In part (a), we have shown that for some sequence $(\tilde{C}_n)$ of nonrandom matrices in $\mathcal{M}_s^d$

$$L_{B_{k_n}}(T\mu)^{k_n} * \varepsilon_{-\tilde{C}_n} \Rightarrow \Upsilon,$$

where Υ has Lévy representation $[C, 0, T\phi]$ for some $C \in \mathcal{M}_s^d$. By using Lévy's Continuity Theorem 1.3.6 this implies

$$\Big(L_{B_{k_n}}(T\mu)^{k_n} * \varepsilon_{-\tilde{C}_n}\Big)^{\widehat{}}(\xi) \to \hat{\Upsilon}(\xi) \tag{10.32}$$

uniformly on compact subsets of $\mathcal{M}_s^d$, where $\hat{\Upsilon}$ denotes the Fourier transform of the measure Υ as a measure on the Euclidean vector space $(\mathcal{M}_s^d, <\cdot,\cdot>)$.

Now, let $C_n = \lambda(n) L_{\lambda(n)^{-E}}(\tilde{C}_{p(n)})$. If $\lambda(n) \to \lambda \in [1, c]$ along a subsequence, then by (10.32) for any $\xi \in \mathcal{M}_s^d$

$$\begin{aligned}
&\Big(L_{\tilde{B}_n}(T\mu)^n * \varepsilon_{-C_n}\Big)^{\widehat{}}(\xi) \\
=&\big((T\mu)^{\widehat{}}(L_{\tilde{B}_n^*}\xi)\big)^n e^{-i<C_n,\xi>} \\
=&\Big((T\mu)^{\widehat{}}\big(L_{B^*_{k_{p(n)}}}(L_{\lambda(n)^{-E^*}}\xi)\big)\Big)^{\lambda(n)k_{p(n)}} e^{-i<\tilde{C}_{p(n)}, L_{\lambda(n)^{-E^*}}\xi>\lambda(n)} \\
=&\bigg(\Big((T\mu)^{\widehat{}}\big(L_{B^*_{k_{p(n)}}}(L_{\lambda(n)^{-E^*}}\xi)\big)\Big)^{k_{p(n)}} e^{-i<\tilde{C}_{p(n)}, L_{\lambda(n)^{-E^*}}\xi>}\bigg)^{\lambda(n)} \\
=&\bigg(\Big(L_{B_{k_{p(n)}}}(T\mu)^{k_{p(n)}} * \varepsilon_{-\tilde{C}_{p(n)}}\Big)^{\widehat{}}\Big(L_{\lambda(n)^{-E^*}}\xi\Big)\bigg)^{\lambda(n)} \\
\to&\Big(\hat{\Upsilon}(L_{\lambda^{-E^*}}\xi)\Big)^{\lambda} \\
=&\Big(L_{\lambda^{-E}}\Upsilon^{\lambda}\Big)^{\widehat{}}(\xi)
\end{aligned}$$

along the same subsequence. Since every subsequence has a further subsequence with this property the result follows. This concludes the proof. □

Corollary 10.2.4. *If* $\mu \in \mathrm{GDOSA}(\nu, c)$ *and* ν *has no normal component, then* (10.26) *holds along with the conclusions of Theorem 10.2.3.*

Proof. In this case, it follows from Theorem 8.2.1 that $\mu \in \mathrm{ROV}_\infty(E, c)$ for some $c > 1$ and some linear operator E whose eigenvalues all have real part greater than $1/2$. □

Remark 10.2.5. It follows from the proof of Theorem 10.2.3 that if the shifts C_{k_n} in (10.27) can be chosen to zero, then the shifts C_n in (10.28) are zero too. In fact, the shifts C_n in (10.28) are connected to those in (10.27) by a linear mapping.

Corollary 10.2.6. *Under the assumptions of Theorem 10.2.3 we can take* $C_n = 0$ *in* (10.27) *and* (10.28) *if every eigenvalue of* E *has real part exceeding* $1/2$. *We can take* $C_{k_n} = k_n B_{k_n}(EXX')B^*_{k_n}$ *in* (10.27) *when every eigenvalue of* E *has real part less than* $1/2$.

Proof. By Theorem 3.2.2, we can take

$$C_{k_n} = k_n \int_{\|A\| \le R} A \; L_{B_{k_n}} T\mu\{dA\} \tag{10.33}$$

where we can take any $R > 0$ as long as $\{\|A\| \le R\}$ is a $T\phi$-continuity set. Given a unit vector $b \in \mathbb{R}^d$ let $B = T(b)$ and use (10.33) to write

$$\begin{aligned} |\langle C_{k_n}, T(b)\rangle| &= \left|\left\langle k_n \int_{\|A\| \le R} A \; L_{B_{k_n}} T\mu\{dA\}, B\right\rangle\right| \\ &\le k_n \int_{\|A\| \le R} |\langle A, B\rangle| L_{B_{k_n}} T\mu\{dA\} \\ &\le k_n \int_{|\langle A, B\rangle| \le R} |\langle A, B\rangle| L_{B_{k_n}} T\mu\{dA\} \\ &= k_n \int_{|\langle x, B^*_{k_n} b\rangle| \le R^{1/2}} \langle x, B^*_{k_n} b\rangle^2 \mu\{dx\} \\ &= k_n r_n^2 U_2(R^{1/2} r_n^{-1}, \theta_n), \end{aligned}$$

where again $B^*_{k_n} b = r_n \theta_n$ with $r_n > 0$ and $\|\theta_n\| = 1$.

Let $1/2 < a_1 < \cdots < a_p$ be the real parts of the eigenvalues of E. Then by Theorems 6.3.4 and 6.3.16 both U_2 and V_0 are uniformly R–O varying. Let $R < 1$. Theorem 6.3.4 yields that for every $\delta > 0$ and all large n

$$\frac{V_0(r_n^{-1}, \theta_n)}{V_0(R^{1/2} r_n^{-1}, \theta_n)} = \frac{V_0(R^{-1/2}(R^{1/2} r_n^{-1}), \theta_n)}{V_0(R^{1/2} r_n^{-1}, \theta_n)} \ge C(R^{-1/2})^{-\frac{1}{a_1} - \delta} = C R^{\frac{1}{2a_1} + \frac{\delta}{2}}$$

where $C > 0$ is a constant. Hence, using Corollary 6.3.9 (b) and (10.31), we obtain that for all large n and some constant $D > 0$

$$\begin{aligned}
&k_n r_n^2 U_2(R^{1/2} r_n^{-1}, \theta_n) \\
=&R \frac{U_2(R^{1/2} r_n^{-1}, \theta_n)}{(r_n^{-1} R^{1/2})^2 V_0(R^{1/2} r_n^{-1}, \theta_n)} \cdot \frac{V_0(R^{1/2} r_n^{-1}, \theta_n)}{V_0(r_n^{-1}, \theta_n)} \cdot k_n V_0(r_n^{-1}, \theta_n) \\
\leq& D R^{1-\frac{1}{2a_1}-\frac{\delta}{2}},
\end{aligned}$$

Then

$$\limsup_{n\to\infty} |<C_{k_n}, T(b)>| \leq C_0 R^{\delta_0}$$

for some constant $C_0 > 0$ and $\delta_0 = 1 - (1/2a_1) - (\delta/2) > 0$ if $\delta > 0$ is chosen small enough since $a_1 > 1/2$. By taking $R > 0$ arbitrary small, we see that the convergence in (10.27) still holds when $C_{k_n} = 0$. But then, by Remark 10.2.5, we can choose $C_n = 0$ in (10.28) too. This concludes the proof of the first assertion of the corollary.

In the other case, we have $1/4 < a_1 < \cdots < a_p < 1/2$. Then, by Theorems 6.3.16 and 6.3.23, for $\eta < 1/a_p$ both U_4 and V_η are uniformly R–O varying. Especially V_2 exists and hence $E<X,\theta>^2$ is finite for every unit vector θ. But since $\theta' E(XX')\theta = E<X,\theta>^2$ if follows that EXX' exists in this case. Use (10.33) to write $C_{k_n} = k_n B_{k_n}(EXX')B_{k_n}^* - I_n$ where

$$I_n = k_n \int_{\|A\|>R} A L_{B_{k_n}} T\mu\{dA\},$$

so that for any unit vector $b \in \mathbb{R}^d$ we have

$$\begin{aligned}
|<I_n, T(b)>| \leq& k_n \int_{\|A\|>R} |<A, T(b)>| L_{B_{k_n}} T\mu\{dA\} \\
=& \quad k_n \int_{|<A,T(b)>|>R} |<A, T(b)>| L_{B_{k_n}} T\mu\{dA\} \\
&+ k_n \int_{\|A\|>R \text{ and } |<A,T(b)>|\leq R} |<A, T(b)>| L_{B_{k_n}} T\mu\{dA\}.
\end{aligned} \tag{10.34}$$

Now, if we write $B_{k_n}^* b = r_n \theta_n$ with $r_n > 0$ and $\|\theta_n\| = 1$ then

$$\begin{aligned}
&k_n \int_{|<A,T(b)>|>R} |<A, T(b)>| L_{B_{k_n}} T\mu\{dA\} = k_n r_n^2 V_2(R^{1/2} r_n^{-1}, \theta_n) \\
&= \frac{V_2(R^{1/2} r_n^{-1}, \theta_n)}{V_2(r_n^{-1}, \theta_n)} \cdot \frac{r_n^{-2} V_2(r_n^{-1}, \theta_n)}{U_4(r_n^{-1}, \theta_n)} \cdot \frac{U_4(r_n^{-1}, \theta_n)}{r_n^{-4} V_0(r_n^{-1}, \theta_n)} \cdot k_n V_0(r_n^{-1}, \theta_n).
\end{aligned}$$

Then Theorem 6.3.23 and Corollary 6.3.9 together with (10.31) yield

$$\limsup_{n\to\infty} k_n \int_{|<A,T(b)>|>R} |<A,T(b)>| L_{B_{k_n}} T\mu\{dA\} \le KR^{-\delta_0} \tag{10.35}$$

for some constant $K > 0$ where $\delta_0 = (1/2a_p) - 1 - (\delta/2) > 0$ if $\delta > 0$ is chosen small enough.

To estimate the other integral on the right-hand side of (10.34) choose $\{b_k\}$ an orthonormal basis of $\mathbb{R}^d$ and note that if $\|y\|^2 > R$ then $<y, b_k>^2 > R/d$ for some $k = 1, \ldots, d$. Hence

$$\begin{aligned}
& k_n \int_{\|A\|>R \text{ and } |<A,T(b)>|\le R} |<A,T(b)>| L_{B_{k_n}} T\mu\{dA\} \\
\le & R k_n \int_{\|A\|>R} L_{B_{k_n}} T\mu\{dA\} \\
= & R k_n \int_{\|B_{k_n} x\|^2 > R} \mu\{dx\} \\
\le & R \sum_{k=1}^{d} k_n \int_{<B_{k_n} x, b_k>^2 > R/d} \mu\{dx\} \\
= & \sum_{k=1}^{d} R k_n V_0(r_{nk}^{-1}\sqrt{R/d}, \theta_{nk}),
\end{aligned}$$

where $B_{k_n}^* b_k = r_{nk}\theta_{nk}$ with $r_{nk} > 0$ and $\|\theta_{nk}\| = 1$. Apply Theorem 6.3.4 and (10.31) to see that each individual summand in the sum above can be bounded for all large n by

$$\begin{aligned}
R k_n V_0(r_{nk}^{-1}\sqrt{R/d}, \theta_{nk}) &= R \frac{V_0(r_{nk}^{-1}\sqrt{R/d}, \theta_{nk})}{V_0(r_{nk}^{-1}, \theta_{nk})} \cdot k_n V_0(r_{nk}^{-1}, \theta_{nk}) \\
&\le KR^{1 - \frac{1}{2a_p} + \frac{\delta}{2}} \\
&= KR^{-\delta_0},
\end{aligned}$$

for some positive constant $K > 0$, where $\delta_0 = (1/2a_p) - 1 - (\delta/2) > 0$ as above. Hence, in view of (10.34) and (10.35) we have shown that for some positive real constant C_1

$$\limsup_{n\to\infty} |<I_n, T(b)>| \le C_1 R^{-\delta_0},$$

where $\delta_0 > 0$. By taking $R > 0$ arbitrary large, we see that the convergence in (10.27) still holds when $C_{k_n} = k_n B_{k_n}(EXX')B_{k_n}^*$. □

After calculating the asymptotic distribution of the normalized sample covariance matrix in Theorem 10.2.3 we now investigate the structure of the limit W in (10.27) or equivalently of the measure Υ in (10.29).

We show that the limit W is operator semistable on a certain subspace of $\mathcal{M}_s^d$ and we compute the exponent. Then we prove that under the assumption that all the real parts of the eigenvalues of E exceed $1/2$, W is almost surely invertible, but we provide an example to show that W need not to be full on $\mathcal{M}_s^d$.

Lemma 10.2.7. *The limit W with distribution Υ in (10.27) and with Lévy representation $[C, 0, T\phi]$ is operator semistable on a certain subspace L of $\mathcal{M}_s^d$ with exponent Ξ, where Ξ is an element of $\mathrm{L}(\mathcal{M}_s^d)$, the tangent space of $\mathrm{GL}(\mathcal{M}_s^d)$ defined by $\Xi(M) = EM + ME^*$. Here E is an exponent of μ. Hence*

$$\Upsilon^c = c^{\Xi}\Upsilon * \delta(B)$$

for some $B \in \mathcal{M}_s^d$.

Proof. Since W is the weak limit of

$$L_{B_{k_n}}\left(\sum_{i=1}^{k_n} T(X_i)\right) - C_{k_n},$$

and since trivially Υ is full on $L = \mathrm{span}\{\mathrm{supp}(\Upsilon)\}$, by Definition 3.3.20 W is a full operator semistable random vector on L with no normal component. But $L_{c^E}(T\phi) = T(c^E\phi) = c \cdot T\phi$ so by Lemma 7.1.6 Υ satisfies

$$\Upsilon^c = L_{c^E}(\Upsilon) * \varepsilon_B$$

for some $B \in \mathcal{M}_s^d$. Since $(L_{t^E})_{t>0}$ is a one parameter subgroup of $\mathrm{GL}(\mathcal{M}_s^d)$ it can be written in the form $t^{\Xi} = \exp(\Xi \log t) = L_{t^E}$ for some element Ξ of $\mathrm{L}(\mathcal{M}_s^d)$ (see, e.g., [186] p. 10–21), where Ξ is the derivative of $\exp(s\Xi)$ at $s = 0$. Since for $M \in \mathcal{M}_s^d$

$$\begin{aligned}
\frac{\exp(s\Xi) - I}{s}(M) &= \frac{L_{\exp(sE)}(M) - M}{s} \\
&= \frac{1}{s}\big(\exp(sE)M\exp(sE^*) - M\big) \\
&= \frac{1}{s}\Big[\big(I + sE + O(s^2)\big)M\big(I + sE^* + O(s^2)\big) - M\Big] \\
&= EM + ME^* + O(s)
\end{aligned}$$

as $s \to 0$ we see that $\Xi(M) = EM + ME^*$. Hence Υ is (c^{Ξ}, c) operator semistable. □

Next, we show that the limit W in (10.27) is almost surely invertible. This requires a simple lemma.

Lemma 10.2.8. $\mathrm{supp}(T\phi) = T\,\mathrm{supp}(\phi)$.

Proof. The proof is essentially identical to the proof of Lemma 1.1.9, using Lemma 10.2.2 (f). □

Theorem 10.2.9. *If μ is* ROV(E, c) *where all the eigenvalues of the exponent E have real parts exceeding $1/2$, then the limiting matrix W in* (10.27) *above is invertible with probability one.*

Proof. Define $L = \text{span}\{\text{supp}(T\phi)\}$ so that L is a linear subspace of $\mathcal{M}_s^d$ and $T\phi$ is full on L. Since W is full and, by Lemma 10.2.7, operator semistable on L, Theorem 7.1.15 shows that W has a Lebesgue density on L, which is a real vector space and hence isomorphic with $\mathbb{R}^m$ for some m.

Define $\mathcal{M}_s^d(+) = \{M \in \mathcal{M}_s^d : \theta' M \theta \geq 0 \text{ for all } \theta \in \mathbb{R}^d\}$, the cone of nonnegative definite matrices, and let $L^+ = L \cap \mathcal{M}_s^d(+)$. Since $T(X_i) \in \mathcal{M}_s^d(+)$ almost surely we have

$$M_n = \frac{1}{n} \sum_{i=1}^{n} T(X_i) \in \mathcal{M}_s^d(+)$$

almost surely and so $k_n B_{k_n} M_{k_n} B_{k_n}^* \in \mathcal{M}_s^d(+)$ almost surely. Note that by corollary 10.2.6 we have $k_n B_{k_n} M_{k_n} B_{k_n}^* \Rightarrow W$. Since $\mathcal{M}_s^d(+)$ is a closed subset of $\mathcal{M}_s^d$ we have

$$P\{W \in \mathcal{M}_s^d(+)\} \geq \limsup_{n \to \infty} P\{k_n B_{k_n} M_{k_n} B_{k_n}^* \in \mathcal{M}_s^d(+)\}$$

by the Portmanteau Theorem 1.2.2 and so $W \in \mathcal{M}_s^d(+)$ almost surely.

Since ϕ is not supported on any proper subspace of $\mathbb{R}^d$ we can choose $\theta_1, \ldots, \theta_d$ linearly independent in $\mathbb{R}^d$ with all $\theta_i \in \text{supp}(\phi)$. Then, by Lemma 10.2.8, $T(\theta_i) \in L^+$ for all $i = 1, \ldots, d$ and $T(\theta_1), \ldots, T(\theta_d)$ are linearly independent in L. If not, then we can write

$$0 = \sum_{i=1}^{d} c_i T(\theta_i)$$

where not all $c_i = 0$. Suppose, e.g., that $c_1 \neq 0$. Then

$$0 = \left(\sum_{i=1}^{d} c_i T(\theta_i) \right) \theta_1 = \sum_{i=1}^{d} c_i \langle \theta_i, \theta_1 \rangle \theta_i$$

but $\langle \theta_1, \theta_1 \rangle \neq 0$, which contradicts the linear independence of $\theta_1, \ldots, \theta_d$.

Lemma 10.2.8 implies that $L = \text{span}\{T\, \text{supp}(\phi)\}$, and so we can choose additional $\theta_i \in \text{supp}(\phi)$ as necessary so that $T(\theta_1), \ldots, T(\theta_m)$ form a basis for L. Note that $m \geq d$ and all $T(\theta_i) \in L^+$. Define $S = \{M \in L^+ : M \text{ singular}\}$ and let $M_0 = T(\theta_1) + \cdots + T(\theta_m)$. For all nonzero $x \in \mathbb{R}^d$ we have

$$x' M_0 x = \sum_{i=1}^{m} x' T(\theta_i) x = \sum_{i=1}^{m} \langle x, \theta_i \rangle^2 > 0$$

since $\theta_1, \ldots, \theta_m$ span $\mathbb{R}^d$, so M_0 is positive definite. Additionally, if $M \in L^+$ and $c > 0$, then $M + cM_0$ is positive definite because $x'(M + cM_0)x =$

$x'Mx + cx'M_0x \geq cx'M_0x > 0$, whenever $x \neq 0$. Now define $V = \{M \in L : <M, M_0> = 0\}$, $V_M = \{M + tM_0 : t \in \mathbb{R}\}$ and $S_M = S \cap V_M$. If both $M + t_1M_0$ and $M + t_2M_0$ were in S_M, then (assuming $t_2 > t_1$) $M + t_2M_0 = M + t_1M_0 + (t_2 - t_1)M_0$ would be positive definite, and hence nonsingular, which is a contradiction. Then S_M can contain at most one point, and so S_M has Lebesgue measure zero in V_M for all $M \in V$. The Fubini Theorem 1.1.17 implies that S has Lebesgue measure zero in L. Since W has a Lebesgue density on L we have $P\{W \in S\} = 0$. Now $P\{W \text{ singular}\} = P\{W \in S\} + P\{W \notin \mathcal{M}_s^d(+)\} = 0 + 0 = 0$, which concludes the proof. □

Recall that a probability measure $\mu \in \mathrm{RVM}_\infty(E)$ if there exists a sequence $(B_n) \subset \mathrm{GL}(\mathbb{R}^d)$ regularly varying with index $(-E)$ such that

$$n(B_n\mu) \to \phi \tag{10.36}$$

where ϕ is $\mathcal{M}$-full. Since such a probability measure μ is also R–O varying, the following results are immediate.

Corollary 10.2.10. *If a probability measure $\mu \in \mathrm{RVM}_\infty(E)$ where every eigenvalue of E has real part exceeding $1/4$ and (10.36) holds then*

(a) *There exist nonrandom $C_n \in \mathcal{M}_s^d$ such that*

$$nB_nM_nB_n^* - C_n \Rightarrow W \tag{10.37}$$

where W is infinitely divisible on $\mathcal{M}_s^d$ with Lévy representation $[C, 0, T\phi]$ for some $C \in \mathcal{M}_s^d$

(b) *If every eigenvalue of E has real part exceeding $1/2$, then we can take $C_n = 0$ in (10.37), and if every eigenvalue of E has real part less than $1/2$, then we can take $C_n = nB_n(EXX')B_n^*$ in (10.37)*

(c) *The limit measure W in (10.37) is operator stable on a certain subspace L of $\mathcal{M}_s^d$ with exponent Ξ, where Ξ is an element of $\mathrm{L}(\mathcal{M}_s^d)$ with $\Xi(M) = EM + ME^*$*

(d) *If every eigenvalue of E has real part exceeding $1/2$, then the limiting matrix W in (10.37) is invertible with probability one*

Remark 10.2.11. If $\mu \in \mathrm{ROV}_\infty(E, c)$ and some eigenvalues of E have real part greater than $1/4$ while others have real part less than $1/4$, then we can use Theorem 8.3.9 to reduce to each separate case. Then the asymptotic distribution of M_n is operator semistable on $\mathcal{M}_s^d$ with both a normal and a nonnormal component.

Remark 10.2.12. We can also use the spectral decomposition of $\mathcal{M}_s^d$ with respect to XX' to sharpen the results of this section. This decomposition of $\mathcal{M}_s^d$ is in terms of the real parts of the eigenvalues of Ξ. If $Ex = \lambda_1 x$ and $Ey = \lambda_2 y$, then it is not hard to check that $\Xi xy' = (\lambda_1 + \lambda_2)xy'$, so the eigenvalues of Ξ are related to those of E. If an eigenvalue of Ξ has real part less than $1/2$, then we can center the corresponding component of M_n to zero expectation, and if an eigenvalue of Ξ has real part greater than $1/2$, then no centering is required.

Remark 10.2.13. Corollary 10.2.10 remains true if we replace M_n by the centered version of the sample covariance matrix defined by

$$G_n = \frac{1}{n}\sum_{i=1}^{n}(X_i - \bar{X}_n)(X_i - \bar{X}_n)' \quad \text{where} \quad \bar{X}_n = \frac{1}{n}\sum_{i=1}^{n} X_i. \tag{10.38}$$

We defer the proof until Theorem 10.6.15, where we consider the more general case of moving averages. Presumably the same is true for μ R–O varying, but this is an open problem.

Example 10.2.14. The limit W in (10.27) is not necessarily full. Suppose we are on $\mathbb{R}^2$ and that $\mu \in \mathrm{RVM}_\infty(E)$ where $E = \alpha^{-1}I$ for some $0 < \alpha < 2$. Then there exists a sequence $(A_n) \subset \mathrm{GL}(\mathbb{R}^2)$ regularly varying with index $(-E)$ and a Lévy measure ϕ on $\mathbb{R}^2$ such that

$$n(A_n\mu) \to \phi$$

where ϕ is $\mathcal{M}$-full. Assume that ϕ is concentrated on the positive coordinate axes with $\phi\{te_i : t > r\} = r^{-\alpha}$ for $i = 1, 2$. Then μ must be full, ϕ is full on $\mathbb{R}^2$, but W is not full. In particular, if we write

$$M = \begin{pmatrix} 0 & 1 \\ 1 & 0 \end{pmatrix} \quad \text{and} \quad T(y) = \begin{pmatrix} y_1^2 & y_1y_2 \\ y_1y_2 & y_2^2 \end{pmatrix}$$

Then $(T\phi)\{N : <M,N> \neq 0\} = \phi\{y : <M,T(y)> \neq 0\} = \phi\{y : y_1y_2 \neq 0\} = 0$ since ϕ is supported on the set $\{y : y_1y_2 = 0\}$. Hence $T\phi$ is supported on the lower dimensional subspace $\{N : <M,N> = 0\}$ and therefore W can not be full. It follows from Lemma 10.2.8 that $W \in L$ almost surely, where $L = \mathrm{span}\{\mathrm{supp}(T\phi)\} = \mathrm{span}\{e_1e_1', e_2e_2'\}$.

10.3 SELF-NORMALIZED SUMS

In this section, we apply the results of Section 10.2 to investigate the behavior of self-normalized sums. Assume that $X, X_1, X_2, \ldots$ are i.i.d. on $\mathbb{R}^d$ with common distribution μ and that μ belongs to the strict generalized domain of semistable attraction of some full operator semistable law ν on $\mathbb{R}^d$ having no normal component. Recall from Definition 9.1.2 that this means, for some sequence (k_n) of natural numbers tending to infinity with $k_{n+1}/k_n \to c > 1$, there exist linear operators A_n such that

$$A_n S_{k_n} \Rightarrow Y,$$

where $S_n = \sum_{i=1}^n X_i$ and Y has distribution ν. In most statistical applications, the norming operators A_n are unknown and it is quite natural to consider the asymptotic behavior of the self-normalized sum $M_{k_n}^{-1/2} S_{k_n}$ instead. This quantity is the multivariable analogue of the t-statistic (without centering).

In view of Corollary 8.2.2, there exists a sequence $(B_n) \subset \mathrm{GL}(\mathbb{R}^d)$ regularly varying with index $(-E)$ such that

$$B_{k_n} S_{k_n} \Rightarrow Y \tag{10.39}$$

where Y is (c^E, c) operator semistable. Furthermore, μ is $\mathrm{ROV}_\infty(E, c)$, where the real parts of the eigenvalues of E exceed $1/2$, and in fact $k_n B_{k_n} \mu \to \phi$ in $\mathcal{M}$, where ϕ is the Lévy measure of ν. Apply Theorem 10.2.3 and Corollary 10.2.6 to see that

$$k_n B_{k_n} M_{k_n} B_{k_n}^* \Rightarrow W \tag{10.40}$$

where W is an operator semistable symmetric matrix with distribution Υ. Additionally, the sequence

$$(nB_n M_n B_n^*)_n \tag{10.41}$$

is relatively compact with limit set contained in $\{W_\lambda : \lambda \in [1, c]\}$ where $\mathcal{L}(W_\lambda) = L_{\lambda^{-E}}(\Upsilon^\lambda)$. Since the Lévy measure of some law $L_{\lambda^{-E}}(\Upsilon^\lambda)$ is $\lambda \cdot T(\lambda^{-E}\phi)$ and $\lambda^{-E}\phi$ is the Lévy measure of some full operator semistable law, it follows as in the proof of Theorem 10.2.9 that every limit W_λ of the sequence in (10.41) is almost surely invertible. By Theorem 10.2.9, we see that W is almost surely invertible. Then weak convergence implies that $P\{k_n B_{k_n} M_{k_n} B_{k_n}^* \text{ is invertible }\} \to 1$ as $n \to \infty$. Hence the probability that M_n^{-1} exists tends to one as $n \to \infty$. To avoid complicated notation, we will assume henceforth, without loss of generality, that M_n is invertible with probability one. Proposition 2.1.20 shows that, since M_n is a nonnegative definite symmetric matrix, it has a unique symmetric square root, which we will denote by $M_n^{1/2}$.

Before we state the main result of this section, we need some preparatory lemmas. We will always assume that (10.39) and hence (10.40) hold.

Lemma 10.3.1. *Suppose that X belongs to the strict generalized domain of semistable attraction of Y, where Y is (c^E, c) operator semistable with no normal component. If* (10.39) *holds for some $k_{n+1}/k_n \to c > 1$ and some $(B_n) \in \mathrm{RV}(-E)$, then $\left(\sqrt{n} B_n M_n^{1/2}\right)_n$ is weakly relatively compact.*

Proof. In view of Prohorov's Theorem 1.2.10, it will suffice to show that $\left(\sqrt{n} B_n M_n^{1/2}\right)_n$ is tight. Note that for $A \in \mathcal{M}_s^d$ we have $\|A\|^2 = \mathrm{trace}(AA^*)$. The Continuous Mapping Theorem 1.2.7 together with (10.41) implies that the sequence $\left(\mathrm{trace}(nB_n M_n B_n^*)\right)_n$ is weakly relatively compact and hence tight. But then for any $\varepsilon > 0$

$$\begin{aligned} P\{\|\sqrt{n}B_n M_n^{1/2}\| > R\} &= P\{\|\sqrt{n}B_n M_n^{1/2}\|^2 > R^2\} \\ &= P\{\operatorname{trace}(nB_n M_n B_n^*) > R^2\} < \varepsilon \end{aligned}$$

for all n and some $R > 0$. The proof is now complete. □

Lemma 10.3.2. *Suppose that X belongs to the strict generalized domain of semistable attraction of Y, where Y is (c^E, c) operator semistable with no normal component. If* (10.39) *and hence* (10.40) *hold for some $k_{n+1}/k_n \to c > 1$ and some $(B_n) \in \mathrm{RV}(-E)$, then*

(a) $\left(B_n S_n, \sqrt{n}B_n M_n^{1/2}\right)_n$ *is weakly relatively compact*

(b) $\left(B_n S_n, nB_n M_n B_n^*\right)_n$ *is weakly relatively compact*

Proof. For $x \in \mathbb{R}^d$ and $M \in \mathcal{M}_s^d$ define $\|(x, M)\| = \sqrt{\|x\|^2 + \|M\|^2}$, a norm on the real vector space $\mathbb{R}^d \oplus \mathcal{M}_s^d$. Then

$$\begin{aligned} &P\{\|(B_n S_n, \sqrt{n}B_n M_n^{1/2})\| > R\} \\ =&P\{\|B_n S_n\|^2 + \|\sqrt{n}B_n M_n^{1/2}\|^2 > R^2\} \\ \leq&P\{\|B_n S_n\| > R/\sqrt{2}\} + P\{\|\sqrt{n}B_n M_n^{1/2}\| > R/\sqrt{2}\} \\ <&\frac{\varepsilon}{2} + \frac{\varepsilon}{2} = \varepsilon \end{aligned}$$

for all n and some $R > 0$ by Theorem 8.3.18 and Lemma 10.3.1. See also Remark 8.3.19 (d). The proof of (b) is similar. □

Lemma 10.3.3. *Suppose that X belongs to the strict generalized domain of semistable attraction of Y, where Y is (c^E, c) operator semistable with no normal component. If* (10.39) *and hence* (10.40) *hold for some $k_{n+1}/k_n \to c > 1$ and some $(B_n) \in \mathrm{RV}(-E)$, then*

$$\left(B_{k_n} S_{k_n}, k_n B_{k_n} M_{k_n} B_{k_n}^*\right) \Rightarrow (Y, W).$$

Proof. We use the standard convergence criteria for triangular arrays, Theorem 3.2.2. If U is a Borel subset of Γ and V a Borel subset of $\mathcal{M}_s^d \setminus \{0\}$ define $\Phi(U \times V) = \phi(U \cap T^{-1}(V))$. If $U \times V$ is a Φ continuity set, then

$$\begin{aligned} &k_n P\{(B_{k_n} X, B_{k_n} X X' B_{k_n}^*) \in U \times V\} \\ =&k_n P\{B_{k_n} X \in U, B_{k_n} X X' B_{k_n}^* \in V\} \\ =&k_n P\{B_{k_n} X \in U, T(B_{k_n} X) \in V\} \\ =&k_n P\{B_{k_n} X \in U, B_{k_n} X \in T^{-1}(V)\} \\ =&k_n P\{B_{k_n} X \in U \cap T^{-1}(V)\} \\ =&k_n (B_{k_n} \mu)(U \cap T^{-1}(V)) \\ \to&\phi(U \cap T^{-1}(V)) = \Phi(U \times V), \end{aligned}$$

so that, in view of Theorem 1.2.5, we have $k_n L_n \Lambda \to \Phi$ where Λ is the distribution of (X, XX') on $\mathbb{R}^d \oplus \mathcal{M}_s^d$ and L_n is the linear operator on $\mathbb{R}^d \oplus \mathcal{M}_s^d$ defined by $L_n(x, M) = (B_{k_n} x, B_{k_n} M B_{k_n}^*)$. Note that the Lévy measure Φ of the joint limit is concentrated on the set $\{(x, xx') : x \in \mathbb{R}^d\}$.

Now in order to establish joint convergence it suffices to show that

$$\lim_{\varepsilon \to 0} \limsup_{n \to \infty} k_n \int_{\|(y,B)\| \le \varepsilon} <(y, B), (x, A)>^2 L_n \Lambda\{dy, dB\} = 0 \qquad (10.42)$$

for all $(x, A) \in \mathbb{R}^d \oplus \mathcal{M}_s^d$. We use the inner product associated with the norm in the proof of Lemma 10.3.2 so that $\mathbb{R}^d \perp \mathcal{M}_s^d$ in $\mathbb{R}^d \oplus \mathcal{M}_s^d$. Then

$$\begin{aligned} &\int_{\|(y,B)\| \le \varepsilon} <(y, B), (x, A)>^2 L_n \Lambda\{dy, dB\} \\ &= \int_{\|y\|^2 + \|B\|^2 \le \varepsilon^2} <y, x>^2 + <B, A>^2 L_n \Lambda\{dy, dB\} \\ &\le \int_{\|y\| \le \varepsilon} <y, x>^2 B_{k_n} \mu\{dy\} + \int_{\|B\| \le \varepsilon} <B, A>^2 L_{B_{k_n}} T\mu\{dB\} \end{aligned}$$

and (10.42) follows from Theorem 3.2.2 along with (10.39) and (10.40). □

Our general result about the asymptotic behavior of self-normalized sums is the following.

Theorem 10.3.4. *Suppose that X belongs to the strict generalized domain of semistable attraction of Y, where Y is (c^E, c) operator semistable with no normal component. If* (10.39) *and hence* (10.40) *hold for some $k_{n+1}/k_n \to c > 1$ and some $(B_n) \in \mathrm{RV}(-E)$ then $\left(\frac{1}{\sqrt{n}} M_n^{-1/2} S_n\right)_n$ is stochastically compact. Especially $\left(\frac{1}{\sqrt{k_n}} M_{k_n}^{-1/2} S_{k_n}\right)_n$ is stochastically compact with all limit points of the form $K^{-1}Y$ where KK^* and W are identically distributed.*

Proof. By Lemma 10.3.2(a), there exists for every subsequence of positive integers a further subsequence (n') along which

$$\left(B_n S_n, \sqrt{n} B_n M_n^{1/2}\right) \Rightarrow (Y_\lambda, K),$$

where by Theorem 8.3.18 we know $\mathcal{L}(Y_\lambda) = \lambda^{-E} \nu^\lambda$ for some $\lambda \in [1, c]$ and K is some random matrix. Then, by an argument similar to those of the proof of Theorem 10.2.3, $n(B_n M_n B_n^*) \Rightarrow W_\lambda$ along the same subsequence (n'), where $\mathcal{L}(W_\lambda) = L_{\lambda^{-E}}(\Upsilon^\lambda)$. On the other hand, $\sqrt{n} B_n M_n^{1/2} \Rightarrow K$ along (n'), so we see that KK^* is identically distributed with W_λ. It follows from an argument similar to the proof of Theorem 10.2.9 that W_λ is almost surely invertible. Hence K must be invertible almost surely and the Continuous Mapping Theorem implies that

$$\frac{1}{\sqrt{n}} M_n^{-1/2} S_n = (\sqrt{n} B_n M_n^{1/2})^{-1} B_n S_n \Rightarrow K^{-1} Y_\lambda$$

along (n'). Note that $K^{-1}Y_\lambda$ is full since K is nonsingular and Y_λ is full.

If we consider the subsequence (k_n) to start with, then $B_{k_n}S_{k_n} \Rightarrow Y$ and $k_nB_{k_n}M_{k_n}B_{k_n}^* \Rightarrow W$ and hence by the same argument as above we conclude that $\left(\frac{1}{\sqrt{k_n}}M_{k_n}^{-1/2}S_{k_n}\right)_n$ is stochastically compact with limit points $K^{-1}Y$, where $\mathcal{L}(KK^*) = \mathcal{L}(W)$. The proof is now complete. □

In the special case of scalar norming, when $B_n = b_nI$ for some regularly varying sequence (b_n) of real numbers, we can give a more detailed description of the asymptotic distribution of the self normalized sum.

Theorem 10.3.5. *Suppose that X belongs to the strict generalized domain of semistable attraction of Y, where Y is (c^E, c) operator semistable with no normal component. If (10.39) and hence (10.40) hold for some $k_{n+1}/k_n \to c > 1$ and some $(B_n) \in \mathrm{RV}(-E)$ with $B_n = b_nI$, then*

$$\frac{1}{\sqrt{k_n}}M_{k_n}^{-1/2}S_{k_n} \Rightarrow W^{-1/2}Y.$$

Moreover the sequence $\left(\frac{1}{\sqrt{n}}M_n^{-1/2}S_n\right)_n$ is stochastically compact with limit set contained in $\left\{W_\lambda^{-1/2}Y_\lambda : \lambda \in [1,c]\right\}$, where $\mathcal{L}(Y_\lambda) = \lambda^{-E}\nu^\lambda$ and $\mathcal{L}(W_\lambda) = L_{\lambda^{-E}}(\Upsilon^\lambda)$.

Proof. In this case, (10.39) becomes $b_{k_n}S_{k_n} \Rightarrow Y$ and (10.40) reduces to $k_nb_{k_n}^2M_{k_n} \Rightarrow W$. Then Lemma 10.3.3 yields

$$\left(b_{k_n}S_{k_n}, k_nb_{k_n}^2M_{k_n}\right) \Rightarrow (Y, W)$$

where W is almost surely invertible by Theorem 10.2.9. Now the Continuous Mapping Theorem implies that

$$\frac{1}{\sqrt{k_n}}M_{k_n}^{-1/2}S_{k_n} = (k_nb_{k_n}^2M_{k_n})^{-1/2}(b_{k_n}S_{k_n}) \Rightarrow W^{-1/2}Y.$$

In the general case, by Theorem 8.3.18, every subsequence contains a further subsequence (n') such that $B_nS_n \Rightarrow Y_\lambda$ along (n') for some $\lambda \in [1,c]$, where $\mathcal{L}(Y_\lambda) = \lambda^{-E}\nu^\lambda$. Then by an argument similar to the proof of Theorem 10.2.3 it follows that $nB_nM_nB_n^* \Rightarrow W_\lambda$ along (n'), where $\mathcal{L}(W_\lambda) = L_{\lambda^{-E}}(\Upsilon^\lambda)$. Now the result follows along the same lines as the special case above. □

Remark 10.3.6. If μ belongs to the strict generalized domain of attraction of some full operator stable law ν on $\mathbb{R}^d$ having no normal component, then $A_nS_n \Rightarrow Y$ where Y has distribution ν. In this case Theorems, 10.3.4 and 10.3.5 apply with $k_n = n$, and the stochastic compactness results are superfluous, since every limit point is the same.

Remark 10.3.7. If μ is a probability distribution on $\mathbb{R}^1$, then M_n reduces to the second sample moment of Definition 10.1.1. Then Theorem 10.3.5 extends Corollary 10.1.8 to the case where $X \in \mathrm{DOSA}(Y)$, extending the result of Logan, Mallows, Rice, and Shepp [89]. Note that in either case, the limiting distribution is the ratio of two dependent (semi)stable random variables.

10.4 TAIL AND MOMENT ESTIMATORS

In this section, we develop estimators for the spectral index function $\alpha(\theta)$ from Definition 8.3.11. Our focus is on the case of a nonnormal limit, where the spectral index function measures the tail thickness. Assume that a probability measure μ belongs to the generalized domain of semistable attraction of some full (c^E, c) operator semistable law ν with no normal component. Then Theorem 8.2.1 implies that μ is $\mathrm{ROV}_\infty(E, c)$ for $c > 1$ where all the real parts of the eigenvalues of the exponent E exceed $1/2$. Recall that $1/\alpha(\theta)$ ranges over the real parts $a_1 < \cdots < a_p$ of the eigenvalues of E, and that when $\alpha(\theta) = 1/a_i$ the absolute moments

$$\int |<x, \theta>|^\rho \mu\{dx\}$$

exist for $\rho < \alpha(\theta)$ and diverge for $\rho > \alpha(\theta)$, see Theorem 8.3.10. This reflects the fact that the tails

$$V_0(r, \theta) = \int_{|<x,\theta>|>r} \mu\{dx\}$$

in the θ direction diminish like $r^{-\alpha(\theta)}$ as $r \to \infty$, see Theorem 6.4.15 (a) (ii). Thus the index function α delivers the real spectrum of E as well as some important information about the tails of μ. In statistical applications, it is important to estimate both the range of $\alpha(\theta)$ and the level sets $\{\theta : \alpha(\theta) = 1/a_i\}$.

Example 10.4.1. This example illustrates the basic problem with tail and moment estimation for multivariable data with heavy tails. The situation is complicated by the fact that the component with the heaviest tail tends to dominate. Suppose that $X = (U, V)$ where U is normal and V is Cauchy, so that EV does not exist. The first marginal U has light tails and hence a finite variance. The projection of X onto any other radial direction has heavy tails, since it has both a normal and a Cauchy component, so that its mean does not exist. Now rotate X so that the normal component lies along the diagonal. Then the projections of X onto the coordinate axes (i.e., the marginal distributions) have heavy tails. If we only consider these marginal distributions, we will only detect the Cauchy-like tails. In order to discover the variations in tail behavior, we must consider every radial direction.

Suppose that $X, X_1, X_2, \ldots$ are independent random vectors with common distribution μ in the generalized domain of semistable attraction of some full operator semistable law ν having no normal component. In view of Corollary 8.2.2, there exist a sequence $(B_n) \subset \mathrm{GL}(\mathbb{R}^d)$ varying regularly with index $(-E)$ and a sequence (k_n) of natural numbers with $k_{n+1}/k_n \to c > 1$ such that

$$B_{k_n}(X_1 + \cdots + X_{k_n}) - b_n \Rightarrow Y \tag{10.43}$$

for some nonrandom vectors b_n where $\mathcal{L}(Y) = \nu$. Apply Theorem 10.2.3 and Corollary 10.2.6 to show that the sample covariance matrix M_n satisfies

$$k_n B_{k_n} M_{k_n} B_{k_n}^* \Rightarrow W \tag{10.44}$$

where W is a random element of $\mathcal{M}_s^d$. The limit W is invertible with probability one by Theorem 10.2.9.

Note that by (10.25) M_n is symmetric and nonnegative definite so that by Proposition 2.1.20 there exists an orthonormal basis of eigenvectors for M_n with nonnegative eigenvalues. Our estimator of the index function $\alpha(\theta)$ is based on the eigenvalues

$$0 \le \lambda_n^{(1)} \le \cdots \le \lambda_n^{(d)}$$

and the associated unit eigenvectors

$$\theta_n^{(1)}, \ldots, \theta_n^{(d)}$$

of the sample covariance matrix M_n.

Let $V_1, \ldots, V_p$ be the spectral decomposition of $\mathbb{R}^d$ with respect to X from Definition 8.3.11, and take Y and (B_n) spectrally compatible with μ in (10.43). Put $d_i = \dim V_i$ and write $E = E_1 \oplus \cdots \oplus E_p$, where every eigenvalue of the $d_i \times d_i$ matrix E_i has real part equal to a_i. Since the subspaces V_i are mutually orthogonal, $EV_i = V_i$ and $E^* V_i = V_i$ for all $i = 1, \ldots, p$. Let π_i denote the orthogonal projection operator onto V_i, and recall from Theorem 8.3.7 that $\pi_i X \in \mathrm{GDOSA}(\pi_i Y, c)$ and $\pi_i Y$ is (c^{E_i}, c) operator semistable on V_i. If we let $P_i = \pi_i + \cdots + \pi_p$, then $P_i X \in \mathrm{GDOSA}(P_i Y, c)$ and $P_i Y$ is operator semistable on $L_i = V_i \oplus \cdots \oplus V_p$ with exponent $E_i \oplus \cdots \oplus E_p$. Similarly, if we let $\bar{P}_i = \pi_1 + \cdots + \pi_i$, then $\bar{P}_i X \in \mathrm{GDOSA}(\bar{P}_i Y, c)$ and $\bar{P}_i Y$ is operator semistable on $\bar{L}_i = V_1 \oplus \cdots \oplus V_i$ with exponent $E_1 \oplus \cdots \oplus E_i$. Recall from Section 6.4 that $\alpha(\theta) = 1/a_i$ if and only if $\theta \in \bar{L}_i \setminus \bar{L}_{i-1}$.

Since $(B_n) \in \mathrm{RV}(-E)$ is spectrally compatible with $-E$, Proposition 4.3.14 shows that the conclusions of Theorem 4.3.1 hold with $L_i = V_i \oplus \cdots \oplus V_p$ for each $i = 1, \ldots, p$. Then for any $\varepsilon > 0$ and any $x \in L_i \setminus L_{i+1}$ we have

$$n^{-a_i - \varepsilon} \le \|B_n x\| \le n^{-a_i + \varepsilon} \tag{10.45}$$

for all large n. Then

$$\frac{\log \|B_n x\|}{\log n} \to -a_i \quad \text{as } n \to \infty \tag{10.46}$$

and since this convergence is uniform on compact subsets of $L_i \setminus L_{i+1}$ we also have

$$\frac{\log \|\pi_i B_n\|}{\log n} \to -a_i \quad \text{as } n \to \infty. \tag{10.47}$$

It follows that

$$\frac{\log \|B_n\|}{\log n} \to -a_1 \quad \text{as } n \to \infty. \tag{10.48}$$

Since $((B_n^*)^{-1}) \in \text{RV}(E^*)$ is spectrally compatible with E^*, a similar argument shows that for any $x \in \bar{L}_i \setminus \bar{L}_{i-1}$ we have

$$n^{a_i - \varepsilon} \le \|(B_n^*)^{-1} x\| \le n^{a_i + \varepsilon} \tag{10.49}$$

for all large n and hence

$$\frac{\log \|(B_n^*)^{-1} x\|}{\log n} \to a_i \quad \text{as } n \to \infty, \tag{10.50}$$

$$\frac{\log \|\pi_i (B_n^*)^{-1}\|}{\log n} \to a_i \quad \text{as } n \to \infty. \tag{10.51}$$

Hence

$$\frac{\log \|(B_n^*)^{-1}\|}{\log n} \to a_p \quad \text{as } n \to \infty. \tag{10.52}$$

Define $T : \mathbb{R}^d \to \mathcal{M}_s^d$ by $T(x) = xx'$ as in Section 10.2, where $\mathcal{M}_s^d$ is the real vector space of $d \times d$ symmetric matrices.

Lemma 10.4.2. *Suppose that $\mu \in \text{GDOSA}(\nu, c)$ where ν has no normal component and* (10.44) *holds. Then for all $\theta \in \Gamma$ the random variable $<W, T(\theta)>$ has a density with respect to Lebesgue measure.*

Proof. Lemma 10.2.7 and Theorem 10.2.9 show that W is full and operator semistable on $L = \text{span}\{\text{supp}(T\phi)\}$, where ϕ is the Lévy measure of ν. Then Theorem 7.1.15 shows that W has a density on L. Since ϕ is $\mathcal{M}$-full on $\mathbb{R}^d$ we can choose $y_1, \dots, y_d$ linearly independent in $\mathbb{R}^d$ with all $y_i \in \text{supp}(\phi)$. Otherwise, $\text{span}\{\text{supp}(\phi)\}$ has dimension less than d and ϕ would not be full. Then Lemma 10.2.8 shows that $T(y_i) \in L$ for all $i = 1, \dots, d$. If $T(\theta)$ is perpendicular to L, then $<T(y_i), T(\theta)> = <y_i, \theta>^2 = 0$ for all $i = 1, \dots, d$, which is a contradiction. Hence $<W, T(\theta)>$ is a one dimensional-marginal of W on L and so it has a density by the Fubini Theorem 1.1.17. □

We are now in position to construct our estimator. We start by showing how to estimate $a_1 < \cdots < a_p$, the real parts of the eigenvalues of E. For $i = 1, \dots, p$ let $b_i = \dim L_i = d_i + \cdots + d_p$ and $\bar{b}_i = \dim \bar{L}_i = d_1 + \cdots + d_i$.

Proposition 10.4.3.

$$\frac{\log(n\lambda_n^{(d)})}{2 \log n} \to a_p \quad \textit{in probability.}$$

Proof. For $\delta > 0$ arbitrary write

$$\begin{aligned}
&P\Big\{\Big|\frac{\log(n\lambda_n^{(d)})}{2\log n} - a_p\Big| > \delta\Big\}\\
\le&P\{\log(n\lambda_n^{(d)}) > 2(a_p+\delta)\log n\} + P\{\log(n\lambda_n^{(d)}) < 2(a_p-\delta)\log n\}\\
=&P\{\lambda_n^{(d)} > n^{2(a_p+\delta)-1}\} + P\{\lambda_n^{(d)} < n^{2(a_p-\delta)-1}\}\\
=&P\Big\{\max_{\|\theta\|=1} <M_n\theta,\theta> > n^{2(a_p+\delta)-1}\Big\} + P\Big\{\max_{\|\theta\|=1} <M_n\theta,\theta> < n^{2(a_p-\delta)-1}\Big\}.
\end{aligned} \tag{10.53}$$

Now choose ρ so that $2a_p < \rho^{-1} < 2(a_p+\delta)$ and note that $E\|X\|^{2\rho} < \infty$ by Corollary 8.2.6. Therefore $E\|T(X)\|^\rho = E\|X\|^{2\rho} < \infty$. Furthermore, since $a_p > 1/2$ we have $\rho < 1$ so that $|x+y|^\rho \le |x|^\rho + |y|^\rho$ for all $x, y \in \mathbb{R}$. Then by Markov's Inequality, Proposition 1.1.28, we can bound the first term on the right hand side of (10.53) by

$$\begin{aligned}
P\Big\{\max_{\|\theta\|=1} <M_n\theta,\theta> > n^{2(a_p+\delta)-1}\Big\} &= P\{\|M_n\| > n^{2(a_p+\delta)-1}\}\\
&= P\Big\{\Big\|\sum_{i=1}^n T(X_i)\Big\| > n^{2(a_p+\delta)}\Big\}\\
&\le P\Big\{\sum_{i=1}^n \|T(X_i)\| > n^{2(a_p+\delta)}\Big\}\\
&\le \left(\frac{1}{n^{2(a_p+\delta)}}\right)^\rho E\Big|\sum_{i=1}^n \|T(X_i)\|\Big|^\rho\\
&\le n^{1-2\rho(a_p+\delta)}E\|T(X)\|^\rho
\end{aligned}$$

which tends to zero as $n \to \infty$ by the choice of ρ.

To estimate the second term on the right-hand side of (10.53) choose $\theta_0 \in \bar{L}_p \setminus \bar{L}_{p-1}$ and write $(B_n^*)^{-1}\theta_0 = r_n\theta_n$, where $r_n > 0$ and $\|\theta_n\| = 1$. By virtue of (10.49), we can choose $\varepsilon < \delta$ so that $n^{a_p-\varepsilon} \le r_n \le n^{a_p+\varepsilon}$ for all large n. Then

$$\begin{aligned}
P\Big\{\max_{\|\theta\|=1} <M_n\theta,\theta> < n^{2(a_p-\delta)-1}\Big\} &\le P\{<M_n\theta_0,\theta_0> < n^{2(a_p-\delta)-1}\}\\
&= P\{<nB_nM_nB_n^*\theta_n,\theta_n> < r_n^{-2}n^{2(a_p-\delta)}\}\\
&\le P\{<nB_nM_nB_n^*\theta_n,\theta_n> < n^{2(\varepsilon-\delta)}\}.
\end{aligned}$$

Since $\|\theta_n\| = 1$ for all n, given any subsequence there exists a further subsequence (n') along which $\theta_{n'} \to \theta$, where $\|\theta\| = 1$. In view of Theorem 10.2.3, there exists a further subsequence $(n'') \subset (n')$ such that $nB_nM_nB_n^* \Rightarrow W_\lambda$ along (n'') for some $\lambda \in [1, c]$, where $\mathcal{L}(W_\lambda)$ has the Lévy representation

$[C_\lambda, 0, \lambda \cdot T(\lambda^{-E}\phi)]$. The Lévy measure $\phi_{(\lambda)} = \lambda \cdot (\lambda^{-E}\phi)$ satisfies $c^E\phi_{(\lambda)} = c \cdot \phi_{(\lambda)}$ by Remark 8.3.19 (b) and hence $<W_\lambda, T(\theta)>$ has a Lebesgue density by Lemma 10.4.2. Then $<nB_nM_nB_n^*\theta_n, \theta_n> = <nB_nM_nB_n^*, T(\theta_n)> \Rightarrow <W_\lambda, T(\theta)>$ along (n'') by the Continuous Mapping Theorem 1.2.8. For any $\varepsilon_1 > 0$, there exists a $\rho > 0$ such that

$$P\{<W_\lambda, T(\theta)> < \rho\} < \varepsilon_1/2$$

since $<W_\lambda, T(\theta)>$ has a Lebesgue density. Choose n_0 such that $(n'')^{2(\varepsilon-\delta)} < \rho$ and $\left|P\{<nB_nM_nB_n^*, T(\theta_n)> < \rho\} - P\{<W_\lambda, T(\theta)> < \rho\}\right| < \varepsilon_1/2$ for all $n'' \geq n_0$. Then

$$\begin{aligned} P\{<nB_nM_nB_n^*, T(\theta_n)> < n^{2(\varepsilon-\delta)}\} &\leq P\{<nB_nM_nB_n^*, T(\theta_n)> < \rho\} \\ &\leq P\{<W_\lambda, T(\theta)> < \rho\} + \varepsilon_1/2 < \varepsilon_1 \end{aligned}$$

for all $n'' \geq n_0$. Since for any subsequence there is a further subsequence along which $P\{<nB_nM_nB_n^*\theta_n, \theta_n> < n^{2(\varepsilon-\delta)}\} \to 0$, this convergence holds along the entire sequence, which concludes the proof. □

Proposition 10.4.4.

$$\frac{\log(n\lambda_n^{(1)})}{2\log n} \to a_1 \quad \textit{in probability.}$$

Proof. In view of Theorem 10.2.9, we can assume without loss of generality that M_n is invertible almost surely. For $\delta > 0$ arbitrary, we have

$$\begin{aligned} &P\left\{\left|\frac{\log(n\lambda_n^{(1)})}{2\log n} - a_1\right| > \delta\right\} \\ \leq &P\{\log(n\lambda_n^{(1)}) > 2(a_1+\delta)\log n\} + P\{\log(n\lambda_n^{(1)}) < 2(a_1-\delta)\log n\} \\ = &P\{\lambda_n^{(1)} > n^{2(a_1+\delta)-1}\} + P\{\lambda_n^{(1)} < n^{2(a_1-\delta)-1}\}. \end{aligned} \tag{10.54}$$

To estimate the first probability on the right-hand side of (10.54), choose a unit vector $\theta_0 \in \bar{L}_1$ and write $(B_n^*)^{-1}\theta_0 = r_n\theta_n$, where $r_n > 0$ and $\|\theta_n\| = 1$ as above. Then, by (10.49), for $\varepsilon < \delta$ we have $n^{a_1-\varepsilon} \leq r_n \leq n^{a_1+\varepsilon}$ for all large n. Therefore

$$\begin{aligned} P\{\lambda_n^{(1)} > n^{2(a_1+\delta)-1}\} &= P\Big\{\min_{\|\theta\|=1} <M_n\theta, \theta> > n^{2(a_1+\delta)-1}\Big\} \\ &\leq P\{<M_n\theta_0, \theta_0> > n^{2(a_1+\delta)-1}\} \\ &= P\{<nB_nM_nB_n^*\theta_n, \theta_n> > r_n^{-2}n^{2(a_1+\delta)}\} \\ &\leq P\{<nB_nM_nB_n^*\theta_n, \theta_n> > n^{2(\delta-\varepsilon)}\}. \end{aligned}$$

Since $||\theta_n|| = 1$ for all n, given any subsequence there exists a further subsequence (n') along which $\theta_{n'} \to \theta$, where $||\theta|| = 1$. In view of Theorem 10.2.3, there exists a further subsequence $(n'') \subset (n')$ and a $\lambda \in [1, c]$ such that $nB_n M_n B_n^* \Rightarrow W_\lambda$ along (n'') and hence, as before, $<nB_n M_n B_n^* \theta_n, \theta_n> = <nB_n M_n B_n^*, T(\theta_n)> \Rightarrow <W_\lambda, T(\theta)>$ by the Continuous Mapping Theorem 1.2.8. Thus the sequence $\left(<nB_n M_n B_n^* \theta_n, \theta_n>\right)_n$ is weakly relatively compact, so by Prohorov's Theorem 1.2.10 this sequence is tight. Then for $\varepsilon_1 > 0$ there exists $R > 0$ such that

$$P\{<nB_n M_n B_n^* \theta_n, \theta_n> > n^{2(\delta-\varepsilon)}\} \leq P\{<nB_n M_n B_n^* \theta_n, \theta_n> > R\} < \varepsilon_1$$

for all $n \geq n_0(\varepsilon_1)$, which shows that $P\{<nB_n M_n B_n^* \theta_n, \theta_n> > n^{2(\delta-\varepsilon)}\} \to 0$ as $n \to \infty$.

Since the smallest eigenvalue of M_n is the reciprocal of the largest eigenvalue of M_n^{-1} we have

$$\begin{aligned}
P\{\lambda_n^{(1)} < n^{2(a_1-\delta)-1}\} &= P\{\frac{1}{\lambda_n^{(1)}} > n^{2(\delta-a_1)+1}\} \\
&= P\Big\{\max_{||\theta||=1} <M_n^{-1}\theta, \theta> > n^{2(\delta-a_1)+1}\Big\} \\
&= P\{\|M_n^{-1}\| > n^{2(\delta-a_1)+1}\} \\
&= P\{\|B_n^*(B_n^*)^{-1} M_n^{-1} B_n^{-1} B_n\| > n^{2(\delta-a_1)+1}\} \\
&\leq P\{\|\frac{1}{n}(B_n^*)^{-1} M_n^{-1} B_n^{-1}\| > ||B_n||^{-2} n^{2(\delta-a_1)}\}.
\end{aligned}$$

It follows from (10.48) that for any $\varepsilon < \delta$ there exists a constant $C > 0$ such that $||B_n|| \leq Cn^{-a_1+\varepsilon}$ for all n. Then for some $K > 0$ we have $||B_n||^{-2} \geq Kn^{2(a_1-\varepsilon)}$ for all n. By Theorem 10.2.3, the sequence $\left(nB_n M_n B_n^*\right)_n$ is weakly relatively compact. Hence every subsequence contains a further subsequence (n') such that $nB_n M_n B_n^* \Rightarrow W_\lambda$ along (n') for some $\lambda \in [1, c]$. It follows from an argument similar to the proof of Theorem 10.2.9 that W_λ is almost surely invertible. Then the Continuous Mapping Theorem 1.2.7 yields

$$\frac{1}{n}(B_n^*)^{-1} M_n^{-1} B_n^{-1} \Rightarrow W_\lambda^{-1} \quad \text{along } (n').$$

Given $\varepsilon_1 > 0$ choose $R > 0$ such that

$$P\Big\{\Big\|\frac{1}{n'}(B_{n'}^*)^{-1} M_{n'}^{-1} B_{n'}^{-1}\Big\| > R\Big\} < \varepsilon_1 \quad \text{for all } n'.$$

Choose n_0 such that $Kn^{2(\delta-\varepsilon)} > R$ for all $n \geq n_0$. Then for all $n' \geq n_0$ we have

$$\begin{aligned}
&P\Big\{\Big\|\frac{1}{n'}(B_{n'}^*)^{-1}M_{n'}^{-1}B_{n'}^{-1}\Big\| > \|B_{n'}\|^{-2}(n')^{2(\delta-a_1)}\Big\}\\
\leq &P\Big\{\Big\|\frac{1}{n'}(B_{n'}^*)^{-1}M_{n'}^{-1}B_{n'}^{-1}\Big\| > K(n')^{2(\delta-\varepsilon)}\Big\}\\
\leq &P\Big\{\Big\|\frac{1}{n'}(B_{n'}^*)^{-1}M_{n'}^{-1}B_{n'}^{-1}\Big\| > R\Big\} < \varepsilon_1
\end{aligned}$$

and so $P\{\lambda_n^{(1)} < n^{2(a_1-\delta)-1}\} \to 0$ along (n'). Since for any subsequence there exists a further subsequence with this property we get

$$P\{\lambda_n^{(1)} < n^{2(a_1-\delta)-1}\} \to 0$$

as $n \to \infty$, which concludes the proof. □

Theorem 10.4.5. *For $i = 1, \ldots, p$ and $\bar{b}_{i-1} < j \leq \bar{b}_i$ we have*

$$\frac{\log(n\lambda_n^{(j)})}{2\log n} \to a_i \quad \textit{in probability.}$$

Proof. Let $\mathcal{C}_j$ denote the collection of all orthogonal projection operators onto subspaces of $\mathbb{R}^d$ with dimension j. The Courant-Fischer Max-Min Theorem (see [150], p. 51) implies that

$$\begin{aligned}
\lambda_n^{(j)} &= \min_{P\in\mathcal{C}_j}\max_{\|\theta\|=1} <PM_nP\theta,\theta>\\
&= \max_{P\in\mathcal{C}_{d-j+1}}\min_{\|\theta\|=1} <PM_nP\theta,\theta>.
\end{aligned} \tag{10.55}$$

Recall that P_iX belongs to the generalized domain of semistable attraction of the operator semistable law P_iY whose exponent $E_i \oplus \cdots \oplus E_p$ has real spectrum $a_i, \ldots, a_p$ and note that $P_iM_nP_i$ is the sample covariance matrix of the data $P_iX_1, \ldots, P_iX_n$. Let λ_n denote the smallest eigenvalue of the matrix $P_iM_nP_i$ and apply Proposition 10.4.4 above to see that

$$\frac{\log(n\lambda_n)}{2\log n} \to a_i \quad \text{in probability.}$$

Similarly, since $\bar{P}_iX$ belongs to the generalized domain of semistable attraction of the operator semistable law $\bar{P}_iY$ whose exponent $E_1 \oplus \cdots \oplus E_i$ has real spectrum $a_1, \ldots, a_i$ we can apply Proposition 10.4.3 above to see that

$$\frac{\log(n\bar{\lambda}_n)}{2\log n} \to a_i \quad \text{in probability,}$$

where $\bar{\lambda}_n$ is the largest eigenvalue of the matrix $\bar{P}_iM_n\bar{P}_i$. Now apply (10.55) to see that

$$\lambda_n \leq \lambda_n^{(\bar{b}_{i-1}+1)} \leq \lambda_n^{(j)} \leq \lambda_n^{(\bar{b}_i)} \leq \bar{\lambda}_n.$$

Now the theorem follows easily. □

After dealing with the asymptotic behavior of the eigenvalues of the sample covariance matrix in Theorem 10.4.5 above we now investigate the convergence of the unit eigenvectors of M_n. It will turn out that if $p \leq 3$ then the eigenvectors converge in probability to the subspaces V_i if the corresponding normalized eigenvalues converge in probability to a_i. To make our results easier to read, we formulate and prove them in terms of random orthogonal projections. Recall the definition of the projections π_i from the beginning of this section.

Proposition 10.4.6. *If $j > \bar{b}_{p-1}$ and $r < p$, then*

$$\pi_r \theta_n^{(j)} \to 0 \quad \textit{in probability.}$$

Proof. Since $M_n \theta_n^{(j)} = \lambda_n^{(j)} \theta_n^{(j)}$ we can write $\pi_r \theta_n^{(j)} = (\pi_r M_n / \lambda_n^{(j)}) \theta_n^{(j)}$, where

$$\begin{aligned} \|\pi_r M_n / \lambda_n^{(j)}\| &= \frac{\|B_n^{-1} B_n \pi_r M_n B_n^* (B_n^*)^{-1}\|}{\lambda_n^{(j)}} \\ &= \frac{\|\pi_r B_n^{-1} B_n M_n B_n^* (B_n^*)^{-1}\|}{\lambda_n^{(j)}} \\ &\leq \frac{\|\pi_r B_n^{-1}\| \|\| n B_n M_n B_n^* \|\| \|(B_n^*)^{-1}\|}{n \lambda_n^{(j)}}. \end{aligned}$$

From Theorem 10.2.3 and the Continuous Mapping Theorem 1.2.7 it follows that $\left(\|nB_n M_n B_n^*\|\right)_n$ is weakly relatively compact and hence by Prohorov's Theorem 1.2.10 this sequence is tight. Also

$$\begin{aligned} &\frac{\log\left(\|\pi_r B_n^{-1}\| \|\|(B_n^*)^{-1}\| / (n\lambda_n^{(j)})\right)}{\log n} \\ =&\frac{\log \|\pi_r B_n^{-1}\|}{\log n} + \frac{\log \|(B_n^*)^{-1}\|}{\log n} - \frac{\log(n\lambda_n^{(j)})}{\log n} \\ \to& a_r + a_p - 2a_p < 0 \quad \text{in probability,} \end{aligned}$$

by (10.51), (10.52) and Theorem 10.4.5. Hence $\pi_r M_n / \lambda_n^{(j)} \to 0$ in probability. Since $\|\theta_n^{(j)}\| = 1$ this implies that

$$\|\pi_r \theta_n^{(j)}\| = \|(\pi_r M_n / \lambda_n^{(j)}) \theta_n^{(j)}\| \leq \|\pi_r M_n / \lambda_n^{(j)}\| \to 0$$

in probability, which proves the proposition. □

Similarly, we have

Proposition 10.4.7. *If $j \leq \bar{b}_1$ and $r > 1$, then*

$$\pi_r \theta_n^{(j)} \to 0 \quad \textit{in probability.}$$

Proof. Since $\theta_n^{(j)} = M_n^{-1}\lambda_n^{(j)}\theta_n^{(j)}$ we have $\pi_r\theta_n^{(j)} = (\pi_r M_n^{-1}\lambda_n^{(j)})\theta_n^{(j)}$, where

$$\begin{aligned}||\pi_r M_n^{-1}\lambda_n^{(j)}|| &= ||B_n^*(B_n^*)^{-1}\pi_r M_n^{-1} B_n^{-1} B_n||\lambda_n^{(j)} \\ &= ||\pi_r B_n^*(B_n^*)^{-1} M_n^{-1} B_n^{-1} B_n||\lambda_n^{(j)} \\ &\le ||\pi_r B_n^*||||\frac{1}{n}(B_n^*)^{-1} M_n^{-1} B_n^{-1}||||B_n||(n\lambda_n^{(j)}).\end{aligned}$$

Now, as in the proof of Proposition 10.4.4 it follows that the sequence $\left(\frac{1}{n}(B_n^*)^{-1}M_n^{-1}B_n^{-1}\right)_n$ is weakly relatively compact and hence by the Continuous Mapping Theorem and Prohorov's Theorem $\left(||\frac{1}{n}(B_n^*)^{-1}M_n^{-1}B_n^{-1}||\right)_n$ is tight. Also, by (10.47), (10.48) and Theorem 10.4.5 we conclude that

$$\begin{aligned}&\frac{\log\left(||\pi_r B_n^*||||B_n||(n\lambda_n^{(j)})\right)}{\log n} \\ &= \frac{\log||\pi_r B_n^*||}{\log n} + \frac{\log||B_n||}{\log n} + \frac{\log(n\lambda_n^{(j)})}{\log n} \\ &\to -a_r - a_1 + 2a_1 < 0 \quad \text{in probability.}\end{aligned}$$

Hence $\pi_r M_n^{-1}\lambda_n^{(j)} \to 0$ in probability. Since $||\theta_n^{(j)}|| = 1$ this implies that

$$||\pi_r\theta_n^{(j)}|| = ||(\pi_r M_n^{-1}\lambda_n^{(j)})\theta_n^{(j)}|| \le ||\pi_r M_n^{-1}\lambda_n^{(j)}|| \to 0$$

in probability, which concludes the proof. □

Theorem 10.4.8. *For $1 \le i \le p$, let*

$$\pi_n^{(i)}(x) = \sum_{j=\bar{b}_{i-1}+1}^{\bar{b}_i} <x, \theta_n^{(j)}> \theta_n^{(j)}.$$

Then $\pi_n^{(1)} \to \pi_1$ and $\pi_n^{(p)} \to \pi_p$ in probability.

Proof. We will show that $\pi_n^{(p)} \to \pi_p$ in probability. The proof that $\pi_n^{(1)} \to \pi_1$ in probability is similar. We use that convergence in probability is equivalent to the fact that for any subsequence there exists a further subsequence along which we have almost sure convergence. For any $j > \bar{b}_{p-1}$ and $r < p$, we have by Proposition 10.4.6 that $\pi_r\theta_n^{(j)} \to 0$ in probability. Since $\pi_1 + \cdots + \pi_p = I$ this implies $||\pi_p\theta_n^{(j)}|| \to 1$ in probability. Then for any subsequence (n_1) there exists a further subsequence $(n_2) \subset (n_1)$ such that $||\pi_p\theta_{n_2}^{(j)}(\omega)|| \to 1$ for all $j > \bar{b}_{p-1}$ and all $\omega \in \Omega_0$ where $P(\Omega_0) = 1$. Then for any fixed $\omega \in \Omega_0$ the sequence $(\pi_p\theta_{n_2}^{(j)}(\omega))_{n_2}$ is relatively compact so given any subsequence $(n_3) \subset (n_2)$ there exists a further subsequence $(n_4) \subset (n_3)$ (depending on ω), such that $\pi_p\theta_{n_4}^{(j)}(\omega) \to \theta_j(\omega)$ for all $j > \bar{b}_{p-1}$. By continuity, we have $||\theta_j(\omega)|| = 1$ and $<\theta_j(\omega), \theta_k(\omega)> = 0$ for $j \ne k$ and $j, k > \bar{b}_{p-1}$. Since $\pi_p = \pi_p^2$

we also have $\pi_p\theta_{n4}^{(j)}(\omega) \to \pi_p\theta_j(\omega)$ so that $\pi_p\theta_j(\omega) = \theta_j(\omega)$, hence $\theta_j(\omega) \in V_p$ and so the vectors $\{\theta_j(\omega) : j = \bar{b}_{p-1}+1, \ldots, \bar{b}_p\}$ form an orthonormal basis for V_p. Then for any $x \in \mathbb{R}^d$ we have

$$\begin{aligned}\pi_{n4}^{(p)}(\omega)(x) &= \sum_{j=\bar{b}_{p-1}+1}^{\bar{b}_p} <x, \theta_{n4}^{(j)}(\omega)>\theta_{n4}^{(j)}(\omega) \\ &\to \sum_{j=\bar{b}_{p-1}+1}^{\bar{b}_p} <x, \theta_j(\omega)>\theta_j(\omega) \\ &= \pi_p(x).\end{aligned}$$

We have shown that for all $\omega \in \Omega_0$ for any subsequence $(n_3) \subset (n_2)$ there exists a further subsequence $(n_4) \subset (n_3)$ such that $\pi_{n4}^{(p)}(\omega) \to \pi_p$. Since the limit neither depends on ω nor on the subsequence (n_4) this means that for all $\omega \in \Omega_0$ we have $\pi_{n2}^{(p)}(\omega) \to \pi_p$. Then for any subsequence (n_1) we have shown that there exists a further subsequence $(n_2) \subset (n_1)$ such that $\pi_{n2}^{(p)} \to \pi_p$ almost surely, so $\pi_n^{(p)} \to \pi_p$ in probability. □

If the spectral decomposition of $\mathbb{R}^d$ with respect to the exponent E has at most three different components we have shown:

Corollary 10.4.9. *If $p \leq 3$, then $\pi_n^{(i)} \to \pi_i$ in probability for $i = 1, \ldots, p$.*

Proof. If $p = 1$ there is nothing to prove, and if $p = 2$ the result is contained in Theorem 10.4.8. If $p = 3$, then by Theorem 10.4.8 we have $\pi_n^{(i)} \to \pi_i$ in probability for $i = 1$ and $i = 3$ and then $\pi_n^{(2)} = I - \pi_n^{(1)} - \pi_n^{(3)} \to I - \pi_1 - \pi_3 = \pi_2$ in probability as well. □

In the cases considered in Corollary 10.4.9, we are able to estimate the index function $\alpha(\theta)$.

Theorem 10.4.10. *Define*

$$\hat{\alpha}_n(\theta) = \frac{2\log n}{\log(n\lambda_n^{(j)})}$$

where j is the largest integer such that $<\theta, \theta_n^{(j)}> \neq 0$. If $p \leq 3$ then $\hat{\alpha}_n \to \alpha$ in the following sense: There exists a sequence of random linear operators I_n on $\mathbb{R}^d$ with $I_n \to I$ in probability such that $\hat{\alpha}_n(I_n\theta) \to \alpha(\theta)$ in probability for all $\theta \neq 0$ in $\mathbb{R}^d$.

Proof. Given $\theta \neq 0$ choose $i = 1, \ldots, p$ so that $\alpha(\theta) = 1/a_i$. This is equivalent to $\theta \in \bar{L}_i \backslash \bar{L}_{i-1}$. Let $\bar{P}_n^{(i)} = \pi_n^{(1)} + \cdots + \pi_n^{(i)}$. Then Corollary 10.4.9 implies that $\bar{P}_n^{(i)} \to \bar{P}_i$ in probability, and so $\theta_n \to \bar{P}_i\theta$ in probability where $\theta_n = \bar{P}_n^{(i)}\theta$.

Since $\theta \in \bar{L}_i$ we have $\bar{P}_i\theta = \theta$, so in fact $\theta_n \to \theta$ in probability. Since both $\pi_n^{(i)} \to \pi_i$ and $\theta_n \to \theta$ in probability and since

$$\begin{aligned}||\pi_n^{(i)}\theta_n - \pi_i\theta|| &\leq ||\pi_n^{(i)}\theta_n - \pi_i\theta_n|| + ||\pi_i\theta_n - \pi_i\theta|| \\ &\leq ||\pi_n^{(i)} - \pi_i||||\theta_n|| + ||\pi_i||||\theta_n - \theta||\end{aligned}$$

we also have $\pi_n^{(i)}\theta_n \to \pi_i\theta$ in probability.

Since $\theta \notin \bar{L}_{i-1}$ we have $\pi_i\theta \neq 0$ and so given any $\varepsilon > 0$ and any $\delta > 0$ (choose $\delta < ||\pi_i\theta||$) we have $P\{\pi_n^{(i)}\theta_n = 0\} \leq P\{||\pi_n^{(i)}\theta_n - \pi_i\theta|| > \delta\} < \varepsilon/2$ for all large n. By construction, we always have $\pi_n^{(r)}\theta_n = 0$ when $r > i$. Theorem 10.4.5 implies that $\log(n\lambda_n^{(j)})/2\log n \to a_i$ in probability for all $\bar{b}_{i-1} < j \leq \bar{b}_i$, and so we also have

$$P\left\{\left|\frac{2\log n}{\log(n\lambda_n^{(j)})} - \frac{1}{a_i}\right| > \delta\right\} < \frac{\varepsilon}{2d_i}$$

for all $\bar{b}_{i-1} < j \leq \bar{b}_i$ and all large n. Then, we have

$$\begin{aligned}&P\{|\hat{\alpha}_n(\theta_n) - \alpha(\theta)| > \delta\} \\ \leq &P\{\pi_n^{(i)}\theta_n = 0\} + P\{\pi_n^{(r)}\theta_n \neq 0 \text{ for some } r > i\} \\ &+ P\left\{\left|\frac{2\log n}{\log(n\lambda_n^{(j)})} - \frac{1}{a_i}\right| > \delta \text{ for some } \bar{b}_{i-1} < j \leq \bar{b}_i\right\} \\ \leq &P\{\pi_n^{(i)}\theta_n = 0\} + 0 + \sum_{j=\bar{b}_{i-1}+1}^{\bar{b}_i} P\left\{\left|\frac{2\log n}{\log(n\lambda_n^{(j)})} - \frac{1}{a_i}\right| > \delta\right\} \\ < &\frac{\varepsilon}{2} + 0 + d_i\frac{\varepsilon}{2d_i} = \varepsilon\end{aligned}$$

for all large n, so that $\hat{\alpha}_n(\theta_n) \to \alpha(\theta)$ in probability. Now let $I_n\theta = \bar{P}_n^{(i)}\theta$ when $\alpha(\theta) = 1/a_i$ so that $\theta_n = I_n\theta$. Since $\theta = I\theta = \bar{P}_i\theta$ for such θ we see that $||I_n - I||$ is bounded above by the maximum of $||\bar{P}_n^{(i)} - \bar{P}_i||$ over $i = 1, \ldots, p$ and hence $I_n \to I$ in probability. The proof is now complete. □

Remark 10.4.11. Since we must estimate both the range of $\alpha(\theta)$ and the level sets $\{\theta : \alpha(\theta) = 1/a_i\}$, the result of Theorem 10.4.10 is the best possible. In fact, it is somewhat surprising that we can detect the lighter tails $\alpha(\theta) = 1/a_i$ for $i < p$ at all, since this tail behavior exists on a set of Lebesgue measure zero on the unit sphere.

Remark 10.4.12. In Remark 10.2.13, we noted that the asymptotics of M_n are identical to those of the centered sample covariance matrix Γ_n given by (10.38). Then all of the results of this section also hold if we replace M_n by Γ_n.

Remark 10.4.13. The proof of Propositions 10.4.6 and 10.4.7 also provide information on the rate of convergence of the eigenvectors. For example, if

$p = 2$, then $n^{a_2 - a_1 - \varepsilon}(\pi_n^{(j)} - \pi_j) \to 0$ in probability for all j, for any $\varepsilon > 0$. In Example 10.4.1, we have $a_1 = 1/2$ for the normal component and $a_2 = 1$ for the Cauchy component, so the rate of convergence is $\sqrt{n}$. The rate of convergence in Theorem 10.4.5 is much slower. Apparently, it is much more difficult to estimate the range of $\alpha(\theta)$ than to estimate the subspaces in the spectral decomposition. For this reason, it may be useful to apply Corollary 10.4.9 together with alternative tail estimation methods. Once we change to the coordinates determined by the eigenvectors of the sample covariance matrix, we can apply any one variable tail estimator to determine the tail thickness for each marginal. See [109] for a recent survey and critique of several one variable tail estimation methods, and [127] for an application of the estimation methods in this section to a problem in finance.

Remark 10.4.14. If X is a random variable, then M_n reduces to the second sample moment of Definition 10.1.1. Then $M_n = \lambda_n^{(d)}$ and Proposition 10.4.3 implies that

$$\hat{a}_n = \frac{\log(nM_n)}{2\log n} \to a \quad \text{in probability} \tag{10.56}$$

where $\alpha = 1/a$ is the index of the semistable limit Y in (10.43). The same holds if we replace M_n by Γ_n, the sample variance. In this case, $E(\hat{a}_n - a)^2 \to 0$ and for some $c_n \to 0$ we have that $2\ln n(\hat{a}_n - a - c_n)$ is stochastically compact with limit set $\{\ln Y^{(\lambda)} : 1 \le \lambda \le c\}$, where $\mathcal{L}(Y^{(\lambda)}) = \lambda^{-2/\alpha}\nu_2^{\lambda}$ for some semistable distribution ν_2 with index $\alpha/2$. If Y is stable with index α, then $2\ln n(\hat{a}_n - a - c_n) \Rightarrow \ln Y_2$ stable with index $\alpha/2$. If Y is normal, then (10.56) holds with $a = 1/2$. See [125] for details, and [4] for an application to hydrology.

10.5 SYMMETRIC K-TENSORS

In Section 10.2, we computed the asymptotic distribution of the sample covariance matrix. In this section, we generalize those arguments to obtain the asymptotics of certain symmetric k-tensors whose underlying distribution is regularly varying. We restrict our attention to regularly varying probability measures. The symmetric k-tensors considered here are also the vector analogue of the sample moments considered in Section 10.1.

Recall from Proposition 6.1.10 that the probability measure μ on $\mathbb{R}^d$ is regularly varying at infinity with exponent E if there exist $(A_n) \in \mathrm{RV}(-E)$ and ϕ $\mathcal{M}$-full such that

$$n(A_n\mu) \to \phi \quad \text{as } n \to \infty. \tag{10.57}$$

Let $a_1 < \cdots < a_p$ denote the real parts of the eigenvalues of E. Theorem 6.3.23 implies that if $\eta < 1/a_p$, then for all $\delta > 0$ there exist constants

$m, r_0 > 0$ such that

$$\frac{V_\eta(\lambda r, \theta)}{V_\eta(r, \theta)} \geq m\lambda^{-1/a_1+\eta-\delta} \tag{10.58}$$

for all $\|\theta\| = 1$, $\lambda \geq 1$ and $r \geq r_0$. Theorem 6.3.16 implies that if $\zeta > 1/a_1$, then for every $\delta > 0$ there exist constants $M, r_1 > 0$ such that

$$\frac{U_\zeta(\lambda r, \theta)}{U_\zeta(r, \theta)} \leq M\lambda^{-1/a_p+\zeta+\delta} \tag{10.59}$$

for all $\|\theta\| = 1$, $\lambda \geq 1$ and $r \geq r_1$. Finally, Corollary 6.3.9 implies that for $\zeta > 1/a_1$ and $\eta < 1/a_p$ there exist positive constants A, B, t_0 such that

$$A \leq \frac{t^{\zeta-\eta} V_\eta(t, \theta)}{U_\zeta(t, \theta)} \leq B \tag{10.60}$$

for all $\|\theta\| = 1$ and $t \geq t_0$.

Let $d \geq 1$ and $k \geq 2$ be natural numbers and let

$$\mathbb{V} = (\underbrace{\mathbb{R}^d \otimes \cdots \otimes \mathbb{R}^d}_{k\text{-times}})_s$$

denote the real vector space of symmetric k-tensors on $\mathbb{R}^d$. If we define $T : \mathbb{R}^d \to \mathbb{V}$; $T(x) = x \otimes \cdots \otimes x = \otimes^k x$, then $\mathbb{V}$ is the linear span of $T(\mathbb{R}^d)$. For $x, y \in \mathbb{R}^d$, we define $<T(x), T(y)> = <\otimes^k x, \otimes^k y> = <x, y>^k$. Extend by linearity to define an inner product $<\cdot,\cdot>$ on $\mathbb{V}$. Then $(\mathbb{V}, <\cdot,\cdot>)$ is an Euclidean vector space and $\|M\| = <M, M>^{1/2}$ is the corresponding norm. For a linear operator A on $\mathbb{R}^d$, let $L_A : \mathbb{V} \to \mathbb{V}$ be the linear extension of $L_A(\otimes^k x) = \otimes^k(Ax)$. For more details on tensor algebra, see, e.g., [45]. The following properties are easily verified:

Lemma 10.5.1. *Let A be an invertible operator on $\mathbb{R}^d$, $x \in \mathbb{R}^d$, $\lambda \in \mathbb{R}$ and $M, N \in \mathbb{V}$. Then*

(a) $L_{A^{-1}} = (L_A)^{-1}$

(b) $L_A(T(x)) = T(Ax)$

(c) $T(\lambda x) = \lambda^k T(x)$

(d) $<L_A(M), N> = <M, L_{A^*}(N)>$, *where A^* denotes the transpose of A*

Definition 10.5.2. Suppose that $X_1, X_2, \ldots, X_n$ are random vectors on $\mathbb{R}^d$. We will call the sum

$$M_n = \sum_{i=1}^{n} \otimes^k X_i$$

the symmetric k-tensor of $X_1, X_2, \ldots, X_n$.

Theorem 10.5.3. *Suppose that $X_1, X_2, \ldots$ are i.i.d. random vectors whose common distribution μ varies regularly at infinity with index E and that (10.57) holds. For any natural number $k \geq 2$ such that $1/(2k) < a_1$, there exist nonrandom $B_n \in \mathbb{V}$ such that*

$$L_{A_n}(M_n) - B_n \Rightarrow W, \tag{10.61}$$

where W is infinitely divisible on $\mathbb{V}$ with Lévy representation $[C, 0, T\phi]$ for some $C \in \mathbb{V}$ and ϕ is the limit measure in (10.57).

Proof. The proof is similar to Theorem 10.2.3, but here we only consider μ regularly varying. Since $\|T(x)\|^2 = <T(x), T(x)> = <x, x>^k = \|x\|^{2k}$ we have

$$\begin{aligned}\int_{\mathbb{V}\setminus\{0\}} \min(\|A\|^2, 1) T\phi\{dA\} &= \int_{0<\|A\|\leq 1} \|A\|^2 T\phi\{dA\} \\ &\quad + T\phi\{A \in \mathbb{V} : \|A\| > 1\} \\ &= \int_{0<\|x\|\leq 1} \|x\|^{2k} \phi\{dx\} + \phi\{x : \|x\| > 1\}.\end{aligned}$$

where the last integral is finite by virtue of Lemma 6.3.11 since $2k > 1/a_1$, hence $T\phi$ is a Lévy measure on $\mathbb{V}$.

We show (10.61) by applying the standard convergence criteria for triangular arrays, Theorem 3.3.8. The first condition

$$nL_{A_n} T\mu \to T\phi \tag{10.62}$$

follows exactly as in Theorem 10.2.3. In order to verify the second condition, an application of Proposition 3.3.11 shows that it is enough to verify

$$\lim_{\varepsilon\to 0} \limsup_{n\to\infty} n \int_{\|A\|\leq\varepsilon} <A, B>^2 L_{A_n} T\mu\{dA\} = 0 \tag{10.63}$$

for all unit vectors $B \in \mathbb{V}$.

Suppose first that $B = T(b) = \otimes^k b$ for some unit vector $b \in \mathbb{R}^d$. If we write $A_n^* b = r_n \theta_n$ for some $r_n > 0$ and $\|\theta_n\| = 1$, we get as in the proof of Theorem 10.2.3 that

$$n \int_{\|A\|\leq\varepsilon} <A, B>^2 L_{A_n} T\mu\{dA\} \leq n r_n^{2k} U_{2k}(r_n^{-1} \varepsilon^{1/k}, \theta_n).$$

Note that since $(A_n) \in \mathrm{RV}(-E)$ and $a_1 > 0$, we have $r_n \to 0$ as $n \to \infty$. Choose $\delta > 0$ such that $2 - (1/k)((1/a_1) + \delta) > 0$. Then, by (10.58) there exists a constant $m > 0$ such that for any fixed $0 < \varepsilon < 1$ there exists a $n_0 \geq 1$ such that for all $n \geq n_0$

$$\begin{aligned}\frac{V_0(\varepsilon^{1/k} r_n^{-1}, \theta_n)}{V_0(r_n^{-1}, \theta_n)} &= \left(\frac{V_0(\varepsilon^{-1/k}(\varepsilon^{1/k} r_n^{-1}), \theta_n)}{V_0(\varepsilon^{1/k} r_n^{-1}, \theta_n)} \right)^{-1} \\ &\leq \left(m \left(\varepsilon^{-1/k}\right)^{-1/a_1 - \delta} \right)^{-1} \\ &= \frac{1}{m} \varepsilon^{(1/k)(-1/a_1 - \delta)}.\end{aligned}$$

Note that n_0 does depend on ε in general.

Choose $n_1 \geq 1$ such that $\varepsilon^{1/k} r_n^{-1} \geq t_0$ for all $n \geq n_1$, where t_0 is as in (10.60). Then for all $n \geq \max(n_0, n_1)$ we get from (10.60)

$$\begin{aligned}&n r_n^{2k} U_{2k}(\varepsilon^{1/k} r_n^{-1}, \theta_n) \\ =&\varepsilon^2 \frac{U_{2k}(\varepsilon^{1/k} r_n^{-1}, \theta_n)}{(\varepsilon^{1/k} r_n^{-1})^{2k} V_0(\varepsilon^{1/k} r_n^{-1}, \theta_n)} \frac{V_0(\varepsilon^{1/k} r_n^{-1}, \theta_n)}{V_0(r_n^{-1}, \theta_n)} n V_0(r_n^{-1}, \theta_n) \\ \leq&\varepsilon^2 \frac{1}{A} \frac{1}{m} \varepsilon^{-\frac{1}{k}(\frac{1}{a_1} + \delta)} n(A_n \mu)\{x : |<x, b>| > 1\}\end{aligned}$$

In view of (10.57) and the Portmanteau Theorem 1.2.2, we get

$$\limsup_{n \to \infty} n(A_n \mu)\{|<x, b>| > 1\} \leq \phi\{x : |<x, b>| \geq 1\} = D < \infty. \tag{10.64}$$

Hence we have shown that for $0 < \varepsilon < 1$

$$\limsup_{n \to \infty} n \int_{\|A\| \leq \varepsilon} <A, B>^2 L_{A_n} T\mu\{dA\} \leq \frac{D}{Am} \varepsilon^{2 - \frac{1}{k}(\frac{1}{a_1} + \delta)}.$$

This implies (10.63) by the choice of δ for $B = T(b)$.

For the general case, choose unit vectors $\{b_k : k = 1, \ldots, m\} \subset \mathbb{R}^d$ such that $\{T(b_k) : k = 1, \ldots, m\}$ spans $\mathbb{V}$. Now, for any $B \in \mathbb{V}$ we can write $B = \sum_{k=1}^m c_k T(b_k)$ and then by the Cauchy-Schwarz inequality

$$\begin{aligned}&n \int_{\|A\| \leq \varepsilon} <A, B>^2 L_{A_n} T\mu\{dA\} \\ &\qquad = \sum_{k,\ell=1}^m c_k c_\ell \cdot n \int_{\|A\| \leq \varepsilon} <A, T(b_k)><A, T(b_\ell)> L_{A_n} T\mu\{dA\} \\ &\qquad \leq \sum_{k,\ell=1}^m |c_k c_\ell| \left(n \int_{\|A\| \leq \varepsilon} <A, T(b_k)>^2 L_{A_n} T\mu\{dA\} \right)^{1/2} \\ &\qquad\qquad \times \left(n \int_{\|A\| \leq \varepsilon} <A, T(b_\ell)>^2 L_{A_n} T\mu\{dA\} \right)^{1/2}.\end{aligned}$$

Since we have already shown (10.63) for $B = T(b)$ the general case of an arbitrary $B \in \mathbb{V}$ follows from the special case considered above. □

In certain cases, we can say more about the possible choice of the centering tensors B_n in (10.61).

Corollary 10.5.4. *Under the assumptions of Theorem 10.5.3, we can take*

$$B_n = \begin{cases} 0 & \text{if } \frac{1}{k} < a_1 \\ nL_{A_n}(E(\otimes^k X)) & \text{if } a_p < \frac{1}{k} \end{cases}$$

where X is a $\mathbb{R}^d$-valued random vector with distribution μ.

Proof. In view of the standard convergence criteria for triangular arrays of random vectors (Theorem 3.3.8), we can take

$$B_n = n \int_{\|A\| \le R} A\; L_{A_n} T\mu\{dA\} \tag{10.65}$$

where $R > 0$ is arbitrary as long as $\{A \in \mathbb{V} : \|A\| = R\}$ has $T\phi$-measure zero, which holds for all but countable many $R > 0$.

Assume first that $1/k < a_1$. Let $b \in \mathbb{R}^d$ be a unit vector and set $B = T(b)$. Then as in the proof of Corollary 10.2.6 it follows from (10.65) that

$$|\langle B_n, T(b)\rangle| \le n r_n^k U_k(r_n^{-1} R^{1/k}, \theta_n)$$

where $A_n^* b = r_n \theta_n$ for some $r_n > 0$ and $\|\theta_n\| = 1$.

Choose $\delta > 0$ such that $1 - \frac{1}{k}(\frac{1}{a_1} + \delta) > 0$. Then, by (10.58) there exists a constant $m > 0$ such that for a fixed $0 < R < 1$ there exists a $n_0 \ge 1$ such that for all $n \ge n_0$

$$\frac{V_0(R^{1/k} r_n^{-1}, \theta_n)}{V_0(r_n^{-1}, \theta_n)} = \left(\frac{V_0(R^{-1/k}(R^{1/k} r_n^{-1}), \theta_n)}{V_0(R^{1/k} r_n^{-1}, \theta_n)} \right)^{-1} \le \frac{1}{m} R^{-\frac{1}{k}(\frac{1}{a_1}+\delta)}.$$

Moreover, if we choose a $n_1 \ge 1$ such that $R^{1/k} r_n^{-1} \ge t_0$ for all $n \ge n_1$, where t_0 is as in (10.60), it follows that for $n \ge \max(n_0, n_1)$

$$\begin{aligned} & n r_n^k U_k(r_n^{-1} R^{1/k}, \theta_n) \\ =& R \frac{U_k(r_n^{-1} R^{1/k}, \theta_n)}{(r_n^{-1} R^{1/k})^k V_0(r_n^{-1} R^{1/k}, \theta_n)} \frac{V_0(r_n^{-1} R^{1/k}, \theta_n)}{V_0(r_n^{-1}, \theta_n)} n V_0(r_n^{-1}, \theta_n) \\ \le& R \frac{1}{A} \frac{1}{m} R^{-\frac{1}{k}(\frac{1}{a_1}+\delta)} n(A_n\mu)\{x : |\langle x, b\rangle| > 1\} \\ \le& C R^{1-\frac{1}{k}(\frac{1}{a_1}+\delta)} \end{aligned}$$

for some constant $C > 0$. Hence we have shown that there exist positive constants C, δ_0 such that for $0 < R < 1$

$$\limsup_{n\to\infty} |\langle B_n, T(b)\rangle| \le C R^{\delta_0}. \tag{10.66}$$

Now, let $B \in \mathbb{V}$ be arbitrary and write $B = \sum_{k=1}^{m} c_k T(b_k)$ for some unit vectors $b_k \in \mathbb{R}^d$. Then for $0 < R < 1$ we get from (10.66)

$$\limsup_{n\to\infty} |<B_n, B>| \leq \sum_{k=1}^{m} |c_k| \limsup_{n\to\infty} |<B_n, T(b_k)>| \leq KR^{\delta_0}$$

for some constant $K > 0$ and δ_0 as above. Since $\delta_0 > 0$ we see that by taking R arbitrary small we can choose $B_n = 0$.

In the other case, we have $a_1 < \cdots < a_p < 1/k$. Then Theorem 6.3.23 implies that $E<X, \theta>^k$ is finite for every $\theta \in \mathbb{R}^d \setminus \{0\}$. Since $E<\otimes^k X, \otimes^k \theta> = E<X, \theta>^k$ it follows that $E(\otimes^k X)$ exists in this case.

Use (10.65) to write $B_n = nL_{A_n}(E(\otimes^k X)) - I_n$ for $R \geq 1$, where

$$I_n = n \int_{||A||>R} A\, L_{A_n} T\mu\{dA\}.$$

Then, for a unit vector $b \in \mathbb{R}^d$ we have

$$\begin{aligned} |<I_n, T(b)>| &\leq n \int_{||A||>R} |<A, T(b)>|\, L_{A_n} T\mu\{dA\} \\ &= n \int_{|<A,T(b)>|>R} |<A, T(b)>|\, L_{A_n} T\mu\{dA\} \\ &\quad + n \int_{||A||>R \text{ and } |<A,T(b)>|\leq R} |<A, T(b)>|\, L_{A_n} T\mu\{dA\}. \end{aligned} \tag{10.67}$$

Now, if we write $A_n^* b = r_n \theta_n$ with $r_n > 0$ and $||\theta_n|| = 1$ as before, then for any $\zeta > 1/a_1$ we obtain

$$\begin{aligned} & n \int_{|<A,T(b)>|>R} |<A, T(b)>|\, L_{A_n} T\mu\{dA\} \\ =& nr_n^k V_k(R^{1/k} r_n^{-1}, \theta_n) \\ =& R^{1-\zeta/k} \frac{(R^{1/k} r_n^{-1})^{\zeta-k} V_k(R^{1/k} r_n^{-1}, \theta_n)}{U_\zeta(R^{1/k} r_n^{-1}, \theta_n)} \cdot \frac{U_\zeta(R^{1/k} r_n^{-1}, \theta_n)}{U_\zeta(r_n^{-1}, \theta_n)} \\ & \times \frac{U_\zeta(r_n^{-1}, \theta_n)}{(r_n^{-1})^\zeta V_0(r_n^{-1}, \theta_n)} \cdot nV_0(r_n^{-1}, \theta_n). \end{aligned}$$

For $R \geq 1$, an application of (10.58)–(10.60) together with (10.64) yields

$$\limsup_{n\to\infty} n \int_{|<A,T(b)>|>R} |<A, T(b)>|\, L_{A_n} T\mu\{dA\} \leq KR^{-\delta_0} \tag{10.68}$$

for some constant $K > 0$ and $\delta_0 = \frac{1}{ka_p} - 1 - \frac{\delta}{k} > 0$ if $\delta > 0$ is chosen small enough.

To estimate the other integral on the right-hand side of (10.67), choose $\{b_l\}$ an orthonormal basis of $\mathbb{R}^d$ and note that if $\|y\|^k > R$ then $|<y, b_l>|^k > R/d^{k/2}$ for some $l = 1, \ldots, d$. Hence

$$\begin{aligned} n & \int_{\|A\|>R \text{ and } |<A,T(b)>|\leq R} |<A, T(b)>| \, L_{A_n} T\mu\{dA\} \\ \leq & Rn \int_{\|A\|>R} L_{A_n} T\mu\{dA\} \\ = & Rn \int_{\|A_n x\|^k > R} \mu\{dx\} \\ \leq & R \sum_{l=1}^{d} n \int_{|<A_n x, b_l>|^k > R/d^{k/2}} \mu\{dx\} \\ = & \sum_{l=1}^{d} RnV_0((R^{1/k}/\sqrt{d}) r_{nl}^{-1}, \theta_{nl}), \end{aligned}$$

where $A_n^* b_l = r_{nl}\theta_{nl}$ with $r_{nl} > 0$ and $\|\theta_{nl}\| = 1$.

Apply (10.58)–(10.60) and (10.64) for some $\zeta > 1/a_1$, to see that each individual summand in the sum above can be bounded for all large n by

$$\begin{aligned} & RnV_0((R^{1/k}/\sqrt{d}) r_{nl}^{-1}, \theta_{nl}) \\ = & d^{\zeta/2} R^{1-\zeta/k} \frac{((R^{1/k}/\sqrt{d}) r_{nl}^{-1})^{\zeta} V_0((R^{1/k}/\sqrt{d}) r_{nl}^{-1}, \theta_{nl})}{U_\zeta((R^{1/k}/\sqrt{d}) r_{nl}^{-1}, \theta_{nl})} \\ & \times \frac{U_\zeta((R^{1/k}/\sqrt{d}) r_{nl}^{-1}, \theta_{nl})}{U_\zeta(r_{nl}^{-1}, \theta_{nl})} \cdot \frac{U_\zeta(r_{nl}^{-1}, \theta_{nl})}{(r_{nl}^{-1})^{\zeta} V_0(r_{nl}^{-1}, \theta_{nl})} \cdot nV_0(r_{nl}^{-1}, \theta_{nl}) \\ \leq & KR^{1 - \frac{1}{ka_p} + \frac{\delta}{k}} = KR^{-\delta_0} \end{aligned}$$

for some positive constant K, where $\delta_0 = \frac{1}{ka_p} - 1 - \frac{\delta}{k} > 0$ as above. Hence, in view of (10.67) and (10.68) we have shown that for some positive real constant C_1 and all $R \geq 1$

$$\limsup_{n \to \infty} |<I_n, T(b)>| \leq C_1 R^{-\delta_0}$$

for every unit vector $b \in \mathbb{R}^d$, where $\delta_0 > 0$.

This implies as before that

$$\limsup_{n \to \infty} |<I_n, B>| \leq C_2 R^{-\delta_0}$$

for every unit vector $B \in \mathbb{V}$ and some constant $C_2 > 0$. By taking $R > 1$ arbitrary large, we see that the convergence in (10.61) still holds when $B_n = nL_{A_n}(E(\otimes^k X))$. The proof is now complete. □

After calculating the asymptotic distribution of the normalized sum of i.i.d. symmetric k-tensors $\otimes^k X_i$ in Theorem 10.5.3, we now investigate the structure of the limit W in (10.61).

Proposition 10.5.5. *Under the assumptions of Theorem 10.5.3, the limit W with Lévy representation $[C, 0, T\phi]$ is a full operator stable law on a certain subspace of $\mathbb{U}$ of $\mathbb{V}$.*

Proof. The Lévy measure $T\phi$ of W satisfies

$$L_{t^E}(T\phi) = T(t^E\phi) = t \cdot T\phi \quad \text{for all } t > 0.$$

Since W has no normal component, Lemma 7.1.6 together with Theorem 7.2.1 imply that W is a full (L_{t^E}) operator stable law on the subspace $\mathbb{U} = \text{span}(\text{supp}(\Psi))$ where Ψ denotes the distribution of W. Note that (L_{t^E}) is a one-parameter subgroup of $\text{GL}(\mathbb{V})$. □

Theorem 10.5.6. *Let W be as in Theorem 10.5.3. Then for all $\theta \in \mathbb{R}^d \setminus \{0\}$, the random variable $<W, \otimes^k\theta>$ has a density with respect to Lebesgue measure on $\mathbb{R}$.*

Proof. Proposition 10.5.5 shows that W is full and operator semistable on $\mathbb{U}$, where $\mathbb{U} = \text{span}(\text{supp}(T\phi))$. Then Theorem 7.1.15 shows that W has a density on $\mathbb{U}$. It follows as in Lemma 10.4.2 that $T\theta \in \mathbb{U}$, and then $<W, T\theta>$ has a density by the Fubini Theorem. □

Remark 10.5.7. Symmetric k-tensors can also be used for tail and moment estimation. If $k = 2$, then the symmetric 2-tensor M_n is the sample covariance matrix. In Section 10.4, we saw that the eigenvalues of M_n can be used to estimate tail thickness in this case. This assumes that $a_1 > 1/2$ so that μ belongs to some nonnormal GDOSA. If μ is regularly varying with $a_1 \leq 1/2$, choose an *even* natural number $k \geq 2$ with $a_1 > 1/k$ and define the random variable

$$\lambda_n = \max_{\|\theta\|=1} <M_n, \otimes^k\theta>.$$

If we let

$$\hat{a}_n = \frac{\log \lambda_n}{k \log n}$$

then an argument similar to Proposition 10.4.3 shows that $\hat{a}_n \to a_p$ in probability. This gives a family of tail estimators. In some cases, it may be useful to compare the results of estimators for different k. Also, since we require $1/(2k) < a_1$ this allows tail estimation for random vectors whose tails are heavy (i.e., they diminish at a power law rate) but too light to be attracted to a nonnormal limit. It may be possible to estimate the spectral index in this way, but this is an open problem.

Remark 10.5.8. It is also possible to obtain a stochastic compactness result for symmetric k-tensors. For any integer $k \geq 1$ such that $1/(2k) < a_1$, an argument similar to Theorem 8.3.20 shows that for all nonzero $\theta \in \mathbb{R}^d$ there exists a sequence (r_n) of positive real numbers tending to zero and a sequence of shifts (s_n) in $\mathbb{R}$ such that

$$\left(r_n^k \sum_{i=1}^{n} <\otimes^k X_i, \otimes^k \theta> - s_n\right)_n \tag{10.69}$$

is stochastically compact with limit set contained in the set

$$\left\{<W, \otimes^k \theta_0> : \|\theta_0\| = 1\right\}$$

where W is the limit in (10.61). The mapping $x \mapsto <\otimes^k x, \otimes^k \theta> = <x, \theta>^k$ is a homogeneous polynomial of degree k in the components of $x \in \mathbb{R}^d$. Hence (10.69) is an i.i.d. sum of certain homogeneous polynomials in the components of the X_i.

10.6 TIME SERIES ANALYSIS

In this section, we establish the basic asymptotic theory for moving averages of random vectors with heavy tails. Moving averages of random vectors with regularly varying tail probabilities are asymptotically operator stable. The sample autocovariance matrix formed from these moving averages is also asymptotically operator stable, as a random element of the vector space of $d \times d$ symmetric matrices.

Assume that $Z, Z_1, Z_2, Z_3, \ldots$ are i.i.d. random vectors on $\mathbb{R}^d$ with common distribution μ. Recall from Proposition 6.1.10 that μ is regularly varying at infinity with exponent E if there exist linear operators $(A_n) \in \mathrm{RV}(-E)$ such that

$$nA_n\mu \to \phi \tag{10.70}$$

where ϕ is $\mathcal{M}$-full. Recall from Definition 3.3.24 that $Z \in \mathrm{GDOA}(Y)$ if

$$A_n(Z_1 + \cdots + Z_n - nb_n) \Rightarrow Y \tag{10.71}$$

for some linear operators A_n and nonrandom centering vectors b_n. Theorem 8.2.10 shows that (10.71) holds with Y nonnormal if and only if μ is regularly varying with exponent E, where the real parts of the eigenvalues of E satisfy $1/2 < a_1 < \cdots < a_p$. In this case, Corollary 8.2.15 shows that $E\|X\|^\rho$ exists if $\rho < 1/a_p$ and is infinite if $\rho > 1/a_p$.

Suppose that (10.70) holds and define the moving average process

$$X_t = \sum_{j=-\infty}^{\infty} C_j Z_{t-j} \tag{10.72}$$

where C_j are $d \times d$ real matrices such that for each j either $C_j = 0$ or else C_j^{-1} exists and $A_n C_j = C_j A_n$ for all n. Our first result shows that the moving average (10.72) is well defined as long as

$$\sum_{j=-\infty}^{\infty} \|C_j\|^{\delta} < \infty \tag{10.73}$$

for some $\delta < 1/a_p$ with $\delta \leq 1$. From now on, we will assume that this is the case, so that the moving average (10.72) is well defined.

Lemma 10.6.1. *Suppose that* $E\|Z\|^{\alpha} < \infty \ \forall \alpha < \alpha_0$ *and that* (10.73) *holds for some* $\delta < \alpha_0$, $\delta \leq 1$. *Then* (10.72) *exists almost surely.*

Proof. First, suppose $\alpha_0 > 1$ so that $E\|Z\| < \infty$. Then $\sum \|C_j\|^{\delta} < \infty$ implies that $\sum \|C_j\| < \infty$ because $\|C_j\|^{\delta} \to 0$ since the series converges, so $\|C_j\| < 1$ for all large n, and for such n we have $\|C_j\| \leq \|C_j\|^{\delta}$. Then we have $\|X_t\| = \|\sum_j C_j Z_{t-j}\| \leq \sum \|C_j Z_{t-j}\| \leq \sum \|C_j\| \cdot \|Z_{t-j}\|$ so that $E\|X_t\| \leq E \sum \|C_j\| \|Z_{t-j}\| = E\|Z\| \sum \|C_j\| < \infty$ by monotone convergence and then X_t exists almost surely.

If $\alpha_0 \leq 1$, then $\sum \|C_j\|^{\delta} < \infty$ for some $\delta < \alpha_0$ and $E\|Z\|^{\delta} < \infty$ as well. Also $|x + y|^{\delta} \leq |x|^{\delta} + |y|^{\delta} \ \forall x, y \geq 0$ so

$$\begin{aligned} E\|X_t\|^{\delta} &= E\Big[\Big\| \sum_j C_j Z_{t-j} \Big\|^{\delta}\Big] \\ &\leq E\Big[\Big(\sum_j \|C_j\| \|Z_{t-j}\|\Big)^{\delta}\Big] \\ &\leq E\Big[\Big(\sum_j \|C_j\|^{\delta} \|Z_{t-j}\|^{\delta}\Big)\Big] \\ &= \sum_j \|C_j\|^{\delta} E\|Z\|^{\delta} < \infty \end{aligned}$$

so as before X_t converges absolutely almost surely. □

Theorem 10.6.2. *Suppose that* X_t *is a moving average defined by* (10.72) *where* $\{Z_n\}$ *are i.i.d. with common distribution* μ *on* $\mathbb{R}^d$. *Suppose that* μ *varies regularly with exponent* E, *where every eigenvalue of* E *has real part exceeding 1/2. Then*

$$A_n(X_1 + \cdots + X_n - n a_n) \Rightarrow U \tag{10.74}$$

where $a_n \in \mathbb{R}^d$, $A_n \in \mathrm{GL}(\mathbb{R}^d)$ *and* U *is full-dimensional and operator stable.*

Recall the spectral decomposition of $\mathbb{R}^d$ with respect to X from Definition 8.3.11. Henceforth we will assume that Y and (A_n) are as in Theorem 8.3.7, so that Y is spectrally compatible with X and (A_n) is regularly varying with index E and respects the direct sum decomposition $\mathbb{R}^d = V_1 \oplus \cdots \oplus V_p$ into mutually orthogonal subspaces. Then

(a) For any nonzero vector $\theta \in V_i$ the marginal absolute moment $E|<Z,\theta>|^\rho$ exists for $\rho < 1/a_i$, and if $a_i > 1/2$ then this moment is infinite for $\rho > 1/a_i$

(b) $\pi_i A_n = A_n \pi_i$ and $\pi_i E = E \pi_i$ where π_i is orthogonal projection onto V_i

(c) every eigenvalue of $\pi_i E$ has real part equal to a_i

(d) for any nonzero vector $x \in \mathbb{R}^d$, we have $\log \|\pi_i A_n x\| / \log n \to -a_i$ (uniformly on compact subsets) and $\log \|\pi_i A_n\| / \log n \to -a_i$

The following lemma shows that the moving average parameter matrices C_j also respect this spectral decomposition.

Lemma 10.6.3. *Suppose that $(A_n) \in \mathrm{RV}(-E)$ is spectrally compatible with $-E$, and that $V_1, \ldots, V_p$ is the spectral decomposition of $\mathbb{R}^d$ with respect to E. If C_j is invertible and $C_j A_n = A_n C_j$ for all n, then $C_j V_i = V_i$ for all $i = 1, \ldots, p$.*

Proof. Let $L_i = V_i \oplus \cdots \oplus V_p$ and $\bar{L}_i = V_1 \oplus \cdots \oplus V_i$. Proposition 4.3.14 shows that for any $x \in L_i \setminus L_{i+1}$ we have $\log \|A_n x\| / \log n \to -a_i$, uniformly on compact subsets of $x \in L_i \setminus L_{i+1}$, and similarly for any $x \in \bar{L}_i \setminus \bar{L}_{i-1}$ we have $\log \|(A_n^*)^{-1} x\| / \log n \to a_i$ (uniformly on compact sets). Since C_j is invertible there exist $c_j > 0$ and $b_j < \infty$ such that $c_j \|x\| \le \|C_j x\| \le b_j \|x\|$ for all x. Then $\log \|C_j A_n x\| / \log n \to -a_i$ if and only if $x \in L_i \setminus L_{i+1}$, and $\log \|A_n C_j x\| / \log n \to -a_i$ if and only if $C_j x \in L_i \setminus L_{i+1}$. Since $A_n C_j = C_j A_n$ this implies that $C_j^{-1}(L_i \setminus L_{i+1}) = L_i \setminus L_{i+1}$, and so $C_j(L_i \setminus L_{i+1}) = L_i \setminus L_{i+1}$. Similarly, we have $\log \|(C_j^*)^{-1}(A_n^*)^{-1} x\| / \log n \to a_i$ if and only if $x \in \bar{L}_i \setminus \bar{L}_{i-1}$ as well as $\log \|(A_n^*)^{-1}(C_j^*)^{-1} x\| / \log n \to a_i$ if and only if $(C_j^*)^{-1} x \in \bar{L}_i \setminus \bar{L}_{i-1}$. Then $(C_j^*)^{-1}(\bar{L}_i \setminus \bar{L}_{i-1}) = \bar{L}_i \setminus \bar{L}_{i-1}$ and since the subspaces V_i are mutually orthogonal, it follows that $C_j(\bar{L}_i \setminus \bar{L}_{i-1}) = \bar{L}_i \setminus \bar{L}_{i-1}$. Then $C_j(L_i \cap \bar{L}_i) = L_i \cap \bar{L}_i$ and $L_i \cap \bar{L}_i = V_i$. □

Lemma 10.6.4. *Suppose that Z is a random vector on $\mathbb{R}^d$ with distribution $\mu \in \mathrm{RVM}_\infty(E)$, and that (10.70) holds with $(A_n) \in \mathrm{RV}(-E)$ spectrally compatible with E. Let $a_1 < \cdots < a_p$ be the real spectrum of E and $V_1, \ldots, V_p$ the spectral decomposition of $\mathbb{R}^d$ with respect to E. Then for any $0 < \eta < 1/a_i < \zeta$ there exists n_0 such that*

(i) $nE|<A_n Z, \theta>|^\zeta I(|<A_n Z, \theta>| \le 1) < K_1$

(ii) $nE|<A_n Z, \theta>|^\zeta I(\|A_n Z\| \le 1) < K_2$

(iii) $nE|<A_n Z, \theta>|^\eta I(|<A_n Z, \theta>| > 1) < K_3$

(iv) $nE|<A_n Z, \theta>|^\eta I(\|A_n Z\| > 1) < K_4$

for all $n \geq n_0$ *and all* $\|\theta\| = 1$ *in* V_i, *where* $K_1, \ldots, K_4$ *are positive reals.*

Proof. Suppose we are given an arbitrary sequence of unit vectors θ_n in V_i, and write $r_n x_n = A_n^* \theta_n$ where $r_n > 0$ and $\|x_n\| = 1$. Then we have

$$\begin{aligned} nE|<A_n Z, \theta_n>|^\zeta I(|<A_n Z, \theta_n>| \leq 1) &= nE|<Z, A_n^* \theta_n>|^\zeta I(|<Z, A_n^* \theta_n>| \leq 1) \\ &= nE|<Z, r_n x_n>|^\zeta I(|<Z, r_n x_n>| \leq 1) \\ &= n r_n^\zeta E|<Z, x_n>|^\zeta I(|<Z, x_n>| \leq r_n^{-1}) \\ &= n r_n^\zeta U_\zeta(r_n^{-1}, x_n) \end{aligned}$$

Since $\zeta > 1/a_i$, Theorem 6.4.15 shows that both $U_\zeta(r, \theta)$ and $V_0(r, \theta)$ are uniformly R–O varying on compact subsets of $\theta \neq 0$ in V_i. Then Corollary 6.3.9 shows that for some m, M, t_0 we have

$$m \leq \frac{t^\zeta V_0(t, \theta)}{U_\zeta(t, \theta)} \leq M \tag{10.75}$$

for all $\|\theta\| = 1$ in V_i and all $t \geq t_0$. Corollary 8.2.12 shows that $\|A_n\| \to 0$, hence for some n_0 we have $\|A_n \theta\| \leq t_0^{-1}$ for all $\|\theta\| = 1$ and all $n \geq n_0$. Then for all $n \geq n_0$ we have

$$\begin{aligned} n r_n^\zeta U_\zeta(r_n^{-1}, x_n) &\leq m^{-1} n V_0(r_n^{-1}, x_n) \\ &= m^{-1} n P(|<Z, x_n>| > r_n^{-1}) \\ &= m^{-1} n P(|<Z, r_n x_n>| > 1) \\ &= m^{-1} n P(|<Z, A_n^* \theta_n>| > 1) \\ &= m^{-1} n P(|<A_n Z, \theta_n>| > 1) \\ &\leq m^{-1} n P(\|A_n Z\| > 1) \end{aligned}$$

as $n \to \infty$, and this upper bound holds independent of our choice of the sequence θ_n. By (10.70), we have $nP(\|A_n Z\| > 1) = nA_n \mu\{z : \|z\| > 1\} \to \phi\{z : \|z\| > 1\}$ (if $\{z : \|z\| > 1\}$ is not a continuity set of ϕ, then we can use the upper bound $nP(\|A_n Z\| > r)$ instead, where $0 < r < 1$). Then the sequence of real numbers $nP(\|A_n Z\| > 1)$ is bounded above by some $K > 0$, and (i) holds with $K_1 = m^{-1} K$. Since

$$nE|<A_n Z, \theta_n>|^\zeta I(\|A_n Z\| \leq 1) \leq nE|<A_n Z, \theta_n>|^\zeta I(|<A_n Z, \theta_n>| \leq 1)$$

we immediately obtain (ii) with $K_2 = K_1$. Now write

$$\begin{aligned} nE|<A_n Z, \theta_n>|^\eta I(|<A_n Z, \theta_n>| > 1) &= nE|<Z, A_n^* \theta_n>|^\eta I(|<Z, A_n^* \theta_n>| > 1) \\ &= nE|<Z, r_n x_n>|^\eta I(|<Z, r_n x_n>| > 1) \\ &= n r_n^\eta E|<Z, x_n>|^\eta I(|<Z, x_n>| > r_n^{-1}) \\ &= n r_n^\eta V_\eta(r_n^{-1}, x_n) \end{aligned}$$

Since $\eta < 1/a_i$, Theorem 6.4.15 shows that $V_\eta(r,\theta)$ is also uniformly R–O varying on compact subsets of $\theta \neq 0$ in V_i. Then Corollary 6.3.9 shows that for some m', M', t_0 we have

$$m' \leq \frac{t^{\zeta-\eta} V_\eta(t,\theta)}{U_\zeta(t,\theta)} \leq M' \tag{10.76}$$

for all $\|\theta\| = 1$ in V_i and all $t \geq t_0$. By choosing t_0 large enough so that both (10.75) and (10.76) hold whenever $t \geq t_0$ and $\|\theta\| = 1$ in V_i, and then choosing $n_0 = n_0(t_0)$ such that $\|A_n\theta\| \leq t_0^{-1}$ for all $\|\theta\| = 1$ whenever $n \geq n_0$, we have for all $n \geq n_0$ that

$$r_n^\eta V_\eta(r_n^{-1}, x_n) \leq M' n r_n^\zeta U_\zeta(r_n^{-1}, x_n)$$

and so (iii) holds with $K_3 = M'K_1$. Finally, write

$$\begin{aligned}
&nE|{<}A_nZ,\theta_n{>}|^\eta I(\|A_nZ\| > 1) \\
=&nE|{<}A_nZ,\theta_n{>}|^\eta I(|{<}A_nZ,\theta_n{>}| > 1) \\
&\quad + nE|{<}A_nZ,\theta_n{>}|^\eta I(\|A_nZ\| > 1 \text{ and } |{<}A_nZ,\theta_n{>}| \leq 1) \\
\leq&nE|{<}A_nZ,\theta_n{>}|^\eta I(|{<}A_nZ,\theta_n{>}| > 1) \\
&\quad + nP(\|A_nZ\| > 1)
\end{aligned}$$

so that (iv) holds with $K_4 = K_3 + K$. □

Proof of Theorem 10.6.2. Suppose that (10.71) holds and define

$$\begin{aligned}
X_t^{(m)} &= \sum_{|j|\leq m} C_j Z_{t-j} \\
a_n^{(m)} &= \sum_{|j|\leq m} C_j b_n \\
U^{(m)} &= \sum_{|j|\leq m} C_j Y
\end{aligned}$$

for $m \geq 1$, where Y is the limit in (10.71). From (10.71) it follows that

$$\left(A_n \sum_{t=1}^{n} (Z_{t-j} - b_n) : |j| \leq m\right) \Rightarrow \Big(Y, \ldots, Y\Big)$$

since we know that the convergence holds for the $j = 0$ component, and since the difference between this component and any other component (at most $2m$ terms) tends to zero in probability. Now, by the Continuous Mapping Theorem 1.2.7 we obtain

$$A_n(X_1^{(m)} + \cdots + X_n^{(m)} - na_n^{(m)}) \Rightarrow U^{(m)}$$

and since $U^{(m)} \to U = \sum_j C_j W$ almost surely it suffices to show (e.g., by Billingsley [17] Theorem 4.2) that

$$\lim_{m\to\infty} \limsup_{n\to\infty} P(\|T\| > \varepsilon) = 0$$

where

$$T = \sum_{|j|>m} \sum_{t=1}^{n} A_n C_j (Z_{t-j} - b_n).$$

For this, it suffices to show that

$$\lim_{m\to\infty} \limsup_{n\to\infty} P(|\langle T, x\rangle| > \varepsilon) = 0 \tag{10.77}$$

for any unit vector $x \in V_i$ where $1 \le i \le p$. Decompose $T = A + B - C$ where

$$\begin{aligned} A &= \sum_{|j|>m} \sum_{t=1}^{n} A_n C_j Z_{t-j} I(\|A_n Z_{t-j}\| \le 1) \\ B &= \sum_{|j|>m} \sum_{t=1}^{n} A_n C_j Z_{t-j} I(\|A_n Z_{t-j}\| > 1) \\ C &= \sum_{|j|>m} n A_n C_j b_n. \end{aligned}$$

First, suppose that $a_i > 1$, so that assertion (ii) of Lemma 10.6.4 holds with $\zeta = 1$. Write $C_j^* x = r_j x_j$ where $r_j > 0$ and $\|x_j\| = 1$. Note that $x_j \in V_i$ by Lemma 10.6.3. Then

$$\begin{aligned} P[|\langle A, x\rangle| > \varepsilon/3] &\le 3\varepsilon^{-1} E|\langle A, x\rangle| \\ &\le 3\varepsilon^{-1} \sum_{|j|>m} nE|\langle A_n C_j Z, x\rangle| I(\|A_n Z\| \le 1) \\ &= 3\varepsilon^{-1} \sum_{|j|>m} nE|\langle C_j A_n Z, x\rangle| I(\|A_n Z\| \le 1) \\ &= 3\varepsilon^{-1} \sum_{|j|>m} nE|\langle A_n Z, C_j^* x\rangle| I(\|A_n Z\| \le 1) \\ &= 3\varepsilon^{-1} \sum_{|j|>m} r_j nE|\langle A_n Z, x_j\rangle| I(\|A_n Z\| \le 1) \\ &\le 3\varepsilon^{-1} K_2 \sum_{|j|>m} \|C_j\| \end{aligned}$$

which tends to zero as $m \to \infty$ in view of the fact that (10.73) holds with $\delta < 1/a_p \le 1/a_i < 1$. By choosing $0 < \delta < 1/a_p$ so that (10.73) holds, we have that assertion (iv) of Lemma 10.6.4 holds with $\eta = \delta$. Note also that

$|x+y|^\delta \le |x|^\delta + |y|^\delta$ since $\delta < 1/a_i < 1$. Then

$$\begin{aligned}
P[|\langle B,x\rangle| > \varepsilon/3] &\le 3\varepsilon^{-\delta} E|\langle B,x\rangle|^\delta \\
&\le 3\varepsilon^{-\delta} \sum_{|j|>m} nE|\langle A_n C_j Z, x\rangle|^\delta I(\|A_n Z\| > 1) \\
&= 3\varepsilon^{-\delta} \sum_{|j|>m} nE|\langle C_j A_n Z, x\rangle|^\delta I(\|A_n Z\| > 1) \\
&= 3\varepsilon^{-\delta} \sum_{|j|>m} nE|\langle A_n Z, C_j^* x\rangle|^\delta I(\|A_n Z\| > 1) \\
&= 3\varepsilon^{-\delta} \sum_{|j|>m} r_j^\delta nE|\langle A_n Z, x_j\rangle|^\delta I(\|A_n Z\| > 1) \\
&\le 3\varepsilon^{-\delta} K_4 \sum_{|j|>m} \|C_j\|^\delta
\end{aligned}$$

which tends to zero as $m \to \infty$ in view of the fact that (10.73) holds. Theorem 3.3.8 shows that we can take $b_n = EZI(\|A_n Z\| \le 1)$ in (10.71). Then for all $n \ge n_0$ we have

$$\begin{aligned}
|\langle C,x\rangle| &\le \sum_{|j|>m} n|\langle A_n C_j b_n, x\rangle| \\
&= \sum_{|j|>m} n|\langle A_n C_j EZI(\|A_n Z\| \le 1), x\rangle| \\
&\le \sum_{|j|>m} nE|\langle A_n C_j Z, x\rangle| I(\|A_n Z\| \le 1) \\
&\le K_2 \sum_{|j|>m} \|C_j\|
\end{aligned}$$

which tends to zero as $m \to \infty$ as in the argument for A. Since

$$\begin{aligned}
P(|\langle T,x\rangle| > \varepsilon) &\le P(|\langle A,x\rangle| > \varepsilon/3) \\
&\quad + P(|\langle B,x\rangle| > \varepsilon/3) + P(|\langle C,x\rangle| > \varepsilon/3)
\end{aligned}$$

it follows that (10.77) holds when x is a unit vector in V_i and $a_i > 1$.

Suppose then that $1/2 < a_i \le 1$ and note that Lemma 10.6.4 (ii) holds with $\zeta = 2$. Since $C = EA$, we have by Chebyshev's Inequality 1.1.30 that

$$\begin{aligned} P[|<A-C,x>| > \varepsilon/2] &\le 2\varepsilon^{-2}\operatorname{Var}(<A,x>) \\ &= 2\varepsilon^{-2}\sum_{|j|>m} n\operatorname{Var}[<A_nC_jZ,x>I(\|A_nZ\|\le 1)] \\ &\le 2\varepsilon^{-2}\sum_{|j|>m} nE[<A_nC_jZ,x>^2 I(\|A_nZ\|\le 1)] \\ &= 2\varepsilon^{-2}\sum_{|j|>m} nr_j^2 E[<A_nZ,x_j>^2 I(\|A_nZ\|\le 1)] \\ &\le 2\varepsilon^{-2}K_2\sum_{|j|>m}\|C_j\|^2 \end{aligned}$$

which tends to zero as $m \to \infty$ in view of the fact that (10.73) holds with $\delta \le 1$. If $1/2 < a_i < 1$, then Lemma 10.6.4 (iv) holds with $\eta = 1$ and so

$$\begin{aligned} P[|<B,x>| > \varepsilon/2] &\le 2\varepsilon^{-1}E|<B,x>| \\ &\le 2\varepsilon^{-1}\sum_{|j|>m} nE|<A_nC_jZ,x>|I(\|A_nZ\| > 1) \\ &= 2\varepsilon^{-1}\sum_{|j|>m} r_j nE|<A_nZ,x_j>|I(\|A_nZ\| > 1) \\ &\le 2\varepsilon^{-1}K_4\sum_{|j|>m}\|C_j\| \end{aligned}$$

which tends to zero as $m \to \infty$ in view of the fact that (10.73) holds with $\delta \le 1$. Finally, if $a_i = 1$, then Lemma 10.6.4 (iv) holds with $\eta = \delta$ where we have $\delta < 1/a_p \le 1/a_i = 1$ such that (10.73) holds. Then argue as before that

$$\begin{aligned} P[|<B,x>| > \varepsilon/2] &\le 2\varepsilon^{-\delta}E|<B,x>|^\delta \\ &\le 2\varepsilon^{-\delta}\sum_{|j|>m} nE|<A_nC_jZ,x>|^\delta I(\|A_nZ\| > 1) \\ &= 2\varepsilon^{-\delta}\sum_{|j|>m} r_j^\delta nE|<A_nZ,x_j>|^\delta I(\|A_nZ\| > 1) \\ &\le 2\varepsilon^{-\delta}K_4\sum_{|j|>m}\|C_j\|^\delta \end{aligned}$$

which tends to zero as $m \to \infty$ in view of the fact that (10.73) holds. Since

$$P(|<T,x>| > \varepsilon) \le P(|<A-C,x>| > \varepsilon/2) + P(|<B,x>| > \varepsilon/2)$$

it follows that (10.77) holds when x is a unit vector in V_i and $a_i \le 1$. $\square$

Remark 10.6.5. The proof of Theorem 10.6.2 shows that we can take the centering constants $a_n = \sum_j C_j b_n$ in (10.74) where b_n are the centering constants from (10.71) above. The operator stable limit $U = \sum_j C_j Y$ in (10.74) is nonnormal with Lévy measure $\sum_j C_j \phi$.

Definition 10.6.6. The *sample covariance matrix at lag* h of the moving average X_t is defined by

$$\hat{\Gamma}_n(h) = \frac{1}{n}\sum_{t=1}^{n} X_t X'_{t+h}. \tag{10.78}$$

Suppose that μ varies regularly with exponent E, and every eigenvalue of E has real part exceeding 1/4. Then Theorem 10.2.3 and Lemma 10.2.7 show that

$$A_n\left(\sum_{i=1}^{n} Z_i Z'_i - B_n\right)A_n^* \Rightarrow W \tag{10.79}$$

where A_n is taken from (10.71), $B_n = EZZ'I(\|A_n Z\| \leq 1)$, and W is a nonnormal operator stable random element of the vector space $\mathcal{M}_s^d$ of symmetric $d \times d$ matrices with real entries.

Theorem 10.6.7. *Suppose that* $\hat{\Gamma}_n(h)$ *is the sample covariance matrix defined by* (10.79) *where* X_t *is the moving average* (10.72) *and* $\{Z_n\}$ *are i.i.d. with common distribution* μ *on* $\mathbb{R}^d$. *Suppose that* μ *varies regularly with exponent* E, *where every eigenvalue of* E *has real part exceeding 1/4. Assume that* $Ex'Z = 0$ *when the mean exists. Then for all* h *we have*

$$nA_n\Big[\hat{\Gamma}_n(h) - \sum_{j=-\infty}^{\infty} C_j B_n C_{j+h}^*\Big]A_n^* \Rightarrow \sum_{j=-\infty}^{\infty} C_j W C_{j+h}^*. \tag{10.80}$$

Proposition 10.6.8. *Assume that* $Ex'Z_n = 0$ *when the mean exists. Then*

$$A_n\left[n\hat{\Gamma}_n(h) - \sum_{t=1}^{n}\sum_{j=-\infty}^{\infty} C_j Z_{t-j} Z'_{t-j} C_{j+h}^*\right]A_n^* \xrightarrow{P} 0.$$

Proof. Let $V_1, \ldots, V_p$ be the spectral decomposition of $\mathbb{R}^d$ with respect to E and $a_1 < \cdots < a_p$ the real spectrum of E. Recall that $V_i \perp V_j$ and that π_i is the orthogonal projection onto V_i. We also assume as before that (10.73) holds for some $\delta < 1/a_p$ with $\delta \leq 1$. It will suffice to show that

$$x'A_n\Big[n\hat{\Gamma}_n(h) - \sum_{t=1}^{n}\sum_{j=-\infty}^{\infty} C_j Z_{t-j} Z'_{t-j} C_{j+h}^*\Big]A_n^* y \xrightarrow{P} 0 \tag{10.81}$$

for all unit vectors $x \in V_i$ and $y \in V_\ell$. Note that

$$\begin{aligned}
n\hat{\Gamma}_n(h) &= \sum_{t=1}^{n} X_t X_{t+h}' \\
&= \sum_{t=1}^{n}\Big(\sum_{j=-\infty}^{\infty} C_j Z_{t-j}\Big)\Big(\sum_{q=-\infty}^{\infty} C_q Z_{t+h-q}\Big)' \\
&= \sum_{t=1}^{n}\Big(\sum_{j=-\infty}^{\infty} C_j Z_{t-j}\Big)\Big(\sum_{k=-\infty}^{\infty} C_{k+h} Z_{t-k}\Big)' \qquad (\text{set } k = q - h) \\
&= \sum_{t=1}^{n}\sum_{j=-\infty}^{\infty}\sum_{k=-\infty}^{\infty} C_j Z_{t-j} Z_{t-k}' C_{k+h}^* \\
&= \sum_{t=1}^{n}\sum_{j=-\infty}^{\infty}\Big(C_j Z_{t-j} Z_{t-j}' C_{j+h}^* + \sum_{k\neq j} C_j Z_{t-j} Z_{t-k}' C_{k+h}^*\Big) \\
&= \sum_{t=1}^{n}\sum_{j=-\infty}^{\infty} C_j Z_{t-j} Z_{t-j}' C_{j+h}^* + \sum_{t=1}^{n}\sum_{j=-\infty}^{\infty}\sum_{k\neq j} C_j Z_{t-j} Z_{t-k}' C_{k+h}^*
\end{aligned}$$

so that (10.81) is equivalent to

$$x' A_n \Big[\sum_{t=1}^{n}\sum_{j=-\infty}^{\infty}\sum_{k\neq j} C_j Z_{t-j} Z_{t-k}' C_{k+h}^*\Big] A_n^* y \xrightarrow{P} 0. \tag{10.82}$$

We now have to consider several cases separately.

Case I: Suppose that $x \in V_i$ and $y \in V_\ell$ are unit vectors with $a_i + a_\ell > 1$. Choose $\delta > 0$ such that $\delta > 1/(a_i + a_\ell)$, $\delta < 1/a_i$, $\delta < 1/a_\ell$, and $\delta \le 1$. Write $C_j^* A_n^* x = r\theta$ where $r > 0$ and $\|\theta\| = 1$. Then $r = \|C_j^* A_n^* x\| = \|C_j^* A_n^* \pi_i x\| = \|C_j^* \pi_i A_n^* x\| \le \|C_j\|\|\pi_i A_n\|$ by Proposition 2.1.3 (c) and Remark 2.1.5 and so

$$\begin{aligned}
E|x' A_n C_j Z_1|^\delta &= E|(C_j^* A_n^* x)' Z_1|^\delta \\
&= E|r\theta' Z_1|^\delta \\
&\le \|\pi_i A_n\|^\delta \|C_j\|^\delta E|\langle Z_1, \theta\rangle|^\delta \\
&= \|\pi_i A_n\|^\delta \|C_j\|^\delta E|\langle Z_1, \pi_i\theta\rangle|^\delta \\
&= \|\pi_i A_n\|^\delta \|C_j\|^\delta E|\langle \pi_i Z_1, \theta\rangle|^\delta \\
&\le \|\pi_i A_n\|^\delta \|C_j\|^\delta E\|\pi_i Z_1\|^\delta
\end{aligned}$$

and likewise

$$E|Z_2' C_{k+h}^* A_n^* y|^\delta \le \|\pi_\ell A_n\|^\delta \|C_{k+h}\|^\delta E\|\pi_\ell Z_2\|^\delta.$$

Then

$$E\Big|x'A_n\Big[\sum_{t=1}^{n}\sum_{j=-\infty}^{\infty}\sum_{k\neq j}C_jZ_{t-j}Z'_{t-k}C^*_{k+h}\Big]A^*_ny\Big|^{\delta}$$

$$= E\Big|\sum_{t=1}^{n}\sum_{j=-\infty}^{\infty}\sum_{k\neq j}x'A_nC_jZ_{t-j}Z'_{t-k}C^*_{k+h}A^*_ny\Big|^{\delta}$$

$$\leq E\Big[\Big(\sum_{t=1}^{n}\sum_{j=-\infty}^{\infty}\sum_{k\neq j}|x'A_nC_jZ_{t-j}Z'_{t-k}C^*_{k+h}A^*_ny|\Big)^{\delta}\Big]$$

$$\leq E\Big[\sum_{t=1}^{n}\sum_{j=-\infty}^{\infty}\sum_{k\neq j}|x'A_nC_jZ_{t-j}Z'_{t-k}C^*_{k+h}A^*_ny|^{\delta}\Big]$$

$$= E\Big[\sum_{t=1}^{n}\sum_{j=-\infty}^{\infty}\sum_{k\neq j}|x'A_nC_jZ_{t-j}|^{\delta}|Z'_{t-k}C^*_{k+h}A^*_ny|^{\delta}\Big]$$

$$= \sum_{t=1}^{n}\sum_{j=-\infty}^{\infty}\sum_{k\neq j}E\Big(|x'A_nC_jZ_{t-j}|^{\delta}|Z'_{t-k}C^*_{k+h}A^*_ny|^{\delta}\Big)$$

$$= \sum_{t=1}^{n}\sum_{j=-\infty}^{\infty}\sum_{k\neq j}E|x'A_nC_jZ_{t-j}|^{\delta}E|Z'_{t-k}C^*_{k+h}A^*_ny|^{\delta}$$

by independence of Z_{t-j}, Z_{t-k}

$$\leq \sum_{t=1}^{n}\sum_{j=-\infty}^{\infty}\sum_{k\neq j}\|\pi_iA_n\|^{\delta}\|C_j\|^{\delta}E\|\pi_iZ\|^{\delta}\cdot\|\pi_\ell A_n\|^{\delta}\|C_{k+h}\|^{\delta}E\|\pi_\ell Z_2\|^{\delta}$$

$$= n\|\pi_iA_n\|^{\delta}\|\pi_\ell A_n\|^{\delta}E\|\pi_iZ\|^{\delta}E\|\pi_\ell Z_2\|^{\delta}\sum_{j=-\infty}^{\infty}\sum_{k\neq j}\|C_j\|^{\delta}\|C_{k+h}\|^{\delta}$$

where $E\|\pi_iZ\|^{\delta}<\infty$ because $\delta<1/a_i$, $E\|\pi_\ell Z_2\|^{\delta}<\infty$ because $\delta<1/a_\ell$, and

$$\sum_{j=-\infty}^{\infty}\sum_{k\neq j}\|C_j\|^{\delta}\|C_{k+h}\|^{\delta}\leq\Big(\sum_{j=-\infty}^{\infty}\|C_j\|^{\delta}\Big)\Big(\sum_{k=-\infty}^{\infty}\|C_{k+h}\|^{\delta}\Big)$$
$$=\Big(\sum_{j=-\infty}^{\infty}\|C_j\|^{\delta}\Big)^2<\infty$$

and furthermore since

$$\frac{\log\|\pi_iA_n\|}{\log n}\to -a_i \quad\text{and}\quad \frac{\log\|\pi_\ell A_n\|}{\log n}\to -a_\ell$$

we have

$$\frac{\log(n\|\pi_iA_n\|^{\delta}\|\pi_\ell A_n\|^{\delta})}{\log n}\to 1-\delta a_i-\delta a_\ell<0$$

since $\delta > 1/(a_i + a_\ell)$. Then $n\|\pi_i A_n\|^\delta \|\pi_\ell A_n\|^\delta \to 0$ and it follows that (10.82) holds in this case.

Case II: Suppose that $x \in V_i$ and $y \in V_\ell$ are unit vectors with $a_i + a_\ell \le 1$. Then $a_i < 1$ and $a_\ell < 1$ so $E\|\pi_i Z\| < \infty$ and $E\|\pi_\ell Z_2\| < \infty$. Then by assumption $Ex'Z = 0$ and $EZ_2'y = 0$. Define $Z_{in} = Z_i I(\|A_n Z_i\| \le 1)$ and $\mu_n = EZ_{in}$ and let

$$\begin{aligned} A_{i,j,n} &= (Z_{in} - \mu_n)(Z_{jn} - \mu_n)' \\ B_{i,j,n} &= Z_{in}\mu_n' + \mu_n Z_{jn}' \\ C_{i,j,n} &= Z_i Z_j' I(\|A_n Z_i\| > 1 \text{ or } \|A_n Z_j\| > 1) \\ D_{i,j,n} &= -\mu_n \mu_n' \end{aligned}$$

so that $EA_{i,j,n} = 0$ for $i \ne j$ by independence and $Z_i Z_j' = A_{i,j,n} + B_{i,j,n} + C_{i,j,n} + D_{i,j,n}$. Then let

$$A = \sum_{t=1}^{n} \sum_{j=-\infty}^{\infty} \sum_{k \ne j} x' A_n C_j A_{t-j,t-k,n} C_{k+h}^* A_n^* y$$

and similarly for B, C, D to obtain

$$x' A_n \Big[\sum_{t=1}^{n} \sum_{j=-\infty}^{\infty} \sum_{k \ne j} C_j Z_{t-j} Z_{t-k}' C_{k+h}^* \Big] A_n^* y = A + B + C + D$$

and so in order to establish (10.82) it suffices to show that each of A, B, C, D tends to zero in probability. Let $\bar{Z}_{in} = Z_{in} - \mu_n$ so that $A_{i,j,n} = \bar{Z}_{in}\bar{Z}_{jn}'$. Since $\bar{Z}_{in}$ and $\bar{Z}_{jn}$ are independent with mean zero for $i \ne j$ it is easy to check that $EA = 0$. Then $\mathrm{Var}(A)$ equals

$$\begin{aligned} & E\Big(\sum_{t=1}^{n} \sum_{j=-\infty}^{\infty} \sum_{k \ne j} x' A_n C_j \bar{Z}_{t-j,n} \bar{Z}_{t-k,n}' C_{k+h}^* A_n^* y \Big)^2 \\ =& E\Big[\Big(\sum_{t_1=1}^{n} \sum_{j_1=-\infty}^{\infty} \sum_{k_1 \ne j_1} x' A_n C_{j_1} \bar{Z}_{t_1-j_1,n} \bar{Z}_{t_1-k_1,n}' C_{k_1+h}^* A_n^* y \Big) \\ & \cdot \Big(\sum_{t_2=1}^{n} \sum_{j_2=-\infty}^{\infty} \sum_{k_2 \ne j_2} x' A_n C_{j_2} \bar{Z}_{t_2-j_2,n} \bar{Z}_{t_2-k_2,n}' C_{k_2+h}^* A_n^* y \Big) \Big] \\ =& \sum_{t_1=1}^{n} \sum_{t_2=1}^{n} \sum_{j_1=-\infty}^{\infty} \sum_{j_2=-\infty}^{\infty} \sum_{k_1 \ne j_1} \sum_{k_2 \ne j_2} E\Big[\langle \bar{Z}_{t_1-j_1,n}, C_{j_1}^* A_n^* x \rangle \\ & \cdot \langle \bar{Z}_{t_1-k_1,n}, C_{k_1+h}^* A_n^* y \rangle \langle \bar{Z}_{t_2-j_2,n}, C_{j_2}^* A_n^* x \rangle \langle \bar{Z}_{t_2-k_2,n}, C_{k_2+h}^* A_n^* y \rangle \Big]. \end{aligned}$$

If any one of $t_1 - j_1$, $t_1 - k_1$, $t_2 - j_2$, or $t_2 - k_2$ is distinct then this term vanishes since $E\bar{Z}_{in} = 0$, using the independence to write the mean of the product as the product of the means. The only nonvanishing terms for a given t_1, j_1, k_1 occur when either

$$\text{(a) } t_1 - j_1 = t_2 - j_2 \text{ and } t_1 - k_1 = t_2 - k_2$$
$$\text{(b) } t_1 - j_1 = t_2 - k_2 \text{ and } t_1 - k_1 = t_2 - j_2.$$

Given t_1, j_1, k_1 for every $t_2 = 1, \ldots, n$ there exists a unique j_2, k_2 satisfying (a) or (b). Indeed, we have

$$\text{(a) } j_2 = (t_2 - t_1) + j_1 \text{ and } k_2 = (t_2 - t_1) + k_1$$
$$\text{(b) } j_2 = (t_2 - t_1) + k_1 \text{ and } k_2 = (t_2 - t_1) + j_1$$

so that $j_2 \neq k_2$ is guaranteed. Now let $p = t_1 - j_1$, $q = t_1 - k_1$, $j = j_1$, $t = t_1$, $k = k_1$, and $\Delta = t_2 - t_1$ so that in case

$$\text{(a) } t_2 - j_2 = p,\ t_2 - k_2 = q,\ j_2 = \Delta + j, \text{ and } k_2 = \Delta + k$$
$$\text{(b) } t_2 - j_2 = q,\ t_2 - k_2 = p,\ j_2 = \Delta + k, \text{ and } k_2 = \Delta + j.$$

Then in case (a) the only nonvanishing terms for a given t, j, k are of the form

$$\begin{aligned}
&E[<\bar{Z}_{pn}, C_j^* A_n^* x><\bar{Z}_{qn}, C_{k+h}^* A_n^* y><\bar{Z}_{pn}, C_{\Delta+j}^* A_n^* x><\bar{Z}_{qn}, C_{\Delta+k+h}^* A_n^* y>] \\
=&E[<\bar{Z}_{pn}, C_j^* A_n^* x><\bar{Z}_{pn}, C_{\Delta+j}^* A_n^* x><\bar{Z}_{qn}, C_{k+h}^* A_n^* y><\bar{Z}_{qn}, C_{\Delta+k+h}^* A_n^* y>] \\
=&E[x' A_n C_j \bar{Z}_{pn} \bar{Z}_{pn}' C_{\Delta+j}^* A_n^* x] E[y' A_n C_{k+h} \bar{Z}_{qn} \bar{Z}_{qn}' C_{\Delta+k+h}^* A_n^* y] \\
=& (x' A_n C_j M_n C_{\Delta+j}^* A_n^* x)(y' A_n C_{k+h} M_n C_{\Delta+k+h}^* A_n^* y)
\end{aligned}$$

where $M_n = E\bar{Z}_{in}\bar{Z}_{in}'$. In case (b) the only nonvanishing terms are of the form

$$\begin{aligned}
&E[<\bar{Z}_{pn}, C_j^* A_n^* x><\bar{Z}_{qn}, C_{k+h}^* A_n^* y><\bar{Z}_{qn}, C_{\Delta+k}^* A_n^* x><\bar{Z}_{pn}, C_{\Delta+j+h}^* A_n^* y>] \\
=&E[<\bar{Z}_{pn}, C_j^* A_n^* x><\bar{Z}_{pn}, C_{\Delta+j+h}^* A_n^* y><\bar{Z}_{qn}, C_{k+h}^* A_n^* y><\bar{Z}_{qn}, C_{\Delta+k}^* A_n^* x>] \\
=& (x' A_n C_j M_n C_{\Delta+j+h}^* A_n^* y)(y' A_n C_{k+h} M_n C_{\Delta+k}^* A_n^* x)
\end{aligned}$$

so that Var(A) equals

$$\begin{aligned}
\sum_{t=1}^{n} \sum_{t+\Delta=1}^{n} \sum_{j=-\infty}^{\infty} \sum_{k \neq j} \Big[&(x' A_n C_j M_n C_{\Delta+j}^* A_n^* x)(y' A_n C_{k+h} M_n C_{\Delta+k+h}^* A_n^* y) \\
&+ (x' A_n C_j M_n C_{\Delta+j+h}^* A_n^* y)(y' A_n C_{k+h} M_n C_{\Delta+k}^* A_n^* x) \Big].
\end{aligned}$$

Now, we need to bound the terms in this sum. Choose unit vectors x_j, x_{nj}, y_j, y_{nj} and positive reals r_j, a_{nj}, ρ_j, b_{nj} so that $C_j^* x = r_j x_j$, $A_n^* x_j = a_{nj} x_{nj}$,

$C_j^* y = \rho_j y_j$, and $A_n^* y_j = b_{nj} y_{nj}$. Then, e.g., we have

$$\begin{aligned} x' A_n C_j M_n C_{\Delta+j}^* A_n^* x &= x' C_j A_n M_n A_n^* C_{\Delta+j}^* x \\ &= (C_j^* x)' A_n M_n A_n^* C_{\Delta+j}^* x \\ &= r_j r_{\Delta+j} x_j' A_n M_n A_n^* x_{\Delta+j} \end{aligned}$$

where

$$\begin{aligned} x_j' A_n M_n A_n^* x_{\Delta+j} &= x_j' A_n E \bar{Z}_{1n} \bar{Z}_{1n}' A_n^* x_{\Delta+j} \\ &= E x_j' A_n \bar{Z}_{1n} \bar{Z}_{1n}' A_n^* x_{\Delta+j} \\ &= E{<}\bar{Z}_{1n}, A_n^* x_j{>}{<}\bar{Z}_{1n}, A_n^* x_{\Delta+j}{>} \\ &\le \sqrt{E{<}\bar{Z}_{1n}, A_n^* x_j{>}^2 E{<}\bar{Z}_{1n}, A_n^* x_{\Delta+j}{>}^2}. \end{aligned}$$

Since $\bar{Z}_{1n} = Z_{1n} - \mu_n$, where $\mu_n = E Z_{1n}$, we have, e.g., that

$$\begin{aligned} E{<}\bar{Z}_{1n}, A_n^* x_j{>}^2 &= E{<}Z_{1n} - \mu_n, A_n^* x_j{>}^2 \\ &= E\Big({<}Z_{1n}, A_n^* x_j{>} - E{<}Z_{1n}, A_n^* x_j{>}\Big)^2 \\ &= \mathrm{Var}\Big({<}Z_{1n}, A_n^* x_j{>}\Big) \\ &\le E[{<}Z_{1n}, A_n^* x_j{>}^2] \end{aligned}$$

so that, calculating in a similar manner for the remaining terms, we have, e.g.,

$$x' A_n C_j M_n C_{\Delta+j}^* A_n^* x \le r_j r_{j+\Delta} \sqrt{E{<}Z_{1n}, A_n^* x_j{>}^2 E{<}Z_{1n}, A_n^* x_{\Delta+j}{>}^2}$$

and then $\mathrm{Var}(A)$ is bounded above by

$$\begin{aligned} &\sum_{t=1}^{n} \sum_{t+\Delta=1}^{n} \sum_{j=-\infty}^{\infty} \sum_{k \neq j} \Bigg(r_j r_{j+\Delta} \Big[E{<}Z_{1n}, A_n^* x_j{>}^2 E{<}Z_{1n}, A_n^* x_{j+\Delta}{>}^2\Big]^{1/2} \\ &\cdot \rho_{k+h} \rho_{k+h+\Delta} \Big[E{<}Z_{1n}, A_n^* y_{k+h}{>}^2 E{<}Z_{1n}, A_n^* y_{k+h+\Delta}{>}^2\Big]^{1/2} \\ &+ r_j \rho_{j+h+\Delta} \Big[E{<}Z_{1n}, A_n^* x_j{>}^2 E{<}Z_{1n}, A_n^* y_{j+h+\Delta}{>}^2\Big]^{1/2} \\ &\cdot \rho_{k+h} r_{k+\Delta} \Big[E{<}Z_{1n}, A_n^* y_{k+h}{>}^2 E{<}Z_{1n}, A_n^* x_{k+\Delta}{>}^2\Big]^{1/2} \Bigg). \end{aligned}$$

Lemma 10.6.9. *For all unit vectors $x \in V_i$, $y \in V_\ell$, for some n_0, for all $n \geq n_0$ and all j, k, h, Δ, for some $K_n > 0$ satisfying $nK_n \to 0$ we have*

$$\begin{aligned}&\Big[E{<}Z_{1n}, A_n^* x_j{>}^2 E{<}Z_{1n}, A_n^* x_{j+\Delta}{>}^2 \\ &\quad \cdot E{<}Z_{1n}, A_n^* y_{k+h}{>}^2 E{<}Z_{1n}, A_n^* y_{k+h+\Delta}{>}^2\Big]^{1/2} \\ &+\Big[E{<}Z_{1n}, A_n^* x_j{>}^2 E{<}Z_{1n}, A_n^* y_{j+h+\Delta}{>}^2 \\ &\quad \cdot E{<}Z_{1n}, A_n^* y_{k+h}{>}^2 E{<}Z_{1n}, A_n^* x_{k+\Delta}{>}^2\Big]^{1/2} \leq K_n.\end{aligned} \tag{10.83}$$

Proof. Suppose that $a_i > 1/2$. Then $a_\ell < 1/2$ so $E\|\pi_i Z_1\|^2 = \infty$ while $E\|\pi_\ell Z_1\|^2 < \infty$. Recall that $\pi_\ell = \pi_\ell^2 = \pi_\ell^*$ for this orthogonal projection operator, $A_n \pi_\ell = \pi_\ell A_n$, and that $|{<}x, \theta{>}| \leq \|x\|$ when θ is a unit vector. Then for all unit vectors $\theta \in V_\ell$ we have

$$\begin{aligned} E{<}Z_{1n}, A_n^* \theta{>}^2 &= E{<}A_n Z_1, \theta{>}^2 I(\|A_n Z_1\| \leq 1) \\ &\leq E{<}A_n Z_1, \theta{>}^2 \\ &= E{<}A_n Z_1, \pi_\ell \theta{>}^2 \\ &= E{<}\pi_\ell A_n \pi_\ell Z_1, \theta{>}^2 \\ &\leq E\|\pi_\ell A_n \pi_\ell Z_1\|^2 \\ &\leq \|\pi_\ell A_n\|^2 E\|\pi_\ell Z_1\|^2. \end{aligned}$$

Since $1/a_i < 2$, we can apply part (ii) of Lemma 10.6.4 with $\zeta = 2$ to get

$$nE{<}Z_{1n}, A_n^* \theta{>}^2 = nE{<}A_n Z_1, \theta{>}^2 I(\|A_n Z_1\| \leq 1) < K_2$$

for all $n \geq n_0$ and all $\|\theta\| = 1$ in V_i. Then (10.83) holds in this case with

$$\begin{aligned} K_n &= 2\Big[(K_2/n)^2 \Big(\|\pi_\ell A_n\|^2 E\|\pi_\ell Z_1\|^2\Big)^2\Big]^{1/2} \\ &= 2(K_2/n)\|\pi_\ell A_n\|^2 E\|\pi_\ell Z_1\|^2 \end{aligned}$$

so that $nK_n \to 0$ since $\|\pi_\ell A_n\| \to 0$. If $a_\ell > 1/2$ so that $a_i < 1/2$, then a similar argument establishes (10.83) with $K_n = 2(K_2/n)\|\pi_i A_n\|^2 E\|\pi_i Z_1\|^2$ and again $nK_n \to 0$. Otherwise, we have $a_i \leq 1/2$ and $a_\ell \leq 1/2$. If both $E\|\pi_i Z_1\|^2$ and $E\|\pi_\ell Z_1\|^2$ exist then (10.83) holds with

$$K_n = 2\|\pi_i A_n\|^2 \|\pi_\ell A_n\|^2 E\|\pi_i Z_1\|^2 E\|\pi_\ell Z_1\|^2$$

and then

$$\frac{\log nK_n}{\log n} \sim \frac{\log(n\|\pi_i A_n\|^2 \|\pi_\ell A_n\|^2)}{\log n} \to 1 - 2(a_i + a_\ell) < 0$$

since $a_i > 1/4$ and $a_\ell > 1/4$ by assumption, so again $nK_n \to 0$. Otherwise, suppose that $a_i \leq 1/2$ and $a_\ell \leq 1/2$ with $E\|\pi_i Z_1\|^2 = \infty$ and $E\|\pi_\ell Z_1\|^2 < \infty$.

Then $a_i = 1/2$ and so we can apply part (ii) of Lemma 10.6.4 with $\zeta = 2+\varepsilon$ for any $\varepsilon > 0$. Observe that for all t and all θ we have

$$\begin{aligned}U_2(t,\theta) &= \int_{|<z,\theta>|\le t} |<z,\theta>|^2 \mu(dz)\\ &= \int_{0\le|<z,\theta>|<1} |<z,\theta>|^2\mu(dz) + \int_{1\le|<z,\theta>|\le t} |<z,\theta>|^2\mu(dz)\\ &\le \int_{0\le|<z,\theta>|<1} |<z,\theta>|^2\mu(dz) + \int_{1\le|<z,\theta>|\le t} |<z,\theta>|^{2+\varepsilon}\mu(dz)\\ &\le 1 + U_{2+\varepsilon}(t,\theta)\end{aligned}$$

and since $U_{2+\varepsilon}(t,\theta) \to \infty$ uniformly on $\|\theta\| = 1$ in V_i, for some t_0 and $C > 0$ we have for all $t \ge t_0$ and all $\|\theta\| = 1$ in V_i that $U_2(t,\theta) \le CU_{2+\varepsilon}(t,\theta)$. As in the proof of Lemma 10.6.4 we can choose n_0 such that $\|A_n\theta\| \le t_0^{-1}$ for all $\|\theta\| = 1$ and all $n \ge n_0$. Enlarge n_0 if necessary so that Lemma 10.6.4 also applies. Then

$$\begin{aligned}nE<Z_{1n}, A_n^* x_j>^2 &= nE<Z_1, A_n^* x_j>^2 I(\|A_n Z_1\| \le 1)\\ &\le n a_{nj}^2 U_2(a_{nj}^{-1}, x_{nj})\\ &\le C a_{nj}^{-\varepsilon} \cdot n a_{nj}^{2+\varepsilon} U_{2+\varepsilon}(a_{nj}^{-1}, x_{nj})\\ &= C a_{nj}^{-\varepsilon} \cdot nE|<A_n Z_1, x_j>|^{2+\varepsilon} I(|<A_n Z_1, x_j>| \le 1)\\ &\le C a_{nj}^{-\varepsilon} \cdot K_1\end{aligned}$$

for all $n \ge n_0 = n_0(t_0)$ and all j. Note also that for all j and all n we have $a_{nj} = \|A_n x_j\| \ge \min\{\|A_n\theta\| : \|\theta\| = 1 \text{ in } V_i\} = 1/\|\pi_i A_n^{-1}\|$ so that $a_{nj}^{-\varepsilon} \le \|\pi_i A_n^{-1}\|^\varepsilon$ for all n and all j. Then (10.83) holds in this case with

$$\begin{aligned}K_n &= \left[\left(n^{-1}K_1\|\pi_i A_n^{-1}\|^\varepsilon\right)^2 \left(\|\pi_\ell A_n\|^2 E\|\pi_\ell Z_1\|^2\right)^2\right]^{1/2}\\ &= n^{-1}K_1\|\pi_i A_n^{-1}\|^\varepsilon \|\pi_\ell A_n\|^2 E\|\pi_\ell Z_1\|^2\end{aligned}$$

so that

$$\frac{\log nK_n}{\log n} \sim \frac{\log(\|\pi_i A_n^{-1}\|^\varepsilon \|\pi_\ell A_n\|^2)}{\log n} \to \varepsilon a_i - 2a_\ell < 0$$

for all $\varepsilon > 0$ sufficiently small. Note that $\varepsilon > 0$ can be chosen independently of x, y, j and so $nK_n \to 0$ here too. Finally, if $a_i \le 1/2$ and $a_\ell \le 1/2$ with $E\|\pi_i Z_1\|^2 < \infty$ and $E\|\pi_\ell Z_1\|^2 = \infty$, then (10.83) holds with $K_n = n^{-1}K_1\|\pi_\ell A_n^{-1}\|^\varepsilon \|\pi_i A_n\|^2 E\|\pi_i Z_1\|^2$ and $\log(nK_n)/\log n \to \varepsilon a_\ell - 2a_i < 0$ for $\varepsilon > 0$ sufficiently small, and again $nK_n \to 0$. □

Returning to the proof of Proposition 10.6.8, we now have that Var(A) is bounded above by

$$
\begin{aligned}
&K_n \sum_{t=1}^{n} \sum_{t+\Delta=1}^{n} \sum_{j=-\infty}^{\infty} \sum_{k\neq j} \left(r_j r_{j+\Delta} \rho_{k+h} \rho_{k+h+\Delta} + r_j \rho_{j+h+\Delta} \rho_{k+h} r_{k+\Delta}\right) \\
&\leq K_n \sum_{t=1}^{n} \sum_{\Delta=1-t}^{n-t} \sum_{j=-\infty}^{\infty} \sum_{k\neq j} \Big(\|C_j\| \|C_{j+\Delta}\| \|C_{k+h}\| \|C_{k+h+\Delta}\| \\
&\qquad\qquad + \|C_j\| \|C_{j+h+\Delta}\| \|C_{k+h}\| \|C_{k+\Delta}\| \Big) \\
&\leq n K_n 2 \Big(\sum_{j=-\infty}^{\infty} \|C_j\| \Big)^4 \to 0
\end{aligned}
$$

and so $A \to 0$ in probability.

Next, we have

$$
B = \sum_{t=1}^{n} \sum_{j=-\infty}^{\infty} \sum_{k\neq j} x' A_n C_j \left[Z_{t-j,n} \mu_n' + \mu_n Z_{t-k,n}' \right] C_{k+h}^* A_n^* y
$$

and we will let

$$
\begin{aligned}
B_{rs}^{jk} &= x' A_n C_j \left[Z_{rn} \mu_n' + \mu_n Z_{sn}' \right] C_k^* A_n^* y \\
&= (C_j^* x)' A_n \left[Z_{rn} \mu_n' + \mu_n Z_{sn}' \right] A_n^* (C_k^* y) \\
&= r_j r_k (A_n^* x_j)' \left[Z_{rn} \mu_n' + \mu_n Z_{sn}' \right] (A_n^* y_k) \\
&= r_j r_k \left[B_r^{(1)} + B_s^{(2)} \right]
\end{aligned}
$$

where

$$
B_r^{(1)} = (A_n^* x_j)' Z_{rn} \mu_n' A_n^* y_k \quad \text{and} \quad B_s^{(2)} = (A_n^* x_j)' \mu_n Z_{sn}' A_n^* y_k
$$

Lemma 10.6.10. *For all unit vectors $x \in V_i$ and $y \in V_\ell$, for some n_0, for all $n \geq n_0$, for all j, k, r, s we have*

$$
E|B_{rs}^{jk}| \leq \|C_j\| \|C_k\| K_n \tag{10.84}
$$

where $nK_n \to 0$.

Proof. Write

$$
\begin{aligned}
E|B_{rs}^{jk}| &= r_j r_k E|B_r^{(1)} + B_s^{(2)}| \\
&\leq \|C_j\| \|C_k\| \Big(E|B_r^{(1)}| + E|B_s^{(2)}| \Big)
\end{aligned}
$$

where

$$
\begin{aligned}
E|B_r^{(1)}| &= E|\langle Z_{1n}, A_n^* x_j\rangle \langle \mu_n, A_n^* y_k\rangle| \\
&= |\langle \mu_n, A_n^* y_k\rangle| E|\langle Z_{1n}, A_n^* x_j\rangle|.
\end{aligned}
$$

Since $a_i < 1$ we have $E\|\pi_i Z_1\| < \infty$. Then we also have $E|<Z_{1n}, A_n^* x_j>| \leq \|\pi_i A_n\| E\|\pi_i Z_1\|$ as in the proof of Lemma 3.2. Apply part (iv) of Lemma 10.6.4 with $\eta = 1$ (so that $\eta < 1/a_\ell$). By using the fact that $E<Z_1, y_k> = 0$, we have for some $n_1 > 0$ that

$$\begin{aligned} n|<\mu_n, A_n^* y_k>| &= n|<EZ_{1n}, A_n^* y_k>| \\ &= n|E<Z_{1n}, A_n^* y_k>| \\ &= n|E<Z_1, A_n^* y_k> I(\|A_n Z_1\| \leq 1)| \\ &= n|E<Z_1, A_n^* y_k> I(\|A_n Z_1\| > 1)| \\ &\leq nE|<A_n Z_1, y_k>| I(\|A_n Z_1\| > 1) < K_4 \end{aligned}$$

for all $n \geq n_1$ and all k. It follows that

$$nE|B_r^{(1)}| \leq K_4 \|\pi_i A_n\| E\|\pi_i Z_1\|$$

for all $n \geq n_1$ and all k. A similar argument (note that $a_\ell < 1$ so that $E\|\pi_\ell Z_1\| < \infty$ and apply part (iv) of Lemma 10.6.4 again with $\eta = 1$, so that $\eta < 1/a_i$) shows that for some $n_2 > 0$ we have

$$nE|B_s^{(2)}| \leq K_4 \|\pi_\ell A_n\| E\|\pi_\ell Z_1\|$$

for all $n \geq n_2$ and all j. By letting $n_0 = \max\{n_1, n_2\}$, we have that (10.84) holds with

$$nK_n = K_4 \Big(\|\pi_i A_n\| E\|\pi_i Z_1\| + \|\pi_\ell A_n\| E\|\pi_\ell Z_1\| \Big)$$

for all j, k, r, s and all $n \geq n_0$. Then $nK_n \to 0$ since $\|A_n\| \to 0$. □

Now we have

$$E|B| \leq \sum_{t=1}^{n} \sum_{j=-\infty}^{\infty} \sum_{k \neq j} \|C_j\| \|C_{k+h}\| K_n \leq nK_n \Big(\sum_{j=-\infty}^{\infty} \|C_j\| \Big)^2 \to 0$$

and so $B \to 0$ in probability. Next, we write

$$\begin{aligned} C_{rs}^{jk} &= x' A_n C_j Z_r Z_s' I(\|A_n Z_r\| > 1 \text{ or } \|A_n Z_s\| > 1) C_k^* A_n^* y \\ &= r_j r_k <Z_r, A_n^* x_j><Z_s, A_n^* y_k> I(\|A_n Z_r\| > 1 \text{ or } \|A_n Z_s\| > 1) \end{aligned}$$

so that

$$\begin{aligned} E|C_{rs}^{jk}| \leq \|C_j\| \|C_k\| \Big(& E|x_j' A_n Z_r Z_s' A_n^* y_k| I(\|A_n Z_r\| > 1) \\ & + E|x_j' A_n Z_r Z_s' A_n^* y_k| I(\|A_n Z_s\| > 1) \Big). \end{aligned}$$

Lemma 10.6.11. *For all unit vectors $x \in V_i$ and $y \in V_\ell$, for some n_0, for all $n \geq n_0$, and all j, k, r, s, we have*

$$E|C_{jk}^{rs}| \leq \|C_j\| \|C_k\| K_n \tag{10.85}$$

where $nK_n \to 0$.

Proof. Choose $b_{nk} > 0$ and $||y_{nk}|| = 1$ such that $A_n^* y_k = b_{nk} y_{nk}$, and write

$$E|x_j' A_n Z_r Z_s' A_n^* y_k| I(||A_n Z_r|| > 1) = E|<Z_s, A_n^* y_k>| \cdot E|<Z_r, A_n^* x_j>| I(||A_n Z_r|| > 1)$$

where $E|<Z_s, A_n^* y_k>| = b_{nk} E|<Z_s, y_{nk}>| \le b_{nk} E||\pi_\ell Z_2|| \le ||\pi_\ell A_n|| E||\pi_\ell Z_2||$ and $nE|<Z_r, A_n^* x_j>| I(||A_n Z_r|| > 1) \le K_4$ and similarly

$$nE|x_j' A_n Z_r Z_s' A_n^* y_k| I(||A_n Z_s|| > 1) \le ||\pi_i A_n|| E||\pi_i A_n|| K_4$$

so that (10.85) holds with

$$nK_n = K_4 \Big(||\pi_i A_n|| E||\pi_i Z_1|| + ||\pi_\ell A_n|| E||\pi_\ell Z_1|| \Big)$$

which is the same K_n as for Lemma 3.3, so $nK_n \to 0$. □

Now we have

$$E|C| \le \sum_{t=1}^{n} \sum_{j=-\infty}^{\infty} \sum_{k \ne j} ||C_j|| ||C_{k+h}|| K_n \le nK_n \Big(\sum_{j=-\infty}^{\infty} ||C_j|| \Big)^2 \to 0$$

and so $C \to 0$ in probability.

Finally, we can write

$$D_{jk} = x' A_n C_j \mu_n \mu_n' C_k^* A_n^* y = r_j r_k <\mu_n, A_n^* x_j><\mu_n, A_n^* y_k>.$$

Lemma 10.6.12. *For all unit vectors $x \in V_i$ and $y \in V_\ell$, for some n_0, for all $n \ge n_0$, and all j, k, we have $|D_{jk}| \le ||C_j|| ||C_k|| K_n$, where $nK_n \to 0$.*

Proof. Recall that for all large n we have $|<\mu_n, A_n^* x_j>| < K_4/n$ from the proof of Lemma 3.3 above. Likewise $|<\mu_n, A_n^* y_k>| \le K_4/n$ so we can take $K_n = (K_4/n)^2$. □

Now we have

$$|D| \le \sum_{t=1}^{n} \sum_{j=-\infty}^{\infty} \sum_{k \ne j} ||C_j|| ||C_{k+h}|| K_n \le nK_n \Big(\sum_{j=-\infty}^{\infty} ||C_j|| \Big)^2 \to 0$$

which completes the proof of Proposition 10.6.8. □

Proposition 10.6.13. *For all h, we have*

$$A_n \Big[\sum_{t=1}^{n} \sum_{j=-\infty}^{\infty} C_j (Z_{t-j} Z_{t-j}' - B_n) C_{j+h}^* \Big] A_n^* \Rightarrow \sum_{j=-\infty}^{\infty} C_j W C_{j+h}^*. \qquad (10.86)$$

Proof. Define

$$
\begin{aligned}
S_n^{(m)} &= \sum_{|j|\le m} C_j\Big(\sum_{t=1}^{n} Z_{t-j}Z'_{t-j}\Big)C^*_{j+h}\\
d_n^{(m)} &= n\sum_{|j|\le m} C_jB_nC^*_{j+h}\\
W^{(m)} &= \sum_{|j|\le m} C_jWC^*_{j+h}
\end{aligned}
$$

for $m \ge 1$. We will now apply (10.79) to show that

$$\left(A_n\Big(\sum_{t=1}^{n} Z_{t-j}Z'_{t-j} - B_n\Big)A_n^* : |j| \le m\right) \Rightarrow \Big(W, \ldots, W\Big). \tag{10.87}$$

Since we know that the convergence holds for the $j = 0$ component, it suffices to show that the difference between this component and any other component tends to zero in probability. This difference consists of at most $2m$ identically distributed terms, any one of which must tend to zero in probability in view of the convergence (10.79) above, and so (10.87) holds. Now by the Continuous Mapping Theorem 1.2.7 we obtain from (10.87) that

$$A_n(S_n^{(m)} - d_n^{(m)})A_n^* \Rightarrow W^{(m)}$$

and since

$$
\begin{aligned}
\sum_{|j|>m} \|C_jWC^*_{j+h}\| &\le \|W\| \sum_{|j|>m} \|C_j\|\|C_{j+h}\|\\
&\le \|W\|\Big(\sum_{|j|>m} \|C_j\|^2\Big)^{1/2}\Big(\sum_{|j|>m} \|C_{j+h}\|^2\Big)^{1/2} \to 0
\end{aligned}
$$

as $m \to \infty$ we also have

$$W^{(m)} \to \sum_{j=-\infty}^{\infty} C_jWC^*_{j+h}$$

almost surely. Then in order to establish (10.86) it suffices to show (e.g., by [17] Theorem 4.2) that

$$\lim_{m\to\infty} \limsup_{n\to\infty} P(\|T\| > \varepsilon) = 0 \tag{10.88}$$

where

$$T = \sum_{|j|>m} \sum_{t=1}^{n} A_nC_j\big(Z_{t-j}Z'_{t-j} - B_n\big)C^*_{j+h}A_n^*.$$

Note that $||T||^2 = \sum_{i,j} |e_i' T e_j|^2$ where $e_1, \ldots, e_d$ is the standard basis for $\mathbb{R}^d$. Then in order to show that (10.88) holds it suffices to show that

$$\lim_{m\to\infty} \limsup_{n\to\infty} P(|x'Ty| > \varepsilon) = 0 \tag{10.89}$$

for all unit vectors $x \in V_i$ and $y \in V_\ell$ where $1 \le i, \ell \le p$. Write $T = A + B - C$ where

$$A = \sum_{|j|>m} \sum_{t=1}^{n} A_n C_j Z_{t-j} Z_{t-j}' C_{j+h}^* A_n^* I(||A_n Z_{t-j}|| \le 1)$$

$$B = \sum_{|j|>m} \sum_{t=1}^{n} A_n C_j Z_{t-j} Z_{t-j}' C_{j+h}^* A_n^* I(||A_n Z_{t-j}|| > 1)$$

$$C = \sum_{|j|>m} \sum_{t=1}^{n} A_n C_j B_n C_{j+h}^* A_n^*.$$

Since $T = A + B - C$ it suffices to show that (10.89) holds with T replaced by A, B, or C. We begin with B. Choose $\delta > 0$ with $\delta < 1/a_p$ and $\delta \le 1$ and let $\delta_1 = \delta/2 < 1$. Use the Markov Inequality 1.1.28 along with the fact that $|x+y|^{\delta_1} \le |x|^{\delta_1} + |y|^{\delta_1}$ to see that $P(|x'By| > \varepsilon)$ is bounded above by

$$\begin{aligned}
&\varepsilon^{-\delta_1} E|x'By|^{\delta_1} \\
&\le \varepsilon^{-\delta_1} \sum_{|j|>m} nE|x'C_j A_n Z_1 Z_1' A_n^* C_{j+h}^* y|^{\delta_1} I(||A_n Z_1|| > 1) \\
&\le \varepsilon^{-\delta_1} \sum_{|j|>m} ||C_j||^{\delta_1} ||C_{j+h}||^{\delta_1} nE|x_j' A_n Z_1 Z_1' A_n^* y_{j+h}|^{\delta_1} I(||A_n Z_1|| > 1) \\
&\le \varepsilon^{-\delta_1} \Big[\sum_{|j|>m} ||C_{j+h}||^{\delta} \Big]^{1/2} \\
&\quad \cdot \Big[\sum_{|j|>m} ||C_j||^{\delta} \Big(nE|x_j' A_n Z_1 Z_1' A_n^* y_{j+h}|^{\delta_1} I(||A_n Z_1|| > 1) \Big)^2 \Big]^{1/2}
\end{aligned}$$

by the Cauchy-Schwarz Inequality 1.1.23. Then in order to show that (10.89) holds with T replaced by B it will suffice to show that for some $K_4 > 0$ we have

$$nE|x_j' A_n Z_1 Z_1' A_n^* y_{j+h}|^{\delta_1} I(||A_n Z_1|| > 1) \le K_4 \tag{10.90}$$

for all j, h and all n large. Apply the Cauchy-Schwarz Inequality again to see that

$$\begin{aligned}
&nE|x_j' A_n Z_1 Z_1' A_n^* y_{j+h}|^{\delta_1} I(\|A_n Z_1\| > 1) \\
=&nE|<A_n Z_1, x_j><A_n Z_1, y_{j+h}>|^{\delta_1} I(\|A_n Z_1\| > 1) \\
\leq&\sqrt{nE|<A_n Z_1, x_j>|^{\delta} I(\|A_n Z_1\| > 1) \cdot nE|<A_n Z_1, y_{j+h}>|^{\delta} I(\|A_n Z_1\| > 1)}
\end{aligned}$$

where $nE|<A_n Z_1, x_j>|^{\delta} I(\|A_n Z_1\| > 1) < K_4$ for all n large and all j by an application of Lemma 10.6.4 part (iv) with $\eta = \delta < 1/a_i$. A similar argument holds for the remaining term, which establishes (10.90), and so (10.89) holds with T replaced by B.

Next, we look at A and C, which requires us to consider different cases. First, suppose that $a_i > 1/2$ and $a_\ell > 1/2$. As in the proof of Proposition 10.6.8 we write $C_j^* x = r_j x_j$, $A_n^* x_j = a_{nj} x_{nj}$, $C_j^* y = \rho_j y_j$, and $A_n^* y_j = b_{nj} y_{nj}$ where x_j, x_{nj}, y_j, y_{nj} are unit vectors and r_j, a_{nj}, ρ_j, b_{nj} are positive reals. By the Markov Inequality, we get that $P(|x'Ay| > \varepsilon)$ is bounded above by

$$\begin{aligned}
&\varepsilon^{-1} E|x'Ay| \\
&\leq \varepsilon^{-1} \sum_{|j|>m} nE|x' C_j A_n Z_1 Z_1' A_n^* C_{j+h}^* y| I(\|A_n Z_1\| \leq 1) \\
&\leq \varepsilon^{-1} \sum_{|j|>m} \|C_j\| \|C_{j+h}\| nE|<Z_1, A_n^* x_j><Z_1, A_n^* y_{j+h}>| I(\|A_n Z_1\| \leq 1).
\end{aligned}$$

Next, we will show that for some $K_2 > 0$ we have

$$nE|<Z_1, A_n^* x_j><Z_1, A_n^* y_{j+h}>| I(\|A_n Z_1\| \leq 1) \leq K_2 \tag{10.91}$$

for all j, h and all n large. By the Cauchy-Schwarz Inequality we have that the left hand side of (10.91) is bounded above by

$$\sqrt{nE<Z_1, A_n^* x_j>^2 I(\|A_n Z_1\| \leq 1) nE<Z_1, A_n^* y_{j+h}>^2 I(\|A_n Z_1\| \leq 1)}$$

where we have $nE<Z_1, A_n^* x_j>^2 I(\|A_n Z_1\| \leq 1) < K_2$ independent of j and n large as in the proof of Lemma 10.6.9, since $a_i > 1/2$ (so that $\zeta = 2 > 1/a_i$ in Lemma 10.6.4 part (ii)). Similarly, $nE<Z_1, A_n^* y_{j+h}>^2 I(\|A_n Z_1\| \leq 1) < K_2$ independent of j and n large by another application of Lemma 10.6.4 part (ii) (note that $a_\ell > 1/2$ so that $\zeta = 2 > 1/a_\ell$). Hence (10.91) holds for all j, h and all n large, and so for n large we have

$$\begin{aligned}
P(|x'Ay| > \varepsilon) &\leq \varepsilon^{-1} K_2 \sum_{|j|>m} \|C_j\| \|C_{j+h}\| \\
&\leq \varepsilon^{-1} K_2 \Big(\sum_{|j|>m} \|C_j\|^2 \Big)^{1/2} \Big(\sum_{|j|>m} \|C_{j+h}\|^2 \Big)^{1/2}
\end{aligned}$$

which tends to zero as $m \to \infty$, and so (10.89) holds with T replaced by A in this case. Since $|x'Cy| = |Ex'Ay|$ we also have that (10.89) holds with T replaced by C in this case. Finally, suppose that either $a_i \le 1/2$ or $a_\ell \le 1/2$ (or both). As in the proof of Proposition 10.6.8 define $Z_{in} = Z_i I(\|A_n Z_i\| \le 1)$ and let $Q_{in} = Z_{in}Z_{in}' - EZ_{in}Z_{in}' = Z_{in}Z_{in}' - B_n$. By the Markov Inequality, we get

$$\begin{aligned} P(|x'(A-C)y| > \varepsilon) &\le \varepsilon^{-2}E[(x'(A-C)y)^2] \\ &= \varepsilon^{-2}E\Big(\sum_{|j|>m}\sum_{t=1}^{n} x'A_nC_jQ_{t-j,n}C_{j+h}^*A_n^*y\Big)^2 \\ &= \varepsilon^{-2}\sum_{|j|>m}\sum_{t=1}^{n}\sum_{|j'|>m}\sum_{t'=1}^{n} E\big(x'A_nC_jQ_{t-j,n}C_{j+h}^*A_n^*y \\ &\qquad \cdot x'A_nC_{j'}Q_{t'-j',n}C_{j'+h}^*A_n^*y\big). \end{aligned}$$

Since Q_{in} are independent with mean zero the only terms that contribute to the sum above are those for which $t - j = t' - j'$. Then the sum above equals

$$\sum_{|j|>m}\sum_{t=1}^{n}\sum_{j'=1-t+j}^{n-t+j} r_j\rho_{j+h}r_{j'}\rho_{j'+h}E\big(x_j'A_nQ_{t-j,n}A_n^*y_{j+h}x_{j'}'A_nQ_{t-j,n}A_n^*y_{j'+h}\big).$$

Next, we will show that for some $K_2 > 0$ we have

$$nE\big(x_j'A_nQ_{1n}A_n^*y_{j+h}x_{j'}'A_nQ_{1n}A_n^*y_{j'+h}\big) \le K_2 \tag{10.92}$$

for all j, j', h and all n large. Write

$$\begin{aligned} x_j'A_nQ_{1n}A_n^*y_k &= x_j'A_nZ_{1n}Z_{1n}'A_n^*y_k - Ex_j'A_nZ_{1n}Z_{1n}'A_n^*y_k \\ &= \langle A_nZ_{1n}, x_j\rangle\langle A_nZ_{1n}, y_k\rangle - E\langle A_nZ_{1n}, x_j\rangle\langle A_nZ_{1n}, y_k\rangle \\ &= \xi_{nj}\beta_{nk} - E\xi_{nj}\beta_{nk} \end{aligned}$$

where $\xi_{nj} = \langle A_nZ_{1n}, x_j\rangle$ and $\beta_{nk} = \langle A_nZ_{1n}, y_k\rangle$. Then

$$\begin{aligned} &E\big(x_j'A_nQ_{1n}A_n^*y_{j+h}x_{j'}'A_nQ_{1n}A_n^*y_{j'+h}\big) \\ =&E\big(\xi_{nj}\beta_{n,j+h} - E\xi_{nj}\beta_{n,j+h}\big)\big(\xi_{nj'}\beta_{n,j'+h} - E\xi_{nj'}\beta_{n,j'+h}\big) \\ =&E\xi_{nj}\beta_{n,j+h}\xi_{nj'}\beta_{n,j'+h} - E\xi_{nj}\beta_{n,j+h}E\xi_{nj'}\beta_{n,j'+h} \\ &- E\xi_{nj}\beta_{n,j+h}E\xi_{nj'}\beta_{n,j'+h} + E\xi_{nj}\beta_{n,j+h}E\xi_{nj'}\beta_{n,j'+h} \\ =&E\xi_{nj}\beta_{n,j+h}\xi_{nj'}\beta_{n,j'+h} - E\xi_{nj}\beta_{n,j+h}E\xi_{nj'}\beta_{n,j'+h} \end{aligned}$$

where

$$\begin{aligned}
&E|\xi_{nj}\beta_{n,j+h}\xi_{nj'}\beta_{n,j'+h}| \\
&\leq \left(E(\xi_{nj}\beta_{n,j+h})^2\right)^{1/2}\left(E(\xi_{nj'}\beta_{n,j'+h})^2\right)^{1/2} \\
&\leq \left(E\xi_{nj}^4\right)^{1/4}\left(E\beta_{n,j+h}^4\right)^{1/4}\left(E\xi_{nj'}^4\right)^{1/4}\left(E\beta_{n,j'+h}^4\right)^{1/4} \\
&= \left(E\xi_{nj}^4 E\beta_{n,j+h}^4 E\xi_{nj'}^4 E\beta_{n,j'+h}^4\right)^{1/4}
\end{aligned}$$

by the Cauchy-Schwarz Inequality, and similarly

$$\begin{aligned}
E|\xi_{nj}\beta_{n,j+h}| &\leq \left(E\xi_{nj}^4 E\beta_{n,j+h}^4\right)^{1/4} \\
E|\xi_{nj'}\beta_{n,j'+h}| &\leq \left(E\xi_{nj'}^4 E\beta_{n,j'+h}^4\right)^{1/4}
\end{aligned}$$

so that together we have

$$\begin{aligned}
&E\left(x_j' A_n Q_{t-j,n} A_n^* y_{j+h} x_{j'}' A_n Q_{t-j,n} A_n^* y_{j'+h}\right) \\
&\leq 2\left(E\xi_{nj}^4 E\beta_{n,j+h}^4 E\xi_{nj'}^4 E\beta_{n,j'+h}^4\right)^{1/4}
\end{aligned}$$

where $nE\xi_{nj}^4 = nE{<}A_n Z_{1n}, x_j{>}^4 = nE{<}A_n Z_1, x_j{>}^4 I(||A_n Z_1|| \leq 1)$ which is bounded above by some $K_2 < \infty$, for all $n \geq n_0$ and all j by an application of part (ii) of Lemma 10.6.4 with $4 = \zeta > 1/a_i$. A similar argument shows that $nE\beta_{nk}^4 < K_2$ for all k and all n large [apply part (ii) of Lemma 10.6.4 again with $4 = \zeta > 1/a_\ell$]. Then (10.92) holds for j, j', h and all n large, and so for all large n we have that $P(|x'(A-C)y| > \varepsilon)$ is bounded above by

$$\begin{aligned}
&\varepsilon^{-2} 2K_2 n^{-1} \sum_{|j|>m} \sum_{t=1}^{n} \sum_{j'=1-t+j}^{n-t+j} ||C_j|| ||C_{j+h}|| ||C_{j'}|| ||C_{j'+h}|| \\
&= \varepsilon^{-2} 2K_2 \sum_{|j|>m} ||C_j|| ||C_{j+h}||\; n^{-1} \sum_{t=1}^{n} \sum_{j'=1-t+j}^{n-t+j} ||C_{j'}|| ||C_{j'+h}|| \\
&\leq \varepsilon^{-2} 2K_2 \Big(\sum_{|j|>m} ||C_j||^2\Big)^{1/2} \Big(\sum_{|j|>m} ||C_{j+h}||^2\Big)^{1/2} \\
&\qquad \cdot \Big(\sum_{j'=-\infty}^{\infty} ||C_{j'}||^2\Big)^{1/2} \Big(\sum_{j'=-\infty}^{\infty} ||C_{j'+h}||^2\Big)^{1/2}
\end{aligned}$$

which tends to zero as $m \to \infty$, and so (10.89) holds with T replaced by $A - C$ in this case. □

Proof of Theorem 10.6.7. By combining Propositions 10.6.8 and 10.6.13, we have

$$
\begin{aligned}
nA_n\Big[\hat{\Gamma}_n(h) - \sum_{j=-\infty}^{\infty} C_j B_n C_{j+h}^*\Big] A_n^* \\
&= A_n\Big[n\hat{\Gamma}_n(h) - \sum_{t=1}^{n}\sum_{j=-\infty}^{\infty} C_j B_n C_{j+h}^*\Big] A_n^* \\
&= A_n\Big[n\hat{\Gamma}_n(h) - \sum_{t=1}^{n}\sum_{j=-\infty}^{\infty} C_j Z_{t-j} Z_{t-j}' C_{j+h}^*\Big] A_n^* \\
&\quad + A_n\Big[\sum_{t=1}^{n}\sum_{j=-\infty}^{\infty} C_j \big(Z_{t-j}Z_{t-j}' - B_n\big) C_{j+h}^*\Big] A_n^* \\
&\Rightarrow 0 + \sum_{j=-\infty}^{\infty} C_j W C_{j+h}^*
\end{aligned}
$$

which concludes the proof. □

Remark 10.6.14. If $a_p < 1/2$, then it follows from Theorem 8.2.14 that EZZ' exists, and consequently the autocovariance matrix $\Gamma(h) = EX_tX_{t+h}'$ is well defined. In this case, Corollary 10.2.6 shows that we can take $B_n = EZZ'$ in (10.79) and then (10.80) becomes

$$
nA_n[\hat{\Gamma}_n(h) - \Gamma(h)]A_n^* \Rightarrow \sum_{j=-\infty}^{\infty} C_j W C_{j+h}^*. \tag{10.93}
$$

Since $A_n \to 0$ slower than $n^{-1/2}$ when $a_p < 1/2$, we also have in this case that $\hat{\Gamma}_n(h) \to \Gamma(h)$ in probability, and (10.93) provides the rate of convergence. If $a_1 > 1/2$, then Corollary 10.2.6 shows that we can take $B_n = 0$ in (10.79) and then (10.80) becomes

$$
nA_n\hat{\Gamma}_n(h)A_n^* \Rightarrow \sum_{j=-\infty}^{\infty} C_j W C_{j+h}^*.
$$

Since $A_n \to 0$ faster than $n^{-1/2}$ when $a_1 > 1/2$, we also have in this case that $\hat{\Gamma}_n(h)$ is not bounded in probability, and (10.93) provides the rate at which it blows up.

In Theorem 10.6.7, the assumption that $Ex'Z = 0$ when the mean exists can be removed if we use the centered version of the sample covariance matrix defined by

$$
\hat{G}_n(h) = \frac{1}{n}\sum_{t=1}^{n}(X_t - \bar{X}_0)(X_{t+h} - \bar{X}_h)' \tag{10.94}
$$

where $\bar{X}_k = \frac{1}{n}\sum_{t=1}^{n} X_{t+k}$ is the sample mean of $X_{k+1}, \ldots, X_{k+n}$ for any $k \geq 0$.

Theorem 10.6.15. *Suppose that $\hat{G}_n(h)$ is the sample autocovariance matrix defined by (10.94), where X_t is the moving average (10.72) and $\{Z_n\}$ are i.i.d. with common distribution μ on $\mathbb{R}^d$. Suppose that μ varies regularly with exponent E, where every eigenvalue of E has real part exceeding 1/4. Then for all h we have*

$$nA_n\Big[\hat{G}_n(h) - \sum_{j=-\infty}^{\infty} C_j B_n C_{j+h}^*\Big]A_n^* \Rightarrow \sum_{j=-\infty}^{\infty} C_j W C_{j+h}^*.$$

Proof. Note that the difference between the two formulas (10.78) and (10.94) can be written in the form

$$\begin{aligned}
&\frac{1}{n}\sum_{t=1}^{n}(X_t - \bar{X}_0)(X_{t+h} - \bar{X}_h)' - \frac{1}{n}\sum_{t=1}^{n} X_t X_{t+h}' \\
&= -\frac{1}{n}\sum_{t=1}^{n} X_t \bar{X}_h' - \frac{1}{n}\sum_{t=1}^{n} \bar{X}_0 X_{t+h}' + \frac{1}{n}\sum_{t=1}^{n} \bar{X}_0 \bar{X}_h' \\
&= -\bar{X}_0 \bar{X}_h'
\end{aligned}$$

and so it suffices to show that $nA_n\bar{X}_0\bar{X}_h'A_n^* \to 0$ in probability. For this, it suffices to show that for all unit vectors $x \in V_i$ and $y \in V_\ell$ we have

$$nx'A_n\bar{X}_0\bar{X}_h'A_n^*y = \sqrt{n}x'A_n\bar{X}_0 \cdot \sqrt{n}y'A_n\bar{X}_h \stackrel{P}{\to} 0$$

and so it suffices to show that $\sqrt{n}x'A_n\bar{X}_0 \to 0$ and $\sqrt{n}y'A_n\bar{X}_h \stackrel{P}{\to} 0$ in probability for any unit vectors $x, y \in V_i$ for any $i = 1, \ldots, p$. Since the proof of both parts is similar, we only consider the first.

If $a_i < 1$, then $E\pi_i Z_t$ exists. Note that (10.94) is not changed if we replace Z_t by $Z_t - E\pi_i Z_t$ so that without loss of generality $E\pi_i Z_t = 0$. Let $S_n = X_1 + \cdots + X_n$ so that $\bar{X}_0 = S_n/n$. In the case $a_i < 1$, Theorem 8.3.7 (d) along with Theorem 8.2.16 (a) shows that we can take $\pi_i b_n = E\pi_i Z_t = 0$ in (10.71) and then we have from (10.74) that $A_n\pi_i S_n = nA_n\pi_i\bar{X}_0 \Rightarrow U$ and so $n^{1/2}A_n\pi_i\bar{X}_0 \to 0$ in probability. It follows that $n^{1/2}x'A_n\bar{X}_0 \to 0$ in probability. If $a_i > 1$, then Theorem 8.3.7 (d) along with Theorem 8.2.16 (b) shows that we can also take $\pi_i b_n = 0$ in (10.71) and it follows as before that $n^{1/2}x'A_n\bar{X}_0 \to 0$ in probability.

If $a_i = 1$, we can apply Lemma 10.6.4 part (ii) with $\zeta = 1 + \varepsilon > 1/a_i$ for any $\varepsilon > 0$. By (10.74), we have $A_n(S_n - na_n) \Rightarrow U$, and so we have $nA_n(\bar{X}_0 - a_n) \Rightarrow U$, and then $nx'A_n(\bar{X}_0 - a_n) \Rightarrow x'U$ by continuous mapping. Then $\sqrt{n}x'A_n(\bar{X}_0 - a_n) \to 0$ in probability and so it suffices to show that $\sqrt{n}x'A_na_n \to 0$, where without loss of generality $a_n = \sum_j C_j b_n$ and $b_n = EZ_1 I(\|A_nZ_1\| \leq 1)$. Argue as in the proof of Lemma 10.6.9 that $U_1(r, \theta) <$

$CU_{1+\varepsilon}(r,\theta)$ for all r large and all θ unit vectors in V_i. Then as in the proof of Lemma 10.6.9 we have (letting $A_n^* x = r_n\theta_n$ with $r_n > 0$ and $\|\theta_n\| = 1$) that

$$\begin{aligned}|nx'A_nb_n| &\leq nE|\langle A_nZ_1,x\rangle| I(\|A_nZ_1\| \leq 1)\\ &\leq C\|\pi_iA_n^{-1}\|^{\varepsilon} nE|\langle A_nZ_1,x\rangle|^{1+\varepsilon} I(|\langle A_nZ_1,x\rangle| \leq 1)\\ &< C\|\pi_iA_n^{-1}\|^{\varepsilon}K_1\end{aligned}$$

for all n large, and so for all large n we have

$$\begin{aligned}|n^{1/2}x'A_na_n| &= n^{-1/2}\Big|\sum_{j=-\infty}^{\infty} nx'A_nC_jb_n\Big|\\ &\leq n^{-1/2}\sum_{j=-\infty}^{\infty}\|C_j\||nx'A_nb_n|\\ &\leq n^{-1/2}C\|\pi_iA_n^{-1}\|^{\varepsilon}K_1\sum_{j=-\infty}^{\infty}\|C_j\|\end{aligned}$$

where the series converges. Since $\log(n^{-1/2}\|\pi_iA_n^{-1}\|^{\varepsilon})/\log n \to -1/2 + a_i\varepsilon < 0$ for small $\varepsilon > 0$ we obtain $n^{1/2}x'A_na_n \to 0$, which concludes the proof. □

Remark 10.6.16. The assumption that $C_jA_n = A_nC_j$ restricts the moving average model. In view of Lemma 10.6.3, this restriction implies that C_j is block diagonal with one block for each spectral component, which rules out interactions between components with different tail behavior. Relaxing this restriction would be useful in some applications, but this is an open problem.

Remark 10.6.17. One important application of stationary vector moving averages is to model periodically stationary scalar moving averages

$$Y_t = \sum_{j=-\infty}^{\infty} \psi_t(j)\varepsilon_{t-j} \tag{10.95}$$

where $\psi_t(j)$ is periodic in t with the same period $d \geq 1$ for all j, and (ε_t) are i.i.d. random variables. If we let $Z_t = (\varepsilon_{td},\ldots,\varepsilon_{(t+1)d-1})'$ and $Y_t = (X_{td},\ldots,X_{(t+1)d-1})'$, then we can rewrite (10.95) in the form (10.72) where C_j is the $d\times d$ matrix with ij entry $\psi_i(td+i-j)$, and we number the rows and columns $0,1,\ldots,d-1$ for ease of notation. If the distribution of (ε_t) is regularly varying, then the distribution μ of (Z_t) is also regularly varying and (10.70) holds with $A_n = a_nI$ for some scalars (scalar-normed R–O variation) and ϕ concentrated on the coordinate axes. Then there is no restriction on the model parameters C_j since scalar multiples of the identity commute with every linear operator. Another important application is to compare scalar moving averages

$$X_{it} = \sum_{j=-\infty}^{\infty} \psi_{ij}\varepsilon_{i,t-j}$$

where (ε_{it}) are i.i.d. random variables for each i and $Z_t = (\varepsilon_{1t}, \ldots, \varepsilon_{dt})'$. Then $X_t = (X_{1t}, \ldots, X_{dt})'$ satisfies (10.72) with $C_j = \operatorname{diag}(\psi_{1j}, \ldots, \psi_{dj})$. If the distribution μ of (Z_n) varies regularly and (10.70) holds with A_n diagonal matrices with respect to some fixed basis for $\mathbb{R}^d$ (vector-normed R–O variation) then there is no restriction on the model parameters, since all diagonal matrices commute.

Remark 10.6.18. If (10.70) and (10.71) hold with $A_n = a_n I$ for some scalars a_n (scalar norming), then we can also obtain the asymptotics of the sample autocorrelations. Write $X_t = X_{1t} + \cdots X_{dt}$ in the standard coordinates. Then the ij entry of the matrix $\hat{G}_n(h)$ is

$$\hat{\gamma}_{ij}(h) = \frac{1}{n}\sum_{t=1}^{n}(X_{it} - \bar{X}_{i,0})(X_{j,t+h} - \bar{X}_{j,h}) \quad \text{where} \quad \bar{X}_{i,k} = \frac{1}{n}\sum_{t=1}^{n} X_{i,t+k}$$

the sample autocovariance of the i and j components at lag h. The asymptotics of the sample autocorrelation of the i and j components at lag h defined by

$$\hat{\rho}_{ij}(h) = \frac{\hat{\gamma}_{ij}(h)}{\sqrt{\hat{\gamma}_{ii}(0)\hat{\gamma}_{jj}(0)}}$$

can be obtained using the Continuous Mapping Theorem in this case. See [3] for the case of periodic moving averages, and [4] for an application to Hydrology.

10.7 NOTES AND COMMENTS

The material in Section 10.1 is taken from Mandrekar and Meerschaert [103]. Theorem 10.1.2 extends results of Tucker [191] and Seidl' [176]. These papers also proved the characteristic function formula in Corollary 10.1.4. Operator stable laws and generalized domains of attraction on $\mathbb{R}^\infty$ are used in Remark 10.1.10. The theory of operator normed limit theorems on infinite dimensional spaces is still developing. Krakowiak [83] and Siegel [185] established the existence of exponents for operator stable laws on Banach spaces. Seigel [184] contains some interesting results on operator normed domains of attraction in Banach spaces. Rubin and Vitale [156] pioneered the modern study of symmetric statistics. Symmetric statistics are also considered in a more general setting by Dynkin and Mandelbaum [37], Avram and Taqqu [7], and Mandrekar and Patterson [102].

Sections 10.2 and 10.3 follow Meerschaert and Scheffler [128]. Since the limiting distribution in Theorem 10.2.3 is not always full dimensional, one might ask whether a different norming would produce a full limit. However, even when X has independent stable marginals their cross-product will not usually belong to any domain of attraction, see Cline [26]. Asymptotics of the sample covariance matrix are also useful in the study of linear regression models, see for example Cline [27]. Self-normalized sums for i.i.d. random variables in the domain of attraction of a nonnormal stable law were considered by Logan, Mallows, Rice and Shepp [89], see Corollary 10.1.8. Multivariate self-normalized sums for i.i.d. random vectors in the generalized domain of attraction of a normal law are considered in Sepanski [178, 179]. When $X \in \mathrm{GDOA}(Y)$ normal, Sepanski [178] shows that $M_n^{-1/2} S_n \Rightarrow Y$ even if $E\|X\|^2 = \infty$. Vu, Maller, and Klass [194] extend this result to certain dependent sequences.

The material in Section 10.4 is from Scheffler [166]. The special case of operator stable limits was considered in Meerschaert and Scheffler [127]. Similar tail estimation methods were used in the one variable case by Meerschaert and Scheffler [125]. Many other tail estimation techniques are available for scalar data, see McCulloch [109] for a survey and critique of several common methods. The proliferation of tail estimates seems to stem from the fact that, even in the one variable case, estimating the regular variation parameter α is a difficult task.

Section 10.5 is taken from Meerschaert and Scheffler [132]. Despite the ubiquity of tensor notation in the physical sciences, probabilists have paid very little attention to the problem of estimating k-tensors. Tensors and tensor products are in fact quite useful in probability and statistics, see for example McCullagh [106], Parthasarathy and Schmidt [145] and Takemura [189]. Random tensors are also considered in Masuyama [105]. The most familiar example is the sample covariance matrix, which is a random 2-tensor. The tensors considered in Section 10.5 are also the vector analogue of sample moments for i.i.d. random variables, and in this sense the results of Section 10.5 extend those of Section 10.1.

Section 10.6 follows Meerschaert and Scheffler [133]. Time series models with heavy tail innovations have been studied by Adams and Goodwin [1], Anderson, et al. [3, 4, 5], Balkema and de Haan [9], Bhansali [15], Brockwell and Davis [22], Hannan and Kanter [54], Klüppelberg and Mikosch [80], Knight [81], Kokoszka and Taqqu [82], and numerous others. Davis and Resnick [32] proved Theorems 10.6.2 and 10.6.7 on $\mathbb{R}^1$ (see also [33, 34]). Davis, et al. [35, 36] extended these results to moving averages of random vectors with scalar-normed regularly varying tails. Time series with heavy tails occur in many areas of application, including electrical engineering [140], hydrology [4], and finance [39, 99, 138]. Estimation of the index α of regular variation has developed into quite a controversy in finance, see for example [72, 90, 107].

11

Self-Similar Stochastic Processes

Self-similarity is a useful physical and mathematical concept. A power function $y = f(x) = x^\rho$ is *self-similar* in the sense that $f(x) = c^{-\rho} f(cx)$ for all $c > 0$. In other words, for any linear rescaling in x there is an appropriate rescaling in y that recovers the original function. A regularly varying function $R : \mathbb{R}^+ \to \mathbb{R}^+$ with index ρ is asymptotically self-similar in the sense that $R(cx) \sim c^\rho R(x)$ as $x \to \infty$ for all $c > 0$. A function $\varphi : \mathbb{R}^d \to [0, \infty)$ is *operator self-similar* if $c \cdot \varphi = c^E \varphi$ for any $c > 0$, and a regularly varying real-valued function on $\mathbb{R}^d$ is asymptotically operator self-similar. The same idea also applies to regularly varying measures. In this chapter, we will extend this important idea to stochastic processes. Scaling properties of natural phenomena are important in many classical physical models. In modern science, the emergence of fractals as a useful physical model has intensified the interest in scaling properties and self-similarity.

11.1 OPERATOR SELF-SIMILAR PROCESSES

A stochastic process on $\mathbb{R}^d$ is *operator self-similar* if any linear rescaling in time produces a new process whose finite dimensional distributions scale back to those of the original process under some affine transformation. The best known example is a standard Brownian motion $Y(t)$, which has the same finite dimensional distributions as the process $c^{-1/2}Y(ct)$ for any $c > 0$. The theory of operator self-similar processes and their generalized domains of attraction extends the theory of operator stable laws. Many of the results from Chapter

7 also extend to this more general setting, and as before, regular variation is quite useful in the statement and proof of some fundamental results.

Definition 11.1.1. A stochastic process $\{Y(t)\}_{t\geq 0}$ on $\mathbb{R}^d$ is *continuous in law* if $Y(t_n) \Rightarrow Y(t)$ whenever $t_n \to t$.

Definition 11.1.2. A stochastic process $\{Y(t)\}_{t\geq 0}$ on $\mathbb{R}^d$ is *operator self-similar* if it is continuous in law and there exists a linear operator E on $\mathbb{R}^d$ and nonrandom vectors $\{d(t)\}_{t\geq 0}$ in $\mathbb{R}^d$ such that

$$\{Y(ct)\} \stackrel{d}{=} \{c^E Y(t) + d(c)\} \quad \text{for all } c > 0 \tag{11.1}$$

where $\stackrel{d}{=}$ means the equality of all finite dimensional marginal distributions of the processes. The linear operator E is called an *exponent* of the operator self-similar process, and we denote the set of all exponents of $\{Y(t)\}$ by $\mathcal{E}(\{Y(t)\})$.

As in the case of domains of attraction, in order to obtain useful results we need to assume that the random vectors $Y(t)$ are not supported on a lower dimensional subspace.

Definition 11.1.3. A stochastic process $\{Y(t)\}$ on $\mathbb{R}^d$ is *proper* if for all $t > 0$ the distribution of $Y(t)$ is full, i.e., if it cannot be supported on any $d-1$ dimensional hyperplane.

Definition 11.1.4. We say that a linear operator $A \in \mathrm{L}(\mathbb{R}^d)$ is a *symmetry* of the stochastic process $\{Y(t)\}$ if

$$\{Y(t)\} \stackrel{d}{=} \{AY(t) + a\}$$

for some $a \in \mathbb{R}^d$. We denote the set of symmetries by $\mathcal{S}(\{Y(t)\})$.

The exponent E of an operator self-similar process is not unique in general, because of possible symmetries. The following result is due to Hudson and Mason [65]. See also Theorems 5.2.13 and 6.1.21.

Theorem 11.1.5. *The collection of symmetries $\mathcal{S}(\{Y(t)\})$ forms a compact group, and for any $E \in \mathcal{E}(\{Y(t)\})$ we have*

$$\mathcal{E}(\{Y(t)\}) = E + T\mathcal{S}(\{Y(t)\})$$

The following result is due to Maejima [95]. See also Theorem 5.2.13 and Theorem 6.1.21.

Theorem 11.1.6. *For some $E_0 \in \mathcal{E}(\{Y(t)\})$ we have $E_0 A = A E_0$ for every $A \in \mathcal{S}(\{Y(t)\})$.*

Corollary 11.1.7. *Every $B \in \mathcal{E}(\{Y(t)\})$ has the same real spectrum.*

Proof. The proof is identical to Theorem 5.2.14, using Theorems 11.1.5 and 11.1.6 in place of Theorem 5.2.13. □

Now we develop the spectral decomposition for these processes. Recall from Definition 2.1.15 that for any linear operator E on $\mathbb{R}^d$ there exists a direct sum decomposition $\mathbb{R}^d = V_1 \oplus \cdots \oplus V_p$ into E-invariant subspaces such that, if we write $E = E_1 \oplus \cdots \oplus E_p$ where $E_i : V_i \to V_i$ then every eigenvalue of E_i has real part equal to a_i and $a_1 < \cdots < a_p$. This is called the spectral decomposition of $\mathbb{R}^d$ with respect to E.

Definition 11.1.8. Suppose that $\{Y(t)\}$ is a proper operator self-similar process on $\mathbb{R}^d$ and that E is an exponent of $\{Y(t)\}$. Let $V_1, \ldots, V_p$ denote the spectral decomposition of $\mathbb{R}^d$ relative to E, and $\pi_i : \mathbb{R}^d \to V_i$ the projections associated with this direct sum decomposition. Write $Y(t) = Y_1(t) + \cdots + Y_p(t)$ for $t \geq 0$ with respect to this direct sum decomposition. Then the stochastic processes $\{Y_i(t)\}$ for $i = 1, \ldots, p$ are called the *spectral decomposition of* $\{Y(t)\}$ *with respect to* E. If every eigenvalue of E has the same real part, then there is only one component in the spectral decomposition. In this case we say that $\{Y(t)\}$ is *spectrally simple.*

Lemma 11.1.9. *Suppose that* $\{Y(t)\}$ *is a proper operator self-similar process on* $\mathbb{R}^d$ *with exponent* E. *Then every* $\{Y_i(t)\}$ *in the spectral decomposition of* $\{Y(t)\}$ *relative to* $E = E_1 \oplus \cdots \oplus E_p$ *is a proper operator self-similar process on* V_i *with exponent* E_i. *Furthermore,* $\{Y_i(t)\}$ *is spectrally simple.*

Proof. Let $\mathbb{R}^d = V_1 \oplus \cdots \oplus V_p$ denote the spectral decomposition of $\mathbb{R}^d$ relative to E, and write $d(t) = d_1(t) + \cdots + d_p(t)$ with respect to this direct sum decomposition. By using (11.1) along with the fact that the subspaces V_i are E-invariant, we have

$$\begin{aligned} Y_1(ct) + \cdots + Y_p(ct) &= Y(ct) \\ &\stackrel{d}{=} c^E Y(t) + d(c) \\ &= \left(c^{E_1} Y_1(t) + d_1(c)\right) + \cdots + \left(c^{E_p} Y_p(t) + d_p(c)\right) \end{aligned}$$

for all $c > 0$. Projection onto V_i yields

$$\{Y_i(ct)\}_{t \geq 0} \stackrel{d}{=} \{c^{E_i} Y_i(t) + d_i(c)\}_{t \geq 0} \qquad \forall c > 0 \tag{11.2}$$

for all $i = 1, 2, \ldots, p$. Moreover $\{Y_i(t)\}$ is continuous in law and proper, so $\{Y_i(t)\}$ is a proper operator self-similar process on V_i with exponent E_i, where the real part of the eigenvalues of the exponent E_i are all equal to a_i. □

We say that $\{Y(t)\}$ is *strictly operator self-similar* if the centering function $d(c)$ in (11.1) is identically zero. Hudson and Mason [65] show that, for any proper operator self-similar process $\{Y(t)\}$, there exists a nonrandom vector $y \in \mathbb{R}^d$ such that $\{Y(t) - y\}$ is strictly operator self-similar. For the remainder of this section, we will assume that $\{Y(t)\}$ is strictly operator self-similar, and this entails no loss of generality.

The fact that $\{Y(t)\}$ is continuous in law at $t = 0$ restricts the real spectrum of the exponent D. Let $a_1 < \ldots < a_p$ denote the real spectrum of the exponent

D. Theorem 3 in Hudson and Mason [65] shows that $a_1 \geq 0$. If $a_1 = 0$, then, by Theorem 4 of [65], $\{Y_1(t)\}$ has constant sample paths. On the other hand, Maejima and Mason [93] show that, if $a_1 > 0$, then $Y(0) = 0$ almost surely. Then $Y(t) \to 0$ in probability as $t \to 0$. The following simple result implies that $Y(t)$ grows faster than $t^{a_1-\varepsilon}$ for t near zero, and slower than $t^{a_p+\varepsilon}$ for t large, for any $\varepsilon > 0$.

Theorem 11.1.10. *Suppose that $\{Y(t)\}_{t\geq 0}$ is a proper strictly operator self-similar process on $\mathbb{R}^d$ with exponent E, and let $0 < a_1 < \ldots < a_p$ denote the real spectrum of the exponent E.*

(a) If $\alpha < a_1$, then $Z(t) = t^{-\alpha}Y(t)$ is a proper strictly operator self-similar process with exponent $E - \alpha I$.

(b) If $\beta > a_p$, then the stochastic process defined by $Z(t) = t^{\beta}Y(1/t)$ for $t > 0$, and $Z(0) = 0$ almost surely, is a proper strictly operator self-similar process with exponent $\beta I - E$.

Proof. If $\alpha < a_1$, then from (11.1) with $d(c) = 0$ we obtain

$$\begin{aligned}\{Z(ct)\} &= \{(ct)^{-\alpha}Y(ct)\}\\ &\stackrel{d}{=} \{c^{-\alpha}t^{-\alpha}c^{E}Y(t)\}\\ &= \{c^{E-\alpha I}t^{-\alpha}Y(t)\}\\ &= \{c^{E-\alpha I}Z(t)\}\end{aligned}$$

for any $c > 0$, so that $\{Y(t)\}$ is strictly operator self-similar with exponent $E - \alpha I$. If $\beta > a_p$, then

$$\begin{aligned}\{Z(ct)\} &= \{(ct)^{\beta}Y(c^{-1}t^{-1})\}\\ &\stackrel{d}{=} \{c^{\beta}t^{\beta}c^{-E}Y(t^{-1})\}\\ &= \{c^{\beta I-E}t^{\beta}Y(1/t)\}\\ &= \{c^{\beta I-E}Z(t)\}\end{aligned}$$

for all $c > 0$ so that $\{Z(t)\}$ is strictly operator self-similar with exponent $\beta I - E$. Note that $c^{\beta I-E} \to 0$ in norm as $c \to 0$ so that $Z(t)$ is continuous in law with $Z(0) = 0$. □

Now we will apply the spectral decomposition to sharpen the results of Theorem 11.1.10. This result implies that the spectral component $X_i(t)$ grows faster than $t^{a_i-\varepsilon}$ for t near zero, and slower than $t^{a_i+\varepsilon}$ for t large, for any $\varepsilon > 0$. Hence the real spectrum of the exponent E characterizes the growth rate of the process.

Corollary 11.1.11. *Suppose that $\{Y(t)\}_{t\geq 0}$ is a proper strictly operator self-similar process with exponent E. Let $V_1 \oplus \cdots \oplus V_p$ be the spectral decomposition of $\mathbb{R}^d$ with respect to E and $0 < a_1 < \cdots < a_p$ the real spectrum of E. Write*

$Y(t) = Y_1(t) + \cdots + Y_p(t)$ and $E = E_1 \oplus \cdots \oplus E_p$ with respect to this direct sum decomposition. Then for $i = 1, \ldots, p$

(a) *If $\alpha < a_i$, then $Z_i(t) = t^{-\alpha} Y_i(t)$ is a proper strictly operator self-similar process with exponent $E_i - \alpha I$.*

(b) *If $\beta > a_i$, then the stochastic process defined by $Z_i(t) = t^{\beta} Y_i(1/t)$ for $t > 0$, and $Z_i(0) = 0$ almost surely, is a proper strictly operator self-similar process with exponent $\beta I - E_i$.*

Proof. Apply Theorem 11.1.10 to the proper strictly operator self-similar component process $\{Y_i(t)\}$, which is spectrally simple with exponent E_i where every eigenvalue of E_i has real part equal to a_i. □

We can also use the spectral decomposition to obtain sharp upper and lower bounds on the growth rate of the norm and radial projections of the process $\{Y(t)\}$. Since these results can also be obtained for stochastic processes in the generalized domain of attraction of a proper operator self-similar process by essentially the same methods, we will defer these results to Section 11.2. Every proper operator self-similar process belongs to its own generalized domain of attraction, and is spectrally compatible with itself. Therefore, sharp upper and lower bounds on $\|Y_i(t)\|$ and $|\langle Y(t), \theta \rangle|$ can be deduced from the results in the next section by letting $\{Y(t)\} = \{X(t)\}$.

Example 11.1.12. Hudson and Mason [65] show that if $\{Y(t)\}$ is a proper stochastic process that is continuous in law and has stationary independent increments with $Y(0) = 0$, then $\{Y(t)\}$ is operator self-similar if and only if $Y(1)$ is strictly operator stable. In this case, every exponent E of $Y(1)$ is also an exponent of $\{Y(t)\}$. We will call this process an *operator Lévy motion*. If $Y(1)$ is normal with mean zero and covariance matrix M, then $Y(t)$ is normal with mean zero and covariance matrix tM. This *Brownian motion* $\{Y(t)\}$ is operator self-similar with

$$\{Y(ct)\} \stackrel{d}{=} c^{1/2}\{Y(t)\} \quad \text{for all } c > 0.$$

When the exponent E of an operator self-similar process is of the form $E = aI$, we also say that $\{Y(t)\}$ is *self-similar* with index a. If $Y(1)$ is operator stable with exponent $E = aI$, then $Y(1)$ is also stable with index $\alpha = 1/a$. This *Lévy motion* $\{Y(t)\}$ is self-similar with index a, the case $a = 1/2$ corresponding to a Brownian motion.

Remark 11.1.13. There is an interesting connection between stable Lévy motions, fractional derivatives, and fractals. Brownian motions and Lévy motions were introduced in Example 11.1.12. When $d > 1$, the sample paths of a Lévy motion $\{Y(t)\}$ with self-similarity index a are random fractals with dimension $\alpha = 1/a$, see the excellent review article of Taylor [190]. In statistical mechanics, particle diffusion is modeled as a Lévy motion $\{Y(t)\}$ where $Y(t)$

denotes the location of a randomly selected particle at time $t \geq 0$. Theorem 7.2.7 shows that $Y(t)$ has a density $P(x,t)$ with respect to Lebesgue measure. The probability densities $P(x,t)$ are also the point-source solution to a fractional diffusion equation, where α is the order of the fractional derivative. The fractional diffusion equation in $\mathbb{R}^1$ is

$$\frac{\partial P(x,t)}{\partial t} = -v\frac{\partial P(x,t)}{\partial x} + qa\frac{\partial^\alpha P(x,t)}{\partial(-x)^\alpha} + pa\frac{\partial^\alpha P(x,t)}{\partial x^\alpha} \tag{11.3}$$

where $v \in \mathbb{R}^1$, $a > 0$, $1 < \alpha \leq 2$, $p \geq 0$, $q \geq 0$ and $p + q = 1$. By using the Fourier transform $\hat{P}(k,t) = \int e^{-i<k,x>} P(x,t)dx$, so that $\hat{P}(-k,t)$ is the characteristic function of $Y(t)$, the fractional derivative $\partial^\alpha P(x,t)/\partial(\pm x)^\alpha$ has Fourier transform $(\pm ik)^\alpha \hat{P}(k,t)$, extending the familiar formula where α is a positive integer, see, e.g., Samko, Kilbas and Marichev [160]. By taking Fourier transforms in (11.3), we obtain

$$\frac{d\hat{P}(k,t)}{dt} = -v(ik)\hat{P}(k,t) + qa(-ik)^\alpha \hat{P}(k,t) + pa(ik)^\alpha \hat{P}(k,t) \tag{11.4}$$

so that $\hat{P}(k,t) = \exp\left(-tv(ik) + tqa(-ik)^\alpha + tpa(ik)^\alpha\right)$ for the point source initial condition $\hat{P}(k,0) \equiv 1$, corresponding to $Y(0) = 0$ almost surely. Then an argument similar to the proof of Theorem 7.3.5 shows that $P(x,t)$ is the density of a stable law $Y(t)$ with distribution $S_\alpha(\sigma, p-q, tv)$ in the notation of Definition 7.3.6, with $\sigma^\alpha = -ta\cos(\pi\alpha/2)$. If $0 < \alpha < 1$, then the same result holds with $a < 0$, so that $\sigma^\alpha = -ta\cos(\pi\alpha/2)$ remains positive. If $\alpha = 2$ the fractional diffusion equation (11.3) reduces to the classical diffusion equation, and $P(x,t)$ is a normal density with mean tv and variance $2ta$. Fractional derivatives are almost as old as ordinary derivatives, see Miller and Ross [137] for a brief history. Benson et al. [12, 14] apply the fractional diffusion equation (11.3) to model the spread of contaminant plumes in underground water. Shlesinger et al. [174] contains several additional applications to physics.

11.2 SCALING LIMITS

Generalized domains of attraction of operator self-similar processes extend the idea of the generalized domain of attraction of an operator stable law. Many of the results from Chapter 8 also extend to this more general setting, and as before, regular variation is quite useful in the statement and proof of some fundamental results.

Definition 11.2.1. We say that a stochastic process $\{X(t)\}_{t\geq 0}$ belongs to the *generalized domain of attraction* of a proper process $\{Y(t)\}_{t\geq 0}$ if there exist invertible linear operators $\{A(t)\}_{t\geq 0}$ on $\mathbb{R}^d$ and nonrandom vectors $\{a(t)\}_{t\geq 0}$ in $\mathbb{R}^d$ such that

$$\{A(c)X(ct) + a(c)\}_{t\geq 0} \stackrel{d}{\Rightarrow} \{Y(t)\}_{t\geq 0} \quad \text{as } c \to \infty \tag{11.5}$$

where $\stackrel{d}{\Rightarrow}$ denotes the convergence of all finite dimensional marginal distributions. In this case we write $\{X(t)\} \in \mathrm{GDOA}(\{Y(t)\})$.

The following result is due to Hudson and Mason [65].

Theorem 11.2.2. *For a proper stochastic process $\{Y(t)\}_{t\geq 0}$ that is continuous in law, $\mathrm{GDOA}(\{Y(t)\}) \neq \emptyset$ if and only if $\{Y(t)\}$ is operator self-similar. A proper operator self-similar process always belongs to its own generalized domain of attraction.*

The following results are due to Maejima [95]. They show that we can always choose the scaling operators in (11.5) to be regularly varying, and they also establish a Convergence of Types result for operator self-similar processes.

Theorem 11.2.3. *Let $\{Y(t)\}$ be a proper operator self-similar process with exponent E. If $\{X(t)\} \in \mathrm{GDOA}(\{Y(t)\})$ and (11.5) holds, then*

$$A(\lambda t)A(t)^{-1} \rightharpoonup \lambda^{-E}\mathcal{S}(\{Y(t)\}) \quad \text{as } t \to \infty \tag{11.6}$$

for all $\lambda > 0$.

Theorem 11.2.4. *Let $\{Y(t)\}$ be a proper operator self-similar process with exponent E. If $\{X(t)\} \in \mathrm{GDOA}(\{Y(t)\})$ and (11.5) holds, then we also have*

$$\{B(c)X(ct) + b(c)\}_{t\geq 0} \stackrel{d}{\Rightarrow} \{Y(t)\}_{t\geq 0} \quad \text{as } c \to \infty \tag{11.7}$$

for some $B(t)$ regularly varying with index $-E$ and some $b(t) \in \mathbb{R}^d$. Furthermore, if (11.5) holds then

$$B(t)A(t)^{-1} \rightharpoonup \mathcal{S}(\{Y(t)\}) \quad \text{as } t \to \infty. \tag{11.8}$$

Recall from Definition 4.3.9 that for $A, B : \mathbb{R}^+ \to \mathrm{GL}(\mathbb{R}^d)$ we write $A \sim B$ if $A(t)B(t)^{-1} \to I$ as $t \to \infty$. This defines an equivalence relation. The following lemma shows that the norming function A in (11.5) can be replaced by any $B \sim A$, without affecting the limit process.

Lemma 11.2.5. *Assume that $\{X(t)\} \in \mathrm{GDOA}(\{Y(t)\})$ and (11.5) holds. Then for any $B \sim A$ there exists a shift function $b : \mathbb{R}^+ \to \mathbb{R}^d$ such that*

$$\{B(c)Y(ct) + b(c)\}_{t\geq 0} \stackrel{d}{\Rightarrow} \{Y(t)\}_{t\geq 0} \quad \text{as } c \to \infty.$$

Proof. Let $b(c) = B(c)A(c)^{-1}a(c)$. Fix any $n \geq 1$ and any $0 \leq t_1 < \cdots < t_n$ and define for $c > 0$ a mapping $\Psi(c) : (\mathbb{R}^d)^n \to (\mathbb{R}^d)^n$ by letting $\Psi(c)(x_1, \ldots, x_n) = (B(c)A(c)^{-1}x_1, \ldots, B(c)A(c)^{-1}x_n)$. Then $\Psi(c)$ is a linear operator and since $B \sim A$ it follows that $\Psi(c) \to Id$ as $c \to \infty$, where Id is the identity on $(\mathbb{R}^d)^n$. Moreover, by assumption $(A(c)X(ct_i) + a(c))_{i=1,\ldots,n} \Rightarrow (Y(t_i))_{i=1,\ldots,n}$ and hence the Convergence of Types Theorem 2.3.17 yields

$$\begin{aligned}(B(c)X(ct_i) + b(c))_{i=1,\ldots,n} &= \Psi(c)(A(c)X(ct_i) + a(c))_{i=1,\ldots,n} \\ &\Rightarrow Id(Y(t_i))_{i=1,\ldots,n} = (Y(t_i))_{i=1,\ldots,n}\end{aligned}$$

which concludes the proof. □

Definition 11.2.6. Suppose that $\{X(t)\} \in \text{GDOA}(\{Y(t)\})$ and (11.5) holds, where $\{Y(t)\}$ is a proper operator self-similar process with exponent E. Let $V_1, \ldots, V_p$ denote the spectral decomposition of $\mathbb{R}^d$ relative to E, and write $X(t) = X_1(t) + \cdots + X_p(t)$ for $t \geq 0$ with respect to this direct sum decomposition. If every V_i is an $A(c)$-invariant subspace for each $c > 0$ then we say that $\{X(t)\}$ is *spectrally compatible* with $\{Y(t)\}$ (or E), and we call the stochastic processes $\{X_i(t)\}$ the *spectral decomposition of* $\{X(t)\}$ *with respect to* E.

Lemma 11.2.7. *Suppose that* $\{X(t)\} \in \text{GDOA}(\{Y(t)\})$ *and* (11.5) *holds. If* $\{X(t)\}$ *is spectrally compatible with* $\{Y(t)\}$ *then* $\{X_i(t)\} \in \text{GDOA}(\{Y_i(t)\})$ *for each* $i = 1, \ldots, p$, *where* $\{Y_1(t)\}, \ldots, \{Y_p(t)\}$ *is the spectral decomposition of* $\{Y(t)\}$ *with respect to* E. *Furthermore, if we write* $A(c) = A_1(c) \oplus \cdots \oplus A_p(c)$ *and* $a(c) = a_1(c) \oplus \cdots \oplus a_p(c)$ *with respect to this direct sum decomposition, then*

$$\{A_i(c)X_i(ct) + a_i(c)\}_{t\geq 0} \stackrel{d}{\Rightarrow} \{Y_i(t)\}_{t\geq 0} \quad \text{as } c \to \infty \tag{11.9}$$

for each $i = 1, \ldots, p$.

Proof. Project (11.5) onto V_i and use the fact that $A(c)$ commutes with this projection. □

Theorem 11.2.8. *If* $\{X(t)\}$ *belongs to some generalized domain of attraction, then there exists some proper operator self-similar process* $\{Y(t)\}$, *some* $E \in \mathcal{E}(\{Y(t)\})$ *and some* $A : \mathbb{R}^+ \to \text{GL}(\mathbb{R}^d)$ *such that*

(a) $\{X(t)\} \in \text{GDOA}(\{Y(t)\})$ *and* (11.5) *holds*

(b) $\{X(t)\}$ *is spectrally compatible with* E

(c) $A(c)$ *varies regularly with index* $-E$

(d) *the subspaces* $V_1, \ldots, V_p$ *in the spectral decomposition of* $\mathbb{R}^d$ *with respect to* E *are mutually orthogonal*

(e) *each of these subspaces are* $A(c)$ *invariant for every* $c > 0$

Proof. Suppose $\{X(t)\}$ belongs to the generalized domain of attraction of some proper operator self-similar process $\{Y'(t)\}$. If E' is any exponent of $\{Y'(t)\}$, then by Theorem 11.2.4 there exists $A'(c)$ regularly varying with index $-E'$ such that

$$\{A'(c)X(ct) + a'(c)\}_{t\geq 0} \stackrel{d}{\Rightarrow} \{Y'(t)\}_{t\geq 0} \quad \text{as } c \to \infty \tag{11.10}$$

for some $a'(c) \in \mathbb{R}^d$. Corollary 4.3.12 along with Theorem 4.2.9 imply that $TA'(c) \sim A(c)$ as $c \to \infty$ for some $T \in \text{GL}(\mathbb{R}^d)$ and some $A(c)$ regularly varying with index $-E = -TE'T^{-1}$ such that (c), (d), and (e) hold. Since $A'(c) \sim T^{-1}A(c)$ it follows from (11.10) in view of Lemma 11.2.5 that

$$\{T^{-1}A(c)X(ct) + a'(c)\}_{t\geq 0} \stackrel{d}{\Rightarrow} \{Y'(t)\}_{t\geq 0} \quad \text{as } c \to \infty \tag{11.11}$$

and then it follows easily that

$$\{A(c)X(ct) + Ta'(c)\}_{t\geq 0} \stackrel{d}{\Rightarrow} \{TY'(t)\}_{t\geq 0} \quad \text{as } c \to \infty$$

which proves that (11.5) holds with $Y(t) = TY'(t)$ and $a(c) = Ta'(c)$. Then (a) holds. Since $E' \in \mathcal{E}(\{Y'(t)\})$ we have from Definition 11.1.2 that

$$\{Y'(ct)\} \stackrel{d}{=} \{c^{E'}Y'(t) + d'(c)\} \quad \text{for all } c > 0 \tag{11.12}$$

for some $d'(t) \in \mathbb{R}^d$. Now use (11.12) along with the fact that $c^E = Tc^{E'}T^{-1}$ to see that

$$\begin{aligned} \{Y(ct)\} &\stackrel{d}{=} \{T(c^{E'}Y'(t) + d'(c))\} \\ &= \{(Tc^{E'}T^{-1})TY(t) + Td'(c))\} \\ &= \{c^E Y(t) + Td'(c))\} \end{aligned}$$

for all $c > 0$. Hence E is an exponent of the proper operator self-similar process $\{Y(t)\}$, and (b) follows immediately. □

Now, we obtain sharp bounds on the growth of the process $\{X(t)\}$ over time. We will obtain bounds on both the norm of the process, and the radial projections.

Theorem 11.2.9. *Suppose that $\{Y(t)\}_{t\geq 0}$ is a proper strictly operator self-similar process on $\mathbb{R}^d$ with exponent E, and that $\{X(t)\} \in \text{GDOA}(\{Y(t)\})$. Let $0 < a_1 < \cdots < a_p$ denote the real spectrum of E. Then*

(a) For any $\beta > a_p$, $P[\|X(t)\| > t^\beta] \to 0$ as $t \to \infty$

(b) If $P[Y(1) = 0] = 0$, then for any $\alpha < a_1$, $P[\|X(t)\| < t^\alpha] \to 0$ as $t \to \infty$

Proof. Apply Theorem 11.2.8 to obtain a limiting process such that $\{Y(t)\}$ and $\{X(t)\}$ are spectrally compatible, and such that (11.5) holds with $A(c)$ regularly varying. Recall that for any proper operator self-similar process $\{Y(t)\}$, there exists a nonrandom vector $y \in \mathbb{R}^d$ such that $\{Y(t) - y\}$ is strictly operator self-similar. Replacing the centering constants $a(c)$ in (11.5) by $a(c) - y$ yields a strictly operator self-similar limiting process. By letting $t = 0$ in (11.5), we see that $A(c)X(0) + a(c) \Rightarrow Y(0)$ where the limit is almost surely equal to zero. Since $a_1 > 0$, it follows from Theorem 4.2.4 that $\|A(c)\| \to 0$. Then $A(c)X(0) \to 0$ in probability, and so we also have $a(c) \to 0$. Hence we obtain the same limit in (11.5) if we take $a(c) = 0$ for all $c > 0$.

Then without loss of generality we can take both $d(c)$ in (11.1) and $a(c)$ in (11.5) to be identically zero.

If $\beta > a_p$ choose $\varepsilon > 0$ so that $\beta - \varepsilon > a_p$. It is easy to check that $t^{\beta-\varepsilon}A(t)$ varies regularly with index $(\beta-\varepsilon)I-E$, where every eigenvalue of $(\beta-\varepsilon)I-E$ has positive real part. Then $t^{\beta-\varepsilon}\|A(t)x\| \to \infty$ as $t \to \infty$ uniformly on $\{x : \|x\| = 1\}$ by Theorem 4.2.4. Choose $t_0 > 0$ such that $t^{\beta-\varepsilon}\|A(t)y\| > 1$ for all $t \geq t_0$ and all $\|y\| = 1$. It follows that $t^{-\varepsilon}\|A(t)y\| > t^{-\beta}\|y\|$ for all $t \geq t_0$ and all $y \neq 0$. Given $\delta > 0$ choose $M > 0$ such that $P[\|Y(1)\| > M] < \delta$ and $P[\|Y(1)\| = M] = 0$. Then for all $t \geq t_0$ for which $t^\varepsilon > M$ we have

$$\begin{aligned} P[\|X(t)\| > t^\beta] &= P[t^{-\beta}\|X(t)\| > 1] \\ &\leq P[t^{-\varepsilon}\|A(t)X(t)\| > 1] \\ &= P[\|A(t)X(t)\| > t^\varepsilon] \\ &\leq P[\|A(t)X(t)\| > M] \\ &\to P[\|Y(1)\| > M] < \delta \end{aligned}$$

as $t \to \infty$. Since $\delta > 0$ can be made arbitrarily small, we see that (a) holds.

If $\alpha < a_1$ choose $\varepsilon > 0$ so that $\alpha + \varepsilon < a_1$. Since $t^{\alpha+\varepsilon}A(t)$ varies regularly with index $(\alpha+\varepsilon)I - E$, where every eigenvalue of $(\alpha+\varepsilon)I - E$ has negative real part, we have $t^{\alpha+\varepsilon}\|A(t)\| \to 0$ as $t \to \infty$ by Theorem 4.2.4. Choose $t_0 > 0$ such that $t^{\alpha+\varepsilon}\|A(t)y\| < 1$ for all $t \geq t_0$ and all $\|y\| = 1$. It follows that $t^\varepsilon\|A(t)y\| < t^{-\alpha}\|y\|$ for all $t \geq t_0$ and all $y \neq 0$. Given $\delta > 0$ choose $m > 0$ such that $P[\|Y(1)\| < m] < \delta$ and $P[\|Y(1)\| = m] = 0$. Then for all $t \geq t_0$ for which $t^{-\varepsilon} < m$ we have

$$\begin{aligned} P[\|X(t)\| < t^\alpha] &= P[t^{-\alpha}\|X(t)\| < 1] \\ &\leq P[t^\varepsilon\|A(t)X(t)\| < 1] \\ &= P[\|A(t)X(t)\| < t^{-\varepsilon}] \\ &\leq P[\|A(t)X(t)\| < m] \\ &\to P[\|Y(1)\| < m] < \delta \end{aligned}$$

as $t \to \infty$. Since $\delta > 0$ can be made arbitrarily small, we see that (b) holds. □

Corollary 11.2.10. *Suppose $\{X(t)\}$ belongs to some generalized domain of attraction. Let $\{Y(t)\}$ and $E \in \mathcal{E}(\{Y(t)\})$ be as in Theorem 11.2.8 and let $\{X_1(t)\}, \ldots, \{X_p(t)\}$ denote the spectral decomposition of $\{X(t)\}$ with respect to the exponent E, where E has real spectrum $a_1 < \cdots < a_p$. Then*

(a) For any $\beta > a_i$, $P[\|X_i(t)\| > t^\beta] \to 0$ as $t \to \infty$

(b) If $P[Y_i(1) = 0] = 0$, then for any $\alpha < a_i$, $P[\|X_i(t)\| < t^\alpha] \to 0$ as $t \to \infty$

Proof. Apply Theorem 11.2.8 along with Lemma 11.2.7 to see that $\{X_i(t)\} \in \text{GDOA}(\{Y_i(t)\})$, and recall that every eigenvalue of E_i has real part equal to $a_i > 0$. Then apply Theorem 11.2.9. □

Remark 11.2.11. Since $Y(ct)$ has the same distribution as $c^E Y(t)$, the condition $P[Y(1) = 0] = 0$ implies that $P[Y(t) = 0] = 0$ for all $t > 0$. It is possible that $Y(1) = 0$ with positive probability. For example, let $\{B(t)\}$ be a standard Gaussian process on $\mathbb{R}^d$ independent of Z Bernoulli, and $Y(t) = ZB(t)$. It is also possible that $Y(t)$ is confined to a lower dimensional hyperplane with positive probability for all $t > 0$. For example, take $B_1(t)$ and $B_2(t)$ independent univariate Brownian motions, independent of Z Bernoulli, and let $Y(t) = uZB_1(t) + v(1 - Z)B_2(t)$ where $u, v \in \mathbb{R}^2$ are linearly independent.

Now we will refine the results of Theorem 11.2.9 and Corollary 11.2.10 to obtain the growth rate of $X(t)$ in any radial direction. We begin with two simple lemmas.

Lemma 11.2.12. *If $X_n \Rightarrow X$ on $\mathbb{R}^d$ and if $\theta_n \to \theta \neq 0$ and $M_n \to \infty$, then*

$$P[|<X_n, \theta_n>| > M_n] \to 0.$$

Proof. It follows from Continuous Mapping Theorem 1.2.8, that as $n \to \infty$ we have $|<X_n, \theta_n>| \Rightarrow |<X, \theta>|$. Hence the sequence $(|<X_n, \theta_n>|)$ is tight, so given $\varepsilon > 0$ there exists an $R > 0$ such that $P[|<X_n, \theta_n>| > R] < \varepsilon$ for all n. Then for all n such that $M_n > R$, we have $P[|<X_n, \theta_n>| > M_n] \leq P[|<X_n, \theta_n>| > R] < \varepsilon$, which concludes the proof. □

Lemma 11.2.13. *If $X_n \Rightarrow X$ on $\mathbb{R}^d$ and $P[<X, \theta> = 0] = 0$, then for any $\theta_n \to \theta \neq 0$ and $\varepsilon_n \to 0$ we have*

$$P[|<X_n, \theta_n>| < \varepsilon_n] \to 0.$$

Proof. Given $\delta > 0$ there exists an $m > 0$ such that $P[|<X, \theta>| < m] < \delta$ and $P[|<X, \theta>| = m] = 0$. Since $\varepsilon_n < m$ for all $n \geq n_0$ and some $n_0 \geq 1$ we conclude that for $n \geq n_0$ by the Portmanteau Theorem 1.2.2 $P[|<X_n, \theta_n>| < \varepsilon_n] \leq P[|<X_n, \theta_n>| < m] \to P[|<X, \theta>| < m] < \delta$, which proves the Lemma since $\delta > 0$ can be made arbitrary small. □

Theorem 11.2.14. *Suppose that $\{Y(t)\}_{t \geq 0}$ is a proper strictly operator self-similar process on $\mathbb{R}^d$ with exponent E, and that $\{X(t)\} \in \text{GDOA}(\{Y(t)\})$. Let $0 < a_1 < \cdots < a_p$ denote the real spectrum of E. Then*

(a) $P[|<X(t), \theta>| > t^\beta] \to 0$ as $t \to \infty$ for any $\theta \neq 0$ and any $\beta > a_p$

(b) If $Y(1)$ has a density, then $P[|<X(t), \theta>| < t^\alpha] \to 0$ as $t \to \infty$ for any $\theta \neq 0$ and any $\alpha < a_1$

Proof. Given $t_n \to \infty$ define $t_n^{-\beta}(A(t_n)^*)^{-1}\theta = r_n \omega_n$ where $r_n > 0$ and $\|\omega_n\| = 1$. Then $r_n \to 0$ by Theorem 4.2.4 since $t^{-\beta}(A(t)^*)^{-1}$ is regularly varying with index $E^* - \beta I$, where every eigenvalue of $E^* - \beta I$ has negative real part. Since ω_n are all unit vectors, for any sequence $t_{n'} \to \infty$ there is a subsequence (n'') along which $\omega_n \to \omega_0$. Then

$$\begin{aligned} P[|<X(t_n),\theta>| > t_n^\beta] &= P[t_n^{-\beta}|<X(t_n),\theta>| > 1] \\ &= P[|<X(t_n),t_n^{-\beta}\theta>| > 1] \\ &= P[|<A(t_n)^{-1}A(t_n)X(t_n),t_n^{-\beta}\theta>| > 1] \\ &= P[|<A(t_n)X(t_n),(A(t_n)^{-1})^*t_n^{-\beta}\theta>| > 1] \\ &= P[|<A(t_n)X(t_n),r_n\omega_n>| > 1] \\ &= P[|<A(t_n)X(t_n),\omega_n>| > r_n^{-1}] \to 0 \end{aligned}$$

by Lemma 11.2.12, since $r_n^{-1} \to \infty$ and $A(t_n)X(t_n) \Rightarrow Y(1)$. This proves (a), since every sequence has a further subsequence with this property.

Now let $t_n^{-\alpha}(A(t_n)^*)^{-1}\theta = r_n\omega_n$, where $r_n > 0$ and $\|\omega_n\| = 1$. Then $r_n \to \infty$ as $t_n \to \infty$ by Theorem 4.2.4, since $t^{-\alpha}(A(t)^*)^{-1}$ is regularly varying with index $E^* - \alpha I$, where every eigenvalue of $E^* - \alpha I$ has positive real part. For any sequence $t_{n'} \to \infty$, there is a subsequence (n'') along which $\omega_n \to \omega_0$. Then $P[|<X(t_n),\theta>| < t_n^\alpha] = P[|<A(t_n)X(t_n),\omega_n>| < r_n^{-1}] \to 0$ by Lemma 11.2.13, since $r_n^{-1} \to 0$ and $A(t_n)X(t_n) \Rightarrow Y(1)$ while $P[<Y(1),\omega_0> = 0] = 0$. Then (b) holds, which concludes the proof. The technical condition that $Y(1)$ has a density is to ensure that $P[|<Y(1),\theta>| = 0] = 0$ for all $\theta \neq 0$. □

Theorem 11.2.15. *Suppose $\{X(t)\}$ belongs to some generalized domain of attraction. Let $\{Y(t)\}$ and $E \in \mathcal{E}(\{Y(t)\})$ be as in Theorem 11.2.8. Define $\alpha(\theta) = a_i^{-1}$ for $\theta \in V_1 \oplus \cdots \oplus V_i \setminus V_1 \oplus \cdots \oplus V_{i-1}$, where $V_1 \oplus \cdots \oplus V_p$ is the spectral decomposition of $\mathbb{R}^d$ with respect to E and $0 < a_1 < \cdots < a_p$ is the real spectrum of E. Then*

(a) $P[|<X(t),\theta>| > t^\beta] \to 0$ as $t \to \infty$ for any $\theta \neq 0$ and any $\beta > \alpha(\theta)^{-1}$

(b) If $Y(1)$ has a density, then $P[|<X(t),\theta>| < t^\alpha] \to 0$ as $t \to \infty$ for any $\theta \neq 0$ and any $\alpha < \alpha(\theta)^{-1}$

Proof. Apply Theorem 11.2.8 to obtain a proper operator self-similar process $\{Y(t)\}$ and an exponent $E \in \mathcal{E}(\{Y(t)\})$ such that (11.5) holds for some $A(c)$ regularly varying with index $-E$, and let $V_1, \ldots, V_p$ denote the spectral decomposition of $\mathbb{R}^d$ with respect to E. Define $L_i = V_1 \oplus \cdots \oplus V_i$ and let $A_1(c) \oplus \cdots \oplus A_p(c)$ be the spectral decomposition of $A(c)$ with respect to E. Then $\bar{A}_i(c) = A_1(c) \oplus \cdots \oplus A_i(c)$ varies regularly on L_i with index $-\bar{E}_i = -E_1 \oplus \cdots \oplus E_i$, so $(\bar{A}_i(c)^*)^{-1}$ varies regularly on L_i with index $\bar{E}_i^*$. Since the eigenvalues of $\bar{E}_i^*$ all have real part less than or equal to a_i, it follows from Theorem 4.2.6 that for any $\rho > a_i$ we have $c^{-\rho}\|(\bar{A}_i(c)^*)^{-1}\theta\| \to 0$ for all $\theta \in L_i$. Since $(A(c)^*)^{-1}$ varies regularly with index E^*, Theorem 4.3.1 implies that for some unique subspace $\tilde{L}_i$ with the same dimension as L_i we have for every $a_i < \rho < a_{i+1}$ that $c^{-\rho}\|(A(c)^*)^{-1}\theta\| \to 0$ if and only if $\theta \in \tilde{L}_i$. Since $(\bar{A}_i(c)^*)^{-1}\theta = (A(c)^*)^{-1}\theta$ for all $\theta \in L_i$, it follows that $L_i = \tilde{L}_i$. Then Theorem 4.3.1 implies that that $t^{-\beta}\|(A(c)^*)^{-1}\theta\| \to 0$ as $c \to \infty$ for any $\theta \in V_1 \oplus \cdots \oplus V_i$ and that $t^{-\alpha}\|(A(t)^*)^{-1}\theta\| \to \infty$ as $t \to \infty$ for any

$\theta \notin V_1 \oplus \cdots \oplus V_{i-1}$. The remainder of the proof is identical to the proof of Theorem 11.2.14. □

Remark 11.2.16. It follows from (11.1) that if $Y(t)$ has a density for some $t > 0$ then it has a density for all $t > 0$. In this case Theorem 11.2.15 implies that for any $\varepsilon > 0$ we have

$$P[t^{\alpha(\theta)^{-1}-\varepsilon} < |<X(t), \theta>| < t^{\alpha(\theta)^{-1}+\varepsilon}] \to 1 \tag{11.13}$$

as $t \to \infty$ for any $\theta \neq 0$. Thus the real spectrum of E governs the growth rate of $X(t)$ in any radial direction. We can also apply these results in the special case $X(t) = Y(t)$, since any proper operator self-similar process belongs to its own generalized domain of attraction. If $Y(t)$ has a density, this gives sharp bounds on the growth rate of the process $\{Y(t)\}$.

Remark 11.2.17. Theorem 11.1.5 shows that the exponent of an operator self-similar process is not unique in general, because of possible symmetries. Although the index function $\alpha(\theta)$ depends on E, it is clear from (11.13) that this index function is unique. Hence the spectral decomposition $V_1, \ldots, V_p$ in Theorem 11.2.8 is uniquely determined by the process $\{X(t)\}$, if $Y(1)$ has a density.

Example 11.2.18. Operator Lévy motions are the scaling limits of random walks. Suppose that $Z_1, Z_2, Z_3, \ldots$ are i.i.d. random vectors on $\mathbb{R}^d$ that belong to the strict generalized domain of attraction of some full strictly operator stable random vector Y with exponent E, so that in view of Theorem 8.1.5 we have $B_n(Z_1 + \cdots Z_n) \Rightarrow Y$ for some $(B_n) \in \mathrm{RV}(-E)$. Recall from Example 11.1.12 that there exists an operator Lévy motion $\{Y(t)\}$ with exponent E such that $Y(1) \approx Y$, where $\approx$ means identically distributed. Define the random walk $X(t) = Z_1 + \cdots + Z_{[t]}$, and let $A(c) = B_{[c]}$. Theorem 4.2.9 shows that $A(c)$ varies regularly with index $-E$. Then the Convergence of Types Theorem 2.3.17 along with the Uniform Convergence Theorem 4.2.1 imply that

$$A(c)X(c) = A(c)A([c])^{-1} \cdot A([c])X(c) \Rightarrow Y(1) \quad \text{as } c \to \infty. \tag{11.14}$$

For any $0 \leq s < t$, it is easy to check that $X(t) - X(s) \approx X([t] - [s])$ and $Y(t) - Y(s) \approx Y(t-s) \approx (t-s)^E Y(1)$. Let $t_c = ([ct] - [cs])/c$ so that $t_c \to (t-s) > 0$ as $c \to \infty$. Then it follows from (11.14) along with Theorems 2.3.17 and 4.2.1 that $A(c)(X(ct) - X(cs)) \approx A(c)X([ct] - [cs]) = A(c)A(ct_c)^{-1} \cdot A(ct_c)X(ct_c) \Rightarrow (t-s)^E Y(1) \approx Y(t) - Y(s)$ as $c \to \infty$. Fix $0 = t_0 < t_1 < \cdots < t_n$ and note that the components of the random vectors $(X(t_i) - X(t_{i-1}))_{i=1,\ldots,n}$ and $(Y(t_i) - Y(t_{i-1}))_{i=1,\ldots,n}$ are independent. Then since $A(c)(X(ct_i) - X(ct_{i-1})) \Rightarrow Y(t_i) - Y(t_{i-1})$ for each $i = 1, \ldots, n$, we also have $(A(c)(X(ct_i) - X(ct_{i-1})))_{i=1,\ldots,n} \Rightarrow (Y(t_i) - Y(t_{i-1}))_{i=1,\ldots,n}$, see, e.g., Billingsley [17] pp. 26–27. Then the Continuous Mapping Theorem 1.2.7 yields $(A(c)X(ct_i))_{i=1,\ldots,n} \Rightarrow (Y(t_i))_{i=1,\ldots,n}$ so that

$$\{A(c)X(ct)\}_{t \geq 0} \overset{d}{\Rightarrow} \{Y(t)\}_{t \geq 0} \quad \text{as } c \to \infty.$$

Remark 11.2.19. Anomalous diffusion models in physics can be justified in terms of scaling limits of random walks, in which the individual particle jumps belong to the domain of attraction of a nonnormal stable law. Remark 11.1.13 introduced the idea, from statistical mechanics, that particle diffusion can be modeled as a Lévy motion $\{Y(t)\}$. Example 11.2.18 shows that this Lévy motion is the scaling limit of a random walk $X(t) = Z_1 + \cdots + Z_{[t]}$, where the (Z_n) represent particle jumps. Schumer et al. [175] use this idea to give a physical derivation of the fractional partial differential equation (11.3) that governs anomalous diffusion in $\mathbb{R}^1$.

11.3 NOTES AND COMMENTS

The study of self-similar stochastic processes was begun by Lamperti [86]. He called these processes semi-stable, but that term has now taken the meaning of Definition 7.4.1. Samorodnitsky and Taqqu [161] contains additional results on self-similar processes, along with a number of useful examples. Hudson and Mason [65] along with Laha and Rohatgi [85] were the first to consider operator self-similar processes. Sections 11.1 and 11.2 are based in part on Meerschaert and Scheffler [129]. Operator Lévy motions were considered in Maejima [94]. Recently, Maejima et al. [96, 97, 98] have begun the study of semi-self-similar processes, which extends Definition 11.1.2 by requiring only that (11.1) hold along a geometric sequence of scales. These processes are related to operator self-similar processes in the same way that operator semistable laws generalize operator stable laws. Hudson and Mason [65] were the first to consider generalized domains of attraction of operator self-similar processes. Fractional diffusion equations on $\mathbb{R}^1$ were derived by Benson et al. [11, 13], Chaves [23], Compte [29, 30], Metzler [136] and Zaslavsky [198]. Some related integral equations were considered by Riesz [154], and connections with stable Lévy motions in $\mathbb{R}^1$ were noted in Bochner [20] and Feller [41]. Janicki and Weron [71] discuss simulation methods for stable Lévy motions, and some interesting connections with chaos theory. Meerschaert, Benson and Bäumer [130, 134] extend the fractional diffusion equation to $\mathbb{R}^d$.

References

1. Adams, G. and C. Goodwin (1995) Parameter estimation for periodic ARMA models. *J. Time Series Anal.* **16**, 127–145.

2. Adler, R., R. Feldman, and M. Taqqu (1998) *A Practical Guide to Heavy Tails.* Birkhäuser, Boston.

3. Anderson, P. and M. Meerschaert (1997) Periodic moving averages of random variables with regularly varying tails. *Ann. Statist.* **25**, 771–785.

4. Anderson, P. and M. Meerschaert (1998) Modeling river flows with heavy tails. *Water Resour. Res.* **34**, 2271–2280.

5. Anderson, P., M. Meerschaert, and A. Vecchia (1999) Innovations algorithm for periodically stationary time series. *Stochastic Process. Appl.* **83**, 149–169.

6. Araujo, A. and E. Giné (1980) *The Central Limit Theorem for Real and Banach Valued Random Variables.* Wiley, New York.

7. Avram, F. and M. Taqqu (1986) Symmetric polynomials of random variables attracted to an infinitely divisible law. *Probab. Theory Relat. Fields* **71**, 491–500.

8. Balkema, A. (1973) Monotone transformations and limit laws, *Mathematical Centre Tracts* **45**, Mathematisch Centrum, Amsterdam.

9. Balkema, A. and L. de Haan (1988) Almost sure continuity of stable moving average processes with index less than one. *Ann. Probab.* **16**, 333–343.

10. Bauer, H. (1972) *Probability Theory and Elements of Measure Theory.* Holt, Rinehart and Winston, New York.

11. Benson, D. (1998) *The Fractional Advection-Dispersion Equation: Development and Application.* Ph.D. Thesis, University of Nevada, Reno.

12. Benson, D., R. Schumer, M. Meerschaert, and S. Wheatcraft (2001) Fractional dispersion, Lévy motions, and the MADE tracer tests. *Transport Porous Med.* **42**, 211–240.

13. Benson, D., S. Wheatcraft, and M. Meerschaert (2000) The fractional-order governing equation of Lévy motion. *Water Resour. Res.* **36**, 1413–1423.

14. Benson, D., S. Wheatcraft, and M. Meerschaert (2000) Application of a fractional advection-dispersion equation. *Water Resour. Res.* **36**, 1403–1412.

15. Bhansali, R. (1993) Estimation of the impulse response coefficients of a linear process with infinite variance. *J. Multivariate Anal.* **45**, 274–290.

16. Billingsley, P. (1966) Convergence of types in k-space. *Z. Wahrsch. Verw. Geb.* **5**, 175–179.

17. Billingsley, P. (1968) *Convergence of Probability Measures.* Wiley, New York.

18. Billingsley, P. (1995) *Probability and Measure.* 3rd Ed., Wiley, New York.

19. Bingham, N., C. Goldie, and J. Teugels (1987) Regular Variation. *Encyclopedia of Mathematics and its Applications* **27**, Cambridge University Press, London.

20. Bochner, S. (1949) Diffusion equation and stochastic processes. *Proc. Nat. Acad. Sci. U.S.A.* **35**, 368–370.

21. Bojanic, R. and E. Seneta (1973) A unified theory of regularly varying sequences. *Math. Z.* **134**, 91–106.

22. Brockwell, P. and R. Davis (1991) *Time Series: Theory and Methods.* 2nd Ed., Springer-Verlag, New York.

23. Chaves, A. (1998) A fractional diffusion equation to describe Lévy flights. *Phys. Lett. A* **239**, 13–16.

24. Chorny, V. (1986) Operator semistable distributions on $\mathbb{R}^d$. *Theory Probab. Appl.* **31**, 703–709.

25. Chover, J. (1966) A law of the iterated logarithm for stable summands. *Proc. Amer. Math. Soc.* **17**, 441–443.

26. Cline, D. (1986) Convolution tails, product tails, and domains of attraction. *Probab. Th. Rel. Fields* **72**, 529–557.

27. Cline, D. (1989) Consistency for least squares regression estimators with infinite variance data. *J. Statist. Plann. Inference* **23**, 163–179.

28. Cohn, P. (1957) Lie groups. *Cambridge Tracts in Mathematics and Mathematical Physics* **46**, Cambridge University Press, New York.

29. Compte, A. (1996) Stochastic foundations of fractional dynamics. *Phys. Rev. E* **53**, 4191–4193.

30. Compte, A. (1997) Continuous time random walks on moving fluids. *Phys. Rev. E* **55,** 6821–6831.

31. Curtis, C. (1974) *Linear Algebra.* 3rd Ed., Allyn and Bacon, Boston.

32. Davis, R. and S. Resnick (1985) Limit theory for moving averages of random variables with regularly varying tail probabilities. *Ann. Probab.* **13**, 179–195.

33. Davis, R. and S. Resnick (1985) More limit theory for the sample correlation function of moving averages. *Stoch. Proc. Appl.* **20**, 257–279.

34. Davis, R. and S. Resnick (1986) Limit theory for the sample covariance and correlation functions of moving averages. *Ann. Statist.* **14**, 533–558.

35. Davis, R., J. Marengo, and S. Resnick (1985) Extremal properties of a class of multivariate moving averages. Proceedings of the 45th session of the International Statistical Institute (Amsterdam, 1985). *Bull. Inst. Internat. Statist.* **51**, (V) 185–192.

36. Davis, R. and J. Marengo (1990) Limit theory for the sample covariance and correlation matrix functions of a class of multivariate linear processes. *Comm. Statist. Stochastic Models* **6**, 483–497.

37. Dynkin, E. and A. Mandelbaum (1983) Symmetric statistics, Poisson point processes, and multiple Wiener integrals. *Ann. Statist.* **11**, 739–745.

38. Embrechts, P., C. Klüppelberg, and T. Mikosch (1997) Modelling extremal events. For insurance and finance. *Applications of Mathematics* **33**, Springer-Verlag, Berlin.

39. Fama, E. (1965) The behavior of stock market prices. *J. Business* **38**, 34–105.

40. Feller, W. (1971) *An Introduction to Probability Theory and Its Applications.* Vol. II, 2nd Ed., Wiley, New York.

41. Feller, W. (1952) On a generalization of Marcel Riesz' potentials and the semi-groups generated by them. *Comm. Sém. Math. Univ. Lund*, Tome Supplementaire, 72–81.

42. Fisz, M. (1954) A generalization of a theorem of Khintchin. *Studia Math.* **14**, 310–313.

43. Giné, E. and M. Hahn (1983) On stability of probability laws with univariate stable marginals. *Z. Wahrsch. Verw. Geb.* **64**, 157–165.

44. Gnedenko, B. and A. Kolmogorov (1968) *Limit Distributions for Sums of Independent Random Variables.* Translated from the Russian, annotated, and revised by K. L. Chung. With appendices by J. L. Doob and P. L. Hsu. Revised edition, Addison-Wesley, Reading, Mass.

45. Greub, W. (1978) *Multilinear Algebra.* 2nd Ed., Springer-Verlag, New York.

46. de Haan, L. (1970) On regular variation and its application to the weak convergence of sample extremes. *Mathematical Centre Tracts* **32**, Mathematisch Centrum, Amsterdam.

47. de Haan, L. and S. Resnick (1979) Derivatives of regularly varying functions in $\mathbb{R}^d$ and domains of attraction of stable distributions. *Stoch. Proc. Appl.* **8**, 349–355.

48. de Haan, L., E. Omey, and S. Resnick (1984) Domains of attraction and regular variation in $\mathbb{R}^d$. *J. Multivariate Anal.* **14**, 17–33.

49. Hahn, M. and M. Klass (1980) Matrix normalization for sums of random vectors in the domain of attraction of the multivariate normal. *Ann. Probab.* **8**, 262–280.

50. Hahn, M. and M. Klass (1980) The generalized domain of attraction of spherically symmetric stable laws on $\mathbb{R}^d$. Probability theory on vector spaces, II (Proc. Second Internat. Conf., Błażejewko, 1979), *Lecture Notes in Math.* **828**, Springer, Berlin, 52–81.

51. Hahn, M. and M. Klass (1985) Affine normability of partial sums of i.i.d. random vectors: A characterization. *Z. Wahrsch. Verw. Geb. 69*, 479–505.

52. Hall, P. (1981) A comedy of errors: the canonical form for a stable characteristic function. *Bull. London Math. Soc.* **13**, 23–27.

53. Hall, P. (1982) On some simple estimates of an exponent of regular variation. *J. Royal Statist. Soc. B* **44** 37–42.

54. Hannan, E. and M. Kanter (1977) Autoregressive processes with infinite variance. *J. Appl. Probability* **14**, 411–415.

55. Hazod, W. and E. Siebert (2000) *Stable Probability Measures on Euclidean Spaces and on Locally Compact Groups.* Kluwer, Amsterdam.

56. Hewitt, E. and K. Ross (1979) *Abstract Harmonic Analysis.* Vol. I., 2nd Ed., Springer-Verlag, New York.

57. Hedye, C. (1967) On large deviation problems for sums of random variables not attracted to the normal law. *Ann. Math. Statist.* **38**, 1575–1578.

58. Heyer, H. (1977) *Probability Measures on Locally Compact Groups.* Springer-Verlag, Berlin.

59. Hill, B. (1975) A simple general approach to inference about the tail of a distribution. *Ann. Statist.* **3** 1163–1173.

60. Hirsch, M. and S. Smale (1974) *Differential Equations, Dynamical Systems and Linear Algebra,* Academic Press, New York.

61. Hoffman, K. and Kunze, R. (1961) *Linear Algebra.* Prentice-Hall, New Jersey.

62. Holmes, J., W. Hudson, and J. D. Mason (1982) Operator Stable Laws: Multiple Exponents and Elliptical Symmetry. *Ann. Probab.* **10**, 602–612.

63. Hosking, J. and J. Wallis (1987) Parameter and quantile estimation for the generalized Pareto distribution. *Technometrics* **29**, 339–349.

64. Hudson, W. (1980) Operator-stable distributions and stable marginals. *J. Multivariate Anal.* **10**, 26–37.

65. Hudson, W. and J. D. Mason (1982) Operator-self-similar processes in a finite-dimensional space. *Trans. Amer. Math. Soc.* **273**, 281–297.

66. Hudson, W., Z. Jurek, and J. Veeh (1986) The symmetry group and exponents of operator stable probability measures. *Ann. Probab.* **14**, 1014–1023.

67. Hudson, W., J. Veeh, and D. Weiner (1988) Moments of distributions attracted to operator-stable laws. *J. Multivariate Anal.* **24**, 1–10.

68. Jacobson, N. (1979) *Lie Algebras.* Dover, New York.

69. Jajte, R. (1977) Semistable probability measures on $\mathbb{R}^N$. *Studia Math.* **61**, 29–39.

70. Jakimiv, A. (1981) Many-dimensional Tauberian theorems and their application to Bellman-Harris branching processes. *Math. Sbornik* **115**, 463–477. (Russian)

71. Janicki, A. and A. Weron (1994) *Simulation and Chaotic Behavior of α-Stable Stochastic Processes.* Marcel Dekker, New York.

72. Jansen, D. and C. de Vries (1991) On the frequency of large stock market returns: Putting booms and busts into perspective. *Rev. Econ. Statist.* **23**, 18–24.

73. Jurek, Z. (1980) Domains of normal attraction of operator-stable measures on Euclidean spaces. *Bull. Acad. Polon. Sci. Sér. Sci. Math.* **28**, 397–409.

74. Jurek, Z. (1984) Polar Coordinates in Banach Spaces. *Bull. Polish Acad. Sci. Math.* **32**, 61–66.

75. Jurek, Z. and J. D. Mason (1993) *Operator-Limit Distributions in Probability Theory.* Wiley, New York.

76. Kanter, M. (1975) Stable densities under change of scale and total variation inequalities. *Ann. Probab.* **3**, 697–707.

77. Karamata, J. (1930) Sur certain "Tauberian theorems" de m.m. Hardy et Littlewood. *Mathematica (Cluj)* **3**, 33–48.

78. Karamata, J. (1930) Sur un mode de croissance régulière des fonctions. *Mathematica (Cluj)* **4**, 38–53.

79. Klosowska, M. (1980) Domain of operator attraction of a Gaussian measure on $\mathbb{R}^N$. *Ann. Soc. Math. Pol., Ser. I, Comment. Math.* **XXII**, 73–80.

80. Klüppelberg, C. and T. Mikosch (1996) The integrated periodogram for stable processes. *Ann. Statist.* **24**, 1855–1879.

81. Knight, K. (1989) Consistency of Akaike's information criterion for infinite variance autoregressive processes. *Ann. Statist.* **17**, 824–840.

82. Kokoszka, P. and M. Taqqu (1994) Infinite variance stable ARMA processes. *J. Time Series Anal.* **115**, 203–220.

83. Krakowiak, W. (1979) Operator stable probability measures on Banach spaces. *Colloq. Math.* **41**, 313–326.

84. Kruglov, V. (1972) On the extension of the class of stable distributions. *Theory Probab. Appl.* **17**, 685–694.

85. Laha, R. and V. Rohatgi (1982) Operator self-similar stochastic processes in $\mathbb{R}^d$. *Stochastic Process. Appl.* **12**, 73–84.

86. Lamperti, J. (1962) Semi-stable stochastic processes. *Trans. Amer. Math. Soc.* **104**, 62–78.

87. Lee, A. (1990) *U-Statistics. Theory and Practice.* Marcel Dekker, New York.

88. Lévy, P. (1954) *Théorie de l'Addition des Variables Aléatoires.* 2nd Ed., Gauthier, Paris.

89. Logan, B., C. Mallows, S. Rice, and L. Shepp (1973) Limit distributions of self-normalized sums. *Ann. Probab.* **1**, 788–809.

90. Loretan, M. and P. Phillips (1994) Testing the covariance stationarity of heavy tailed time series. *J. Empirical Finance* **1**, 211–248.

91. Łuczak, A. (1981) Operator semistable probability measures on R^N. *Colloq. Math.* **45**, 287–300.

92. Łuczak, A. (1984) Elliptical symmetry and characterization of operator-stable and operator semi-stable measures. *Ann. Prob.* **12**, 1217–1223.

93. Maejima, M. and J. D. Mason (1994) Operator-self-similar stable processes. *Stoch. Proc. Appl.* **54**, 139–163.

94. Maejima, M. (1995) Operator-stable processes and operator fractional stable motions. Dedicated to the memory of Jerzy Neyman. *Probab. Math. Statist.* **15**, 449–460.

95. Maejima, M. (1998) Norming operators for operator-self-similar processes. Stochastic processes and related topics, *Trends Math.*, Birkhäuser, Boston, 287–295.

96. Maejima, M., K. Sato, and T. Watanabe (1999) Operator semi-selfdecomposability, (C, Q)-decomposability and related nested classes. *Tokyo J. Math.* **22**, 473–509.

97. Maejima, M., K. Sato, and T. Watanabe (1999) Exponents of semi-self-similar processes. *Yokohama Math. J.* **47**, 93–102.

98. Maejima, M. and K. Sato (1999) Semi-selfsimilar processes. *J. Theoret. Probab.* **12**, 347–373.

99. Mandelbrot, B. (1963) The variation of certain speculative prices. *J. Business* **36**, 394–419.

100. Mandelbrot, B. (1982) *The Fractal Geometry of Nature.* W. H. Freeman, San Francisco.

101. Mandrekar, V. (1981) Domain of attraction problem on Banach spaces: a survey. *Lecture Notes in Math.* **860**, Springer, New York, 285–290.

102. Mandrekar, V. and R. Patterson (1993) Limit theorems for symmetric statistics of exchangeable random variables. *Statist. Probab. Lett.* **17**, 157–161.

103. Mandrekar, V. and M. Meeerschaert (1994) Sample moments and symmetric statistics. Stochastic Analysis on Infinite Dimensional Spaces, *Pitman Res. Notes Math. Ser.* **310**, 197–210.

104. Marcus, D. (1983) Non-stable laws with all projections stable. *Z. Wahrsch. Verw. Geb.* **64**, 139–156.

105. Masuyama, M. (1942) The Bienaymé-Tchebycheff inequality for Hermitic tensors. *Proc. Phys.-Math. Soc. Japan* **24**, (3) 409–411.

106. McCullagh, P. (1987) *Tensor methods in statistics.* Chapman & Hall, London.

107. McCulloch, J. (1996) Financial applications of stable distributions. *Statistical Methods in Finance: Handbook of Statistics* **14**, G. Madfala and C. R. Rao, Eds., Elsevier, Amsterdam, 393–425.

108. McCulloch, J. (1996) On the parametrization of the afocal stable distributions. *Bull. London Math. Soc.* **28**, 651–655.

109. McCulloch, J. (1997) Measuring tail thickness to estimate the stable index α: A critique. *J. Business Econ. Statist.* **15**, 74–81.

110. Meerschaert, M. (1984) *Multivariable Domains of Attraction and Regular Variation*, Ph.D. Thesis, University of Michigan, Ann Arbor.

111. Meerschaert, M. (1986) Domains of attraction of nonnormal operator-stable laws. *J. Multivariate Anal.* **19**, 342–347.

112. Meerschaert, M. (1986) Regular variation and domains of attraction in $\mathbb{R}^k$. *Stat. Probab. Lett.* **4**, 43–45.

113. Meerschaert, M. (1988) Regular variation in $\mathbb{R}^k$. *Proc. Amer. Math. Soc.* **102**, 341–348.

114. Meerschaert, M. (1990) Moments of random vectors which belong to some domain of normal attraction. *Ann. Probab.* **18**, 870–876.

115. Meerschaert, M. (1991) Regular variation in $\mathbb{R}^k$ and vector-normed domains of attraction. *Stat. Prob. Lett.* **11**, 287–289.

116. Meerschaert, M. (1991) Spectral decomposition for generalized domains of attraction. *Ann. Probab.* **19**, 875–892.

117. Meerschaert, M. and J. Veeh (1993) The structure of the exponents and symmetries of an operator stable measure. *J. Theoret. Probab.* **6**, 713–726.

118. Meerschaert, M. (1993) Regular Variation and Generalized Domains of Attraction in $\mathbb{R}^k$. *Statist. Probab. Lett.* **18**, 233–239.

119. Meerschaert, M. and J. Veeh (1995) Symmetry groups in d-space. *Statist. Probab. Lett.* **22**, 1–6.

120. Meerschaert, M. (1994) Norming Operators for Generalized Domains of Attraction. *J. Theoret. Probab.* **7**, 793–798.

121. Meerschaert, M. and H. P. Scheffler (1996) Generalized domains of semistable attraction of nonnormal laws. *J. Applied Anal.* **2**, 203–215.

122. Meerschaert, M. and H. P. Scheffler (1997) Spectral decomposition for generalized domains of semistable attraction. *J. Theoret. Probab.* **10**, 51–71.

123. Meerschaert, M. and H. P. Scheffler (1997) The structure of generalized domains of semistable attraction. *Statist. Probab. Lett.* **33**, 367–372.

124. Meerschaert, M. and H. P. Scheffler (1998) Multivariable regular variation of functions and measures. *J. Appl. Anal.* **5**, 77–98.

125. Meerschaert, M. and H. P. Scheffler (1998) A simple robust estimation method for the thickness of heavy tails. *J. Statis. Plann. Inf.* **71**, 19–34.

126. Meerschaert, M. and H. P. Scheffler (1998) Convergence of semitypes. *Publ. Math. Debrecen* **53**, 119–131.

127. Meerschaert, M. and H. P. Scheffler (1999) Moment estimator for random vectors with heavy tails. *J. Multivariate Anal.* **71**, 145–159.

128. Meerschaert, M. and H. P. Scheffler (1999) Sample covariance matrix for random vectors with heavy tails. *J. Theoret. Probab.* **12**, 821–838.

129. Meerschaert, M. and H. P. Scheffler (1999) Spectral decomposition for operator-self-similar processes and their generalized domains of attraction. *Stoch. Proc. Appl.* **84**, 71–80.

130. Meerschaert, M., D. Benson, and B. Bäumer (1999) Multidimensional advection and fractional dispersion. *Phys. Rev. E* **59**, 5026–5028.

131. Meerschaert, M. and H. P. Scheffler (1999) One dimensional marginals of operator stable laws and their domains of attraction. *Publ. Math. Debrecen* **55**, 487–499.

132. Meerschaert, M. and H. P. Scheffler (2000) Limit laws for symmetric k-tensors of regularly varying measures. *J. Multivariate Anal.* **73**, 241–261.

133. Meerschaert, M. and H. P. Scheffler (2000) Moving averages of random vectors with regularly varying tails. *J. Time Series Anal.* **21**, 297–328.

134. Meerschaert, M., D. Benson, and B. Bäumer (2001) Operator Lévy motion and multiscaling anomalous diffusion. *Physical Review E* **63**, 1112–1117.

135. Mejzler, D. (1973) On a certain class of infinitely divisible distributions. *Israel J. Math.* **16**, 1–19.

136. Metzler, R., J. Klafter, and I. Sokolov (1998) Anomalous transport in external fields: Continuous time random walks and fractional diffusion equations extended. *Phys. Rev. E* **58**, 1621–1633.

137. Miller, K. and B. Ross (1993) *An Introduction to Fractional Calculus and Fractional Differential Equations.* Wiley, New York.

138. Mittnik, S. and S. Rachev (1999) *Asset and Option Pricing with Alternative Stable Models.* Wiley, New York.

139. Neveu, J. (1976) Processes ponctuels. *Lecture Notes in Mathematics* **598**, Springer-Verlag, Berlin.

140. Nikias, C. and M. Shao (1995) *Signal Processing with Alpha Stable Distributions and Applications.* Wiley, New York.

141. Nobel, S. (1991) Limit theorems for probability measures on simply connected nilpotent Lie groups. *J. Theoret. Probab.* **4**, 261–284.

142. Nolan, J., Panorska A. and J. H. McCulloch (2001) Estimation of stable spectral measures, *Math. Comput. Modelling*, to appear.

143. Nolan, J. and A. Panorska (1997) Data analysis for heavy tailed multivariate samples. Heavy tails and highly volatile phenomena. *Comm. Statist. Stochastic Models* **13**, 687–702.

144. Nolan, J. (1997) Numerical calculation of stable densities and distribution functions. Heavy tails and highly volatile phenomena. *Comm. Statist. Stochastic Models* **13**, 759–774.

145. Parthasarathy, K. and K. Schmidt (1972) Positive definite kernels, continuous tensor products, and central limit theorems of probability theory. *Lecture Notes in Math.* **272**, Springer-Verlag, Berlin.

146. Phuc, H. (1994) Domains of operator semi-attraction of operator semi-stable probability measures, preprint.

147. Pillai, R. (1971) Semi stable laws as limit distributions. *Ann. Math. Statist.* **42**, 780–783.

148. Pontryagin, L. (1962) *Ordinary Differential Equations.* Translated from the Russian by Leonas Kacinskas and Walter B. Counts. Addison-Wesley, Reading, Mass.

149. Price, J. (1977) Lie Groups and Compact Groups. *London Math. Soc. Lecture Notes* **25**, Cambridge University Press, London.

150. Rao, C.R. (1965) *Linear Statistical Inference and Its Applications.* Wiley, New York.

151. Resnick, S. and P. Greenwood (1979) A bivariate characterization and domains of attraction. *J. Multivariate Anal.* **9**, 206–221.

152. Resnick, S. (1986) Point processes, regular variation and weak convergence. *Adv. Appl. Prob.* **18**, 66–138.

153. Resnick, S. (1987) *Extreme Values, Regular Variation, and Point Processes.* Springer-Verlag, New York.

154. Riesz, M. (1949) L'intégrale de Riemann-Liouville et le problème de Cauchy. (French) *Acta Math.* **81**, 1–223.

155. Royden, H. (1988) *Real Analysis.* 3rd Ed., Macmillan, New York.

156. Rubin, H. and R. Vitale (1980) Asymptotic distribution of symmetric statistics. *Ann. Statist.* **8**, 165–170.

157. Rudin, W. (1987) *Real and Complex Analysis.* 3rd Ed., McGraw-Hill, New York.

158. Rvačeva, E. (1962) On domains of attraction of multidimensional distributions. *Select. Transl. Math. Stat. Prob.* **2**, American Math. Soc., Providence, Rhode Island, 183–205.

159. Sakovich, G. (1950) A single form for the conditions for attraction to stable laws. *Theory Probab. Appl.* **1**, 322–325.

160. Samko, S., A. Kilbas, and O. Marichev (1993) *Fractional Integrals and Derivatives: Theory and Applications.* Gordon and Breach, London.

161. Samorodnitsky, G. and M. Taqqu (1994) *Stable non-Gaussian Random Processes.* Chapman & Hall, New York.

162. Sato, K. (1999) Lévy processes and infinitely divisible distributions. Translated from the 1990 Japanese original. Revised by the author. *Cambridge Studies in Advanced Mathematics* **68**, Cambridge University Press.

163. Scheffler, H. P. (1992) *Anziehungsbereiche stabiler Wahrscheinlichkeitsmaße auf stratifizierbaren Lie-Gruppen (Domains of Attraction of Stable Probability Measures on Stratified Lie Groups).* Ph.D. Thesis, University of Dortmund, Germany.

164. Scheffler, H. P. (1994) Domains of semi-stable attraction of nonnormal semi-stable laws. *J. Multivariate Anal.* **51**, 432–444.

165. Scheffler, H. P. (1995) Moments of measures attracted to operator semistable laws. *Statist. Probab. Lett.* **24**, 187–192.

166. Scheffler, H. P. (1997) *Multivariable R–O variation and Generalized Domains of Semistable Attraction.* Habilitation Thesis, University of Dortmund, Germany.

167. Scheffler, H. P. (1997) Large deviations and law of the iterated logarithm for generalized domains of attraction. *Probab. Math. Statist.* **17**, 65–77.

168. Scheffler, H. P. (1998) A law of the iterated logarithm for semistable laws on vector spaces and homogeneous groups. Stability problems for stochastic models, Part 1 (Moscow, 1996). *J. Math. Sci. (New York)* **89**, 1545–1552.

169. Scheffler, H. P. (1999) On estimation of the spectral measure of certain nonnormal operator stable laws. *Statist. Probab. Lett.* **43**, 385–392.

170. Scheffler, H. P. (2000) A law of the iterated logarithm for heavy-tailed random vectors. *Probab. Theory Related Fields* **116**, 257–271.

171. Scheffler, H. P. (2000) Norming operators for generalized domains of semistable attraction. *Publ. Math. Debrecen*, to appear.

172. Scheffler, H. P. (2000) Multivariate R–O varying measures. Part I: Uniform bounds. *Proc. London Math. Soc.* **81** (3), 231–256.

173. Scheffler, H. P. (2000) Multivariate R–O varying measures. Part II: Individual bounds, preprint.

174. Shlesinger, M. G. Zaslavsky, and U. Frisch, Eds. (1995) Lévy flights and related topics in physics. Proceedings of the International Workshop held in Nice, June 27–30, 1994. *Lecture Notes in Physics*, **450**, Springer-Verlag, Berlin.

175. Schumer, R., D. Benson, M. Meerschaert, and S. Wheatcraft (2001) Eulerian derivation of the fractional advection-dispersion equation. *J. Contam. Hydrol.* **38**, 69–88.

176. Seidl', L. (1986) On the limit distributions of random symmetric functions. *Theory Probab. Appl.* **31**, 590–603.

177. Seneta, E. (1976) Regularly Varying Functions. *Lecture Notes in Math.* **508**, Springer-Verlag, Berlin.

178. Sepanski, S. (1994) Asymptotics for multivariate t-statistic and Hotelling's T^2-statistic under infinite second moments via bootstrapping. *J. Multivariate Anal.* **49**, 41–54.

179. Sepanski, S. (1996) Asymptotics for multivariate t-statistic for random vectors in the generalized domain of attraction of the multivariate normal law. *Statist. Probab. Lett.* **30**, 179–188.

180. Serfling, R. (1980) *Approximation Theorems of Mathematical Statistics.* Wiley, New York.

181. Sharpe, M. (1969) Operator-stable probability distributions on vector groups. *Trans. Amer. Math. Soc.* **136**, 51–65.

182. Shimuzu, R. (1970) On the domain of partial attraction of semi-stable distributions. *Ann. Inst. Statist. Math.* **22**, 245–255.

183. Siebert, E. (1982) Continuous convolution semigroups integrating a submultiplicative function. *Manus. Math.* **37**, 383–391.

184. Siegel, G. (1992) On the class of operator stable distributions in a separable Banach space. *Probab. Math. Statist.* **13**, 33–37.

185. Siegel, G. (1990) Exponents of operator stable distributions in Banach space. *Probability theory and mathematical statistics* **II** (Vilnius, 1989), 437–445, "Mokslas", Vilnius.

186. Spivak, M. (1979) *A Comprehensive Introduction to Differential Geometry.* Vol. I, 2nd Ed., Publish or Perish, Wilmington, Del.

187. Stadtmüller, U. and R. Trautner (1981) Tauberian theorems for Laplace transforms in dimension $D > 1$. *J. Reine Angew. Math.* **323**, 127–138.

188. Stam, A. (1977) Regular variation in $\mathbb{R}^d_+$ and the Abel-Tauber theorem. *Technical Report T. W. 189,* Mathematisch Instituut Rijksuniversiteit Groningen, The Netherlands.

189. Takemura, A. (1983) Tensor analysis of ANOVA decomposition. *J. Amer. Statist. Assoc.* **78**, 894–900.

190. Taylor, S. (1986) The measure theory of random fractals. *Math. Proc. Cambridge Philos. Soc.* **100**, 383–406.

191. Tucker, H. (1980) Joint limit laws of sample moments of a symmetric distribution. *Ann. Probab.* **8**, 991–998.

192. Uchaikin, V. and V. Zolotarev (1999) *Chance and Stability: Stable Distributions and Their Applications*, VSP, Utrecht.

193. Vasudeva, R. (1984) Chover's law of the iterated logarithm and weak convergence. *Acta Math. Hung* **44**, 215–221.

194. Vu, H., R. Maller, and M. Klass (1996) On the studentization of random vectors. *J. Multivariate Anal.* **57**, 142–155.

195. D. Weiner (1985) A maximal law of the iterated logarithm for operator-normalized stochastically compact partial sums of i.i.d. random vectors. Probability in Banach spaces V, *Lecture Notes in Math.* **1153**, 426–439.

196. Weissman, I. (1976) A note on Bojanic-Seneta theory of regularly varying sequences. *Math. Z.* **151**, 29–30.

197. Weissman, I. (1976) On convergence of types and processes in Euclidean space. *Z. Wahrsch. Verw. Gebiete* **37**, 35–41.

198. Zaslavsky, G. (1994) Fractional kinetic equation for Hamiltonian chaos. Chaotic advection, tracer dynamics and turbulent dispersion. *Phys. D* **76**, 110–122.

199. Zolotarev, V. (1986) One-dimensional stable distributions. Translated from the Russian by H. H. McFaden. Translation edited by Ben Silver. *Translations of Mathematical Monographs* **65**, Amer. Math. Soc., Providence, Rhode Island.

Index

WILEY SERIES IN PROBABILITY AND STATISTICS
ESTABLISHED BY WALTER A. SHEWHART AND SAMUEL S. WILKS

Probability and Statistics Section

*ANDERSON · The Statistical Analysis of Time Series
ARNOLD, BALAKRISHNAN, and NAGARAJA · A First Course in Order Statistics
ARNOLD, BALAKRISHNAN, and NAGARAJA · Records
BACCELLI, COHEN, OLSDER, and QUADRAT · Synchronization and Linearity: An Algebra for Discrete Event Systems
BARNETT · Comparative Statistical Inference, *Third Edition*
BASILEVSKY · Statistical Factor Analysis and Related Methods: Theory and Applications
BERNARDO and SMITH · Bayesian Statistical Concepts and Theory
BILLINGSLEY · Convergence of Probability Measures, *Second Edition*
BOROVKOV · Asymptotic Methods in Queuing Theory
BOROVKOV · Ergodicity and Stability of Stochastic Processes
BRANDT, FRANKEN, and LISEK · Stationary Stochastic Models
CAINES · Linear Stochastic Systems
CAIROLI and DALANG · Sequential Stochastic Optimization
CONSTANTINE · Combinatorial Theory and Statistical Design
COOK · Regression Graphics
COVER and THOMAS · Elements of Information Theory
CSÖRGŐ and HORVÁTH · Weighted Approximations in Probability Statistics
CSÖRGŐ and HORVÁTH · Limit Theorems in Change Point Analysis
*DANIEL · Fitting Equations to Data: Computer Analysis of Multifactor Data, *Second Edition*
DETTE and STUDDEN · The Theory of Canonical Moments with Applications in Statistics, Probability, and Analysis
DEY and MUKERJEE · Fractional Factorial Plans
*DOOB · Stochastic Processes
DRYDEN and MARDIA · Statistical Shape Analysis
DUPUIS and ELLIS · A Weak Convergence Approach to the Theory of Large Deviations
ETHIER and KURTZ · Markov Processes: Characterization and Convergence
FELLER · An Introduction to Probability Theory and Its Applications, Volume I, *Third Edition,* Revised; Volume II, *Second Edition*
FULLER · Introduction to Statistical Time Series, *Second Edition*
FULLER · Measurement Error Models
GHOSH, MUKHOPADHYAY, and SEN · Sequential Estimation
GIFI · Nonlinear Multivariate Analysis
GUTTORP · Statistical Inference for Branching Processes
HALL · Introduction to the Theory of Coverage Processes
HAMPEL · Robust Statistics: The Approach Based on Influence Functions
HANNAN and DEISTLER · The Statistical Theory of Linear Systems
HUBER · Robust Statistics

*Now available in a lower priced paperback edition in the Wiley Classics Library.

Probability and Statistics (Continued)

HUSKOVA, BERAN, and DUPAC · Collected Works of Jaroslav Hajek—with Commentary
IMAN and CONOVER · A Modern Approach to Statistics
JUREK and MASON · Operator-Limit Distributions in Probability Theory
KASS and VOS · Geometrical Foundations of Asymptotic Inference
KAUFMAN and ROUSSEEUW · Finding Groups in Data: An Introduction to Cluster Analysis
KELLY · Probability, Statistics, and Optimization
KENDALL, BARDEN, CARNE, and LE · Shape and Shape Theory
LINDVALL · Lectures on the Coupling Method
MANTON, WOODBURY, and TOLLEY · Statistical Applications Using Fuzzy Sets
MEERSCHAERT and SCHEFFLER · Limit Distributions for Sums of Independent Random Vectors: Heavy Tails in Theory and Practice
MORGENTHALER and TUKEY · Configural Polysampling: A Route to Practical Robustness
MUIRHEAD · Aspects of Multivariate Statistical Theory
OLIVER and SMITH · Influence Diagrams, Belief Nets and Decision Analysis
*PARZEN · Modern Probability Theory and Its Applications
PEÑA, TIAO, and TSAY · A Course in Time Series Analysis
PRESS · Bayesian Statistics: Principles, Models, and Applications
PUKELSHEIM · Optimal Experimental Design
RAO · Asymptotic Theory of Statistical Inference
RAO · Linear Statistical Inference and Its Applications, *Second Edition*
RAO and SHANBHAG · Choquet-Deny Type Functional Equations with Applications to Stochastic Models
ROBERTSON, WRIGHT, and DYKSTRA · Order Restricted Statistical Inference
ROGERS and WILLIAMS · Diffusions, Markov Processes, and Martingales, Volume I: Foundations, *Second Edition;* Volume II: Îto Calculus
RUBINSTEIN and SHAPIRO · Discrete Event Systems: Sensitivity Analysis and Stochastic Optimization by the Score Function Method
RUZSA and SZEKELY · Algebraic Probability Theory
SCHEFFE · The Analysis of Variance
SEBER · Linear Regression Analysis
SEBER · Multivariate Observations
SEBER and WILD · Nonlinear Regression
SERFLING · Approximation Theorems of Mathematical Statistics
SHORACK and WELLNER · Empirical Processes with Applications to Statistics
SMALL and McLEISH · Hilbert Space Methods in Probability and Statistical Inference
STAPLETON · Linear Statistical Models
STAUDTE and SHEATHER · Robust Estimation and Testing
STOYANOV · Counterexamples in Probability
TANAKA · Time Series Analysis: Nonstationary and Noninvertible Distribution Theory
THOMPSON and SEBER · Adaptive Sampling
WELSH · Aspects of Statistical Inference
WHITTAKER · Graphical Models in Applied Multivariate Statistics
YANG · The Construction Theory of Denumerable Markov Processes

Applied Probability and Statistics Section

ABRAHAM and LEDOLTER · Statistical Methods for Forecasting
AGRESTI · Analysis of Ordinal Categorical Data

*Now available in a lower priced paperback edition in the Wiley Classics Library.

Applied Probability and Statistics (Continued)

AGRESTI · Categorical Data Analysis

ANDERSON, AUQUIER, HAUCK, OAKES, VANDAELE, and WEISBERG · Statistical Methods for Comparative Studies

*ARTHANARI and DODGE · Mathematical Programming in Statistics

ASMUSSEN · Applied Probability and Queues

*BAILEY · The Elements of Stochastic Processes with Applications to the Natural Sciences

BARNETT and LEWIS · Outliers in Statistical Data, *Third Edition*

BARTHOLOMEW, FORBES, and McLEAN · Statistical Techniques for Manpower Planning, *Second Edition*

BASU and RIGDON · Statistical Methods for the Reliability of Repairable Systems

BATES and WATTS · Nonlinear Regression Analysis and Its Applications

BECHHOFER, SANTNER, and GOLDSMAN · Design and Analysis of Experiments for Statistical Selection, Screening, and Multiple Comparisons

BELSLEY · Conditioning Diagnostics: Collinearity and Weak Data in Regression

BELSLEY, KUH, and WELSCH · Regression Diagnostics: Identifying Influential Data and Sources of Collinearity

BHAT · Elements of Applied Stochastic Processes, *Second Edition*

BHATTACHARYA and WAYMIRE · Stochastic Processes with Applications

BIRKES and DODGE · Alternative Methods of Regression

BLISCHKE AND MURTHY · Reliability: Modeling, Prediction, and Optimization

BLOOMFIELD · Fourier Analysis of Time Series: An Introduction, *Second Edition*

BOLLEN · Structural Equations with Latent Variables

BOULEAU · Numerical Methods for Stochastic Processes

BOX · Bayesian Inference in Statistical Analysis

BOX and DRAPER · Empirical Model-Building and Response Surfaces

*BOX and DRAPER · Evolutionary Operation: A Statistical Method for Process Improvement

BUCKLEW · Large Deviation Techniques in Decision, Simulation, and Estimation

BUNKE and BUNKE · Nonlinear Regression, Functional Relations and Robust Methods: Statistical Methods of Model Building

CHATTERJEE and HADI · Sensitivity Analysis in Linear Regression

CHERNICK · Bootstrap Methods: A Practitioner's Guide

CHILÈS and DELFINER · Geostatistics: Modeling Spatial Uncertainty

CLARKE and DISNEY · Probability and Random Processes: A First Course with Applications, *Second Edition*

*COCHRAN and COX · Experimental Designs, *Second Edition*

CONOVER · Practical Nonparametric Statistics, *Second Edition*

CORNELL · Experiments with Mixtures, Designs, Models, and the Analysis of Mixture Data, *Second Edition*

*COX · Planning of Experiments

CRESSIE · Statistics for Spatial Data, *Revised Edition*

DANIEL · Applications of Statistics to Industrial Experimentation

DAVID · Order Statistics, *Second Edition*

*DEGROOT, FIENBERG, and KADANE · Statistics and the Law

DODGE · Alternative Methods of Regression

DOWDY and WEARDEN · Statistics for Research, *Second Edition*

GALLANT · Nonlinear Statistical Models

GLASSERMAN and YAO · Monotone Structure in Discrete-Event Systems

GNANADESIKAN · Methods for Statistical Data Analysis of Multivariate Observations, *Second Edition*

GOLDSTEIN and LEWIS · Assessment: Problems, Development, and Statistical Issues

GREENWOOD and NIKULIN · A Guide to Chi-Squared Testing

*Now available in a lower priced paperback edition in the Wiley Classics Library.

Applied Probability and Statistics (Continued)

*HAHN · Statistical Models in Engineering
HAHN and MEEKER · Statistical Intervals: A Guide for Practitioners
HAND · Construction and Assessment of Classification Rules
HAND · Discrimination and Classification
HEIBERGER · Computation for the Analysis of Designed Experiments
HEDAYAT and SINHA · Design and Inference in Finite Population Sampling
HINKELMAN and KEMPTHORNE: · Design and Analysis of Experiments, Volume 1: Introduction to Experimental Design
HOAGLIN, MOSTELLER, and TUKEY · Exploratory Approach to Analysis of Variance
HOAGLIN, MOSTELLER, and TUKEY · Exploring Data Tables, Trends and Shapes
HOAGLIN, MOSTELLER, and TUKEY · Understanding Robust and Exploratory Data Analysis
HOCHBERG and TAMHANE · Multiple Comparison Procedures
HOCKING · Methods and Applications of Linear Models: Regression and the Analysis of Variables
HOGG and KLUGMAN · Loss Distributions
HOSMER and LEMESHOW · Applied Logistic Regression, *Second Edition*
HØYLAND and RAUSAND · System Reliability Theory: Models and Statistical Methods
HUBERTY · Applied Discriminant Analysis
JACKSON · A User's Guide to Principle Components
JOHN · Statistical Methods in Engineering and Quality Assurance
JOHNSON · Multivariate Statistical Simulation
JOHNSON and KOTZ · Distributions in Statistics
JOHNSON, KOTZ, and BALAKRISHNAN · Continuous Univariate Distributions, Volume 1, *Second Edition*
JOHNSON, KOTZ, and BALAKRISHNAN · Continuous Univariate Distributions, Volume 2, *Second Edition*
JOHNSON, KOTZ, and BALAKRISHNAN · Discrete Multivariate Distributions
JOHNSON, KOTZ, and KEMP · Univariate Discrete Distributions, *Second Edition*
JUREČKOVÁ and SEN · Robust Statistical Procedures: Aymptotics and Interrelations
KADANE AND SCHUM · A Probabilistic Analysis of the Sacco and Vanzetti Evidence
KELLY · Reversability and Stochastic Networks
KHURI, MATHEW, and SINHA · Statistical Tests for Mixed Linear Models
KLUGMAN, PANJER, and WILLMOT · Loss Models: From Data to Decisions
KLUGMAN, PANJER, and WILLMOT · Solutions Manual to Accompany Loss Models: From Data to Decisions
KOTZ, BALAKRISHNAN, and JOHNSON · Continuous Multivariate Distributions, Volume 1, *Second Edition*
KOVALENKO, KUZNETZOV, and PEGG · Mathematical Theory of Reliability of Time-Dependent Systems with Practical Applications
LAD · Operational Subjective Statistical Methods: A Mathematical, Philosophical, and Historical Introduction
LEPAGE and BILLARD · Exploring the Limits of Bootstrap
LINHART and ZUCCHINI · Model Selection
LITTLE and RUBIN · Statistical Analysis with Missing Data
LLOYD · The Statistical Analysis of Categorical Data
MAGNUS and NEUDECKER · Matrix Differential Calculus with Applications in Statistics and Econometrics, *Revised Edition*
MANN, SCHAFER, and SINGPURWALLA · Methods for Statistical Analysis of Reliability and Life Data
McLACHLAN · Discriminant Analysis and Statistical Pattern Recognition
McLACHLAN and KRISHNAN · The EM Algorithm and Extensions
McLACHLAN and PEEL · Finite Mixture Models

*Now available in a lower priced paperback edition in the Wiley Classics Library.

Applied Probability and Statistics (Continued)

MEEKER and ESCOBAR · Statistical Methods for Reliability Data
MYERS and MONTGOMERY · Response Surface Methodology: Process and Product in Optimization Using Designed Experiments
NELSON · Accelerated Testing, Statistical Models, Test Plans, and Data Analyses
NELSON · Applied Life Data Analysis
OCHI · Applied Probability and Stochastic Processes in Engineering and Physical Sciences
OKABE, BOOTS, and SUGIHARA · Spatial Tesselations: Concepts and Applications of Voronoi Diagrams
PANKRATZ · Forecasting with Dynamic Regression Models
PANKRATZ · Forecasting with Univariate Box-Jenkins Models: Concepts and Cases
PORT · Theoretical Probability for Applications
POURAHMADI · Foundations of Time Series Analysis and Prediction Theory
PUTERMAN · Markov Decision Processes: Discrete Stochastic Dynamic Programming
RACHEV · Probability Metrics and the Stability of Stochastic Models
RÉNYI · A Diary on Information Theory
RIPLEY · Spatial Statistics
RIPLEY · Stochastic Simulation
ROUSSEEUW and LEROY · Robust Regression and Outlier Detection
RUBIN · Multiple Imputation for Nonresponse in Surveys
RUBINSTEIN · Simulation and the Monte Carlo Method
RUBINSTEIN and MELAMED · Modern Simulation and Modeling
RYAN · Statistical Methods for Quality Improvement, *Second Edition*
SCHIMEK · Smoothing and Regression: Approaches, Computation, and Application
SCHUSS · Theory and Applications of Stochastic Differential Equations
SCOTT · Multivariate Density Estimation: Theory, Practice, and Visualization
*SEARLE · Linear Models
SEARLE · Linear Models for Unbalanced Data
SEARLE, CASELLA, and McCULLOCH · Variance Components
SENNOTT · Stochastic Dynamic Programming and the Control of Queueing Systems
STOYAN, KENDALL, and MECKE · Stochastic Geometry and Its Applications, *Second Edition*
STOYAN and STOYAN · Fractals, Random Shapes and Point Fields: Methods of Geometrical Statistics
THOMPSON · Empirical Model Building
THOMPSON · Sampling
THOMPSON · Simulation: A Modeler's Approach
TIJMS · Stochastic Modeling and Analysis: A Computational Approach
TIJMS · Stochastic Models: An Algorithmic Approach
TITTERINGTON, SMITH, and MAKOV · Statistical Analysis of Finite Mixture Distributions
UPTON and FINGLETON · Spatial Data Analysis by Example, Volume 1: Point Pattern and Quantitative Data
UPTON and FINGLETON · Spatial Data Analysis by Example, Volume II: Categorical and Directional Data
VAN RIJCKEVORSEL and DE LEEUW · Component and Correspondence Analysis
VIDAKOVIC · Statistical Modeling by Wavelets
WEISBERG · Applied Linear Regression, *Second Edition*
WESTFALL and YOUNG · Resampling-Based Multiple Testing: Examples and Methods for *p*-Value Adjustment
WHITTLE · Systems in Stochastic Equilibrium
*ZELLNER · An Introduction to Bayesian Inference in Econometrics

*Now available in a lower priced paperback edition in the Wiley Classics Library.

Biostatistics Section

ARMITAGE and DAVID (editors) · Advances in Biometry
BROWN and HOLLANDER · Statistics: A Biomedical Introduction
CHOW and LIU · Design and Analysis of Clinical Trials: Concepts and Methodologies
DUNN and CLARK · Applied Statistics: Analysis of Variance and Regression, *Second Edition*
*ELANDT-JOHNSON and JOHNSON · Survival Models and Data Analysis
*FLEISS · The Design and Analysis of Clinical Experiments
FLEISS · Statistical Methods for Rates and Proportions, *Second Edition*
FLEMING and HARRINGTON · Counting Processes and Survival Analysis
KADANE · Bayesian Methods and Ethics in a Clinical Trial Design
KALBFLEISCH and PRENTICE · The Statistical Analysis of Failure Time Data
LACHIN · Biostatistical Methods: The Assessment of Relative Risks
LANGE, RYAN, BILLARD, BRILLINGER, CONQUEST, and GREENHOUSE · Case Studies in Biometry
LAWLESS · Statistical Models and Methods for Lifetime Data
LEE · Statistical Methods for Survival Data Analysis, *Second Edition*
MALLER and ZHOU · Survival Analysis with Long Term Survivors
McNEIL · Epidemiological Research Methods
McFADDEN · Management of Data in Clinical Trials
*MILLER · Survival Analysis, *Second Edition*
PIANTADOSI · Clinical Trials: A Methodologic Perspective
WOODING · Planning Pharmaceutical Clinical Trials: Basic Statistical Principles
WOOLSON · Statistical Methods for the Analysis of Biomedical Data

Financial Engineering Section

HUNT and KENNEDY · Financial Derivatives in Theory and Practice
ROLSKI, SCHMIDLI, SCHMIDT, and TEUGELS · Stochastic Processes for Insurance and Finance
SHAFER and VOVK · Probability and Finance: It's Only a Game!

Texts, References, and Pocketbooks Section

AGRESTI · An Introduction to Categorical Data Analysis
ANDĚL · Mathematics of Chance
ANDERSON · An Introduction to Multivariate Statistical Analysis, *Second Edition*
ANDERSON and LOYNES · The Teaching of Practical Statistics
ARMITAGE and COLTON · Encyclopedia of Biostatistics: Volumes 1 to 6 with Index
BARTOSZYNSKI and NIEWIADOMSKA-BUGAJ · Probability and Statistical Inference
BENDAT and PIERSOL · Random Data: Analysis and Measurement Procedures, *Third Edition*
BERRY, CHALONER, and GEWEKE · Bayesian Analysis in Statistics and Econometrics: Essays in Honor of Arnold Zellner
BHATTACHARYA and JOHNSON · Statistical Concepts and Methods
BILLINGSLEY · Probability and Measure, *Second Edition*
BOX · R. A. Fisher, the Life of a Scientist
BOX, HUNTER, and HUNTER · Statistics for Experimenters: An Introduction to Design, Data Analysis, and Model Building

*Now available in a lower priced paperback edition in the Wiley Classics Library.

Texts, References, and Pocketbooks (Continued)

BOX and LUCEÑO · Statistical Control by Monitoring and Feedback Adjustment
CHATTERJEE and PRICE · Regression Analysis by Example, *Third Edition*
COOK and WEISBERG · Applied Regression Including Computing and Graphics
COOK and WEISBERG · An Introduction to Regression Graphics
COX · A Handbook of Introductory Statistical Methods
DANIEL · Biostatistics: A Foundation for Analysis in the Health Sciences, *Sixth Edition*
DILLON and GOLDSTEIN · Multivariate Analysis: Methods and Applications
*DODGE and ROMIG · Sampling Inspection Tables, *Second Edition*
DRAPER and SMITH · Applied Regression Analysis, *Third Edition*
DUDEWICZ and MISHRA · Modern Mathematical Statistics
DUNN and CLARK · Basic Statistics: A Primer for the Biomedical Sciences, *Third Edition*
EVANS, HASTINGS, and PEACOCK · Statistical Distributions, *Third Edition*
FISHER and VAN BELLE · Biostatistics: A Methodology for the Health Sciences
FREEMAN and SMITH · Aspects of Uncertainty: A Tribute to D. V. Lindley
GROSS and HARRIS · Fundamentals of Queueing Theory, *Third Edition*
HALD · A History of Probability and Statistics and their Applications Before 1750
HALD · A History of Mathematical Statistics from 1750 to 1930
HELLER · MACSYMA for Statisticians
HOEL · Introduction to Mathematical Statistics, *Fifth Edition*
HOLLANDER and WOLFE · Nonparametric Statistical Methods, *Second Edition*
HOSMER and LEMESHOW · Applied Logistic Regression, *Second Edition*
HOSMER and LEMESHOW · Applied Survival Analysis: Regression Modeling of Time to Event Data
JOHNSON and BALAKRISHNAN · Advances in the Theory and Practice of Statistics: A Volume in Honor of Samuel Kotz
JOHNSON and KOTZ (editors) · Leading Personalities in Statistical Sciences: From the Seventeenth Century to the Present
JUDGE, GRIFFITHS, HILL, LÜTKEPOHL, and LEE · The Theory and Practice of Econometrics, *Second Edition*
KHURI · Advanced Calculus with Applications in Statistics
KOTZ and JOHNSON (editors) · Encyclopedia of Statistical Sciences: Volumes 1 to 9 with Index
KOTZ and JOHNSON (editors) · Encyclopedia of Statistical Sciences: Supplement Volume
KOTZ, READ, and BANKS (editors) · Encyclopedia of Statistical Sciences: Update Volume 1
KOTZ, READ, and BANKS (editors) · Encyclopedia of Statistical Sciences: Update Volume 2
LAMPERTI · Probability: A Survey of the Mathematical Theory, *Second Edition*
LARSON · Introduction to Probability Theory and Statistical Inference, *Third Edition*
LE · Applied Categorical Data Analysis
LE · Applied Survival Analysis
MALLOWS · Design, Data, and Analysis by Some Friends of Cuthbert Daniel
MARDIA · The Art of Statistical Science: A Tribute to G. S. Watson
MASON, GUNST, and HESS · Statistical Design and Analysis of Experiments with Applications to Engineering and Science
McCULLOCH and SEARLE · Generalized, Linear, and Mixed Models
MONTGOMERY, PECK, and VINING · Introduction to Linear Regression Analysis, *Third Edition*
MURRAY · X-STAT 2.0 Statistical Experimentation, Design Data Analysis, and Nonlinear Optimization
PRESS and TANUR · The Subjectivity of Scientists and the Bayesian Approach

*Now available in a lower priced paperback edition in the Wiley Classics Library.

Texts, References, and Pocketbooks (Continued)

PURI, VILAPLANA, and WERTZ · New Perspectives in Theoretical and Applied Statistics
RENCHER · Linear Models in Statistics
RENCHER · Methods of Multivariate Analysis
RENCHER · Multivariate Statistical Inference with Applications
ROSS · Introduction to Probability and Statistics for Engineers and Scientists
ROHATGI and SALEH · An Introduction to Probability and Statistics, *Second Edition*
RYAN · Modern Regression Methods
SCHOTT · Matrix Analysis for Statistics
SEARLE · Matrix Algebra Useful for Statistics
STYAN · The Collected Papers of T. W. Anderson: 1943–1985
TIAO, BISGAARD, HILL, PEÑA, and STIGLER (editors) · Box on Quality and Discovery: with Design, Control, and Robustness
TIERNEY · LISP-STAT: An Object-Oriented Environment for Statistical Computing and Dynamic Graphics
WONNACOTT and WONNACOTT · Econometrics, *Second Edition*
WU and HAMADA · Experiments: Planning, Analysis, and Parameter Design Optimization

JWS/SAS Co-Publications Section

FREUND and LITTELL · SAS System for Regression, *Third Edition*
KHATTREE and NAIK · Applied Multivariate Statistics with SAS Software, *Second Edition*
KHATTREE and NAIK · Multivariate Data Reduction and Discrimination with SAS Software

WILEY SERIES IN PROBABILITY AND STATISTICS

ESTABLISHED BY WALTER A. SHEWHART AND SAMUEL S. WILKS

Editors
Robert M. Groves, Graham Kalton, J. N. K. Rao, Norbert Schwarz, Christopher Skinner

Survey Methodology Section

BIEMER, GROVES, LYBERG, MATHIOWETZ, and SUDMAN · Measurement Errors in Surveys
COCHRAN · Sampling Techniques, *Third Edition*
COUPER, BAKER, BETHLEHEM, CLARK, MARTIN, NICHOLLS, and O'REILLY (editors) · Computer Assisted Survey Information Collection
COX, BINDER, CHINNAPPA, CHRISTIANSON, COLLEDGE, and KOTT (editors) · Business Survey Methods
*DEMING · Sample Design in Business Research
DILLMAN · Mail and Telephone Surveys: The Total Design Method, *Second Edition*
DILLMAN · Mail and Internet Surveys: The Tailored Design Method
GROVES and COUPER · Nonresponse in Household Interview Surveys

*Now available in a lower priced paperback edition in the Wiley Classics Library.

Survey Methodology (Continued)

GROVES · Survey Errors and Survey Costs

GROVES, BIEMER, LYBERG, MASSEY, NICHOLLS, and WAKSBERG · Telephone Survey Methodology

*HANSEN, HURWITZ, and MADOW · Sample Survey Methods and Theory, Volume I: Methods and Applications

*HANSEN, HURWITZ, and MADOW · Sample Survey Methods and Theory, Volume II: Theory

KISH · Statistical Design for Research

*KISH · Survey Sampling

KORN and GRAUBARD · Analysis of Health Surveys

LESSLER and KALSBEEK · Nonsampling Error in Surveys

LEVY and LEMESHOW · Sampling of Populations: Methods and Applications, *Third Edition*

LYBERG, BIEMER, COLLINS, de LEEUW, DIPPO, SCHWARZ, TREWIN (editors) · Survey Measurement and Process Quality

SIRKEN, HERRMANN, SCHECHTER, SCHWARZ, TANUR, and TOURANGEAU (editors) · Cognition and Survey Research

VALLIANT, DORFMAN, and ROYALL · Finite Population Sampling and Inference: A Prediction Approach